Structure and Reactivity
An Introduction to
Organic Chemistry
Book B

First Edition, Second Printing

Structure and Reactivity: An Introduction to Organic Chemistry is a 4-book series that provides a student-centered approach to the first-year undergraduate course in Organic Chemistry.

The Study Guide for this textbook is a free, openly available resource that has been created and maintained by the author. The Study Guide contains:

- The answer key to all the questions in the books: Practice Questions within the chapters, Chapter Questions at the end of the chapters, and Exam Questions after every block of 3–4 chapters that align with the undergraduate course at the University of Michigan (Ann Arbor).

- Advice to students: 8 Essays on Teaching, Testing, and Learning

- Course materials from the University of Michigan's first-year undergraduate course in Organic Chemistry (CHEM 210, Books A & B; and CHEM 215, Books C & D), located under pull-down menus and blocked off according to the exam content. Materials include: lecture recordings, a picture of the board from the lecture, a "Problem of the Day" (POD) and its solution, and a weekly detailed explanation of a question or two related to the week's lessons ("Thinking in Blue" or TIB).

The Study Guide is located at the following URL: https://sites.lsa.umich.edu/studyguide/

Structure and Reactivity: An Introduction to Organic Chemistry

Book B
First Edition, Second Printing
Brian P. Coppola

Copyright © *by* Brian P. Coppola

All rights reserved. No part of this book may be reproduced or transmitted in any form or by any means, electronic or mechanical, including photocopying, recording or by any information storage and retrieval system, without written permission from the author and publisher.

Printed in the United States of America
10 9 8 7 6 5 4 3
ISBN: 978-1-64565-275-5

Van-Griner Learning
Cincinnati, Ohio
www.van-griner.com

CoppolaB 65-275-5 Sp23
330572-338403

Preface to
Book B
Introduction to Reactivity

Structure and Reactivity: An Introduction to Organic Chemistry (Book B) is the second volume of two that make up a standard first-semester organic chemistry course.

This volume introduces the fundamental reaction mechanisms that form the base set for continued study. We begin with nucleophilic substitution and elimination reactions, featured in Chapter 7, involving the chemistry of polar sigma bonds. And then we continue with electrophilic addition reactions, in Chapters 8 and 9, which involve the reactions of pi bonds. Chapter 10 is the capstone topic and the last chapter for the first term and introduces the first example of a class of reactions requiring a combination of these earlier fundamental types (an addition-elimination sequence).

In addition to the chapters, Book B also includes four appendices. The first one of these is an extensive review of Introductory Chemistry, meant to support and remind students about key topics that are built upon in Organic Chemistry. The other three appendices feature rules of nomenclature, including both names and stereochemical labels.

Two complete sets of practice Exam Questions are included. The first of these sets appears after Chapter 9, where the third examination in our course typically falls. And then after Chapter 10, a question set that is cumulative for Chapters 1–10, which is the final examination in our first-term Organic Chemistry course.

Books C and D for **Structure and Reactivity** are used in the second-term course.

Brian P. Coppola
15 July 2021

Contents of Book B: Introduction to Reactivity

CHAPTER 7

Substitution and Elimination Reactions of Polar Sigma Bonds

Setting the Stage: Predicting Outcomes (Hypotheses) Versus Explaining Observations ... 517

7.1 Possible Outcomes in Substitution and Elimination Reactions of Polar Covalent Bonds
  A. Substitution and Elimination Reactions: Definition of Terms ... 519
  B. Substitution and Elimination Reactions: Possible Outcomes (Connectivity and Stereochemistry) ... 520
  C. Substitution Reactions: Possible Mechanistic Pathways ... 525
  D. Elimination Reactions: Possible Mechanistic Pathways ... 528

PRACTICE QUESTIONS ... 534

7.2 Application of Substitution and Elimination Mechanisms
  A. A Unified View of S_N2 and E2 Mechanisms ... 539
  B. A Unified View of S_N1 and E1 Mechanisms ... 540
  C. The E1cb Mechanism ... 540
  D. Extending the Definition of Leaving Groups ... 541

PRACTICE QUESTIONS ... 548

7.3 A Predictive Model for S_N/E Reactions
  A. Explaining Outcomes Versus Predicting Outcomes ... 552
  B. Developing a Predictive Model for S_N/E Reactions ... 553
  C. Applying a Predictive Model for S_N/E Reactions ... 564
  D. Transformations and Pragmatic Choices in Using S_N/E Reactions in Practice ... 569

PRACTICE QUESTIONS ... 576

7.4 Rate and Reactivity in S_N/E Reactions
  A. Leaving Group Ability ... 582
  B. Solvent Effects ... 583
  C. Nucleophilicity ... 584
  D. Heteroatom Electrophiles ... 585

PRACTICE QUESTIONS ... 587

Reflections About Science: Die Verdammten Experiments—Models and the Limits of Predictions ... 589

Summary ... 591

CHAPTER QUESTIONS ... 592

CHAPTER 8

Electrophilic Addition I: Brønsted Acids

Setting the Stage: Making the Most of the Unexpected ... 611

8.1 Electrophilic Addition of Brønsted Acids
  A. Regioselectivity in Addition Reactions ... 612
  B. Stereoselectivity in Addition Reactions ... 614
  C. Electrophilic Addition of Strong Brønsted Acids: Revisited ... 617
  D. Electrophilic Addition of Weak Brønsted Acids: Revisited ... 620

PRACTICE QUESTIONS ... 623

8.2 Carbocation Chemistry
  A. Origins and Fates of Carbocations ... 627
  B. Carbocation Rearrangements Change Connectivity ... 627
  C. Ring Contraction and Expansion Reactions ... 631
  D. Rearrangements of Carbocations with β-Bonded Heteroatoms ... 634

PRACTICE QUESTIONS ... 636

8.3 Hydration via Hydroboration-Oxidation
  A. Transformation: Anti-Markovnikov Addition of Water ... 641
  B. Stereochemistry and Regioselectivity of the Hydroboration Reaction Mechanism ... 645
  C. Oxidation of Carbon-Boron Bonds ... 648
  D. Synthesis: Here's the Answer, What's the Question? ... 649

PRACTICE QUESTIONS ...

Reflections About Science: Synthesis and Why We Need Options ... 659

Summary ... 661

CHAPTER QUESTIONS ... 662

CHAPTER 9

Electrophilic Addition II: Halogenation, Oxidation, and Reduction

Setting the Stage: Saturated, Partially Hydrogenated, Brominated, and Polyunsaturated ... 677

9.1 Halogenation and Other Reactions of the Halonium Ion Intermediate
  A. Transformation: Anti Addition of Halogens ... 678
  B. Stereoselectivity via the Halonium Ion Intermediate ... 680
  C. Halohydrins and Related Regioselective Reactions of the Halonium Ion Intermediate ... 683
  D. Intramolecular and Other Mixed-Atom Reactions ... 687

PRACTICE QUESTIONS ... 690

9.2 Oxidation Reactions
  A. Peroxyacid Reactions ... 694
  B. Syn Dihydroxylation Reactions ... 697
  C. Ozonolysis Reactions ... 700
  D. Atom-Swap Analogies ... 708

PRACTICE QUESTIONS ... 710

9.3 Reduction Reactions
  A. Catalytic Hydrogenation with Transition Metal Catalysts ... 713
  B. (Z)-Alkenes from Catalytic Hydrogenation with Poisoned Catalysts ... 716
  C. (E)-Alkenes from Dissolving Metal Reduction ... 717
  D. Multistep Transformations and Synthesis ... 719

PRACTICE QUESTIONS ... 725

Reflections About Science: The Power of Mechanistic Organic Chemistry ... 730

Summary ... 732

CHAPTER QUESTIONS ... 733

EXAM QUESTIONS (Chapters 7–9) ... 755

CHAPTER 10

Aromaticity and Electrophilic Aromatic Substitution Reactions

Setting the Stage: Word Meanings Change ... 787

10.1 Aromaticity
  A. Benzene and the Aromatic Sextet ... 790
  B. Hückel Aromaticity (4N+2 Rule) and "Antiaromatic" Compounds ... 791
  C. Cations, Anions, and Heterocyclic Compounds ... 802
  D. Chemical and Physical Properties of Aromatic Compounds ... 808

PRACTICE QUESTIONS ... 811

10.2 Electrophilic Aromatic Substitution
   A. Halogenation, Sulfonation, and Nitration Reactions ... 815
   B. Friedel-Crafts Alkylation and Acylation Reactions ... 819
   C. Regioselectivity of Electrophilic Aromatic Substitution Reactions (Directing Effect) ... 822
   D. Relative Rates of Electrophilic Aromatic Substitution Reactions (Activation) ... 830

PRACTICE QUESTIONS ... 839

Reflections About Science: Occam's Razor (Don't Invoke More Than You Need) ... 843

Summary ... 844

CHAPTER QUESTIONS ... 845

EXAM QUESTIONS (Chapters 1–10) ... 859

APPENDIX 1

Useful Expectations from General Chemistry

Preface ... 907

Introduction ... 909

A1.1 Atoms, Ions, and Molecules
   A. Atoms and the Periodic Table ... 910
   B. Molecular Structure: Valence Bond Model ... 912
   C. The Quantum World ... 919

PRACTICE QUESTIONS ... 925

A1.2 Periodic Trends
   A. Electronegativity and Covalent Bond Polarity ... 928
   B. Atomic Size and Bond Length ... 930
   C. Oxidation-Reduction ... 931

PRACTICE QUESTIONS ... 925

A1.3 Chemical and Physical Properties
   A. Rate (Kinetics) and Stability (Thermodynamics) ... 934
   B. Equilibrium (K_{EQ}), Enthalpy (ΔH), and Entropy (ΔS) ... 936
   C. Phase Changes and Mixing ... 940

PRACTICE QUESTIONS ... 951

APPENDIX 2

Nomenclature of Organic Compounds I: Alkanes, Halides, and Alcohols

Introduction ... 957

A2.1 Acyclic Saturated Hydrocarbons
   A. Longest Carbon Chain (Root Name) ... 960
   B. Identifying Substituents and Numbering the Root Name ... 961
   C. Assembling the Name ... 962
   D. Three More Rules and the Four Allowed Branched Alkyl Group Names ... 963

PRACTICE QUESTIONS ... 967

A2.2 Cyclic Compounds, Halides, and Alcohols
   A. Cyclic Hydrocarbons ... 969
   B. Alkyl and Aryl Halides ... 972
   C. Alcohols ... 972
   D. Priorities When Numbering the Root Name ... 972

PRACTICE QUESTIONS ... 974

APPENDIX 3

Nomenclature of Organic Compounds II: Stereoisomers

Introduction ... 979

A3.1 Prioritizing Substituent Groups
   A. Cahn-Ingold-Prelog (CIP) Rules ... 980
   B. Point of Difference Comparisons ... 980
   C. Nonbonding Electron Pairs ... 983
   D. Duplicate and Phantom Atoms for Multiple Bonds ... 983

PRACTICE QUESTIONS ... 988

A3.2 Assigning Geometrical Labels
   A. Cis/Trans: Relative Stereochemistry ... 992
   B. Cis/Trans: Geometrical Label ... 992
   C. E/Z and R/S Geometrical Labels ... 993
   D. r/s and R_a/S_a Geometrical Labels ... 995

PRACTICE QUESTIONS ... 1002

APPENDIX 4

Nomenclature of Organic Compounds III: Alkoxy Groups, Alkenes, and Alkynes

A4 Alkoxy Groups, Alkenes, and Alkynes
   A. Alkoxyl Group Substituents and Ethers ... 1007
   B. Chain Suffix: Alkenes and Alkynes ... 1008
   C. Prioritizing Substituent Suffixes, Chain Suffixes, and Substituent Prefixes ... 1009
   D. Geometrical Labels in Nomenclature ... 1010

PRACTICE QUESTIONS ... 1012

Cover image copyright 2021, Brian P. Coppola. Commissioned from Ella Maru Studio (https://scientific-illustrations.com).

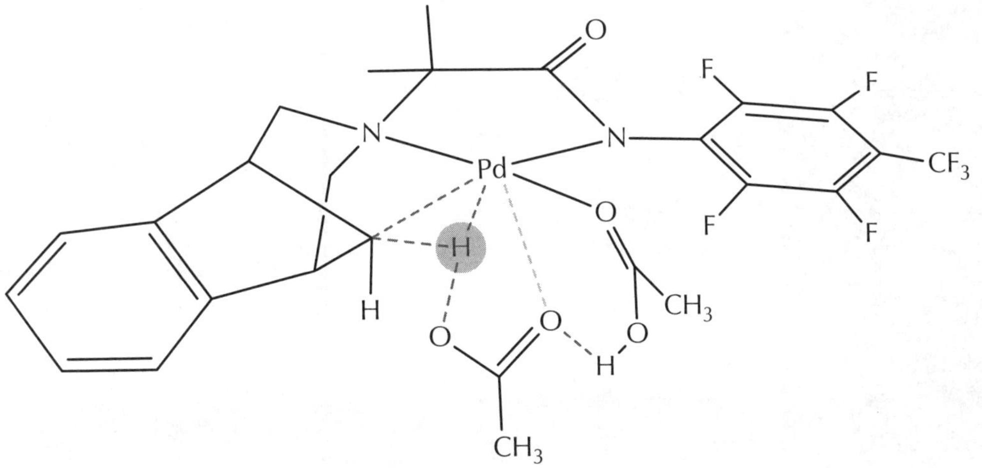

Book B

On the Cover

Hydrocarbon C-H bonds are generally unreactive. Yet, chemists would like to have reliable ways to specifically select a C-H bond from a hydrocarbon and carry out reactions that would transform it into a more reactive C-N, C-O, C-S, and C-halogen bond. This general area, called "C-H activation," has exploded since the 1980s with the discovery of transition metals that can promote these transformations.

In the example presented here, a palladium atom is situated to insert itself into a specific C-H bond thanks to its spatial geometry. The illustration represents a view of the transition state structure for the insertion mechanism, starting from the C-H bond and ending with the C-Pd bond, which can be subsequently replaced by many different atoms, including B, O, S, N, F, Cl, Br, and I.

Research on C-H activation is carried out by the research group led by Professor Melanie S. Sanford at the University of Michigan. Quantum mechanical calculations on these systems are carried out collaboratively with Professor Paul M. Zimmerman and his students, also at the University of Michigan.

Ellen Y. Aguilera, Melanie S. Sanford, "Palladium-Mediated C$_\gamma$-H Functionalization of Alicyclic Amines" *Angewandte Chemie International Edition*, **2021**, *60*, 11227.

Amanda L. Dewyer, Paul M. Zimmerman, "Simulated Mechanism for Palladium-Catalyzed, Directed γ-Arylation of Piperidine" *ACS Catalysis*, **2017**, *7*, 5466.

CHAPTER 7
Substitution and Elimination Reactions of Polar Sigma Bonds

Setting the Stage: Predicting Outcomes (Hypotheses) Versus Explaining Observations

7.1 Possible Outcomes in Substitution and Elimination Reactions of Polar Covalent Bonds
  A. Substitution and Elimination Reactions: Definition of Terms
  B. Substitution and Elimination Reactions: Possible Outcomes (Connectivity and Stereochemistry)
  C. Substitution Reactions: Possible Mechanistic Pathways
  D. Elimination Reactions: Possible Mechanistic Pathways

PRACTICE QUESTIONS

7.2 Application of Substitution and Elimination Mechanisms
  A. A Unified View of S_N2 and E2 Mechanisms
  B. A Unified View of S_N1 and E1 Mechanisms
  C. The E1cb Mechanism
  D. Extending the Definition of Leaving Groups

PRACTICE QUESTIONS

7.3 A Predictive Model for S_N/E Reactions
  A. Explaining Outcomes Versus Predicting Outcomes
  B. Developing a Predictive Model for S_N/E Reactions
  C. Applying a Predictive Model for S_N/E Reactions
  D. Transformations and Pragmatic Choices in Using S_N/E Reactions in Practice

PRACTICE QUESTIONS

7.4 Rate and Reactivity in S_N/E Reactions
  A. Leaving Group Ability
  B. Solvent Effects
  C. Nucleophilicity
  D. Heteroatom Electrophiles

PRACTICE QUESTIONS

Reflections About Science: Die Verdammten Experiments—Models and the Limits of Predictions

Summary

CHAPTER QUESTIONS

CHAPTER 7

Substitution and Elimination Reactions of Polar Sigma Bonds

Setting the Stage: Predicting Outcomes (Hypotheses) Versus Explaining Observations

The process we call "science" is built upon a couple of important ideas: (1) to explain the observable phenomena of the Natural World using the consistency of evidence; and (2) to make as many predictions as possible about new phenomena that are built upon past evidence, and particularly predictions that might be differentiated with new evidence gained through experiments. While different areas of science all operate the same way, each of them (chemistry, physics, biology, etc.) has developed its own focus. Interdisciplinary work benefits because diverse perspectives are brought to solving a problem. Chemistry, speaking broadly, is concerned with the science of matter and its transformations. And quite delightfully, the word for "chemistry" in the standard Chinese language is "化学" (huàxué), which means the study (学) of change (化).

In science, it is common for explanations and theories to change, sometimes radically, while the underlying phenomena and observations of it remain quite constant. Borrowing from physics for a moment: An observable phenomenon, such as gravity, remains consistent over time, but our explanation for gravity is constantly changing, and hopefully getting better as more experiments (and observations) are made. Four hundred years ago, according to one of his students, Galileo famously asserted that Earth's gravity was acting equally on all masses, and that two spheres of different mass dropped from the Pisa Tower were falling with the same acceleration and would reach the ground at the same time (as opposed to acting unequally on different masses, with the heavier one hitting the ground sooner). In a terrific nod to history, American astronaut David Scott conducted a dramatic version of Galileo's experiment while standing on the airless moon in 1971, simultaneously letting go of a feather and a rock hammer in a scientific demonstration broadcast to a worldwide audience (you should check out that video). Our best explanation for gravity, as an action between objects drawn together because of their masses, is incomplete and still under active research. But what we do understand is exceptionally useful, and allows us to throw a ball into a hoop and place a satellite into orbit with predictable certainty.

The theory of gravity is a mature predictive model. The things about gravity that are unknown do not prevent us from making quite accurate predictions about how physical objects will behave.

The number of phenomena we experience in everyday life is substantial, and we readily create predictive models for phenomena that we have not experienced. For instance, it is unlikely you have ever lowered an inflated balloon, with a little water in it, onto the flame of a lit candle. Yet, for most people (and particularly children), the collective human experience with balloons, water, and flame leads to a quite confident prediction that the balloon will burst and the water will quench the flame. Interestingly, however, the balloon does not burst. And this is the sort of observation (called a counterintuitive event), going against a quite reasonable prediction derived from prior experience, which drives curiosity and scientific inquiry.

Chemical reaction phenomena are more complex than lowering a water balloon onto a flame. Trillions of trillions of trillions (of trillions) of molecules, usually as heterogeneous mixtures and with unknown amounts of unknowable impurities, will almost always have many competing reaction pathways that are partially dictated by the actions of the experimentalist (such as temperature, time, concentration, relative concentration, cleanliness of the glassware, age and history of the compounds, how and where substances were stored, and so on). Consequently, a remarkably few chemical reactions occur naturally in high yield and in high purity, particularly the first time you try them. And while it is possible to optimize experimental conditions for a single reaction of specific compounds, even what appears to be a subtle change in molecular structure in one of the reagents, including its crystal form, can give a radically different outcome.

Experienced chemists are accustomed to a high degree of uncertainty when it comes to making predictions about the outcomes from a new experiment. Predictive models for some reactions are quite good, which might mean that you can anticipate 75% of the time what might happen to 60–80% of the molecules. These are made-up numbers, meant to illustrate the idea that the prediction about what will actually happen in a new or unfamiliar chemical reaction is limited. You are sure that one thing is going to happen (the balloon breaks and the water puts out the flame), and something quite different actually happens. This different result does not mean that your prediction was irrational (that the balloon would break is a reasonable prediction… and without the water, it does break). The results mean that nature is complex, and that we should continually learn from nature, experimentally, as a way to deepen our understanding.

Uncertainty does not invalidate a predictive model. In chemistry, many new examples of reactions will, in fact, give exactly what you expect. Experimentally, the yield will not be 100%, the product will not be pure, and changing the conditions of the experiment might completely change the outcome. But predictive models, and their limitations, are how we make decisions about which experiments should be carried out, and how they might be carried out, to produce some desired result. And when you have a new experimental result, it simply does not matter that it was not the predicted outcome (the balloon did not break), an explanation needs to account for the observation.

These twin principles, using a theory to make predictions about the outcome from new experiments, and to explain new observations of phenomena, are at the heart of any science, as are curiosity and skepticism. If you were surprised by what I said about the balloon, and even better if you do not believe me, I can only urge you to try it out for yourself. Or did you already stop reading, a few paragraphs back, and try it (or perhaps, tried to find a video of it)? *Experiment!* Add water to an uninflated balloon, and then inflate the balloon and tie it off. Light a candle (put it in a bowl if you are cautious about the result). Then, lower the balloon onto the top of the flame. Better yet: Do it together with a group of people and talk about the result and why it happens. In this case, don't watch a video… *do it!*

Starting with this chapter, we shift our focus from mainly looking at topics in molecular structure (connectivity, stereochemistry, conformations) to that of chemical reactivity, and how structure influences the outcome from reactions.

In this chapter, we will look at a set of competing reaction pathways and the explanations for how structure and reaction conditions can affect the outcome. We will develop a predictive model that, while imperfect, allows you to anticipate, in the absence of experimental information, what is most likely to happen in millions of reactions that have not necessarily ever been carried out. Even if such a model is only accurate 75% of the time, and the yields are not 100%, the model is still powerful in its ability to predict so many outcomes. Every now and then, though, that damned balloon does not break when you place it over the flame, even when you think it ought to, and that simply means there is something new to learn from nature.

7.1 Possible Outcomes in Substitution and Elimination Reactions of Polar Covalent Bonds

A. Substitution and Elimination Reactions: Definition of Terms

In Section 4.1, four fundamental types of reactions were introduced: complexation-decomplexation (forming and breaking sigma bonds), substitution (S: sigma bond to sigma bond exchange), addition (A: one pi bond to two sigma bonds), and elimination (E: two sigma bonds to one pi bond). A few representative examples are shown in Figure 0701.

Figure 0701
Representative examples of four fundamental reaction types.

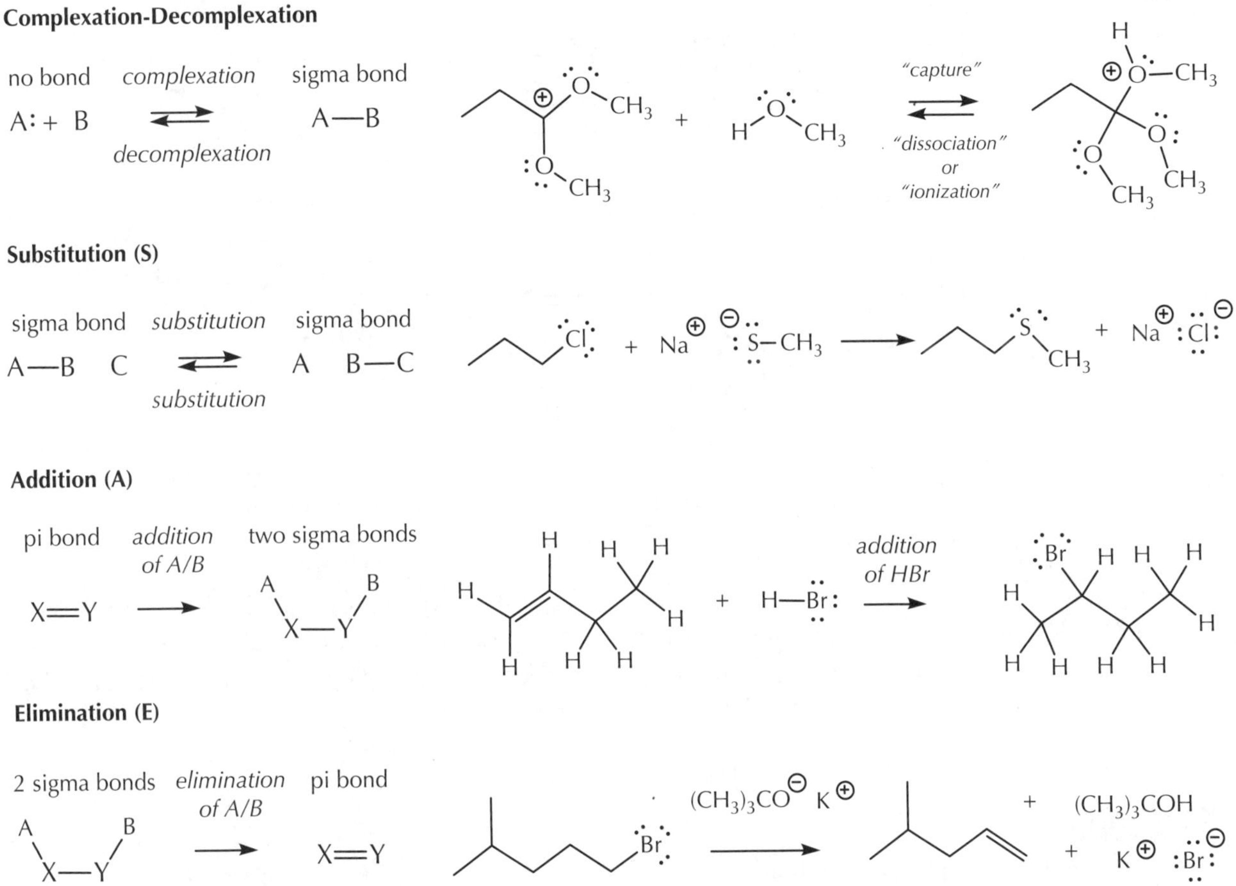

Complexation-Decomplexation

no bond   complexation   sigma bond
A: + B   ⇌   A—B
         decomplexation

Substitution (S)

sigma bond   substitution   sigma bond
A—B C   ⇌   A   B—C
           substitution

Addition (A)

pi bond   addition   two sigma bonds
          of A/B
X=Y   →   A   B
          X—Y

Elimination (E)

2 sigma bonds   elimination   pi bond
               of A/B
A   B   →   X=Y
 \ /
  X—Y

Some details about two of these four reaction types have been introduced already. The complexation-decomplexation reaction (Lewis acid-base reactions), involving open shell atoms and their reactivity, were presented in Sections 1.4B and 3.1C. And the last two sections in Chapter 4 (Sections 4.3 and 4.4) provided the mechanisms for the electrophilic addition (AE) reactions of strong and weak Brønsted acids with double and triple bonds.

In this chapter, the more detailed introductions to substitution (S) and elimination (E) reactions are combined because, as you will see, these reactions can be in competition with one another. This competition creates complexity in the predictive model you will learn because variations in both the structure and properties of the molecules can affect the direction of the outcome, in addition to the specific reaction conditions that are used. Navigating this complexity is an important point that will be revisited a number of times.

B. Substitution and Elimination Reactions: Possible Outcomes (Connectivity and Stereochemistry)

A type of substitution reaction at carbon atoms was introduced in Section 4.2C as an analogy with Brønsted acid-base reactions, which are substitution reactions at hydrogen atoms. From that section: a substitution reaction at carbon is predicted between a Lewis base, as the nucleophile, and unhindered sp³ carbon atoms attached to a halogen atom, as the electrophile. The two reaction mechanisms were described comparably, in which the bond-forming and bond-breaking events occur simultaneously in a molecular collision (Figure 0702).

Figure 0702

Analogous collisional substitution reactions at hydrogen and carbon atoms.

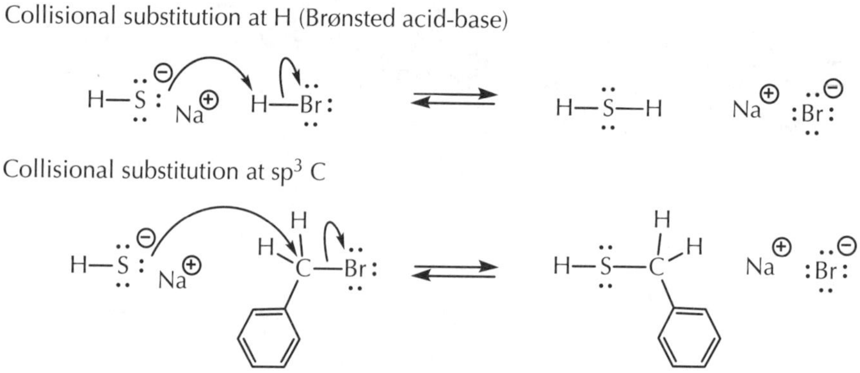

There are mechanisms for substitution (S) reactions other than direct collision, and so we will begin this more detailed look by stepping back for a moment to build a more comprehensive picture of these reactions.

To begin, we will still stick with the carbon atom of any sp³ carbon-halogen bonds (C-X) as the electrophilic site. There are other carbon-electronegative atom groups that are predictably observed to give these same reactions, and so this reactivity will be extended from the halogens, later, in Section 7.2D. The nucleophilic partner is unchanged, that being a Lewis base (LB:).

The observed outcome defines the reaction (Figure 0703).

Figure 0703

Substitution and elimination reactions of sp³ C-X bonds using a Lewis base.

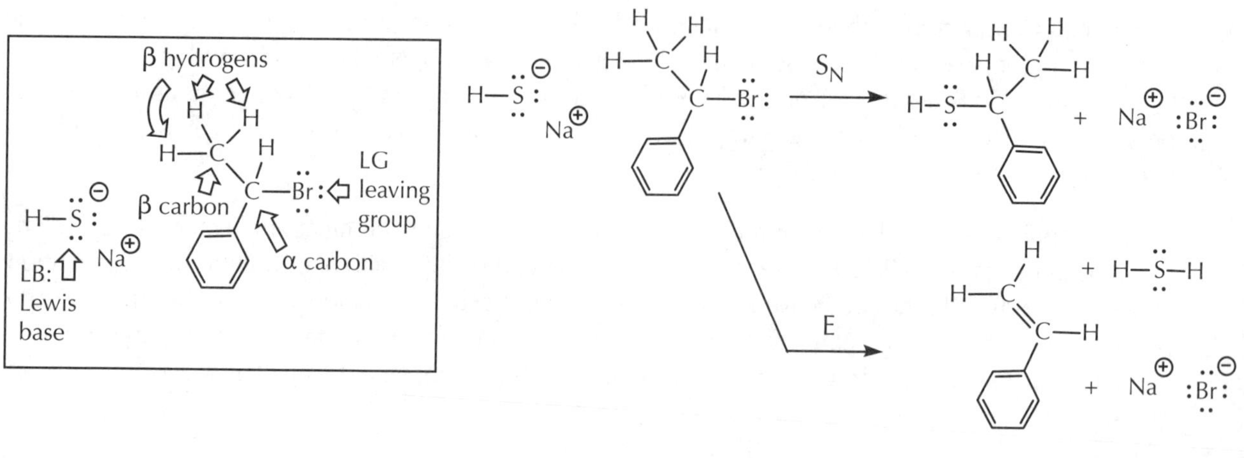

7.1 Possible Outcomes in Substitution and Elimination Reactions of Polar Covalent Bonds

In a substitution (S), the C-X bond is broken, a halide ion forms, and a new sigma bond between the carbon and the Lewis basic partner has formed. These types of substitutions are called nucleophilic substitution reactions and they are symbolized as S_N.

In an elimination (E) reaction, the "C-X" bond is also broken, a halide ion is also formed, but a new pi bond has formed between the carbon that carried the X group (called the *alpha* carbon with respect to the X group) and one of the carbons to which it is attached (called a *beta* carbon), through the deprotonation reaction of a hydrogen atom on the beta carbon (called a *beta* hydrogen). In the case of the elimination reaction, the Lewis basic partner is acting as a Brønsted base and carries out the deprotonation. No mechanisms are implied in Figure 0703, only the observable change in structure.

In nucleophilic substitution (S_N) reactions, and in elimination (E) reactions, the "X" group in the "C-X" bond is classified as a "leaving group" (sometimes generically represented as "LG"). This terminology is self-explanatory: The leaving group is the group that leaves during substitution and elimination reactions.

The carbon atom of the "C-X" bond is familiarly classified according to its degree of substitution (methyl, primary, secondary, and tertiary, depending on whether there are 0-3 alkyl groups attached to the carbon, respectively). When the group attached to the "C-X" carbon is a benzene ring, the carbon is called benzylic; when the group attached to the "C-X" carbon is a double bond, the carbon is called allylic; and when the group attached to the "C-X" carbon is a triple bond, the carbon is called propargylic. Finally, when at least one of the groups attached to the "C-X" carbon is a heteroatom (N, O, S, etc.), then the carbon is merely classified as a heteroatom-substituted carbon. The formal degree of substitution classifications (primary, secondary…) cannot be used meaningfully in these cases, as it only applies to cases where there are only hydrogen and carbon atoms substituent groups being counted. Examples of compounds containing sp^3 carbon-halogen bonds are shown in Figure 0704, along with the classification of the *alpha* carbon atom of the "C-X" bond and the number of *beta* hydrogen atoms there are with respect to the leaving (X) group.

Figure 0704

Classification of the *alpha* carbon and the *beta* hydrogen count for C-X bonds.

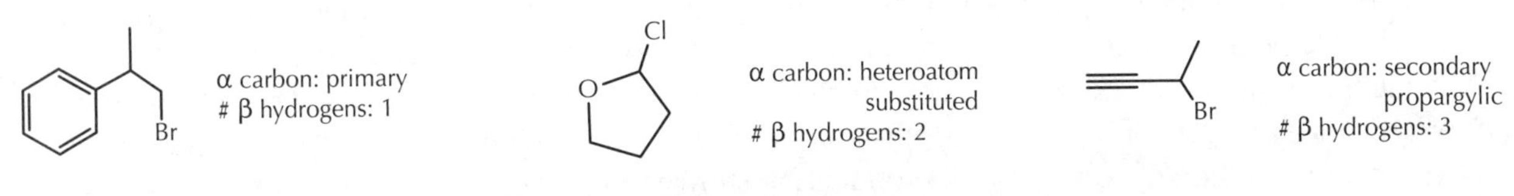

α carbon: secondary
β hydrogens: 4

(Ph)₃CBr

α carbon: tertiary
benzylic
β hydrogens: 0

α carbon: heteroatom
substituted
β hydrogens: 3

α carbon: secondary
allylic
β hydrogens: 2

α carbon: secondary
benzylic
β hydrogens: 3

α carbon: tertiary
β hydrogens: 6

α carbon: primary
β hydrogens: 1

α carbon: heteroatom
substituted
β hydrogens: 2

α carbon: secondary
propargylic
β hydrogens: 3

Stereochemical outcome was not a concern when substitution reactions were introduced in Chapter 4; now it is. If the *alpha* carbon in a nucleophilic substitution reaction is a stereocenter, then there are two (and only two) possible outcomes: the (R)- and/or (S)- geometry. There are two possible stereochemical outcomes, called retention and inversion of configuration. These terms refer to the absolute geometrical arrangements of the groups, not the label.

Retention of configuration is when the Lewis basic partner ends up occupying the same spatial position that the leaving group had.

Inversion of configuration is when the Lewis basic partner ends up in the configuration that is not the retention of configuration (if this position was the only stereocenter in the molecule, then it would be the enantiomer of the retention production; if there were other stereocenters, then it would be the diastereomer at that site).

Examples of substitution reactions with retention and inversion of configuration are shown in Figure 0705 (elimination reaction products are not included).

Figure 0705

Products of retention and inversion of stereochemistry in S_N reactions.

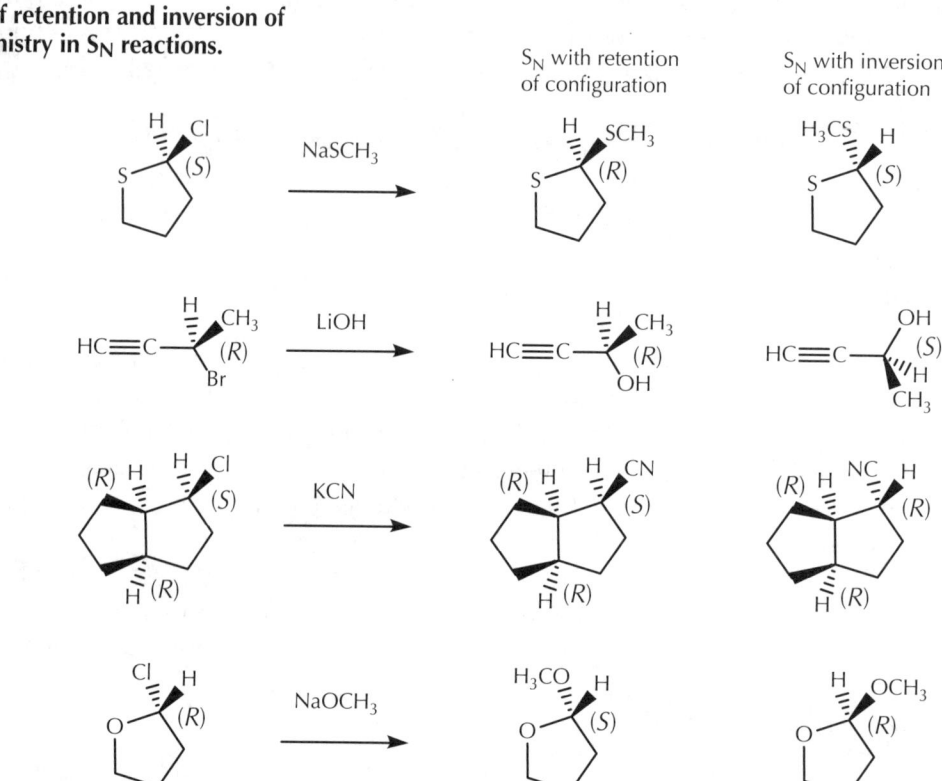

While it is true that retention of configuration will often mean that an (R)-geometry in the starting compound becomes an (R)-geometry in the product, and that inversion of configuration will convert an (R)-geometry to an (S)-geometry, this is not how the terms retention and inversion are defined. Recall that the (R)- and (S)-labels are defined by the relative priorities of the attached groups, and this means that it is quite possible to see cases where an (R)-geometry in a starting material undergoes inversion of configuration to give an (R)-geometry in the product, because the labels are derived from the IUPAC priority rules. Examples of nucleophilic substitution (S_N) reactions with the geometrical labels of the retention and inversion of configuration products are shown in Figure 0705, including those where retention, for instance, ends up with the same and opposite stereochemical label as the starting material.

Elimination (E) reactions can also produce stereoisomeric products if the newly formed double bond has the possibility for (E)- and (Z)-diastereomers. The criteria for identifying possible (E)- and (Z)-diastereomers are unchanged, and need to be applied to the potential double bond formed when a leaving group and a hydrogen atom are lost, on a case-by-case basis from each set of *beta* hydrogen atoms. The examples

7.1 Possible Outcomes in Substitution and Elimination Reactions of Polar Covalent Bonds **523**

in Figure 0706 highlight the *beta* hydrogen atoms with respect to each leaving group, and then consider whether the elimination (E) reaction can produce possible (*E*)- and (*Z*)-diastereomers (substitution reaction products are ignored).

Figure 0706
Possible (*E*)- and (*Z*)-diastereomers in products resulting from E reactions.

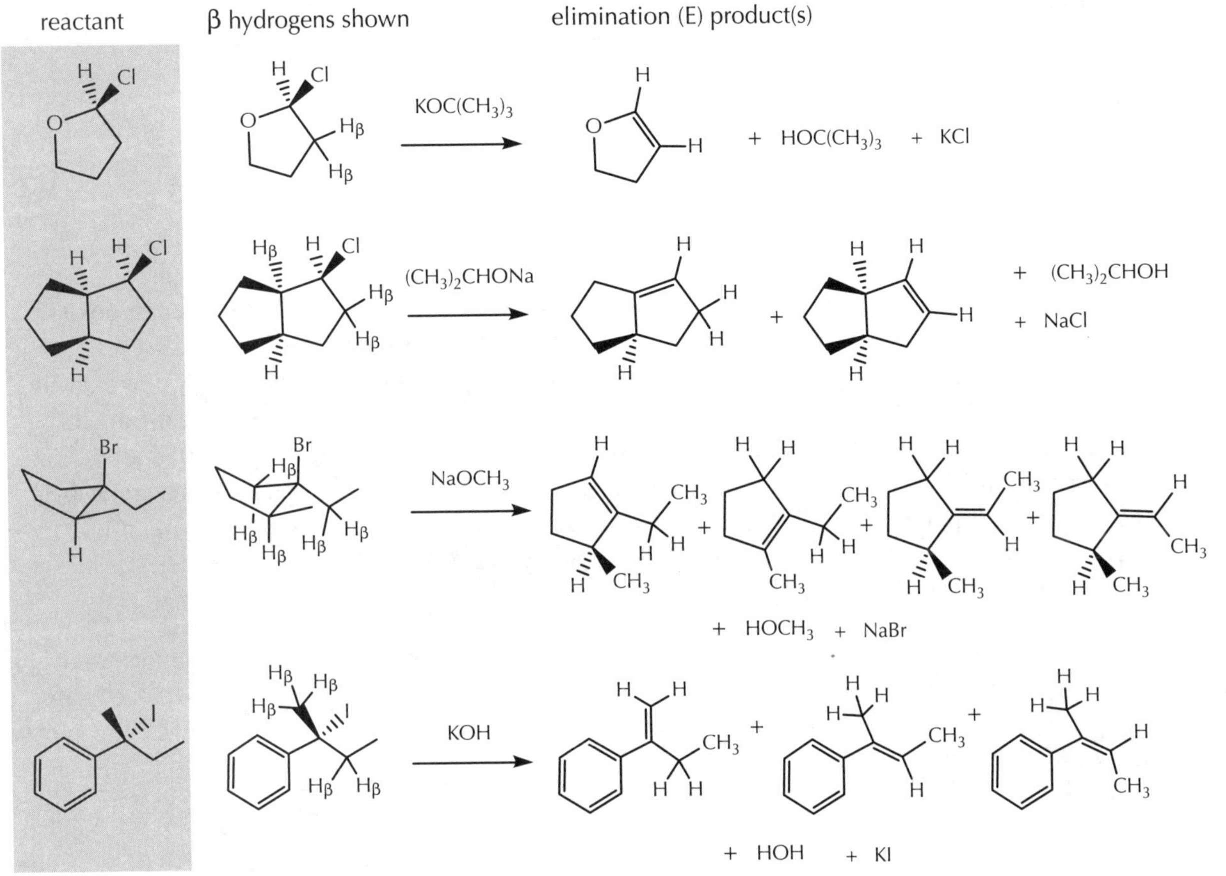

Before looking at examples in which the full array of nucleophilic substitution (S_N) and elimination (E) reaction products are predicted, there are two important features about the geometry of double bonds in small rings that you need to remember, because it can constrain the number of possible elimination reaction products that you can predict.

The first of these reminders is embedded in the first three examples in Figure 0706. Initially introduced in Figure 0518, a double bond inside a small ring can only exist in one of its (*E*)- or (*Z*)- geometries. In all three of these examples, the double bond contained within the 5-membered ring can only have the one geometry that is shown. Whereas, in the third and fourth examples, when the possibility of (*E*)- and (*Z*)- diastereomers exists outside of the ring, both are shown.

The second reminder is the general exclusion of linear (sp) atoms in small rings. This topic was an issue that needed to be considered when deciding which electrons could be delocalized (see Figure 0224). Doing an elimination reaction in which two double bonds will share a central, sp atom is not possible in a small ring. A comparison between an acyclic molecule and its cyclic counterpart illustrates this limitation when predicting the possible outcomes from elimination (E) reactions (Figure 0707 on the next page).

Figure 0707
Possible products from E reactions of cyclic compounds may be limited.

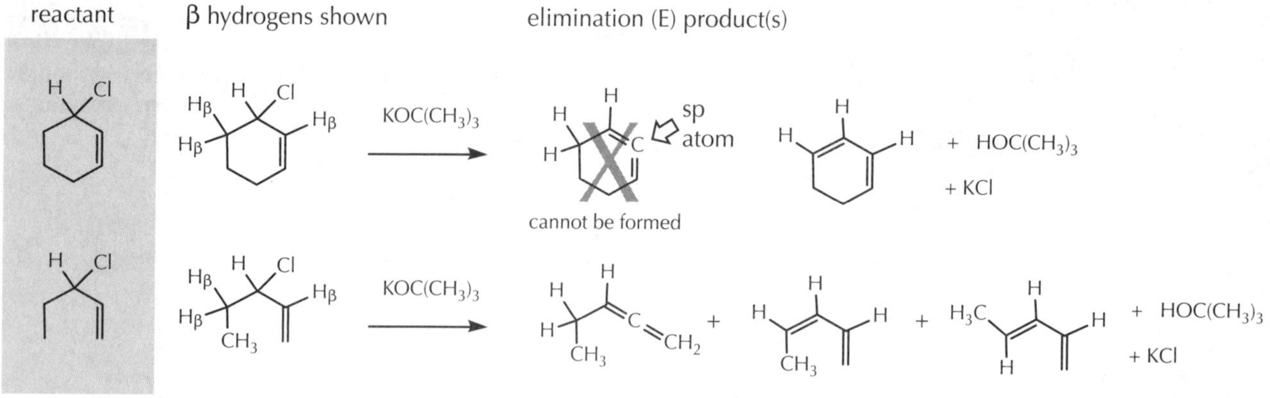

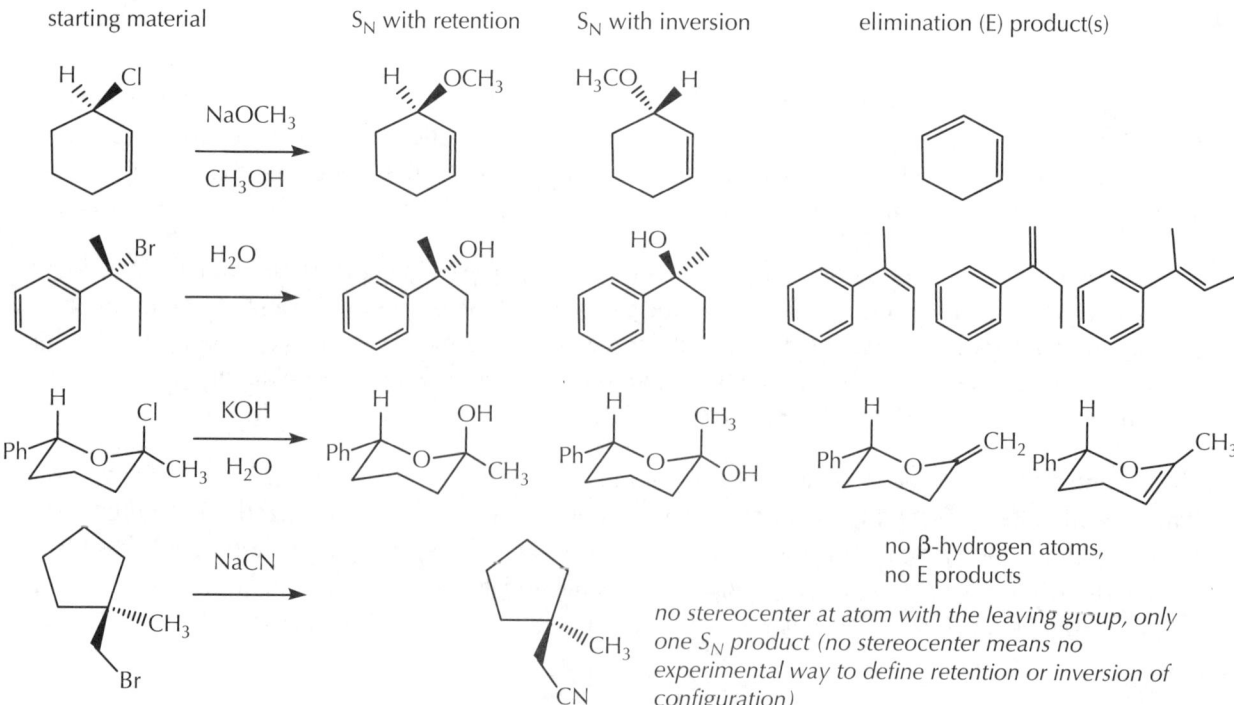

For any given combination of an sp³ "C-X" bond and a Lewis base, there are a finite number of possible nucleophilic substitution (S_N) and elimination (E) products that can be predicted. If the "C-X" bond is part of a stereocenter, then there are two possible substitution products (S_N with inversion, and S_N with retention); if the "C-X" bond is not part of a stereocenter, then there is only one possible S_N product. The number of elimination products depends on the number of different *beta* hydrogen atoms, as well as whether or not both (*E*)- and (*Z*)-diastereomers are possible. At the low end, a compound with no *beta* hydrogen atoms will have no possible elimination reaction products. At the other extreme, a tertiary "C-X" with three different alkyl groups, all of which have possible (*E*)- and (*Z*)-diastereomers, gives the maximum number of elimination products, namely, six.

The examples in Figure 0708 show a range of outcomes from the question: Including stereoisomers, what is the full inventory of possible nucleophilic substitution (S_N) and elimination reaction products in each case?

Figure 0708
Possible products from S_N and E reactions.

Halogen atoms are not the only predictable leaving groups, and in Section 7.2D we will extend the list of leaving groups that you need to recognize. However, because the concept of a leaving group can be generalized, only one new experimental observation of seeing something behave as a leaving group is all you need to make predictions about a large number of new examples that would incorporate that atom or group of atoms. "Leaving group," in other words, is an idea that can be applied to anything that behaves accordingly. For instance, you are provided (Figure 0709) with the observation that the hydroxyl group in methanol (CH_3OH) reacts with hydrochloric acid (HCl) to give chloromethane (CH_3Cl).

Figure 0709

Predicting products from S_N and E reactions based upon a new precedent.

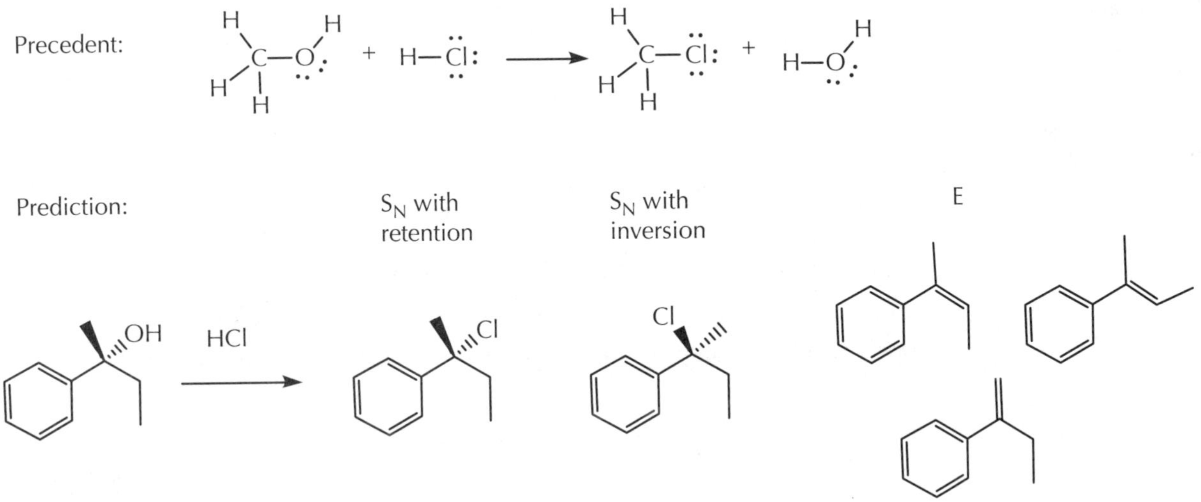

The first conclusion is that this reaction is a nucleophilic substitution, where the hydroxyl group has acted as a leaving group and chloride is the Lewis base. This observation becomes a precedent. In the absence of other experimental information, one prediction (hypothesis) you can make is that other hydroxyl groups on sp^3 carbon atoms might also participate in substitution reactions with HCl, that stereoisomers would be possible if the hydroxyl group was part of a stereocenter, and that elimination reactions would be possible if there were *beta* hydrogen atoms with respect to the hydroxyl group. The hydroxyl group in the second example is at a stereocenter, and so S_N reactions with retention and inversion of configuration are possible, and there are five β hydrogens, which means elimination reactions are possible. The full inventory of potential reaction products is shown for this example. Moreover, products can be predicted for any combination of an sp^3 C-OH with added HCl, and then the experiment carried out to see what happens.

C. Substitution Reactions: Possible Mechanistic Pathways

A reaction mechanism is the story about a reaction in which we define and explain three things: (1) which bonds changed (broken and formed), (2) the timing of those bond-change events (what happened first, second, third…), and (3) the relative rates of the bond-change events. These are all of the features we include in an energy diagram.

Because a substitution reaction, by definition, involves the exchange of one sigma bond for another, the mechanistic questions are how the two bonds changed, the order in which they changed, and the relative ease with which they changed.

In the simplest formulation of a nucleophilic substitution (S_N) at an sp^3 "C-X" bond with a Lewis base (LB:), there are two events: The C-X bond breaks and the C-LB bond forms. With only two bond-change events, there are only three possible pathways along which these two changes can happen, namely:

(S_N Pathway A): in 2 steps: the C-X bond breaks after the LB-C bond forms
(S_N Pathway B): in 2 steps: the C-X bonds breaks before the LB-C bond forms
(S_N Pathway C): in 1 step: the C-X bond breaks at the same time the LB-C bond forms

Using the most generic symbolism for the electrophile ("C-X") and the nucleophile (LB:), the curved arrow representations for these three options are presented in Figure 0710, along with the implied intermediate, if there is one, based on these descriptions. Geometrical considerations are being ignored for the moment, and so no particular information about stereochemistry is implied by these drawings.

Figure 0710

Three possible mechanistic pathways to consider for S_N at sp³ C-X bonds.

S_N Pathway A: in 2 steps: the C-X bond breaks after the LB-C bond forms (octet violation at C)

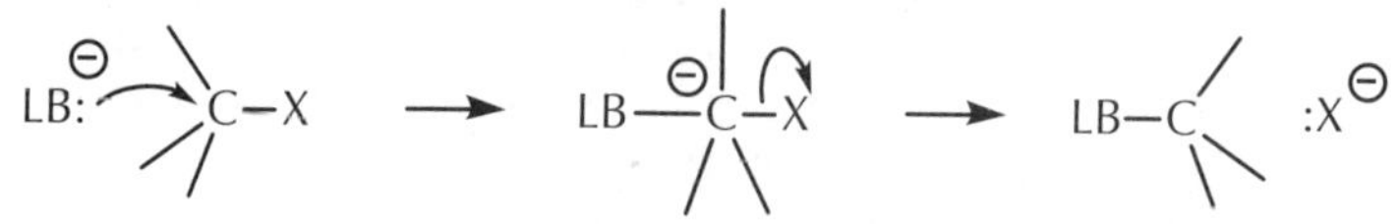

S_N Pathway B: in 2 steps: the C-X bonds breaks before the LB-C bond forms

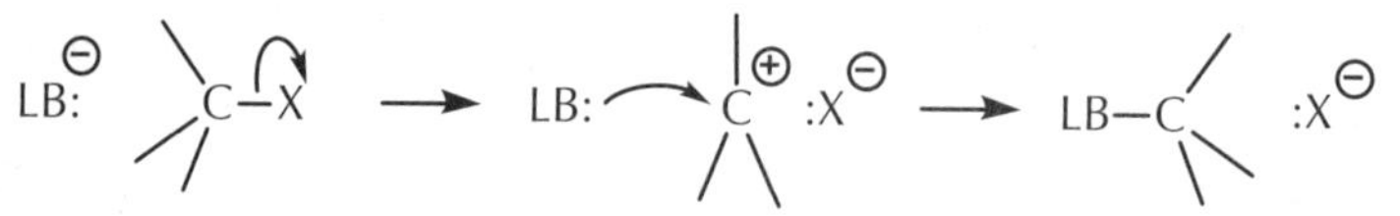

S_N Pathway C: in 1 step: the C-X bond breaks at the same time the LB-C bond forms

One of these three mechanisms, Pathway A, can be readily rejected on the fundamental principle that the carbon of the "C-X" bond would have to have a hypervalent octet violation of 10 bonding electrons around it. This leaves a 2-step mechanism (Pathway B), in which the "C-X" bond breaks to give an open shell carbocation intermediate, and this intermediate is then subsequently captured by the Lewis base; and a 1-step mechanism (Pathway C), in which a collision between the Lewis base and the carbon of the "C-X" bond occurs, simultaneously forming the LB-C bond and breaking the C-X bond.

When there is more than one step, there is another question: Which step is faster? This means that there are two versions of Pathway B, depending on which of the two steps is the slower, rate-determining step. Either the bond-breaking step is the slow step, with the highest absolute peak of the two activation energies, or the bond-forming step is the slow step. Pathway C has only one step. Energy diagrams can be used to represent these mechanistic differences. Figure 0711 shows the energy diagrams for the two versions for Pathway B and for Pathway C.

The first version of Pathway B, where the first step is rate-determining, is well understood. The formation of the high energy, open shell carbocation intermediate is the slow step, and the observed rate at which the product forms will only depend on the concentration of the "C-X" partner and not on the concentration of the Lewis base, which only participates in the second, faster step. A reaction in which the overall rate only depends on one of the molecular components is called a unimolecular reaction, and so this mechanistic option is called a unimolecular nucleophilic substitution pathway, and it is abbreviated as S_N1.

The second version of Pathway B, where the second step is rate-determining, is not a situation that is normally studied. In this option, the carbocation intermediate forms more rapidly than its capture, and the rate at which the product forms will depend on a variety of factors, making the direct study of the reaction more complex than in the first version. The mechanistic pathway is still Pathway B, and the

7.1 Possible Outcomes in Substitution and Elimination Reactions of Polar Covalent Bonds 527

Figure 0711

Unimolecular and bimolecular options for S_N at sp^3 C-X bonds.

S_N Pathway B: in 2 steps (2 options)
the C-X bond breaks before the LB-C bond forms

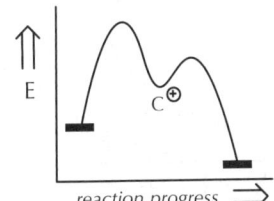

S_N Pathway C: in 1 step (1 option)
the C-X bond breaks at the same time the LB-C bond forms

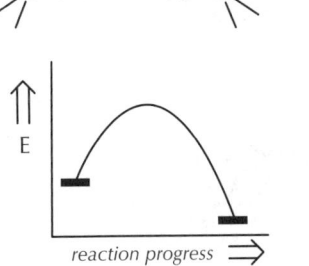

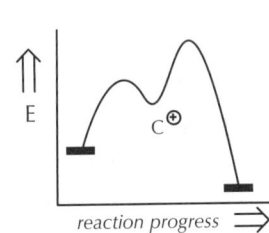

S_N Pathway B
2 steps: step 1 rate-determining

RATE = k[C-X]

S_N1 (rate proportional to [C-X])

S_N Pathway B
2 steps: step 2 rate-determining

Rate studies are complex

S_N Pathway C
1 step

RATE = k[C-X][:LB]

S_N2 (rate proportional to [C-X] & [:LB])

same as the S_N1, above: The "C-X" bonds breaks to give an open shell carbocation intermediate, and this intermediate is captured by the Lewis basic partner.

In Pathway C, a single mechanistic step involves a collision between the electrophilic "C-X" partner and the Lewis base. The overall rate will depend on the number of collisions that can occur, and the rate of formation of the product will depend on the concentration of both reactive partners. This mechanistic option is called a bimolecular nucleophilic substitution pathway, and it is abbreviated as S_N2.

As a matter of fact, all of these mechanistic options are available for every pairing of sp^3 C-X bonds with Lewis bases. What makes the chemistry interesting is that these pathways can vary greatly in their relative energies of activation. Depending on the particular pairing of reaction partners and the experimental conditions under which they are combined, the outcome can strongly favor one of these pathways or the relative rates might be comparable, so the pathways would be competing. You cannot actually know what is happening unless you have experimental evidence, but after many thousands of reactions have been performed, trends appear and a reasonably reliable predictive model has to be created, which we will develop in Section 7.3.

After years of study, the following generalizations can be made about the most commonly observed versions of these reactions.

In a theoretically ideal unimolecular nucleophilic substitution (S_N1) mechanism between a Lewis base and an sp^3 "C-X" partner whose carbon is a stereocenter:

    (a) the first step is rate-determining, and forms an sp^2 open shell, carbocation intermediate
    (b) a stereoisomeric mixture results because both sides of the sp^2 carbon are accessible

In a theoretically ideal bimolecular nucleophilic substitution (S_N2) mechanism between a Lewis base and an sp^3 "C-X" partner whose carbon is a stereocenter:

    (a) every productive collision results in an inversion of configuration at the stereocenter
    (b) the collision has a 180° trajectory with an sp^2 carbon at the transition state

The combination of (R)-1-bromo-1-phenylethane and sodium methoxide (NaOCH₃) in methanol is one of those reactions in which these two mechanistic pathways are known to compete, so the head-to-head comparison of these two idealized mechanisms, and their outcomes, is illustrated in Figure 0712.

Figure 0712

S_N1 and S_N2 comparison for (R)-1-bromo-1-phenylethane and sodium methoxide.

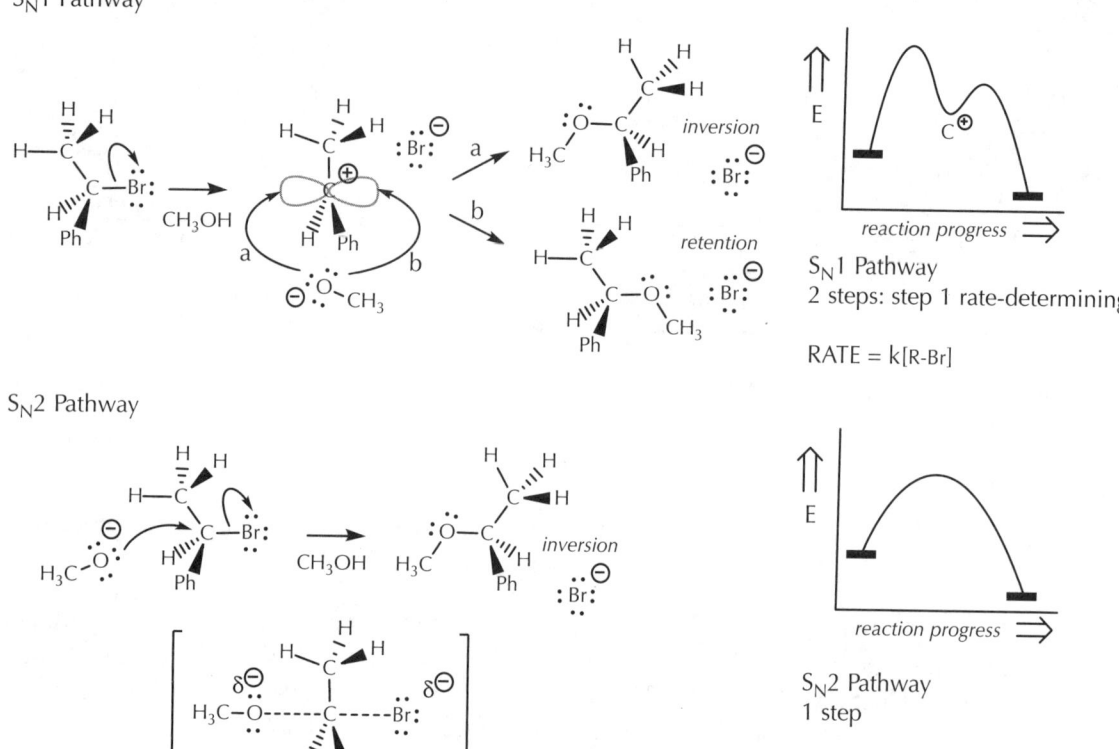

D. Elimination Reactions: Possible Mechanistic Pathways

In the simplest formulation of elimination (E) involving an sp³ "C-X" bond and one of its *beta* hydrogen atoms with a Lewis base (LB:), there are three events: The C-X bond breaks, the C-H is deprotonated by the Lewis base, acting as a Brønsted base, and a pi bond forms. With three bond-change events, there are multiple possible pathways along which these can happen. If we exclude, based on the previous section, the options where the atoms exceed their closed shell electron configurations, there are three possible pathways:

(E Pathway A): in 2 steps: the *beta* C-H is deprotonated by the Lewis base to give the conjugate base of the starting material, then the pi bond forms as the C-X bond breaks

(E Pathway B): in 2 steps: the *beta* C-H is deprotonated by the Lewis base after the C-X bond has been broken to give an intermediate carbocation, and the pi bond forms

(E Pathway C): in 1 step: the *beta* C-H is deprotonated by the Lewis base as the C-X bond breaks, concurrently forming the pi bond

Using the most generic symbolism for the electrophile and the nucleophile, the curved arrow versions of these three options are presented in Figure 0713, along with any implied intermediate, based on the description. Geometrical considerations are being ignored for the moment, and so no particular information about stereochemistry is implied by these drawings.

7.1 Possible Outcomes in Substitution and Elimination Reactions of Polar Covalent Bonds

E Pathway A: in 2 steps: the β C-H is deprotonated by the Lewis base to give the conjugate base of the starting material, then the pi bond forms as the C-X bond breaks

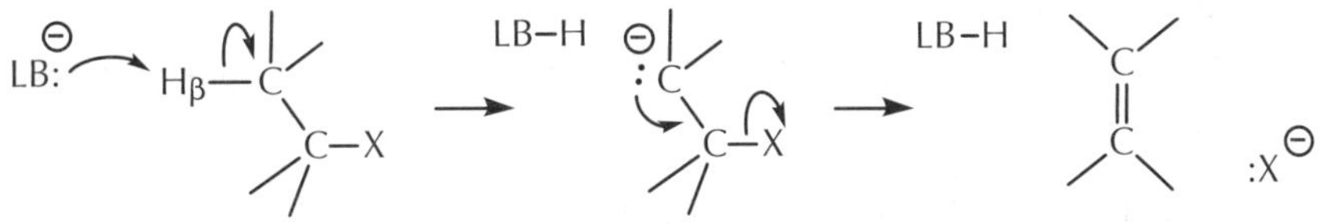

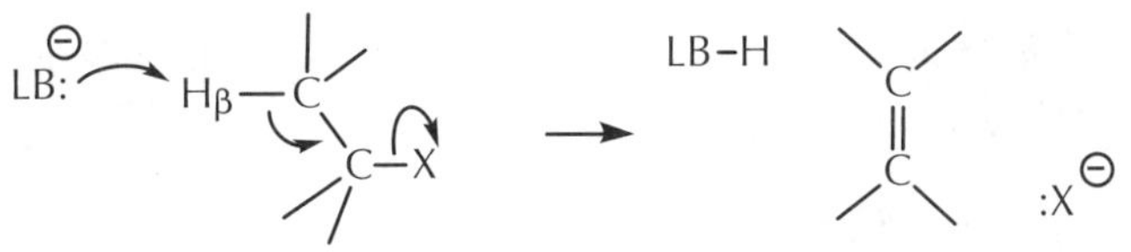

Figure 0713

Three possible mechanistic pathways to consider for E with sp³ C-X bonds.

E Pathway B: in 2 steps: the β C-H is deprotonated by the Lewis base after the C-X bond has been broken to give an intermediate carbocation, and the pi bond forms

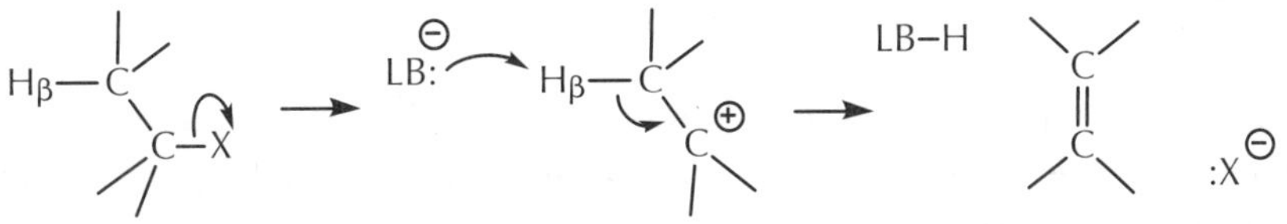

E Pathway C: in 1 step: the β C-H is deprotonated by the Lewis base as the C-X bond breaks, concurrently forming the pi bond

As with the nucleophilic substitution reactions, energy diagrams for the three options can be constructed, along with the implications from the rate experiments. In the 2-step pathways (Pathways A and B), the comparable question exists about which step is rate-determining, which means there are a total of five options to consider. And as with the substitution reactions, the experimental studies narrow the options down to those that are most commonly observed. As a result, we end up with three most common elimination reaction mechanisms for the reactions of sp³ C-X bonds that have *beta* hydrogen atoms. Figure 0714 shows the energy diagrams alongside the mechanisms.

Figure 0714

Unimolecular and bimolecular options for E with sp³ C-X bonds.

E Pathway A: in 2 steps: the β C-H is deprotonated by the Lewis base to give the conjugate base; then the pi bond forms along with the C-X bond breaking

E Pathway B: in 2 steps: the C-X bond breaks to give a carbocation; then the pi bond forms as β C-H bond is deprotonated by the Lewis base

E Pathway C: in 1 step: deprotonation of the β C-H bond is accompanied by the formation of the pi bond and the breaking of the C-X bond

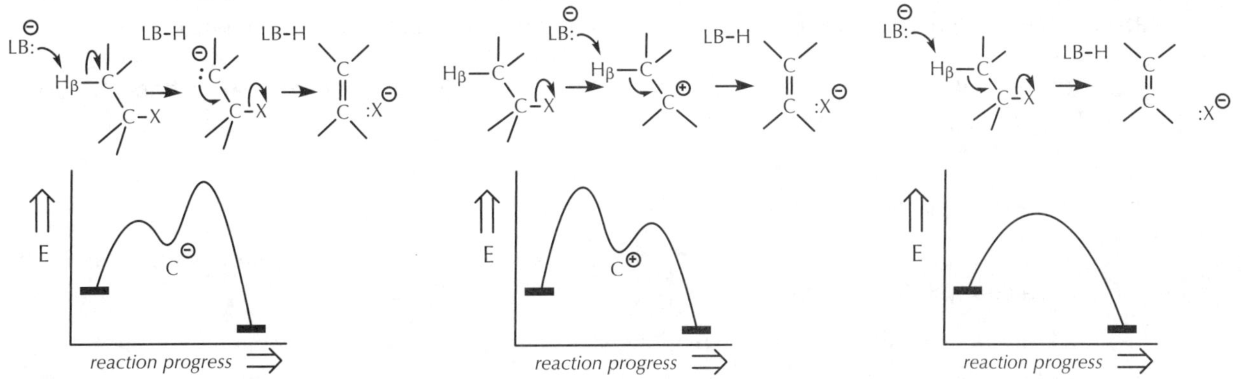

E Pathway A
2 steps: step 2 is often rate-determining

Rate studies are complex

E1cb ("E1 conjugate base")

E Pathway B
2 steps: step 1 rate-determining

RATE = k[C-X]

E1 (rate proportional to [C-X])

E Pathway C
1 step

RATE = k[C-X][:LB]

E2 (rate proportional to [C-X] & [:LB])

In Pathway A, the first step (deprotonation) is usually fast, and the second step (breaking the C-X bond) is rate-determining. The rate at which the products form depends on a variety of factors. Because mechanistic pathway is defined by going through the conjugate base of the starting material, and it is abbreviated as E1cb (unimolecular elimination through the conjugate base). The actual reaction rate equations are complicated and not included here.

In Pathway B, the first step is rate-determining, and it is the loss of the "X" group, which creates the same carbocation intermediate that formed in the S_N1 mechanism by breaking the C-X bond. As in the S_N1 pathway, the rate at which the product forms will only depend on the concentration of the "C-X" partner, so this is the unimolecular elimination pathway, and called E1.

In Pathway C, a single mechanistic step involves all of the bonding changes happening at the same time, during a single bimolecular collision. Analogously with the S_N2 mechanism, the overall rate will depend on the number of collisions that can occur, and the rate of formation of the product will depend on the concentration of both reactive partners. This mechanistic pathway is therefore called the bimolecular elimination, and it is abbreviated as E2.

Not surprisingly, what was true for the substitution pathways is true for the elimination pathways. This paragraph, from earlier, is completely unchanged:

> "All of these mechanistic options are available for every pairing of an sp^3 C-X bond with a Lewis base. What makes the chemistry interesting is that these pathways can vary greatly in their relative energies of activation, so that, depending on the particular pairing of reaction partners and the experimental conditions under which they are combined, the outcome can strongly favor one of these pathways, strongly favor the other, or the relative rates might be comparable so they would be competing. You cannot actually know what is happening unless you have experimental evidence, but after many thousands of reactions have been performed, trends appear and a predictive model can be created, which we will do in Section 7.3."

And again, after years of study, the following generalizations can be made about the most commonly observed mechanistic pathways and stereochemical outcomes:

E1cb: In the ideal unimolecular elimination conjugate base (E1cb) mechanism, in a structure where the new pi bond has both (E)- and (Z)- stereoisomeric outcomes:

(a) deprotonating the *beta* hydrogen (step 1) is faster than the loss of the "X" group
(b) a stereoisomeric mixture results from the carbanion (conjugate base) intermediate

E1: In the ideal unimolecular elimination (E1) mechanism, in a structure where the new pi bond has both (E)- and (Z)- stereoisomeric outcomes:

(a) deprotonating the beta hydrogen (step 2) is faster than the loss of the "X" group
(b) a stereoisomeric mixture results from the carbocation intermediate

E2: In the ideal bimolecular elimination (E2) mechanism, in a structure where the new pi bond has both (E)- and (Z)- stereoisomeric outcomes:

(a) only productive collisions, in which all events can happen, result in forming a pi bond
(b) the elimination rate is favored when the bonds line up in the transition state: the rate is fastest when the beta hydrogen and the C-X bond are in an anti (180°) relationship; an eclipsed (0°) relationship is next fastest; slower rate for other angles, with no productive reaction when the *beta* hydrogen and the C-X bond are at 90°

7.1 Possible Outcomes in Substitution and Elimination Reactions of Polar Covalent Bonds

The same combination of (R)-1-bromo-1-phenylethane and sodium methoxide (NaOCH₃) in methanol used to illustrate a specific S_N1 and S_N2 comparison (Figure 0712) is also observed to give elimination products in this multifaceted competition, where E1 and E2 are competing with one another alongside the S_N1 and S_N2 competition. E1cb is rarer, and requires some special structural features, so it will be considered separately in Section 7.2C. The side-by-side comparison of the two idealized E1 and E2 mechanisms, and their outcomes, is illustrated in Figure 0715.

Figure 0715

E1 and E2 comparison for (R)-1-bromo-1-phenylethane and sodium methoxide.

E1 Pathway

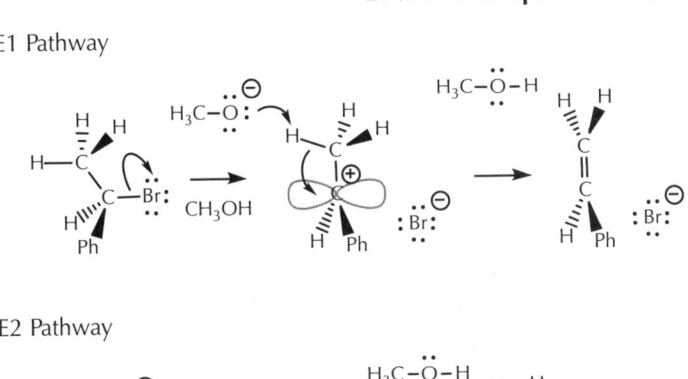

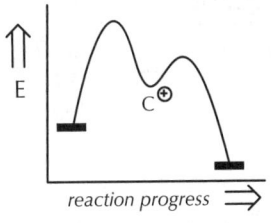

E1 Pathway
2 steps: step 1 rate-determining

RATE = k[R-Br]

E2 Pathway

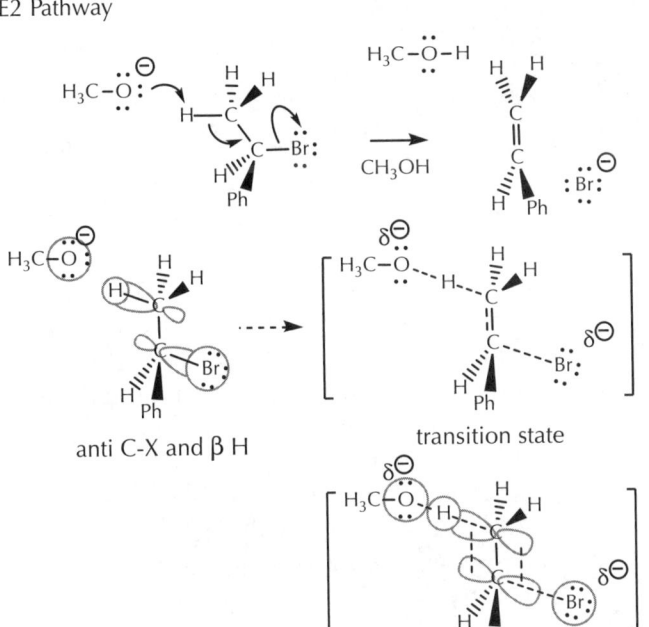

anti C-X and β H   transition state

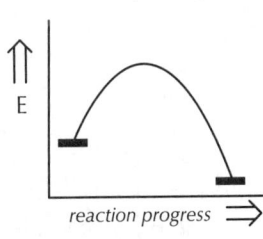

E2 Pathway
1 step

RATE = k[R-Br][CH₃O⁻]

Without experimental results, or a predictive model to help you sort out the factors that might favor one of these mechanistic pathways over the other, you can do two useful things at this point.

First, prediction: You could predict, without regard to mechanism, the full array of substitution and elimination reaction products that might form from any combination of a compound with an sp³ C-X bond plus a Lewis base.

Second, application: If directed, you could apply any of these four mechanisms (S_N1, S_N2, E1, E2) to tell what subset of the full array might form.

One implication from the second statement is quite important to understand. Applying any one (or more) of the specific mechanisms will not predict any products that are not part of the full array of possible products. At the most, the outcome might be a prediction for all of the possible outcomes, but it can never be more than that.

The following example illustrates the principles of "prediction" and "application" described in the previous paragraphs.

Under a certain set of experimental conditions, (1S, 2R)-1-chloro-1,2-diphenylpropane is treated with sodium methoxide in methanol, and the array of observed products is, as in the previous specific example, formed through a set of competing substitution and elimination reaction pathways.

All possible products are formed (Figure 0716).

Figure 0716

Full array of S_N and E from (1S, 2R)-1-chloro-1,2-diphenylpropane and sodium methoxide.

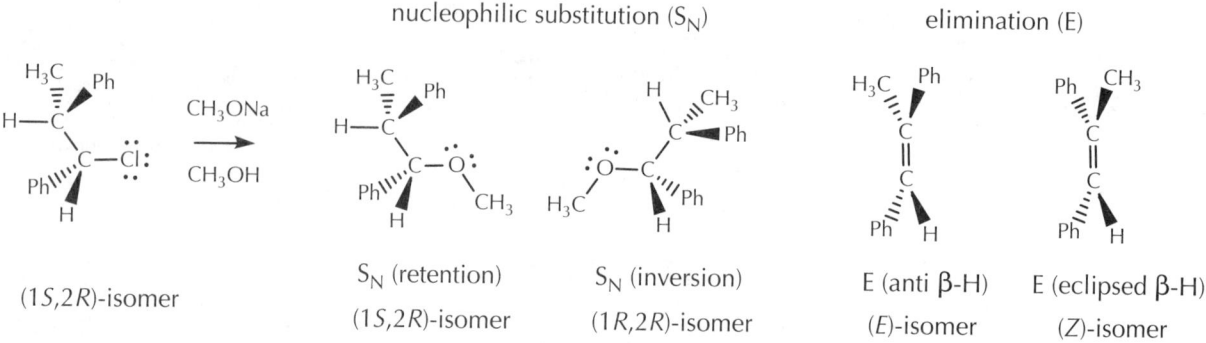

There are four outcomes.

The "C-X" bond (C-Cl in this case) is part of a stereocenter, and so both S_N with retention and S_N with inversion are possible. Because the molecule has two stereocenters in it, these two products are diastereomers. There is only one beta hydrogen atom, and the elimination product that can form does have possible (E)- and (Z)- stereoisomerism, and so there are two potential elimination reaction outcomes, giving a total of four possible substitution and elimination reaction products.

As you might expect, this is not a terribly practical experimental outcome, because we would rather see a reaction that gives as close to one reaction product as possible, but the example illustrates the principle that a specific substitution or elimination mechanistic pathway will not uncover any product that was not anticipated as part of the full array of possible products.

If we now change the conditions of the question, and ask, for each of the four specific mechanistic pathways, S_N1, E1, S_N2, and E2, what is the predicted result from this same combination of reagents. The answer, in all four cases, must be a subset of the molecules in Figure 0716, because that represents the full array of outcomes regardless of mechanism.

As illustrated in Figure 0717, the features of any one of these specific mechanistic pathways can be applied to these starting materials and you only predict a subset of the molecules from among the full array.

7.1 Possible Outcomes in Substitution and Elimination Reactions of Polar Covalent Bonds

Figure 0717

Subset of the full array of S_N and E from (1*S*, 2*R*)-1-chloro-1,2-diphenylpropane and sodium methoxide predicted from the specific S_N1, E1, S_N2, and E2 mechanisms.

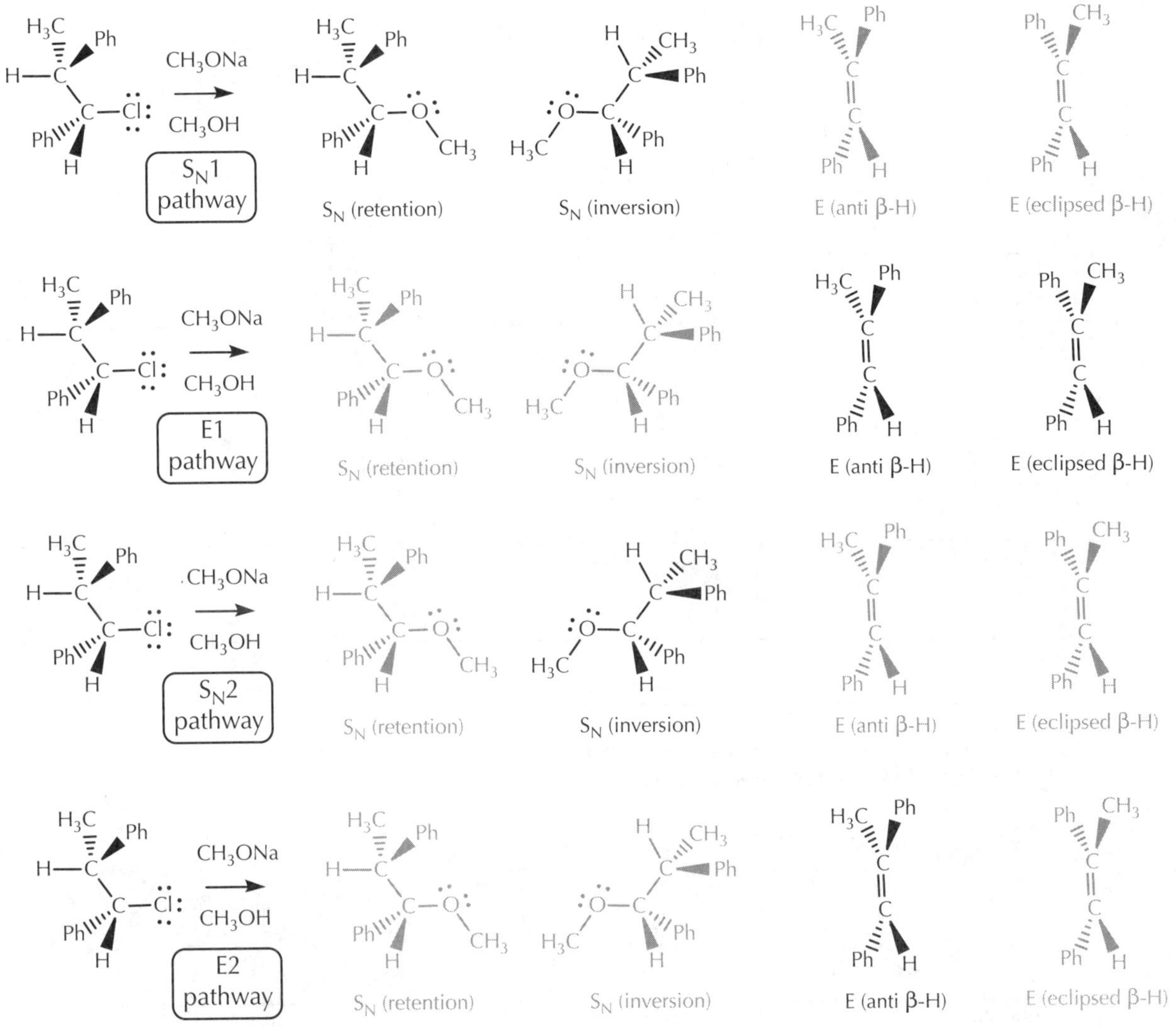

PRACTICE QUESTIONS

7.01 Each of the following molecules contains an sp³ carbon-halogen bond. Classify the electrophilic carbon (1°, 2°, 3°, heteroatom substituted, and allylic, benzylic, or propargylic, as appropriate), and identify the number of β-hydrogen atoms with respect to the leaving group.

(a)

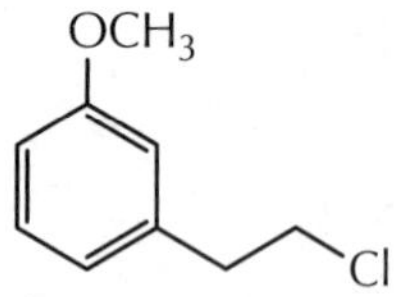

classification:

number of β-hydrogens:

(b)

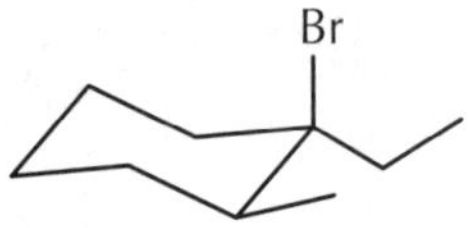

classification:

number of β-hydrogens:

(c)

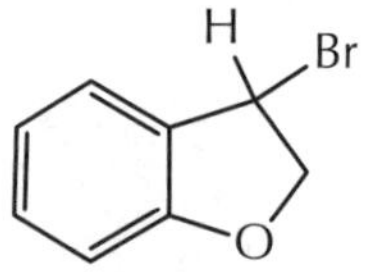

classification:

number of β-hydrogens:

(d)

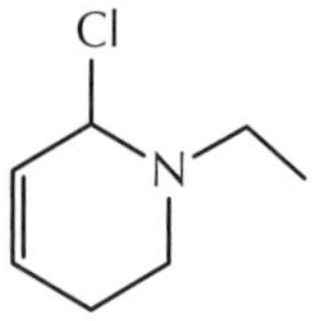

classification:

number of β-hydrogens:

(e)

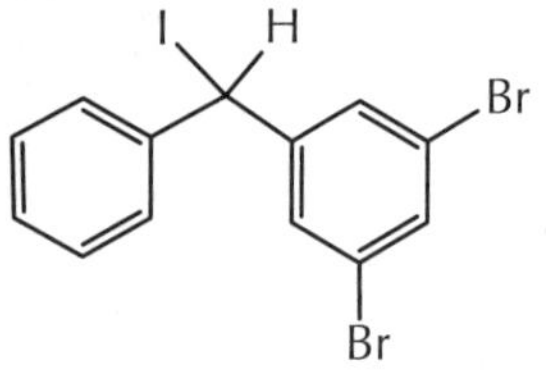

classification:

number of β-hydrogens:

(f)

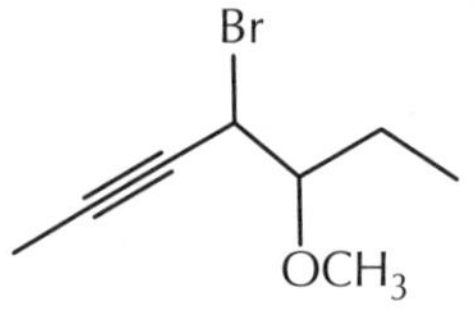

classification:

number of β-hydrogens:

7.02 Including stereoisomers, as well as restrictions on the geometry of double bonds, tell how many substitution and elimination reaction products are possible from each of the following compounds.

(a)

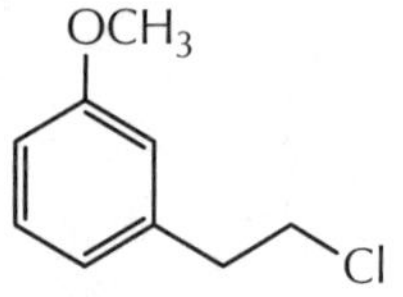

substitution:

elimination:

(b)

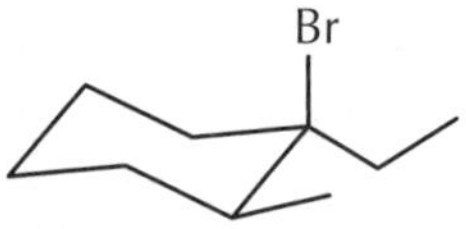

substitution:

elimination:

(c)

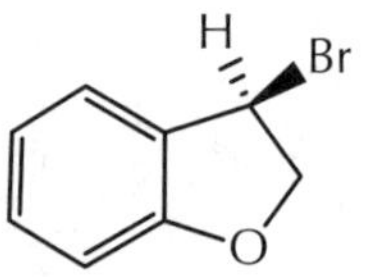

substitution:

elimination:

(d)

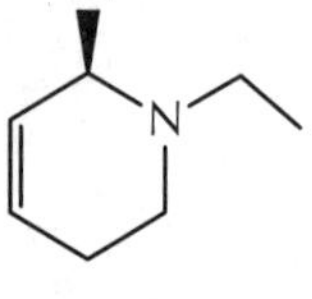

substitution:

elimination:

(e)

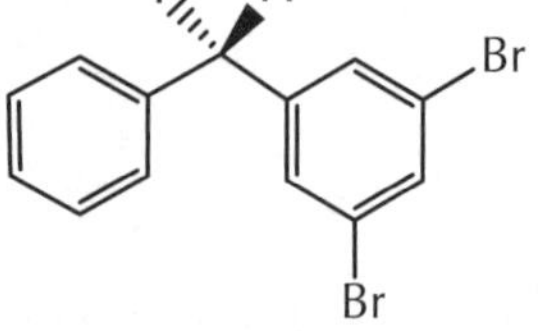

substitution:

elimination:

(f)

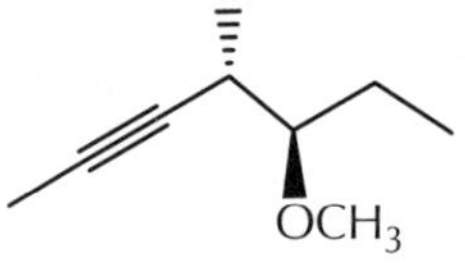

substitution:

elimination:

7.03 Provide the structures for the full array of substitution and elimination reaction products, including stereoisomers, that are possible in each of the following reactions.

(a) 3-methoxyphenethyl chloride + NaCN

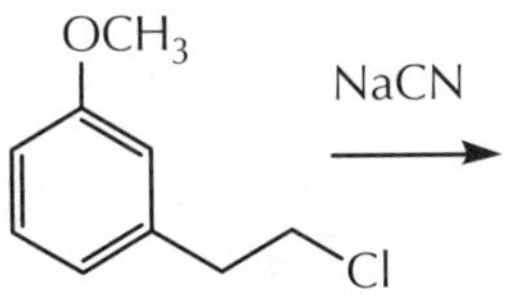

(b) 1-bromo-1-ethylcyclohexane (or similar) + KOH

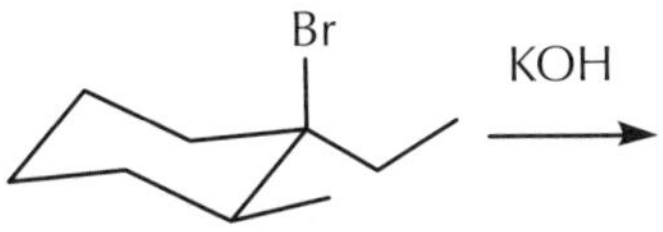

(c) (3-bromo-2,3-dihydrobenzofuran, with H and Br shown as wedge/dash) + KSCH$_2$CH$_3$

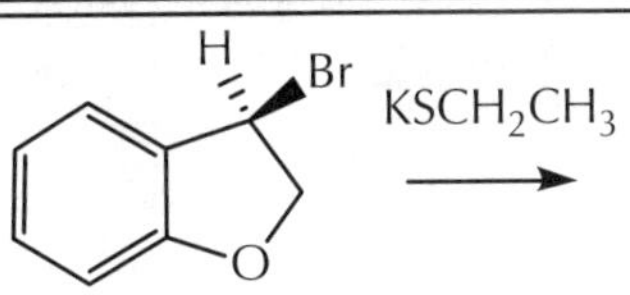

(d) (3-bromo-4-methoxy-hex-1-yne with stereochemistry: Br on dashed wedge, OCH$_3$ on dashed wedge) + NaOCH$_3$

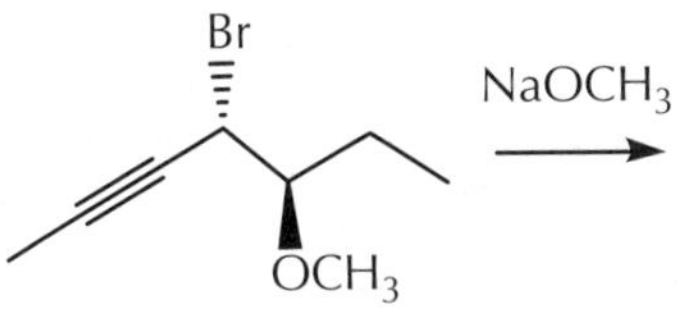

(e) 5-chloro-2(5H)-furanone (Cl on wedge) + LiBr

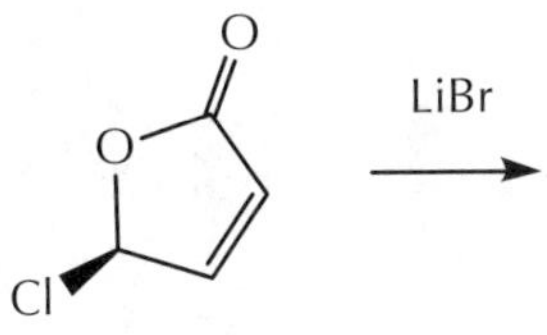

536 CHAPTER 7 Substitution and Elimination Reactions of Polar Sigma Bonds

7.04 The three examples, just below, exhibit groups of atoms that can behave as leaving groups in substitution (and elimination) reactions, exactly the same as halogen atoms. Given these simple precedents, provide the full array of substitution and elimination reaction products that are anticipated in each of these chemical reactions.

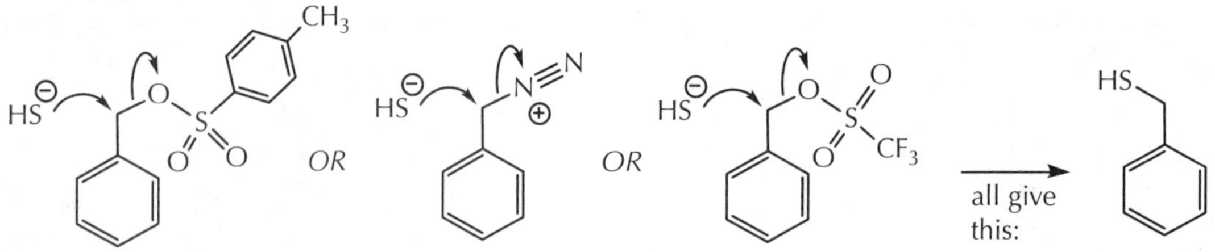

(a) [structure of tetrahydropyran with N₂⁺ leaving group] + LiCN →

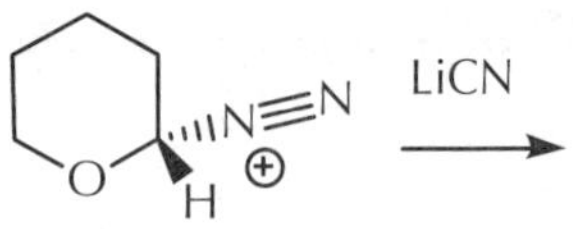

(b) [sec-butyl triflate structure] + H-C(=O)-ONa →

(c) [tetralin-2-yl tosylate structure] + NaSCH₂CH₃ →

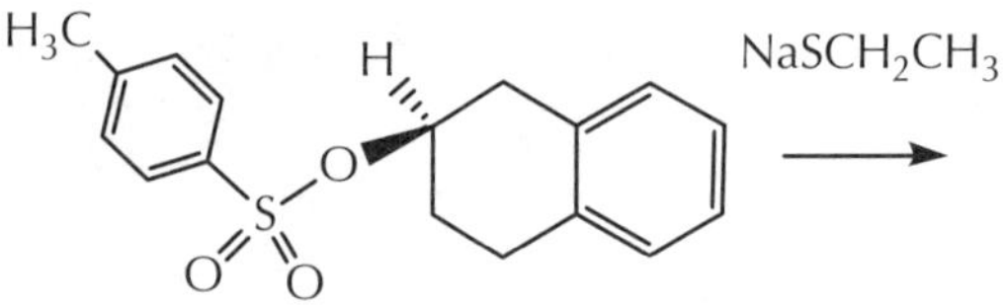

(d) [alanine with diazonium substituent, H₃C, H, N₂⁺, COOH] + KBr →

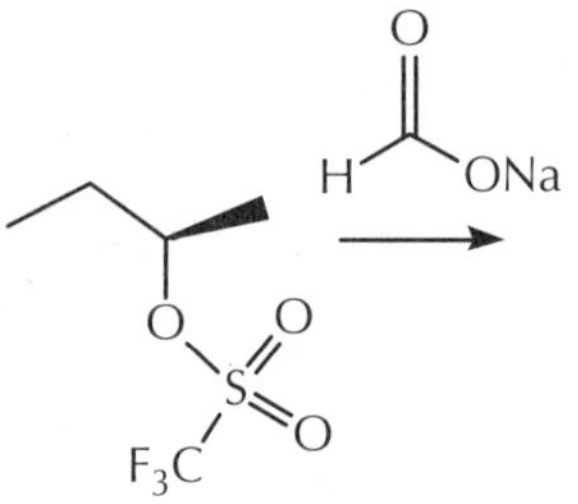

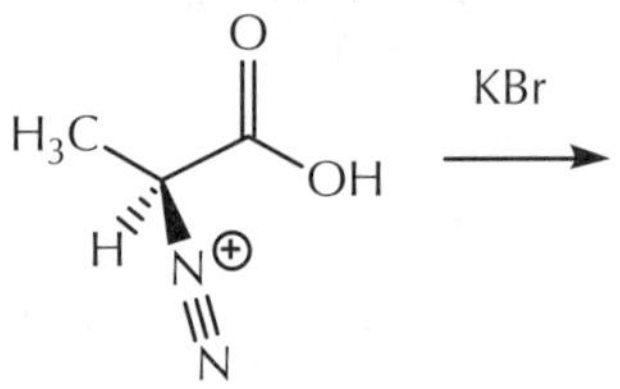

7.05 For each of the following examples, use the specified mechanistic pathway to tell which of the possible substitution and/or elimination reaction products is expected to result. Show the curved arrow mechanism for the reaction. Stereochemistry should be included as needed, and in drawing the mechanisms, any critical three-dimensional relationships between the molecules should be included (e.g., showing a collisional pathway for S_N2 that is appropriate to the inversion of configuration).

(a) Ph-CH(H)(CH₃)-Br + KSH, S_N2

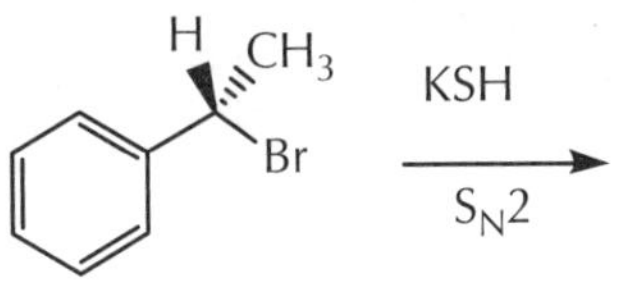

(b) H₃C-C(Cl)(Ph)(CH₂CH₃) + NaN₃, S_N1

(c) 2-chlorotetrahydrofuran + NaOH, S_N2 & E2

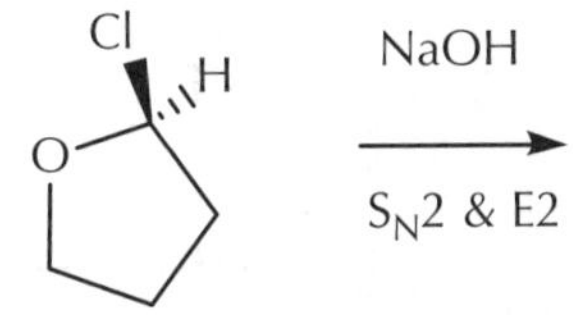

(d) 2-chlorotetrahydrofuran + NaOH, S_N1 & E1

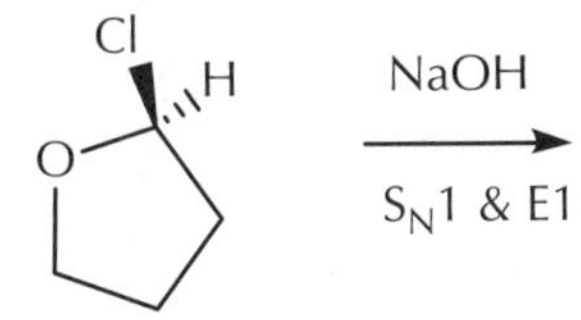

(e) 9-(chloromethyl)fluorene + NH₃, E1cb

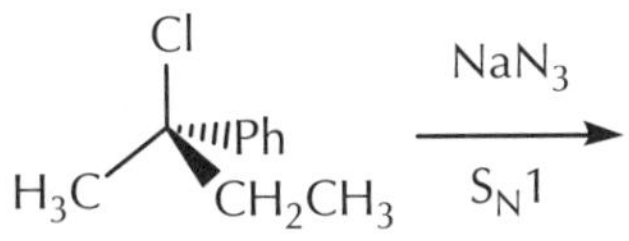

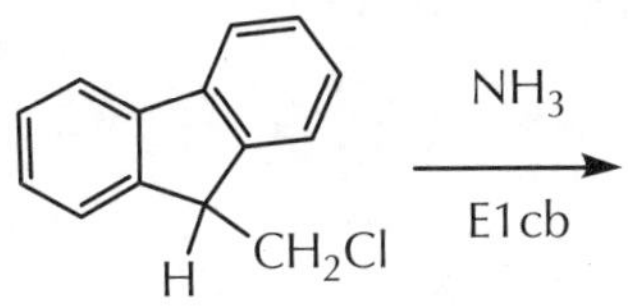

538 **CHAPTER 7** Substitution and Elimination Reactions of Polar Sigma Bonds

7.06 Each of the following reactions gives a total of four possible different substitution and elimination reaction products. Draw the four products and then identify which pathway (or pathways) could be responsible for the product's formation.

(a)

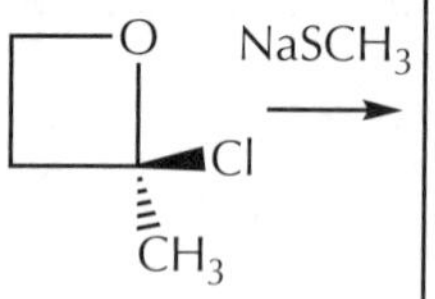

[Structure: oxetane with Cl and CH₃ substituents] + NaSCH₃ →

S_N2 E2 S_N1 E1 S_N2 E2 S_N1 E1 S_N2 E2 S_N1 E1 S_N2 E2 S_N1 E1
☐ ☐ ☐ ☐   ☐ ☐ ☐ ☐   ☐ ☐ ☐ ☐   ☐ ☐ ☐ ☐

(b)

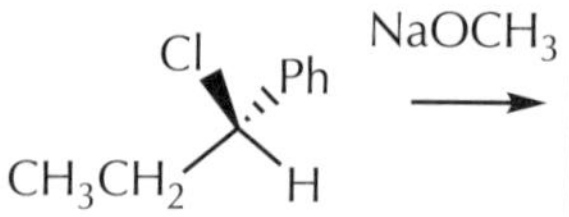

[Structure: CH₃CH₂-C(Cl)(Ph)(H)] + NaOCH₃ →

S_N2 E2 S_N1 E1 S_N2 E2 S_N1 E1 S_N2 E2 S_N1 E1 S_N2 E2 S_N1 E1
☐ ☐ ☐ ☐   ☐ ☐ ☐ ☐   ☐ ☐ ☐ ☐   ☐ ☐ ☐ ☐

(c)

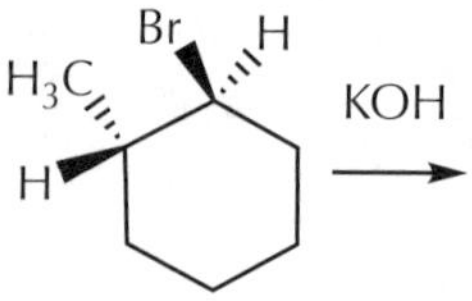

[Structure: cyclohexane with Br and CH₃ substituents] + KOH →

S_N2 E2 S_N1 E1 S_N2 E2 S_N1 E1 S_N2 E2 S_N1 E1 S_N2 E2 S_N1 E1
☐ ☐ ☐ ☐   ☐ ☐ ☐ ☐   ☐ ☐ ☐ ☐   ☐ ☐ ☐ ☐

Draw the two chair conformations for the starting material in part (c). One of them will be involved in the fastest E2 reaction. Select which conformation that will be true for and then, in the space below it, show the transition state for the E2 reaction. Under the other conformation, show the transition state for the S_N2 reaction.

conformation for fastest E2	

7.2 Application of Substitution and Elimination Mechanisms

A. A Unified View of S_N2 and E2 Mechanisms

The bimolecular nucleophilic substitution and elimination mechanisms share a number of attributes. In both cases:

(a) all of the bonding changes take place in one step
(b) there are no high energy intermediate cations or anions in the reactions
(c) the best trajectory puts the electrons that break the "C-X" bond 180° away from the X group
(d) there is stereochemical preference in the outcome

The reason for these shared features can be understood in Figure 0718. There is an implicit competition in ways for the Lewis base to donate electrons to the "C-X" bond, whose polarity is motivating the reaction: a choice for the Lewis base to either collide directly with the carbon of the "C-X" bond or to deprotonate the anti beta hydrogen (β-hydrogen) and send the electrons from the C-H bond towards the "C-X" bond. Either way, a pair of electrons is heading towards the polar "C-X" bond and breaking it, favoring the formation of a new sigma bond (S_N2) or a pi bond (E2).

These mechanisms are both single-step pathways with no high-energy intermediates. The incoming electron pair, in both cases, is being fed directly to the electrophilic carbon atom and preferentially 180° away from the "X" group. The stereochemical outcome is predictable because of the highly organized arrangements proposed for these mechanistic pathways. All of the bonds are breaking and forming at the same time: In the S_N2 pathway, this results in the inversion of stereochemical configuration; and in the E2 pathway, this preferentially results in whichever pi bond geometry forms when the "X" group and the β-hydrogen are anti (180°) with respect to each other.

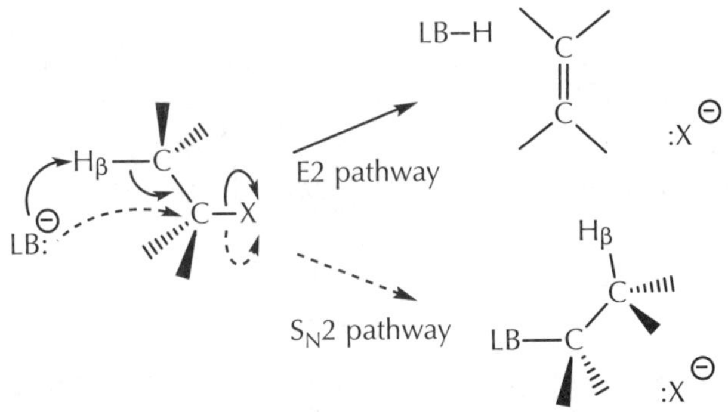

Figure 0718
Direct competition between of S_N2 and E2 mechanistic pathways.

As a general principle, if the conditions for getting a bimolecular reaction pathway exist, the bimolecular pathways will tend to occur faster than the unimolecular pathways, all of which involve high energy intermediates. Learning that criterion is a key part of the predictive model for these reactions. Thus, analyzing of a set of reaction conditions for a potential bimolecular reaction (S_N2, E2) will come before considering the unimolecular ones (S_N1, E1). Said another way: Your ability to predict a unimolecular reaction pathway (S_N1, E1) will come after you have excluded the potential for the bimolecular ones (S_N2, E2).

When you have a predictive model to work with, and you exclude the bimolecular reactions as an option, then it will be time to examine the reaction conditions for a possible unimolecular reaction as the prediction. The unimolecular reaction mechanisms are in an even more direct competition than their bimolecular counterparts.

B. A Unified View of S$_N$1 and E1 Mechanisms

A reminder here that all of the substitution and elimination reactions have the same starting point: an sp^3 "C–X" bond and a Lewis base. If a bimolecular pathway is predicted, then one or both of the one-step collisions (S$_N$2 and/or E2) will eject the leaving ("X") group from the structure.

If the unimolecular pathways (S$_N$1 and/or E1) are predicted, the difference from the bimolecular reactions is that the leaving ("X") group simply leaves first. It might seem a little strange that a sigma bond can simply break to give a couple of ions, but under the right conditions, the bond energy can be overcome and the bond breaks. Again, the discussion about what those conditions are will come a bit later. For now, understand that the "C–X" bond breaks, giving an open shell carbocation and a halide ion. As shown in Figure 0719, the competition is now to close the open shell of this carbocation intermediate. Closing the open shell can be accomplished by the Lewis base directly bonding with the carbon (S$_N$1 pathway, where the sp^2 carbocation can be attacked on either side), or by deprotonation of a β-hydrogen to the carbocation (E1 pathway).

Figure 0719

Direct competition between S$_N$1 and E1 mechanistic pathways.

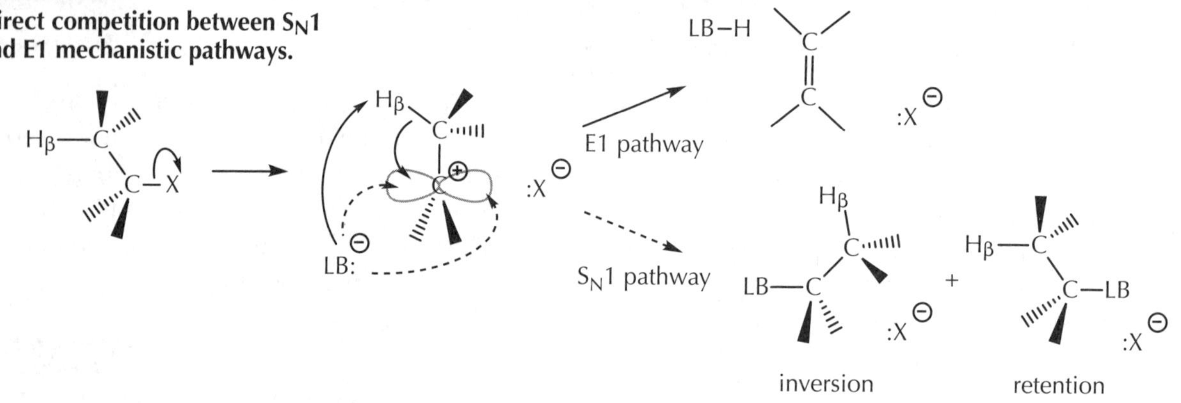

Note that the bond-change events for the unimolecular mechanisms are exactly the same as in the bimolecular mechanisms, and that the difference is in the timing of the changes. In both S$_N$1 and S$_N$2, a sigma bond breaks and another one forms. In the S$_N$1 pathway, there are two steps, where the bond break happens first, followed by the new bond forming. In the S$_N$2 pathway, the same two events happen, but they occur simultaneously.

C. The E1cb Mechanism

As illustrated in Figure 0719, the most common unimolecular elimination mechanisms, S$_N$1 and E1, derive from the rate-determining loss of the "X" group from the "C–X" bond to give an open shell carbocation intermediate. The competition between the Lewis base combining directly with the carbocation versus deprotonation of one of the carbocation's β-hydrogen atoms is the difference between the S$_N$1 and E1 pathways, respectively.

Generally, the β-hydrogen atom is not acidic enough, or the Lewis base is not a strong enough Brønsted base, for deprotonation to take place before the carbocation forms. Once the carbocation forms, however, the pK$_a$ value of any β-hydrogen atoms drops substantially, and even exceptionally weak bases can carry out the fast deprotonation reaction in the second step.

Sometimes, when the β-hydrogen atom is relatively acidic, its deprotonation reaction can end up happening before the leaving ("X") group is lost. The loss of the "X" group is usually still the rate-determining step, but the loss is happening as the second step, after the conjugate base of the original molecule has formed (Figure 0720).

Figure 0720
Direct comparison of E1cb and E1 mechanistic pathways.

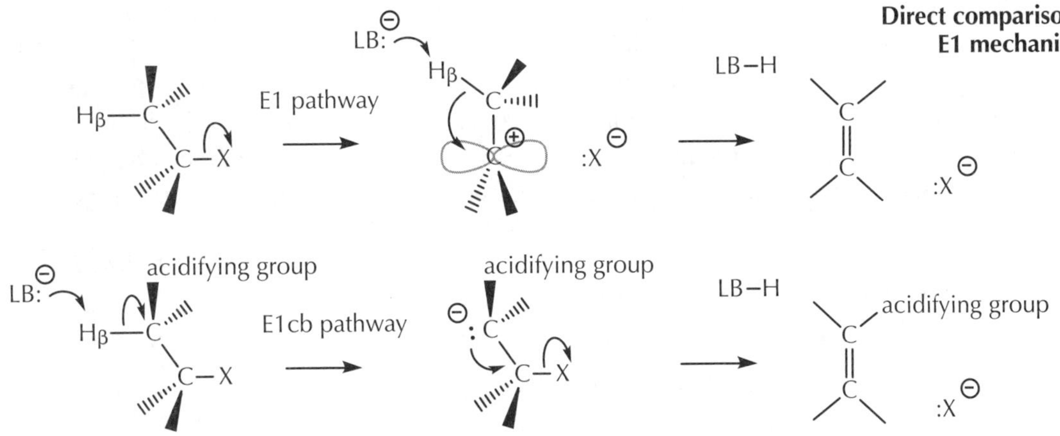

This version of the elimination mechanism, called E1cb, is often difficult to predict without evidence or prior knowledge of a closely related example. In this text, we will assume that the E1 and E2 mechanisms are operating unless specific instructions about a reaction have been provided in which the E1cb pathway is known to be operating. Knowing when to use E1cb will be included explicitly in the discussion, and it will not be included in the general predictive model.

D. Extending the Definition of Leaving Groups

Up until now, substitution and elimination reactions have been represented for sp^3 carbon-halogen bonds as the electrophilic partner, with the halogen atom playing the role of the leaving group. As stated explicitly in an earlier example (Figure 0709), there are other groups that can function as leaving groups in these reactions. In this section, some ways in which the hydroxyl group can be used as a leaving group will be used as examples of how the definition of a leaving group extends beyond only halogen atoms.

In general, the hydroxyl group itself does not predictably act as a leaving group. There are two reasons for this observation. First, the carbon-oxygen bond (C-OH) is more covalent than the carbon-halogen bonds (C-Cl, C-Br, and C-I), and so it is more difficult to break. Second, the OH bond of the hydroxyl group is relatively acidic (pK_a ~16), so in the presence of Lewis bases that are even relatively weak Brønsted bases, the O-H bond would be more likely to break in an acid-base reaction than having the C-O bond break in a substitution or elimination reaction.

The analysis above suggests the two structural features that need to be addressed for an oxygen-atom leaving group: A process is used that weakens the C-O bond, and the deprotonation reaction of the O-H bond is removed as a competing process.

Historically, there have been two most common strategies for transforming a hydroxyl group into a good leaving group, that is, one that participates in substitution and elimination reactions. In both cases, these strategies were not the result of intentionally searching for ways to make a good leaving group, but rather these are examples of how some existing experimental results—sometimes even ones that were originally undesirable—can be exploited to a new purpose. These two strategies are summarized below.

Strategy 1: Sulfonate esters (C-OH is converted to C-OSO$_2$R)
General use is wide: observed to behave as "C-LG" for S_N2, E2, S_N1, and E1

Strategy 2: Protonated hydroxyl (C-OH is converted to C-OH$_2^+$)
General use is limited: Observed to behave as "C-LG" for S_N2 and S_N1 where C-OH becomes C-halogen bond, and in some specific E1 reactions

542 CHAPTER 7 Substitution and Elimination Reactions of Polar Sigma Bonds

The first of these two strategies is quite general: The group called "sulfonate ester" behaves, chemically speaking, just like a halogen atom. Experimentally, hydroxyl groups can be transformed into sulfonate esters in one simple experimental step. There are a few different sulfonate esters that have been commonly used for this purpose, and the transformations and experimental conditions for three representative examples are shown in Figure 0721. The reagents and the sulfonate esters have commonly used abbreviations, which can make redrawing the structures faster and more convenient (although the abbreviations hide the molecular structure from you, you will get accustomed to recognizing them through practice).

Figure 0721
Preparation of three different sulfonate esters from an alcohol.

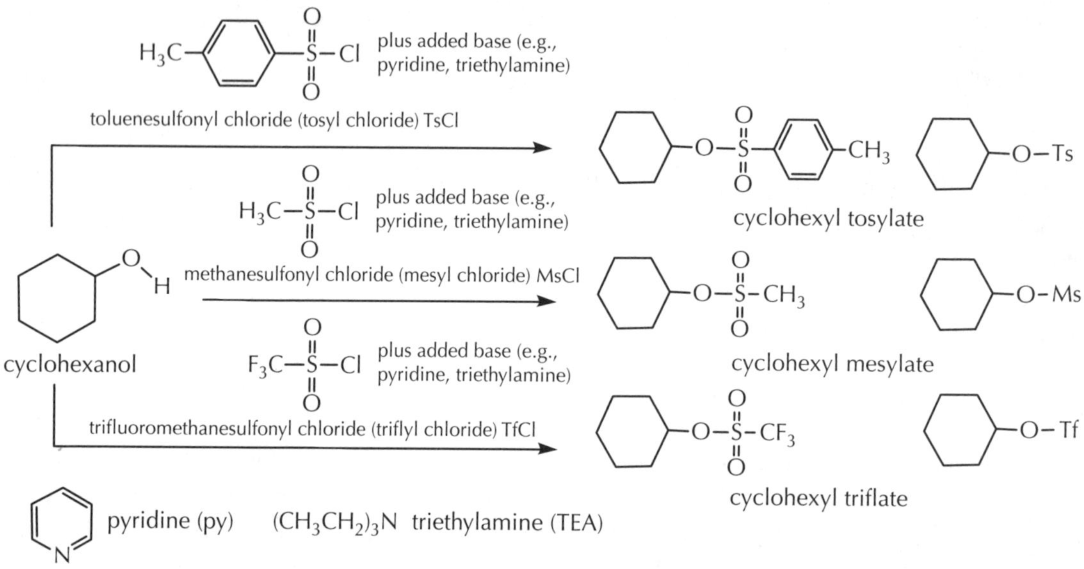

toluenesulfonyl chloride (tosyl chloride) TsCl — plus added base (e.g., pyridine, triethylamine) → cyclohexyl tosylate

methanesulfonyl chloride (mesyl chloride) MsCl — plus added base (e.g., pyridine, triethylamine) → cyclohexyl mesylate

trifluoromethanesulfonyl chloride (triflyl chloride) TfCl — plus added base (e.g., pyridine, triethylamine) → cyclohexyl triflate

cyclohexanol

pyridine (py) $(CH_3CH_2)_3N$ triethylamine (TEA)

There is no uniform mechanism for how the sulfonate esters are formed; they can differ on a case-by-case basis, depending on the structure and the reaction conditions. Indeed, if you end up consulting online resources and other text resources, you will immediately encounter variation for how the mechanism for how the most common of these sulfonate esters, the toluenesulfonate (or tosylate), is presented. The overall result is the same, but the steps for how to get there are presented according to different stories (mechanisms).

Acknowledging the history of how this reaction has been presented, Figure 0722 includes the two most common versions of the mechanism for forming tosylates when pyridine is used as a base, neither of which is consistent with experimental evidence…but you will see them. The version that is more consistent with experimental evidence is shown third.

All three pathways get to the same place, namely, the sulfonate ester. There is simply no consensus among those who teach organic chemistry about which of these mechanisms should be included in the introductory course. Consult with your instructors to see which version you might be held responsible for in your own course.

Returning to the point of this section: When a hydroxyl group is transformed into a sulfonate ester, the sulfonate ester will behave, predictably, the same way that a halogen atom does in substitution and elimination reactions. Sulfonate esters are good oxygen atom leaving groups.

For example, the exercise in Figure 0716 showed how to construct the full array of possible S_N and E products derived from (1S, 2R)-1-chloro-1,2-diphenylpropane plus sodium methoxide. The analogous alcohol, where a hydroxyl group is in the spot occupied by the chloro group, would not participate in substitution or elimination reactions of the hydroxyl group. Instead, sodium methoxide would just reversibly deprotonate the hydroxyl group of (1S, 2R)-1,2-diphenylpropan-1-ol (Figure 0723) to give its conjugate base.

7.2 Application of Substitution and Elimination Mechanisms

Figure 0722

Three different mechanisms for showing the formation of tosylates in pyridine.

Experimental result

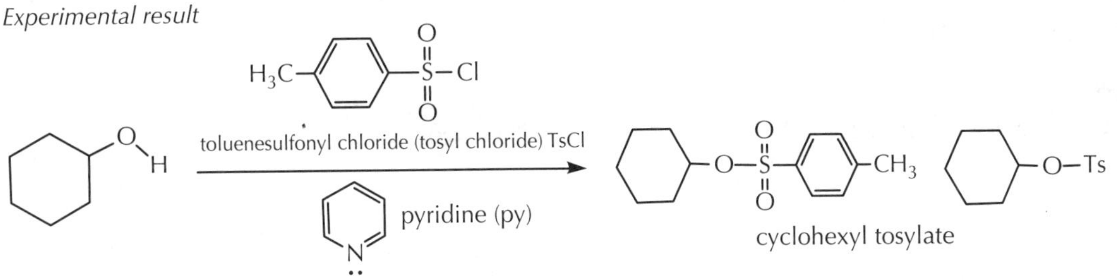

cyclohexanol + toluenesulfonyl chloride (tosyl chloride) TsCl, pyridine (py) → cyclohexyl tosylate

Mechanism 1 (addition, elimination, deprotonation): less favored

addition → elimination → deprotonation

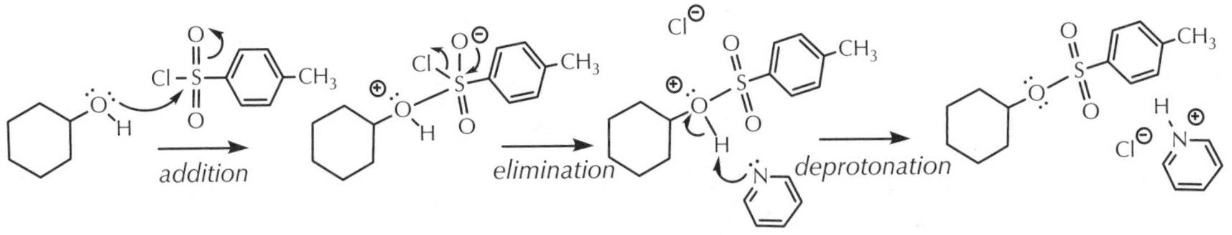

Mechanism 2 (substitution, deprotonation): less favored

substitution → deprotonation

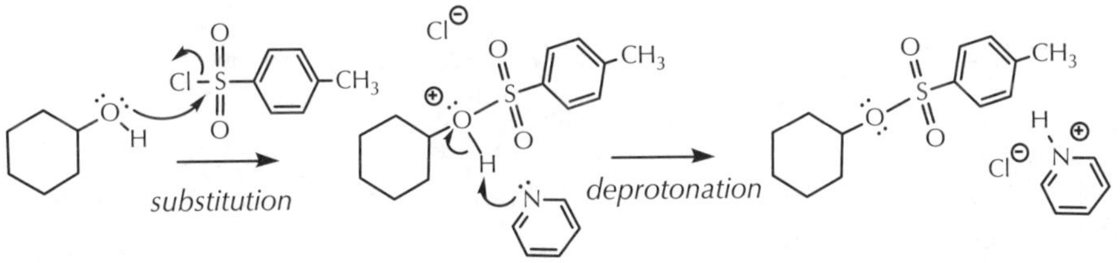

Mechanism 3 (substitution, substitution, deprotonation): most consistent with experiment

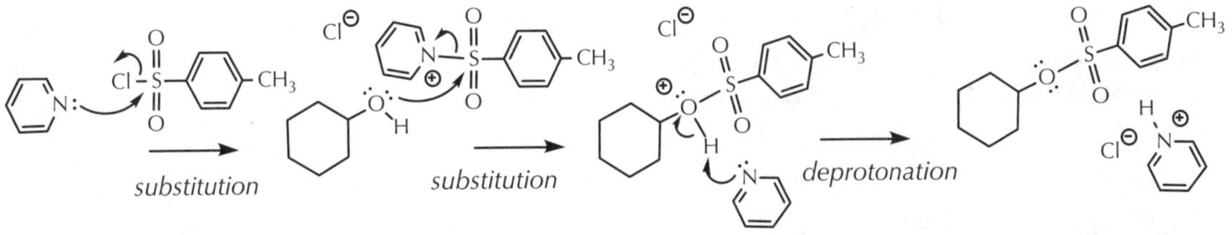

substitution → substitution → deprotonation

Figure 0723

The full array of S_N and E products from (1S, 2R)-1-chloro-1,2-diphenylpropane and sodium methoxide are not observed with (1S, 2R)-1,2-diphenylpropan-1-ol.

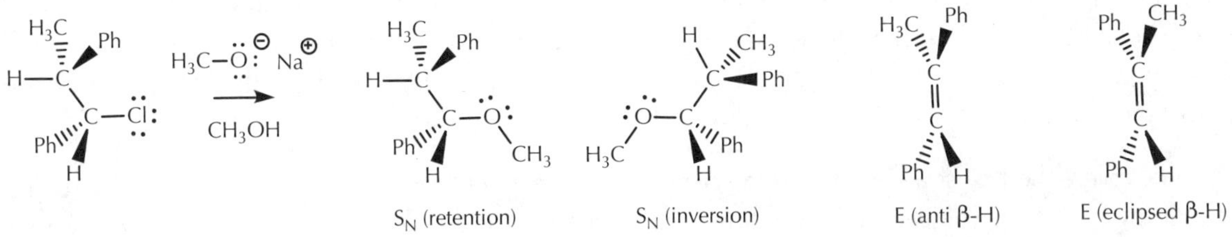

S_N (retention) S_N (inversion)   E (anti β-H)   E (eclipsed β-H)

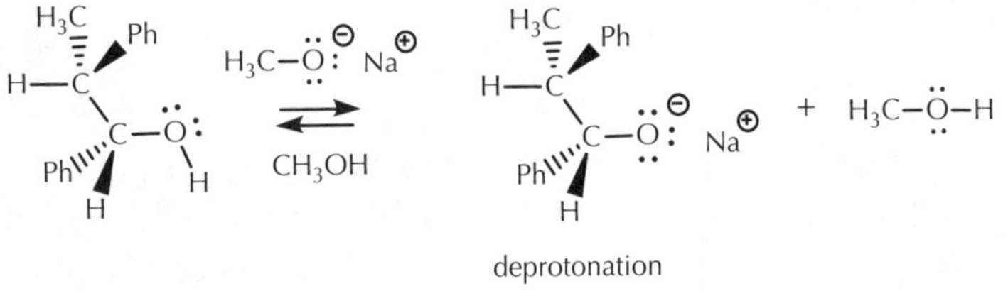

deprotonation

However, if the hydroxyl group in (1*S*, 2*R*)-1,2-diphenylpropan-1-ol is first transformed into the tosylate, and the tosylate behaves as a leaving group, then the full array of substitution and elimination products anticipated in the reaction with sodium methoxide is exactly the same as when the leaving group was the chloro group (Figure 0724).

Figure 0724

The full array of S$_N$ and E products from the tosylate derived from (1*S*, 2*R*)-1,2-diphenylpropan-1-ol plus sodium methoxide.

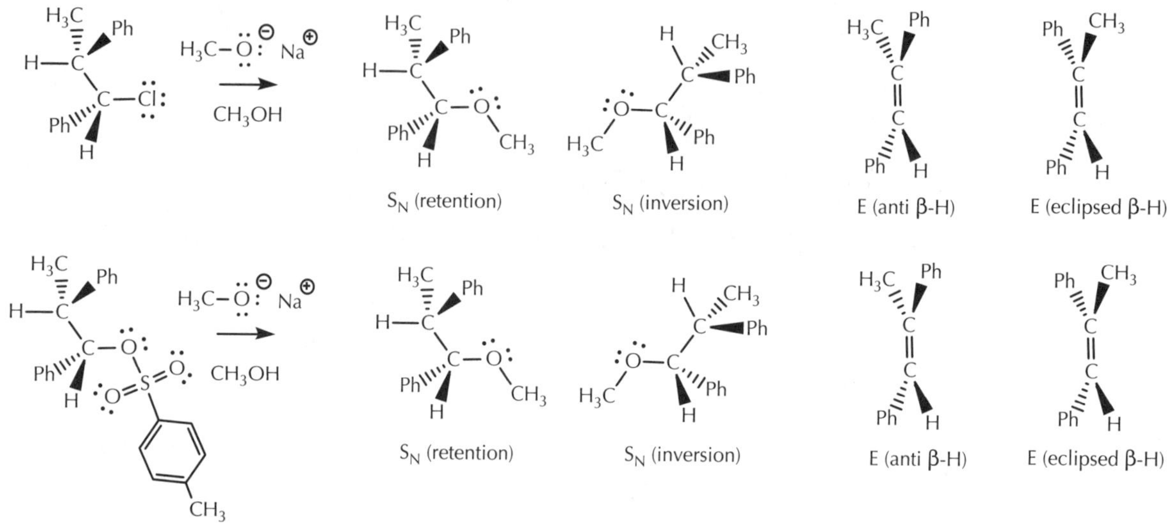

S$_N$ (retention)   S$_N$ (inversion)   E (anti β-H)   E (eclipsed β-H)

S$_N$ (retention)   S$_N$ (inversion)   E (anti β-H)   E (eclipsed β-H)

This exercise was to identify the full set of possible outcomes, not make a prediction about which specific one(s) might or might not form. Experimentally, the distribution of products will depend on all aspects of the reaction conditions, including which leaving group is used.

Why are these sulfonate esters good leaving groups in the first place? A compelling answer to this question can be found in Brønsted acid-base chemistry. The Brønsted acid-base reaction is an S$_N$2 substitution reaction that takes place at a hydrogen atom, and the conjugate base is released as a leaving group. See Figure 0725 for four examples of H-O bonds that are broken and have different conjugate bases released. In thinking about water as a Brønsted acid, one of the hydrogen atoms is the site of the substitution, and (for water) a hydroxyl group needs to behave as a leaving group.

As you move down the list of examples in Figure 0725, from water to acetic acid, to hydronium ion, and to sulfuric acid, the acidity increases. The H-O bond is weakening (more acidic), and the leaving group (conjugate base) is getting more and more likely to leave. Seen from the perspective of the acid form, the oxygen atom is getting progressively more positive; seen from the perspective of the leaving group, the electron pair that is released from the OH bond is getting progressively more stable. When you compare the structure of a tosylate with that of sulfuric acid, the C-O bond that breaks in the tosylate mirrors the reactivity of sulfuric acid, that is, the acidity of its H-O bond.

Tosylates were the first strategy for making oxygen atom leaving groups. The second strategy listed at the beginning of this section is much more limited in scope than the sulfonate esters:

Strategy 2:   Protonated hydroxyl (C-OH is converted to C-OH$_2$$^+$)
General use is limited: Observed to behave as "C-LG" for S$_N$2 and S$_N$1 where C-OH becomes C-halogen bond, and in some specific E1 reactions

As illustrated in Figure 0725, the positively charged oxygen atom in the hydronium ion is a better leaving group in Brønsted acid-base chemistry compared with the hydroxyl group in water. Exactly the same principle applies when a carbon is attached to a positively charged oxygen (C-OH$_2$$^+$) relative to its un-

7.2 Application of Substitution and Elimination Mechanisms 545

Figure 0725

Leaving group ability for sulfonate esters mirrors the reactivity of sulfuric acid.

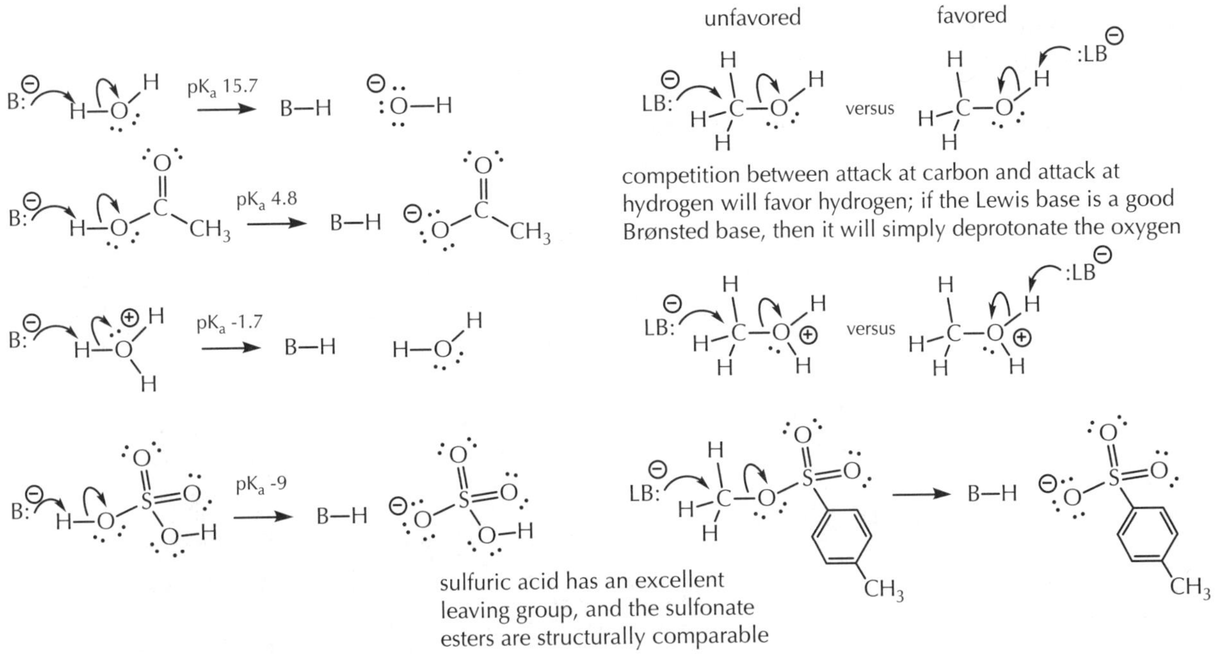

[Reaction scheme, left column: B: attacks H-O-H, pK$_a$ 15.7 → B-H + :O-H]

[B: attacks H-O-C(=O)CH$_3$, pK$_a$ 4.8 → B-H + :O-C(=O)CH$_3$]

[B: attacks H-O(+)(H)-H, pK$_a$ -1.7 → B-H + H-O-H]

[B: attacks H-O-S(=O)(=O)-O-H, pK$_a$ -9 → B-H + :O-S(=O)(=O)-O-H]

sulfuric acid has an excellent leaving group, and the sulfonate esters are structurally comparable

[Right column, top: unfavored vs favored — LB: attacks carbon of H-C(H)(H)-O-H versus LB: attacks H of H-C(H)(H)-O-H]

competition between attack at carbon and attack at hydrogen will favor hydrogen; if the Lewis base is a good Brønsted base, then it will simply deprotonate the oxygen

[Middle right: LB: attacks carbon of H-C(H)(H)-O(+)(H)-H versus LB: attacks H of H-C(H)(H)-O(+)(H)-H]

[Bottom right: LB attacks C of H-C(H)(H)-O-S(=O)(=O)-C$_6$H$_4$-CH$_3$ → B-H + :O-S(=O)(=O)-C$_6$H$_4$-CH$_3$ (tosylate)]

charged (C-OH) counterpart, as shown in the middle of the right-hand column in Figure 0725. Although the protonated hydroxyl group (positively charged oxygen atom) is a better leaving group than its uncharged counterpart, in the example about it, the protons on the C-OH$_2^+$ group have a quite low pK$_a$ value. Therefore, a good Brønsted base would simply deprotonate the C-OH$_2^+$ group. To accomplish a substitution or elimination reaction, and have the protonated hydroxyl group behave as a leaving group, the Lewis basic partner needs to be an exceptionally poor Brønsted base (e.g., a halide ion).

Consequently, the protonated hydroxyl group (-OH$_2^+$) is not as generally useful as sulfonate esters for transforming a hydroxyl group into a leaving group, and its use as a strategy is limited to carefully defined cases. In addition, the strong acid conditions needed to protonate a hydroxyl group can often be incompatible with other functional groups in a complex molecular structure, so the protonated hydroxyl group tends to be used in fairly simple structures.

Four examples of using the protonated hydroxyl group as a leaving group are presented in Figures 0726–0729. It is critically important to remind you, again, that you are not yet predicting which mechanism is operating (that comes next), but rather you are being given the experimental result, to which you

Figure 0726

Transformation of R-OH to R-X using H-X (experimentally an S$_N$2 example).

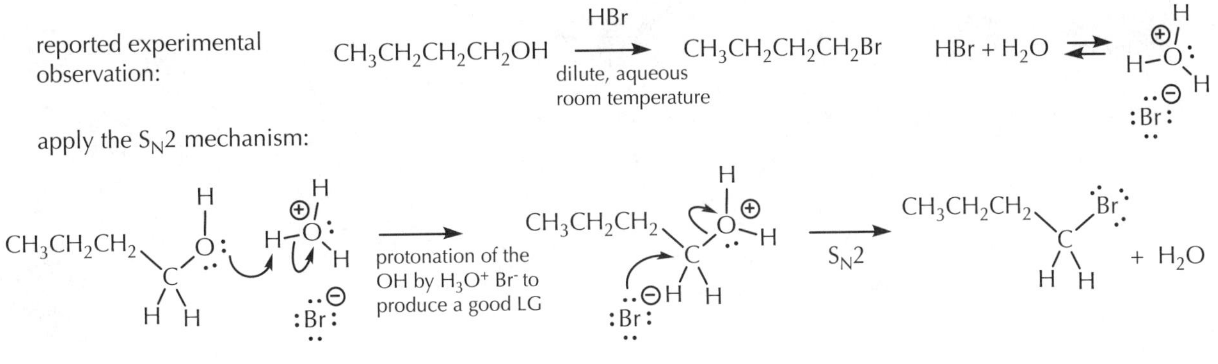

reported experimental observation:

$$CH_3CH_2CH_2CH_2OH \xrightarrow[\text{room temperature}]{\text{HBr, dilute, aqueous}} CH_3CH_2CH_2CH_2Br$$

$$HBr + H_2O \rightleftharpoons H-O(+)(H)-H + :Br:^-$$

apply the S$_N$2 mechanism:

[Mechanism: CH$_3$CH$_2$CH$_2$-C(H)(H)-O-H + H-O(+)(H)(H), :Br:$^-$ → protonation of the OH by H$_3$O$^+$ Br$^-$ to produce a good LG → CH$_3$CH$_2$CH$_2$-C(H)(H)-O(+)(H)-H with :Br:$^-$ attacking → S$_N$2 → CH$_3$CH$_2$CH$_2$-C(H)(H)-Br + H$_2$O]

546 CHAPTER 7 Substitution and Elimination Reactions of Polar Sigma Bonds

can apply the general mechanism. These four examples are the last ones before developing the predictive model, and so one last opportunity to see that once you have the information about which mechanistic pathway is operating, you can then apply it, as a general principle, to any specific case. This identification step derives from a predictive model, and the predictive model, in turn, derives from hundreds and hundreds of experimental results.

Description for Figure 0726: Experimentally, when a solution of 1-butanol is shaken vigorously for just a few minutes with an aqueous solution of hydrobromic acid, a reaction takes place and a layer of nearly pure 1-bromobutane is observed. If you are directed to show the transformation as an application of the S_N2 mechanism, then you need to (a) turn the hydroxyl group into a good leaving group and (b) show the 1-step mechanism. Recall too that an aqueous solution of hydrobromic acid (pK_a -8) is really a solution of hydronium bromide, and so hydronium (pK_a -2) is shown as the Brønsted acid that protonates the hydroxyl group of 1-butanol. The reason that this substitution reaction works is that bromide ion is a quite poor Brønsted base, and it does not favorably deprotonate the protonated hydroxyl group (-OH_2^+). Instead, at some rate, the S_N2 mechanism operates (because this was part of the given information) and 1-bromobutane is formed.

Figure 0727

Transformation of R-OH to R-X using H-X (experimentally an S_N1 example).

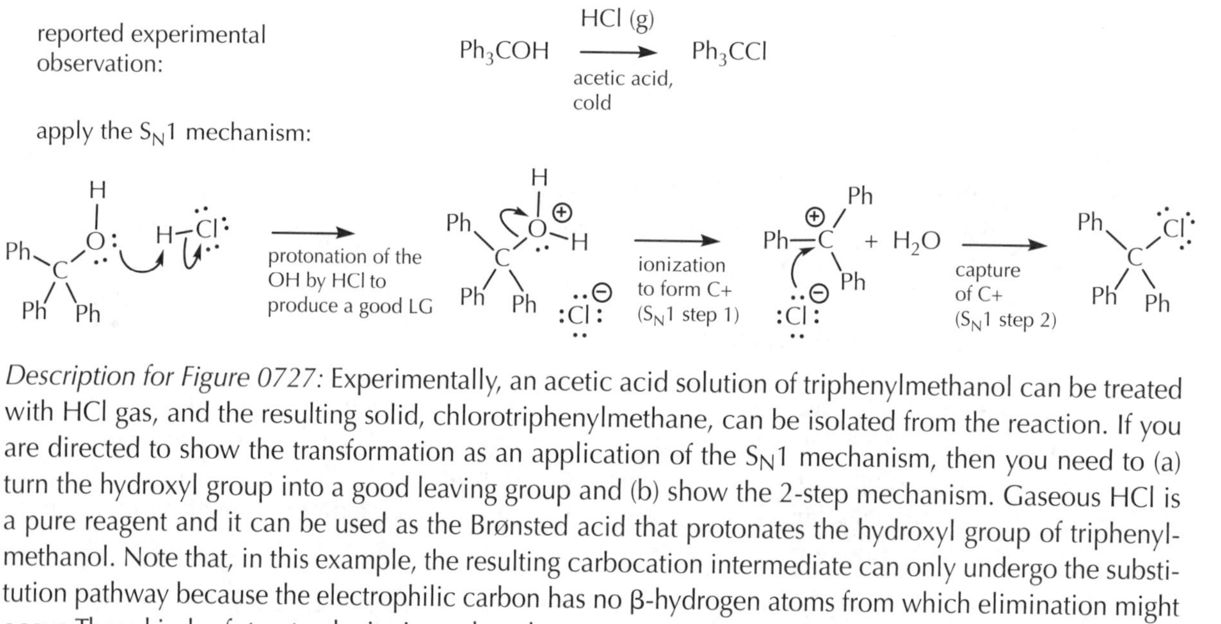

reported experimental observation:

$Ph_3COH \xrightarrow[\text{acetic acid, cold}]{\text{HCl (g)}} Ph_3CCl$

apply the S_N1 mechanism:

[Mechanism diagram showing: protonation of the OH by HCl to produce a good LG → ionization to form C+ (S_N1 step 1) → capture of C+ (S_N1 step 2)]

Description for Figure 0727: Experimentally, an acetic acid solution of triphenylmethanol can be treated with HCl gas, and the resulting solid, chlorotriphenylmethane, can be isolated from the reaction. If you are directed to show the transformation as an application of the S_N1 mechanism, then you need to (a) turn the hydroxyl group into a good leaving group and (b) show the 2-step mechanism. Gaseous HCl is a pure reagent and it can be used as the Brønsted acid that protonates the hydroxyl group of triphenylmethanol. Note that, in this example, the resulting carbocation intermediate can only undergo the substitution pathway because the electrophilic carbon has no β-hydrogen atoms from which elimination might occur. These kinds of structural criteria, such as the presence or absence of β-hydrogen atoms, will become an automatic part of how to decide what might be predicted in any given set of reaction conditions.

Figure 0728

Transformation of R-OH to an alkene (experimentally an E1 example).

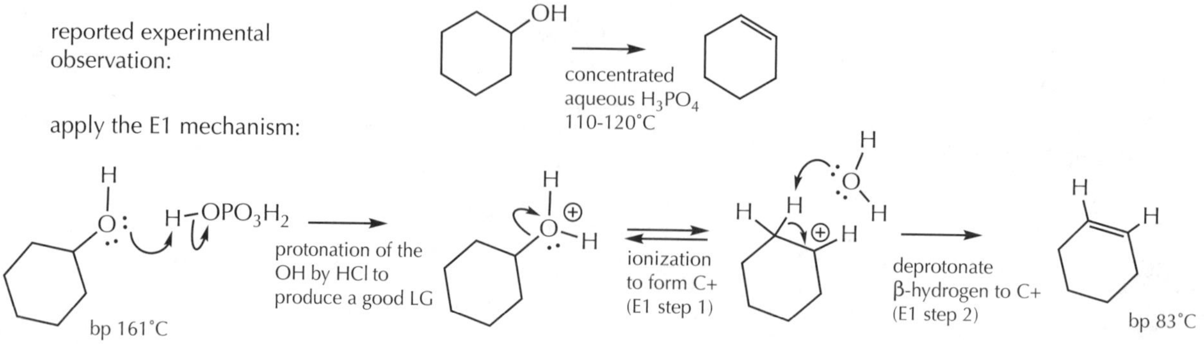

reported experimental observation:

[Cyclohexanol] $\xrightarrow[\text{110-120°C}]{\text{concentrated aqueous } H_3PO_4}$ [cyclohexene]

apply the E1 mechanism:

[Mechanism diagram showing: protonation of the OH by HCl to produce a good LG → ionization to form C+ (E1 step 1) → deprotonate β-hydrogen to C+ (E1 step 2)]

bp 161°C   bp 83°C

Description for Figure 0728: Experimentally, a pure sample of cyclohexanol (bp 161°C) is combined with an 85% (by weight) solution of phosphoric acid in water. The mixture is heated to about 110–120°C, which boils both the water and the observed product, cyclohexene (bp 83°C), out from the reaction vessel. If you are directed to show the transformation as an E1 mechanism, then you need to (a) turn the hydroxyl group into a good leaving group and (b) show the 2-step mechanism. Although the carbocation can be captured by the water in the reaction, that process is simply the reverse of the reaction that formed the carbocation in the first place, and so the substitution pathway is reversible and redundant. The elimination pathway is made irreversible by shifting the equilibrium after the β-hydrogen deprotonation step to form the low boiling elimination product, cyclohexene, and removing it from the reaction (and application of Le Châtelier's Principle).

Figure 0729

Transformation of R-OH to R-OCH$_3$ (experimentally an S$_N$1 example).

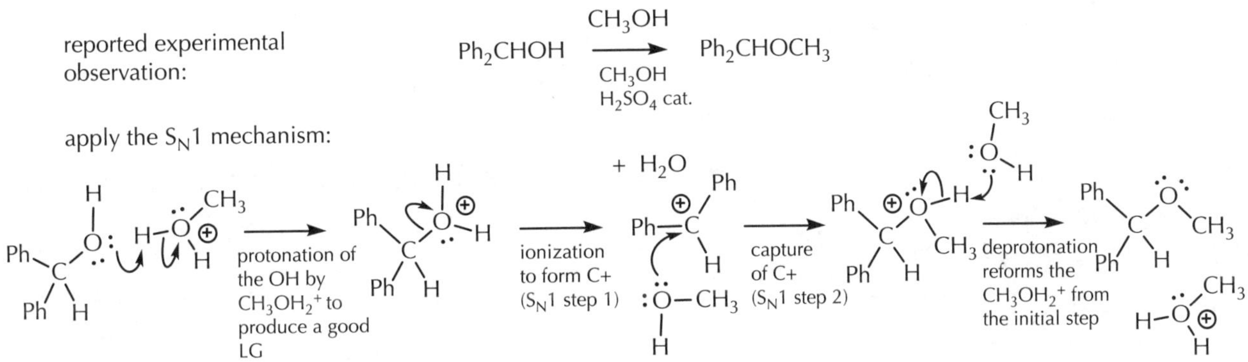

Description for Figure 0729: Experimentally, when a solution of diphenylmethanol (Ph$_2$CHOH) in methanol (solvent) is warmed with a few drops of concentrated sulfuric acid, a reaction takes place within minutes and methoxydiphenylmethane (Ph$_2$CHOCH$_3$) is observed. If you are directed to show the transformation as an application of the S$_N$1 mechanism, then you need to (a) turn the hydroxyl group into a good leaving group and (b) show the mechanism. Sulfuric acid will react instantly with methanol, and so the conjugate acid of methanol is the actual Brønsted acid catalyst that protonates the hydroxyl group of diphenylmethanol. Once the carbocation forms, it is captured by methanol to give the conjugate acid of the observed product, exactly as in the electrophilic addition reaction of weak acids to pi bonds (Section 4.4). And as in those addition reactions, a deprotonation step is needed to complete the mechanism.

PRACTICE QUESTIONS

7.07 Complete the following reactions according to the mechanistic guideline provided with each one of them. If more than one stereoisomer is anticipated, draw all of them. The mechanism does not need to be drawn.

(a) [bicyclic structure with H, H, H, Br substituents] $\xrightarrow{\text{KCN}}_{S_N2}$

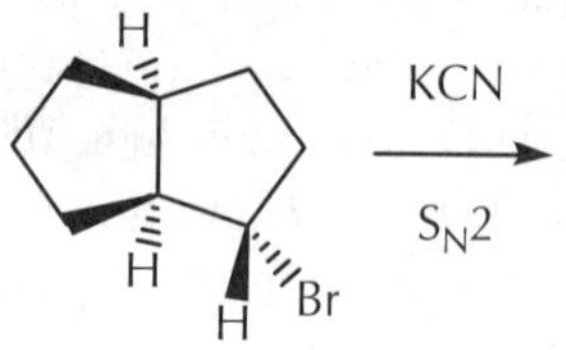

(b) [PhCH(OSO₂CH₃)CH₂CH₃ with H stereochemistry] $\xrightarrow{\text{NaO-iPr}}_{E2}$

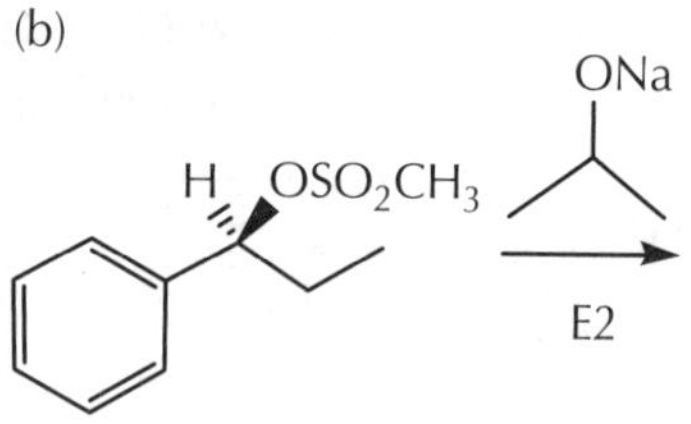

(c) [tetrahydropyran with H and OH at C2] $\xrightarrow{\text{HCl}}_{S_N1}$

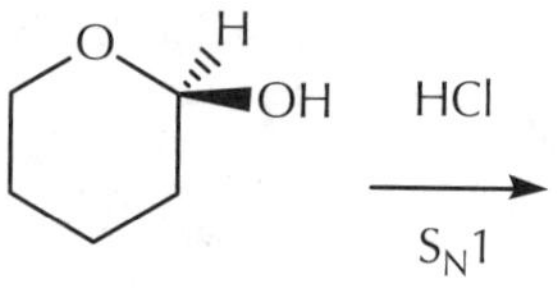

(d) [F₃C-SO₂-O-C(CH₃)(CH₂CH₃)(Ph)] $\xrightarrow{H_3PO_4}_{E1}$

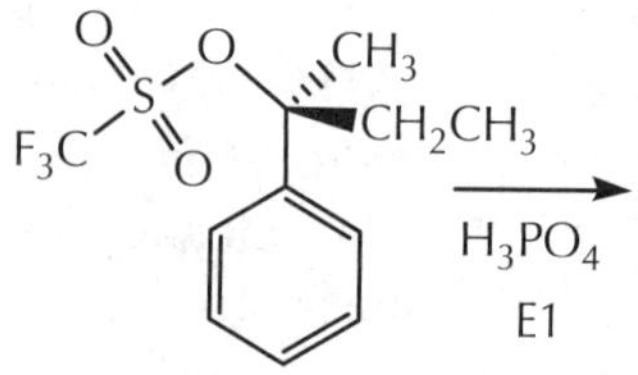

(e) [methylenedioxy-benzofuran derivative with CH₂Br and H] $\xrightarrow{\text{NaSCH}_2\text{Ph}}_{S_N2}$

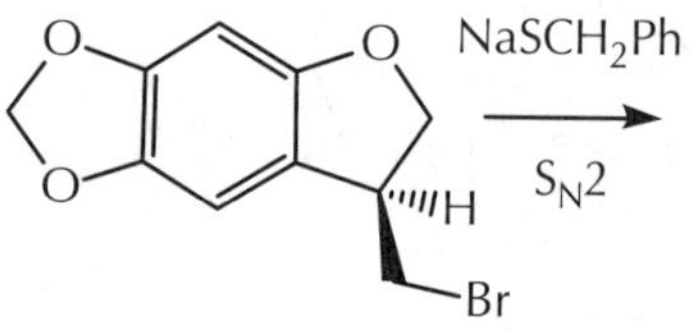

(f) [methyl p-toluenesulfonate] $\xrightarrow{(CH_3)_3N}_{S_N2}$

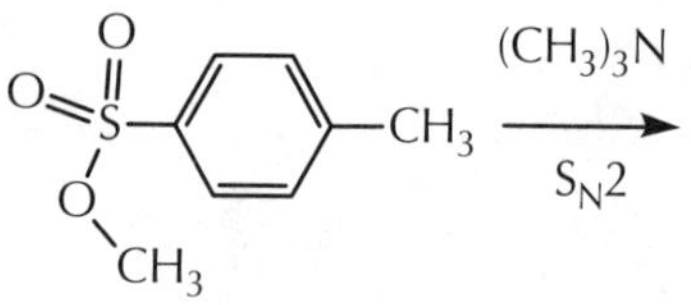

7.2 Application of Substitution and Elimination Mechanisms

7.08 Complete the following reaction schemes according to the information provided. The mechanistic pathway is provided where it is required.

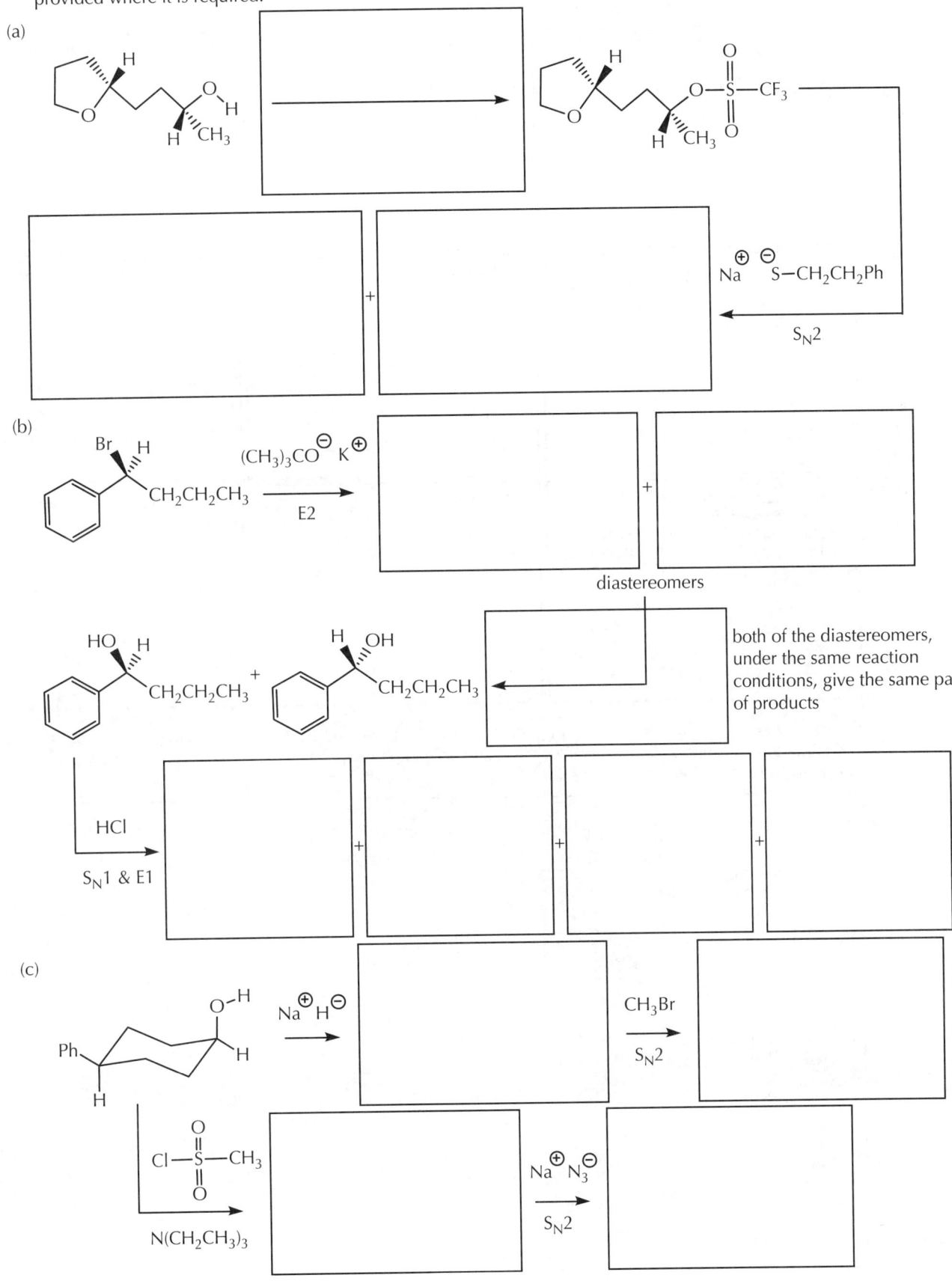

7.09 The mechanistic pathway is provided for each of the following reactions. Apply this information by predicting the product(s) in the first space. Then draw the curved arrow mechanism and the transition for each transformation in the second and third spaces, respectively.

(a)

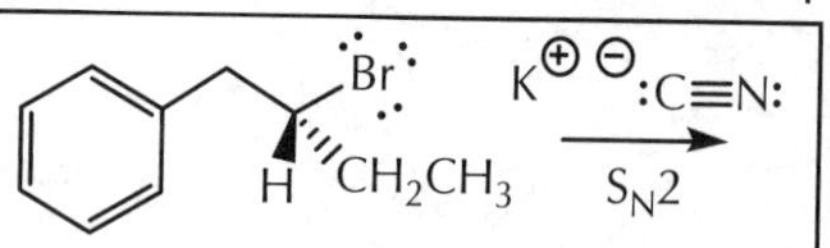

	mechanism	transition state (proper geometry)

PhCH$_2$–C*H(Br)(CH$_2$CH$_3$) + K$^+$ $^-$:C≡N: → (S$_N$2)

(b)

	mechanism	transition state (proper geometry)

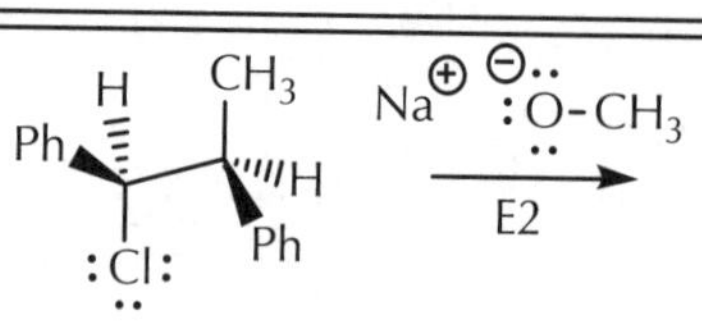

Ph,H / CH$_3$,Ph stereochemistry with :Cl: + Na$^+$ $^-$:O–CH$_3$ → (E2)

(c)

	mechanism	transition state (proper geometry)

cyclohexane with :Br:, H (axial), CH$_3$, H + Na$^+$ $^-$:S–CH$_3$ → (S$_N$2)

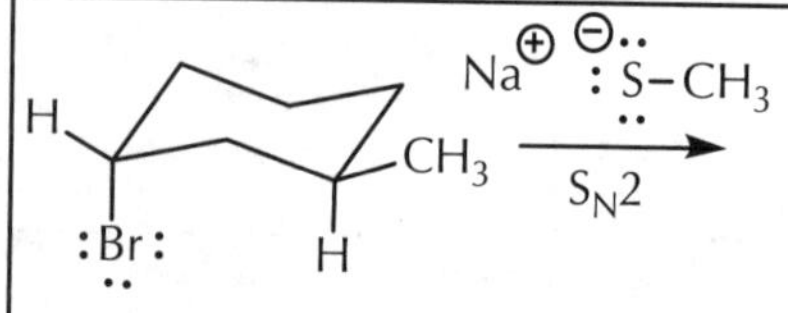

7.10 The mechanistic pathway for the following reaction is S$_N$1/E1, which results in the full array of substitution and elimination products. Apply this information by predicting the products and drawing the curved arrow mechanisms for their formation.

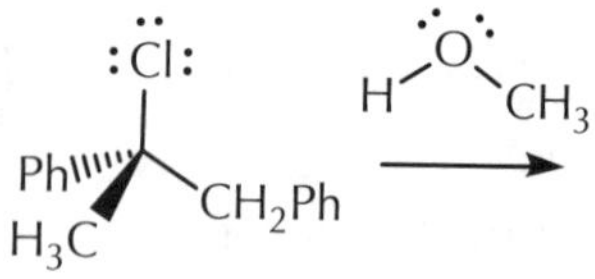

Ph\\\\\\–C(:Cl:)(CH$_3$)(CH$_2$Ph) + H–Ö–CH$_3$ →

7.11 In the following reactions, the hydroxyl groups are converted into leaving groups by protonation, and the resulting conjugate acid immediately undergoes the specified reaction pathway. Predict the product or products that are expected, according to this information. Then provide the full, curved arrow mechanism.

(a) 2-phenyl-2-hydroxytetrahydrofuran + HCl → 2 products (S_N1)

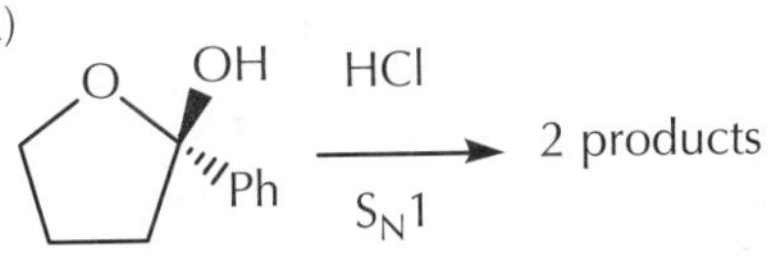

(b) 2-phenyl-2-methyl-propan-1-ol (with stereochemistry) + HBr → 1 product (S_N2)

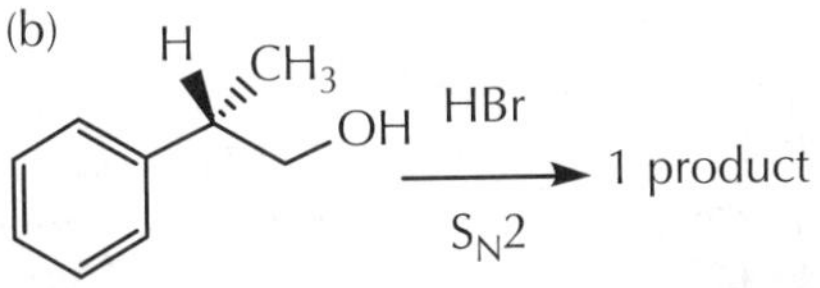

(c) 1-benzyl-cyclohexanol + H_3PO_4 → 2 products (E1)

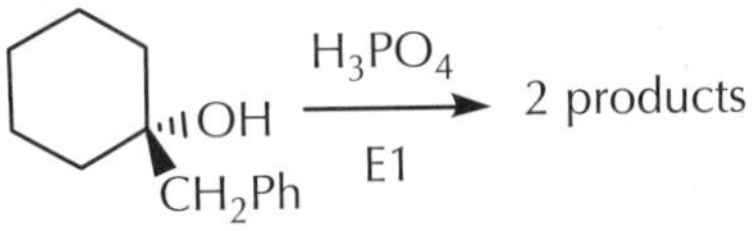

(d) 1-benzyl-cyclohexanol + CH_3OH / H_2SO_4 → 1 product (S_N1)

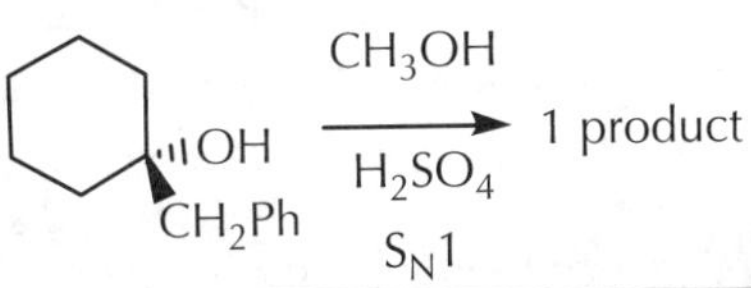

7.3 A Predictive Model for S_N/E Reactions

A. Explaining Outcomes Versus Predicting Outcomes

In the "Setting the Stage" essay that begins this chapter, I was anticipating this section of the text.

So far, you have seen three topics:

(1) the four most common nucleophilic substitution and elimination reaction mechanisms that can be used to explain the reactions of Lewis bases with molecules containing sp^3 carbon-leaving group bonds; these are S_N2, E2, S_N1, and E1;

(2) given molecules that match these criteria (the combination of Lewis bases with molecules containing sp^3 carbon-leaving group bonds), you can generate the full array of possible products, including stereoisomers, that could result; and

(3) given molecules that match these criteria (the combination of Lewis bases with molecules containing sp^3 carbon-leaving group bonds), and by knowing the experimentally determined mechanistic pathway by which they react, you can draw what subset of the full array of possible products will form and provide the curved arrow mechanism for the process.

These three topics all involve explanations of observed results without making a selection from among the possible outcomes unless the information was explicitly provided. What is missing, here, is a way to look at molecules that match these criteria and predict which of the four mechanistic pathways, if any, is the most likely outcome, in the absence of any experimental information. A predictive model, in other words, would deliver a really useful bit of information: You could look at any combination of Lewis bases with molecules containing sp^3 carbon-leaving group bonds and declare which of the mechanistic pathways would be the most likely, or if different pathways might compete.

If you knew that a certain combination was predicted to most likely follow an S_N2 pathway, then you would know it formed the S_N product with inversion of stereochemical configuration. When you have that bit of information, you can not only draw the product(s), but you can also draw the mechanism according to that pathway, as in Figures 0726–0729.

A predictive model for substitution and elimination will not tell you the identity of the products, but it will tell you which pathway is most likely. You then need to apply that pathway to the given set of molecules to propose the structural outcome and the detailed mechanism by which the process took place.

A predictive model is an algorithm to use when you do not have the experimental results. It is based upon as many experimental results as possible, and it will be as consistent with as many experimental results as possible. In a real chemistry setting, however, no predictive model will be perfect. Chemical reactions invariably give mixtures, including the competition between the expected pathways as well as unanticipated pathways that might form byproducts and/or decomposition products.

A predictive model might anticipate the major outcome, but never its percentage with respect to other products and certainly never its yield. What appears to be subtle variation in structure or reaction conditions can give dramatically different outcomes, and so a predictive model is never perfect.

All those warnings aside, the major product(s) from a lot of organic chemical reactions is predictable a great many times, enough to warrant the development of a set of questions and criteria for making a best prediction from among which of the four nucleophilic substitution and elimination reaction pathways is anticipated for a given reaction.

B. Developing a Predictive Model for S$_N$/E Reactions

After years of study, a strong correlation exists between the specific substitution and/or elimination outcome that is observed and the details of the reaction: the given combination of a molecule containing an sp^3 carbon-leaving group bond (electrophile), a Lewis basic partner (nucleophile), and the reaction conditions. The questions that need to be answered for predicting an outcome involve these three features:

(a) the structure of the electrophile:
   Is there a leaving group attached to an sp^3 carbon atom?
   What is the degree of substitution at the carbon?
   Does the carbon have attached hetereoatoms, or is it a benzylic, allylic, or propargylic position?
   Are there β-hydrogens?

(b) the structure of the Lewis base:
   Is it known to be a good electron donor (nucleophile) for S$_N$ reactions?
   What is its Brønsted basicity towards the deprotonation needed for E reactions?
   Is it hindered (bulky) around the Lewis basic atom?

(c) the reaction conditions:
   Is the solvent polar or nonpolar?
   Is it also capable of being a hydrogen bond donor?
   Are there necessary catalysts present?
   Does the temperature permit the reaction to occur at a reasonable rate?

Chemists have created many different ways to navigate the decision-making needed to predict the outcome for potential substitution and elimination reactions. The key aspect to all of them is that the decision-making algorithm is not a simple "if-then" binary decision involving one criterion. That is, no one of these factors, above, can be looked at in isolation from, or to the exclusion of, the others.

Binary decision-making is easy: If you see a red traffic light, you stop. You do not need other information to make the decision to stop. Obviously, even this algorithm is not perfect; after all, you might be driving an emergency response vehicle and you get to cautiously go through a red light. But you get the idea. In fact, the typical traffic lights are used in nonbinary ways by combining the color of the light with whether or not it is blinking. You can proceed through a red light, too, provided that it is blinking, and you have stopped to check whether it is safe to proceed. This is an example of multiconditional decision-making ("the color of the light" combined with "blinking versus steady").

In predicting which substitution or elimination mechanistic pathway a given set of molecules might preferentially undergo, you have a multiconditional decision based upon the three features listed above, before you can know what to propose. Figure 0730 provides a unified overview for how these reaction pathways fit together, plus the driving questions and categories used for decision-making, all deriving from a Lewis base (LB:) combining with a molecule that has an sp^3 carbon-leaving group bond (C-LG).

Figure 0730

Unified overview and decision-making categories for S_N and E reactions.

Stage 1: Is one of the bimolecular pathways predicted?

prediction criteria:
- step 1: Is there both a Lewis basic partner? and
  Is there, or is there the potential for, an sp^3 carbon-leaving group bond?
- step 2: Characterize the electrophile
- step 3: Characterize the Lewis basic partner.
- step 4: Decide if a bimolecular reaction is predicted, and if so, which one (S_N2 or E2).

If S_N2 or E2 is predicted, then apply the appropriate mechanistic pathway to the molecules.

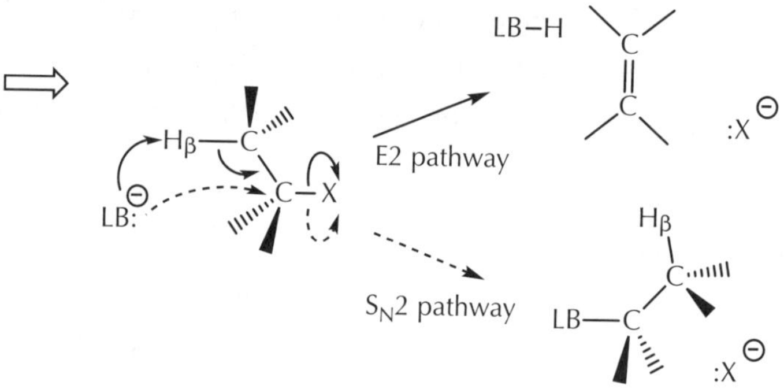

If no bimolecular reaction is predicted, then check for unimolecular.

Stage 2: Is one of the unimolecular pathways predicted?

prediction criteria:
- step 1: Is no bimolecular reaction expected for an eligible electrophile?
- step 2: Is there the potential for a good carbocation (resonance > 3° >> 2°)?
- step 3: Is there a polar reaction medium that can support ions?
- step 4: If all three conditions are met, then S_N1/E1 is predicted; if not, no S_N/E predicted.

If S_N1/E1 is predicted, then apply the appropriate mechanistic pathway to the molecules. If not, then no substitution or elimination is predicted.

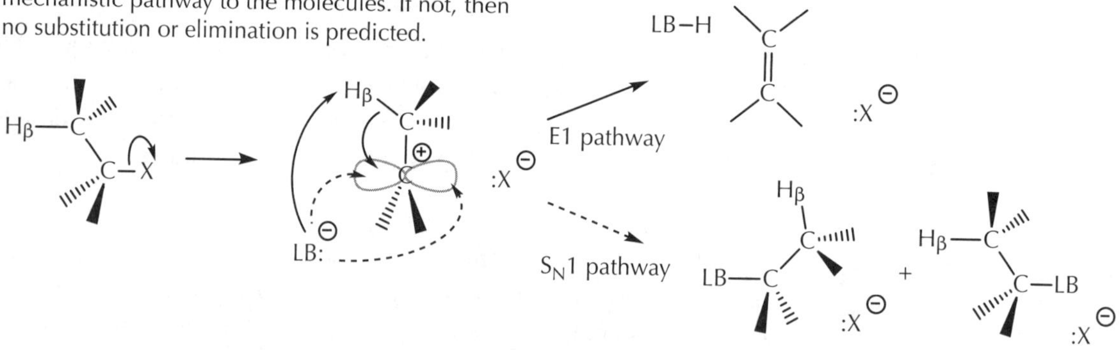

S_N2: The S_N2 pathway is intrinsically the best, because a more stable sigma bond results from a less stable one, and no high energy intermediate is formed; but the S_N2 pathway also has the most simultaneous demands on it: The carbon bearing the leaving group needs to be unhindered, the Lewis base needs to be unhindered, and the Brønsted basicity of the Lewis base cannot be overwhelmingly high (otherwise deprotonation will compete with the substitution pathway).

E2: As the carbon gets more hindered, the activation energy for the S_N2 pathway goes up, and the E2 pathway can begin to dominate as long as the Lewis base has the Brønsted basicity to do the deprotonation.

S_N1/E1: As the carbon bearing the leaving group gets too hindered for direct collision (i.e., the S_N2 activation energy is too high), and if the E2 deprotonation cannot take place because the Brønsted basicity is too low, then there is the possibility of simply losing the leaving group and forming a carbocation intermediate. For this to happen, however, the carbocation needs to be relatively stable and the reaction medium (the solvent) needs to support the formation of ions. Once a carbocation forms, the default assumption, in the absence of other information, is that the S_N1 and E1 pathways are competing and products from both are likely.

In this text, we advocate for the 2-stage decision-making process outlined above:

Stage 1: evaluate the reaction for its potential to follow a bimolecular reaction pathway (S_N2, E2); if one the bimolecular pathways is predicted, then you are done.

Stage 2: only if a bimolecular reaction is not predicted, then evaluate the reaction for its potential to follow a unimolecular reaction pathway (S_N1, E1).

In general, because the bimolecular reaction pathways do not involve a high energy carbocation intermediate, the bimolecular reactions, when the conditions are appropriate, will be favored over their unimolecular counterparts. The two stages are an ordered series of diagnostic questions under the two driving questions: Is one of the bimolecular pathways predicted? And, if not, then is one of the unimolecular pathways predicted?

Here is the brief overview of the prediction criteria, organized by the practical steps and information that will constitute our model.

Stage 1: Is one of the bimolecular pathways predicted?
        Is there both a Lewis basic partner? and

    step i: Is there, or is there the potential for, an sp^3 carbon-leaving group bond?
           Leaving groups: halogens, sulfonate esters, protonated hydroxyl

    step ii: Characterize the electrophile.
           Degree of substitution (hydrocarbon): 1°, 2°, 3° or heteroatom
           Possible pi resonance stabilization in the position: allylic, benzylic, propargylic

    step iii: Characterize the Lewis basic partner.
           Good e-donor/weak base: anion with conj. acid pK_a < 15, uncharged sp^3 N, S, P
           Good e-donor/moderate base: anion with conj. acid pK_a 15-30
           Good e-donor/strong base: conj. acid pK_a > 30 (hindered, conj. acid pK_a >10)
           Poor e-donor/weak base = none of the above

    step iv: Decide if a bimolecular reaction is predicted, and if so, which one.
           See prediction outcomes (Figure 0731)

Stage 2: Is one of the unimolecular pathways predicted?
        Is no bimolecular reaction expected for an eligible electrophile?

    step v: Is there the potential for a good carbocation?
           resonance > 3° >> 2°
        Is there a polar reaction medium that can support ions?
           polar protic (water, alcohols, acids); polar aprotic (DMSO, DMF, HMPA)

    step vi: If all three conditions are met, then S_N1/E1 is predicted; if not, no SN/E predicted

A graphic summary follows, and then a more detailed discussion about each step. And finally, a set of specific examples will be used to illustrate how the steps in this process can be used to actually make predictions about a reaction pathway.

The following concise graphic captures the decision-making steps advocated here for predicting whether or not substitution or elimination reactions are predicted and, if so, what the most likely outcome is in the absence of experimental results.

i. Is there (or is there the potential for) an sp³ carbon-leaving group?	yes		yes, potential		no	iv. Is a bimolecular reaction predicted? If so, which one?		
ii. If so, what is the structural category for the carbon atom?	1°	2°	3°	heteroatom-substituted		S_N2	E2	no bimolecular
	allylic	benzylic		propargylic		v. If no bimolecular, then is there:		
iii. If there is a Lewis base, what is its classification?	anion, conj. acid pK_a <15 uncharged sp³ N, S, or P	anion, conj. acid pK_a 15-30	conj. acid pK_a > 30, hindered base conj. acid pK_a > 10	poor electron donor (not in the other categories)		good carbocation possible		polar solvent
						vi. Is S_N1/E1 predicted?		
						yes		no

Stage 1: Deciding if a bimolecular reaction is expected.

step i: Is there a pair of molecules that could give substitution and/or elimination reactions? Two partners are needed: a Lewis base (LB) and a molecule with an sp³ carbon-leaving group (C-LG), where a leaving group exists (carbon-halogen bond or a sulfonate ester) or the potential for a C-LG to exist (ROH plus reagents for making a sulfonate ester or ROH and a strong acid catalyst).

step ii: Characterize the carbon atom of the C-LG electrophile based upon the kind of substituents it has: (a) is it 1°, 2°, 3°, or heteroatom-substituted, and (b) is there pi resonance stabilization possible if the carbon was a carbocation (i.e., is the C an allylic position, a benzylic position, or a propargylic position?)

step iii: Characterize the Lewis base (LB) according to how good an electron donor it is and its Brønsted basicity. There are four categories for the LB:
    a. A good e-donor that is a weak base: anions with conjugate acids whose pK_a values are less than 15, or Lewis bases with uncharged, sp³ N, S, or P atoms
    b. A good e-donor that is a moderate base: anions with conjugate acids whose pK_a values are between 15–30
    c. A good e-donor that is a strong base: any LB with a conjugate acid pK_a value greater than 30, or LBs that are hindered near the donor atom and have conjugate acids whose pK_a values are greater than 10
    d. A poor e-donor and weak base: a LB that is not in one of the above categories

step iv. Use the characterization of the electrophile and the Lewis base to decide if a bimolecular reaction is predicted, and if so, which one. Here is the multiconditional decision-making described earlier.

If you have a pair of eligible molecules (step i), then the combination of characterizing the electrophile (step ii) and the Lewis base (step iii) will either point to one of the bimolecular reactions as a predicted outcome or that no bimolecular reaction is predicted (Figure 0731).

And, if no bimolecular reaction is predicted in Stage 1, then Stage 2 (is there a unimolecular reaction?) is considered.

7.3 A Predictive Model for S$_N$/E Reactions

Figure 0731

Predicting bimolecular reaction pathways (S$_N$2 and E2)

	category a good e donor weak base anion, c.a. pK$_a$ <15 uncharged sp^3 N/S/P	category b good e donor moderate base anion, c.a. pK$_a$ ~15-30	category c good e donor strong base c.a. pK$_a$ > 30 hindered: c.a. pK$_a$ > 10	category d poor e donor weak base
Lewis base → sp^3 C-LG ↓				
1°C	S$_N$2	S$_N$2	E2 (no β-H: S$_N$2)	no bimolecular predicted
2°C	S$_N$2	E2 (no β-H: S$_N$2)	E2 (β-H)	
3°C	no S$_N$2; no E2	E2 (β-H)	E2 (β-H)	

The three criteria for the electrophile are along the left-hand column (degree of substitution), and the four criteria for the Lewis base are along the top. Matching the criteria from any given pair of molecules gives the recommended prediction.

There are overall trends in these predictions that make intrinsic sense.

The main 3x3 grid is consistent with the unified view in Figure 0730. In the upper left-hand trio, the carbon of the sp^3 C-LG is unhindered and the Lewis base is a good electron donor but not a strong Brønsted base. This is an ideal situation for predicting the S$_N$2 pathway.

The lower right-hand trio is just as ideal for predicting the E2 pathway: The electrophilic carbon is hindered, and the Lewis base is a strong Brønsted base and may itself be hindered (Figure 0732). The indication of "β-H" is a reminder that you cannot get an elimination reaction without β-hydrogen atoms.

Figure 0732

Four typical hindered or bulky bases.

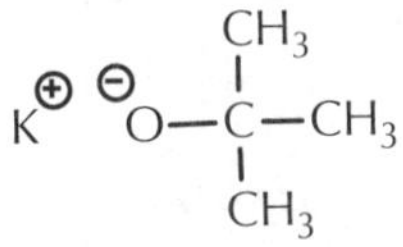

potassium *tert*-butoxide
" KOBu^t " or "KOtBu"
(conj. acid pK$_a$ = 16.5)

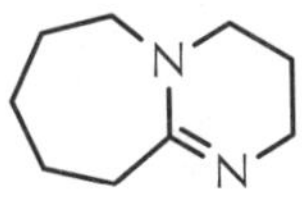

diazabicycloundecane
" DBU "
(conj. acid pK$_a$ = 13.5)

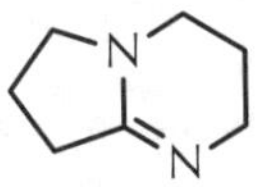

diazabicyclononane
" DBN "
(conj. acid pK$_a$ = 13)

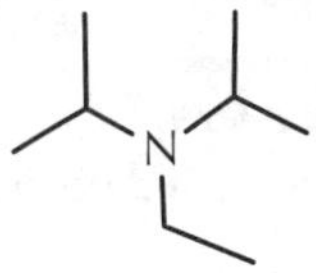

diisopropylethylamine
" Hünig's base " or DIPEA
(conj. acid pK$_a$ = 10.75)

The lower left corner is important. The electrophilic carbon is hindered, no S$_N$2 is anticipated; and if the Lewis base is a poor Brønsted base, no E2 is anticipated. In that corner, the predicted outcome is for no bimolecular pathway (note that this does not mean "no reaction" ... merely that no bimolecular reaction is anticipated).

The middle and upper right corners are slightly complicated. If the electrophilic carbon is unhindered and has no β-hydrogen atoms, then even a strong Brønsted base can follow the S$_N$2 pathway. But if there are β-hydrogen atoms, then this strong base strength of the Lewis base will carry out the deprotonation (E2) pathway before it can collide with the electrophilic carbon atom (collisions with the protons, sitting on the exterior surface of the molecule, happen much more often than collisions with the carbons, sitting more on the interior of the molecule).

Note that in this model the Lewis base needs to be classified as a good electron donor before its participation in a bimolecular reaction pathway can be predicted. So, the characterization of the Lewis base into categories a, b, or c are the "good electron donor" ones, while Lewis bases in category d are not predicted to follow one of the bimolecular reaction pathways.

If no bimolecular reaction is predicted, then we move to Stage 2.

Stage 2: Is one of the unimolecular pathways predicted?

step v: Three criteria need to be met to predict the formation of a carbocation.

The *first criterion* is that no bimolecular reaction was anticipated in Stage 1.

The *second criterion* is that there is the potential of a stable carbocation. The criteria are the same as they were previously: If the leaving group leaves, a good carbocation will have resonance-stabilization (heteroatom and/or pi resonance), or if not, the tertiary (3°) is much better than secondary (2°), which is not anticipated to form without some experimental evidence, and both are significantly better than primary (1°), which is not anticipated to form at all.

The *third criterion* is that the reaction medium (solvent) is polar and can support the formation of ions. Solvent polarity is a complicated topic, so we are going to use some simplified guidelines for predicting whether a reaction medium can support the formation of ions when if the sp³ C-LG bond breaks. We will classify solvents in three ways (Figure 0733):

(a) polar, protic: these are highly polar solvents that are also good hydrogen bond donors; the typical examples are water (H_2O), the low molecular weight alcohols such as methanol (CH_3OH) and ethanol (CH_3CH_2OH), and the low molecular weight acids such as formic acid (HCO_2H) and acetic acid (CH_3CO_2H)

(b) polar, aprotic: these are highly polar solvents that lack hydrogen bond donors; the typical examples are dimethylformamide [DMF: $(CH_3)_2NCHO$], dimethylsulfoxide [DMSO: $(CH_3)_2SO$], and hexamethylphosphoramide [HMPA: $((CH_3)_2N)_3PO$].

(c) nonpolar: there are many organic solvents that do not support the formation of ions; the typical examples are simple hydrocarbons (pentane, hexane, cyclohexane, toluene) and chlorocarbons (dichloromethane: CH_2Cl_2; chloroform: $CHCl_3$; carbon tetrachloride: CCl_4).

Figure 0733

Polar protic and polar aprotic solvents support carbocation formation.

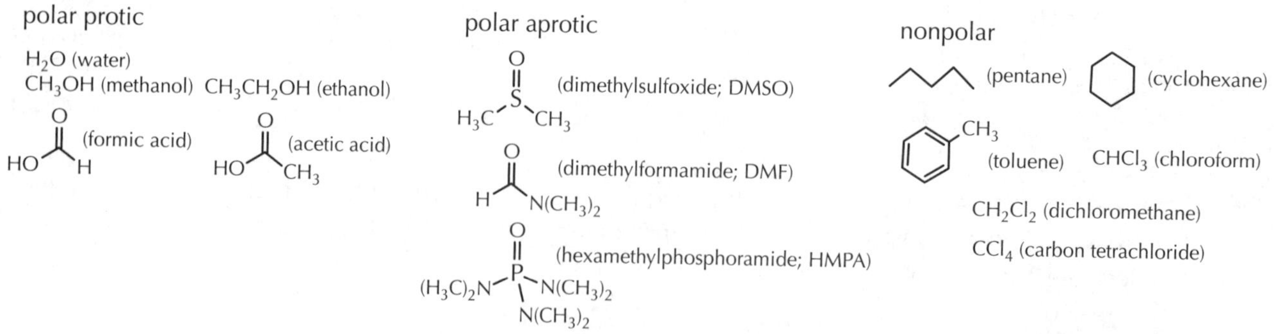

polar protic
H_2O (water)
CH_3OH (methanol) CH_3CH_2OH (ethanol)
HO—CHO (formic acid)   HO—C(O)—CH_3 (acetic acid)

polar aprotic
H_3C—S(O)—CH_3 (dimethylsulfoxide; DMSO)
H—C(O)—$N(CH_3)_2$ (dimethylformamide; DMF)
$(H_3C)_2N$—P(O)(—$N(CH_3)_2$)—$N(CH_3)_2$ (hexamethylphosphoramide; HMPA)

nonpolar
(pentane)   (cyclohexane)
(toluene) $CHCl_3$ (chloroform)
CH_2Cl_2 (dichloromethane)
CCl_4 (carbon tetrachloride)

step vi: If all three conditions for forming a carbocation are met, then $S_N1/E1$ is predicted; if not, then no S_N/E reactions are predicted. The default assumption is that the carbocation is a reactive intermediate and S_N1 and E1 is in competition, producing a mixture of all possible substitution and elimination products. If there are no β hydrogens, then no elimination is possible. If other experimental evidence is consistent with selectivity in the $S_N1/E1$ competition, then go with the experimental results (always).

7.3 A Predictive Model for S_N/E Reactions

The predictive model is obviously a lot of information to keep track of! Time for some practice.

In the rest of this section, we will take a look at six different chemical reactions and step through the decision-making process using these criteria and the decision-making guidelines. It is worth pointing out that this process is only aiming to provide one of four possible recommendations (predictions) for the combination of a Lewis base with any molecule with an sp³ C-LG bond: S_N2, E2, S_N1/E1, or no expected substitution/elimination.

We are exclusively focused on the prediction. Once we have the prediction, then (as already demonstrated in prior examples) the anticipated product(s) can be drawn. Six reactions (A–F) are shown in Figure 0734.

Figure 0734

Six chemical reactions for predicting S_N and E reactions.

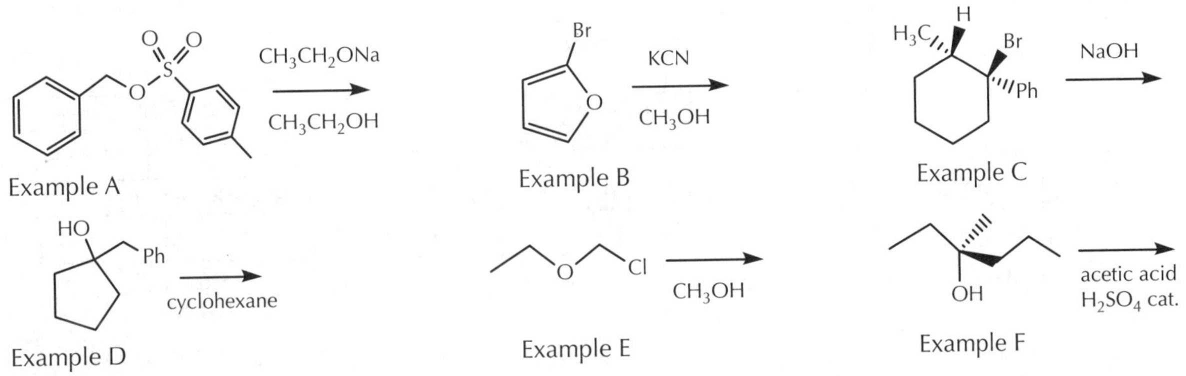

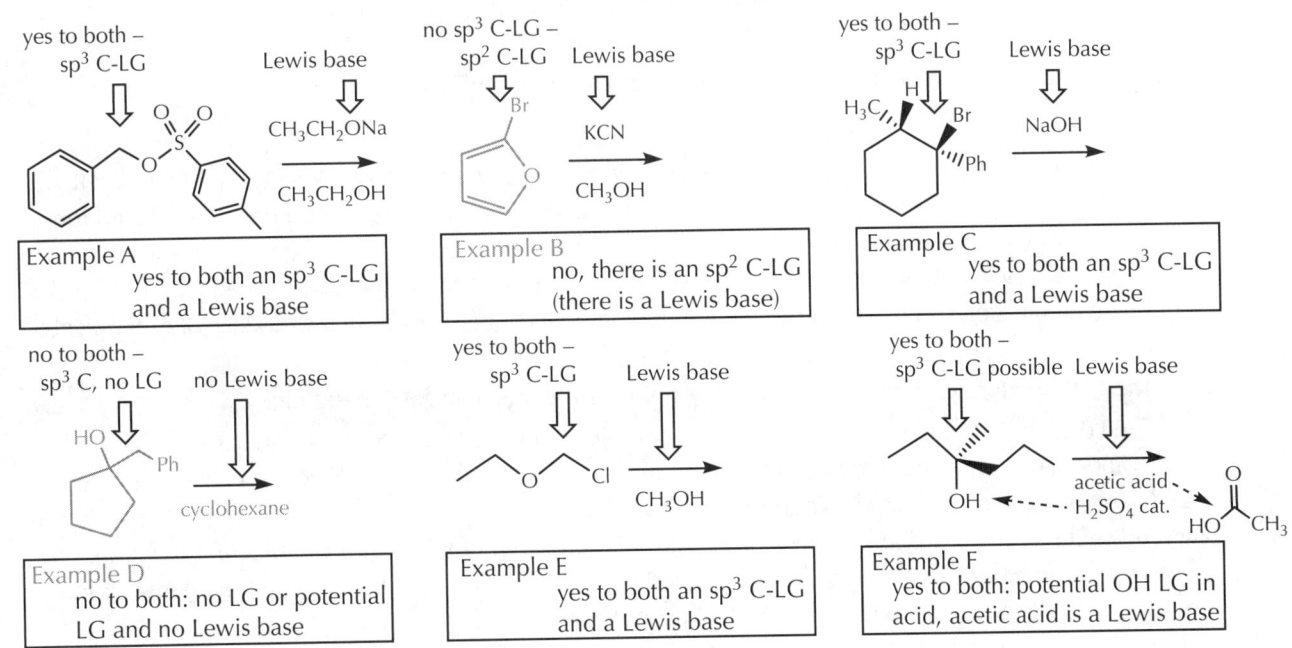

Figure 0735

Six chemical reactions to examine for S_N and E reactions: B and D are excluded.

The first step is to decide if the baseline conditions are satisfied for analyzing these reagents for the possibility of a bimolecular nucleophilic substitution or elimination reaction, as both reactive partners (an electrophilic sp³ C-LG bond and a nucleophilic Lewis base) must be present (Figure 0735).

560 **CHAPTER 7** Substitution and Elimination Reactions of Polar Sigma Bonds

With Examples B and D out of consideration for substitution/elimination reactions, we will now look at the remaining four, starting with the analysis of Example A (Figure 0736), followed by three others in Figures 0737–0739.

Figure 0736

Six chemical reactions to examine for S_N and E reactions: Example A is S_N2.

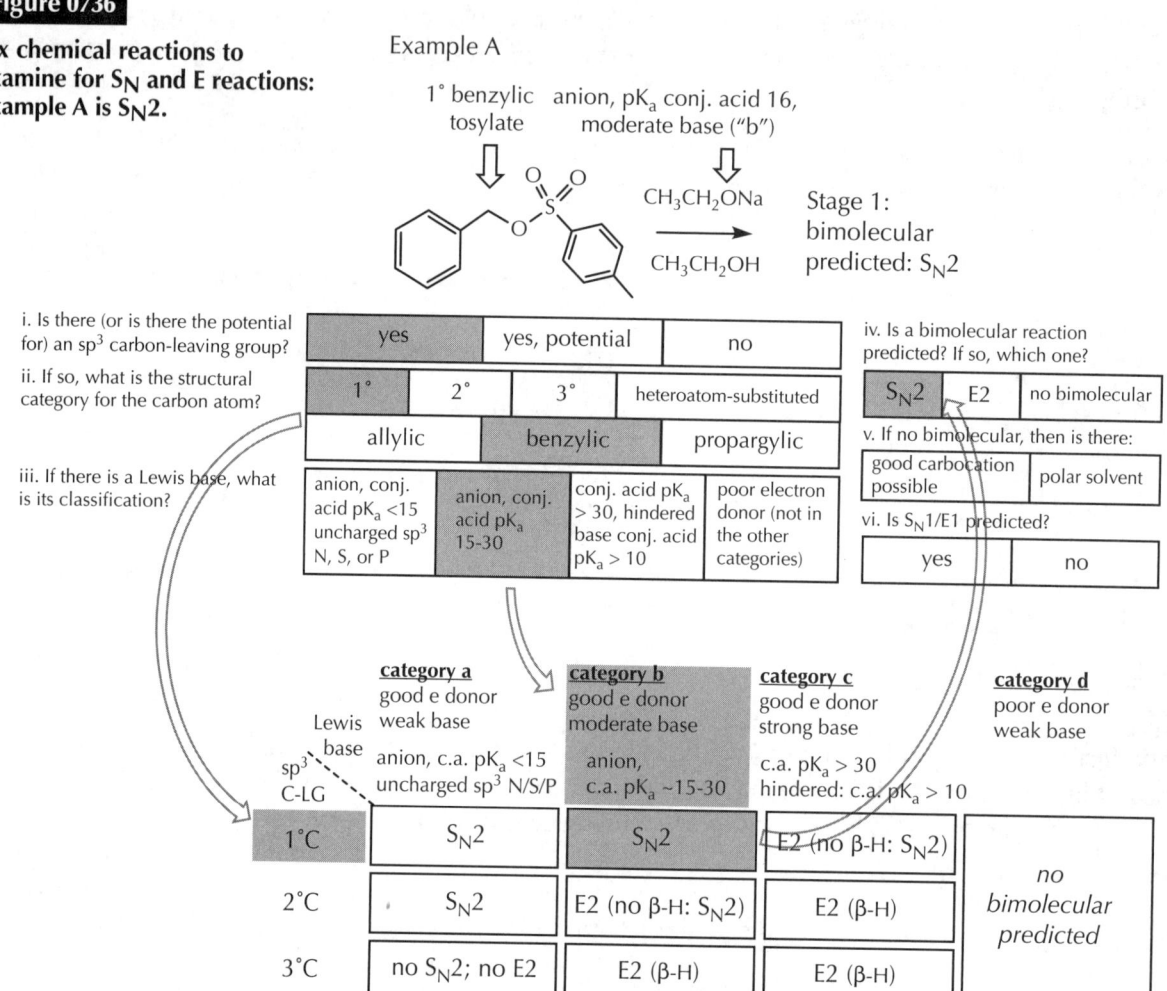

We will step through the decision-making for Example A in detail.

Example A:

(i) As established in Figure 0735, there is an sp^3 C with a leaving group (tosylate) attached to it, and there is a Lewis basic partner (ethoxide), so this reaction is up for consideration as a possible substitution/elimination; note that you could take only this information and draw all of the possible substitution and elimination outcomes, but the purpose of this analysis is to decide which, if any, of the substitution/elimination pathways is predicted for this combination of reagents, in the absence of experimental results.

(ii) The carbon atom with the leaving group on it is primary (1°) and it is benzylic.

(iii) The Lewis base (ethoxide) is an anion. The conjugate acid of ethoxide is ethanol, with its pK_a value of about 16; this places ethoxide in the category of being a good electron donor that is also a moderate base for substitution and elimination reactions.

With these items answered about these molecules, the question of whether a bimolecular reaction, and which one, can be answered. Moving to the prediction guidelines, the intersection between a primary compound and a moderate base predicts the S_N2 outcome.

(iv) S_N2 is predicted, and there is no need to proceed any further.

Earlier in the chapter, you saw examples where the mechanistic pathway was specified, and so with that information you can apply the concept to the case. We are keeping the prediction of the pathway separate from the application of the pathway to emphasize strongly that these are quite separate and distinct activities. That said, just as a reminder for now, predicting "S_N2" as the pathway means that you need to draw the result from substitution with inversion of configuration by applying those concepts to the molecules of the example.

In this section, we will step through and summarize the predictions for the remaining examples, and in the next section, take those predictions and apply the concepts to draw the product(s).

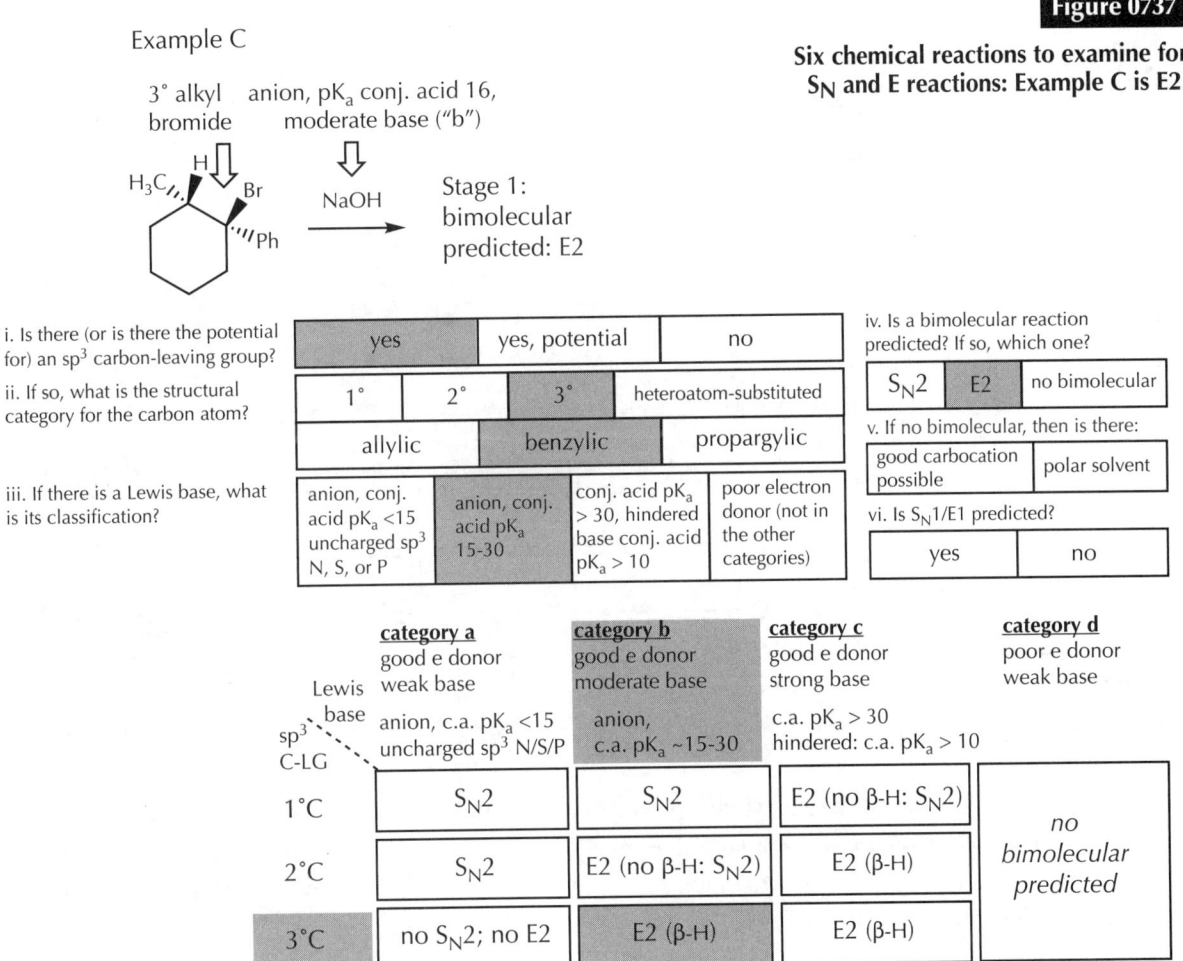

Figure 0737

Six chemical reactions to examine for S_N and E reactions: Example C is E2.

Example C:
(i) There is an sp^3 C with a leaving group (bromide) attached to it, and there is a Lewis basic partner (hydroxide).
(ii) The carbon atom with the leaving group on it is tertiary (3°) and it is benzylic.
(iii) The Lewis base (hydroxide) is an anion. The conjugate acid of hydroxide is water, with its pK_a value of about 16; this places hydroxide in the category of being a good electron donor that is also a moderate base for substitution and elimination reactions.

With these items answered about these molecules, the question of whether a bimolecular reaction, and which one, can be answered. Moving to the prediction guidelines, the intersection between a tertiary compound and a moderate base predicts the E2 outcome.

(iv) E2 is predicted, and there is no need to proceed any further.

562 CHAPTER 7 Substitution and Elimination Reactions of Polar Sigma Bonds

Figure 0738

Six chemical reactions to examine for S_N and E reactions: Example E is S_N1.

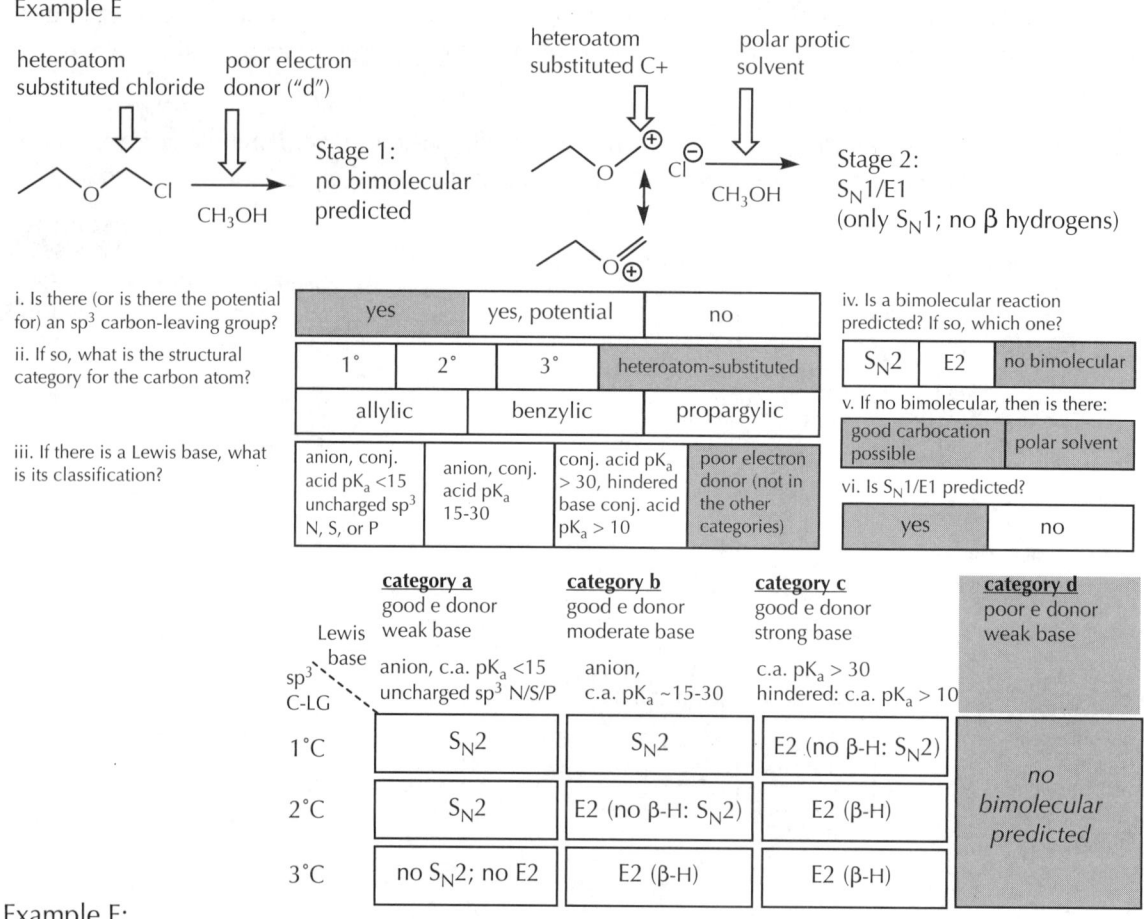

Example E:
- (i) There is an sp³ C with a leaving group (chloride) attached to it, and there is a Lewis basic partner (methanol).
- (ii) The carbon atom with the leaving group on it is heteroatom-substituted.
- (iii) The Lewis base (methanol) is not an anion, and its conjugate acid has a pK_a value of about -2; this places methanol in the category of being a poor electron donor that is also a poor base for substitution and elimination reactions.

With these items answered about these molecules, the question of whether a bimolecular reaction, and which one, can be answered. Moving to the prediction guidelines, any time you have a poor electron donor, no bimolecular reaction is predicted.

- (iv) So, in this example, no bimolecular reaction is predicted, and the possibility for a unimolecular reaction is now explored.
- (v) The first criterion of three is met: no predicted bimolecular reaction. The second criterion is whether a good carbocation would result if the leaving group ionized; the answer here is yes, because a heteroatom-substituted carbocation has a highly significant, closed shell resonance contributor. The third criterion is whether the reaction solvent is polar and can support the formation of these ions. Methanol, which is the prospective Lewis base, is also a polar protic solvent, and so the third criterion is also met.
- (vi) With all three criteria met, $S_N1/E1$ is predicted. In Example E, it is immediately noted that the leaving group has no β hydrogens, so elimination products are not even an option. Thus only S_N1 is predicted.

7.3 A Predictive Model for S$_N$/E Reactions

Figure 0739

Six chemical reactions to examine for S$_N$ and E reactions: Example F is S$_N$1/E1.

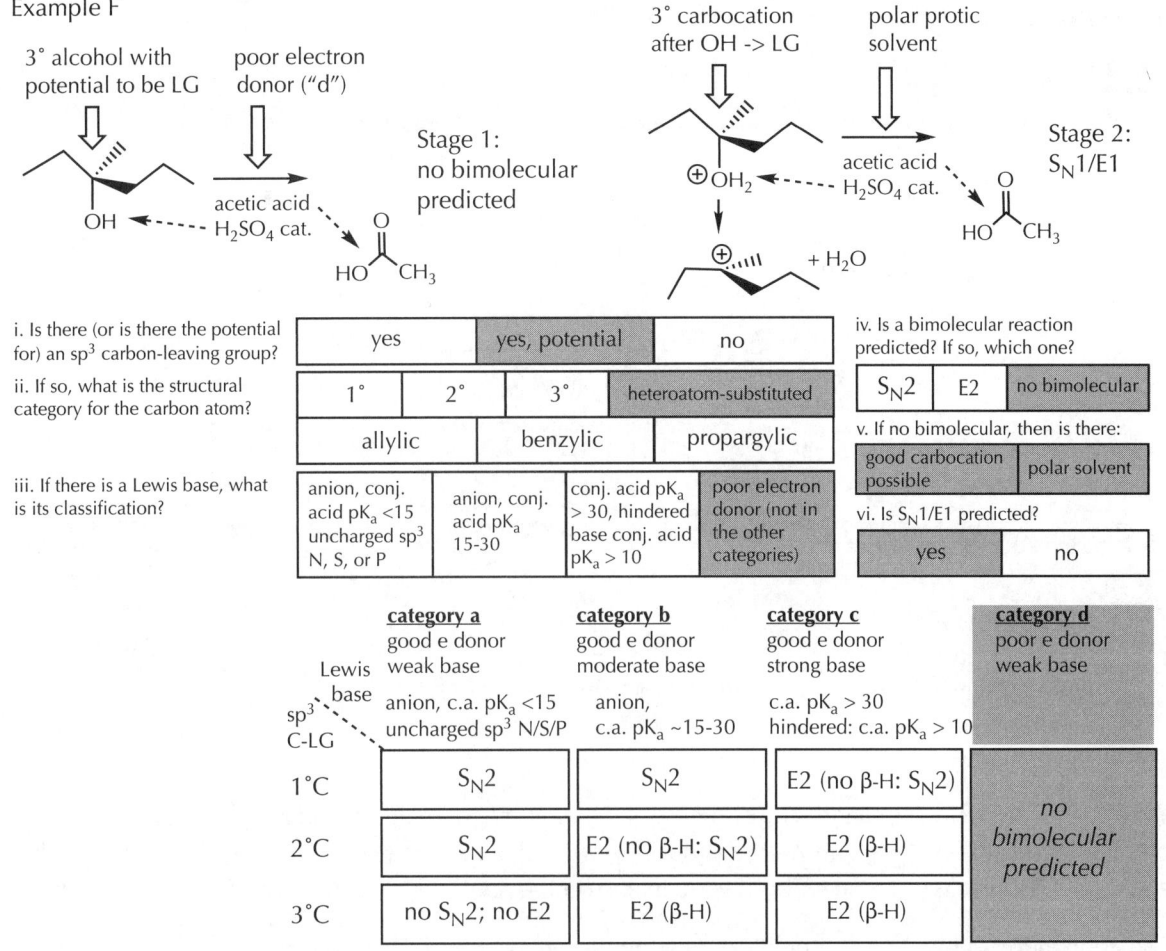

Example F:
(i) There is an sp^3 C with the possibility of a leaving group attached to it because the hydroxyl group is in the presence of a strong acid (H$_2$SO$_4$), and that could give the protonated hydroxyl group; and there is a Lewis basic partner (acetic acid).
(ii) The carbon atom with the leaving group on it is tertiary (3°).
(iii) The Lewis base (acetic acid) is not an anion, and its conjugate acid has a pK$_a$ value of about -5; this places acetic acid in the category of being a poor electron donor that is also a poor base for substitution and elimination reactions.

With these items answered about these molecules, the question of whether a bimolecular reaction, and which one, can be answered. Moving to the prediction guidelines, any time you have a poor electron donor, no bimolecular reaction is predicted.

(iv) So, in this example, no bimolecular reaction is predicted, and the possibility for a unimolecular reaction is now explored.
(v) The first criterion of three is met: no predicted bimolecular reaction. The second criterion is whether a good carbocation would result if the leaving group ionized; the answer here is yes, because a tertiary carbocation could result. The third criterion is whether the reaction solvent is polar and can support the formation of these ions. Acetic acid, which is the prospective Lewis base, is also a polar protic solvent, and so the third criterion is also met.
(vi) With all three criteria met, S$_N$1/E1 is predicted.

Going back to Figure 0734, in which the original six examples were presented, the work of the predictive model is only to deliver one of four recommendations: Is this combination of reagents more likely predicted to follow the S$_N$2 pathway, the E2 pathway, the S$_N$1/E1 pathway, or none of these? That's it. The results from using the predictive model are summarized in Figure 0740.

Figure 0740

Six chemical reactions to examine for S$_N$ and E reactions: Summary of predictions.

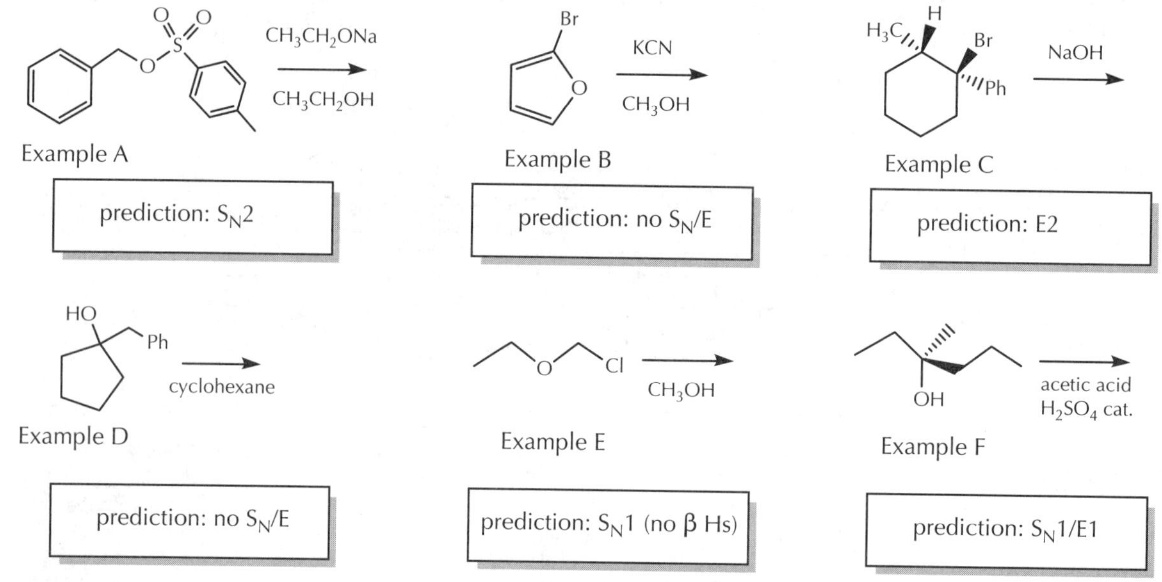

Example A — prediction: S$_N$2

Example B — prediction: no S$_N$/E

Example C — prediction: E2

Example D — prediction: no S$_N$/E

Example E — prediction: S$_N$1 (no β Hs)

Example F — prediction: S$_N$1/E1

With these conclusions in hand, you can now, independently of making the prediction, apply these recommendations to the molecules in question. The predictive model does not tell you what the products are, it only tells you which concept you need to answer the question. You still need to generate the answer.

The application of the recommendations to examples A, C, E, and F is illustrated in the next section.

C. Applying a Predictive Model for S$_N$/E Reactions

Using the diagnosis and treatment analogy, the work so far on these six examples has been only diagnosis.

Each of the six examples has had its symptoms noted: Is there a leaving group present? If so, is it attached to an sp^3 atom? How crowded is the atom? Is there a Lewis base and what is its character? From these questions, a diagnosis results: S$_N$2, no reaction, and so on.

Now we return to the earlier exercise: What if someone told you that the reaction was S$_N$1/E1 or E2? You have already done those kinds of "treatments" when the diagnosis was done for you. The only difference now is that you have the skills to do the diagnosis yourself.

Example A from Figure 0740 is predicted (diagnosed) to follow the S$_N$2 pathway.

Recall that the structure of the S$_N$2 product can be anticipated by using the definition of "nucleophilic substitution" combined with the identification of the sp^3 C-LG bond and the Lewis base. Indeed, the electrophilic carbon is not a stereocenter, and so there is only one substitution product; and there are no β-hydrogens, so elimination is not even an option. Structurally, there is only one S$_N$ product here, even if the question was "what are all possible products?" In this example, the outcome from S$_N$ looks the same, regardless of the mechanistic pathway.

If the mechanism needs to be shown, or any question related to the mechanism is asked, then the identification of "S$_N$2" is still critical. On the next four pages, the treatment protocols for the predicted substitution and/or elimination reaction will be applied to the four examples in Figure 0740 predicted to undergo one or both of these reactions.

Figure 0741 shows the original Example A, the prediction of the mechanistic pathway, the drawing of the product (simply derived from the definition of substitution), the mechanism, and the transition state (in which the inversion of configuration geometry must be illustrated, even if it is not observable, experimentally).

Figure 0741

Example A is predicted to be an S$_N$2 pathway.

Example A

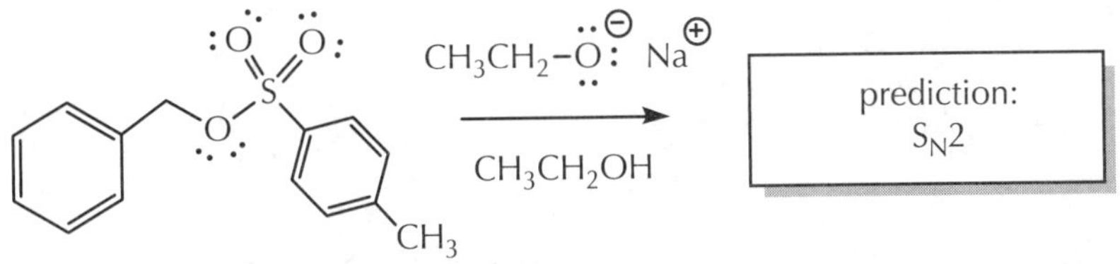

prediction: S$_N$2

structural outcome based on "S$_N$2" prediction

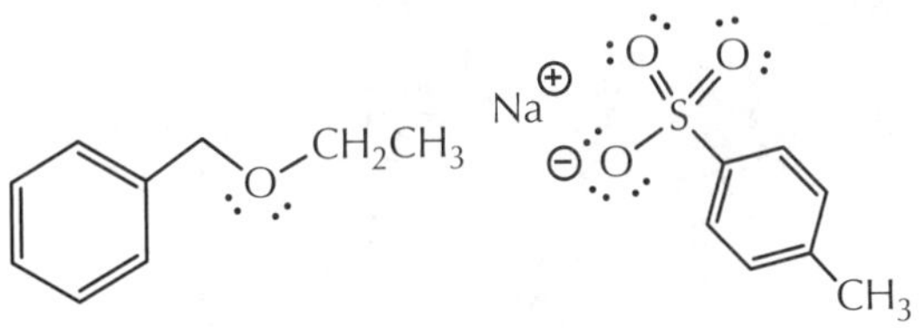

Of note:

Because there is no stereocenter, there is only one possible S$_N$ product that can be drawn, regardless of the mechanism. And with no β-hydrogens in the structure, there is no competition with E.

S$_N$2 mechanism (showing inversion trajectory)

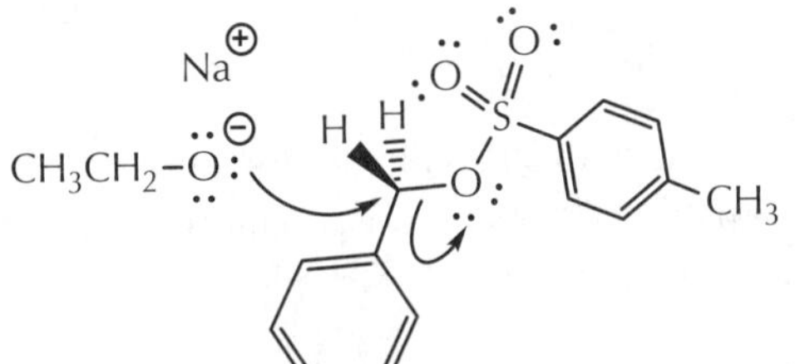

Of note:

An S$_N$2 mechanism is a concept about a collisional reaction and the anticipated trajectory. Even when there is no stereocenter to observe the inversion of configuration, the inversion pathway is occurring and it is a good habit to draw it that way.

transition state for S$_N$2

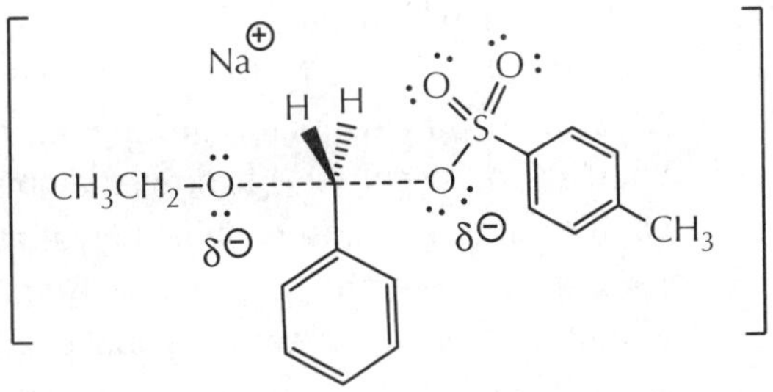

Of note:

Following directly from the statement above: If the drawing of the transition state is required, then the proper geometry is an integral part of that representation.

A reminder here that Example B from Figure 0740 was excluded for not meeting the structural criteria for substitution/elimination (the carbon attached to the prospective leaving group was sp^2).

The next illustration is Example C.

Example C is predicted to follow the E2 pathway. As a tertiary compound, competition with the S_N2 pathway is not an option. There are three β-hydrogen atoms, two in one direction on the ring and one in the other, so there are two different elimination options to consider. Again, using the definition of "elimination," the two possible products (C1 and C2) can be drawn (Figure 0742).

Figure 0742

Example C is predicted to be an E2 pathway.

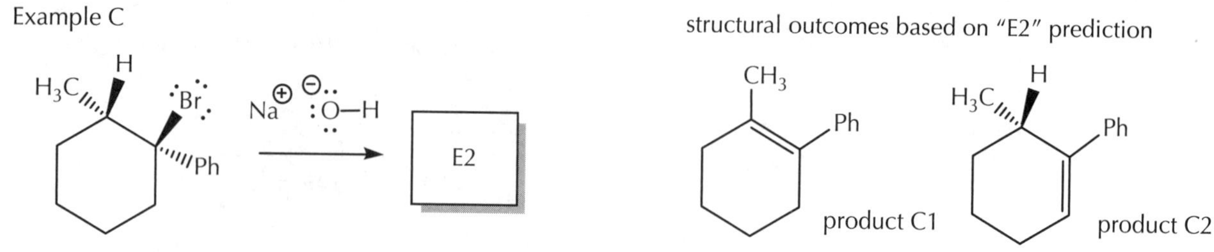

Chair conformations for Example C, mechanism for fastest E2 (anti β-H to LG) and its transition state

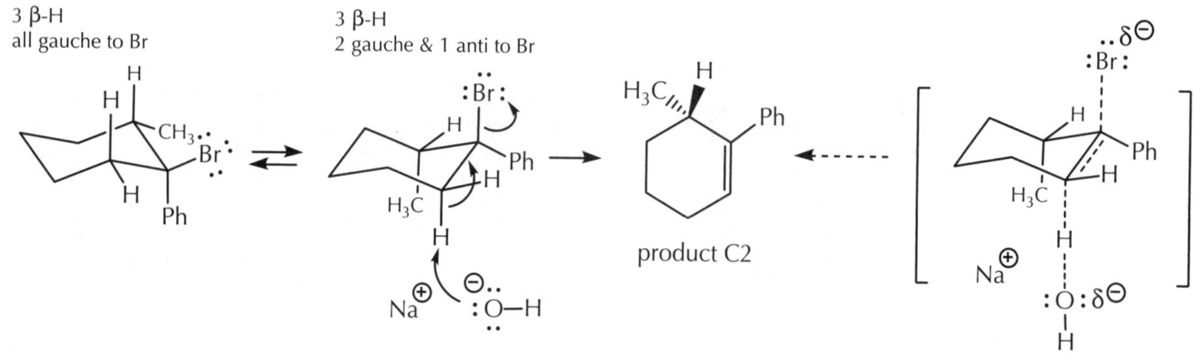

Figure 0742 shows the original example C, the prediction of the mechanistic pathway, the drawing of the products C1 and C2 (derived from the definition of elimination), the two chair conformations, the mechanism for what is expected to be the fastest reaction, and the transition state.

Notice how the predictions for "S_N2" and "E2" were not in any way providing the detailed answers for these reactions, but only the mechanistic pathway that needed to be applied to produce the answers.

As in S_N2, the E2 mechanism has a strong spatial bias: The E2 mechanism is fastest when the leaving group and the β-hydrogen atom are anti. There is a hint that the product in which the double bond forms away from the methyl group (product C2) will form faster than the product with the methyl group on the double bond (product C1), because the β-hydrogen sitting on the carbon with the methyl group is cis to the leaving group, and there is no way for it to end up anti. Whereas, in the pair of β-hydrogen atoms, one is cis and the other is trans. Assessing the prospect for this anti geometry is most convincingly illustrated when the two chair conformations of this molecule are drawn. In the conformation with the bromo group in the axial position, one of the β-hydrogen atoms is anti to it, and the other two are gauche and would give an E2 reaction at a slower rate than the axial one. And when the bromo group is in the equatorial position, it is gauche with respect to all three β-hydrogen atoms. The rate of elimination of the one β-hydrogen atom that is anti with respect to the axial leaving group is expected to be much faster than any of the other possible β-hydrogen atoms, and so product C2 is anticipated as the major product.

Example D from Figure 0740 was excluded for not meeting the substitution/elimination criteria, and Example E is predicted to follow the $S_N1/E1$ pathway.

As with Example A, the electrophilic carbon in Example E is not a stereocenter, and there are no β-hydrogen atoms. There is only one nucleophilic substitution product possible (Figure 0743).

Figure 0743

Example E is predicted to be an S_N1 pathway.

Example E

structural outcome based on "S_N1" prediction

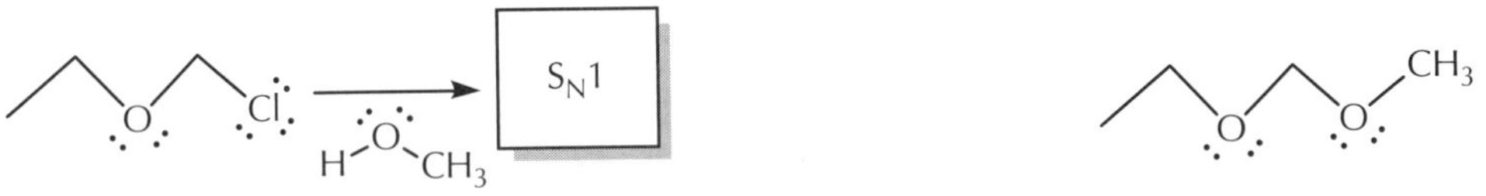

S_N1 mechanism: a choice between two formalisms, the "ionization" (one arrow) and "assisted ionization" (two arrows) drawings, can be used to show the formation of the intermediate carbocation, depending on which of its resonance contributors is desired

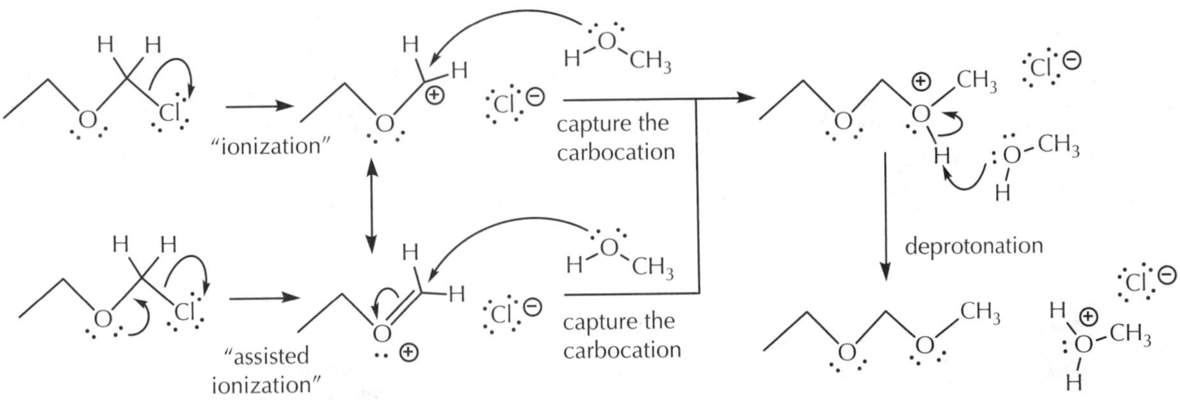

As in the previous reactions using methanol (CH_3OH) as a nucleophile as well as the solvent, its OH group will end up deprotonated in the final product, and so methanol is the source of the methoxy group (CH_3O) in the structure of the product.

The mechanism in Figure 0743 is depicted in one of two ways, depending on which of the resonance contributors for the carbocation you wish to show. Here is it important to remember that there is only one carbocation, and it can be shown with either the open shell resonance contributor or the closed shell resonance contributor.

Most experienced chemists prefer drawing the curved arrows that lead directly to the drawing of the closed shell resonance form because it is the more significant contributor to the actual structure. When the two-arrow mechanism is drawn, where the nonbonding electrons on the oxygen are shown moving in to form the CO pi bond at the same time the chloride leaves, it is called the "assisted ionization" version of the mechanism. Alternatively, the one-arrow loss of chloride ion to give the drawing of the open shell contributor is simply called the "ionization" version of the mechanism. This is a good time to remember that these are not alternative pathways, there is only one carbocation, and these are merely curved-arrow formalisms for showing which of the resonance contributors of the carbocation that the person writing the mechanism wishes to show.

568 CHAPTER 7 Substitution and Elimination Reactions of Polar Sigma Bonds

The last example is Example F, which is also predicted to follow the $S_N1/E1$ pathway.

As in Example E, the full array of substitution and elimination reaction products is predicted if no limiting information is provided (Figure 0744). In this example, there are no limits.

Figure 0744

Example F is predicted to be an $S_N1/E1$ pathway.

Example F

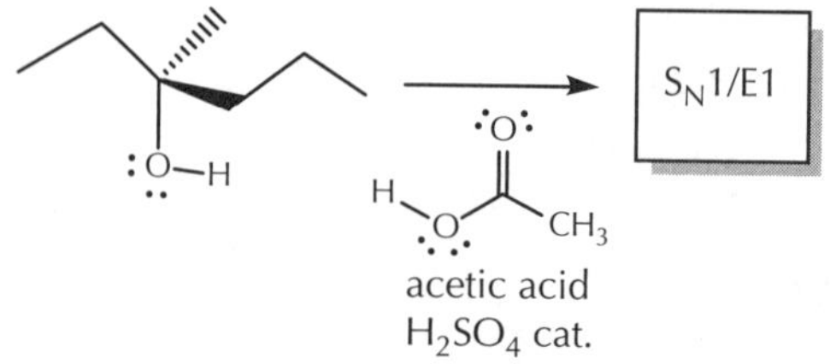

acetic acid
H_2SO_4 cat.

structural outcome based on "$S_N1/E1$" prediction (all possible outcomes)

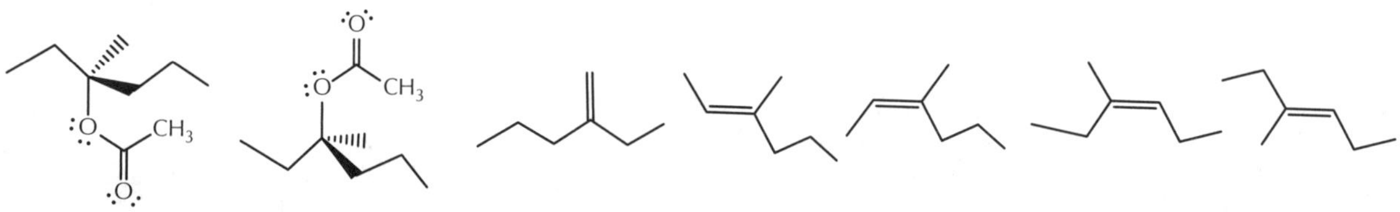

mechanism: protonation to give a LG, ionization to the sp^2 carbocation, followed by its capture/deprotonation (S_N1) or its deprotonation (E1)

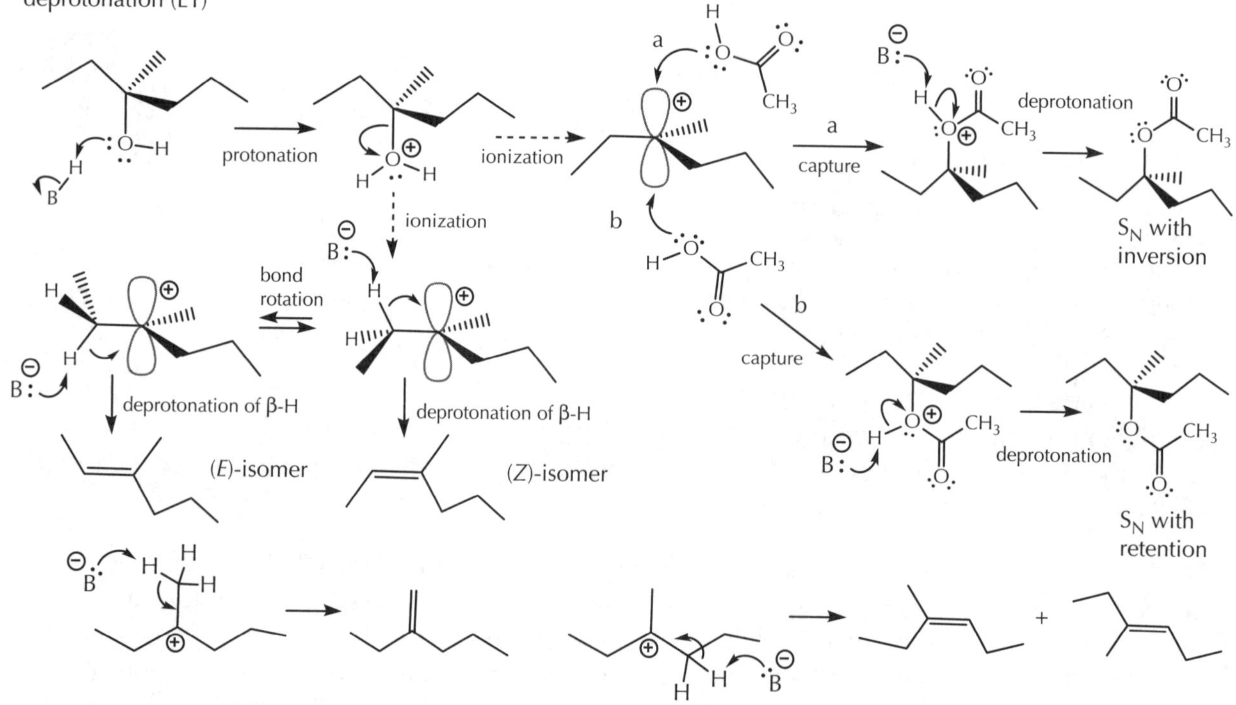

(E)-isomer

(Z)-isomer

S_N with inversion

S_N with retention

In Example F, the leaving group is located on a stereocenter. Both substitution products (inversion and retention) are possible, and five elimination reaction products are possible because there are 7 β-hydrogen atoms on three different alkyl groups, where (E)- and (Z)-diastereomers are possible from two of the three substituent groups. The exercise of predicting all possible S_N and E outcomes is the exact same process as from earlier in the chapter.

D. Transformations and Pragmatic Choices in Using S$_N$/E Reactions in Practice

Much earlier in the text, in Section 1.2, the three principles used, historically, in the early analyses of chemical reactions were introduced: atoms are conserved, the main group elements are well-behaved, and reactions are rarely explosions. A selection of balanced equations was provided in Figure 0105, including the one that is now repeated at the top of Figure 0745.

Figure 0745
Drawing inferences about transformations based upon reaction chemistry.

$$C_6H_{11}Cl + NaOH \longrightarrow C_6H_{10} + H_2O + NaCl$$

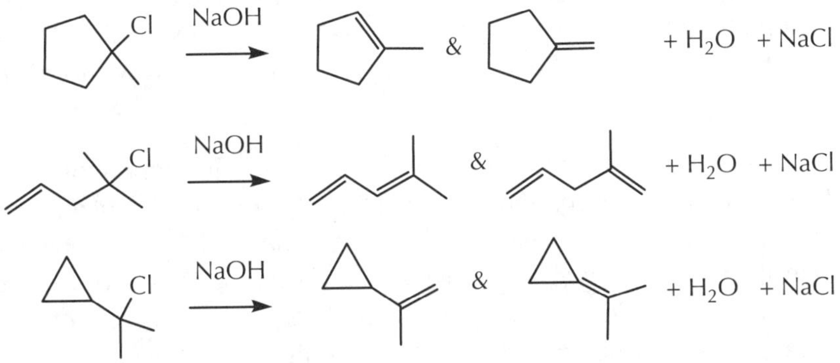

Armed with the knowledge about typical organic reaction chemistry, this balanced equation takes on deeper meaning. In this reaction, looking only at the molecular formulas, an organic chloride ($C_6H_{11}Cl$) with one unit of unsaturation reacts with sodium hydroxide (NaOH). The organic product (C_6H_{10}) shows the loss of HCl and is accompanied by water and sodium chloride. The reaction has the characteristics of an elimination reaction. A molecule of HCl has been lost from $C_6H_{11}Cl$ to give C_6H_{10}. The chlorine atom, which presumably started out covalently bonded to a carbon, has been lost as a chloride ion (in NaCl), typical of leaving group behavior. And the hydroxide ion has been protonated to give water, and deprotonation is the other part of an elimination reaction. The implied elimination reaction would most likely have been an E2 mechanism, as the C-Cl is a potential leaving group bond and the sodium hydroxide is a good electron donor (moderate base).

There is no other evidence in the first equation about the structure of any molecules, but the predictive model can narrow down the options. If this is an E2 reaction, the carbon of the C-Cl bond is likely to be an sp^3 atom. The reaction only shows the elimination product, so that carbon might well be tertiary (3°), which would preclude any competing S$_N$2 product. Three possible reactions (of many), consistent with the presumption about this being an E2 reaction, are proposed.

Another equation from Figure 0105 is shown in Figure 0746.

Figure 0746
Drawing inferences about transformations based upon reaction chemistry.

$$C_5H_{12}O + HCl \longrightarrow C_5H_{11}Cl + H_2O$$

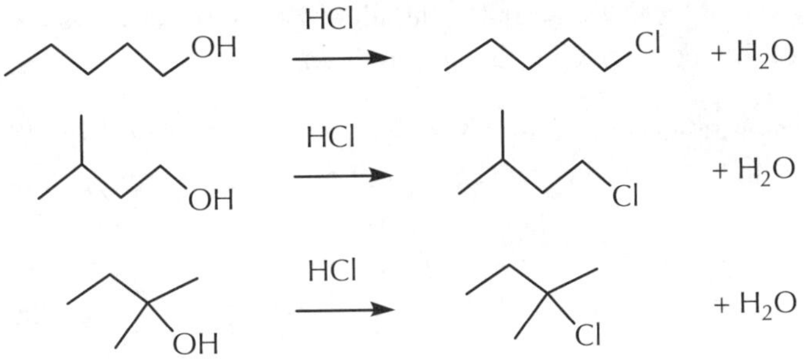

570 CHAPTER 7 Substitution and Elimination Reactions of Polar Sigma Bonds

Water is also formed here, but the pathway is different. The saturated organic molecule loses "OH" and gains "Cl," which is the hallmark of a substitution reaction. Although a hydroxyl group is not, itself, a good leaving group, its protonation by HCl can be followed by its replacement by a halide ion.

The protonated hydroxyl group can be used in both S_N1 and S_N2 reactions, depending on the structure and the reaction conditions. If this substitution reaction occurs along the S_N2 pathway, then the sp^3 carbon of the C–OH bond would best be primary (1°), while if the reaction happens along the S_N1 pathway, that carbon would likely be tertiary (3°). As in the previous example, there is no information to decide which molecules these actually are, although a few suggestions are included.

The important point here is that the predictive model also defines the characteristics of these S_N/E reaction pathways, and the information can be used for more than just predicting which pathway might be followed by a set of reagents. The reaction chemistry embedded in that predictive model allows for many different types of analyses.

Chemical transformations (how to get from "here" to "there") where you have specific information about the structures of the starting and ending points, is as common a way to think about reaction chemistry as predicting the products from combining reagents. In a real-world situation, there might be a molecule that someone wants for some practical reason, such as testing it for its biomedical properties. If you cannot buy that molecule, then it needs to be made, or, the most commonly used verb: synthesized. A chemist would search for molecules that are as close in structure to the desired substance, and then ask: Can I get from here (what I have) to there (what I want)?

Figure 0747 illustrates this type of transformation question.

Figure 0747
Solving a substitution transformation question using reaction chemistry.

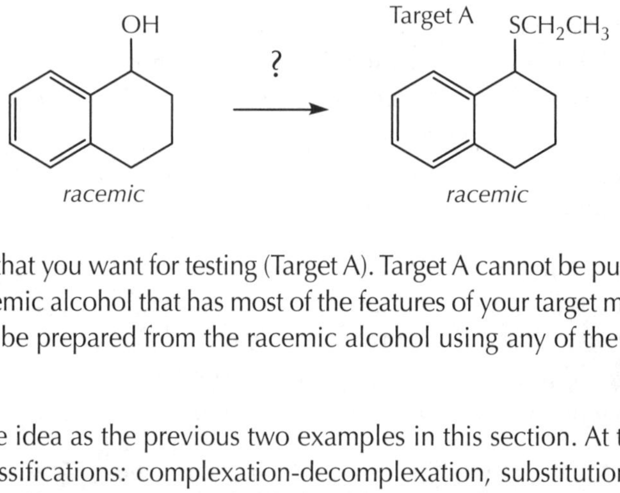

Imagine that there is a targeted substance that you want for testing (Target A). Target A cannot be purchased, but you find a commercial source of a racemic alcohol that has most of the features of your target molecule. The question is: Can this target molecule be prepared from the racemic alcohol using any of the types of reactions you have seen so far?

The answer to this question uses the same idea as the previous two examples in this section. At the most general level, you have four reaction classifications: complexation-decomplexation, substitution, elimination, and addition. And you have a limited but growing set of specific reactions at this point: Lewis acid-base reactions, Brønsted acid-base reactions, electrophilic addition reactions of strong and weak Brønsted acids, and nucleophilic substitution (S_N1, S_N2) and elimination (E1, E2) reactions. If the transformation matches the connectivity change for one of these four general classifications, then a more specific analysis can follow to see if it might be one of the more specific reaction pathways.

The transformation from the alcohol starting point to Target A is easily classified as a substitution reaction. A sigma-bonded sulfur substituent group has replaced the sigma bonded hydroxyl group. Does this substitution match the features required for an S_N1 or S_N2 pathway, or is it some type of substitution you have not yet seen? All of the questions from the predictive model apply, but now you are not predicting the outcome. You are deciding whether you can get from one of these molecules to the other, and if so, how?

7.3 A Predictive Model for S$_N$/E Reactions

Is there, or is there the potential for, an sp³ C-LG bond?

Yes. The hydroxyl group is not a leaving group, but it has the potential to be one by doing the reaction that forms one of the sulfonate esters. As a proposal for what might work, you would have the choice of any of them.

Is the character of the electrophilic carbon amenable to S$_N$1, S$_N$2, both or neither?

Yes. The sp³ carbon where the substitution takes place is 2° benzylic. Some secondary compounds can undergo S$_N$2, so that pathway is possible. As a benzylic atom, a resonance-stabilized carbocation would form, so the S$_N$1 pathway is possible, also.

What is the character of the Lewis base that is needed for SN2?

For a secondary compound, a Lewis base that is a good electron donor but a poor Brønsted base is predicted to follow the S$_N$2 pathway. The "CH$_3$CH$_2$S" group could be delivered as its anion using, e.g., CH$_3$CH$_2$SNa; its conjugate acid has a pK$_a$ value of about 10, so it fits into that category of Lewis base. In fact, the conjugate acid, CH$_3$CH$_2$SH, as an uncharged, sp³ sulfur atom, is also predicted to be a Lewis base in this same category, and amenable to S$_N$2.

What is the character of the Lewis base that is needed for S$_N$1?

The bimolecular reactions are all possible for secondary compounds, and so the Lewis base would need to have failed to predict an S$_N$2 reaction. Both the uncharged and anionic form of the sulfur substituent are predicted to give the S$_N$2 pathway, so even under conditions where the carbocation is encouraged to form (polar solvent, heated to speed up the ionization), the S$_N$2 pathway can still operate. From a purely predictive standpoint, the S$_N$1 pathway is not anticipated to be favored when an S$_N$2 reaction is possible because the S$_N$1 pathway goes through a high energy carbocation intermediate.

Is the stereochemical result consistent with S$_N$1, S$_N$2, both or neither?

Both. The S$_N$2 pathway gives inversion of configuration. Because the starting alcohol is racemic, both of its enantiomers are present, which means that both enantiomers will be present when whatever leaving group is formed. Each of the enantiomers from the racemic mixture would undergo inversion of configuration, so the racemic mixture of the Target A molecules would be expected. The S$_N$1 pathway would produce an achiral carbocation through the loss of the only stereocenter (i.e., both enantiomers would give the exact same carbocation), so a racemic mixture would be expected from that pathway, also, as the carbocation would form each enantiomer at the same rate.

The conclusion here is that two different S$_N$2 reactions could be used to carry out this transformation (Figure 0748).

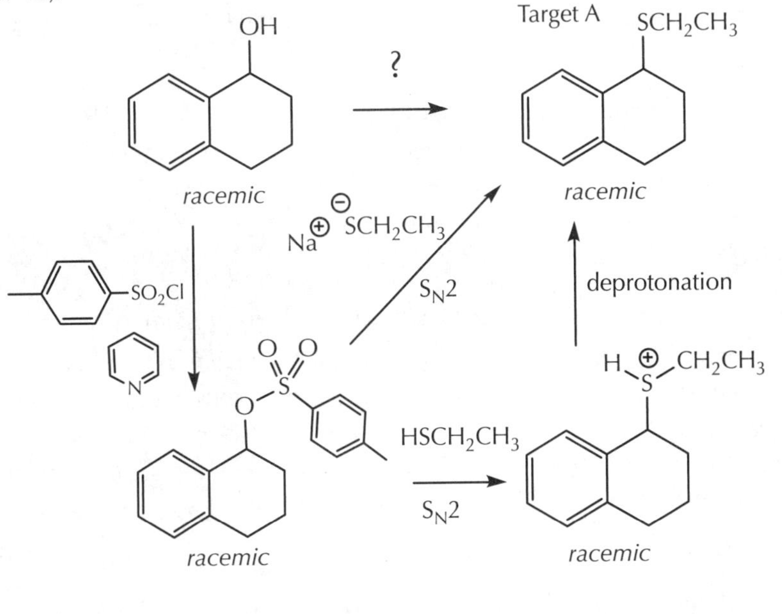

Figure 0748

Options for preparing Target A from the racemic alcohol.

The transformation would be a two- or three-step process. In both cases, the hydroxyl group needs to be changed into a proper leaving group. Then, in one case, the reagent bearing the sulfur anion can be used, directly. In the other case, the uncharged sulfur is used. The latter case would be predicted to produce the cationic sulfur intermediate, which would then need to be deprotonated to give the target compound. The cationic sulfur is expected to have a quite low pK_a value (more acidic than hydronium ion), and so this deprotonation reaction is easy to do. Experimentally, there are many different choices for how that step might get carried out, and so just indicating "base" or "adjust pH to neutral" is typically adequate.

The same starting material might lead to other compounds than Target A. For example, Target B could have been the desired outcome (Figure 0749). Target B cannot be purchased, either, but you see that its structure is also close to the commercially available alcohol used in the previous example. The question is the same: Can this second target molecule be prepared from the racemic alcohol using any of the types of reactions you have seen so far?

Figure 0749

Solving an elimination transformation question using reaction chemistry.

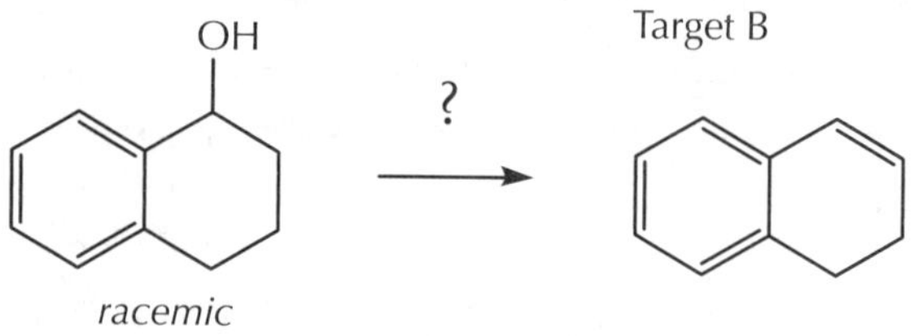

The transformation from the alcohol starting point to Target B is easily classified as an elimination reaction. Water has been removed from the structure through the loss of the hydroxyl group and a β-hydrogen, leaving a new pi bond behind. Does this elimination match the features required for an E1 or E2 pathway, or is it some type of elimination you have not yet seen?

Is there, or is there the potential for, an sp^3 C-LG bond?

Yes. The first question was already answered in the previous example.

Is the character of the electrophilic carbon amenable to E1, E2, both or neither?

Both. A secondary C-LG can give the E2 pathway. And as in the previous example, the secondary benzylic carbon can form a good carbocation, so E1 is possible.

What is the character of the Lewis base that is needed for E2?

A Lewis base that is also a moderate or strong Brønsted base is predicted to follow the E2 pathway with secondary compounds.

What is the character of the Lewis base that is needed for E1?

As long as the Lewis base failed to predictably give one of the bimolecular reactions, the conditions to form the carbocation could be used. The problem here is selectivity. Once the carbocation forms, controlling the E1 versus S_N1 outcome is difficult to predict, and so the specification to "only" form Target B would be more difficult to predict.

Is the regiochemical and stereochemical result consistent with E1, E2, both or neither?

Both. Because it is in a small ring, only the (Z)-configuration for the double bond is possible. Regioisomers are not a concern, either, because there are only β-hydrogen atoms at one site relative to the leaving group, and it is where the double bond needs to form. The E1 pathway has no stereochemical guidelines, as deprotonation of the carbocation is fast. One of the β-hydrogen atoms is *trans* to the leaving group, and so the favored anti relationship for a fast E2 reaction is possible.

The conclusion here is that both E1 and E2 reaction pathways could be used to carry out this transformation (Figure 0750), but the E2 would be the far better choice because the E1 pathway would be competing with the S$_N$1 pathway. The specification to "only" form the elimination product is more confidently predicted with the E2 pathway.

Figure 0750

Options for preparing Target B from the racemic alcohol.

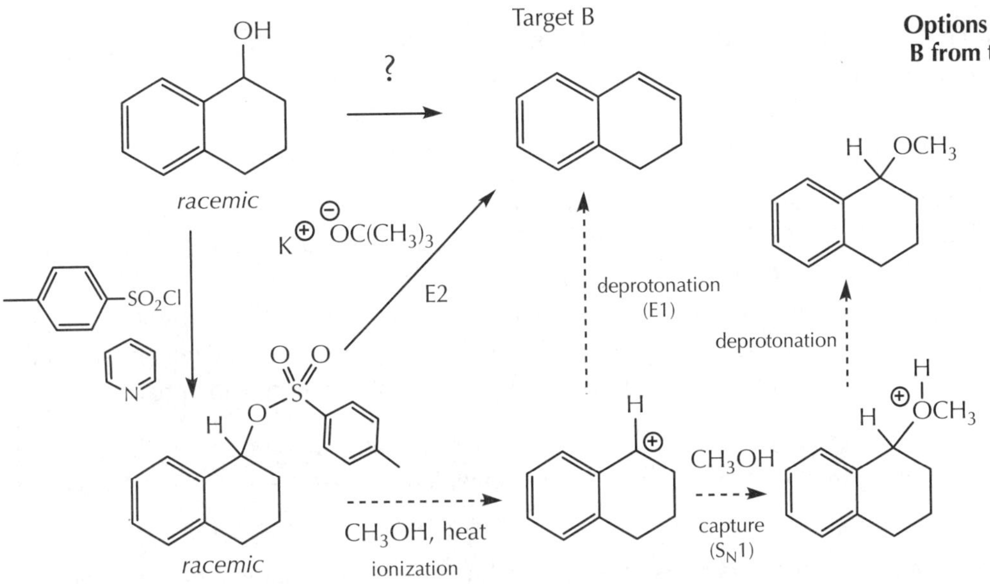

The E2 pathway takes two steps. As in making Target A, the hydroxyl group needs to be changed into a proper leaving group. There are then many different bases that might be used because the secondary compound favors the E2 pathway with both moderate and strong bases. Here is a case where chemists would make a pragmatic choice: Why not use one of the strong and hindered bases that favors the E2 pathway over the S$_N$2 pathway as much as possible? Reagents such as potassium *tert*-butoxide or DBU are often informally thought of as "E2 bases" because they favor the high E2 > S$_N$2 selectivity. You should consult Figure 0732 for four of the typical hindered or bulky bases used in elimination reactions.

In this regard, the choice of reagents used in other questions can also be a clue for the type of reaction taking place. When you see potassium *tert*-butoxide or DBU being used as a reagent, it is worth checking right away to see if an E2 reaction is being promoted.

The last transformation example in this section is shown in Figure 0751.

Figure 0751

A multistep transformation or a new reaction?

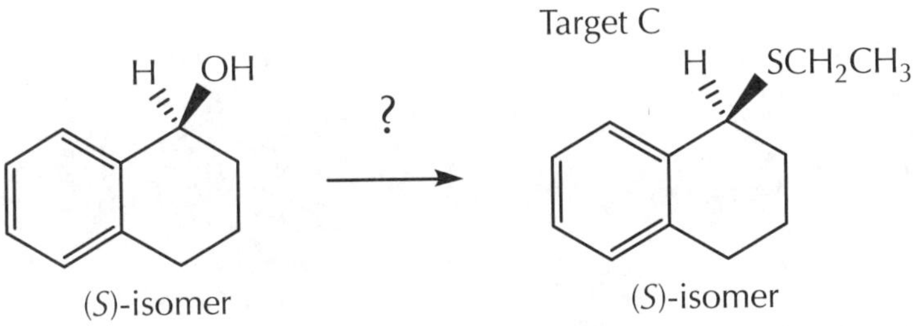

As with the transformation to give Target A, earlier, the change from C-OH to the C-SCH$_2$CH$_3$ is easily classified as a substitution reaction. Taking the mental inventory: This is also a secondary benzylic compound, the hydroxyl group can be changed into a proper leaving group, and the sulfur anion would be an appropriate Lewis base for substitution. But there is an interesting puzzle here: The stereochemistry of the transformation is only the retention of configuration S_N product. This outcome does not match the prediction from the S_N2 reaction (inversion) or the S_N1 reaction (racemic, in this case). As a general principle, when you are faced with a transformation that does not fit one of your prescribed pathways, either it is a new reaction, or it can be accomplished through a combination of the existing pathways.

This transformation is challenging to solve. You have no information to help sort out whether this is a new reaction or a combination of existing ones. And if it is a combination, you cannot tell how many of the previous reactions might need to be combined, which increases the number of options substantially. There is no algorithm to use here. Solving this sort of question requires a bit of playfulness and flexibility.

One substantial clue here is the stereochemical outcome. The result is unfamiliar, but the idea that some of these reactions have stereochemical outcomes that are stricter than the others is useful. And the transformation has most of the features of nucleophilic substitution, so that is a good place to be thinking, also. The other clue (just to be pragmatic again) is that you are a student in a course, learning this subject for the first time. If the answer here is that this transformation is a new reaction, it would be fundamentally unsolvable by a beginning student without a great deal of additional information being provided.

Sometimes the answer to a question comes through insight, where you appear to be sorting through multiple possible solutions and you have a little "aha" moment in which you see the answer. This strategy is tough to count on!

In multistep organic chemistry transformations, there is a deliberate strategy that you can use: working backwards. You have the structure of the product. You can always ask the question, "where might this have come from?" and then, with its potential predecessor in hand, ask, "can I link that predecessor to my starting compound? Figure 0752 outlines this strategy.

Figure 0752

Solving a multistep transformation question using reaction chemistry: the backwards thinking strategy.

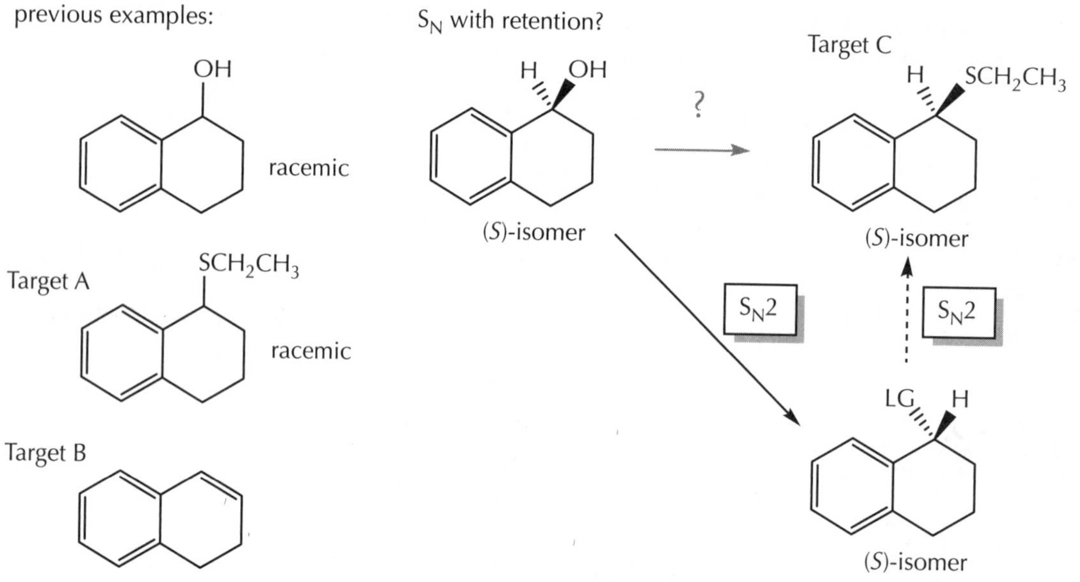

If we ignore the alcohol starting material for a moment, and focus only on the product, it looks like Target A. The molecule contains a sulfur substituent at the stereocenter of a secondary benzylic compound. Target A was the product of an S_N2 reaction. Without specifying the identity of the leaving group (for now), we can propose the type of starting material that would be needed for doing this reaction and the reagent that would be needed. This answers the first question, above, "where might this have come from?"

7.3 A Predictive Model for S$_N$/E Reactions

The second question you ask of yourself is a new transformation problem, "can I link the starting alcohol to this proposed predecessor molecule?"

The relationship between the starting alcohol and predecessor molecule with "LG" looks like a typical S$_N$2 reaction with inversion of configuration. At this point, the number of different groups that you know of as leaving groups is limited to sulfonate esters and halogen atoms, so it is easy to examine these two options explicitly. The two leaving group options (sulfonate ester, halogen) create two new transformation questions: "can the stereoisomer of the starting alcohol be transformed into a tosylate with inversion of configuration?" And, "can the stereoisomer of the starting alcohol be transformed into a halide with inversion of configuration?"

The answer to the second question is an immediate yes. Starting with the tosylate of the hydroxyl group, as before, Lewis bases that are good electron donors but poor Brønsted bases are predicted to follow the S$_N$2 pathway. In this case, a source of chloride ion satisfies this condition; it is an anion whose conjugate acid's pK$_a$ value is quite low.

The summary solution here is called a double inversion strategy (Figure 0753). Because S$_N$2 reactions give inversion of stereochemical configuration, two S$_N$2 reactions in a row give the retention of configuration outcome. The first S$_N$2 reaction is carried out with a Lewis base (chloride) that, in turn, can behave as a leaving group in the second S$_N$2 reaction.

Figure 0753

The double inversion strategy.

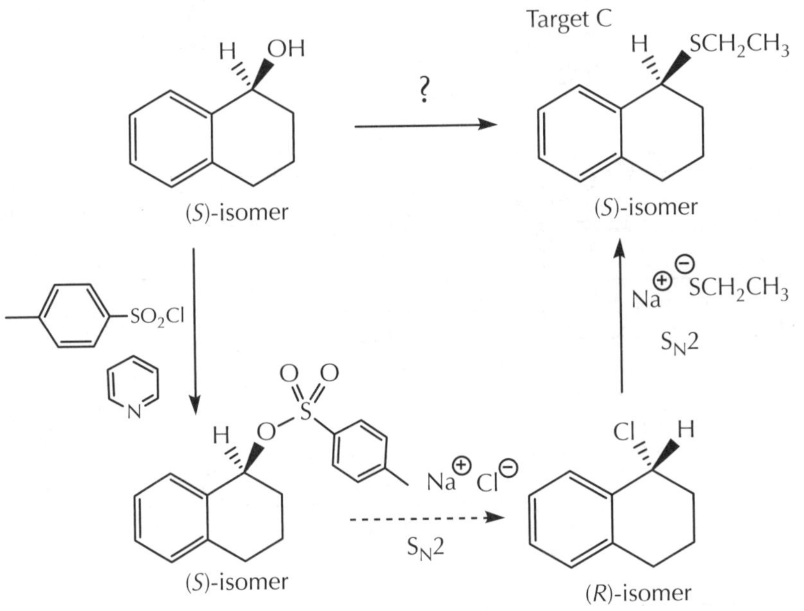

Here, you see organic chemistry getting sophisticated and clever. The double inversion strategy is a way to expand the scope of nucleophilic substitution reactions. Now you have a way to get the retention of configuration product from molecules that undergo two sequential S$_N$2 reactions.

576 CHAPTER 7 Substitution and Elimination Reactions of Polar Sigma Bonds

PRACTICE QUESTIONS

7.12 For each of the following examples, move through the questions of the predictive model for substitution and elimination reactions. Apply the criteria and anticipate which pathway is predicted. Then, apply the predicted pathway to the example and draw the anticipated product(s).

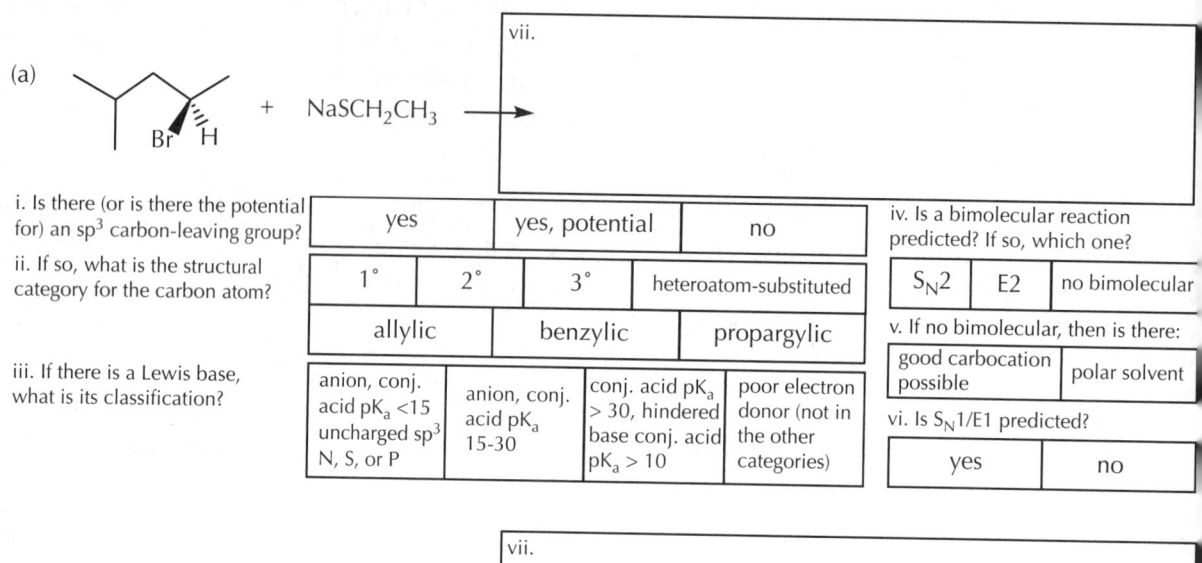

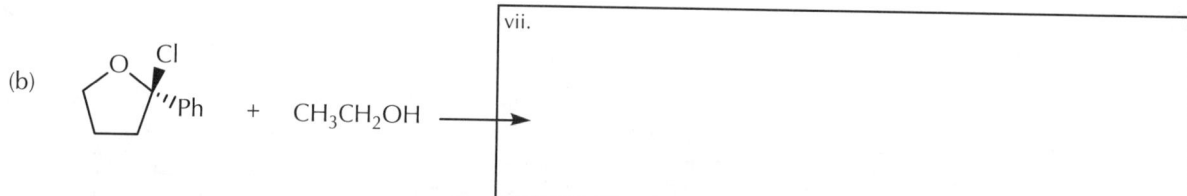

(c)

i. Is there (or is there the potential for) an sp³ carbon-leaving group? **yes** | yes, potential | no
ii. If so, what is the structural category for the carbon atom? 1° | **2°** | 3° | heteroatom-substituted
allylic | benzylic | propargylic
iii. If there is a Lewis base, what is its classification? anion, conj. acid pKₐ <15 uncharged sp³ N, S, or P | anion, conj. acid pKₐ 15-30 | conj. acid pKₐ > 30, hindered base conj. acid pKₐ > 10 | poor electron donor (not in the other categories)

iv. Is a bimolecular reaction predicted? If so, which one? S_N2 | **E2** | no bimolecular
v. If no bimolecular, then is there: good carbocation possible | polar solvent
vi. Is S_N1/E1 predicted? yes | no

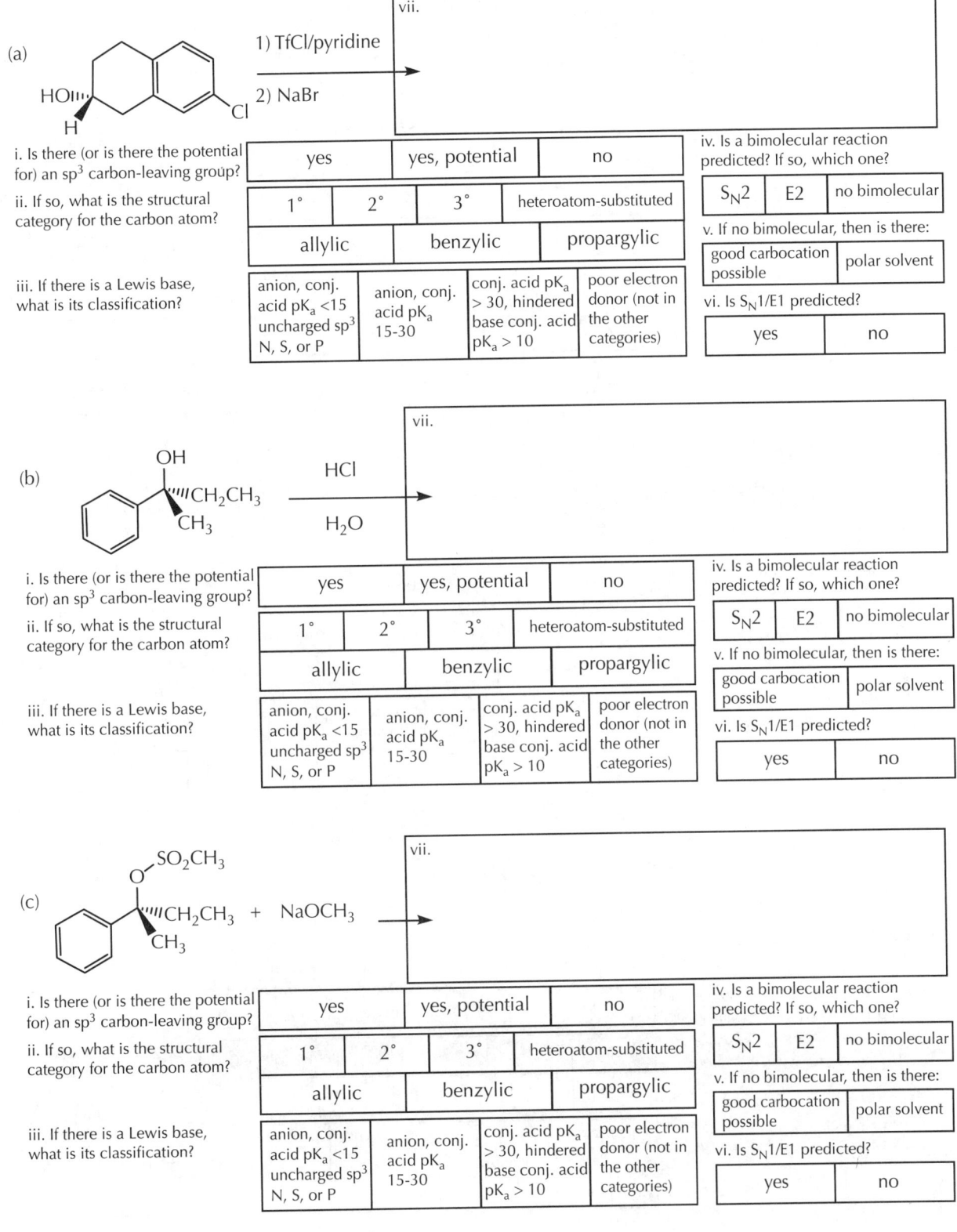

578 CHAPTER 7 Substitution and Elimination Reactions of Polar Sigma Bonds

7.14 For each of the following examples, move through the questions of the predictive model for substitution and elimination reactions. Apply the criteria and anticipate which pathway is predicted. Then, apply the predicted pathway to example and draw the anticipated product(s).

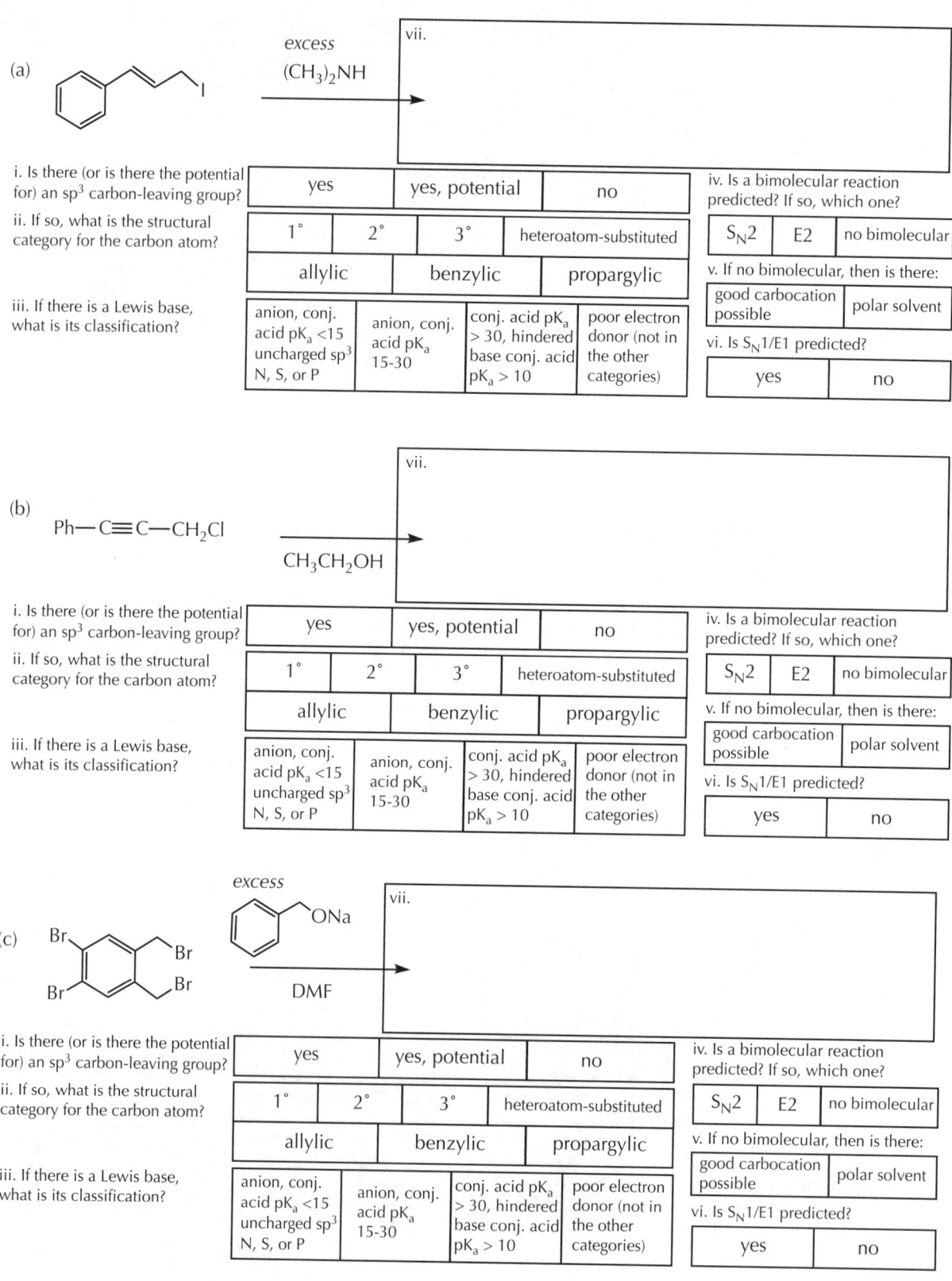

7.15 For each of the following examples, anticipate which substitution and/or elimination pathway is predicted. Then, apply the predicted pathway to the example and draw the anticipated product(s).

(a)

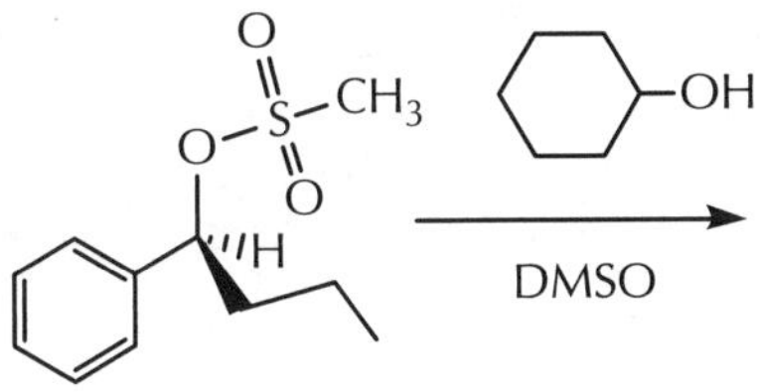

(b)

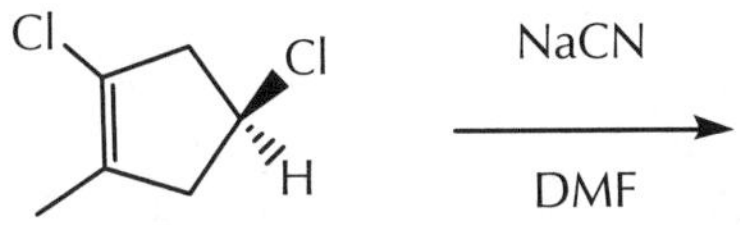

(c)

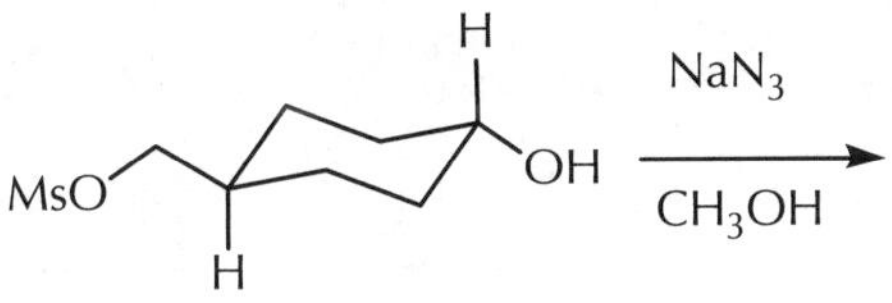

(d)

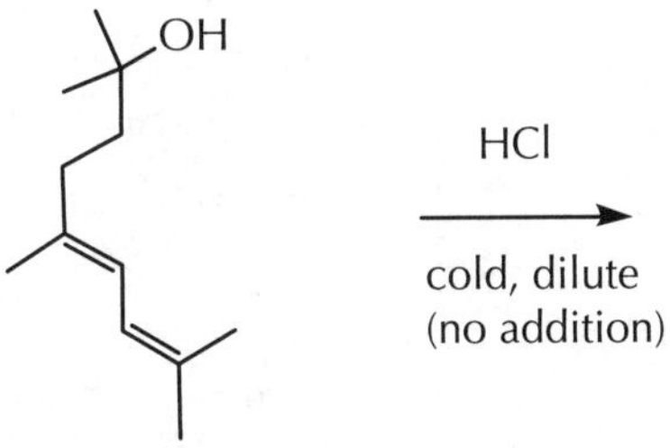

(e)

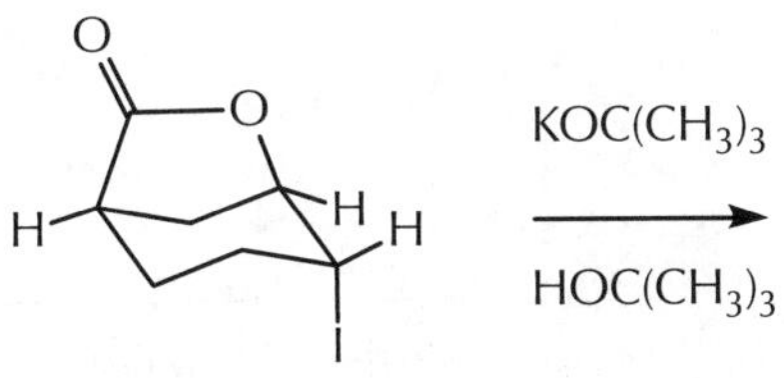

(f)

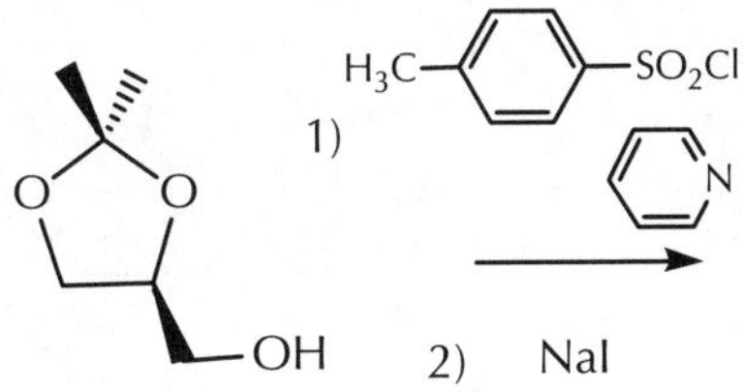

580 CHAPTER 7 Substitution and Elimination Reactions of Polar Sigma Bonds

7.16 Provide the reaction conditions needed to carry out the following transformations. If more than one experimental step is needed, number them appropriately.

(a)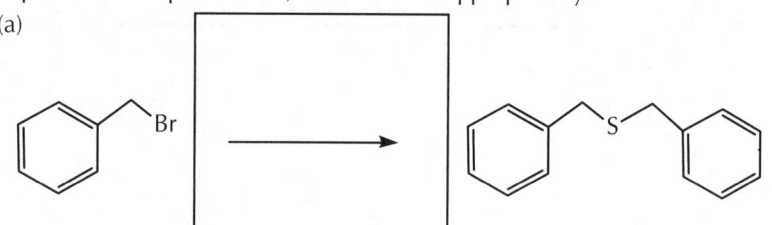

what is the anticipated mechanism and why?

(b)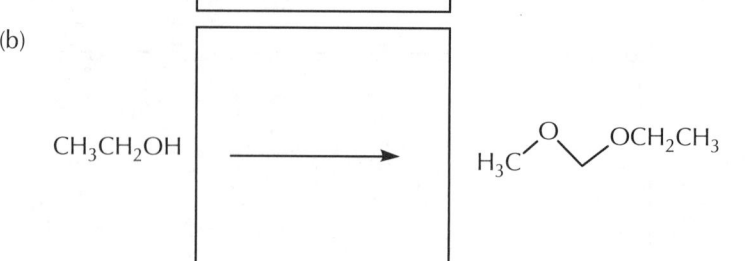

what is the anticipated mechanism and why?

(c)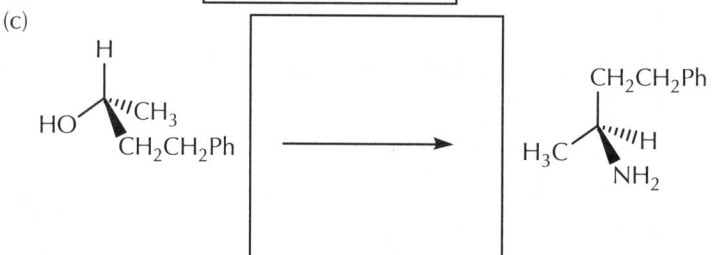

what is the anticipated mechanism and why?

(d)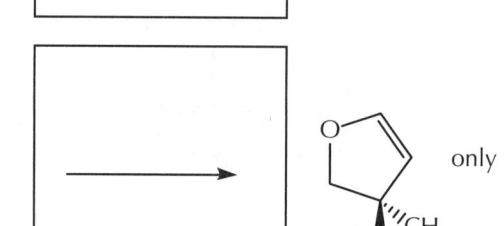

what is the anticipated mechanism and why?

(e)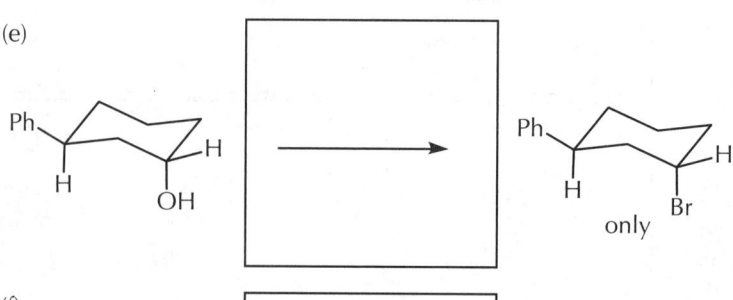

what is the anticipated mechanism and why?

(f)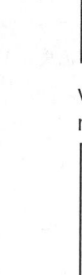

what is the anticipated mechanism and why?

7.17 For the following three reactions, provide (a) the curved arrow mechanism for each step, if there is more than one, and (b) the transition state structure for each step, including geometrical features.

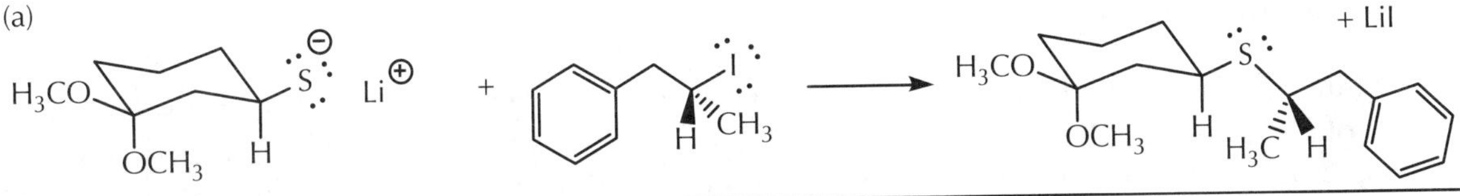

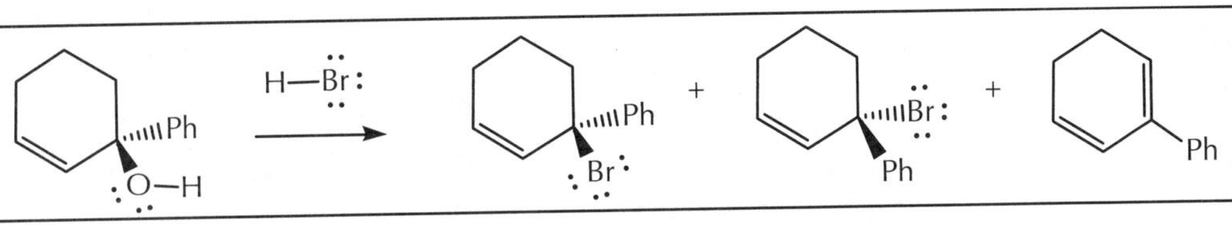

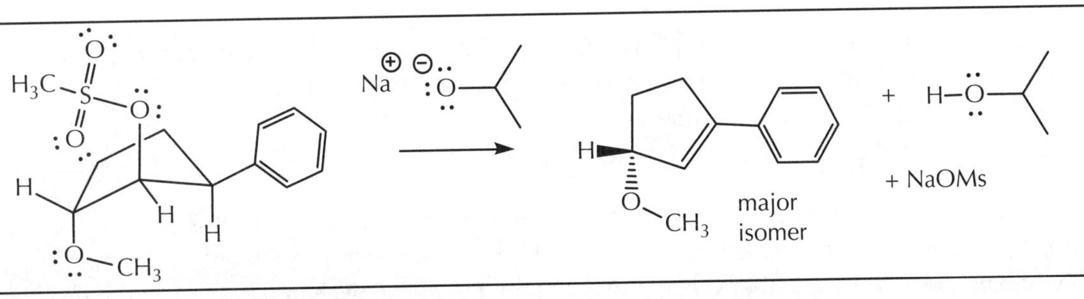

7.4 Rate and Reactivity in S_N/E Reactions

A. Leaving Group Ability

During the 1940s and 1950s, many different rate experiments were carried out to get insight into the details of these mechanistic pathways. Three of the properties identified through rate experiments were (1) leaving group ability, (2) solvent effects, and (3) nucleophilicity. The first one, leaving group ability, refers to the observation that, when all else was held constant, different leaving groups resulted in reactions with different rates. From our modern perspective, this is sensible, as the loss of the leaving group is involved in the rate-determining step for all of these mechanistic pathways. The property "leaving group ability" is described on a range good/better (faster rate) to poor/poorer (slower rate).

Although there is variation in individual cases, the generalized experimental results are given in Figure 0754 for a collection of reactions whose rates have been determined.

Figure 0754
Leaving group ability.

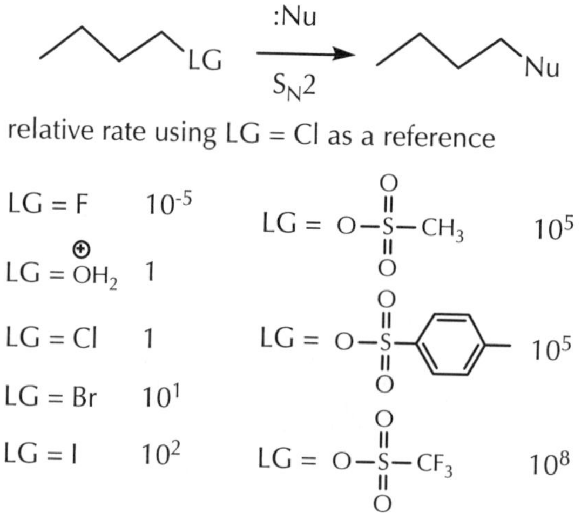

relative rate using LG = Cl as a reference

LG = F	10^{-5}	LG = $O-S(=O)_2-CH_3$	10^5
LG = $\overset{\oplus}{O}H_2$	1		
LG = Cl	1	LG = $O-S(=O)_2-C_6H_4$	10^5
LG = Br	10^1		
LG = I	10^2	LG = $O-S(=O)_2-CF_3$	10^8

The leaving group ability does not change the prediction about which reaction is expected: Whether it leaves faster or more slowly, a leaving group is still a leaving group. This means that only questions about relative rate can be asked about this property, not which reaction is taking place. In general, the leaving group trends you see in Figure 0754 are true:

(a) The sulfonate esters are better (faster) leaving groups than protonated hydroxyl and the halogens.
(b) Triflates are better (faster) than tosylates and mesylates, which are equivalent.
(c) Protonated hydroxyl is about the same as the halogens.
(d) Within the halogen group, iodides are faster than bromides, which are faster than chlorides.

There are resonance contributors for the sulfonate esters in which the oxygen atom at the electrophilic carbon is positively charged, which makes the C-O bond that much more highly polarized and more readily broken. Seen from the other side of the equation, the negative charge formed when the sulfonate leaving group leaves is more stabilized than on the halides, thanks to delocalization, and the equilibrium constant for losing these groups is highly favorable. The electron-withdrawing effect of the fluorine atoms in the triflate enhances both of these factors, and so it is the best of these sulfonate ester leaving groups.

Within the halides, the same trend observed for the Brønsted acidity of the HX acids is seen. The atom size difference for the C-I bond is greater than for C-Br, which is, in turn, greater than for C-Cl. The intrinsic difference in the covalent bond strength makes the C-I bond easier to break, and so on down the line.

Fluorine atoms have been studied, and although the bond is polar, the bond strength is also high (C and F are in the same row, are similar in size, and form a strong bond), and the relative rate, on this scale, is 10^{-5}, or 100,000 times slower than for the corresponding molecule with a chlorine atom.

To reiterate: Leaving group ability does not influence your prediction about the reaction outcome, but rather the relative rate of that reaction across a range of leaving groups. From a practical standpoint, in the laboratory, it might mean that if you were trying to carry out a reaction using a chlorine atom as a leaving group and the reaction was being sluggish, you might try the molecule with a bromine atom as a leaving group, instead.

B. Solvent Effects

The selection of a solvent can have a strong influence on the outcome of a chemical reaction. As you have seen, a polar (protic or aprotic) solvent is one of the three criteria for predicting the unimolecular pathways (S_N1/E1). The ions that form when the bond breaks (ionizes) need to be stabilized by solvation.

How does the solvent assist a bond in breaking? As the C-LG bond is undergoing its normal bond vibration, it is lengthening and shortening. As it lengthens, the bond becomes more polarized due to the electronegativity difference between the atoms in the bond. In a polar solvent, the increased magnitude of the partial charges during the stretch can be stabilized, making it easier (during heating) for some of the bonds to have enough energy to break. And at higher temperatures, there is a greater average energy to promote the bond breaking events. Without the stabilization of a polar solvent for these ions, they would likely just snap back together to relieve the buildup of charge, just as a sodium ion that might detach from its ionic NaCl lattice would quickly rejoin the lattice if faced with only a nonpolar solvent. Figure 0755 illustrates how the normal bond vibration process can lead to bond breaking (ionization) in a polar solvent. The experimentally observed values for one example are included. In the methanol/dioxane mixture, methanol is the more polar solvent. As the percentage of methanol increases, the overall polarity increases for the solution, and the corresponding S_N1 rate increases.

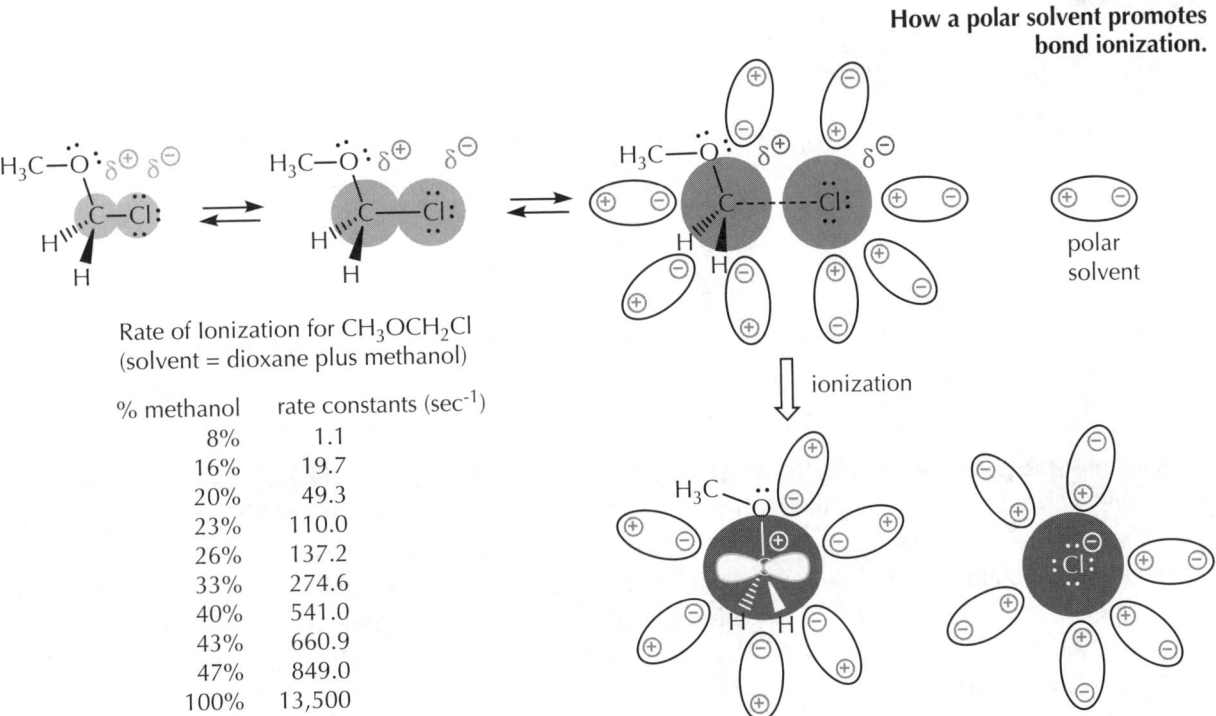

Figure 0755

How a polar solvent promotes bond ionization.

Rate of Ionization for CH_3OCH_2Cl
(solvent = dioxane plus methanol)

% methanol	rate constants (sec^{-1})
8%	1.1
16%	19.7
20%	49.3
23%	110.0
26%	137.2
33%	274.6
40%	541.0
43%	660.9
47%	849.0
100%	13,500

A second solvent effect is the difference that hydrogen bonding makes when comparing the polar protic and polar aprotic solvents, particularly in the reactions of ionic Lewis bases. As a rule, the reactions of Lewis bases in a polar protic solvent are significantly slower than the same reaction carried out in a polar aprotic solvent. The hydrogen-bonded Lewis base is stabilized relative to its non-hydrogen-bonded counterpart, so the distance to the transition state is greater. Also, the hydrogen-bonded Lewis base is effectively larger, with solvent molecules clinging to it. Lastly, some of these solvent molecules need to be removed (desolvation) to make the nucleophile available for reaction. None of these effects are present with a polar aprotic solvent, and the ionic Lewis bases are often referred to as "naked ions" in these solvents because they lack these three features associated with hydrogen bonding.

The generalization is summarized in Figure 0756, along with a specific example that compares the relative S_N2 rates for sodium bromide (NaBr) with iodomethane (CH_3I) in the polar protic methanol as solvent versus the polar aprotic dimethylformamide (DMF).

Figure 0756

Rate effects of polar protic versus polar aprotic solvents on Lewis bases.

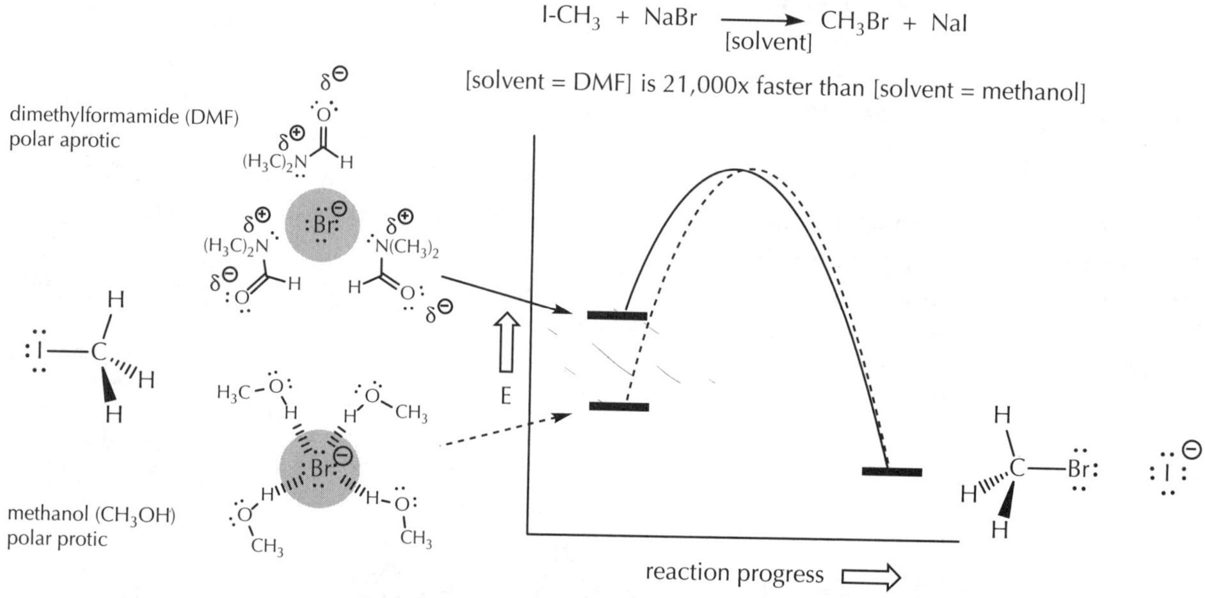

$$I\text{-}CH_3 + NaBr \xrightarrow{[solvent]} CH_3Br + NaI$$

[solvent = DMF] is 21,000x faster than [solvent = methanol]

C. Nucleophilicity

Another rate difference observed in S_N2 reactions is based upon the identity of the Lewis base and whether it is charged or uncharged. These rate differences are referred to as the intrinsic nucleophilicity of the Lewis base. A faster rate is interpreted as being a better or stronger nucleophile.

There are some parallels between this kinetic property (nucleophilicity) and the thermodynamic property of Brønsted basicity, as each of them relates to the ability to donate electrons for forming a new bond. Certainly, in S_N2 reactions, the more Brønsted basic version of a Lewis base (e.g., hydroxide ion) is faster, a better nucleophile, than its less basic version (e.g., water). And for comparable compounds in a row of the periodic table, the stronger Brønsted base is also the faster nucleophile (e.g., ammonia, NH3, and amines, RNH_2, R_2NH, R_3N, are faster than water, OH_2, and alcohols, ROH). The same trends are observed for compounds of the third row elements, particularly those of sulfur and phosphorus (e.g., $(CH_3)_3P$ and CH_3CH_3SNa are both faster, more nucleophilic, than H_2S).

Comparisons within a column of the periodic table are less uniform, and there can be a strong solvent effect that depends on whether the Lewis base is charged or uncharged. Three generalizations are typically made.

(1) For uncharged Lewis bases in the same column of the periodic table, the larger atom is faster (more nucleophilic); its electrons are less tightly held by the nucleus, and they can more readily undergo the distortion needed to form a new bond. This property of the larger atom's electrons is called polarizability. For example, H_2S is a better nucleophilic than H_2O, and $(CH_3)_3P$ is a better nucleophile than $(CH_3)_3N$.

(2) For charged Lewis bases in the same column of the periodic table, the larger atom is faster (more nucleophilic) in protic solvents; a larger, less basic atom is not as strongly hydrogen bonded as the smaller, more basic atom. In water or methanol, NaI is more nucleophilic than NaBr, which is more nucleophilic than NaCl.

(3) For charged Lewis bases in the same column of the periodic table, the smaller atom is faster (more nucleophilic) in aprotic solvents; without the hydrogen bonding to slow it down, the smaller, more basic atoms are faster than their larger counterparts. In dimethylformamide (DMF) or dimethylsulfoxide (DMSO), for instance, NaCl is more nucleophilic than NaBr, which is more nucleophilic than NaI.

Figure 0757 summarizes these trends. The rates for the chloride/bromide/iodide comparisons in methanol versus dimethylformamide (DMF) are reported on the same scale as a reminder that the solvent effect is also operating here. Despite the order of reactivity, note that all three of the rates of the "naked ions" in DMF are significantly faster than all of the rates in methanol, where hydrogen bonding occurs.

$$I-CH_3 + Nu: \xrightarrow{[solvent]} CH_3Nu + NaI$$

identity of Nu	[CH_3OH] rate relative to Nu = NaBr in CH_3OH	[DMF] rate relative to Nu = NaBr in CH_3OH
H_2O, ROH	1.7×10^{-5}	
NaCl	4.2×10^{-1} slower	4.3×10^5 faster
H_2S, R_2S	2.1×10^{-1}	
NaBr	1	2.1×10^4
NaOCH$_3$	3.3	
$(CH_3)_3N$	5.3 faster	slower
NaI	4.2×10^2	6.3×10^3
NaSCH$_3$	1.7×10^3	
$(CH_3)_3P$	1.3×10^3	

Figure 0757
Summary for nucleophilicity comparisons.

D. Heteroatom Electrophiles

Carbon atoms are not the only sites where substitution and elimination reactions can take place. Many different tetrahedral atoms with an attached leaving group can undergo these reactions.

Earlier in this chapter (Figure 0722), two of the mechanisms for forming a tosylate were shown as an S_N2 reaction at a tetrahedral sulfur atom, and this is repeated in Figure 0758 on the next page, along with a few other known examples of substitution and elimination reactions where there are heteroatom electrophiles. There can be a lot of variation in these reactions, so no good general statement can be made. The purpose of this short section is to remind you that the fundamental chemistry of many different atoms is being introduced here, not just the chemistry of carbon.

Figure 0758

Examples of substitution and elimination reactions using heteroatom electrophiles.

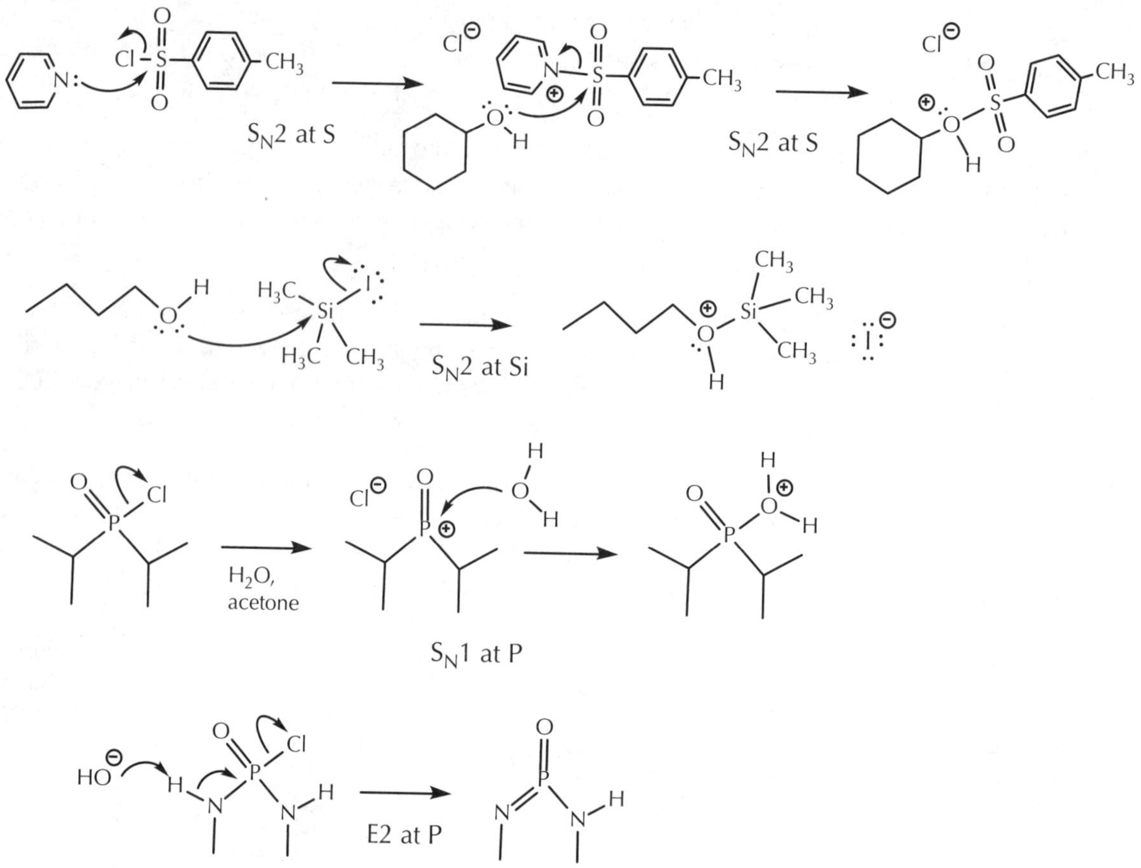

7.4 Rate and Reactivity in S_N/E Reactions

PRACTICE QUESTIONS

7.18 For each of the following pairs of reactions, tell which one is anticipated to be faster, and explain briefly the reason why.

(a)
- A. CH₃CH₂CH₂Br + NaCN →(DMSO) CH₃CH₂CH₂CN + NaBr
- B. (CH₃)₂CHBr + NaCN →(DMSO) (CH₃)₂CHCN + NaBr

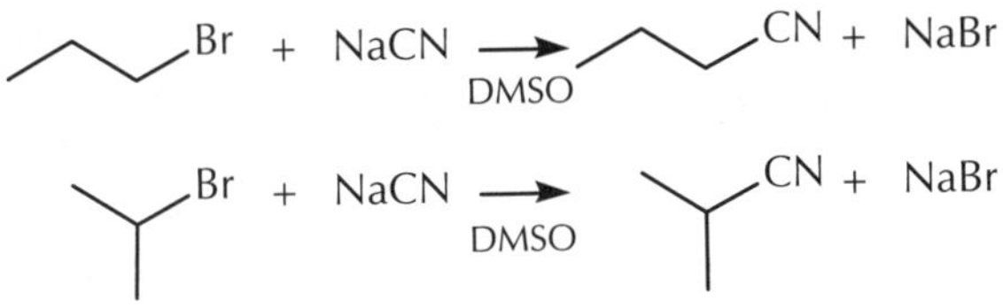

(b)
- A. CH₃CH₂CH₂Br + KOCH₃ →(H₂O) CH₃CH₂CH₂OCH₃ + KBr
- B. CH₃CH₂CH₂Br + KSCH₃ →(H₂O) CH₃CH₂CH₂SCH₃ + KBr

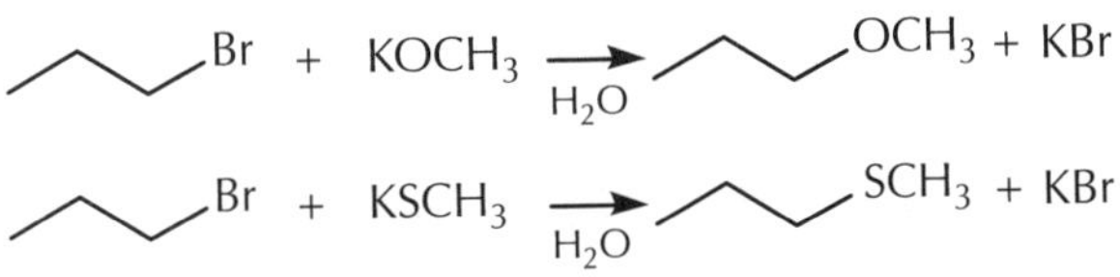

(c)
- A. CH₃CH₂CH₂CH₂Br →(H₂O) CH₃CH₂CH₂CH₂OH + HBr
- B. (CH₃)₃CBr →(H₂O) (CH₃)₃COH + HBr

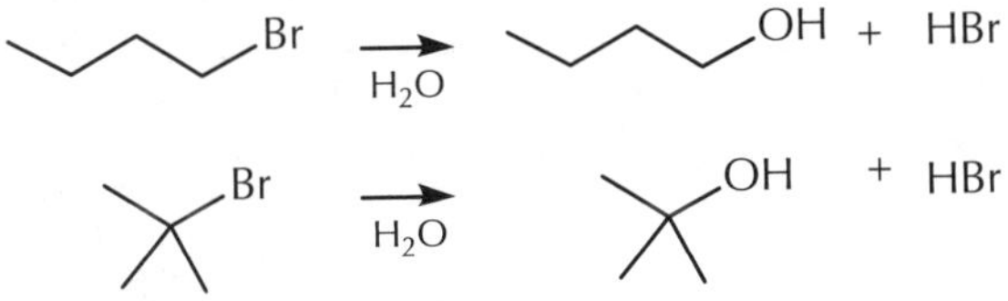

(d)
- A. (CH₃)₂CHBr + NaN₃ → (CH₃)₂CHN₃ + NaBr
- B. (CH₃)₂CHI + NaN₃ → (CH₃)₂CHN₃ + NaI

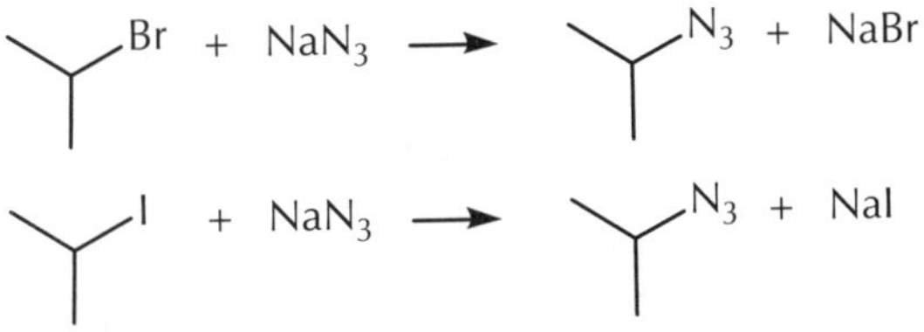

(e)
- A. CH₃CH₂CH₂Cl + NaCN →(DMSO) CH₃CH₂CH₂CN + NaCl
- B. CH₃CH₂CH₂Cl + NaCN →(H₂O) CH₃CH₂CH₂CN + NaCl

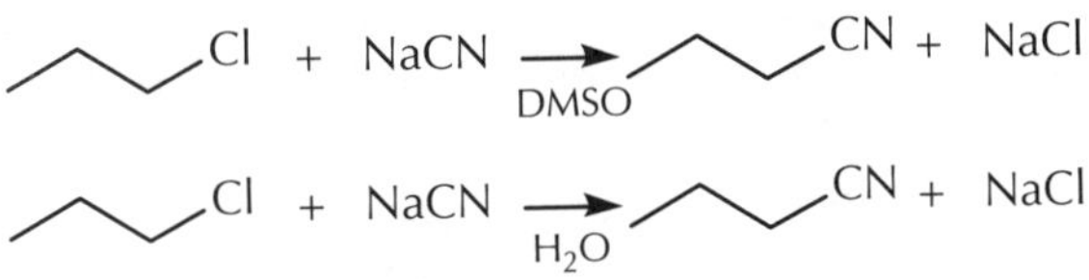

(f)
- A. PhCH₂CH₂I + P(CH₃)₃ → PhCH₂CH₂P⁺(CH₃)₃ I⁻
- B. PhCH₂CH₂I + S(CH₃)₂ → PhCH₂CH₂S⁺(CH₃)₂ I⁻

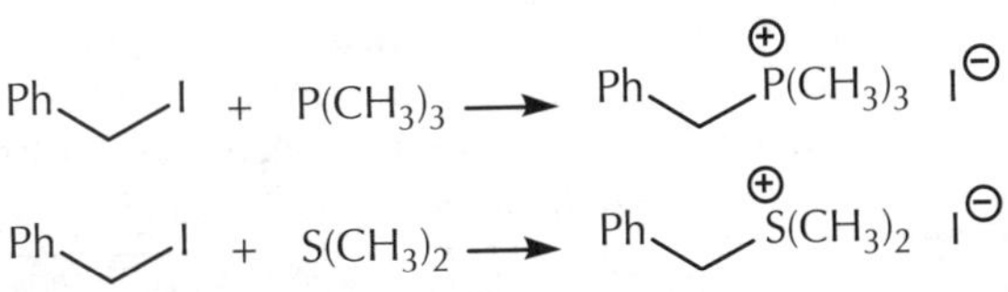

(g)
- A. CH₃CH₂CH₂OMs + NaSH → CH₃CH₂CH₂SH + NaOMs
- B. CH₃CH₂CH₂OTf + NaSH → CH₃CH₂CH₂SH + NaOTf

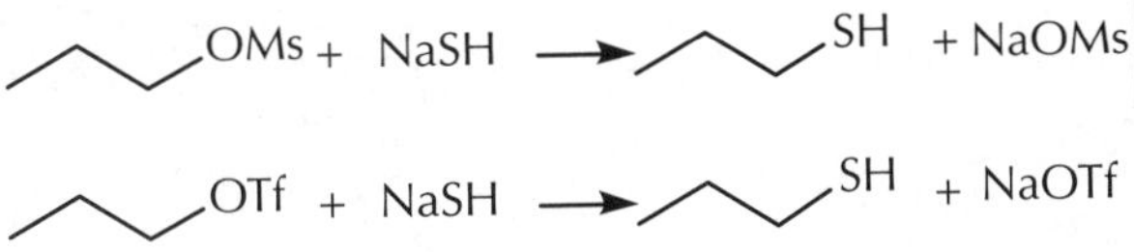

7.19 When the following three compounds are heated in dioxane-water, under exactly the same experimental conditions, there is a significant difference in the observed reaction rate. Using 1-chlorobutane as the reference, one of these two others, chloroethoxymethane, reacts 10^9 times faster, while the other, 1-chloro-2-methyoxyethane, reacts at 1/5 (0.20) of the rate. Explain this difference in the observed unimolecular substitution rates.

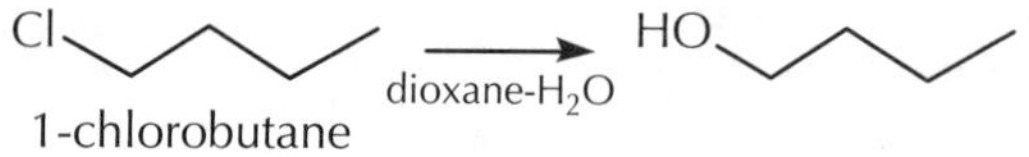

Cl~~~ →(dioxane-H₂O) HO~~~
1-chlorobutane

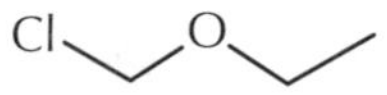

Cl~O~
chloroethoxymethane
10^9 faster than 1-chlorobutane

Cl~~O~
1-chloro-2-methoxyethane
0.20 of the rate of 1-chlorobutane

7.20 When benzyl bromide (PhCH₂Br) reacts with the cyanate ion, the observed product is derived from reaction at the nitrogen atom end of the ion. Under exactly the same conditions, the product resulting from the thiocyanate ion is derived from reaction at the sulfur atom end of the ion. Draw out the structures for the two ions and provide the reaction mechanisms. Give an explanation for these observations that addresses the selectivity observed for each ion.

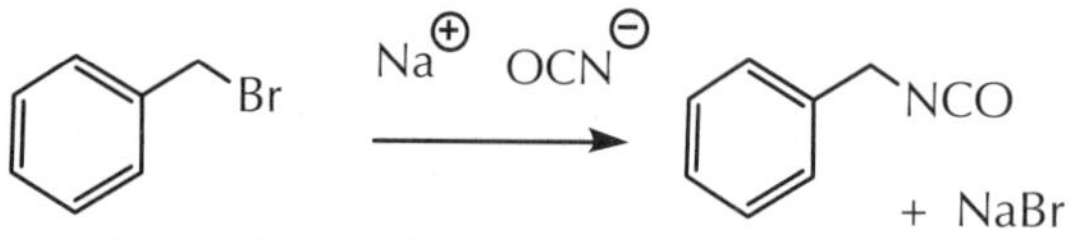

benzyl bromide + Na⁺ OCN⁻ (sodium cyanate) → PhCH₂-NCO + NaBr

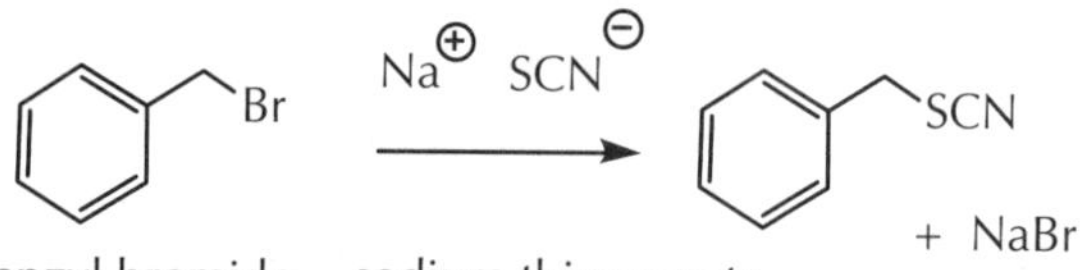

benzyl bromide + Na⁺ SCN⁻ (sodium thiocyanate) → PhCH₂-SCN + NaBr

7.21 Solutions of (S)-2-bromopentane in ethanol have specific rotations of [α] +31° at room temperature. If you take a solution of (S)-2-bromopentane and add some sodium bromide to it, the optical rotation of the solution diminishes over the time of a few hours until the solution is optically inactive and remains that way. If the amount of added sodium bromide is increased, the rate at which the solution becomes optically inactive increases. Explain what is happening in these experiments and why the solution becomes optically inactive.

Reflections About Science: Die Verdammten Experiments—Models and the Limits of Predictions

One definition of science is the process by which we create reliable explanatory and predictive models about observable natural phenomena.

In this chapter, the interlocking relationship between explanation and prediction is key. Observations of chemical reactions started long ago. Representations using molecular structures to catalog the changes began in the early 1800s, and familiar terms such as "substitution," "elimination," and "addition" were applied as classifications for the observed changes. By the mid-20th century, the bonding model featuring subatomic particles was widely accepted, as was the underlying idea (quantum mechanics) behind sigma and pi bonds.

In the 1930s and 1940s, clever experiments in which relative rate, isotopic distribution, and stereochemical outcomes were measured began to provide a set of unified explanations for a hundred years of experimental results. As they evolve, any set of explanations naturally diverge, and new experiments are needed to help sort out which explanations are better than others. In this way, models sit at the center of our thinking: explaining the experiments that have been carried out, and predicting the results from those that have not.

When the results from an experiment contradict a prediction, and the result is affirmed through repeated experiments to ensure the observation is real, then the model needs to evolve to incorporate this new information without invalidating everything else at the same time. Or, sometimes, the model is just unable to be modified or refined because of inconsistencies that are too severe, and we have to "head back to the drawing board" and rethink the whole thing.

I want to point out two things here.

First, more than ever, science is an evolving process. Around the world, more humans than at any time in history are producing new scientific knowledge, and so the rate of change is likely to be the highest it has ever been. This fact gives rise to great progress. But as good as things are, our predictive models (which are better than ever) are still imperfect. Models are simply a program, a form of an algorithm, and the goodness of the output is only as good as the understanding of the people who wrote the program to all of the factors and their relative balance. Predictive models are imperfect, and that is just fine. They are generally better today than they were yesterday, and they will be better tomorrow.

Second, our strong reliance on reproducible and accurate experimental results to make models is a surprisingly recent feature of the culture of science. Humans have always wanted to provide explanations for the phenomena of the world, from sunrise and sunset to life and death. For a couple of millennia, at least, the prevailing idea was that the human brain possessed pure philosophical thinking (theory), and if the observations of the world contradicted the theory, then the problem was with the world, not with the thinking.

You can see a terrific example of this from the early history of modern chemistry.

One of the early organic chemistry theorists was Justus von Liebig (1803–1873), from Germany. In the early 19th century, small-town family chemistry businesses prospered along the Rhine River in Germany. Unrefined coal had been moving from the Black Forest out to industrialized England for many years, and those barges, upon their return, included coal tar, which was an industrial waste. The Germans developed modern organic chemistry by learning how to extract useful materials (medicines and dyes) from the coal tar. A young genius, Liebig was an exceptional educator, and helped develop the first teaching laboratory that was used to better train Germans in the methods of coal tar chemistry. He also championed the practice of writing down and publishing the results from experiments in a form that could then be used in teaching as reference materials; this is the beginning of the modern scientific journal.

The Swedish chemist, Jöns Jacob Berzelius (1779–1848) maintained a seemingly inexhaustible written correspondence with many of his contemporaries, exchanging and working out ideas, including with Liebig. Their correspondence was preserved. As I noted in the essay at the end of Chapter 2, Liebig writes to Berzelius in 1834:

Die schönsten Theorien werden durch die verdammten Versuche über den Haufen geworfen, es ist gar keine Freude mehr Chemiker zu sein.

And looking again at this translation is insightful (and delightful).

The most beautiful theories are thrown into the heap by these damned experiments, it is no fun at all to be a chemist any more.

Here we have a sense of frustration from a theorist (in the classical sense, his personal "beautiful" ideas about the way things run and why) and the apparent demolition of those ideas by experimental facts (*die verdammten Versuche*... the damned experiments). In context, I have been told by native speakers familiar with this era that this particular "into the heap" expression (*über den Haufen*) likely referred to the pile of excrement associated with an outhouse.

We all get enamored with the beautiful theories that we invent to explain what we see around us. There is nothing at all wrong with speculation and mad hypothesis... just do not confuse them with any approximation of reality. To borrow from Sir Isaac Newton:

> "I have not as yet been able to discover the reason for these properties of gravity from phenomena, and I do not feign hypotheses. For whatever is not deduced from the phenomena must be called a hypothesis; and hypotheses, whether metaphysical or physical, or based on occult qualities, or mechanical, have no place in experimental philosophy. In this philosophy particular propositions are inferred from the phenomena, and afterwards rendered general by induction."
>
> Isaac Newton, *Philosophiae Naturalis Principia Mathematica*, General Scholium, 3E (1726); from the 1999 translation by Cohen & Whitman, University of California Press, 943.

Historians have debated the translation and meaning of the phrase at the end of the first sentence for years ("I do not feign hypotheses" from the Latin *Hypotheses non fingo*). Some have argued that he meant that he does not believe in speculation whatsoever. Some early translations of this were "I frame no hypotheses," which was interpreted as permission to speculate, but not to mistake or confuse personal beliefs (beautiful theories) with defensible theories.

Friedrich August Kekulé (1829–1896) gets the credit for tetravalent carbon (1857), the resulting explosion in the development of organic chemistry, and published an 1865 paper on the cyclic structure of benzene that reportedly came to him in a dream. Over 150 years after Newton, he wrote about the role of speculation and the caution to be sure to be testing one's ideas.

Lernen wir träumen, meine Herren, dann finden wir vielleicht die Wahrheit:
    'Und wer nicht denkt,
    Dem wird sie geschenkt,
    Er hat sie ohne Sorgen' —
aber hüten wir uns, unsere Träume zu veröffentlichen, ehe sie durch den wachenden Verstand geprüft worden sind.

Let's learn to dream, gentlemen, then we may find the truth:
   And to those who do not think,
   It will be given,
   They'll have no worries—
But let us beware of publishing our dreams before they have been tested by the waking understanding.

Kekulé at Benzolfest, published in *Berichte*, **1890**, *23*, 1302.

It is okay to dream big; just do not confuse the way you would like the world to be with the way the world is.

Summary

The combination of molecule with an sp^3 carbon-leaving group bond with a Lewis basic electron donor can result in nucleophilic substitution and/or elimination reaction products. A range of mechanisms can account for these reaction products, differing by which bonds change and the relative timing of the bonding changes.

In substitution reactions, the Lewis base, acting as a nucleophile, can directly displace the leaving group in one step (S_N2), giving inversion of stereochemical configuration. Or, the Lewis base can combine with a carbocation that forms after the initial departure of the leaving group, in a two-step process (S_N1), in which a mixture of stereoisomers can form.

In elimination reactions, the Lewis base acts as a Brønsted base. The direct displacement of the leaving group can happen in one step when the base deprotonates a β-hydrogen atom, releasing an electron pair that ejects the leaving group as a new pi bond forms (E2). An anti relationship between the leaving group and the β-hydrogen atom gives the fastest E2 reaction. Or, the deprotonation of the β-hydrogen atom can occur on the carbocation that forms after the initial departure of the leaving group in a two-step process (E1). Less commonly, the deprotonation of a β-hydrogen atom with a low pK_a value can occur in a separate step, preceding the departure of the leaving group and formation of the pi bond (E1cb).

In the absence of experimental information, a useful predictive model can guide the anticipation of the major pathway from among these choices. After establishing that there is an sp^3 carbon-leaving group bond, or its potential, the balance of multiple factors leads to a prediction. Initially, the structure of the electrophile and the classification of the Lewis base can be used to predict whether one of the bimolecular reactions (S_N2 or E2) is expected. If these are not anticipated, and the structure of the electrophile and the reactions conditions favor forming a carbocation, then the unimolecular reactions (S_N1, E1) are expected.

Differences in reaction rates have been used to infer a variety of properties in these reactions, including leaving group ability, solvent effects, and the nucleophilicity of the Lewis base.

CHAPTER 7 Substitution and Elimination Reactions of Polar Sigma Bonds

CHAPTER QUESTIONS

7.22 When the molecule shown below is treated with NaOCH$_3$, it undergoes a rapid E2 elimination of HF to give an alkene with a single stereochemical configuration (*J Org Chem*, **1999**, *64*, 7768).

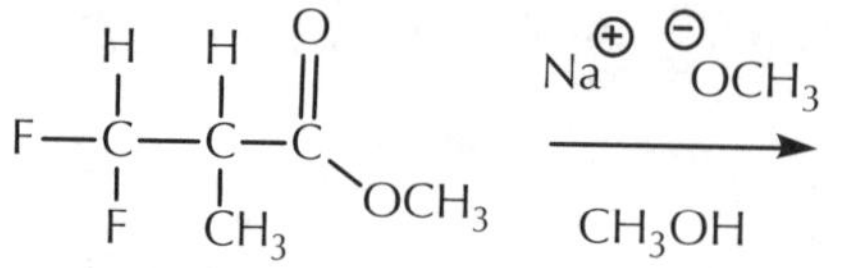

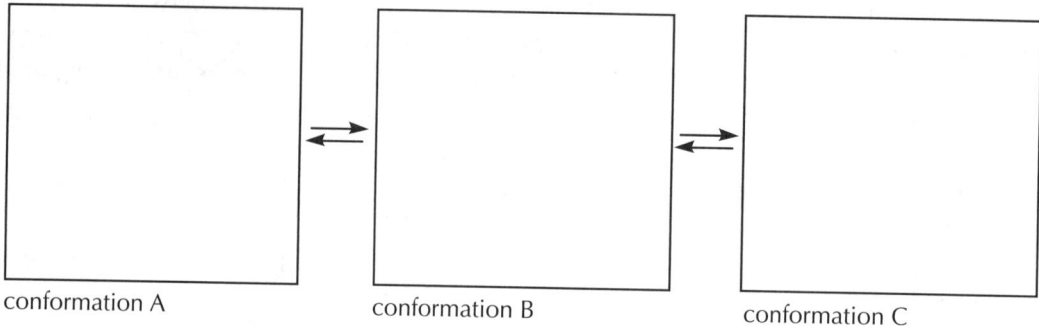

- loss of HF
- E2 mechanism
- one alkene stereoisomer

(a) Analysis of Newman projections is a useful way to understand this reaction. Draw the Newman projections for the three staggered conformations of the C-C bond involved in the E2 reaction (choose the point of view where the CHF$_2$ group is the front atom).

conformation A ⇌ conformation B ⇌ conformation C

(b) Two of the staggered conformations can give E2 reactions much faster than the third one. Which two are these, and why?

Faster E2 from:	Why are these two faster than the other one?
(circle 2) A B C	

(c) Of the two staggered conformations that can give faster E2 reactions, one of them is a more stable form than the other, so it dominates the equilibrium. This more stable form is the one proposed to give rise to the single, observed alkene product. Use this conformation to predict and draw clearly the structure of the alkene resulting from this reaction.

Draw the E2 mechanism (no stereochemistry implied)	Draw the transition state for E2 reaction based on your curved arrows (no stereochemistry needed)

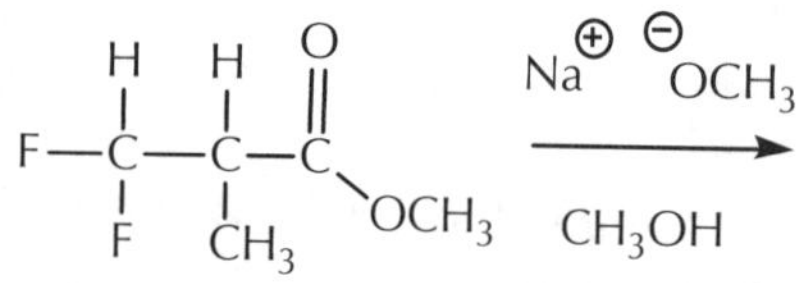

Which of the 2 conformations you selected in part (b) is more stable? Put the letter here:	Draw the E2 product derived from this conformation.

7.23 E2 reactions are observed when menthyl chloride and neomenthyl chloride are combined with sodium ethoxide (NaOCH$_2$CH$_3$). The elimination reaction of neomenthyl chloride is observed to happen about 200 times faster than that of menthyl chloride. Comparing the most stable chair conformation of these two molecules is used to explain this observed rate difference.

(i) Draw the most stable chair conformation for menthyl chloride and neomenthyl chloride. Draw the hydrogen atoms for each substituent-bearing cyclohexane ring carbon.

most stable chair for menthyl chloride	most stable chair for neomenthyl chloride

(ii) Based on the most stable chair conformers drawn above, explain the observed difference in rates for this pair of E2 reactions.

(iii) Draw all the products observed from the reaction of menthyl chloride with sodium ethoxide.

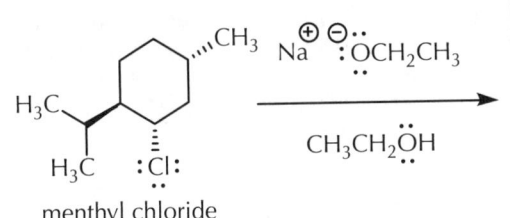

all reaction products (i.e., the balanced equation)

7.24 For the reaction shown below, answer the following questions.

(i) draw the organic product

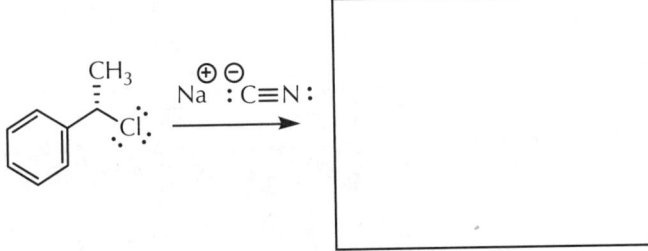

(ii) The reaction rate will depend on the concentration(s) of:

(circle one)
electrophile   nucleophile   both

(iii) Complete the energy diagram, including any intermediates.

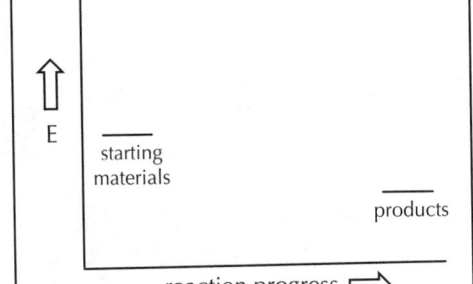

7.25 Answer the questions about the following generic reaction scheme:

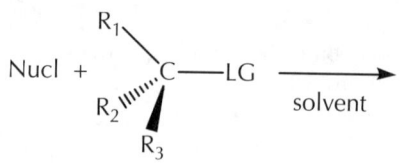

	Nucl	R_1	R_2	R_3	LG	solvent
I.	$Cl^\ominus$	-CH_3	-CH_2CH_3	-H	Br	CH_3OH
II.	$Cl^\ominus$	-CH_3	-CH_2CH_3	-CH_3	Br	CH_3OH
III.	$Br^\ominus$	-CH_3	-CH_2CH_3	-H	Br	CH_3OH
IV.	$Br^\ominus$	-CH_3	-CH_2CH_3	-H	I	CH_3OH
V	$Br^\ominus$	-CH_3	-CH_2CH_3	-H	I	$(CH_3)_2S=O$
VI.	$Cl^\ominus$	-CH_3	-$CH(CH_3)_2$	-H	Br	CH_3OH

Which of the following S_N2 reactions is faster (circle one from each compared pair)?

(a) I or VI (b) II or III (c) V or III (d) V or IV

(e) I or III (f) I or IV (g) III or IV (h) I or II

7.26 The following molecule can undergo a number of E2 reactions with CH_3CH_2ONa/CH_3CH_2OH. Two of these E2 products are formed much faster than the others. What are the structures of the two E2 products that are anticipated to be formed at the fastest rates (chair forms are not required)?

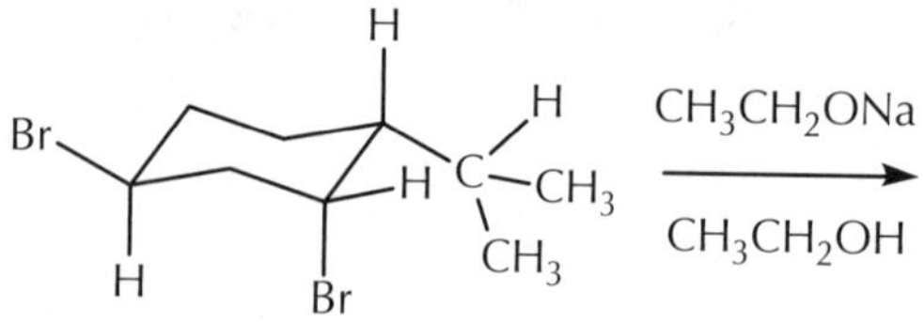

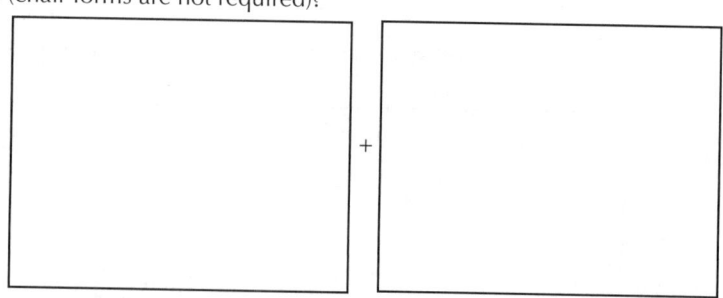

7.27 When the following optically active compound is treated with strong Brønsted bases (such as NaH), its conjugate base undergoes an intramolecular ("within the same molecule") reaction. Experimental data about the product is provided. Deduce its structure (including stereochemistry).

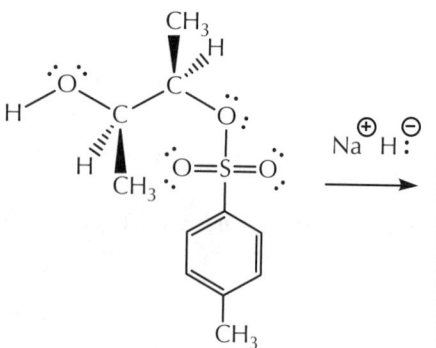

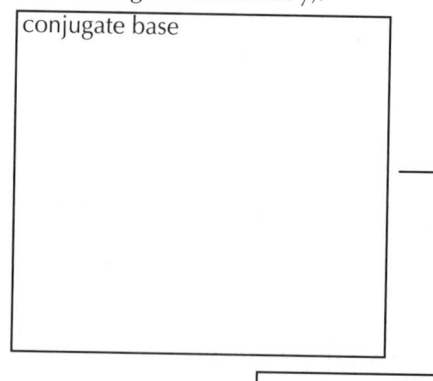

conjugate base undergoes an intramolecular reaction

Data on the observed product:

(a) C_4H_8O

(b) optically inactive

(c) ^{1}H-NMR: 2 signals

(d) ^{13}C-NMR: 2 signals

+ H_2 + NaOTs

7.28 The following reaction gives a mixture of products (*J Org Chem*, **1993**, *58*, 1672). The major outcome results from an S_N2 reaction, while the minor outcome is the (*E*)-stereoisomer of a product resulting from an E2 reaction.

(i) Given this information, draw the two products, including stereochemistry.

(ii) Draw the 3D transition state structure leading to the S_N2 product (lines, dashes, wedges, dotted lines for partial bonds, nonbonding electrons, and any needed partial charges).

(iii) Draw the Newman projection for the conformation of the starting material leading to the E2 product (use the perspective that places the leaving group on the front carbon of the Newman projection).

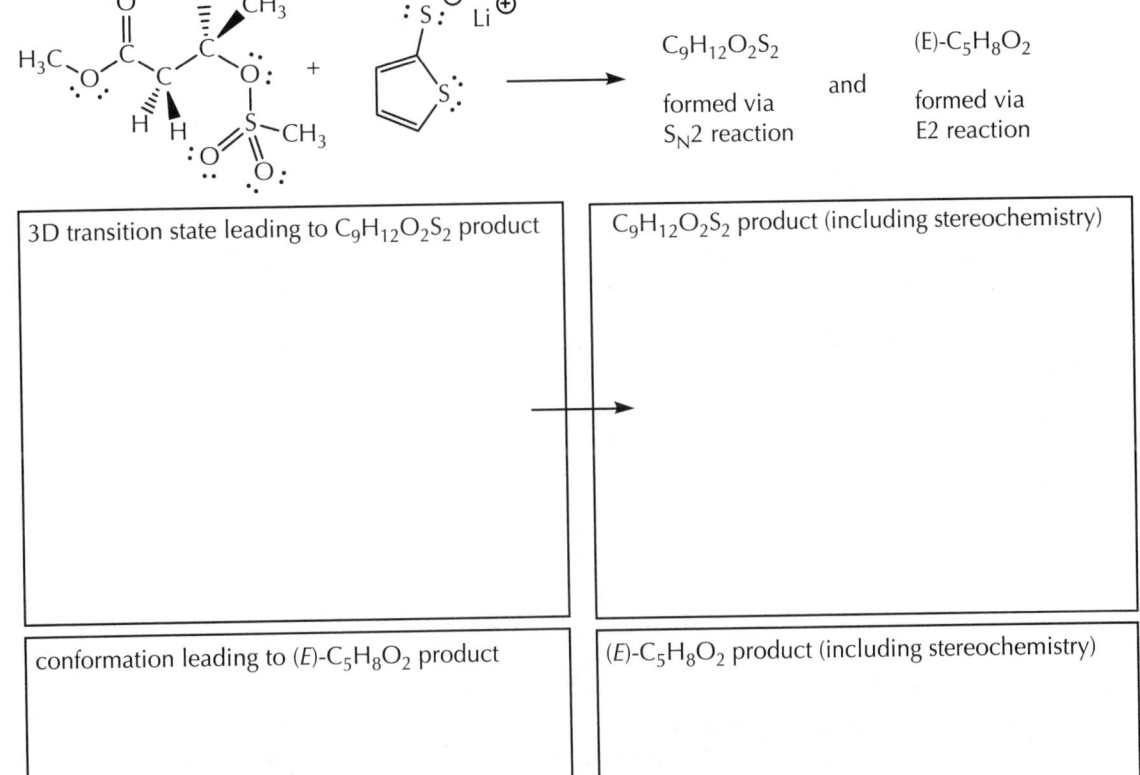

(iv) Provide the structures of the compound and the reagents that can be used to prepare the starting material used in the reaction, above.

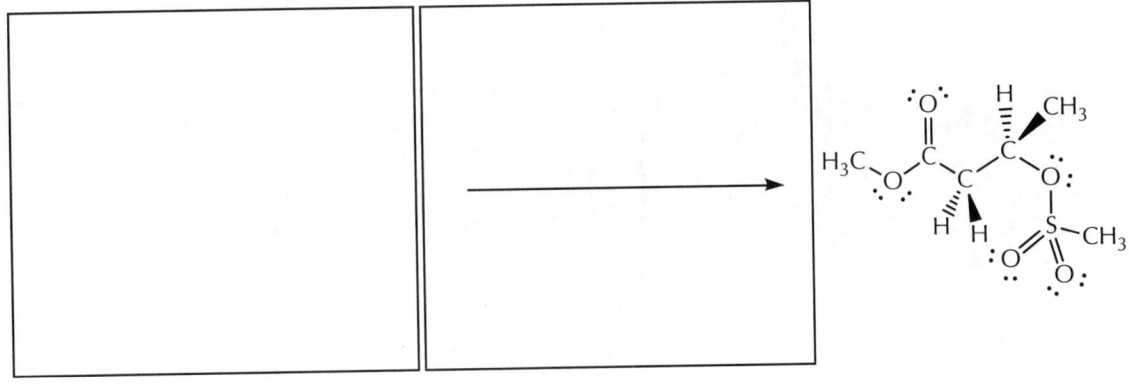

7.29 Provide the reagents for this transformation. If more than one step is needed, be sure to number the steps as necessary.

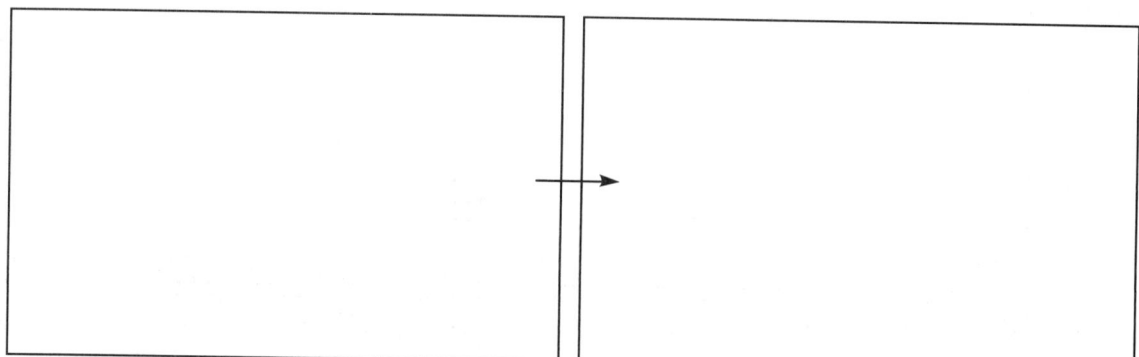

7.30 The reaction shown below leads to a single, kinetically favored bimolecular elimination product.

(a) Provide a chair conformational drawing for the starting material that favors the fastest E2 mechanistic pathway.
(b) Draw the complete curved arrow mechanism for this kinetically favored bimolecular elimination.
(c) Draw the product.

7.31 Predict all products; draw them in the answer spaces, as indicated.

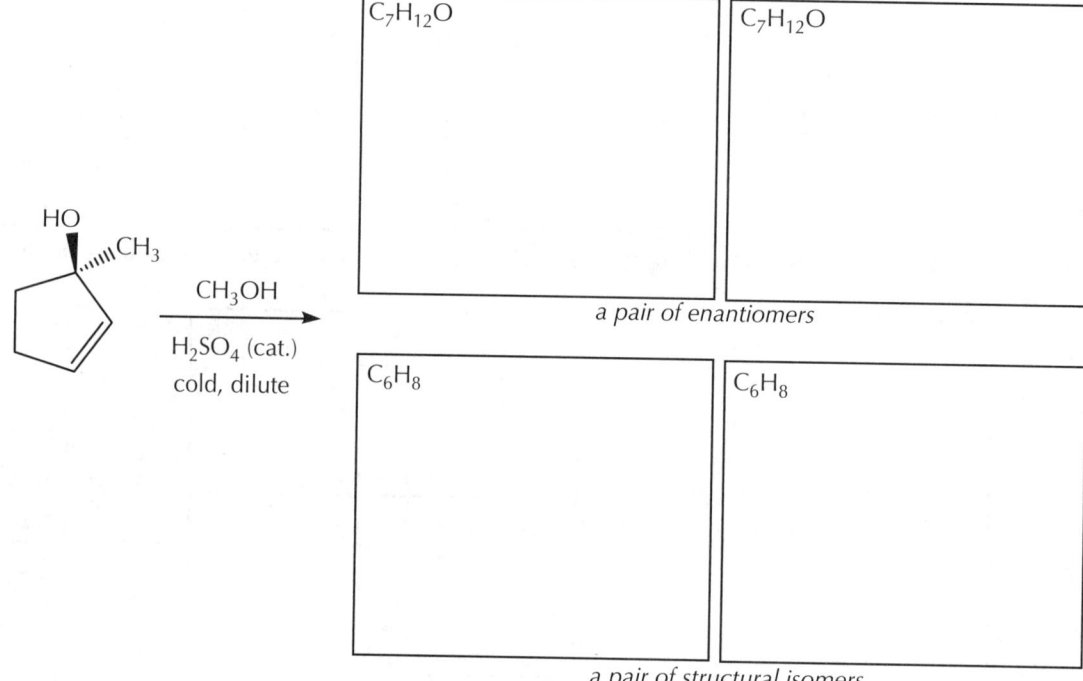

7.32 In 1971, a substitution/elimination study of compound A was published (*J Am Chem Soc*, **1971**, *93*, 4753). Provide the information required to answer the following questions about the chemistry of compound A.

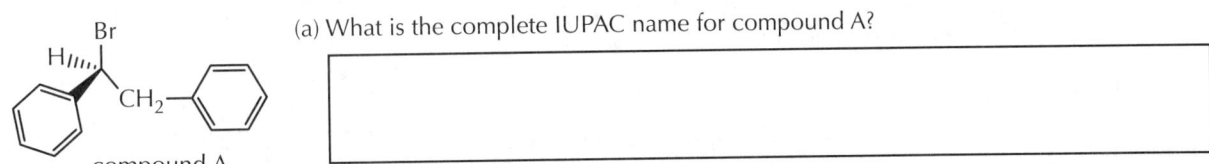

compound A

(a) What is the complete IUPAC name for compound A?

(b) Complete the following reactions. If a mixture of stereoisomers is expected or required, then draw one of them and write "plus enantiomer" or "plus diastereomer" as anticipated.

(i)

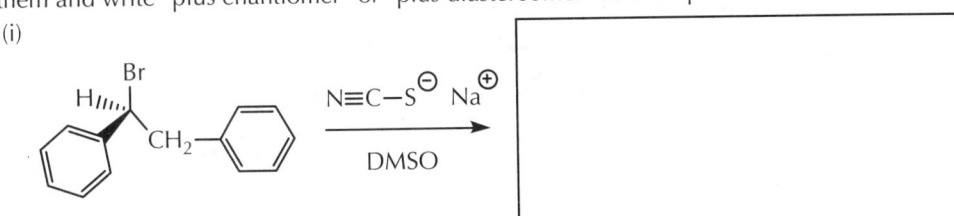

(ii)

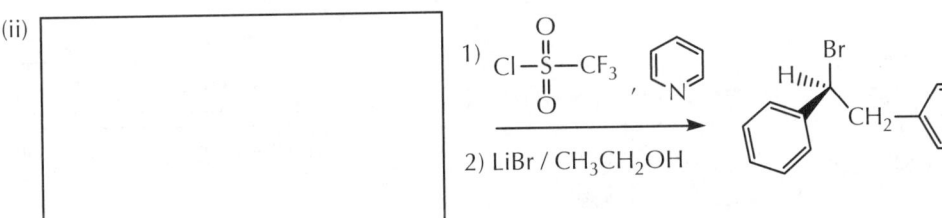

(iii)

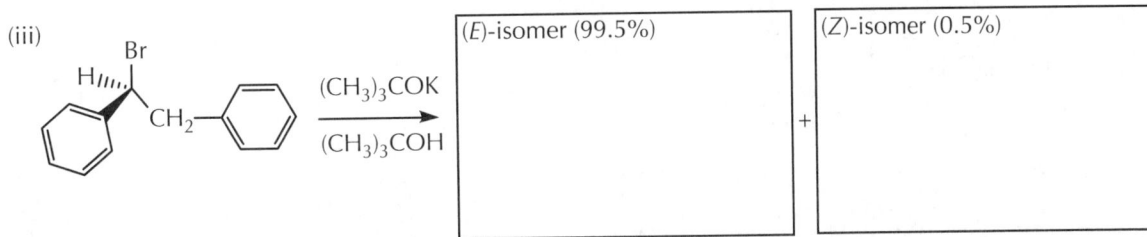

(*E*)-isomer (99.5%) + (*Z*)-isomer (0.5%)

Draw a clear 3D (line/dash/wedge) representation for the conformation that leads to the major product [i.e., the (*E*)-isomer] in this reaction.

Then show the curved arrow mechanism for the reaction.

Using line/dash/wedge notation, along with dotted lines for partial bonds and indicating any required full or partial charges, draw a clear 3D representation for the transition state structure that leads to the major product [i.e., the (*E*)-isomer] in this reaction.

598 CHAPTER 7 Substitution and Elimination Reactions of Polar Sigma Bonds

7.33 In each of the following cases, one of the two reaction conditions (condition 1 versus condition 2) can be used to perform the indicated transformation, and the other one does not work. Decide which condition is expected to work. Then, for the condition that does not give this product, draw the actual product expected to form. If no reaction is expected under the other condition, write "no reaction."

(a) *J Am Chem Soc*, **2000**, *122*, 5666.

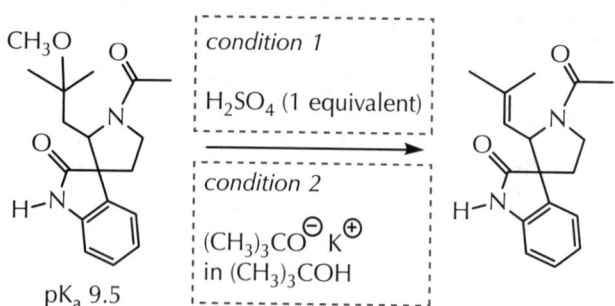

| Which condition works? | 1 | 2 |

What happens under the other condition?

(b) *ACS Med Chem Lett*, **2015**, *6*, 1004.

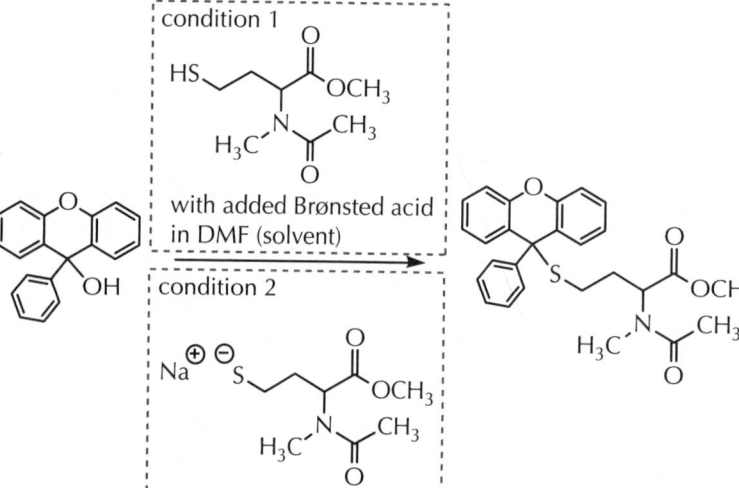

| Which condition works? | 1 | 2 |

What happens under the other condition?

7.34 Rank the following according to the criteria given.
 (a) the number of different possible elimination reaction products.

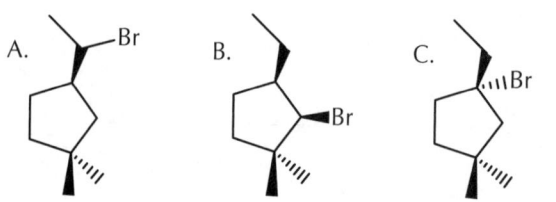

greatest number ☐ > ☐ > ☐ lowest number

(b) the anticipated rate of reaction with CH_3CH_2SNa.

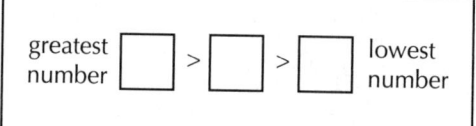

fastest ☐ > ☐ > ☐ slowest

7.35 Compound **N** (see below) was used as a key intermediate in the synthesis of a pharmaceutical that is being examined for the treatment of respiratory diseases. Compound **N** was prepared by treating compound **M** with a catalytic amount of H₂SO₄. Provide a complete curved arrow mechanism for the conversion of **M** to **N**. You may abbreviate any Brønsted acid you use for the mechanism as "H–B" and its conjugate base as "B: ⊖."

7.36 Provide the missing structures and reagents in the following reaction sequence.

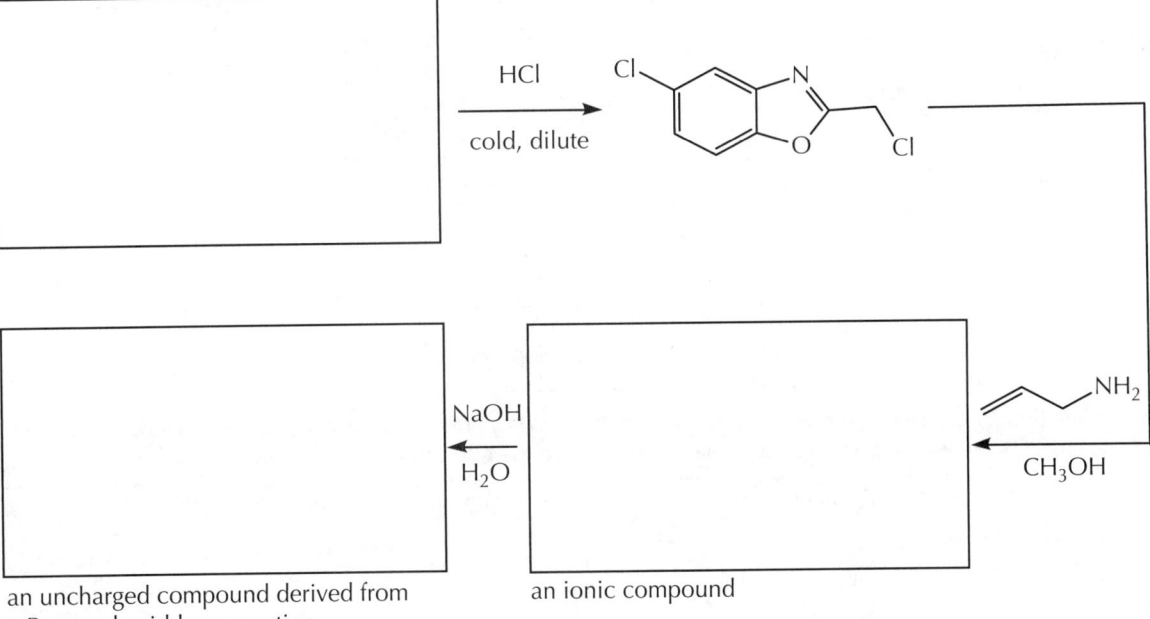

an uncharged compound derived from a Brønsted acid-base reaction

an ionic compound

7.37
When the reaction below is carried out in the lab, the products obtained indicate that both S_N1 and S_N2 pathways are occurring at about the same rate. Provide the structure of the product which indicates that some unimolecular (S_N1) substitution occurred. Briefly explain why the product you drew supports the conclusion that S_N1 was occurring.

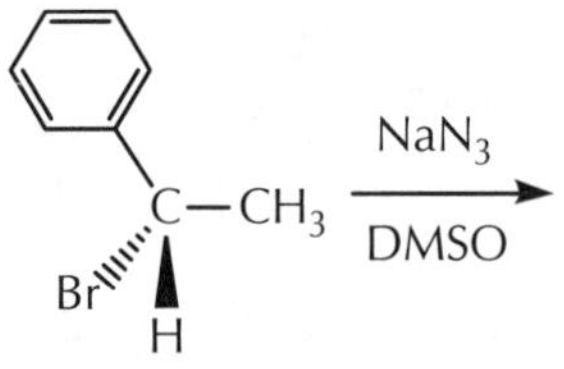

Which one of the products indicates that S_N1 was occurring?	Why does this particular product indicate S_N1?

7.38
Compounds **A** and **B** (see below) each undergo a bimolecular elimination reaction when treated with a strong, hindered base. In both cases, only a single elimination product is formed. Provide the requested information.

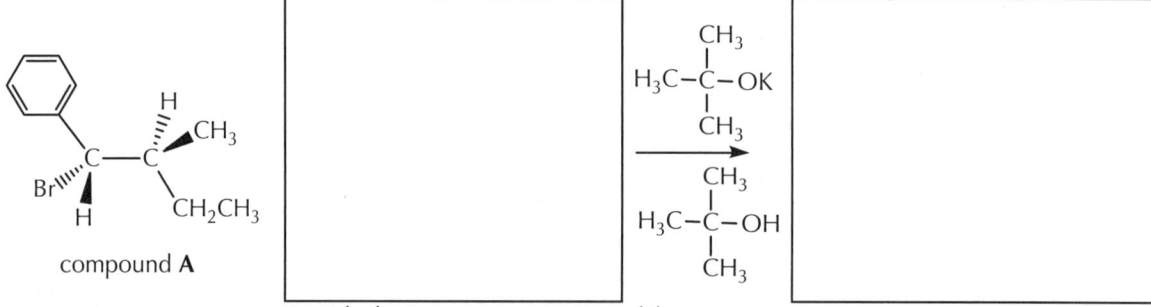

compound **A**

Provide the Newman projection of the conformation of **A** which undergoes a fast E2 reaction. The carbon with the phenyl group should be in front.

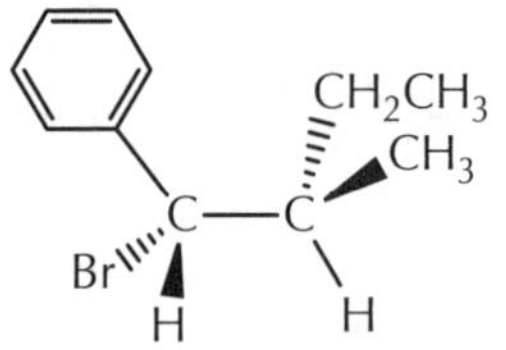

compound **B**

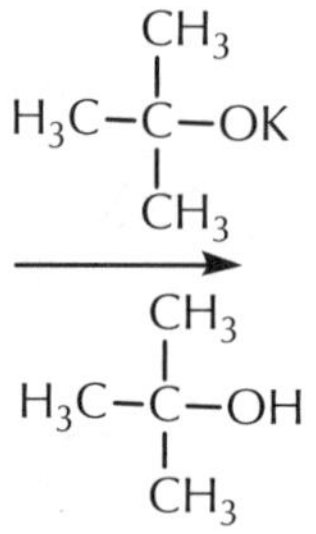

Provide the 3D (line/dash/wedge) drawing of the conformation of **B** which undergoes a fast E2 reaction.

The elimination reactions involving compound **A** and compound **B** occur at drastically different rates. Which occurs faster? And why? You should make reference to the conformational drawings, above.

Which compound undergoes a faster E2 reaction?   compound **A**   compound **B**   Why?

7.39 Ethyl eicosapentaenoic acid ("E-EPA") is a cardiovascular disease drug that works by lowering triglyceride levels (it is marketed under trade names: Vascepa, Epadel, and EPAX). E-EPA can be prepared using some simple chemistry from EPA (an omega-3 fatty acid) in two experimental steps. In the first step, EPA reacts with base; in the second step, the product from the first step reacts with chloroethane (in DMSO) to form E-EPA. Provide the complete, stepwise curved arrow mechanisms for these reactions. Include all nonbonding electron pairs and charges, as needed, in your structures.

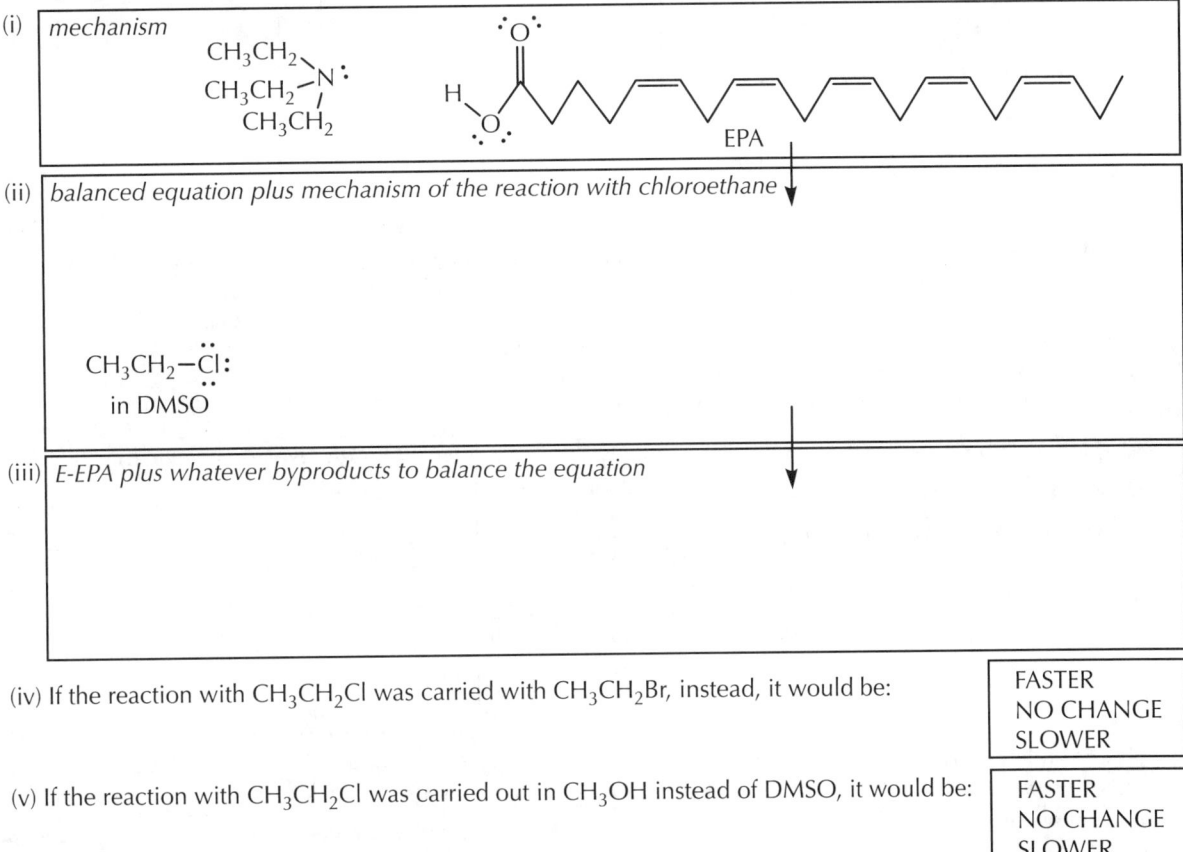

(iv) If the reaction with CH_3CH_2Cl was carried with CH_3CH_2Br, instead, it would be: FASTER / NO CHANGE / SLOWER

(v) If the reaction with CH_3CH_2Cl was carried out in CH_3OH instead of DMSO, it would be: FASTER / NO CHANGE / SLOWER

7.40 (R)-1-Chloro-1-ethoxyethane was treated with a variety of nucleophiles in order to investigate the kinetics of substitution reactions. For each of the reactions below, draw the product(s) expected from a substitution reaction, and predict whether the rate of the reaction would increase, decrease, or be unchanged if the concentration of the nucleophile was increased.

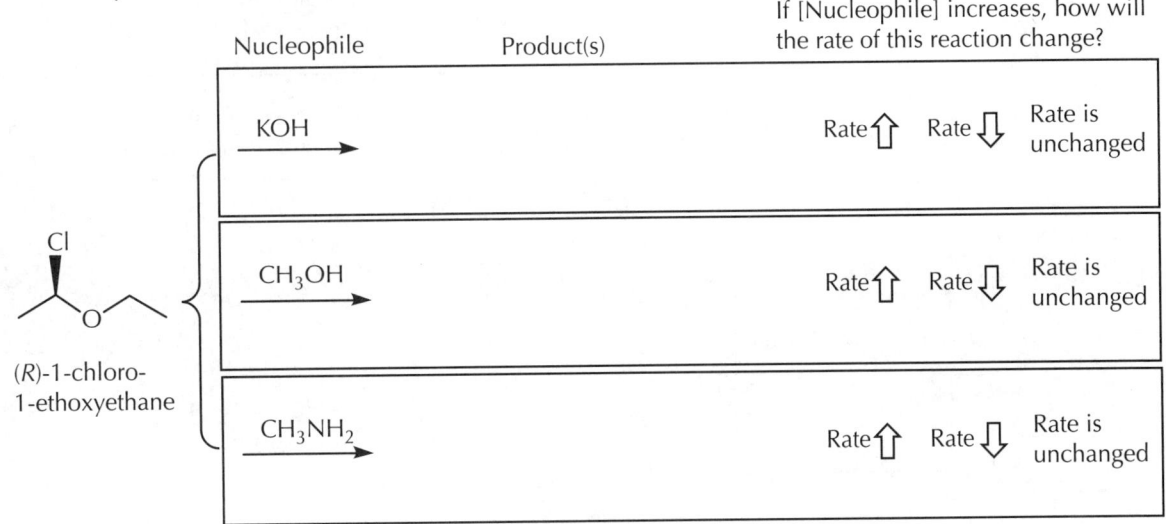

7.41 Compound **P** undergoes reaction with potassium methoxide in methanol to give a mixture of a single S_N2 product (as the major outcome) and a single E2 product (as the minor outcome).

(a) Draw these anticipated major and minor products.

major product (S_N2)	minor product (E2)

compound **P** + 1 equiv. CH$_3$OK, CH$_3$OH, 0.05 M, 25°C, 2 hours

(b) In an attempt to increase the proportion of the elimination product, the experimentalists shifted to potassium tert-butoxide, figuring that the rate of the substitution reaction would be slowed down much more than the elimination reaction. However, the resulting reaction mixture was EXACTLY the same two products, and in the same proportion! Draw the key chemical reaction that explains why changing to this stronger base, potassium tert-butoxide (pK_a conjugate acid = 19), ends up with a reaction mixture that is identical to that using potassium methoxide.

compound **P** + 1 equiv. (CH$_3$)$_3$COK, CH$_3$OH, 0.05 M, 25°C, 2 hours — exactly the same outcome as observed in part (a) .... not just the ratio, but also the structures!

Draw the key reaction needed to explain result.

7.42 An enthusiastic student made the following proposal to prepare compound **R**. Unfortunately, none of the expected product was observed. As the expert consultant, you took one look at this plan and realized the problem!

(a) What products (balanced equation) formed, instead?

PhCH$_2$CH$_2$I + CH$_3$CH$_2$CH$_2^{\ominus}$ Li$^{\oplus}$ → PhCH$_2$CH$_2$CH$_2$CH$_2$CH$_3$ + Li$^{\oplus}$ I$^{\ominus}$

compound **R**
not observed

(b) You're clever: you propose an alternative sequence that begins with this transformation. How is this carried out and why doesn't it give the same outcome as the reaction that was proposed and did not work?

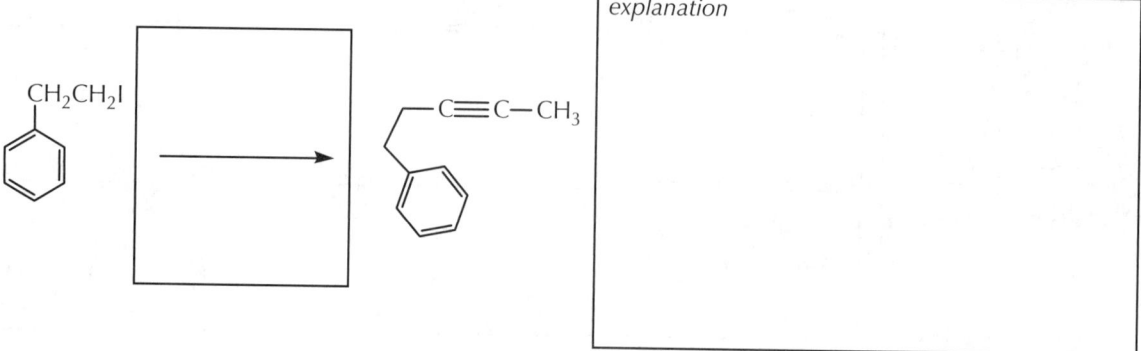

explanation

7.43 Answer the following questions and provide a brief rationale for your choice.

(a) Which of the following gives the fastest E2 reaction?

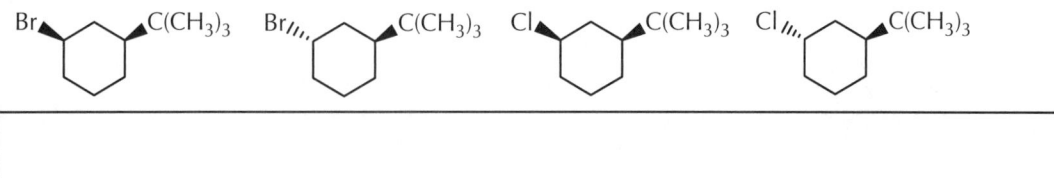

(b) Which of the following gives the fastest S$_N$1 reaction?

Ph$_2$C(CH$_3$)Br   PhCH$_2$Br   PhC(CH$_3$)$_2$Br   PhCH(CH$_3$)Br

(c) Which of the following gives the greatest number of different possible E reaction products?

(d) Which of the following gives the fastest S$_N$2 reaction with bromomethane?

CH$_3$SNa/CH$_3$OH   CH$_3$OH/DMSO   CH$_3$SNa/DMSO   CH$_3$ONa/CH$_3$OH

7.44 The following reagents, when combined, give the observed outcome. Provide a complete, curved arrow mechanism that explains the observation. Note the following: There are exactly two mechanistic steps, and the first one is a Brønsted acid/base reaction. No other reagents are required and none should be shown.

7.45 For the following transformations, the starting materials are chiral and only the (S)-isomer was used (as shown). Draw in the expected product(s) and answer the questions about changing the reaction conditions.

(a)

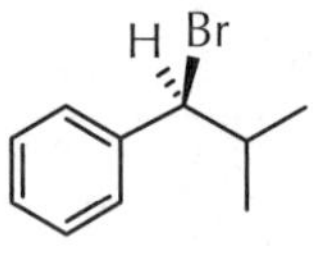

compound **A** + HCO₂⁻K⁺ / HCO₂H →

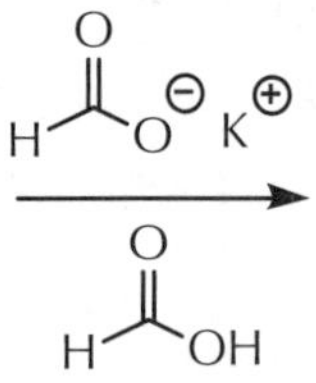

(b) Assuming that the same reaction mechanism occurs (with identical or analogous product(s) forming); then:

(i) If **A** was changed to [structure with H, Br on stereocenter bonded to phenyl and methyl] the reaction rate would: increase decrease remain the same
brief explanation:

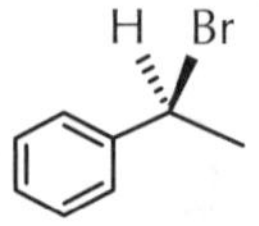

(ii) If **A** was changed to [structure with H, Cl on stereocenter bonded to phenyl and methyl] the reaction rate would: increase decrease remain the same
brief explanation:

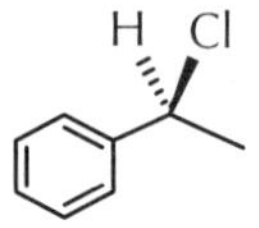

(iii) If the concentration of **A** doubled, the reaction rate would: increase decrease remain the same
brief explanation:

(c)

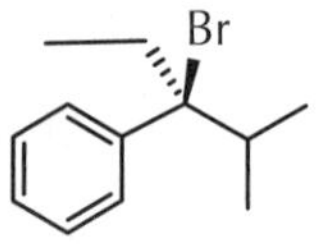

compound **B** + HCO₂⁻K⁺ / HCO₂H → Draw the product(s) with the following formula: $C_{13}H_{18}O_2$.

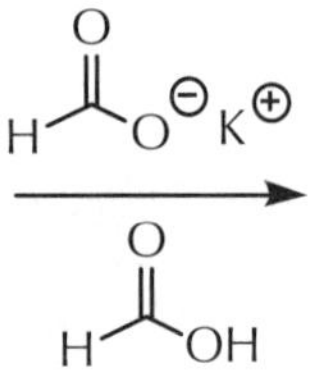

(d) Assuming that the same reaction mechanism occurs (with identical or analogous product(s) forming); then:

(i) If **B** was changed to  the reaction rate would: increase decrease remain the same
brief explanation:

(ii) If the HCO₂⁻K⁺ concentration doubled, the reaction rate would: increase decrease remain the same
brief explanation:

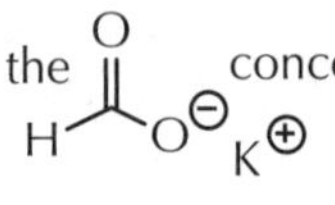

(iii) If the solvent was changed to CCl₄, the reaction rate would: increase decrease remain the same
brief explanation:

7.46 When compound **A** is treated with methanol (CH₃OH) and a small amount of acid (HCl), the transformation to compound **B** is observed (*Org Lett*, **1999**, *1*, 1392).

(a) What is the mechanistic classification for this reaction? Consider this carefully because it affects the entire remainder of the answer.

(i) S_N1   (ii) S_N2   (iii) E1   (iv) E2   (v) electrophilic addition

(b) Provide a complete, stepwise, curved arrow mechanism for this transformation. You may use "H–B" for a generic acid and "B:⊖" for its conjugate base.

(c) The following set of reaction conditions does not result in the formation of compound **B**.

These reaction conditions do result in a reaction, however! What two isomeric compounds would form instead of molecule **B**?

these two products are structural isomers of each other

7.47 When compound **X** is dissolved in methanol (CH₃OH), the following mixture of products is observed.

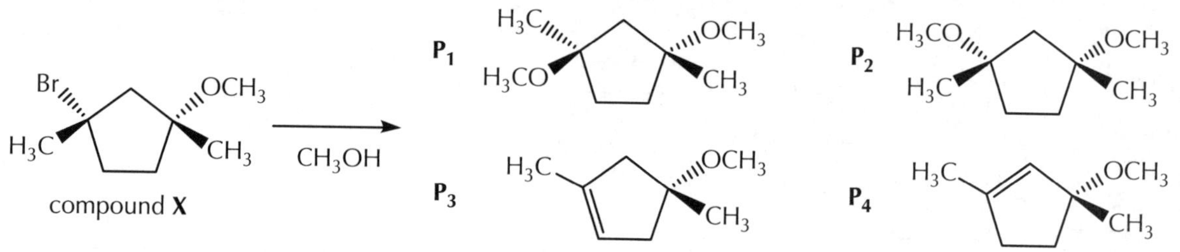

(a) Provide a complete, stepwise mechanism for the formation of products **P₁** and **P₃**. You may use "H-B" and "B:⊖" for any Brønsted acid or base you might need.

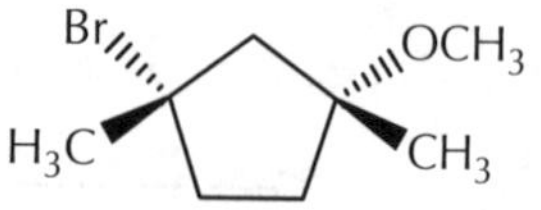

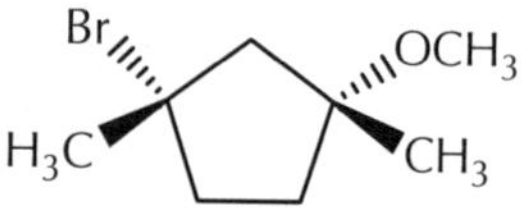

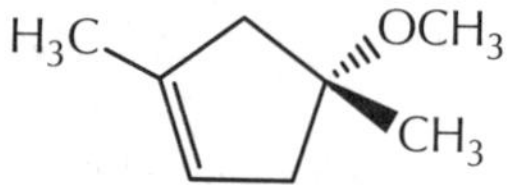

(b) The "-OCH₃" group is named using the prefix designation "methoxy." Including any necessary stereochemical label, what is the name of compound **P₄**?

(c) A fifth compound (**P₅**) was also formed as a minor product. What is its structure?

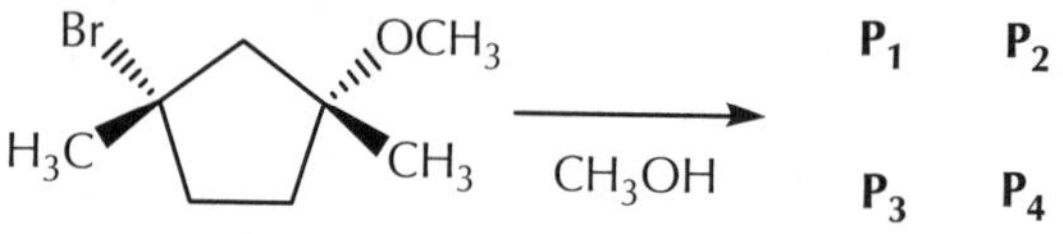

7.4 Chapter Questions

7.48 Compound **P** undergoes a reaction with formic acid (HCO$_2$H) to provide a single, uncharged product.

(a) Draw the structure of this product.

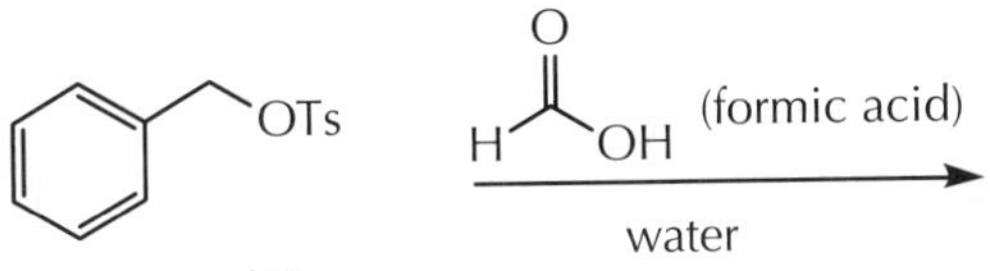

[compound **P**: benzyl OTs] + H-C(=O)-OH (formic acid), water →

(b) The transformation of compound **P** is predicted to be what type of reaction?

| S$_N$1 reaction | S$_N$2 reaction | E1 reaction | E2 reaction | electrophilic addition reaction |
| reduction reaction | complexation/decomplexation reaction | | Brønsted acid/base reaction |

(c) Indicate the relative rates for reactions of compounds **R**, **S**, **T**, and **U** with formic acid (analogous to the reaction of **P**, shown above) by placing the appropriate numbers in the boxes below

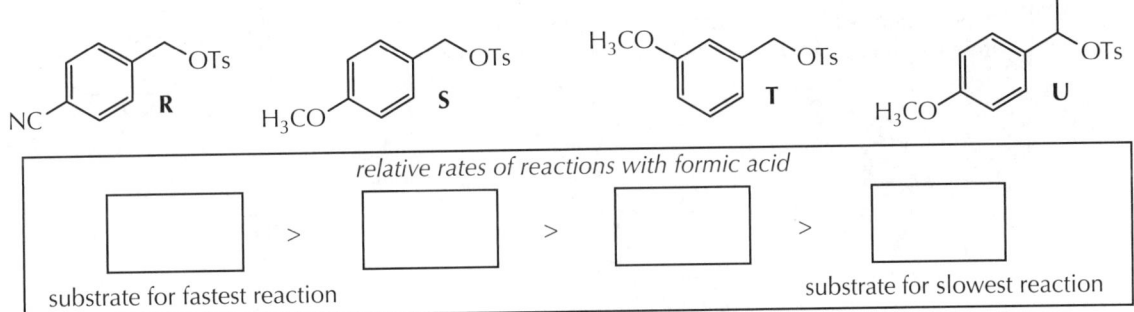

relative rates of reactions with formic acid

☐ > ☐ > ☐ > ☐

substrate for fastest reaction .... substrate for slowest reaction

(d) Which of the following additives or changes to reaction conditions would increase the rate of the reaction between compound **P** and formic acid without leading to a change in the mechanism of the reaction (circle all that apply)?

| pyridine | heating | NaOH | Increased concentration of formic acid | cooling |

7.49 Provide the single substitution product formed in each of the following selective transformations.

(a)

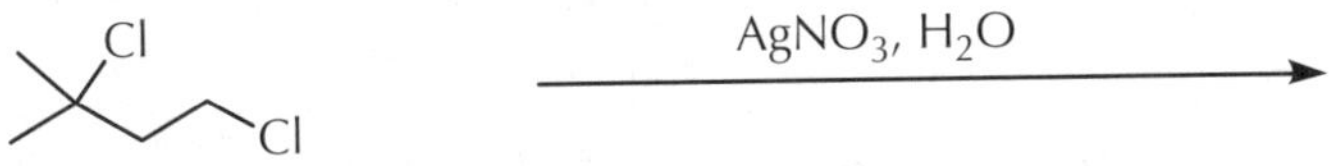

[neopentyl-like dichloride] —AgNO$_3$, H$_2$O→

(b)

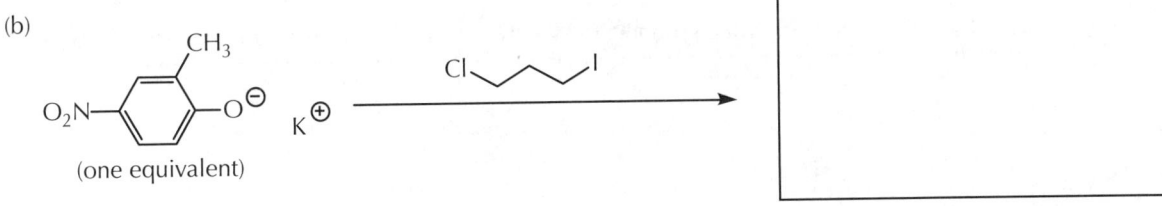

(one equivalent)

7.50 The following questions relate to a study about the regioselectivity in E2 elimination reactions.

(a) Provide the single major product formed in this reaction of compound **M**.

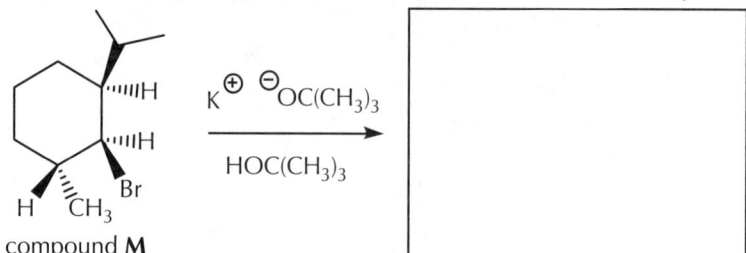

This major product is formed as:

A single chiral molecule

A single achiral molecule

A mixture of enantiomers

A mixture of diastereomers

(b) What is the reason for the observed selectivity in the reaction of compound **M**?

(c) Compound **N** is a stereoisomer of compound **M**. It gives two structurally isomeric products at relatively comparable rates. What are the two products?

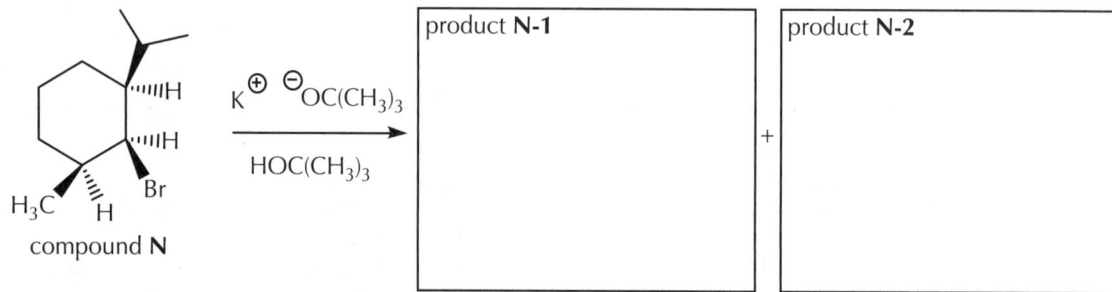

product **N-1**

product **N-2**

+

(d) When the reaction on compound **N** is carried out with sodium methoxide (NaOCH$_3$) in methanol (CH$_3$OH) as the base, the same two products form (products **N-1** and **N-2**), but the ratio of the two products is quite different than in part (c), where potassium *tert*-butoxide was used as the base.

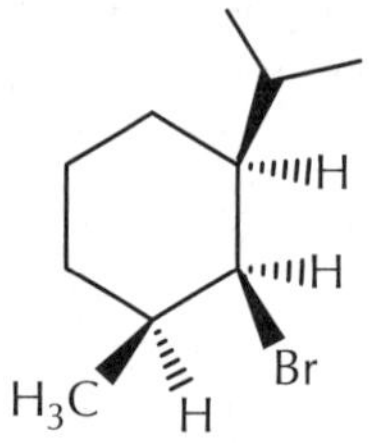

product **N-1** + product **N-2**

less selectivity observed when compared with using KOC(CH$_3$)$_3$

Experimentally, the reaction shown in part (c) is more selective, that is, the ratio of the major product to the minor product is greater when using potassium *tert*-butoxide than when using sodium methoxide.

Based on this information, which product (**N-1** or **N-2**) is the major product in part (c), and why?

7.51 The following alkyl chlorides undergo first-order nucleophilic substitution by water (solvolysis) with dramatically different rates. The first-order reaction rates were measured for the following alkyl chlorides using water-dioxane mixtures as the solvent system.

		relative rate
chloroethoxymethane	Cl—CH$_2$OCH$_2$CH$_3$	1 x 10^9
1-chlorobutane	Cl—CH$_2$CH$_2$CH$_2$CH$_3$	1
1-chloro1-methoxyethane	Cl—CH$_2$OCH$_2$CH$_3$	0.2

(a) Write the mechanism for the rate-determining step of the reaction of chloroethoxymethane.

(b) Using words and drawings, explain the relative rates for these three compounds.

7.52 The specific rotation of (S)-2-bromopentane in ethanol is [α]$_D$ +31°. The optical activity of a solution of (S)-2-bromopentane with an observed rotation (α$_{obs}$) of +20° is monitored, and it remains unchanged until sodium bromide is added to the solution.

Experiment A:
When a 0.10 M solution of sodium bromide is added to a solution of (S)-2-bromopentane, the initially observed rotation gradually decreases until the solution is optically inactive (shown to the right). Analysis of the solution at any point along the way indicates that only sodium bromide and 2-bromopentane are present.

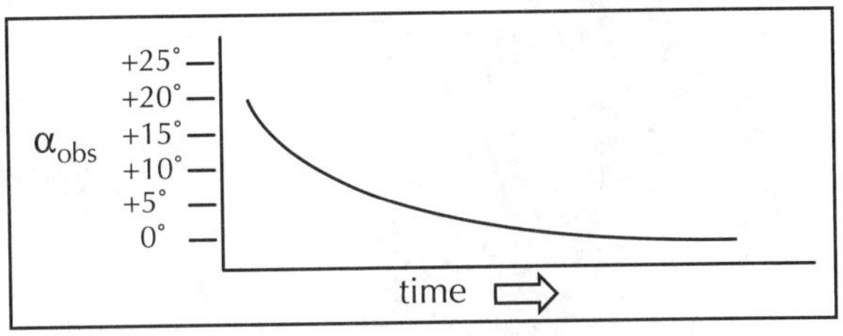

Experiment B:
A 0.25 M solution of sodium bromide is used. No other experimental conditions are changed.

(a) On the graph above, add the curve for the change in observed rotation over time that is anticipated from Experiment B.
(b) Using words and drawings, explain the original experimental result and the rationale behind your prediction about Experiment B.

(c) Only one of these statements explains why the solution becomes optically inactive. Which one is it?
☐ All the starting material is irreversibly converted to the product.
☐ One-half of the starting material is irreversibly converted to the product.
☐ The starting material is reversibly converted to the product.

CHAPTER 8
Electrophilic Addition I: Brønsted Acids

Setting the Stage: Making the Most of the Unexpected

8.1 Electrophilic Addition of Brønsted Acids
    A. Regioselectivity in Addition Reactions
    B. Stereoselectivity in Addition Reactions
    C. Electrophilic Addition of Strong Brønsted Acids: Revisited
    D. Electrophilic Addition of Weak Brønsted Acids: Revisited

PRACTICE QUESTIONS

8.2 Carbocation Chemistry
    A. Origins and Fates of Carbocations
    B. Carbocation Rearrangements Change Connectivity
    C. Ring Contraction and Expansion Reactions
    D. Rearrangements of Carbocations with β-Bonded Heteroatoms

PRACTICE QUESTIONS

8.3 Hydration via Hydroboration-Oxidation
    A. Transformation: Anti-Markovnikov Addition of Water
    B. Stereochemistry and Regioselectivity of the Hydroboration Reaction Mechanism
    C. Oxidation of Carbon-Boron Bonds
    D. Synthesis: Here's the Answer, What's the Question?

PRACTICE QUESTIONS

Reflections About Science: Synthesis and Why We Need Options

Summary

CHAPTER QUESTIONS

CHAPTER 8

Electrophilic Addition I: Brønsted Acids

Setting the Stage: Making the Most of the Unexpected

In an 1854 lecture, the noted microbiologist and chemist Louis Pasteur declared:

Dans les champs de l'observation le hasard ne favorise que les esprits préparés.

"In the fields of observation chance favors only the prepared mind."

Our understanding of the world is a work in progress, and scientists often make observations that do not match their predictions. An unexpected result is not rejected as "wrong" simply because it was not what you wanted. Often, that result ends up being an interesting insight into the original topic, or the solution to a completely different problem or area of study. Pasteur's idea is that a mind should be prepared for—to be aware of—a variety of problems that need to be solved, or how an unexpected result from one area might be used to do something useful in another area.

In 1878, a chemist working at Johns Hopkins University forgot to wash his hands before lunch. He noticed that everything he was eating tasted quite sweet and tracked it down to the compounds he was working with in the lab. He was lucky these substances were not toxic! One of these compounds, which was patented and named Saccharin, is 300–400 times as sweet as sugar, and was the first nonnutritive artificial sweetener produced for the mass market.

In 1942, another chemist was experimenting with different plastics to create clear gun sights to be used for rifles during World War II. One of these compounds, rejected for being too sticky, was examined again nine years later while trying to develop some heat-resistant coatings for glass. The substance was spread between two glass lenses as part of a routine laboratory test, and the lenses ended up permanently stuck together. The lenses were ruined, but the material ended up being marketed as Super Glue.

In 1946, an engineer working on a project to improve radar technology was testing some new, high-powered vacuum tubes, called magnetrons, when he noticed that the snack he had in his pocket, a peanut cluster chocolate bar, had melted into "a gooey, sticky mess." Suspecting that the magnetron was the culprit, the engineer then placed an egg under it, and within a few moments the egg exploded. The next day, he brought some dried corn kernels to work, and everyone was soon snacking on popcorn. A year later, the first commercial microwave oven was on the market.

In 1953, a chemist at the 3M Company was working on a project to create a form of rubber that would not degrade upon exposure to jet fuels. Some of the material she was working on had dropped on her shoes, and she noticed that those spots remained clean while the rest of her shoe became soiled from wear. The stain-resistant material was developed and sold as Scotchguard.

In 1968, another 3M researcher created a "low-tack" adhesive that could be used to hold papers together, but which also allowed them to be separated without any damage. After a few years of trying to think of an application for this material, a colleague was looking for a way to bookmark the pages of a church hymnal; the idea was born. In 1979, Post-It Notes became available to purchase.

CHAPTER 8 Electrophilic Addition I: Brønsted Acids

These stories involve unexpected results, unrelated to the original research objectives, which were turned into popular commercial products. Only a few research projects result in billion dollar industries, but every day, scientists, from research students to seasoned professionals, are doing experiments where there are unexpected results. Sometimes the observation is discarded. Other times, the prepared mind sees potential in the unexpected result.

Until now, this text has presented organic chemistry as quite generalizable. For instance, as long as you can identify the sp³ carbon-leaving group bond and characterize the Lewis base, you can look at thousands of new and unfamiliar experiments and make predictions about the outcomes. With this chapter, we begin to move into areas of organic chemistry where predictions about what the reagents can do are not so general, and where the outcomes are tied strongly to the specific identity of the compounds. In many of these cases, unexpected results have ended up being useful.

8.1 Electrophilic Addition of Brønsted Acids

A. Regioselectivity in Addition Reactions

In Section 4.3C, the relationship between the stability of carbocations and their rates of formation was explained, resulting in a mechanistic basis for the contemporary version of the Markovnikov Rule in electrophilic addition reactions: Given the choice of different carbocation intermediates, the more stable one forms faster, and gives rise to the major product.

Although Markovnikov's original generalization described regioselectivity based only upon differences in degree of substitution, the contemporary version of this rule uses mechanism to more broadly consider the stability of potential carbocation intermediates. The modern application encompasses Markovnikov's original observations, and also more generally applies to carbocations stabilized by resonance, a factor that is more significant than only the degree of substitution.

The regioselectivity observed in a typical reaction is shown in Figure 0801.

Figure 0801

Regioselective electrophilic addition of HBr.

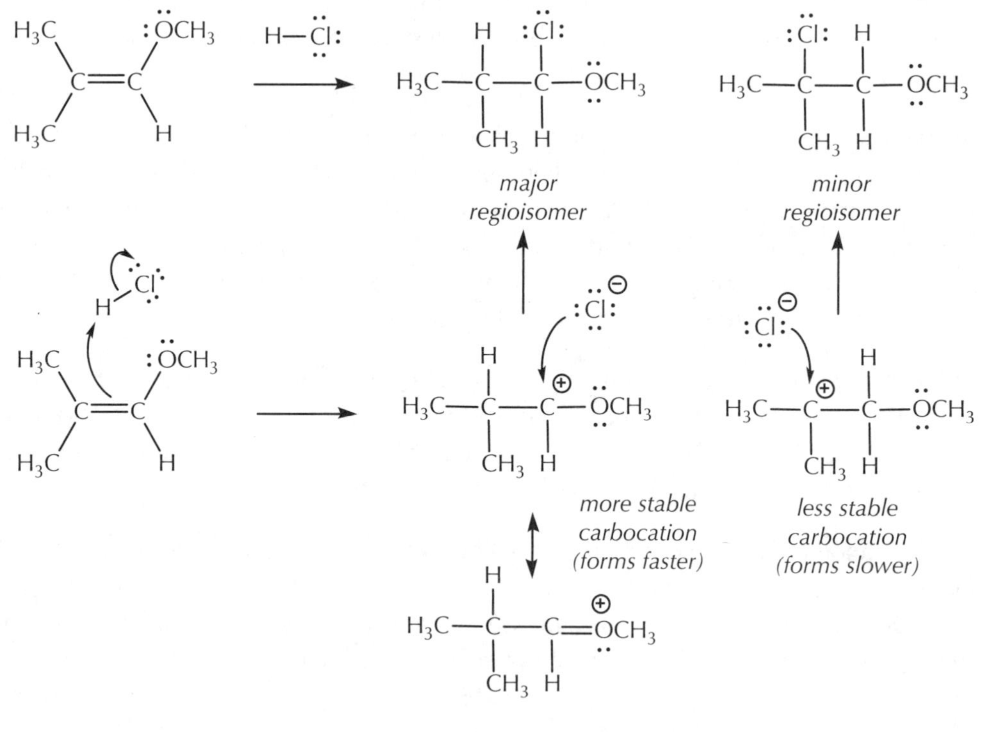

The formation of the more stable, resonance-stabilized heteroatom-substituted carbocation is faster than the tertiary carbocation.

Molecules are not psychic, so it is reasonable to question this explanation by asking: How do the molecules know where they are going? Two explanations answer this question. First, the double bond and the acid are both polarized, and would have partial charges according to differences in electronegativity (in the HCl) and because of the effect of resonance contributors (in the double bond). As these molecules approach one another, the combination of opposite partial charges between the two reaction partners would be favored. Second, as the bonds begin to break and to form, the different transition state structures differ in stability, leading to different activation energies.

The better relative orientation in the collision between the starting materials, in which the opposite charges are aligned, and then the more stable transition state structure, both lead to the more stable carbocation (Figure 0802). These factors contribute to the faster rate of formation of the more stable intermediate.

Figure 0802

Mechanistic features leading to regioselective addition reactions of Brønsted acids.

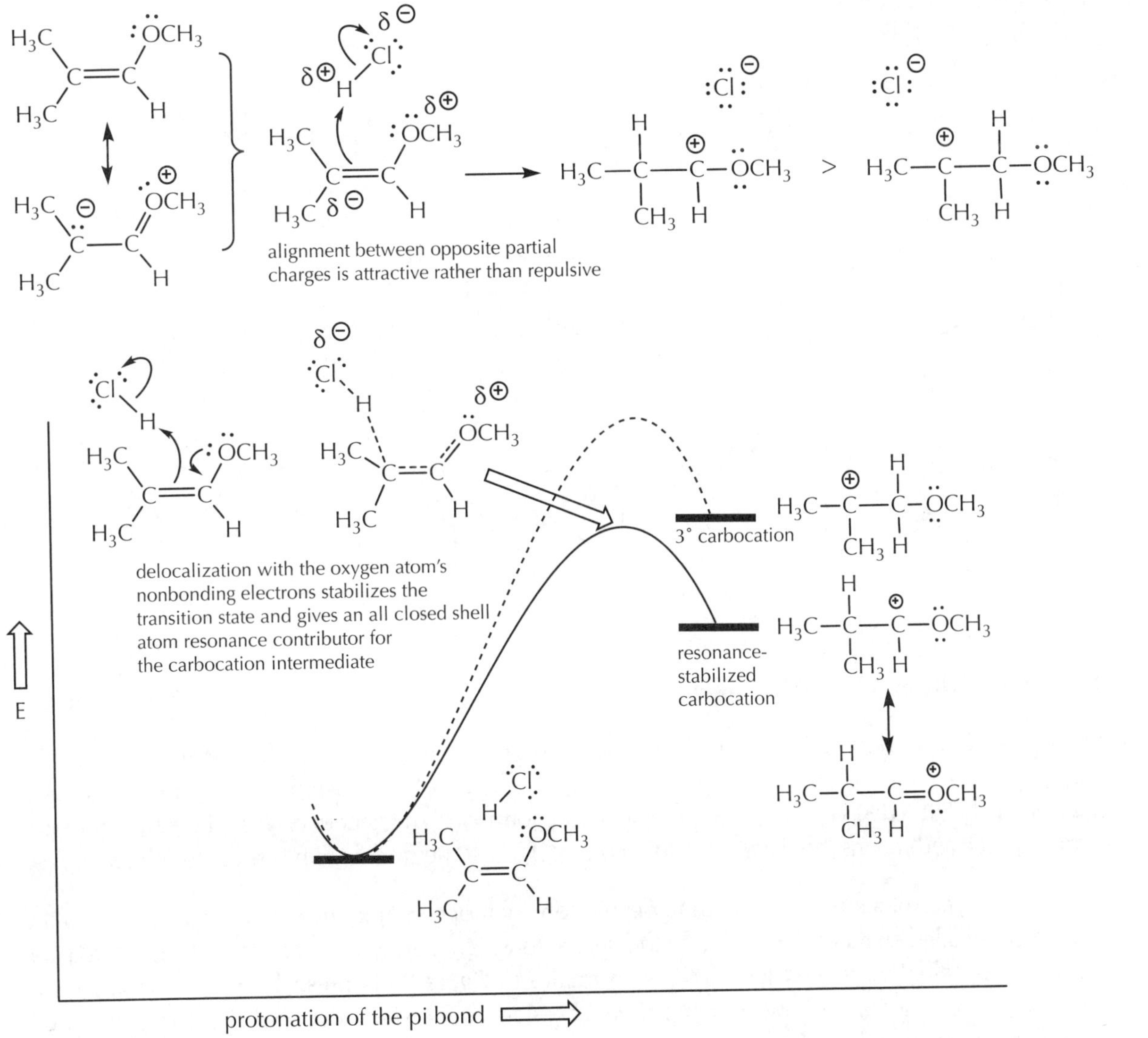

In the electrophilic addition reactions of Brønsted acids, the experimentally determined regioselectivity is rarely completely selective; that is, the result is unlikely to be 100:0 in favor of the major regioisomer. The regioselectivity represented by the major product in electrophilic addition reactions, following the Markovnikov Rule (the more stable carbocation forms faster), is referred to as the "Markovnikov product." The other, minor regioisomer in these reactions is referred to as the "anti-Markovnikov product." Some examples of regioselective reactions are shown in Figure 0803, along with these labels.

Figure 0803

Regioselective electrophilic addition reactions (Markovnikov and anti-Markovnikov products).

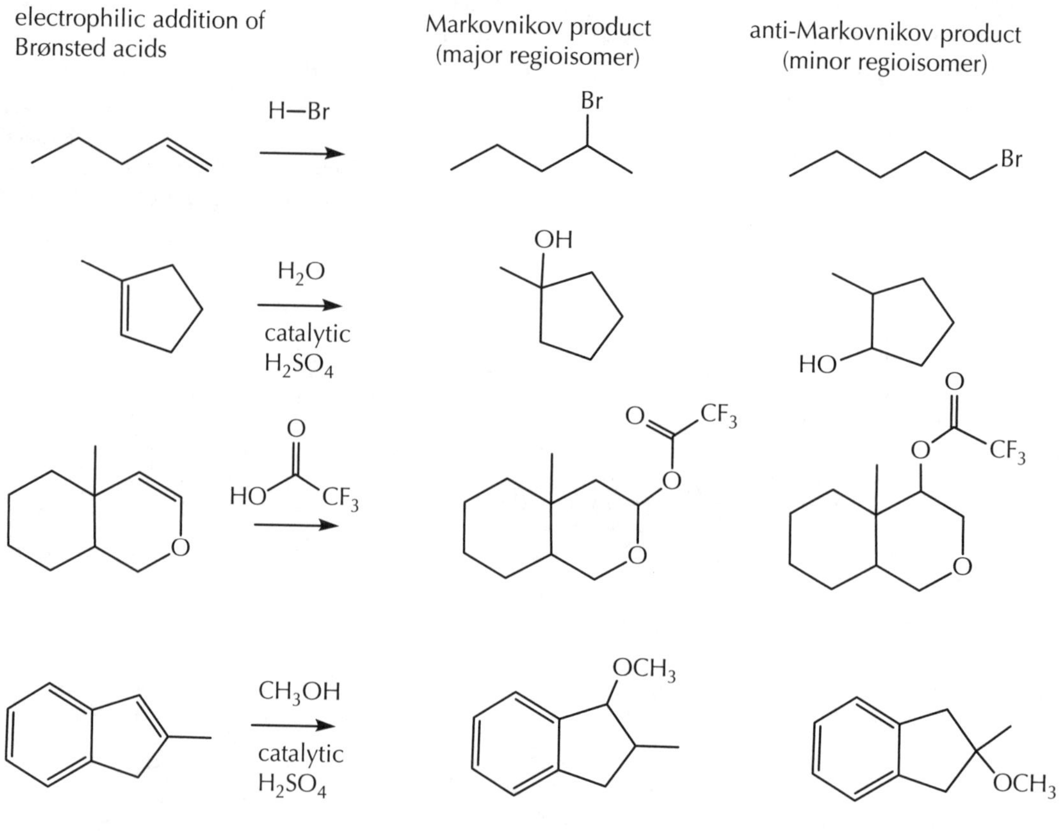

B. Stereoselectivity in Addition Reactions

Addition reactions of double bonds create two new sp³ atoms, which means there is the potential for forming up to two new tetrahedral stereocenters and a mixture of stereoisomers. Just as with the term "regioselectivity" (the degree to which one regioisomer forms with respect to others), the term "stereoselectivity" exists, and means the degree to which one stereoiosomer forms with respect to others.

There are two types of stereoselectivity: the degree to which one enantiomer forms with respect to the other, which is called enantioselectivity; and the degree to which diastereomers form with respect to each other, which is called diastereoselectivity. The examples in Figure 0804 present the connectivity result for some addition reactions followed by the stereochemical analysis of all possible outcomes from that connectivity, including where the assessment of stereoselectivity would be relevant.

8.1 Electrophilic Addition of Brønsted Acids

Figure 0804
Analysis of addition reactions for relevant stereoselectivity.

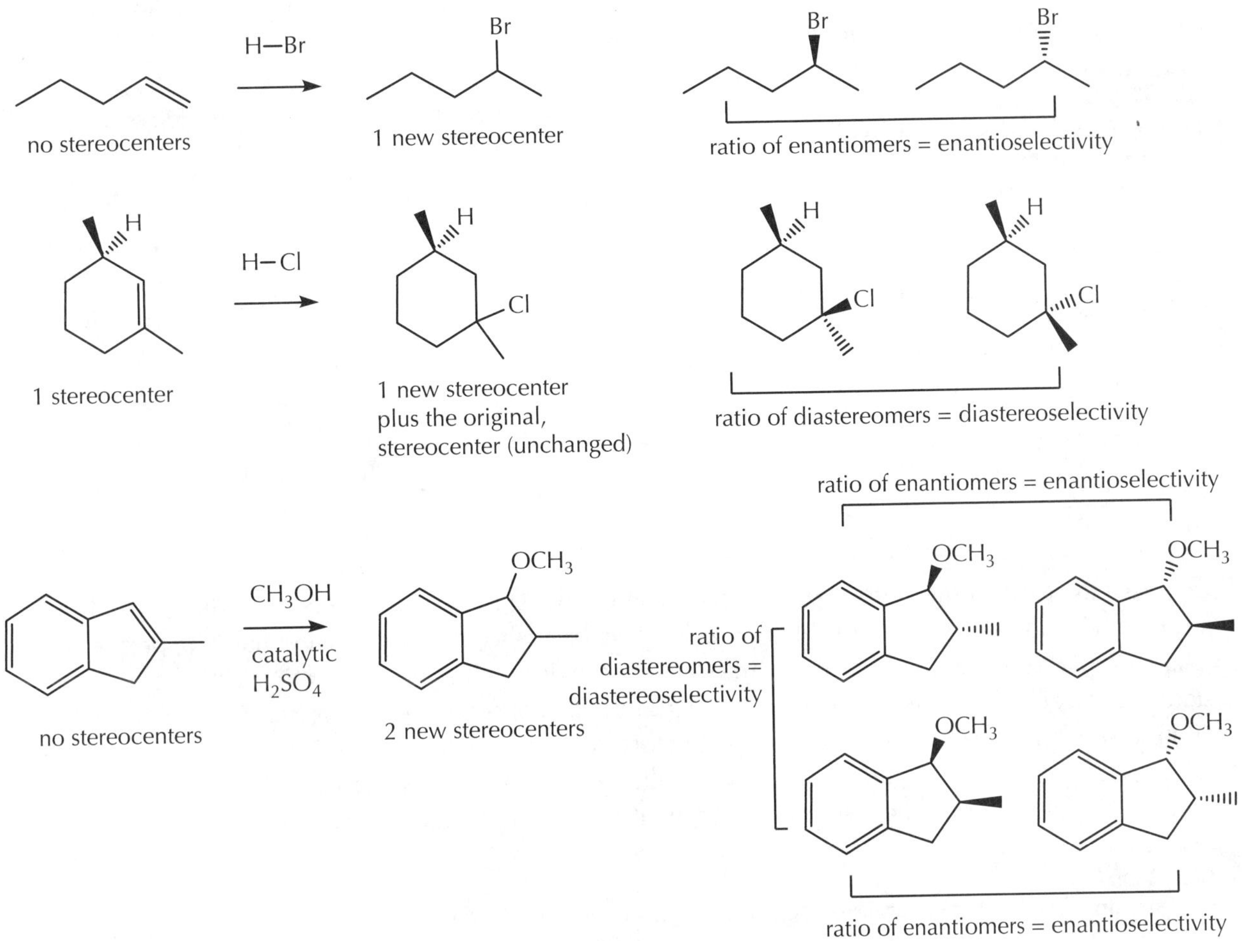

In an addition reaction to a double bond, regardless of the mechanism, there are a total of four different relationships that the two newly formed sigma bonds might have with respect to the original double bond. These four orientations are represented abstractly in Figure 0805.

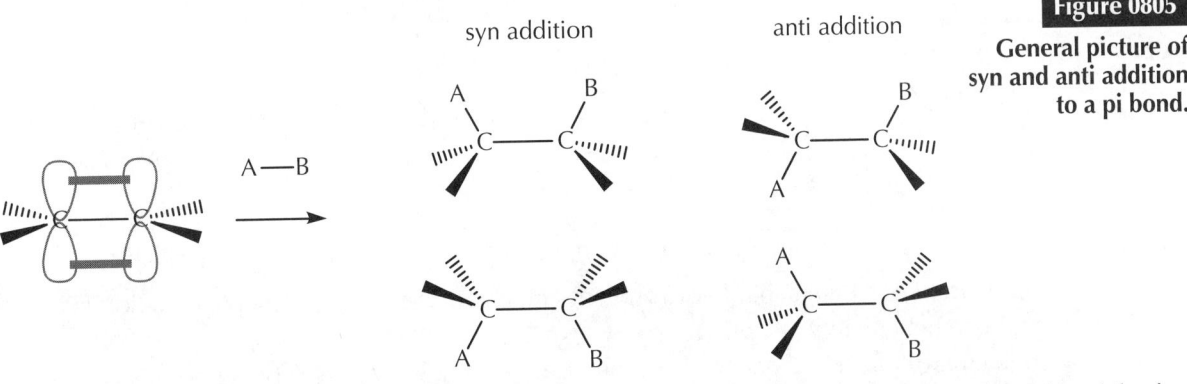

Figure 0805
General picture of syn and anti addition to a pi bond.

A hypothetical addition of the atoms from compound A-B to the same face of the original pi bond (both atoms from above or both from below) means that the two new sigma bonds are initially formed on the same side, and this is called a "syn" (same-sided) addition. Alternatively, the two new sigma bonds could end up forming on the opposite faces of the original pi bond (one above, one below), in which case it is called an "anti" addition.

Returning to one of the examples from Figure 0804, where a ring provides a convenient point of reference for visualizing stereochemical relationships. The previously shown set of four stereoisomers are labeled according to whether the two groups provided in the addition reaction represent a syn addition product or an anti addition product (Figure 0806).

Figure 0806
A specific example of visualizing syn and anti addition to a pi bond.

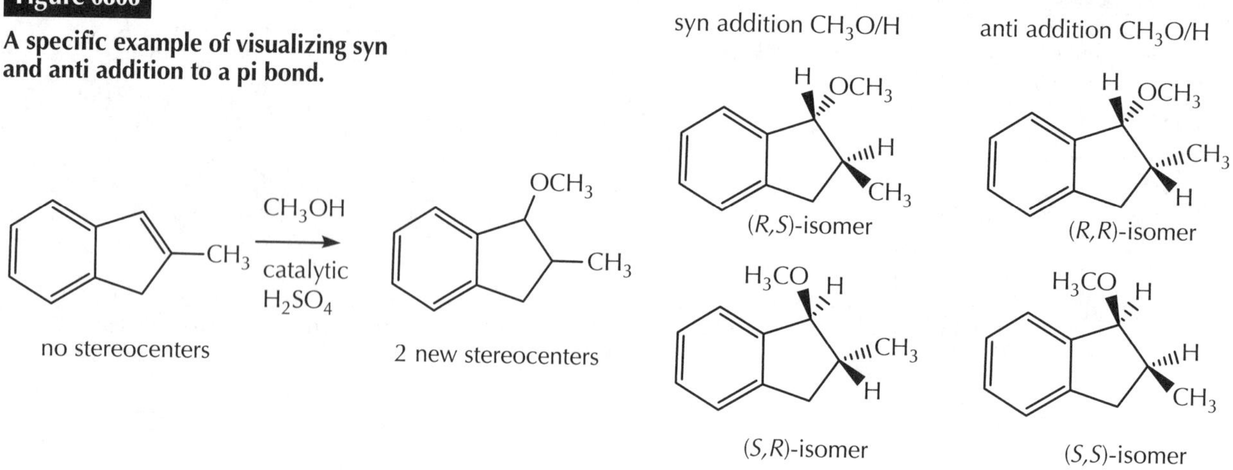

syn addition CH$_3$O/H — (R,S)-isomer; (S,R)-isomer

anti addition CH$_3$O/H — (R,R)-isomer; (S,S)-isomer

Language alert: The term "anti" is used in two different ways, so pay attention to this difference. In regioselectivity, the term anti-Markovnikov regioisomer refers to the connectivity that is opposite from what is expected when applying the Markovnikov Rule. In stereoselectivity, anti addition refers to the stereochemical result where two new sigma bonds form on opposite faces of a pi bond. A comprehensive comparison of when and where the various structural terms used in regioselectivity and stereoselectivity is shown in Figure 0807. In this example, the full array of eight possible products is represented, combining syn and anti stereochemistry with Markovnikov and anti-Markovnikov regiochemistry.

Figure 0807
A comparison of regiochemical and stereochemical terms with "anti" in them.

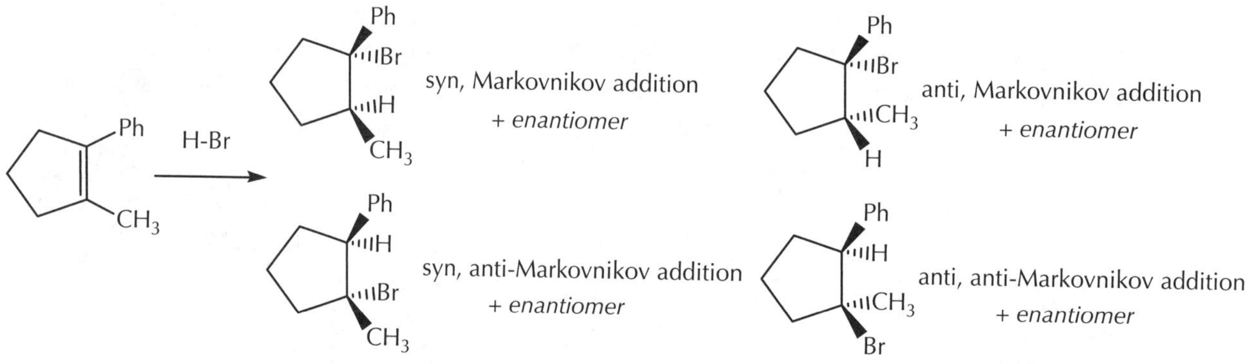

- syn, Markovnikov addition + *enantiomer*
- anti, Markovnikov addition + *enantiomer*
- syn, anti-Markovnikov addition + *enantiomer*
- anti, anti-Markovnikov addition + *enantiomer*

Four general ideas from stereochemistry apply when looking at stereoselectivity as a topic:

(1) Reactions carried out in an achiral environment will create racemic mixtures of enantiomers, resulting in no net chirality.
(2) A racemic (50:50) mixture of enantiomers means no enantioselectivity is being observed or predicted.
(3) If stereocenters already exist in the molecule to which an addition takes place, then diastereomers result when new stereocenters are formed.
(4) Diastereoselectivity ranges from low to high and is often unpredictable unless the mechanism dictates the outcome.

C. Electrophilic Addition of Strong Brønsted Acids: Revisited

In this section, and in the next, we will revisit the electrophilic addition reactions of strong and weak Brønsted acids that were introduced in Sections 4.3 and 4.4, respectively, prior to the chapters on stereochemical and conformational analysis. This reintroduction will now blend stereochemical and conformational topics into the outcomes from these addition reactions.

In the electrophilic addition reactions of Brønsted acids, the guideline for whether an acid is predicted to be strong enough to protonate a pi bond is a pK_a value of 5. This guideline is not a law, but simply a useful recommendation in the absence of experimental information. Depending on the structure and reactivity of the exact reagents, and the experimental conditions under which they are being used, an acid might need to be stronger or a weaker one might well work.

Trifluoroacetic acid has a pK_a value of 0.3 and readily adds to carbon-carbon double bonds and triple bonds (Figure 0808).

Figure 0808
Electrophilic addition reactions of trifluoroacetic acid.

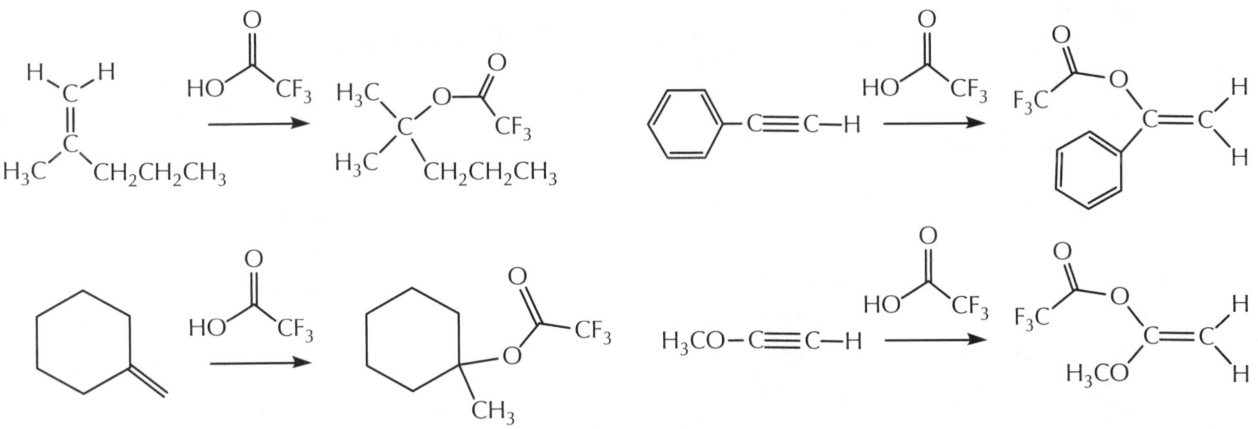

The reaction is highly regioselective when the two prospective carbocation intermediates have different stabilities. In the 2-step reaction mechanism (protonation, followed by capture of the intermediate carbocation), the two new sigma bonds are formed independently of one another.

Because the new sigma bonds in these addition reactions are formed independently of one another, the general stereochemical outcome is a mixture of all of the possible stereoisomers. As illustrated in Figure 0809 (on the next page), when the protonation reaction produces a new stereocenter, both stereochemical configurations are possible because the pi bond can be protonated on either of its faces.

If there is no other stereocenter in the molecule, as in this example, then a pair of enantiomeric carbocations results in the first step. Once the pi bond is broken, the remaining sigma bond from the double bond is free to rotate, so there is no memory of the original stereochemical relationships between the groups that existed in the starting material. Then, when the carbocation is captured by the nucleophile in the second step of the mechanism, it can also be attacked at either side of the planar sp^2 atom, in this case creating a second stereocenter and a pair of diastereomers. The distribution of diastereomers from each of the enantiomeric carbocations is the same, and so a pair of racemic, diastereomeric products results. The enantioselectivity is predictable (they are racemic mixtures, so there is none); the diastereoselectivity would have to be determined experimentally.

618   CHAPTER 8 Electrophilic Addition I: Brønsted Acids

Figure 0809

Stereochemical analysis: electrophilic addition reaction of trifluoroacetic acid.

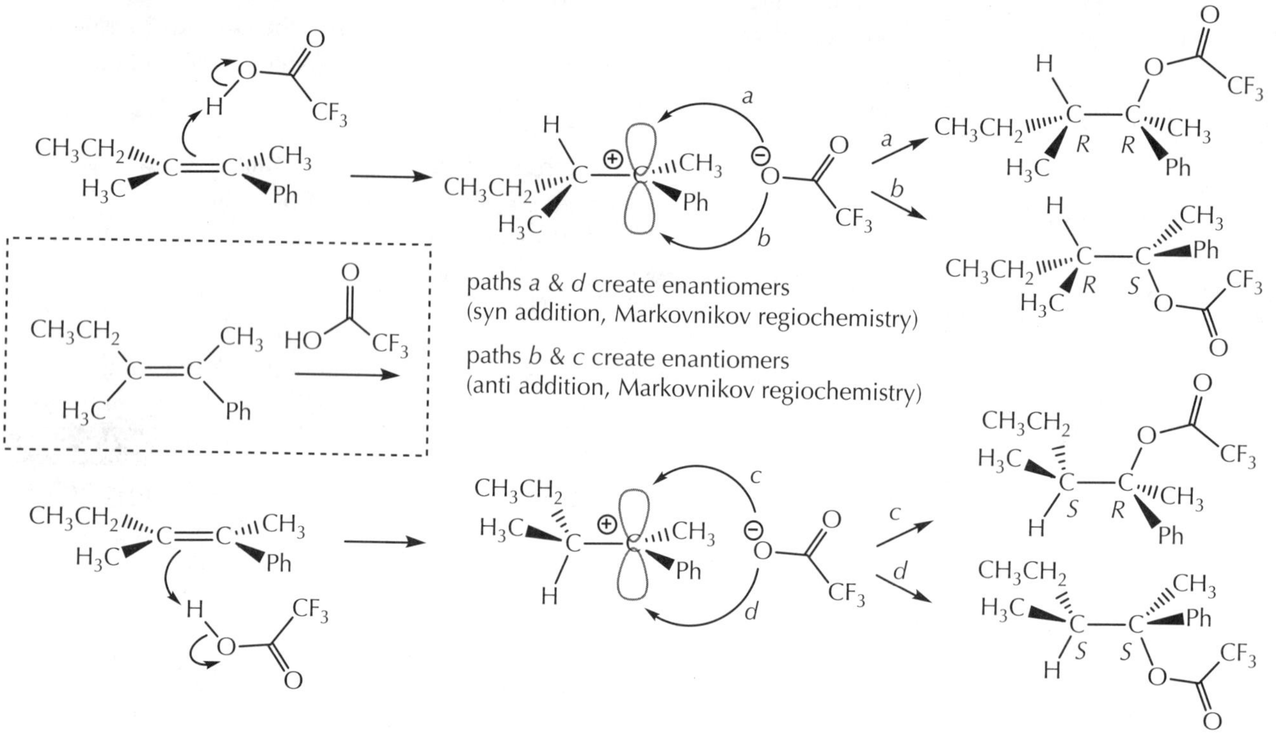

paths *a* & *d* create enantiomers
(syn addition, Markovnikov regiochemistry)

paths *b* & *c* create enantiomers
(anti addition, Markovnikov regiochemistry)

The definition of which products have resulted from a syn versus anti addition does not depend on the accident of which conformation is drawn, but rather on using the fixed stereochemistry of the starting material as a point of reference. Figure 0810 contains examples where the addition reaction products are shown, and then whether the product can be assessed as having resulted from a syn addition pathway, an anti addition pathway, or both.

In the addition reactions of cyclic compounds that form two new stereocenters, such as in the first example, the relative relationship between all of the atoms and groups can be readily seen in the drawings of the products. The ring bonds cannot rotate, and so the syn/anti addition outcome is visible. In the second, third, and fourth examples, no stereoisomers are formed and the syn versus anti addition cannot be determined experimentally. This does not imply any difference in the mechanism, and we would hypothesize that all three of these examples result from a combination of syn plus anti addition because that is the nature of the electrophilic addition mechanism; it is simply that there is not enough information in the structures, in the form of stereoisomers, to detect this.

The last example in Figure 0810 is one that takes a little more work to solve. The central sigma bond is free to rotate, which means that the conformation shown does not necessarily give a fast reading of whether the "H" and "OCH$_3$" groups from the methanol added syn or anti. A common strategy to determine the stereochemical result is redrawing the molecule into a conformational picture where the relative positions of the four substituent groups derived from the double bond are the same in the product as they are in the starting material. Then, a direct visual inspection can be made to assess the syn versus anti relationship of the groups that were added. As shown, the given structure of the product needs to be redrawn into an eclipsed conformation for the four substituent groups to align as they appear on the original double bond. Based upon that alignment, the "H" and "OCH$_3$" groups would have added in a syn manner to get the product that was shown. There are three other stereoisomers that form in that reaction: the enantiomer of the one that is shown, plus its diastereomers that resulted from an anti addition.

8.1 Electrophilic Addition of Brønsted Acids

Figure 0810

Stereochemical analysis: assessing syn versus anti addition reaction products.

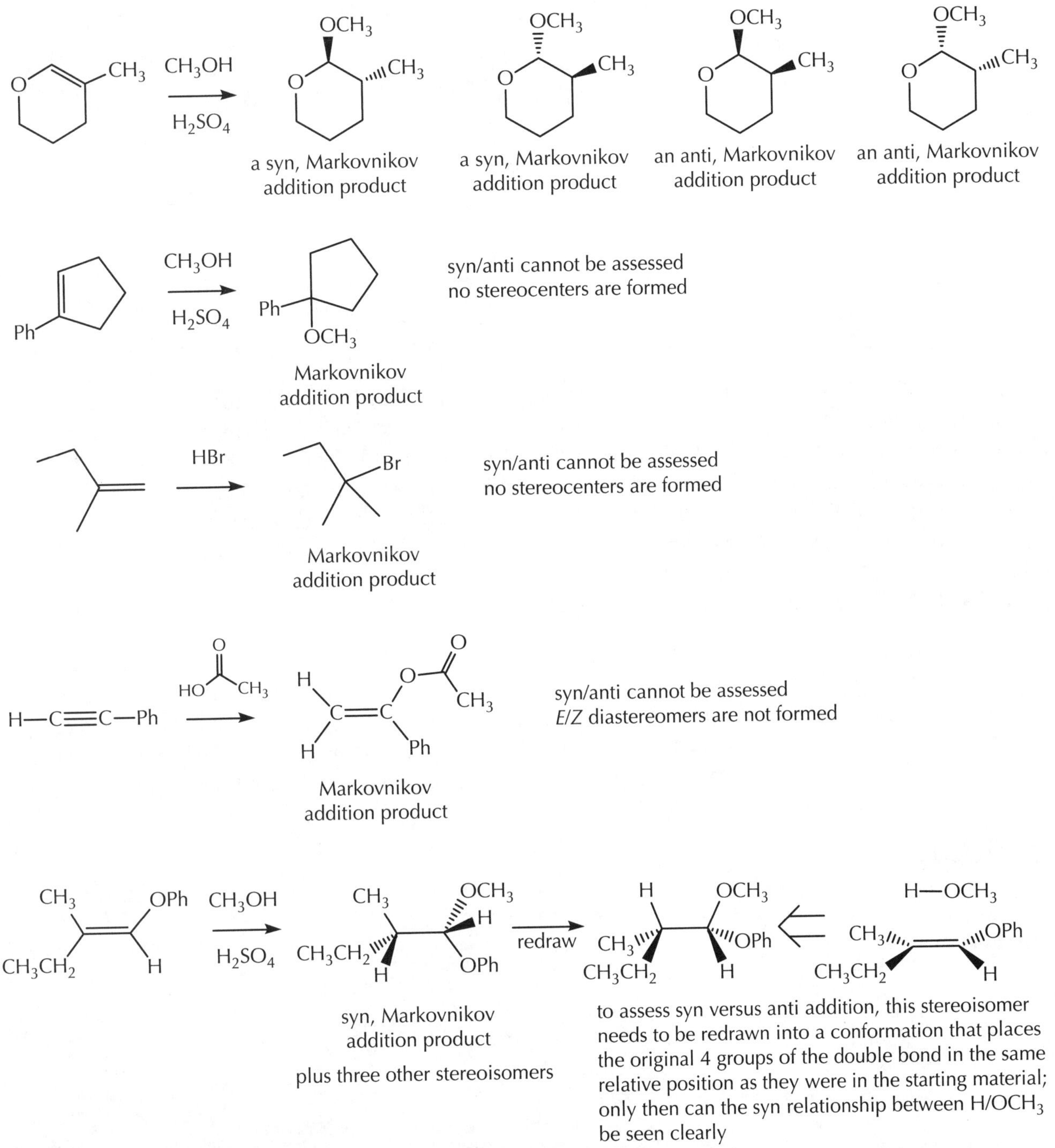

a syn, Markovnikov addition product

a syn, Markovnikov addition product

an anti, Markovnikov addition product

an anti, Markovnikov addition product

Markovnikov addition product

syn/anti cannot be assessed
no stereocenters are formed

Markovnikov addition product

syn/anti cannot be assessed
no stereocenters are formed

Markovnikov addition product

syn/anti cannot be assessed
E/Z diastereomers are not formed

syn, Markovnikov addition product

plus three other stereoisomers

to assess syn versus anti addition, this stereoisomer needs to be redrawn into a conformation that places the original 4 groups of the double bond in the same relative position as they were in the starting material; only then can the syn relationship between H/OCH$_3$ be seen clearly

Electrophilic addition reactions of strong Brønsted acids are also observed to occur with the carbon-carbon triple bonds of alkynes (Figure 0811), as shown on the next page.

Both of the pi bonds in a triple bond can undergo addition, in sequence, and so it is necessary to indicate whether the product results from adding one or two equivalents of the acid. The regioselectivity of the addition reaction(s) follows the Markovnikov Rule, and the possibility of stereoselectivity exists if stereoisomers might be formed. In the regioselective addition reaction of HBr with 1-phenylpropyne, protonation

620 CHAPTER 8 Electrophilic Addition I: Brønsted Acids

Figure 0811

Addition of strong Brønsted acids to triple bonds.

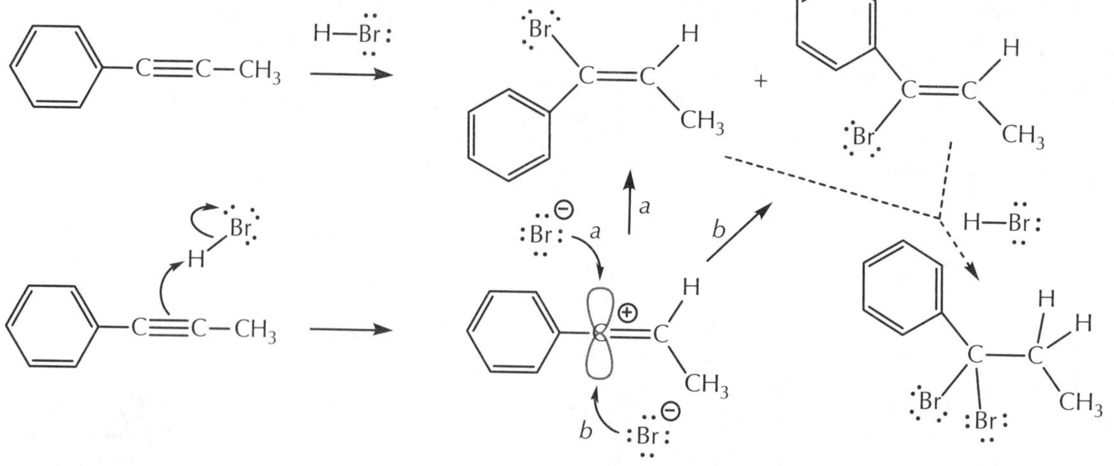

of the triple bond gives a resonance-stabilized sp carbocation intermediate. The carbocation intermediate can be captured by the bromide ion to give both the (E)- and (Z)-diastereomers of the mono-addition product (1-bromo-1-phenylpropene). Under some experimental conditions, these alkene diastereomers might be the observed products. Under other conditions, a second equivalent of HBr could add. The regioselectivity of the second addition reaction also follows the Markovnikov Rule. No stereocenters are formed, so there is only one product, regardless of which of the mono-addition (E)- or (Z)- diastereomers you start from.

D. Electrophilic Addition of Weak Brønsted Acids: Revisited

The electrophilic addition of weak Brønsted acids ($pK_a > 5$) requires a strong acid catalyst. Typically, the strong acid will first be deprotonated by something (the weak acid or the solvent, if different), and the resulting conjugate acid is used to protonate the pi bond undergoing addition. The original weak acid captures the resulting carbocation, and this cationic intermediate is deprotonated to give the addition reaction product(s). The addition reaction of methanol (a weak Brønsted acid) to a double bond, in the presence of sulfuric acid, is shown in Figure 0812.

Figure 0812

Stereochemical analysis: acid-catalyzed electrophilic addition reaction of methanol.

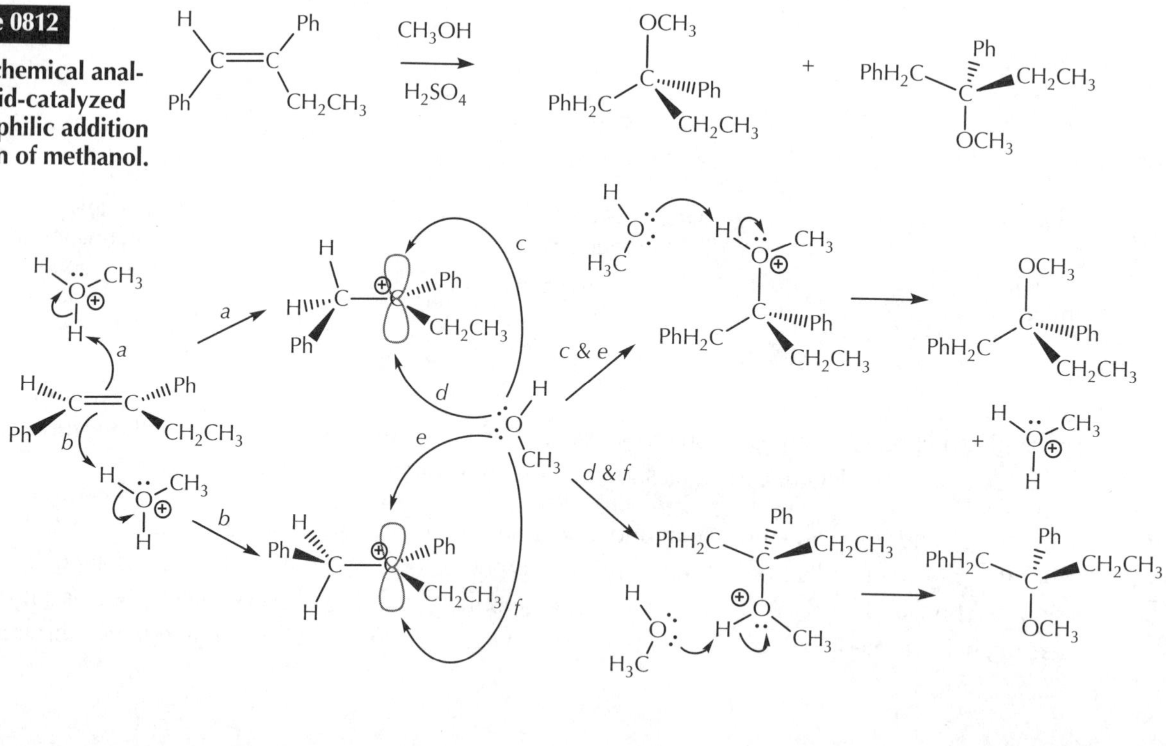

The stereochemical analysis, in this case, predicts forming the product as a racemic mixture because only one new stereocenter is created. Both the syn and anti addition pathways give the same products because the protonation step does not create a new stereocenter (i.e., pathways a and b in Figure 0812 form the same, achiral carbocation intermediate, and so pathways c and e are identical, as are pathways d and f). Only the pair of enantiomers results.

As before, we assume that the mechanisms are operating no differently in those molecules where we cannot experimentally see the full stereochemical outcome as those in which we can. For instance, the stereochemical course of the reaction in Figure 0812 could be studied experimentally by using the compound where one of the ^{1}H hydrogen atoms has been replaced by a deuterium (^{2}H, D) atom, in which case the reaction product would be revealed as coming from both the syn and anti mechanistic pathways (Figure 0813).

Figure 0813

Stereochemical analysis: experimentally monitoring the acid-catalyzed electrophilic addition reaction of methanol.

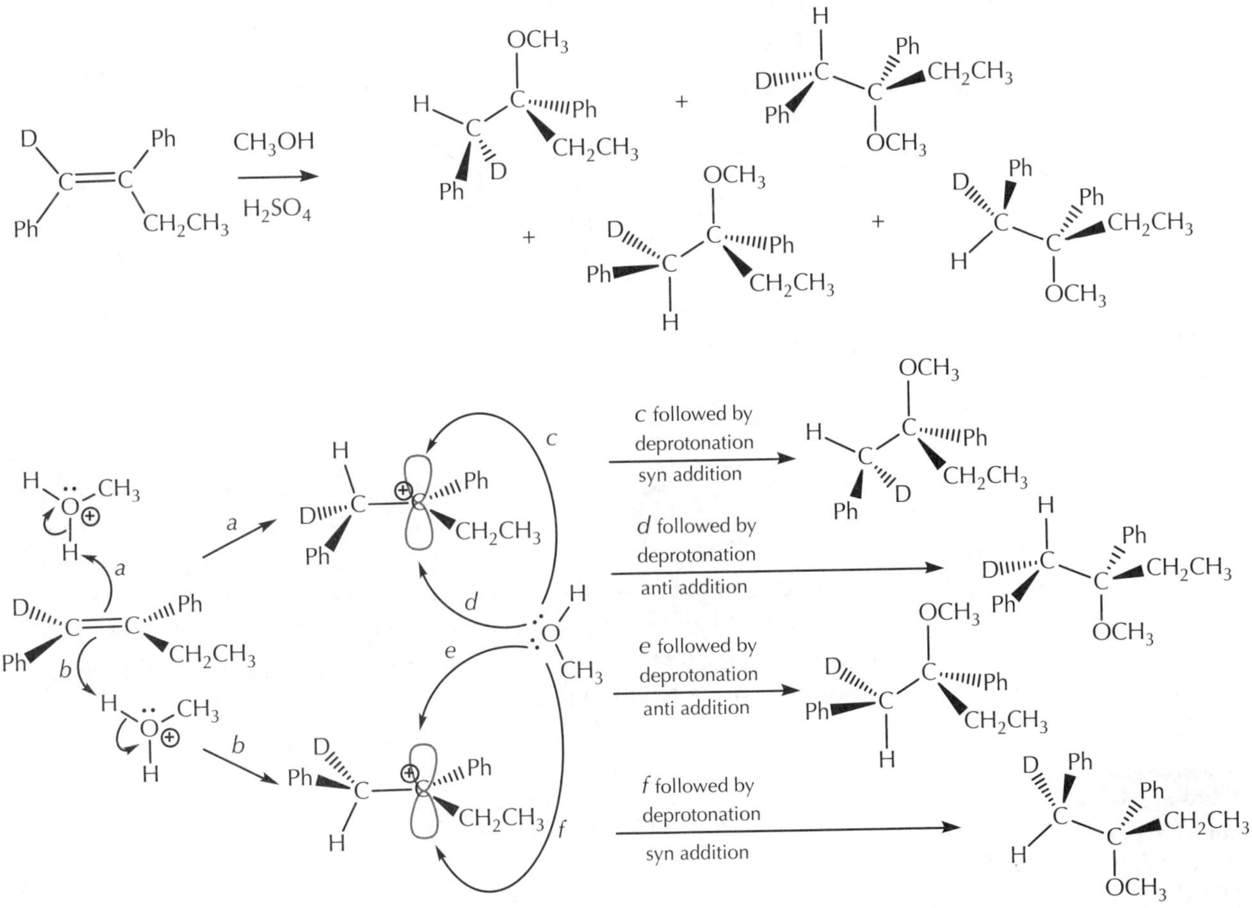

As with their strong acid counterparts, the addition of weak Brønsted acids to triple bonds can happen once (one equivalent of acid, mono-addition) or twice (two equivalents or excess, double addition). The reaction mechanism with triple bonds is the same as it is with double bonds, and regioselectivity follows the Markovnikov Rule (Figure 0814 on the next page).

Figure 0814

Addition of weak Brønsted acids to triple bonds.

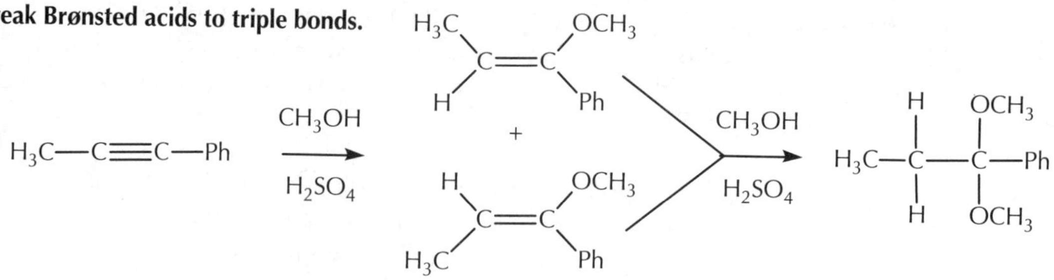

$$H_3C-C \equiv C-Ph \xrightarrow[H_2SO_4]{CH_3OH} \left[\begin{array}{c} H_3C \diagdown \diagup OCH_3 \\ C=C \\ H \diagup \diagdown Ph \end{array} + \begin{array}{c} H \diagdown \diagup OCH_3 \\ C=C \\ H_3C \diagup \diagdown Ph \end{array} \right] \xrightarrow[H_2SO_4]{CH_3OH} \begin{array}{c} H \quad OCH_3 \\ | \quad | \\ H_3C-C-C-Ph \\ | \quad | \\ H \quad OCH_3 \end{array}$$

The addition of water, which qualifies as a weak acid, to triple bonds gives an unexpected structural outcome. Only one equivalent of water is added, and the structure does not look like the anticipated product from an addition reaction (Figure 0815), although it is a structural isomer of the anticipated product.

Figure 0815

Unanticipated result from the addition of water to triple bonds.

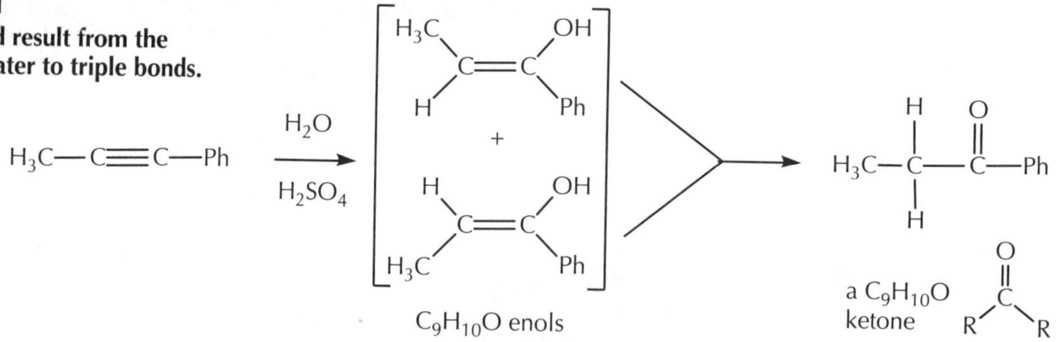

$$H_3C-C \equiv C-Ph \xrightarrow[H_2SO_4]{H_2O} \left[\begin{array}{c} H_3C \diagdown \diagup OH \\ C=C \\ H \diagup \diagdown Ph \end{array} + \begin{array}{c} H \diagdown \diagup OH \\ C=C \\ H_3C \diagup \diagdown Ph \end{array} \right] \longrightarrow \begin{array}{c} H \quad O \\ | \quad \| \\ H_3C-C-C-Ph \\ | \\ H \end{array}$$

$C_9H_{10}O$ enols

a $C_9H_{10}O$ ketone $\quad R\diagdown C(=O) \diagup R$

The functional group of the anticipated (but unobserved) product is called an enol, which is a double bond with a hydroxyl group substituent. The functional group of the observed product is called a ketone, in which the carbon atom of a carbon-oxygen double bond (C=O) is, in turn, attached to two other hydrocarbon ("R") groups.

The chemistry of the enol functional group. The enol product, expected from the addition reaction of water to a triple bond, as shown in Figure 0815, does form. However, except for a few cases, an enol functional group undergoes a rapid and favorable equilibrium process to give the structural isomer with the carbon-oxygen double bond (C=O). The sum of bond energy differences between the two structures (C=C and O-H in the enol, C=O and C-H in the ketone) generally favors the ketone, and this connectivity is the observed product for that reason. The mechanism for the isomerization of an enol functional group to a ketone is shown in Figure 0816.

Figure 0816

Isomerization of an enol to a ketone.

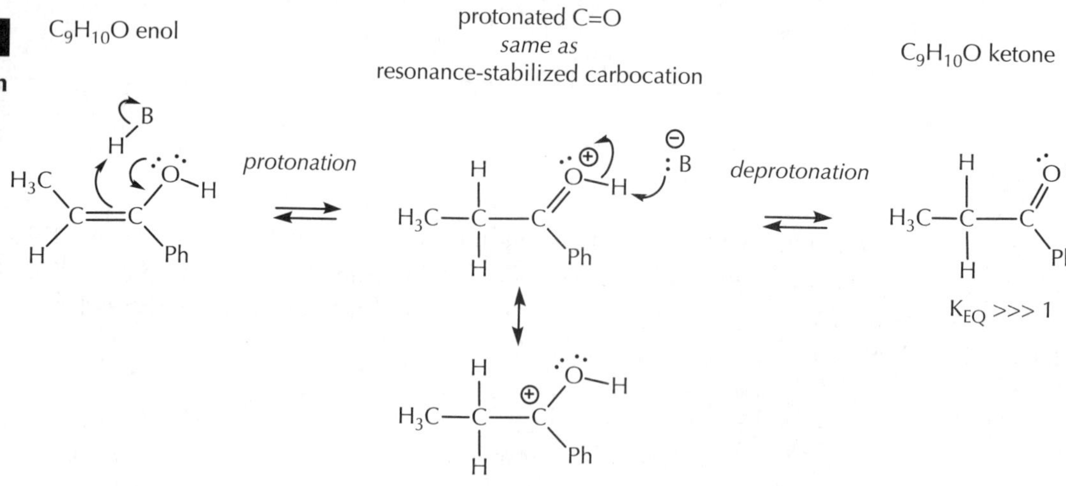

$C_9H_{10}O$ enol $\quad\rightleftharpoons\quad$ protonated C=O same as resonance-stabilized carbocation $\quad\rightleftharpoons\quad$ $C_9H_{10}O$ ketone

$K_{EQ} >>> 1$

8.1 Electrophilic Addition of Brønsted Acids

PRACTICE QUESTIONS

8.01 For each of the following addition reactions of Brønsted acids, one of the possible products is shown. Identify the regioselectivity (Markovnikov, anti-Markovnikov, cannot tell) and the stereoselectivity (syn addition, anti addition, cannot tell) of the specific outcome shown.

(a) regioselectivity: _____ stereoselectivity: _____

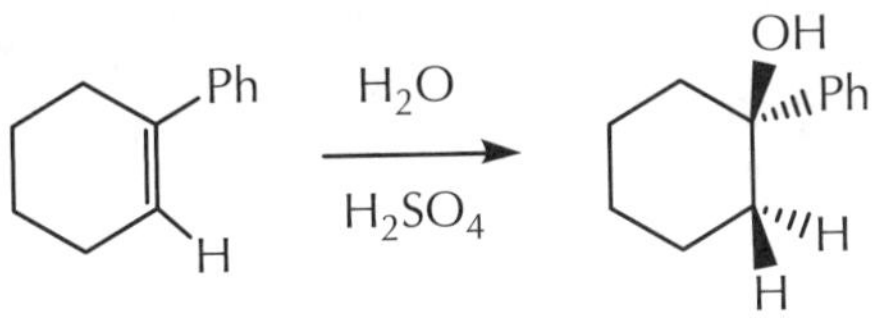

(b) regioselectivity: _____ stereoselectivity: _____

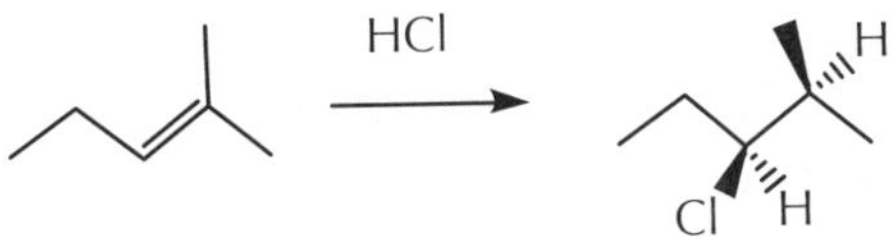

(c) regioselectivity: _____ stereoselectivity: _____

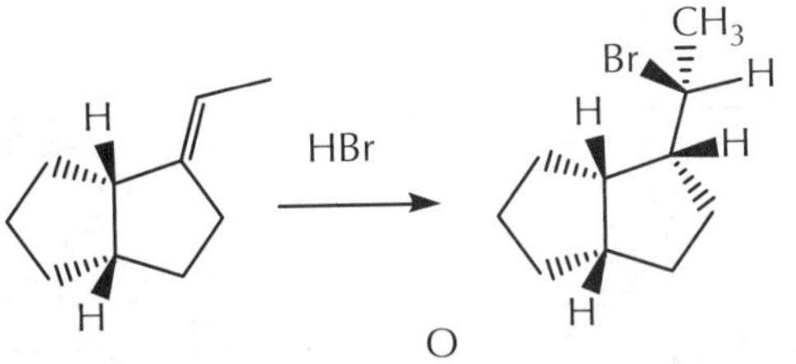

(d) regioselectivity: _____ stereoselectivity: _____

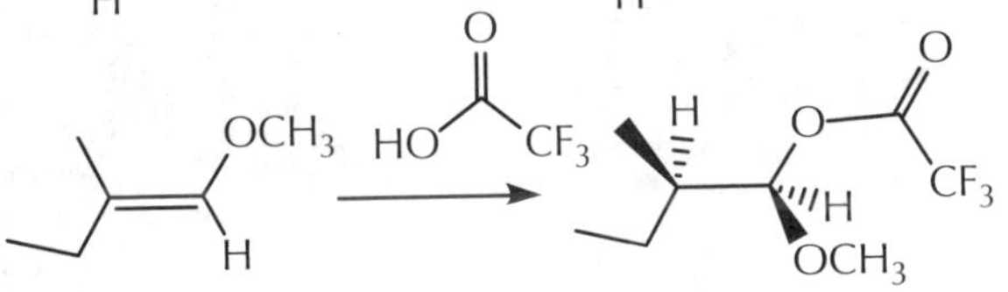

(e) regioselectivity: _____ stereoselectivity: _____

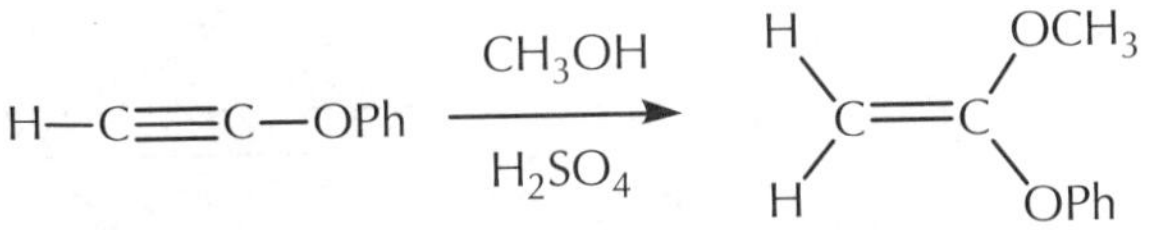

(f) regioselectivity: _____ stereoselectivity: _____

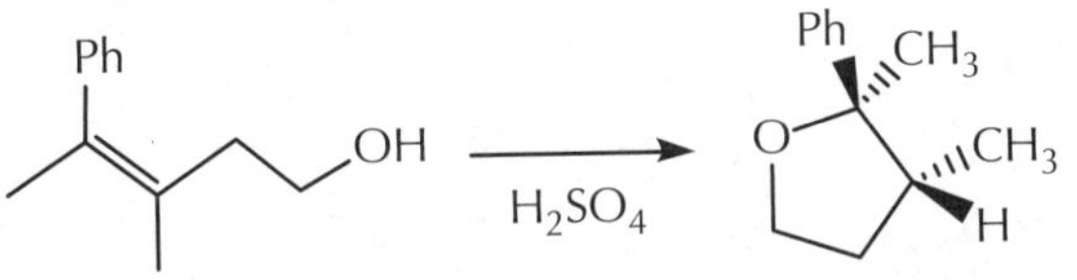

8.02 Draw both of the stereoisomers resulting from the syn, anti-Markovnikov addition of water to each of these compounds.

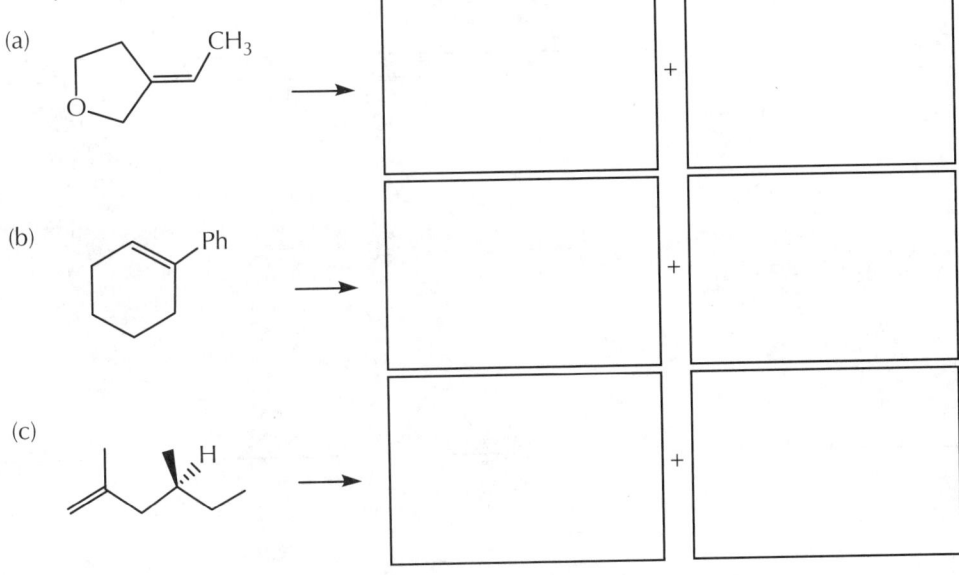

8.03 Provide the full set of stereoisomeric products from the following electrophilic addition reactions of strong and weak Brønsted acids. Show only the products with the Markovnikov regioselectivity. For each product that you draw, indicate whether the stereochemical outcome is from a syn addition, an anti addition, or that it cannot be determined from the structure.

(a) [cyclopentene with –OCH₃ and –CH₂CH₃ substituents] + HBr →

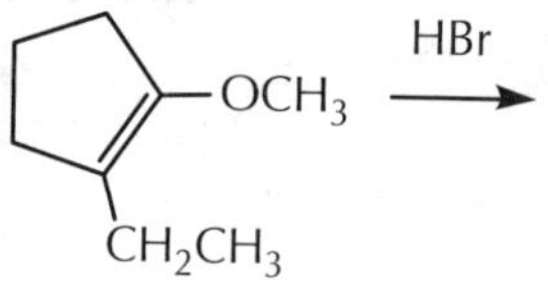

(b) $CH_3-C{\equiv}C-CH_2Ph$ + HO−C(=O)−CHCl₂, 1 equivalent →

(c) [2-methyl-2-butene] + CH₃CH₂OH / H₂SO₄ →

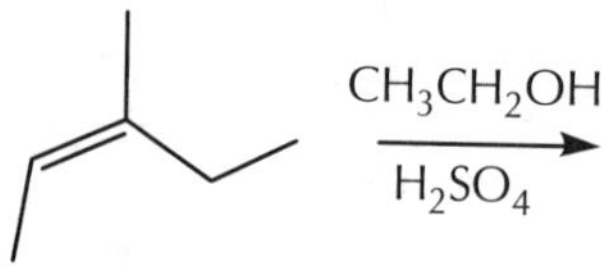

(d) [methylenecyclohexane with wedge stereochemistry] + cyclopentanol / H₂SO₄ →

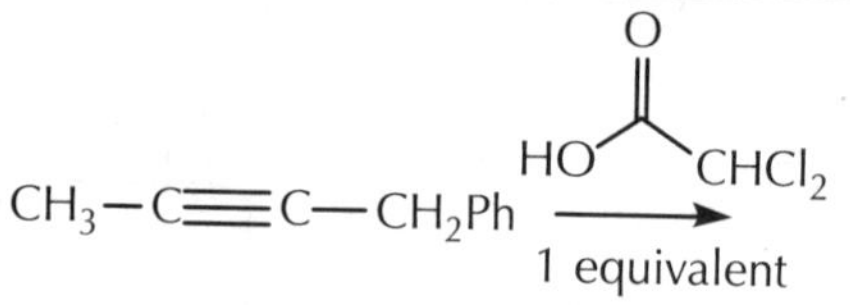

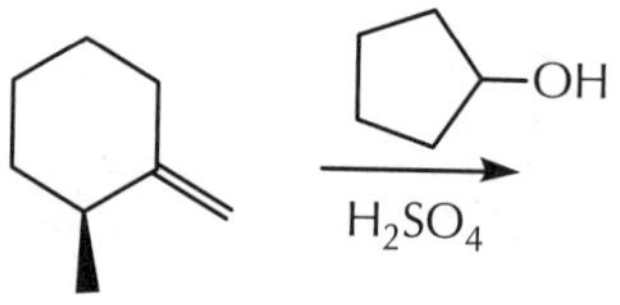

8.04 The following electrophilic addition reactions of Brønsted acids result in mixtures of stereoisomers. In each example, only one specific product is drawn along with the indication about whether there are also others. Provide the complete, curved arrow mechanism for each example.

(a)

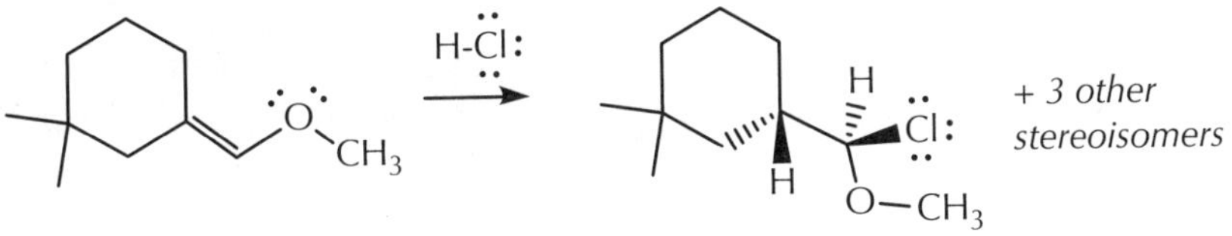

(b)

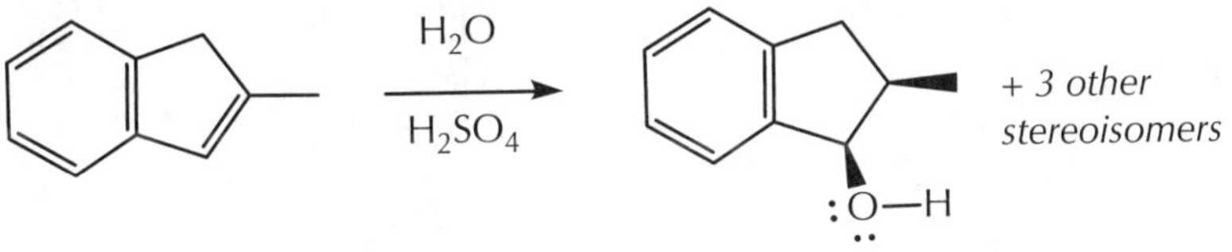

(c)

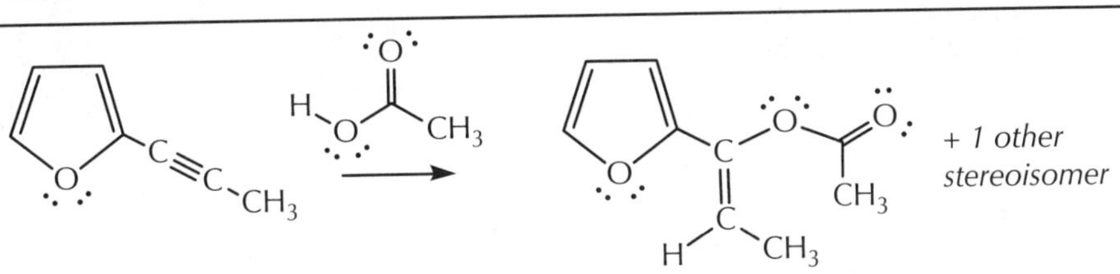

626 CHAPTER 8 Electrophilic Addition I: Brønsted Acids

8.05 Complete the following reactions by providing the missing component. If an acid catalyst is needed, or more than one equivalent, then be sure to include that information. If specific information about the stereochemistry or other structural features is indicated, then be sure your answer is consistent.

(a) [ ] + ~~~OH →(H₂SO₄) [benzofuran derivative with OCH₂CH₂CH₂CH₂CH₃ and CH₃] + enantiomer

^{1}H-NMR shows 6 sp² CH atoms

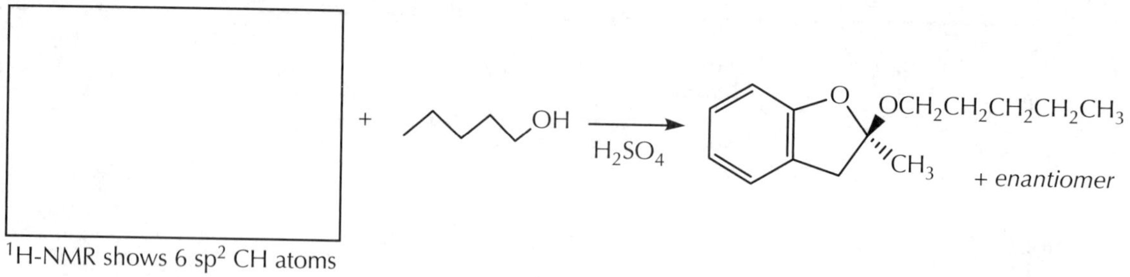

(b) [H₃CO-cyclohexane with =CH₂] + [ ] → [H₃CO-cyclohexane with Br and CH₃] + diastereomer

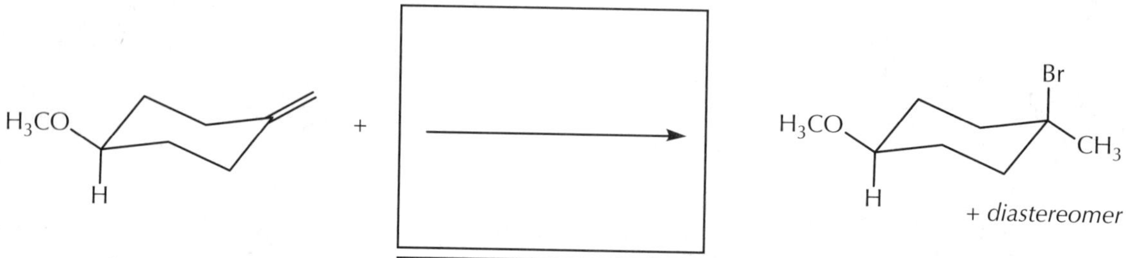

(c) [cyclopentyl-CH₂-C≡C-H] + [ ] → [cyclopentyl-CH₂-C(=O)-CH₃]

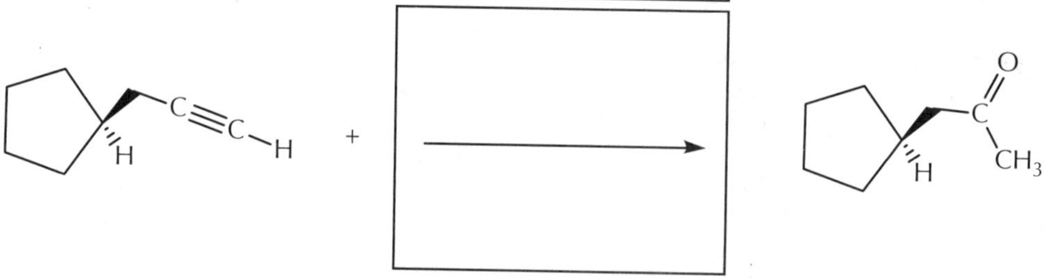

(d) [ ] →(H₂SO₄) [tetrahydrofuran with CH₃ and gem-dimethyl]

a chiral molecule with 3 methyl groups

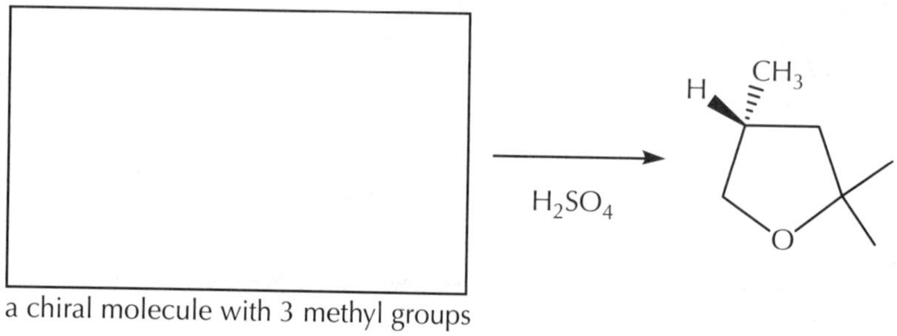

(e) [naphthyl-C≡C-H] →(2 equivalents CH₃OH, H₂SO₄) [ ]

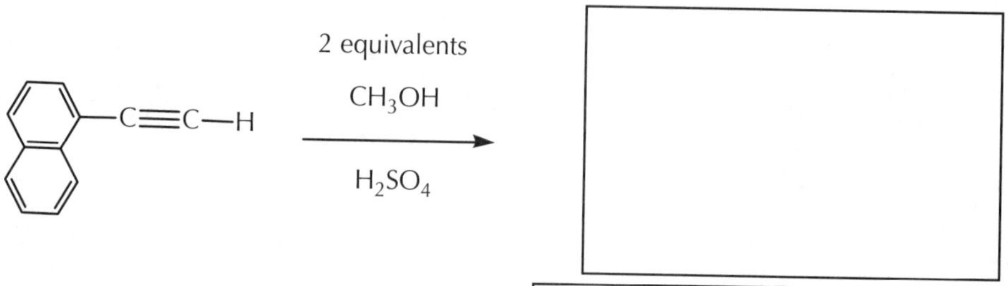

(f) [bicyclic compound with H, H₃C, CH₃, Ph] + HO-C(=O)-CF₃ → [ ]

syn, Markovnikov addition; methyl groups end up *trans*

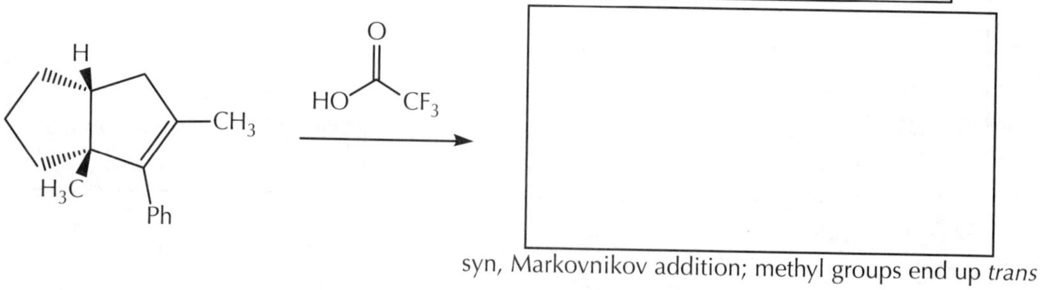

8.2 Carbocation Chemistry

A. Origins and Fates of Carbocations

Carbocations are reactive intermediates that are proposed during the electrophilic addition reactions of strong or weak Brønsted acids to pi bonds, as well as in the S_N1 and E1 reactions. Taken together, these reactions provide a complete look at the origins (formation) and fates (reaction outcomes) of carbocations. Figure 0817 gives a look at how these ideas are connected.

Figure 0817

Origins and fates of carbocations.

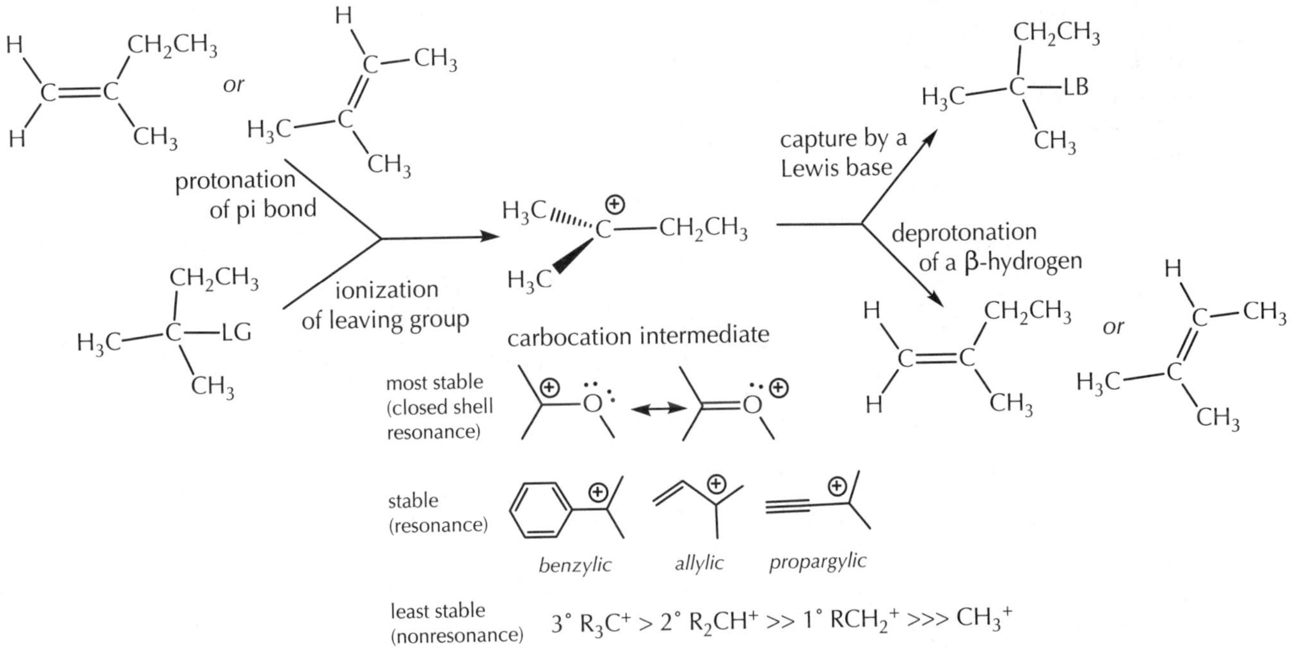

In your studies so far, carbocations can result from (a) the protonation reaction of pi bonds (addition reaction) or (b) from the loss of a leaving group in an ionization reaction (S_N1/E1). The two reaction outcomes are (a) capture to form a new sigma bond (addition, S_N1) or (b) deprotonation of a β-hydrogen atom (E1). Figure 0817 also includes a reminder about relative carbocation stability, which correlates with their relative rates of formation. A more stable carbocation forms faster than a less stable one. Based upon the examples presented so far in the text, heteroatom substituted carbocations are most stable because they have significant resonance contributors in which all of the atoms are closed shell; carbocations with pi delocalization (allyl, benzyl, propargyl) are the next most stable, also due to delocalization; then the simple alkyl substituted carbocations are ranked based their degree of substitution (3° > 2° >> 1° >>> methyl).

B. Carbocation Rearrangements Change Connectivity

Evidence for carbocations goes back to the 1890s, but only in retrospect. A modern picture of atoms did not exist then. This time period is still 10 years before electrons are discovered, 20–30 years before protons are discovered, and about 40 years before a model of atoms based upon these particles was widely accepted.

By the 1920s, reaction outcomes were being reliably observed in which the connectivity of the carbon framework was changing, and these were called "shifts" or "rearrangements." For instance, the electrophilic addition of water to 3,3-dimethyl-1-butene does not produce very much of the 3,3-dimethyl-2-butanol anticipated from the addition reaction, but rather 2,3-dimethyl-2-butanol (Figure 0818, on next page), where the original connectivity of the carbon atom framework has changed, or rearranged.

Figure 0818

Unexpected result from the addition of water to 3,3-dimethyl-1-butene.

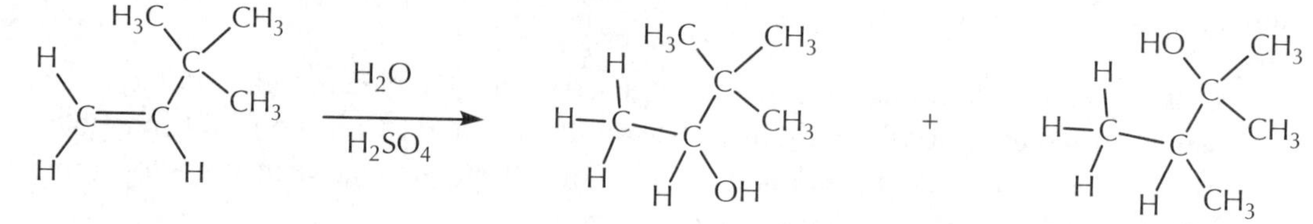

3,3-dimethyl-1-butene → (H₂O, H₂SO₄) → 3,3-dimethyl-2-butanol (expected; minimal or not observed product) + 2,3-dimethyl-2-butanol (unexpected; major observed product)

Figure 0819

General picture for 1,2-rearrangements of carbocations.

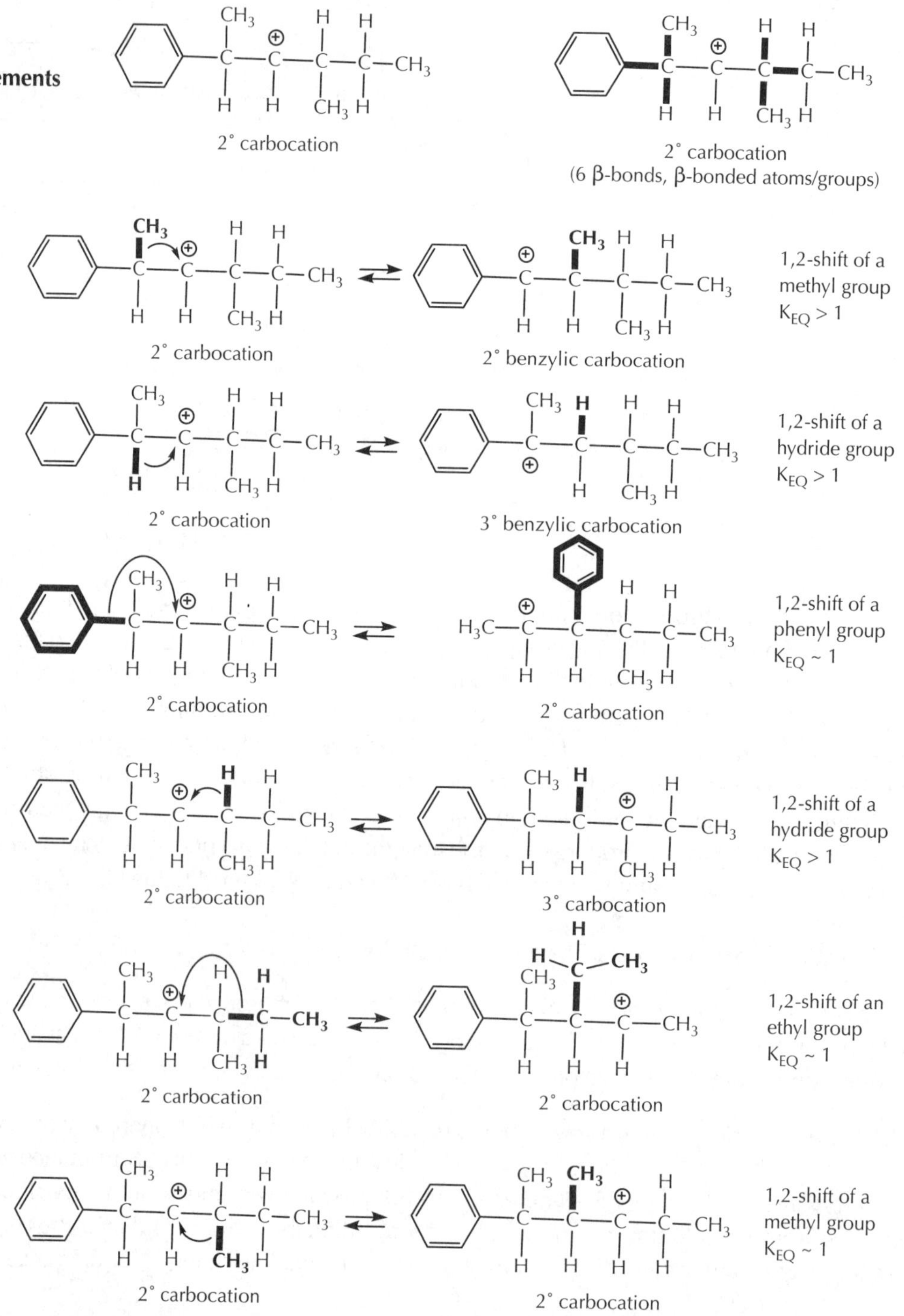

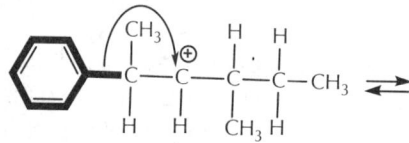

2° carbocation → 2° carbocation (6 β-bonds, β-bonded atoms/groups)

2° carbocation ⇌ 2° benzylic carbocation — 1,2-shift of a methyl group, $K_{EQ} > 1$

2° carbocation ⇌ 3° benzylic carbocation — 1,2-shift of a hydride group, $K_{EQ} > 1$

2° carbocation ⇌ 2° carbocation — 1,2-shift of a phenyl group, $K_{EQ} \sim 1$

2° carbocation ⇌ 3° carbocation — 1,2-shift of a hydride group, $K_{EQ} > 1$

2° carbocation ⇌ 2° carbocation — 1,2-shift of an ethyl group, $K_{EQ} \sim 1$

2° carbocation ⇌ 2° carbocation — 1,2-shift of a methyl group, $K_{EQ} \sim 1$

8.2 Carbocation Chemistry 629

By the 1940s and 1950s, these changes in connectivity were explained using a general property of carbocations called carbocation rearrangement. In general, any carbocation that forms can undergo 1,2-rearrangements, also called 1,2-shifts, where the electrons and atom(s) from any β-bond with respect to the carbocation migrates to the location of the open shell carbon, forming a new carbocation at the position it moved from (Figure 0819).

These movements are fast, with rates of at least 10^{10}/sec, faster than molecular collisions, even at extremely low temperature. All of the possible equilibria involving these shifts are taking place, which means the molecule usually ends up in whatever structure corresponds to the most stable possible carbocation for that set of atoms.

Following Markovnikov's Rule, we predict that the protonation of 3,3-dimethyl-1-butene will give the more stable 2° carbocation faster than the 1° carbocation (Figure 0820).

Figure 0820

Rearrangements (1,2-shifts) in the addition of water to 3,3-dimethyl-1-butene.

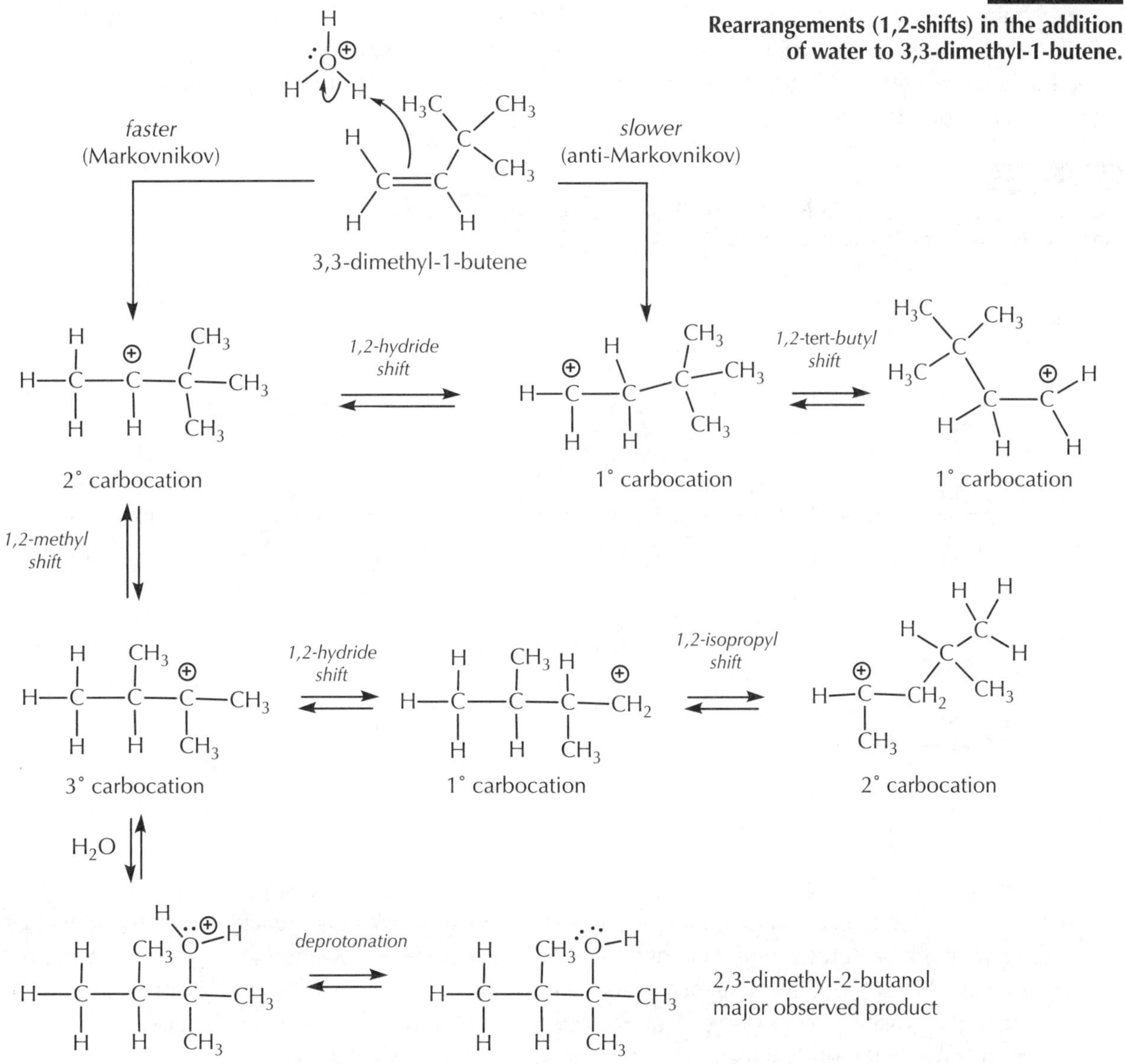

Of the two initially formed carbocations, the 2° carbocation has 6 β-bonds and the 1° carbocation has 3 β-bonds. In theory, all of these β-bonds (that is, the electrons and the appended atoms) can migrate in a 1,2-shift. Linguistically, the "1,2-" label used here is completely unrelated to the carbon numbering system

CHAPTER 8 Electrophilic Addition I: Brønsted Acids

in nomenclature. The "1,2" designation indicates that the action is restricted to a pair of adjacent atoms: the atom on which the carbocation is located (the alpha atom) and the bonds on any of the carbons that are attached to carbocation (the beta atoms, whose bonds are the beta bonds with respect to the carbocation center).

All of the possible rearrangement equilibria are taking place, although not all of them are shown in Figure 0820. Some are favorable ($K_{EQ} > 1$), some are not ($K_{EQ} < 1$), and given enough time, more of the molecules will end up in the most stable possible destination, from which the observed product results. When the first-formed carbocation is the most stable one, the reaction outcome is typically derived from that one. But as you might expect, the more atoms there are, and the less stable the initially formed carbocation is, the more difficult this outcome is to predict.

Experimentally, carbocation rearrangements can be a nuisance, leading to unexpected products and/or reactions that produce multiple outcomes. For carbocations derived from small molecules (3–5 carbons), the rearrangement outcome is predictable because there is one obviously most stable arrangement (Figure 0821). That said, even the extent to which rearrangement products are observed can depend significantly on the exact reaction conditions, so even in simple cases the distribution of products is a matter of experimental observation.

Figure 0821

Rearrangement outcomes in carbocations with 3–5 carbons are more predictable than in larger molecules.

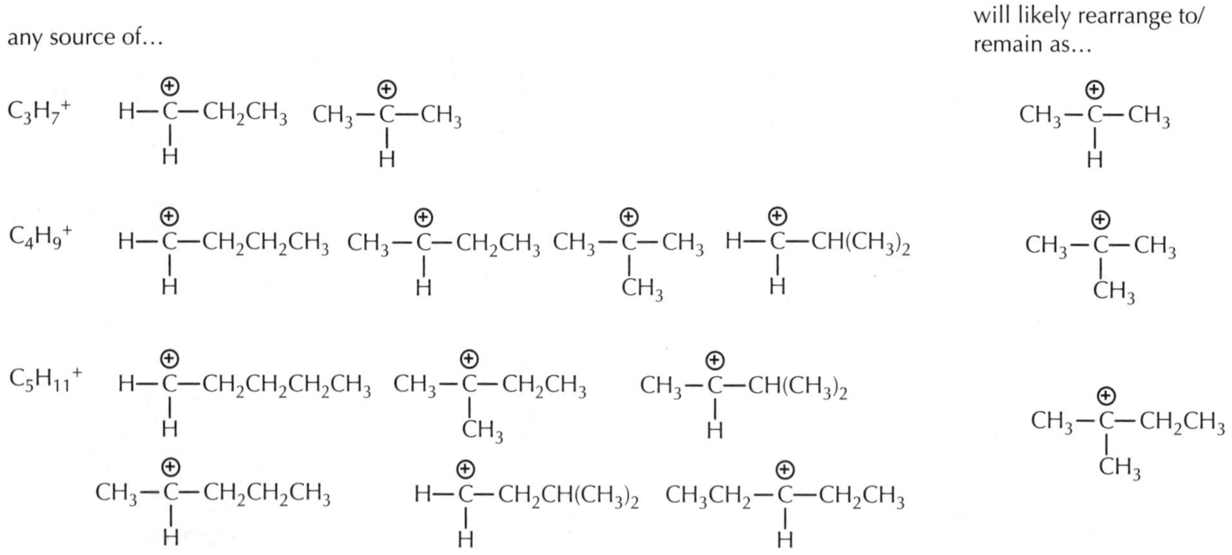

any source of...		will likely rearrange to/remain as...
$C_3H_7^+$	$H-\overset{+}{\underset{H}{C}}-CH_2CH_3$ $CH_3-\overset{+}{\underset{H}{C}}-CH_3$	$CH_3-\overset{+}{\underset{H}{C}}-CH_3$
$C_4H_9^+$	$H-\overset{+}{\underset{H}{C}}-CH_2CH_2CH_3$ $CH_3-\overset{+}{\underset{H}{C}}-CH_2CH_3$ $CH_3-\overset{+}{\underset{CH_3}{C}}-CH_3$ $H-\overset{+}{\underset{H}{C}}-CH(CH_3)_2$	$CH_3-\overset{+}{\underset{CH_3}{C}}-CH_3$
$C_5H_{11}^+$	$H-\overset{+}{\underset{H}{C}}-CH_2CH_2CH_2CH_3$ $CH_3-\overset{+}{\underset{CH_3}{C}}-CH_2CH_3$ $CH_3-\overset{+}{\underset{H}{C}}-CH(CH_3)_2$ $CH_3-\overset{+}{\underset{H}{C}}-CH_2CH_2CH_3$ $H-\overset{+}{\underset{H}{C}}-CH_2CH(CH_3)_2$ $CH_3CH_2-\overset{+}{\underset{H}{C}}-CH_2CH_3$	$CH_3-\overset{+}{\underset{CH_3}{C}}-CH_2CH_3$

In this text, you will not be expected to anticipate when carbocation rearrangements will occur. Instead, you need to be able to use carbocation rearrangements in explanations for reactions where the hydrocarbon framework has undergone a change in its connectivity, which is indicative of carbocation rearrangements. Working out the sequence of rearrangements is not a linear thought process, so playing around with the possible rearrangements at each step along the way may be needed before making the connection between the starting material and the product. As described in the previous section, you can be sure that a carbocation will form by one of the anticipated pathways (protonation, ionization), and that its final fate will be deprotonation or capture by a nucleophile. In between, rearrangements may or may not be taking place.

Figure 0822 shows a few examples of reactions where carbocation rearrangements are required to explain the results. In each case, note that changes in the connectivity of the original hydrocarbon structure provide a clue that rearrangements are taking place.

Figure 0822

Examples of reactions involving carbocation rearrangements.

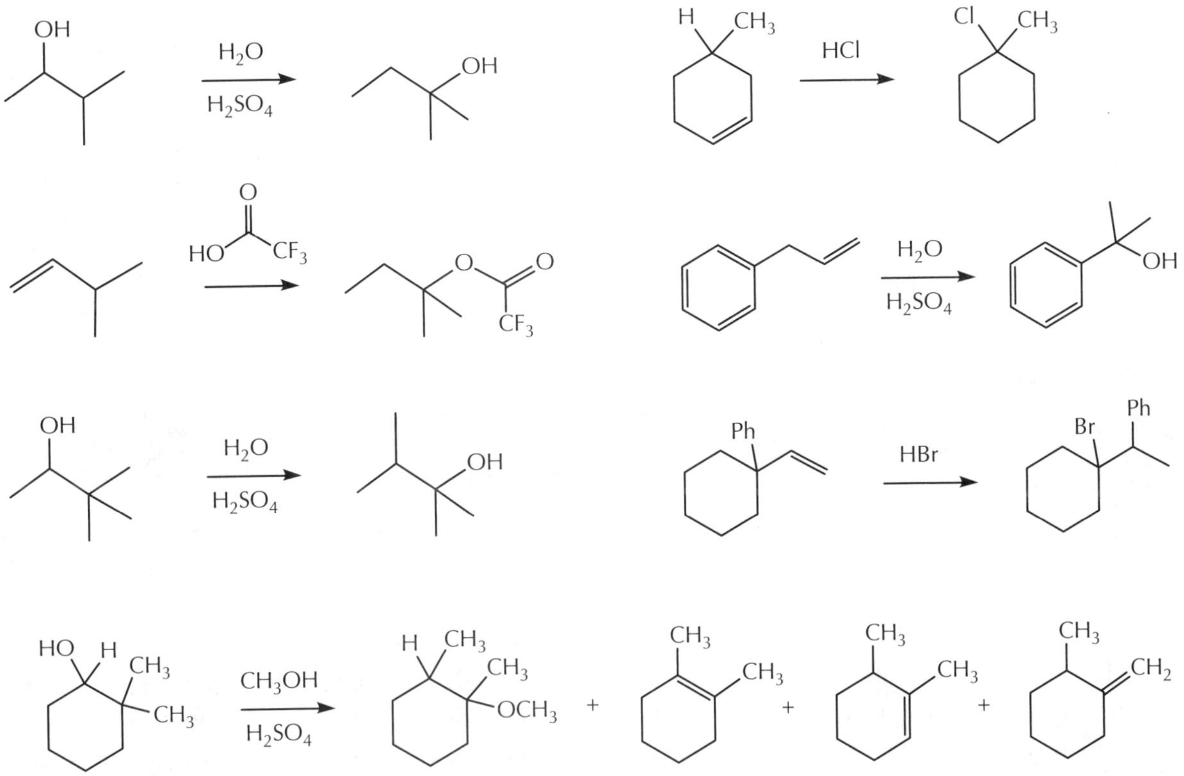

C. Ring Contraction and Expansion Reactions

You may be questioning the presence of unstabilized 1° and 2° carbocations in the previous section on rearrangements. In the discussions about $S_N1/E1$ reactions, we were using reaction conditions where these less stable carbocations would not form readily (i.e., the cases where no $S_N1/E1$ is predicted if the carbocation is an unstabilized 1° or 2°). Under harsh reaction conditions, those unstabilized carbocations can form, but they universally undergo rearrangement reactions to give mixtures of products derived from more stable carbocations, so claiming that they form and give predictably productive outcomes is not generally true.

Experimentally, the main reason to avoid forming those less stable carbocations is their unpredictable nature, although, as shown in Figure 0821, some of the less stable carbocations derived from small groups of atoms have been used productively. Also, the chemists who wish to study those less stable carbocations, and their fates, were motivated to create reliable ways to form them. In particular, when strong aluminum trihalide Lewis acids (AlX$_3$) are heated with a corresponding sp³ alkyl halide (RX), the resulting Lewis acid-base complex turns the halogen atom into an excellent, cationic leaving group, good enough to predictably be ionized from both primary (1°) or secondary (2°) carbons atoms that do not form stabilized carbocations (Figure 0823).

632 CHAPTER 8 Electrophilic Addition I: Brønsted Acids

Figure 0823

Strong Lewis acids promote carbocation formation.

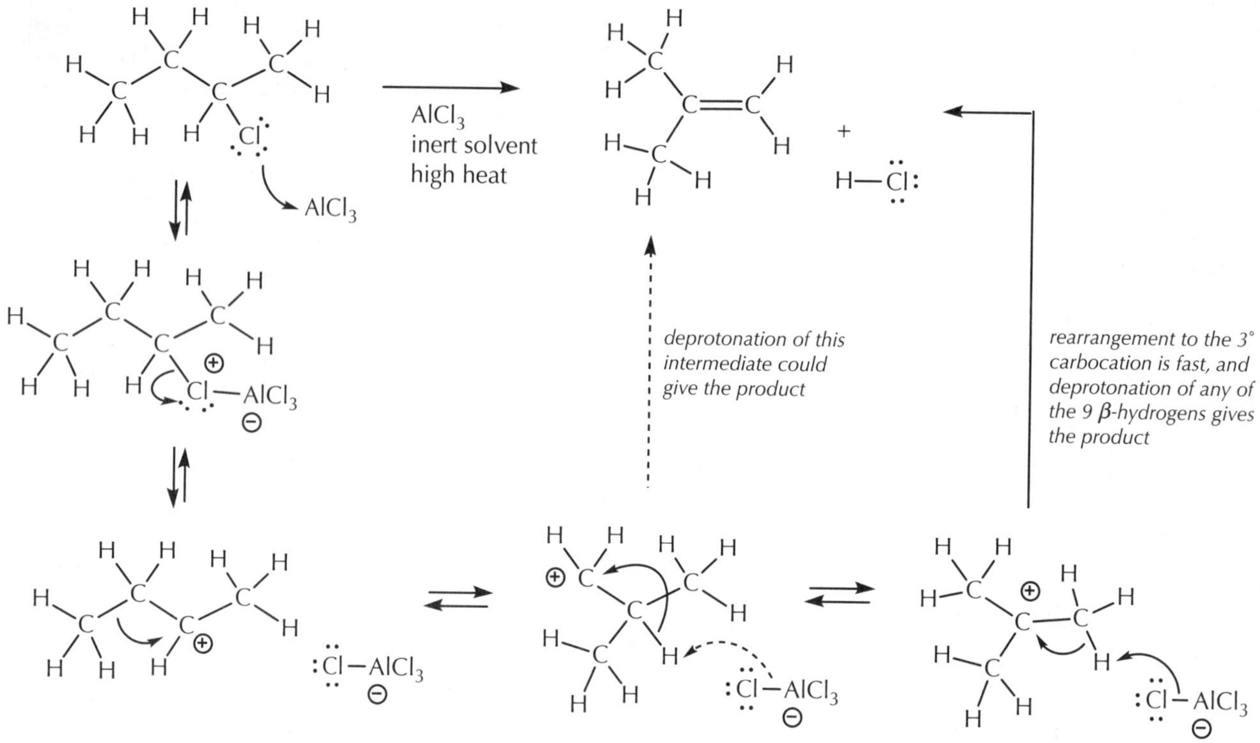

deprotonation of this intermediate could give the product

rearrangement to the 3° carbocation is fast, and deprotonation of any of the 9 β-hydrogens gives the product

When sp³ alkyl halides (RX) are heated with a corresponding aluminum trihalide Lewis acids (AlX₃), products derived from carbocations are observed. And if the halide is on a simple primary (1°) or secondary (2°) carbon, the carbocations invariably undergo rearrangements. For example, heating 2-chlorobutane with aluminum trichloride, in an inert solvent, results in the formation of gaseous mixture of 2-methylpropene and HCl. As stated earlier, the process of working out a mechanism is not linear, and benefits from exploring the options until you can see the connection you need. The key moment is when you see when the connectivity of the carbon atoms matches that of the observed product. After that, it is often shifting bonds with hydrogen atoms, capturing or deprotonating the carbocation that leads to the structure of the product.

When a simple alkyl group moves with its electron pair, it is called a 1,2-alkyl shift (e.g., 1,2-methyl shift; 1,2-ethyl shift; 1,2-phenyl shift). Because a hydrogen atom with an electron pair is called a hydride ion, those rearrangements are called 1,2-hydride shifts.

Sometimes, the β-bond is part of a ring. If the 1,2-shift of that ring bond decreases the size of a ring, then that is called a ring contraction. Alternatively, if the 1,2-shift of the ring bond increases the size of a ring, then that is called a ring expansion.

Figure 0824 gives an example of an observed ring contraction. Among other products, chlorocyclohexane, when heated with aluminum trichloride, gives 1-methylcyclopentene.

A Lewis acid-base reaction between chlorocyclohexane and aluminum trichloride forms a strong complex. Ionization of the positively charged chlorine atom group, which is now an excellent leaving group, gives a relatively unstable 2° carbocation. The 2° carbocation has 6 β-bonds, four of which are carbon-hydrogen atoms and two of which are carbon-carbon bonds of the ring. We presume that the 1,2-hydride shifts are taking place rapidly, and with a K_{EQ} of exactly 1 because the newly formed carbocation is iden-

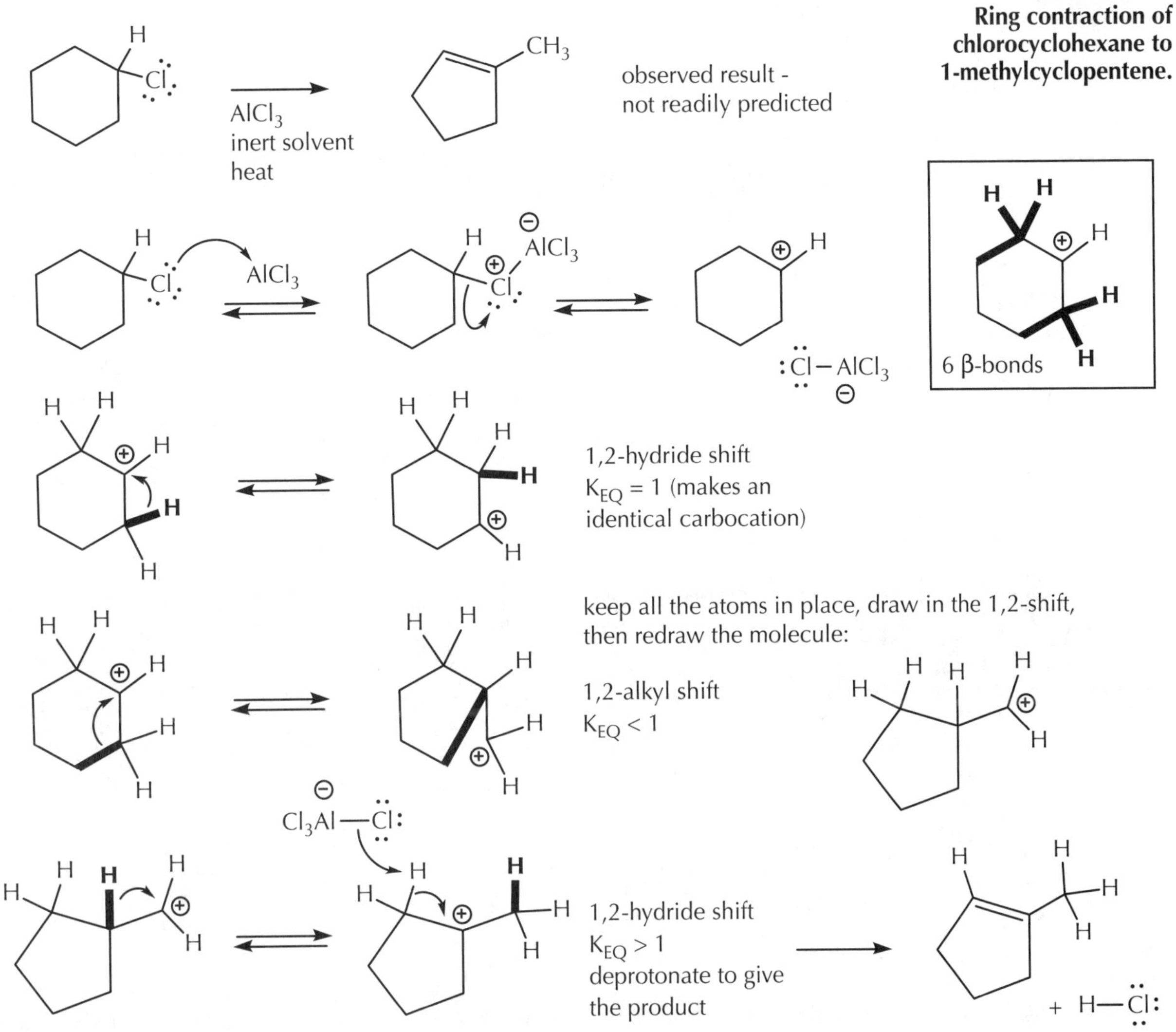

8.2 Carbocation Chemistry

Figure 0824
Ring contraction of chlorocyclohexane to 1-methylcyclopentene.

tical to the one it comes from. At some rate, the ring bond can also shift. A 1° carbocation results, and the K_{EQ} is less than 1—but it is not 0.

Experimentally, we know that the ring contracts, and the 1,2-alkyl shift of the ring bond gives the connectivity of the carbon framework in the observed product. It is not possible to think of a reaction that connects the 1° carbocation to the observed product in one step, but the formation of the 5-membered ring is a step in the right direction. The 1° carbocation has 3 β-bonds, two of which are carbon-carbon ring bonds and one of which is a carbon-hydrogen bond. If either of the carbon-carbon bonds migrate, the original 6-membered ring reforms, and this is presumably taking place at some rate. If the 1,2-hydride shift happens, then we are one step closer to the product's structure because a methyl group results, and this is a favorable K_{EQ} because of the resulting 3° carbocation. The 3° carbocation has 9 β-bonds, but another migration is not necessary to consider because a deprotonation of the 3° carbocation results in the structure of the given product.

Sometimes, forming a carbocation can result in a ring expansion reaction (Figure 0825). To reiterate: Reactions involving carbocation rearrangements are challenging to predict without added experimental information, and so these examples are not predictions, and you have no guidelines for that. Instead, you should expect information, often the structure of the product, as your starting point.

Figure 0825
Ring expansion starting from a cyclobutane compound.

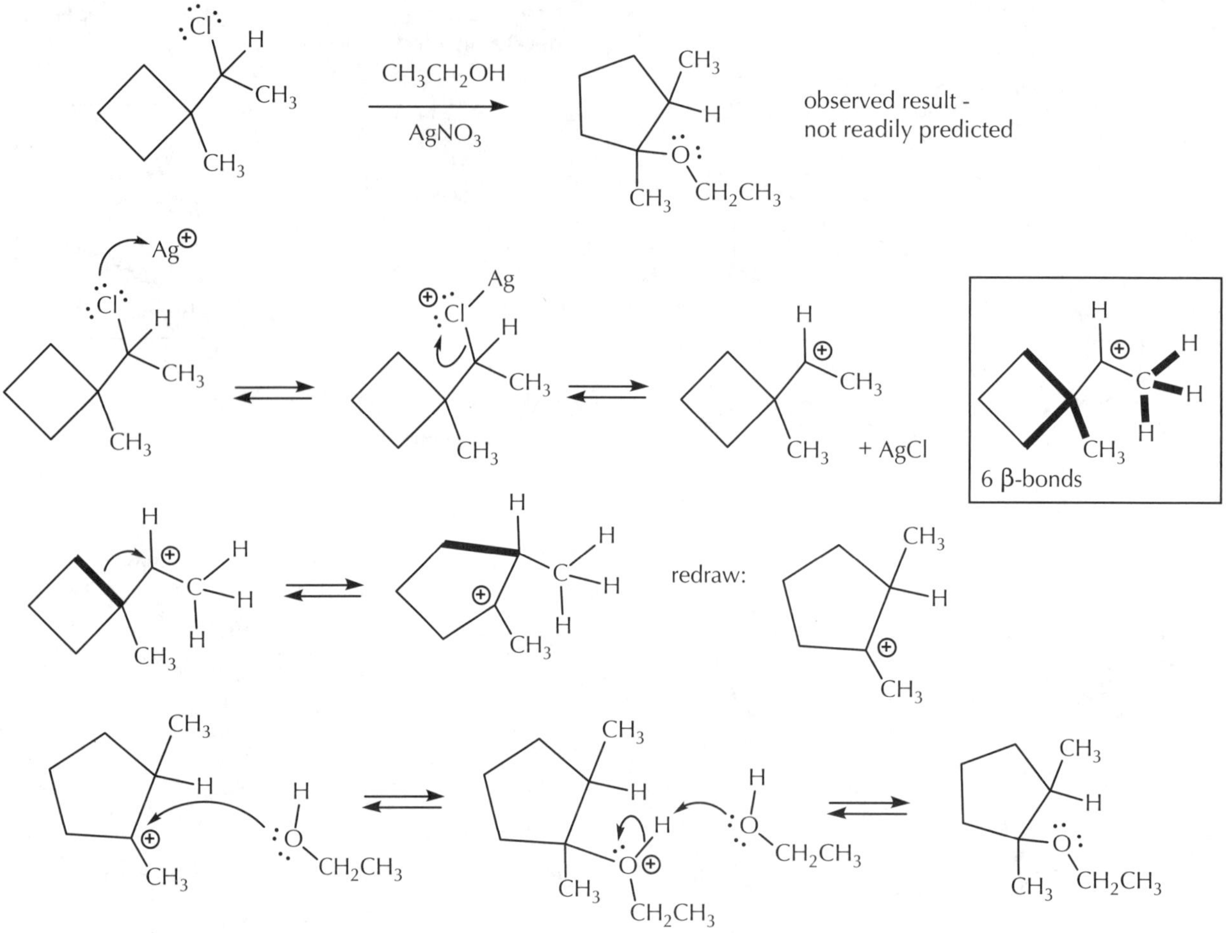

The chloro-substituted compound derived from 1-ethyl-1-methylcyclobutane is treated with silver chloride in ethanol. Like aluminum trihalides, silver (I) ions are also observed to be excellent Lewis acids for halide ions, and so the 2° carbocation can be formed. This 2° carbocation has 6 β-bonds: three CH bonds on the adjoining methyl group, two CC ring bonds, and the carbon-methyl bond. Keeping the observed product in mind, a 1,2-alkyl shift of one of the ring bonds results in a 5-membered ring and a new 3° carbocation. This 3° carbocation has all of the features of the hydrocarbon skeleton of the product, so all that remains is capturing the carbocation with ethanol followed by deprotonation to give the observed outcome.

D. Rearrangements of Carbocations with β-Bonded Heteroatoms

If one of the β-bonds is connected to a heteroatom, the migration of the other β-bonds to that carbon is favorable (Figure 0826).

Figure 0826
Carbocation rearrangements with heteroatoms on the β-bond.

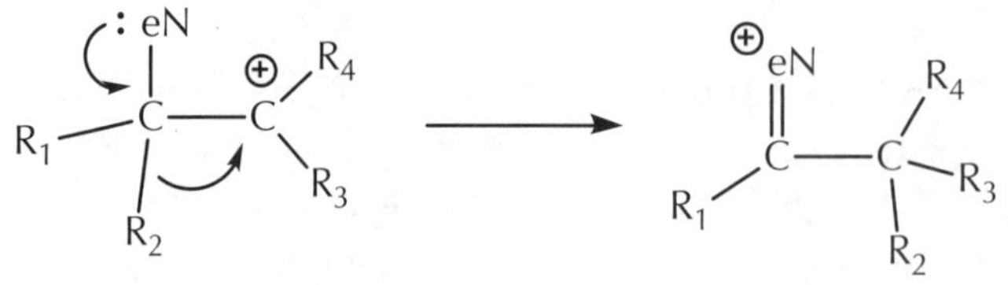

eN = some electronegative heteroatom with at least one nbe pair (e.g., N, O, S)

Molecules with hydroxyl groups on adjacent carbons are called "vicinal diols" or "1,2-diols." When they are treated with Brønsted acids (Figure 0827), a generally predictable rearrangement takes place.

Figure 0827

Heteroatoms on the β-bond: pinacol rearrangement of a vicinal diol.

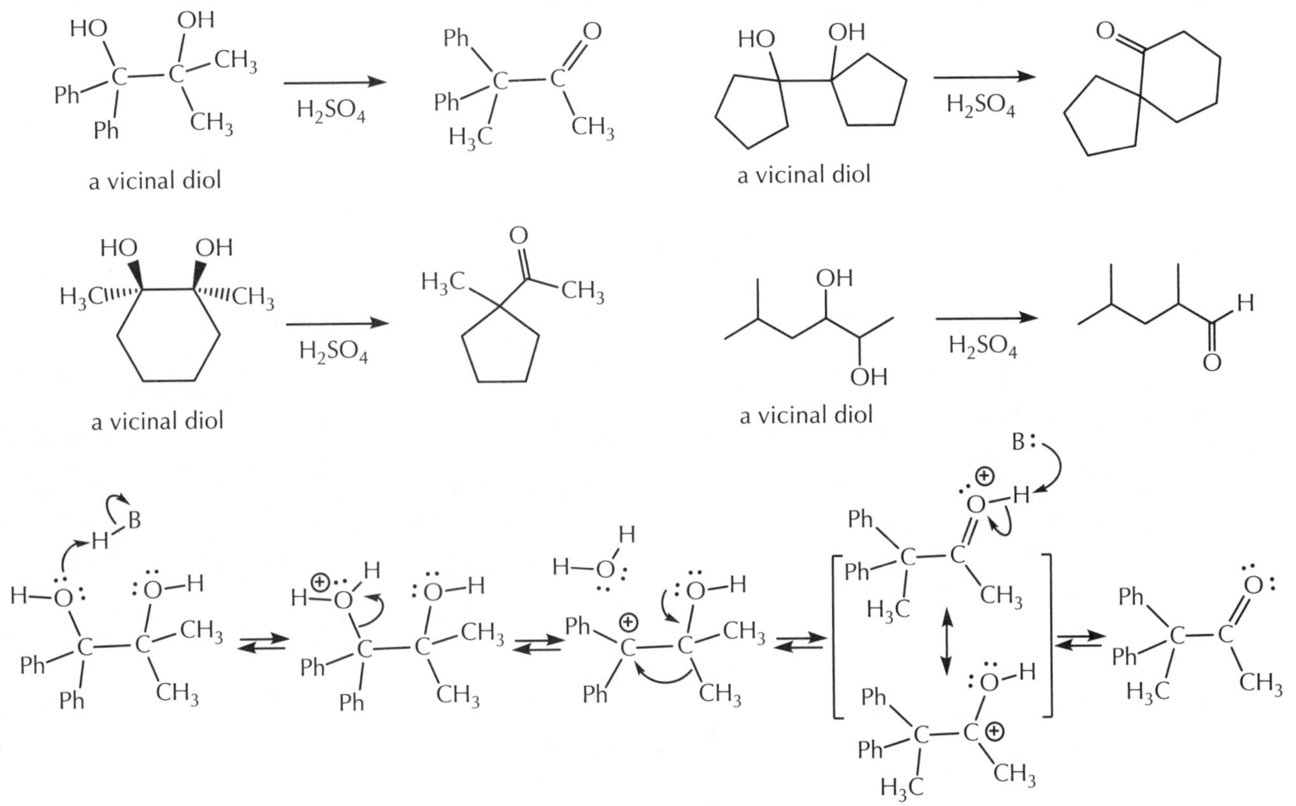

The 1,2-shift that originates from the carbon where the heteroatom is located creates a new, resonance-stabilized carbocation (a protonated ketone). In the case where the heteroatom comes from a hydroxyl group, a deprotonation reaction of the oxygen atom will be quite favorable, resulting in the observed ketone product. These vicinal diol rearrangements are known as pinacol rearrangements and are named after the first molecule where this rearrangement was observed and studied.

Other methods for forming carbocations, such as ionization of carbon-halogen bonds promoted by Ag(I) or protonation of a pi bond, have been observed to give the analogous rearrangement reactions. A few examples of these are shown in Figure 0828. Regardless of how the carbocation is formed, getting a double bond to form with the β-bonded heteroatom is a strong driving force for rearrangement reactions.

Figure 0828

Heteroatoms on the β-bond: additional carbocation sources.

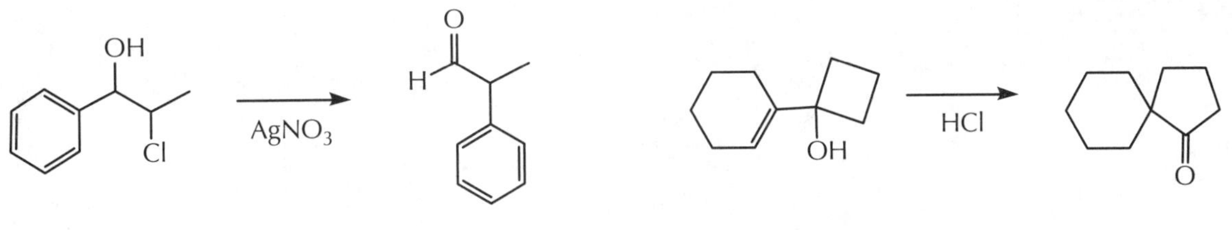

636 CHAPTER 8 Electrophilic Addition I: Brønsted Acids

PRACTICE QUESTIONS

8.06 There are many carbocation rearrangement reactions shown in Figure 0820. Provide the curved arrow mechanism for each of the steps indicated with the letters a–n. In those cases where the same structure is being used to show two mechanisms, label the arrows to indicate which arrow corresponds to which step.

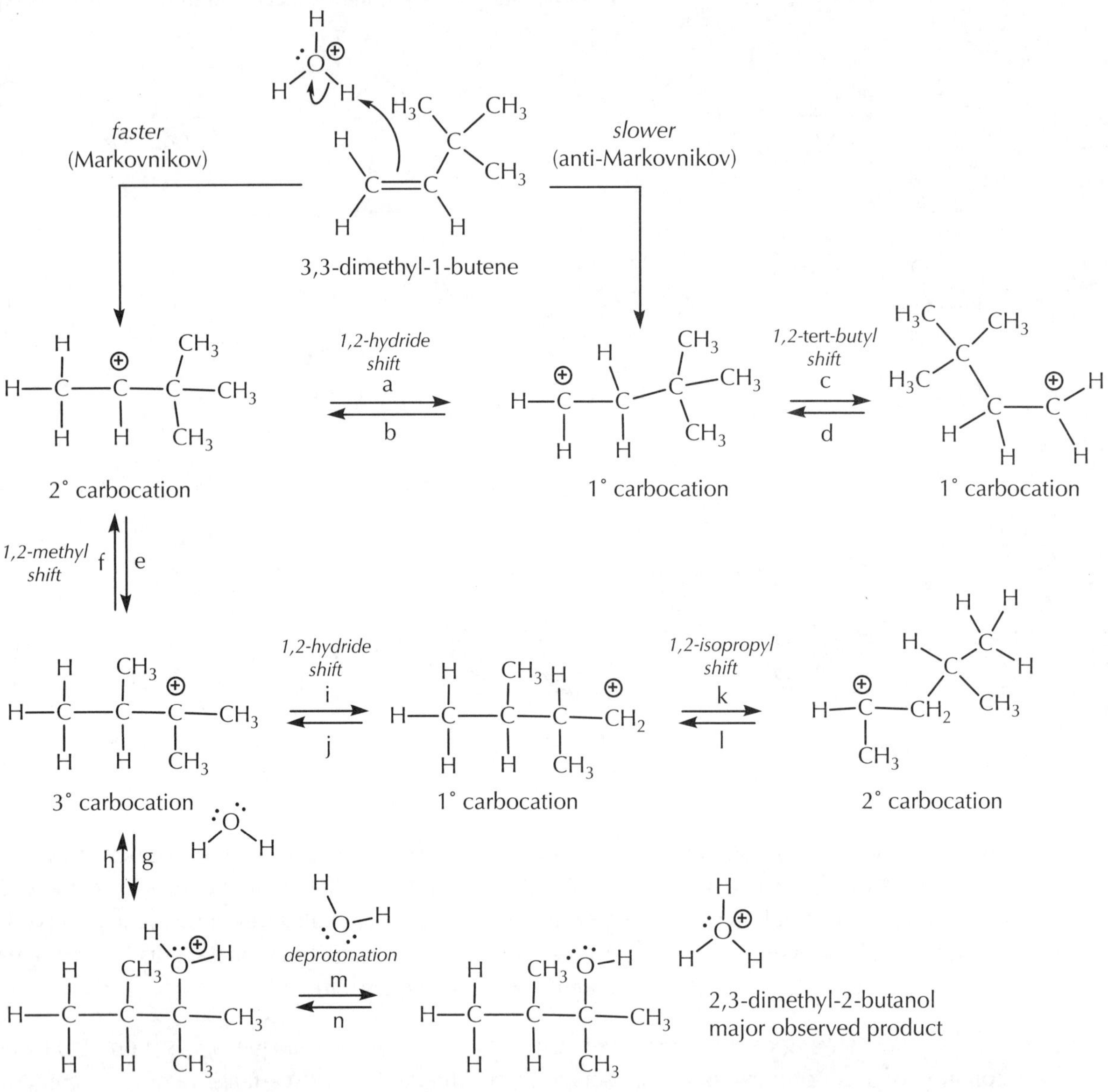

One of the other products in this reaction is drawn below, and one of the carbocations, above, can be used to explain its formation, after two additional rearrangement steps. Which of the carbocations can lead to this product and what is the mechanism for getting from the carbocation to the product?

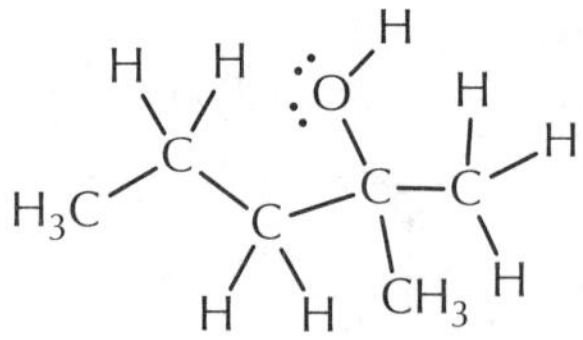

8.07 When 2,2-dimethylcyclohexan-1-ol is treated with methanol (CH$_3$OH) and catalytic sulfuric acid, the following distribution of products is observed. Provide the complete, curved arrow mechanism for this reaction, the products from which come through a single, common intermediate.

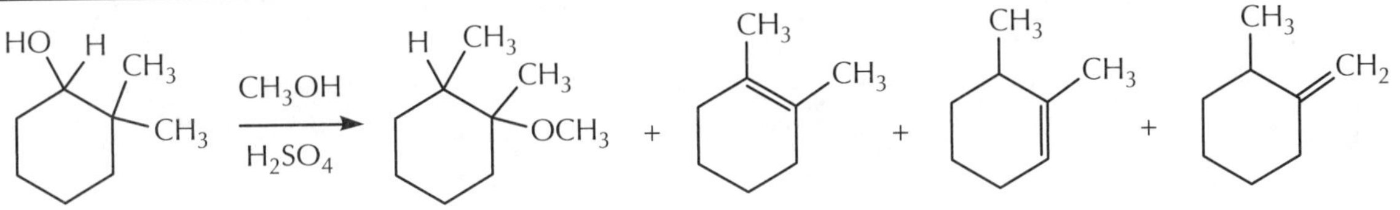

Small amount of the following products are also observed. What is their mechanism of formation?

8.08 These reactions involve carbocation rearrangements. Provide the curved arrow mechanism for each of these examples.

(a) 3-methylbutan-2-ol $\xrightarrow[H_2SO_4]{H_2O}$ 2-methylbutan-2-ol

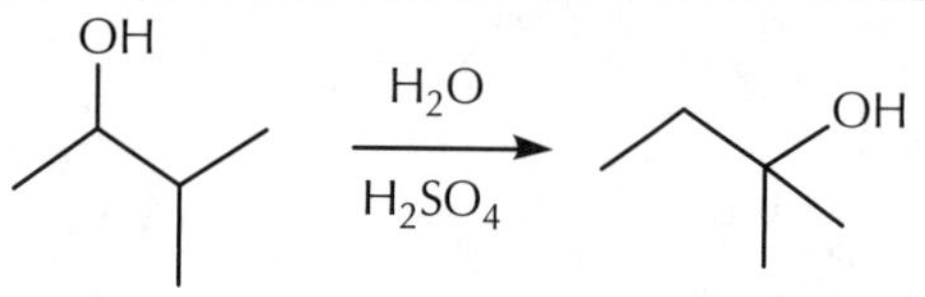

(b) 3-methylbut-1-ene + HO-C(=O)-CF$_3$ → 2-methylbut-2-yl trifluoroacetate

(c) 1-methylcyclohex-3-ene (with H, CH$_3$) $\xrightarrow{HCl}$ 1-chloro-1-methylcyclohexane (with Cl, CH$_3$)

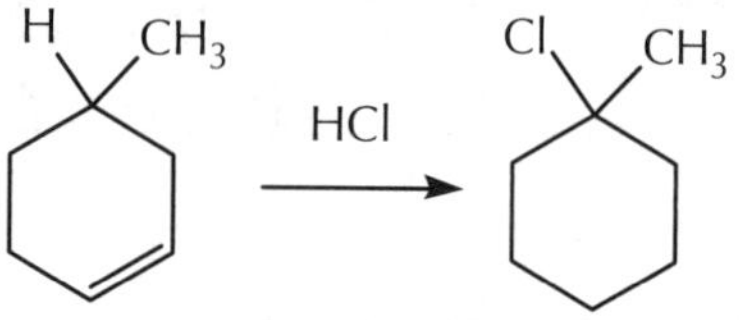

(d) allylbenzene $\xrightarrow[H_2SO_4]{H_2O}$ 2-phenylpropan-2-ol

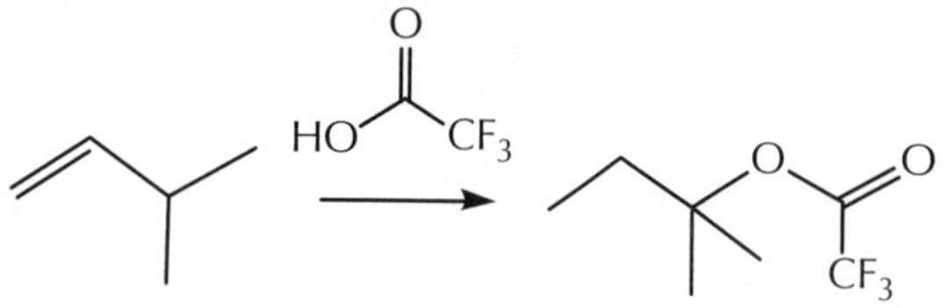

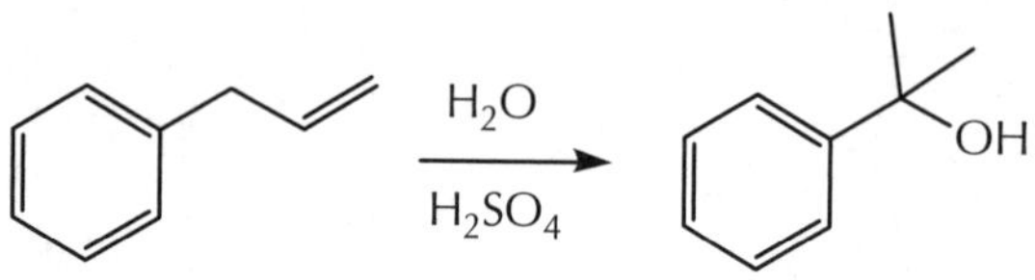

8.09 The following reactions involve carbocation rearrangements in which there is a contraction or expansion of a ring. Provide the complete, curved arrow mechanism for these reactions.

(a)

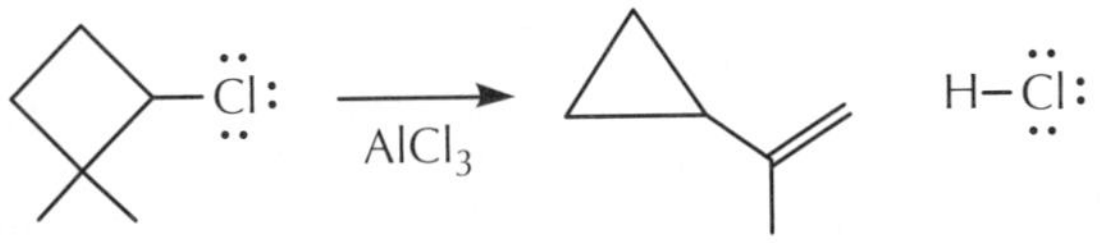

(b)

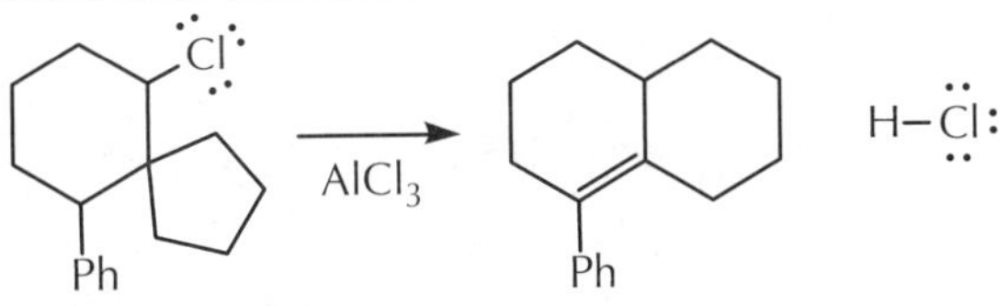

(c)

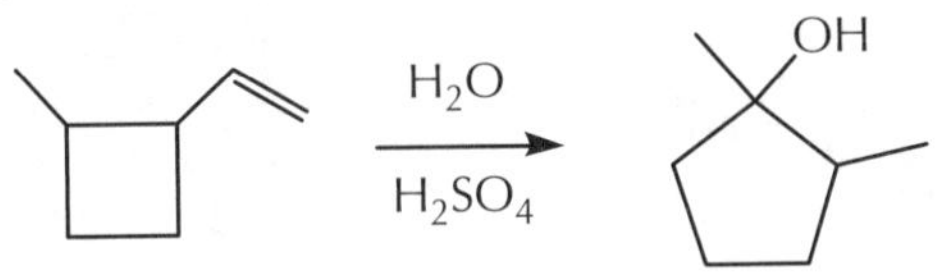

(d)

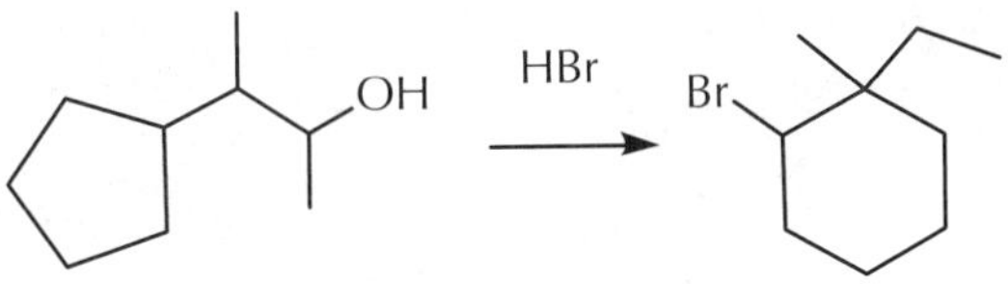

8.10 The following reactions involve carbocation rearrangements in which nearby heteroatoms are involved. Provide the complete, curved arrow mechanism for these reactions.

(a)

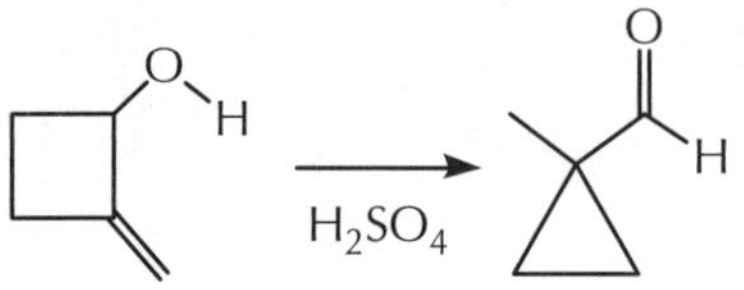

(b)

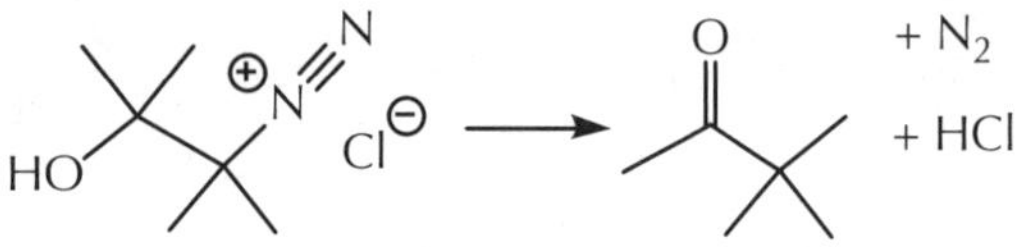

(c)

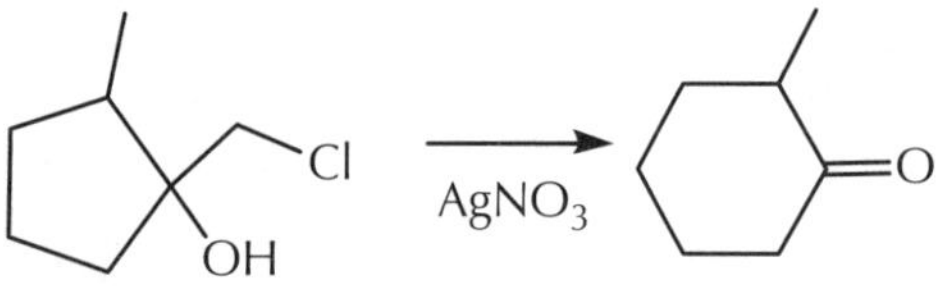

(d)

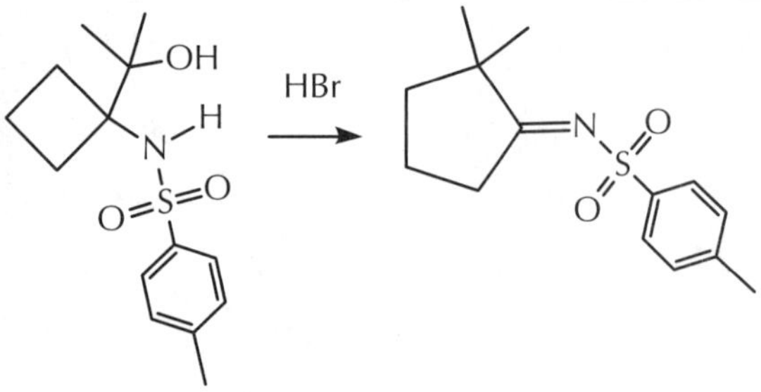

8.3 Hydration via Hydroboration-Oxidation

A. Transformation: Anti-Markovnikov Addition of Water

In electrophilic addition reactions of double and triple bonds, water is a weak acid. This chemistry was discussed in Section 8.1D. In both cases, the acid catalyzed reaction follows the Markovnikov Rule, forming the more stable carbocation intermediate faster than the less stable one. When the addition reaction of water (hydration) is carried out on triple bonds, the initially formed enol product undergoes a favorable equilibration to give a structural isomer possessing a carbon-oxygen double bond (Figure 0829).

Figure 0829

Revisiting the acid-catalyzed electrophilic Markovnikov addition reaction of water (hydration) of double and triple bonds.

alkene hydration (addition of water -> alcohol)

alkyne hydration (addition of water -> enol -> C=O product)

an enol undergoes a favorable equilibrium to give a C=O containing isomer

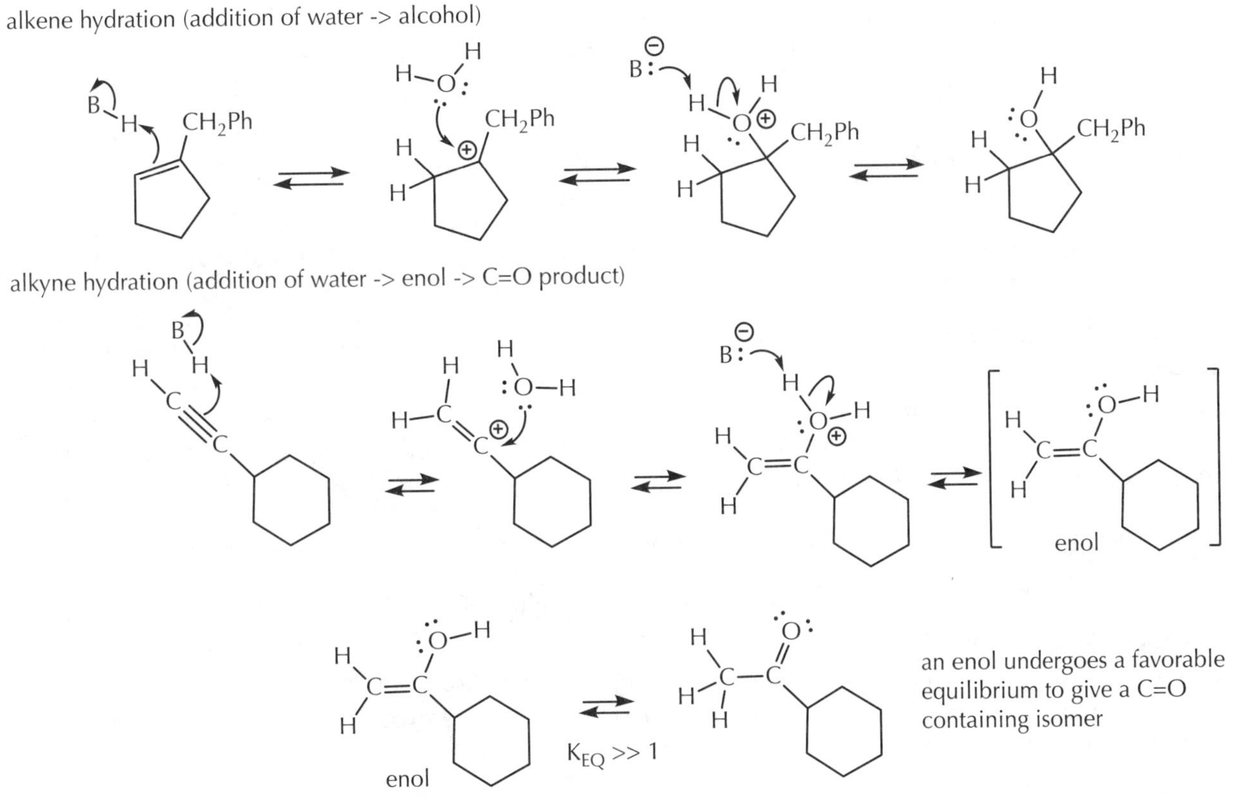

Chemists want to create flexible options for constructing new molecules. If carrying out a reaction with the Markovnikov regioselectivity is possible, then are there also ways to carry out a reaction with the anti-Markovnikov regioselectivity? This question could be the basis of a research project. An anti-Markovnikov result might also be an unexpected or new observation, while looking for something else, which was noted by a "prepared mind" (see the essay at the start of this chapter).

The addition of water to double and triple bonds is an example where the flexibility exists to carry out the reaction with the Markovnikov regioselectivity or the anti-Markovnikov regioselectivity. Without specifying how it might take place, the concept "anti-Markovnikov addition of water" can be applied to any example where the Markovnikov regioselectivity can be observed. Figure 0830 (on the next page) gives the Markovnikov and anti-Markovnikov counterparts for the addition of water to some double and triple bonds.

642 CHAPTER 8 Electrophilic Addition I: Brønsted Acids

Figure 0830
Outcomes from Markovnikov and anti-Markovnikov addition of water.

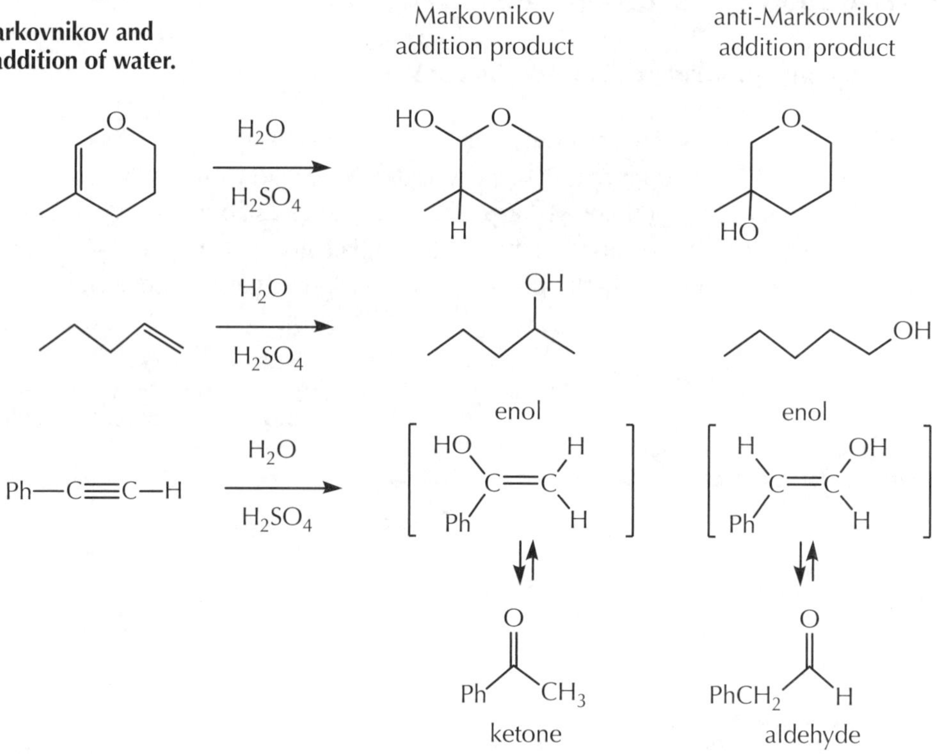

For the addition to triple bonds, as it did in the acid catalyzed additions of water, an initially formed enol undergoes a favorable equilibration to a form with a carbon-oxygen double bond. When the resulting structure has two hydrocarbon groups on the carbon of the C=O, then that functional group was called a ketone; if there is only one hydrocarbon group on the carbon of the C=O, as well as a hydrogen atom, the resulting functional group is called an aldehyde.

But the question remains: How can the anti-Markovnikov addition of water be carried out?

The answer to this question cannot be deduced from any of the chemistry presented so far in this text. And, historically, it was not deduced, either. This transformation was an unexpected observation, derived from a combination of two reactions: (a) hydroboration of a pi bond, followed by (b) oxidation of a C-B bond to C-OH. A representative set of reaction conditions for these experimentally observed transformations is shown in Figure 0831.

Figure 0831
Anti-Markovnikov addition of water via hydroboration-oxidation.

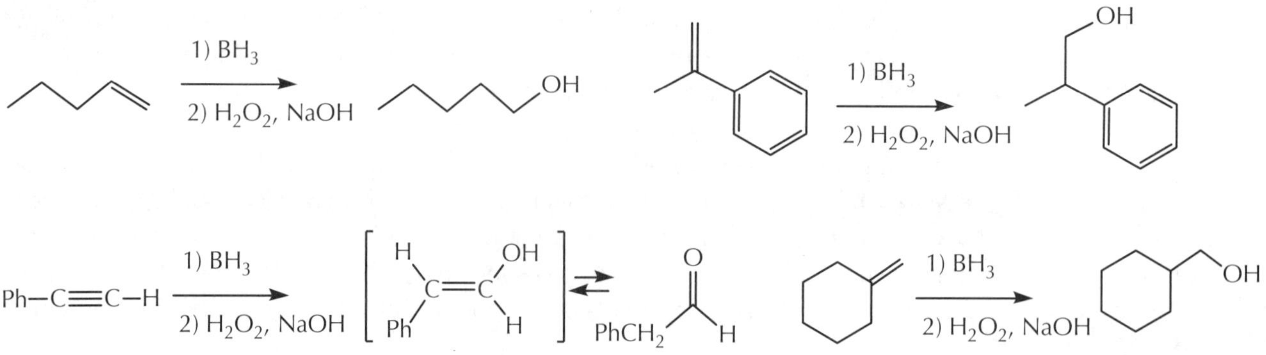

8.3 Hydration via Hydroboration-Oxidation

As the name implies, "hydroboration" is an addition reaction of hydrogen and boron. Just about any molecule with an uncharged B-H bond will participate in this addition reaction. Figure 0831 presents the simplest source of B-H bonds, namely, BH_3, which is the strong Lewis acid known as borane.

In its pure form, borane exists in a favorable equilibrium with a dimeric form called diborane (B_2H_6). When diborane is dissolved in the common organic solvent, tetrahydrofuran (THF), the BH_3 is captured in a Lewis acid-base complexation reaction (Figure 0832), and sometimes used in that form. Many different dialkylboranes (R_2BH) have been prepared, over the years, all of which are used to provide a BH bond in addition reactions. Some of these dialkylboranes, along with their names, are also included in Figure 0832.

Figure 0832

Borane, diborane, and some dialkylboranes.

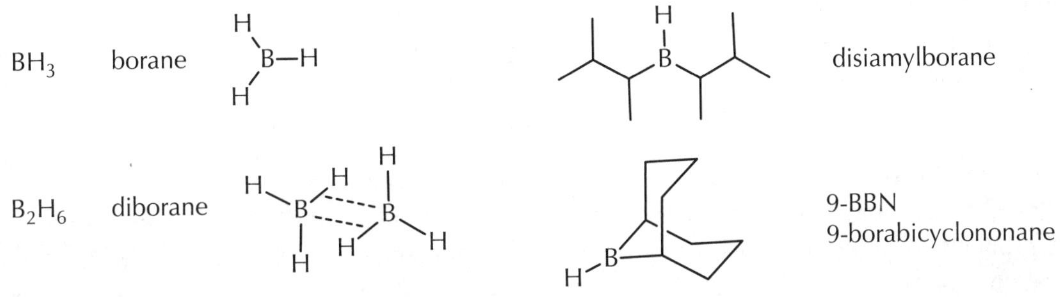

The hydroboration-oxidation reaction sequence results in the anti-Markovnikov (regioselectivity) addition of water to double and triple bonds. In cases where two new stereocenters are formed, the elements of water (H/OH) are observed to result from a syn (stereochemistry) addition.

The correlation between identifying a "hydroboration-oxidation reaction" and the outcome "syn, anti-Markovnikov addition of water" is a statement of experimental fact. As an applied concept, it means that the idea can be used to predict the structure of the products from carrying out the reaction on 1-phenylcyclohexene (Figure 0833).

Figure 0833

Hydroboration-oxidation gives the syn, anti-Markovnikov addition of water.

the reaction conditions "hydroboration-oxidation" define an outome:

syn, anti-Markovnikov addition of water

defining what syn addition means

defining what addition of water means

defining what syn, anti-Markovnikov addition of water means

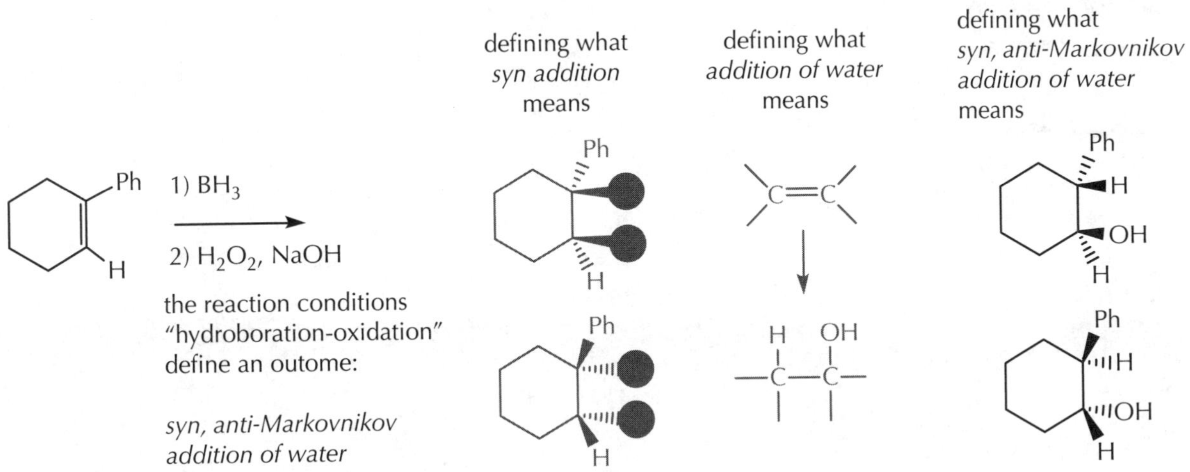

644 CHAPTER 8 Electrophilic Addition I: Brønsted Acids

Figure 0833 breaks down each part of the "syn, anti-Markovnikov addition of water" concept, as it applies to the reaction of 1-phenylcyclohexene.

"Syn addition" means that the two new sigma bonds are formed on the same face of the pi bond (either both above or both below).

"Addition of water" means that a pi bond has been broken and H and OH are now bonded.

"Anti-Markovnikov addition of water" means the opposite regioselectivity from what you would expect from the acid catalyzed electrophilic addition reaction: For the anti-Markovnikov product, the H atom actually ends up on the carbon atom that would have been the more stable carbocation. The lingering questions are how and why?

Understanding the anti-Markovnikov outcome requires the mechanisms, and these are presented in detail in the next two sections. The point being made here is that this particular scientific area started with a set of unexpected experimental observations, which, after being replicated in many different molecules, turned into a general concept that was worth exploring.

In learning some parts of organic chemistry, staring at the reagents does not allow you to figure out what is happening. The experimentalists who discovered many of the transformations you will see had no idea that those products would form, either. This means that it is sometimes simply useful to recognize that the reaction conditions, such as "hydroboration-oxidation reaction," stand for the "syn, anti-Markovnikov addition of water." That latter concept, then, can be applied to any molecule (Figure 0834).

Figure 0834

Predicting the products from hydroboration-oxidation reactions by applying the concept "syn, anti-Markovnikov addition of water."

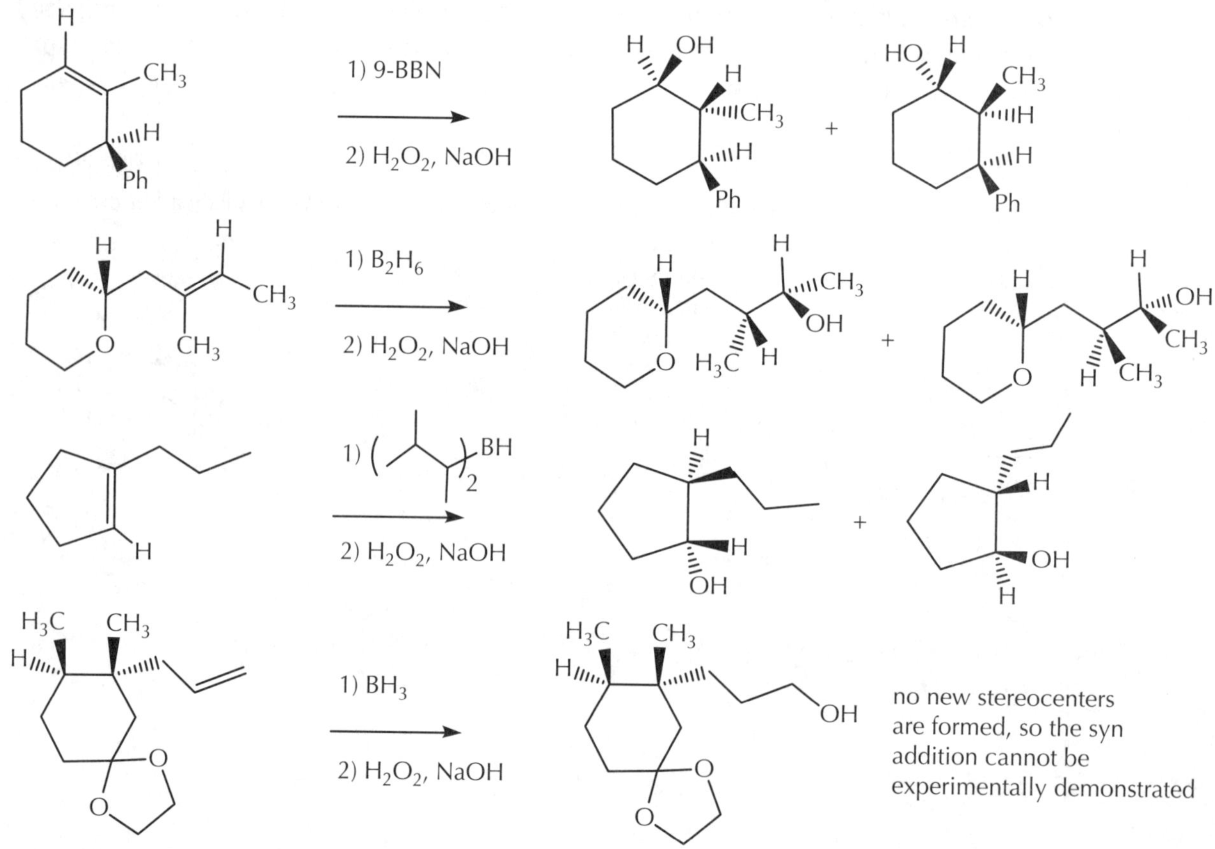

8.3 Hydration via Hydroboration-Oxidation

Having a conceptual organizer such as "syn, anti-Markovnikov addition of water" is also applicable in other reaction contexts. In particular, if you recognize the structural relationship in a transformation as a "syn, anti-Markovnikov addition of water," then you know that the "hydroboration-oxidation reaction" is what connects the starting material and the product (Figure 0835).

Figure 0835

Recognizing transformations that require the hydroboration-oxidation reaction.

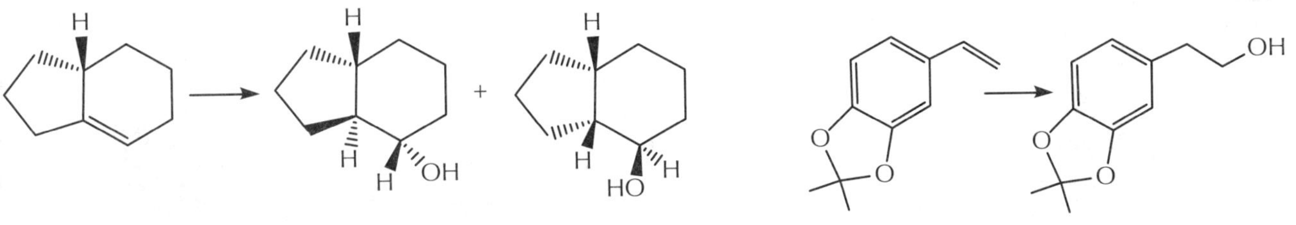

B. Stereochemistry and Regioselectivity of the Hydroboration Reaction Mechanism

Molecules with uncharged B-H bonds react with carbon-carbon pi bonds in a one-step reaction called hydroboration. The open shell, sp² boron atom, with its empty p orbital, is a strong Lewis acid, and it can interact with electron-rich pi bonds as potential electron donors (Lewis base). In the resulting reaction, a 4-centered transition state is proposed, where both the pi bond and the B-H bond are breaking as two new sigma bonds (C-B and C-H) are forming at carbon atoms of the double bond. The two-arrow mechanism (Figure 0836) accounts for all four bonding changes occurring at the same time.

Figure 0836

Mechanism and stereochemistry of the hydroboration reaction.

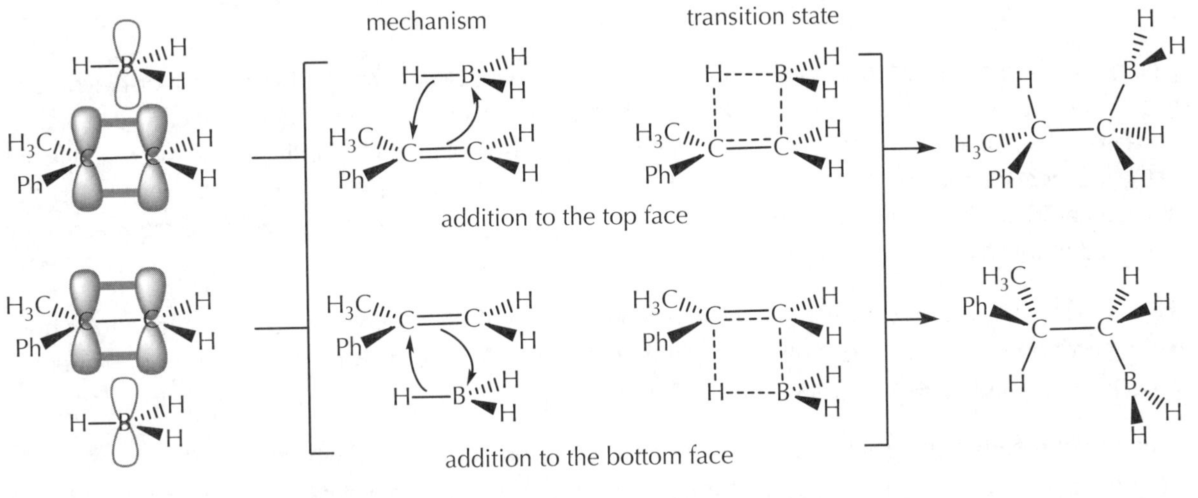

Even this mechanism is not easily predictable. Historically, all of the early experimental results were consistent with a syn addition of the B-H bond, which is why the 4-centered transition state was proposed as the mechanism. With all four bonding changes happening simultaneously, the spatial constraints were used to explain why the new C-H and C-B bonds ended up adding with the syn stereoselectivity. As usual, addition to the pi bond can occur from either of its faces.

CHAPTER 8 Electrophilic Addition I: Brønsted Acids

If the reagent is borane itself (BH₃), then all three B-H bonds can, in theory, end up adding to a pi bond. The exact outcome depends on the reaction conditions. Alternatively, if the borane reagent has only one B-H bond, such as a dialkylborane, the reaction occurs with an exact 1:1 ratio.

Both electronic and steric reasons are used to explain the regioselectivity of the hydroboration reaction (Figure 0837).

Figure 0837

Electronic and steric models for explaining regioselectivity of hydroboration.

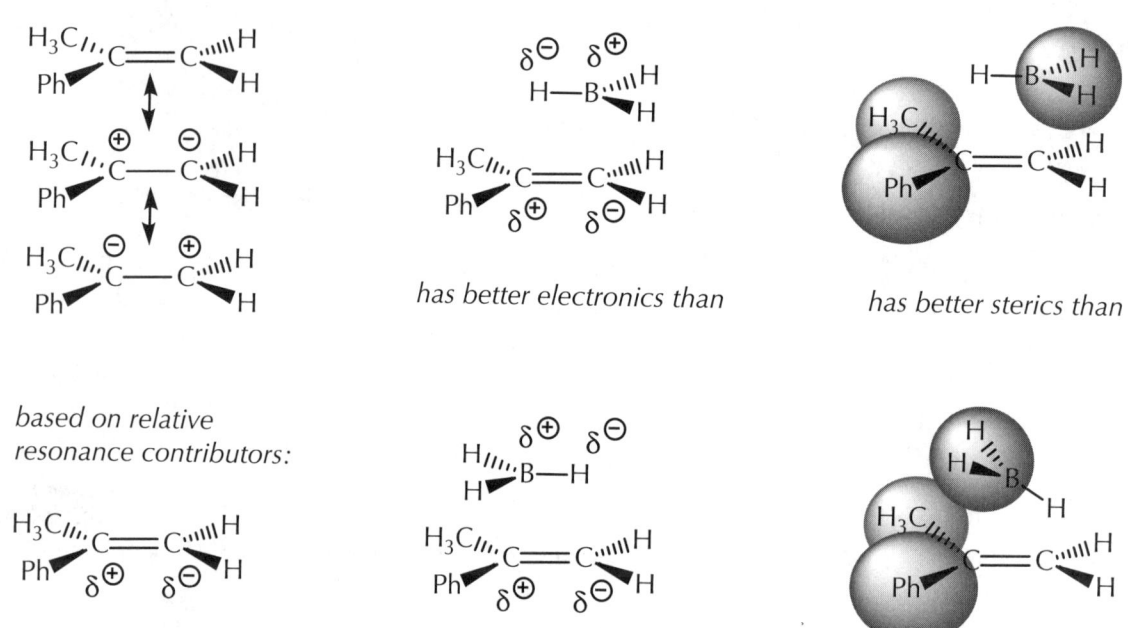

Electronically, the B-H bond is polarized with its electron density towards the hydrogen atom (hydrogen is more electronegative than boron), and the open shell boron atom is a Lewis acid that can accept electrons from the pi bond. The unsymmetrical double bond is also polarized according to the more significant of the minor resonance contributors, which places more positive charge at the carbon that can better stabilize a carbocation. The favorable orientation of these two dipoles is consistent with the regioselectivity observed in this reaction.

Steric effects also favor the observed regioselectivity, as the collision between the molecules allows both of the new bonds to form more easily when the more hindered end of the B-H bond (the boron) is oriented away from the more hindered end of the double bond.

Lining up the partial charges for the most favorable collision is reinforced when imagining the pathway towards getting to the transition state. Starting to form the new C-B bond can be more advanced than forming the new C-H bond as the transition state is approached because donating electrons to the open shell boron atom is favorable. The partial positive change on the more substituted and resonance-stabilized carbon is more stable than in the alternative orientation (Figure 0838) and keeping the larger ends of the two molecules apart, rather than together, is also more stable.

8.3 Hydration via Hydroboration-Oxidation

Figure 0838

Electronic and steric models in the two hydroboration transition states.

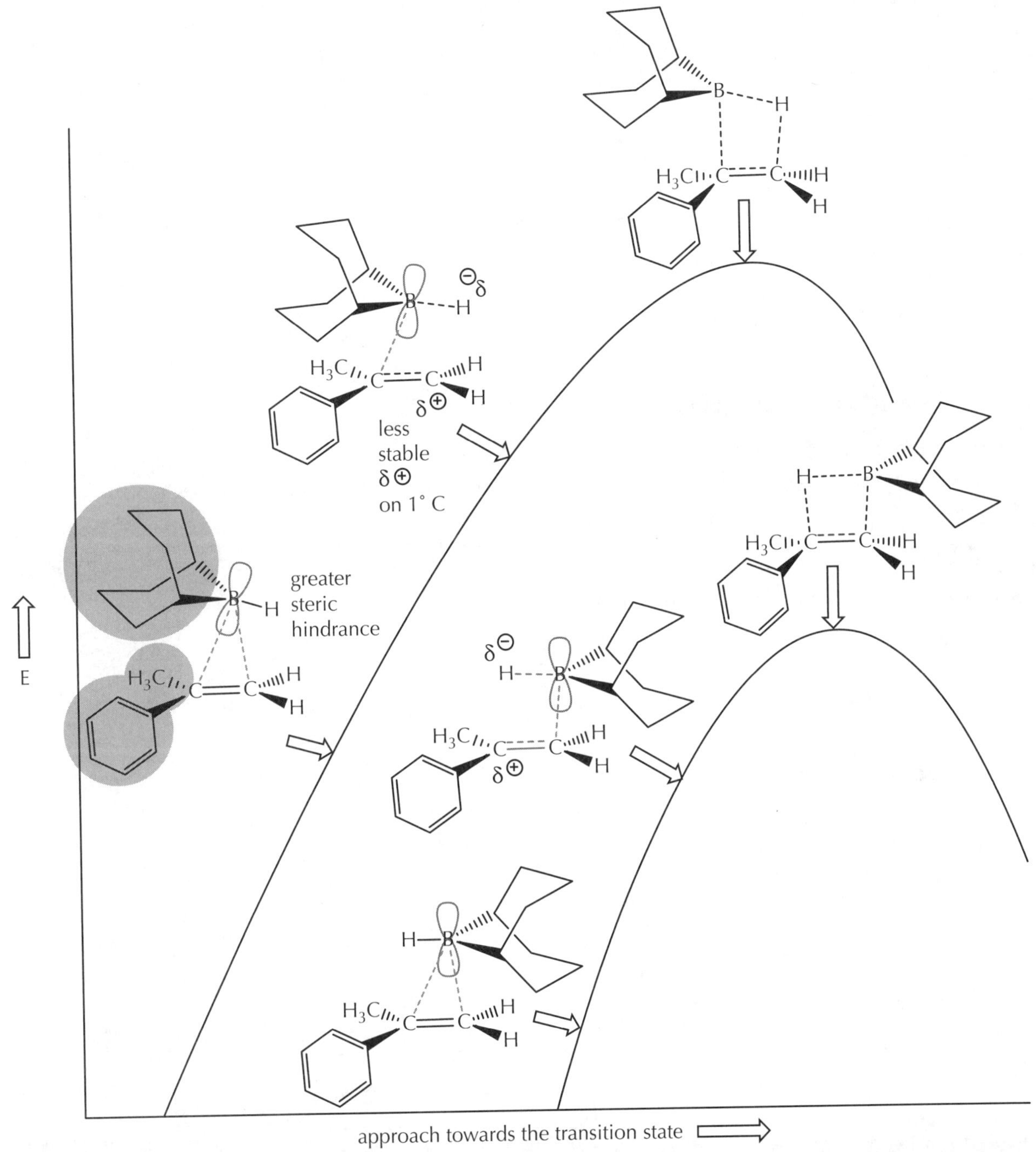

less stable δ⊕ on 1° C

greater steric hindrance

approach towards the transition state ⇨

As the boron end of the B-H gets more sterically hindered, as in the series of dialkylborane reagents (R_2BH) shown in Figure 0839, the regioselectivity in the hydroboration reaction is observed to be higher, which is consistent with the steric effect described above.

Figure 0839

Hydroboration-oxidation regioselectivity observed with increasingly hindered dialkylboranes.

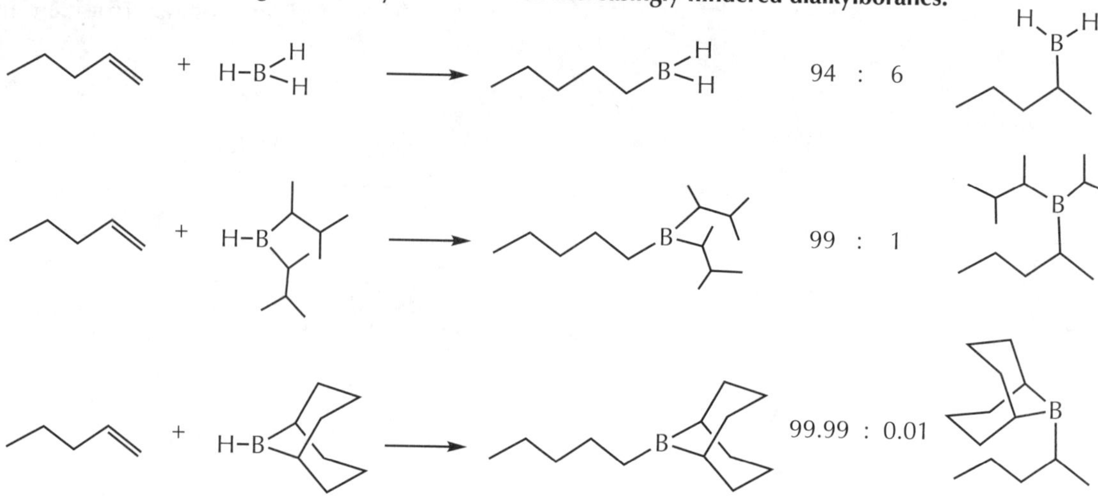

C. Oxidation of Carbon-Boron Bonds

The oxidation reaction of carbon-boron (C-B) bonds to give a carbon-hydroxyl group (C-OH) is a specific set of reaction conditions developed for the hydroboration-oxidation reaction. The mechanism for this reaction is presented in Figure 0840.

Figure 0840

Oxidation mechanism for C-B bonds in hydroboration-oxidation reactions.

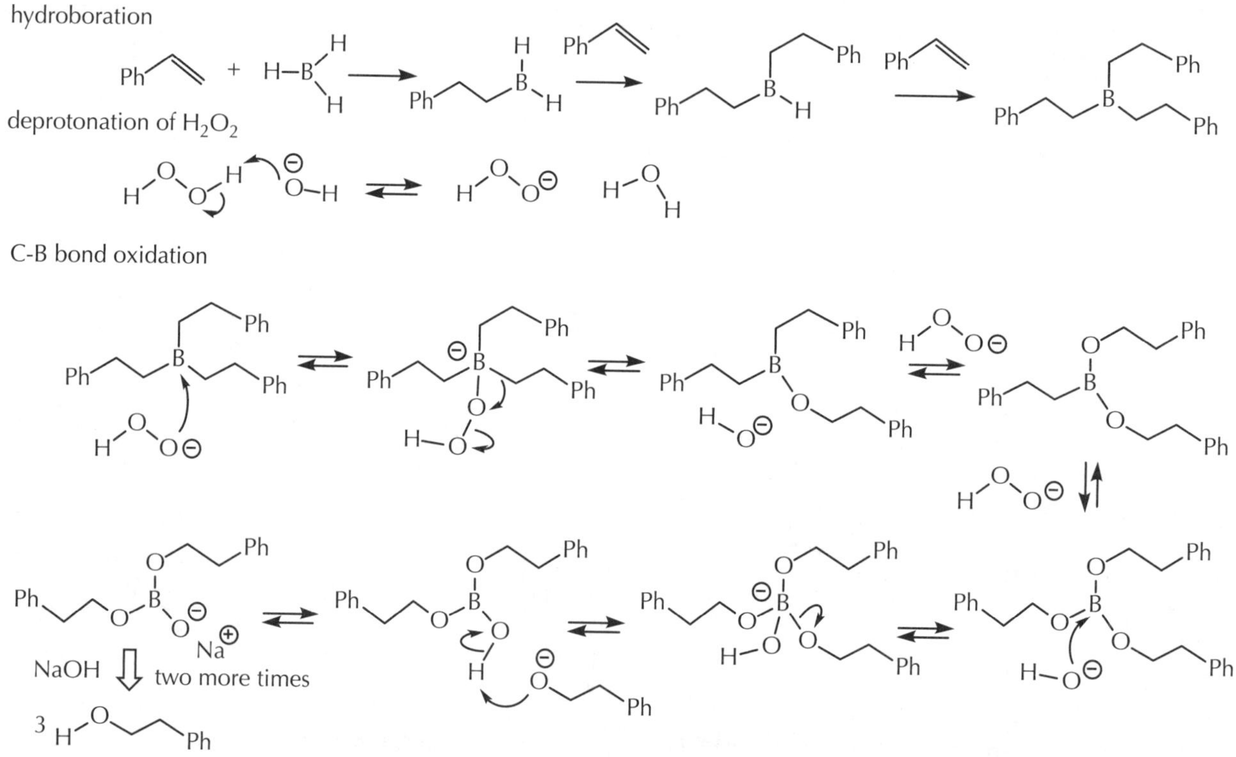

In this example, all three of the B-H bonds in borane (BH_3) have added to three phenylethene molecules, in the hydroboration reaction, to give $(PhCH_2CH_2)_3B$. This represents the hydroboration step, which is followed by the oxidation step. All three C-B bonds will become C-OH bonds.

8.3 Hydration via Hydroboration-Oxidation

In basic solutions, hydrogen peroxide (H_2O_2, $pK_a = 11.7$) is deprotonated to give the sodium hydroperoxide anion (NaOOH). The hydroperoxide anion undergoes a Lewis acid-base complexation reaction with the open shell boron atom.

In the next step, one of the alkyl groups on the anionic boron atom migrates onto the adjacent oxygen atom in a 1,2-shift, breaking the oxygen-oxygen bond and releasing hydroxide ion. This is the key step that makes the C-O bond from the C-B bond.

Finally, under the conditions of excess base, hydroxide (or some other Lewis base) can cause an association-dissociation reaction, releasing the newly formed C-O group, which can then be protonated in the solution by some Brønsted acid (water is shown as the proton source in Figure 0840).

In this example, under conditions of excess peroxide, this reaction will occur two more times, oxidizing all three of the C-B bonds in the original hydroboration product, $(PhCH_2CH_2)_3B$.

D. Synthesis: Here's the Answer, What's the Question?

Chemical reactions are typically thought about in three different ways.

Predict the products. The first of these ways is most familiar to you: Given the starting materials, predict the outcome. Answering such a question relies on properly identifying the classification of the reaction (addition, substitution, etc.) and then applying that classification to the specific compounds. Some familiar examples are presented in Figure 0841.

Figure 0841

Chemical reactions formatted as "predict the products" (A + B -> ?)

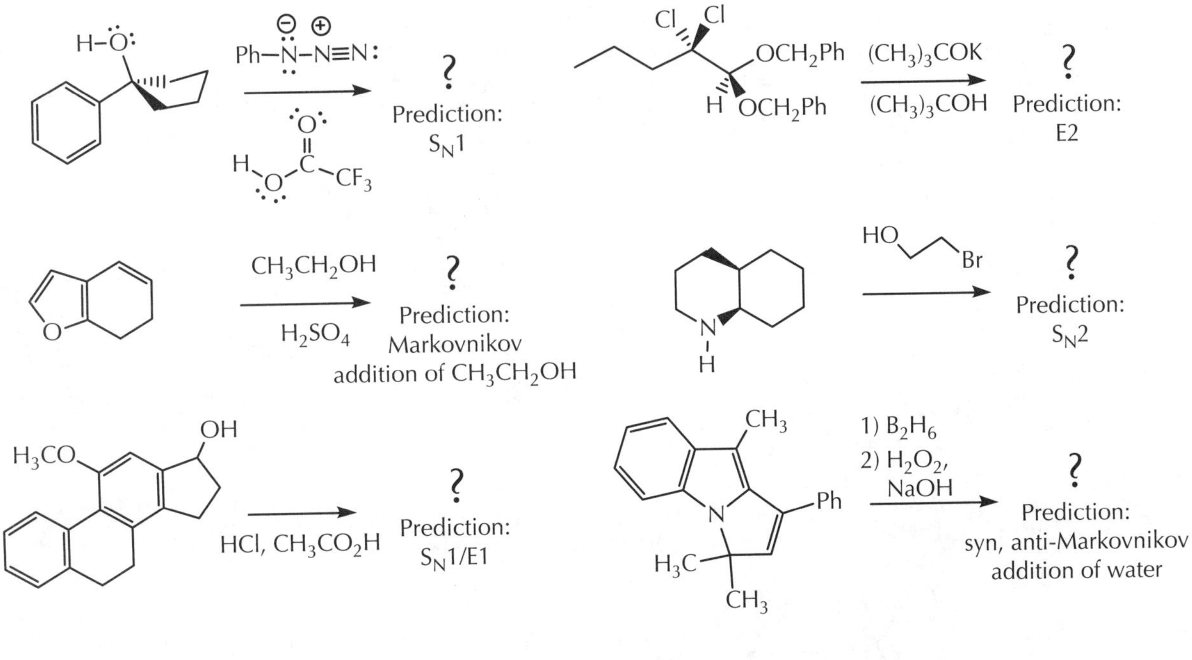

650 CHAPTER 8 Electrophilic Addition I: Brønsted Acids

Note how the example with the hydroboration-oxidation reaction relies on identifying and associating those conditions with "syn, anti-Markovnikov addition of water" and then applying that idea to the molecule in question, whereas the others rely on identifying a more generic class of reactions and applying that general idea (e.g., S$_N$2). The products are not shown. At this point, you should be sure you can readily draw (or imagine) the structural outcomes.

Transformations. A second common format for thinking about chemical reactions is called a transformation. Here, you have the starting material and the product, and the set of reagents and/or reaction conditions are to be inferred from that information. The same underlying reaction classifications as above still hold, but you are now viewing them from the relationship of how the starting and end points are related (Figure 0842). And, in comparison with the format used in Figure 0841, not only does the relationship "syn, anti-Markovnikov addition of water" need to be identified, but the hydroboration-oxidation conditions must be recalled as the way to carry this out.

Figure 0842

Chemical reactions formatted as "transformations" (A + ? -> C).

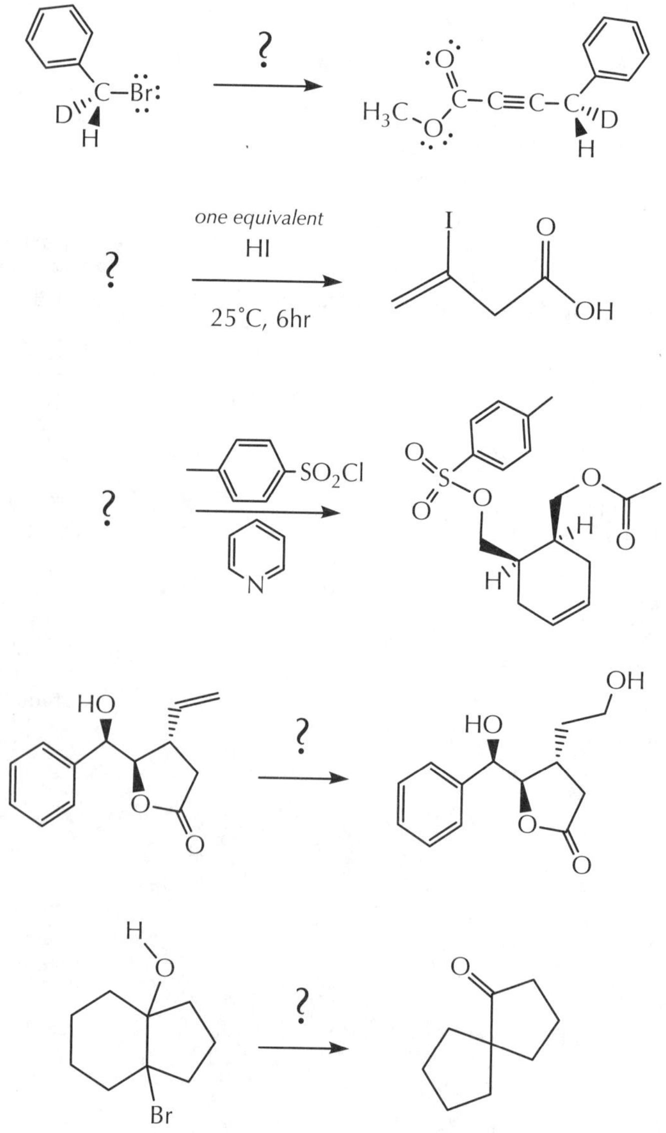

Analysis: S$_N$2 reaction (via an sp carbanion nucleophile)

Analysis: Markovnikov addition of HI to a triple bond

Analysis: Preparation of a sulfonate ester leaving group from a hydroxyl group

Analysis: anti-Markovnikov addition of water (via hydroboration-oxidation)

Analysis: rearrangement reaction (via a carbocation generated by heating the molecule in, e.g., a polar solvent)

Synthesis. The third format used for thinking about chemical reactions is called synthesis. In this format, you are only presented with the structure of the product (what you might otherwise think of as the answer to a question), and you need to think about what kind of reactions... and what starting material... and what reaction conditions might yield this molecule as a product.

This way of thinking about chemistry is more challenging than the others because you have less explicit information to work from, and you need to rely on recalling previous examples according to the type of product they formed. Synthesis is a divergent and opened-ended format, in that there are always multiple reasonable answers for how to prepare any given compound. Often, the only way to resolve any debate about which way might be better than another is by carrying out the experiments and comparing the results.

In a typical example, shown in Figure 0843, the analysis reveals how significantly the subject matter you are learning can build upon itself.

Figure 0843

A chemical reaction formatted as "synthesis" (? + ? -> C).

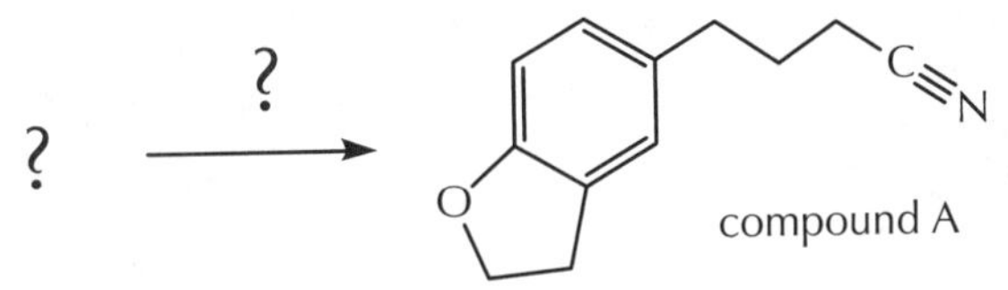

Your prior knowledge really matters here, because there is so little to go on in a synthesis question. Looking at the structure of compound A, the only thing you have to rely upon is your previous experience with reactions, and to analyze this structure as being a product from your increasing list of options (Lewis acid-base, Brønsted acid-base, electrophilic addition of strong and weak Brønsted acids, S_N2, E2, S_N1, E1, and hydroboration-oxidation).

This topic highlights a frustrating challenge for learning organic chemistry that was talked about in detail in some earlier essays: The information in learning organic chemistry builds on itself. Unlike some other introductory science subjects, we have not been restarting each chapter with new topics that are independent of those that came before. Instead, as is hopefully quite transparent in simply looking at this one synthesis question, answering something posed in this format is not possible with a good command of the prior knowledge, not to mention the strategies needed to answer such a question even when you have that command.

Before you get too anxious, this particular kind of question (synthesis) is not going to be developed in any depth during this text. The topic is interesting and important to science, but it can be looked at with varying levels of detail. The analysis for how you might think about answering this question, using your prior knowledge, is outlined in Figure 0844 (on the next page).

The best clues for answering this synthesis question are recalling the examples in which cyanide ion was used as a nucleophile for S_N2 reactions, and then also the structural requirements for S_N2 reactions. Following the idea of an S_N2 reaction with a source of cyanide ion, there are 18 common (and correct) answers, from your experience, that all vary slightly in their literal representation. There are six usual leaving groups (the halides and the sulfonate esters) that might each be combined with one of three different common sources for cyanide ion (LiCN, NaCN, KCN).

Understanding the generality of these answers is also critical for you because an answer key will only show one of these 18 combinations, so you will need to analyze that example and decide whether your solution is equivalent or not.

Figure 0844

Analysis of chemical reaction formatted as "synthesis" (? + ? -> C).

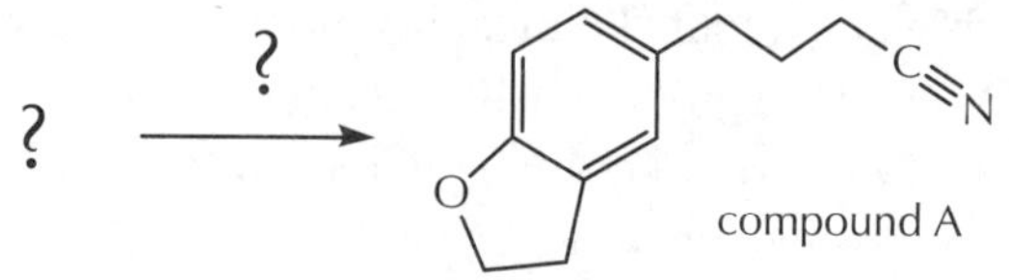

Analysis: At this point, some reactions and reagents need to have become familiar, and they need to be recalled, because they become the tools for answering more complex questions. For example, seeing the cyano group (CN) in compound A needs to inspire the recollection of S_N2 reactions in which cyanide ion (CN^-) was the nucleophile.

Thus, in compound A, if the carbon to which the CN group is attached is an unhindered sp^3 atom capable of having undergone an S_N2 reaction, then compound A could be the product from that type of reaction.

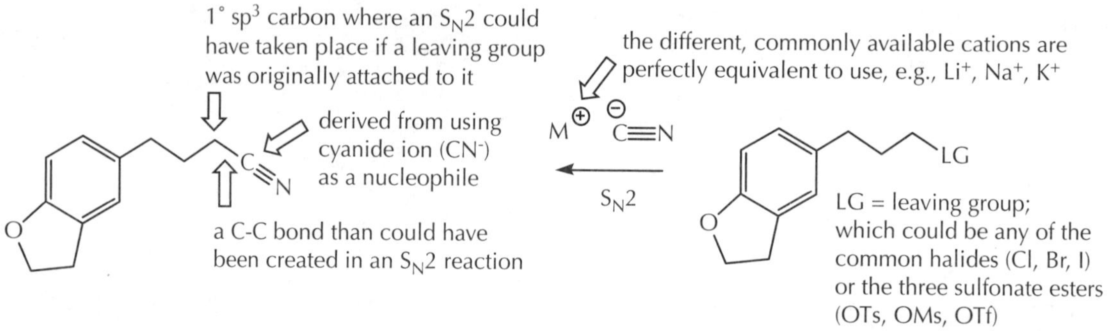

- 1° sp^3 carbon where an S_N2 could have taken place if a leaving group was originally attached to it
- derived from using cyanide ion (CN^-) as a nucleophile
- a C-C bond than could have been created in an S_N2 reaction
- the different, commonly available cations are perfectly equivalent to use, e.g., Li^+, Na^+, K^+
- LG = leaving group; which could be any of the common halides (Cl, Br, I) or the three sulfonate esters (OTs, OMs, OTf)

With six potential leaving groups, there are six possible answers to the question ... and all of them are correct. When combined with the three common ionic cyanides (LiCN, NaCN, KCN), a set of 18 literally different but equivalent solutions might be proposed.

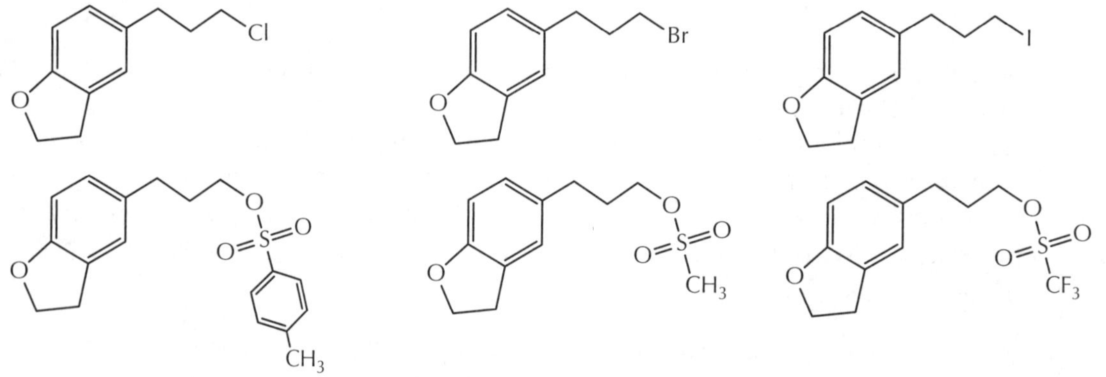

As you might anticipate, the analytical process of synthesis can continue. Each of the six molecules with a leaving group in it shown in Figure 0844 is now just its own new synthesis question: Where did they come from (i.e., what reaction results in these molecules as a product)? And so on... and so on...

Figure 0845 shows how interconnected these molecule structures are, using only the reactions that you have seen to date. Naturally, as the number of reactions you learn increases, so will the options.

Synthesis is one of the most critically important contributions that chemistry brings to humanity because it answers the question "how do we make important substances that we cannot otherwise gets our hands on?"

For nearly the entire history of humanity, bacterial infections have been one of the most significant health threats that existed. Although there are about 2,000 years of history with folk medicines that we now understand to have antibiotic activity, systematic treatments for many bacterial diseases did not exist until the early 20th century. Organoarsenic compounds, first identified in the 1880s, were developed as a

Figure 0845
A further analysis of a synthesis question.

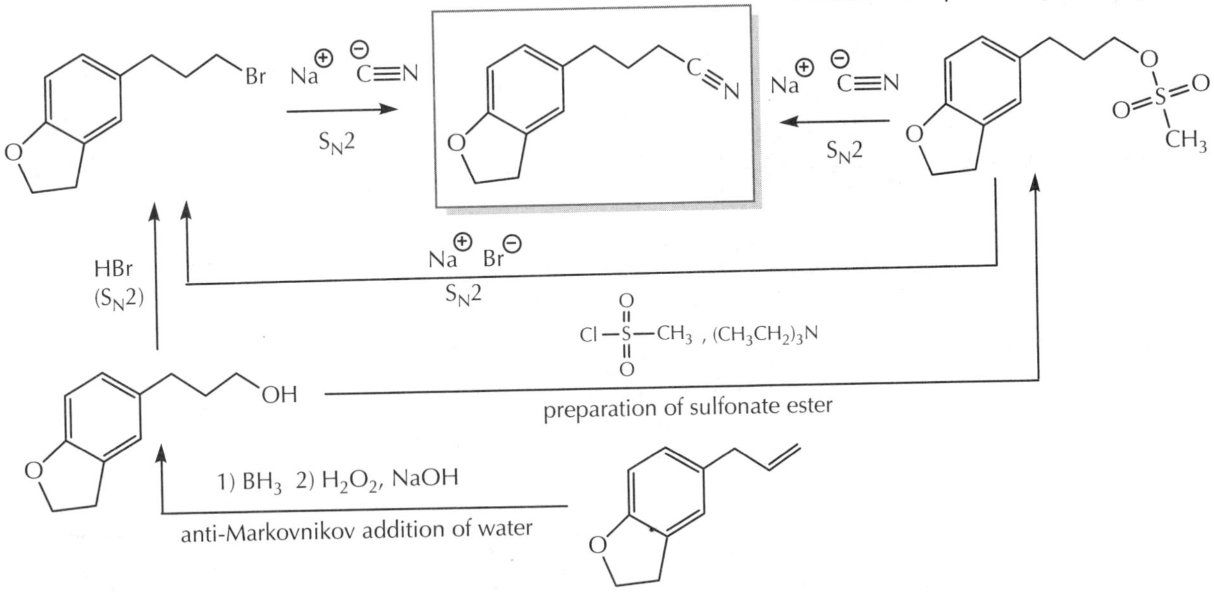

pharmaceutical treatment for syphilis in 1910. Syphilis was responsible for multiple "pox" epidemics, historically, and it still infects about 0.5% of adults on our planet every year.

Arsenic compounds are quite toxic, and new agents were discovered as the concepts of pharmaceutical action and testing matured. The discovery and development of naturally occurring antibiotics, particularly the noteworthy penicillin, whose use became widespread in 1945 after its success in World War II, started to increase the repertoire of compounds we could use for treating infections.

We humans have not won the battle with bacteria, but we have used chemical synthesis to change the battle for both good and bad.

Bacteria are much more adept at evolving than we are. We might eliminate the majority of a bacterial strain, but we cannot eliminate the mutants that survive because something in their genetics allows them to be immune to the effects of an antibiotic.

By wiping out the dominant strain with an antibiotic, we can allow a mutant strain to end up thriving when it might not have otherwise. This creates the ongoing battle between chemical synthesis and bacterial evolution. Nature only provided one molecular structure: penicillin G. But by learning its structure and how it functions, we have synthesized thousands and thousands of compounds that never existed in nature, to test them for their potential antibiotic action.

The fight against our other major foe, viruses, was clearly highlighted during the COVID pandemic that started in 2020. The rapid vaccine response and development of antiviral agents are all directly attributable to synthetic chemistry.

The synthesis of new compounds, combined with the detailed knowledge of the mechanisms of biological action, have allowed us to stay ahead of the rapid evolution of the organisms that infect us. And although we are also directly contributing to their evolution, we are not faced with the same widespread health risks from bacterial infections that faced humans quite significantly less than 200 years ago.

Medicine is not the only area to benefit from synthetic chemistry. New materials, starting with durable and lightweight plastics, coatings, and fabrics, have also revolutionized life on our planet, both for better and for worse.

PRACTICE QUESTIONS

8.11 Provide the necessary structure for the starting material or products in the following hydroboration-oxidation reactions. If stereoisomers are expected, draw them.

(a)

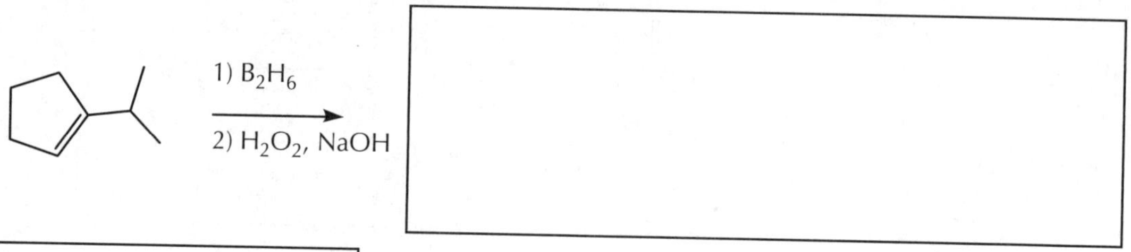

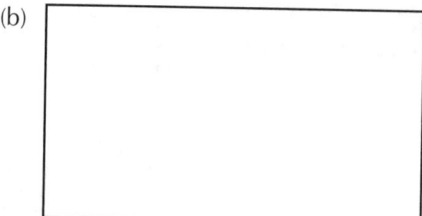

(b)

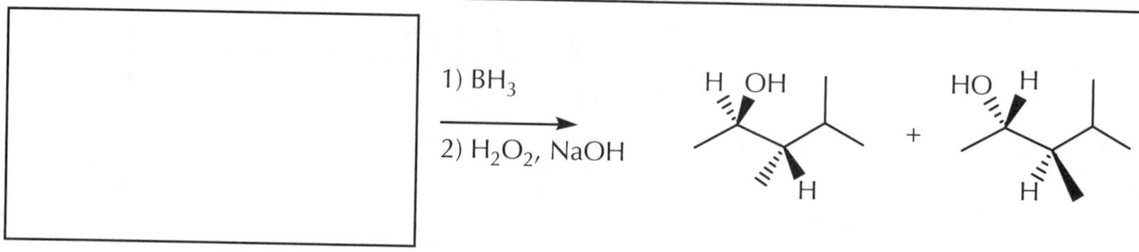

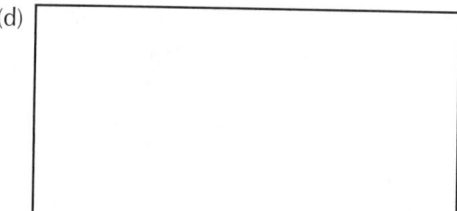

(c)

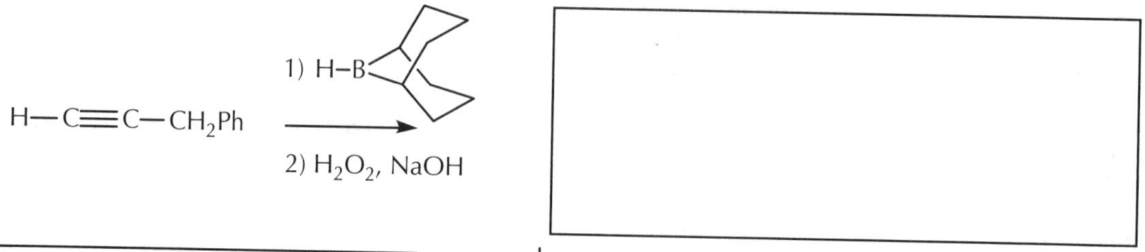

(d)

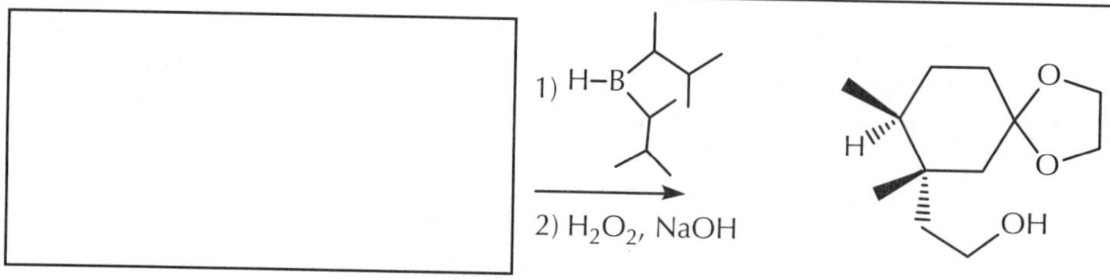

(e)

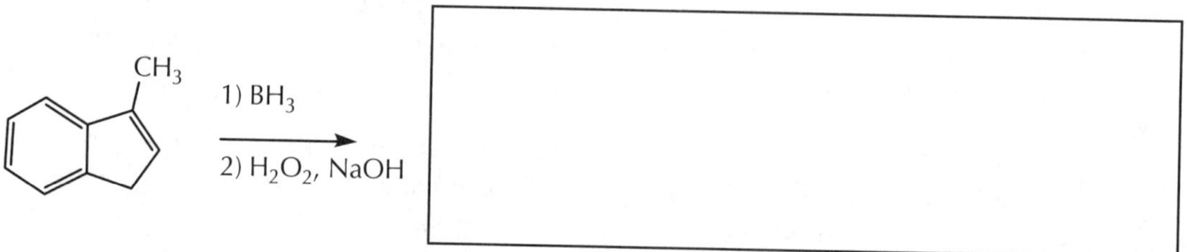

(f)

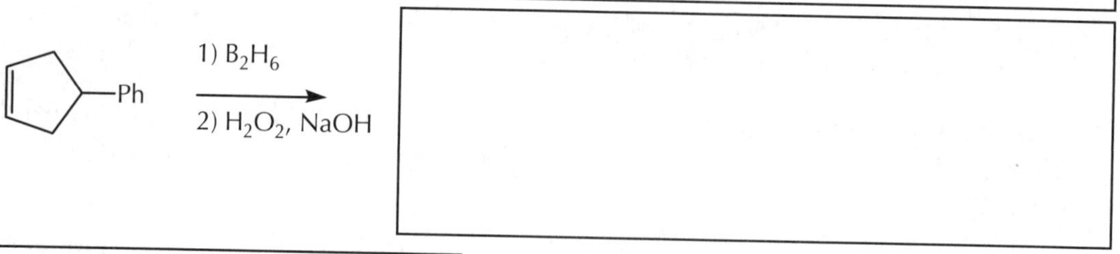

(g)

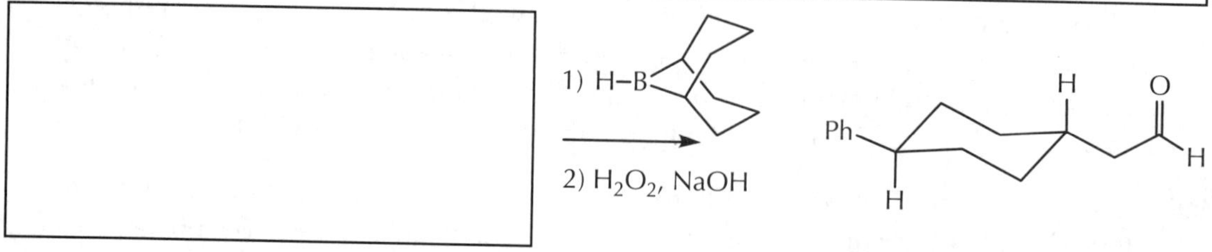

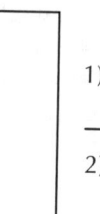

8.12 Use the information present to classify the reaction. Include the reaction (e.g., acid-base) and/or mechanistic classification (e.g., S_N1), as well as the regiochemistry (e.g., Markovnikov; when relevant), and the expected stereochemical outcome (e.g., syn, when relevant).

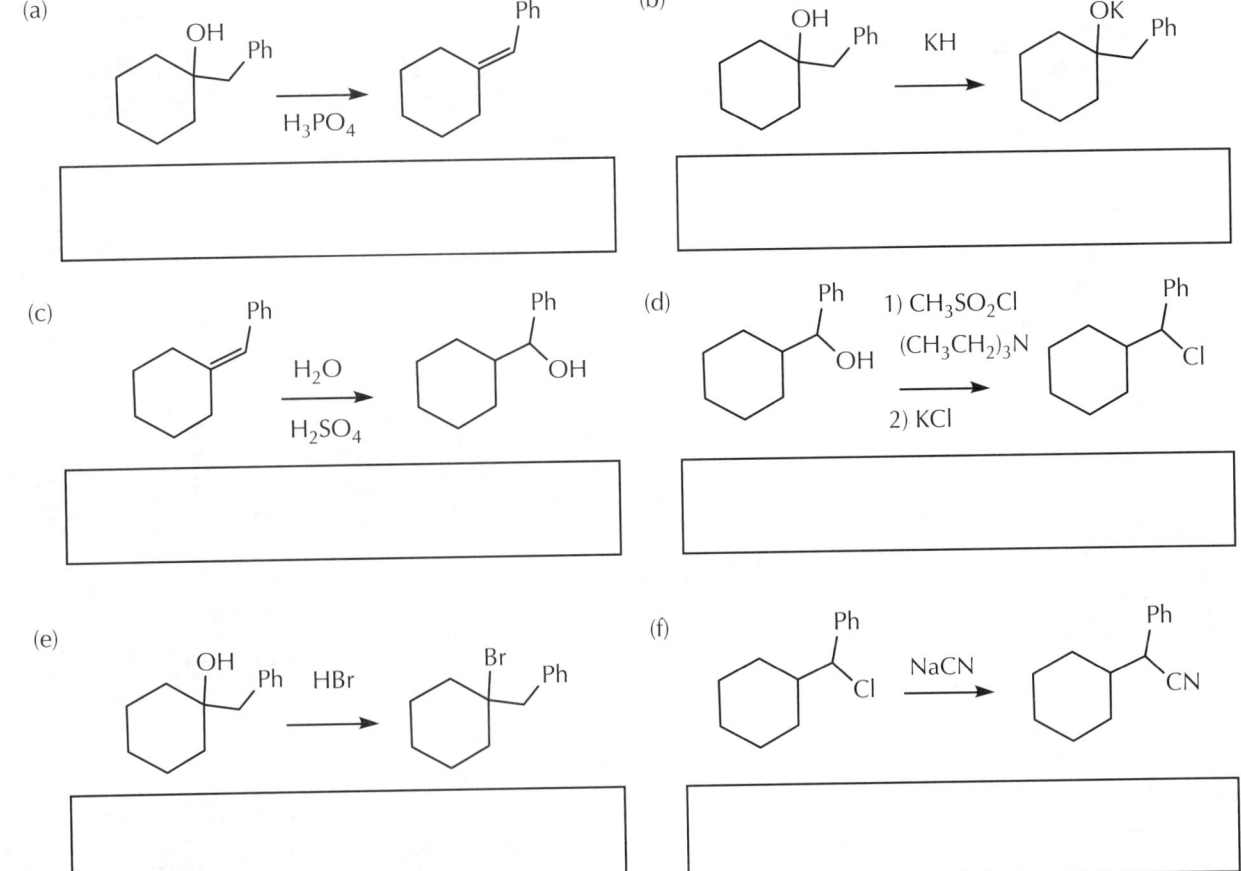

8.13 Provide the reagents needed to carry out the following transformations. You should explicitly classify the reaction, first, as the way to guide your solution.

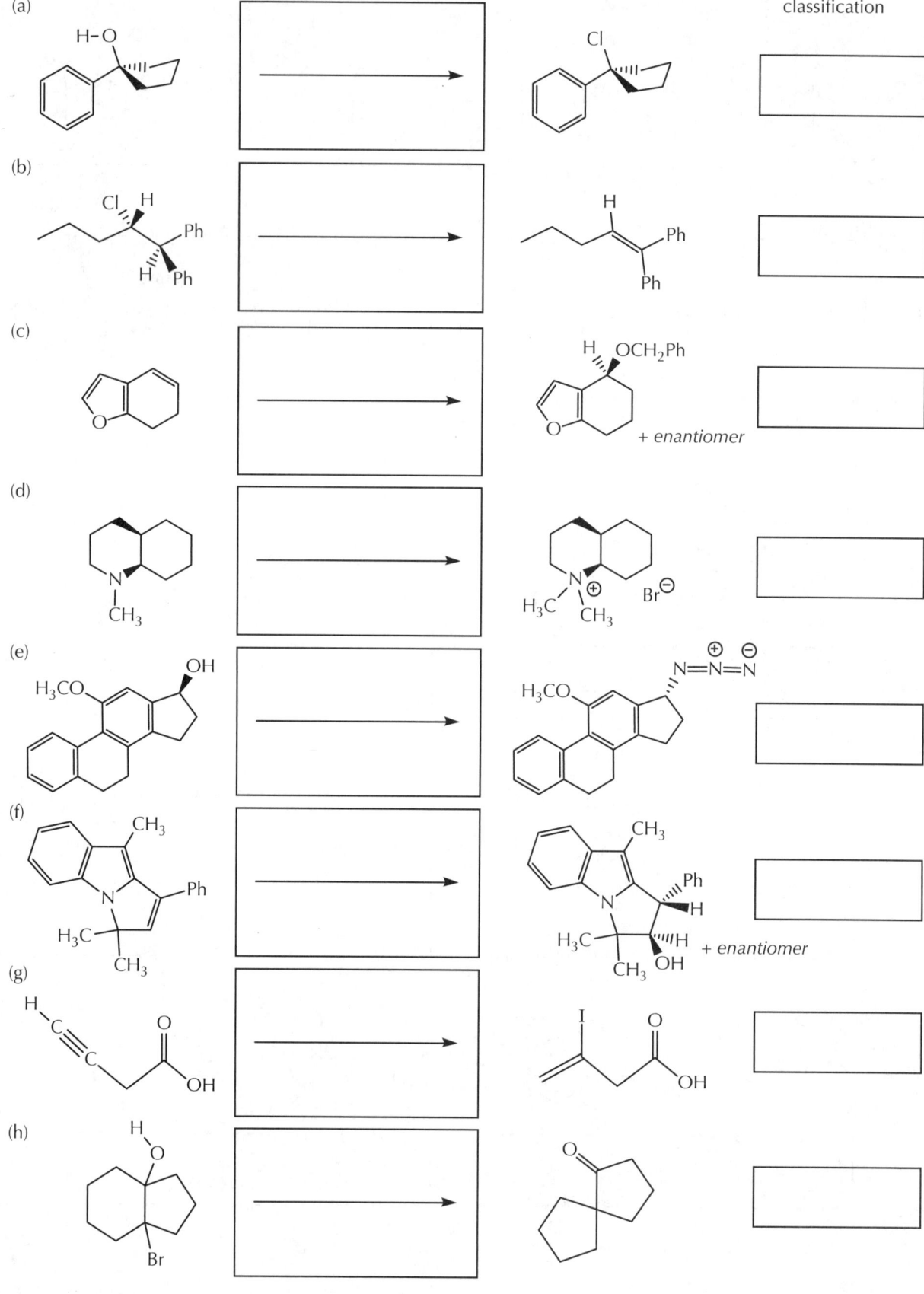

8.3 Hydration via Hydroboration-Oxidation 657

8.14 Provide the reagents needed to carry out the following transformations in the following synthesis. You should explicitly classify the reaction, first, as the way to guide your solution.

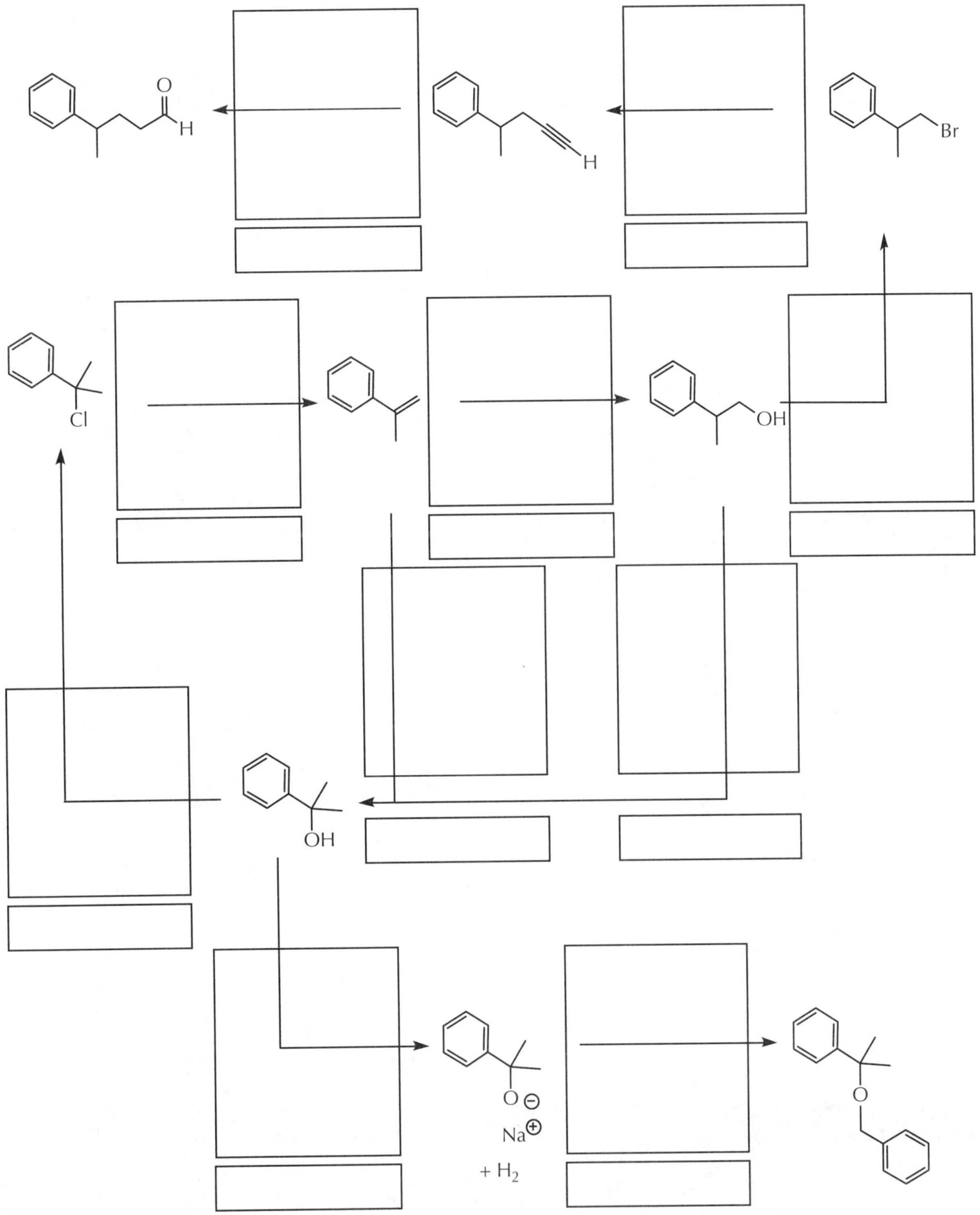

8.15 Complete the following reactions.

(a) PhCH(CH₃)CH₂C≡CH + H₂O / H₂SO₄ →

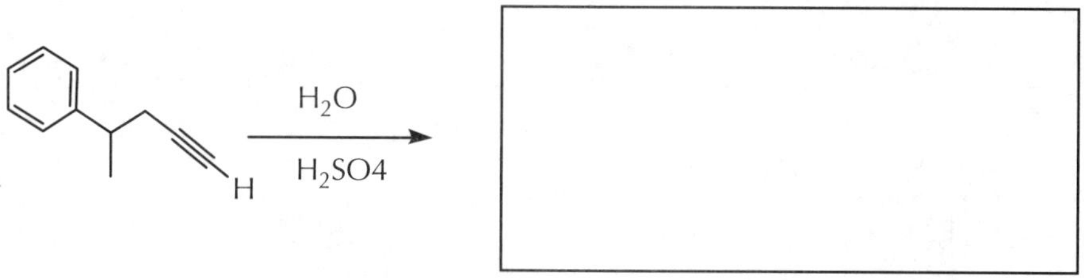

(b) PhC(CH₃)=CH₂ + CH₃SH / H₂SO₄ →

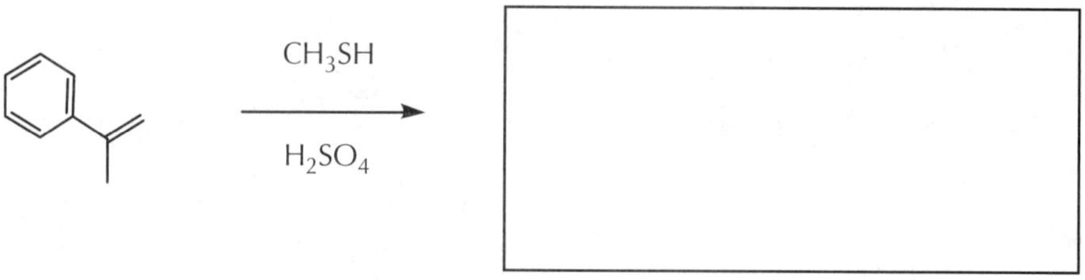

(c) PhCH(CH₃)CH₂OH 1) Na⁺H⁻ 2) PhCH(CH₃)CH₂Br →

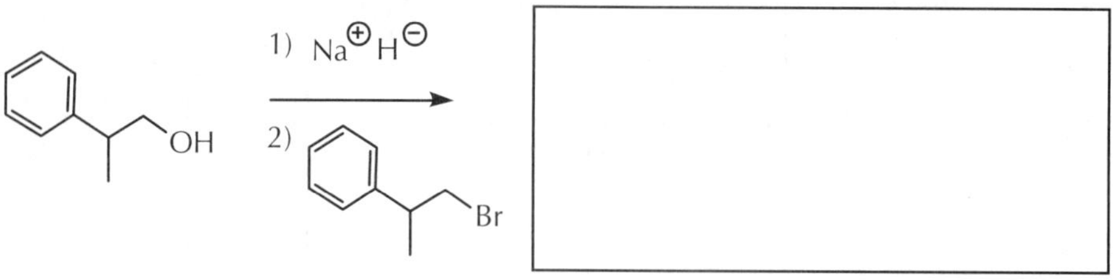

(d) PhC(CH₃)₂OH 1) Na⁺H⁻ 2) PhC(CH₃)₂Cl →

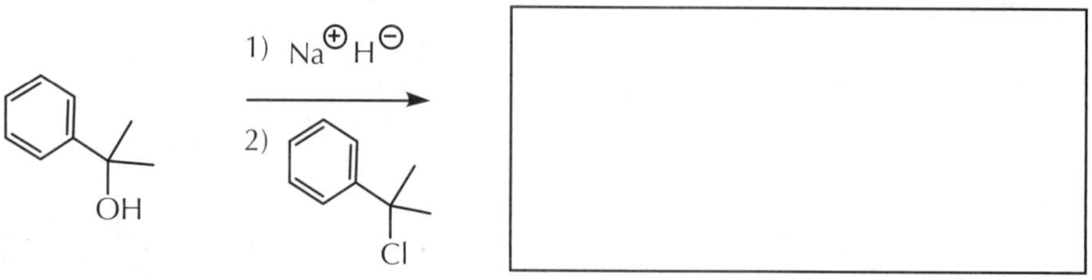

(e) PhCH(CH₃)C≡CH + 2 equivalents HBr →

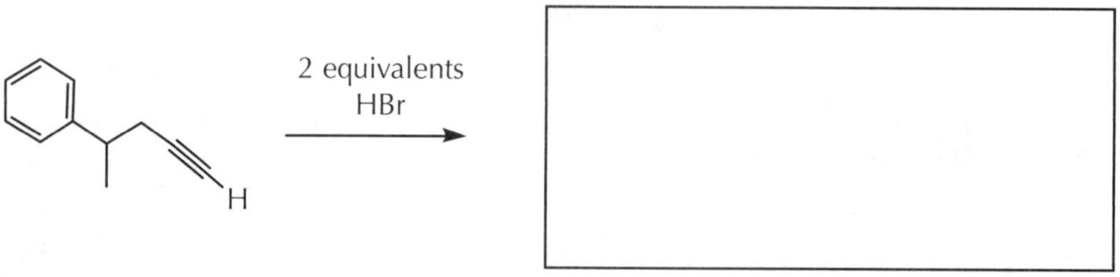

(f) PhC(CH₃)=CH₂ + F₃C−C(=O)−O−H →

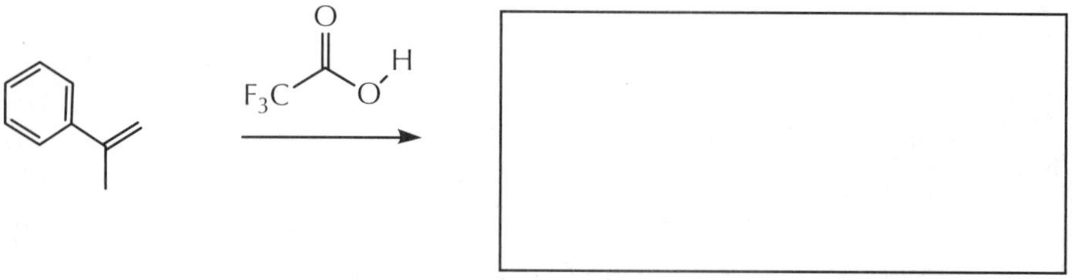

(g) PhCH(CH₃)CH₂C≡CH 1) Na⁺H⁻ 2) PhCH(CH₃)CH₂Br →

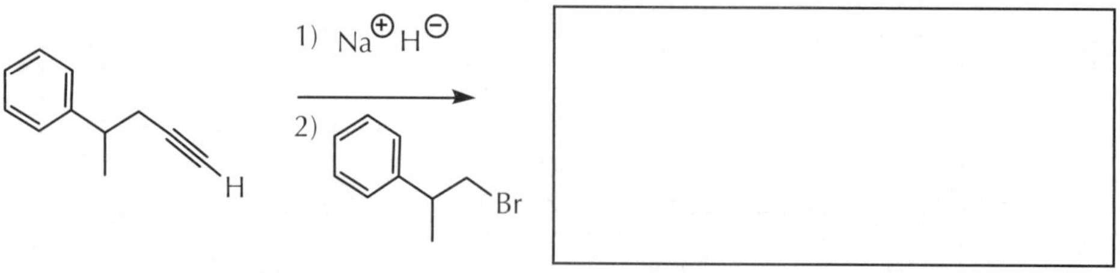

Reflections About Science: Synthesis and Why We Need Options

Even without the formal knowledge of modern chemistry—the structure and reactivity of molecules—humans have been using practical chemical knowledge since the dawn of time. From naturally occurring, plant-based medicines to transforming materials such as metals, people have been doing chemistry. Willow bark, for example, has been recorded as a treatment for pain for at least 3,500 years. Through the early 1800s, apothecaries were people from whom such botanical treatments could be obtained.

With the rise of organic chemistry in Germany in the early to mid-1800s, the apothecary started to be more recognizable as a chemist, as local manufacturing of drug (and dye) compounds became more widespread. This time period is the beginning of the pharmaceutical industry.

Willow bark extract was well known, in the 1800s, for its medical effects, and a great deal of effort in the early years of organic chemistry was devoted to trying to isolate and identify the active chemical ingredient in medically active substances such as these.

In the 1820s and 1830s, chemists isolated a compound they named salicin from the willow bark extracts. An even more potent compound, salicylic acid, was also isolated, and its sodium salt (the conjugate base, sodium salicylate) was used as an effective medicine for relief of pain and fever.

The process of extraction and purification was difficult, and by the 1860s, a synthesis of sodium salicylate was devised from the conjugate base of phenol (PhONa), a coal tar extract, when it was heated under pressure with carbon dioxide.

Sodium salicylate was not tolerated well by many people, and so the search was on for other compounds that could be used. New chemical manufacturing companies, mainly built upon the dye industry, were located along the lower Rhine river, where cargo barges carried coal tar as a byproduct from the refining of coal used in manufacturing steel. In 1897, a chemist at one of these businesses (Frederick Bayer & Company), developed an efficient way to prepare acetylsalicylic acid, whose identity had been known since 1853, but which had not been examined for its medical properties. The mass production of acetylsalicylic acid (Bayer Aspirin) began in 1899 (Figure 0846).

Aspirin is not a naturally occurring substance, and so it is classified as a synthetic material. Inspired by the salicin in the extracts of the natural willow bark, the efficient and large-scale manufacturing of salicylic acid relies on a process that was intentionally designed to achieve a specific structural outcome, that is, a synthesis. The acetyl derivative of salicylic acid was found while seeking to improve the medicinal properties of salicylic acid, and was one of many, many different compounds prepared for testing.

Once aspirin was identified as a target of interest, its synthesis was refined and optimized to maximize the yield and purity. Having multiple options for preparing compounds provides chemists with more ways to be successful in the search for an optimal preparation of an important target molecule.

660 CHAPTER 8 Electrophilic Addition I: Brønsted Acids

Figure 0846

A brief overview of the history of aspirin.

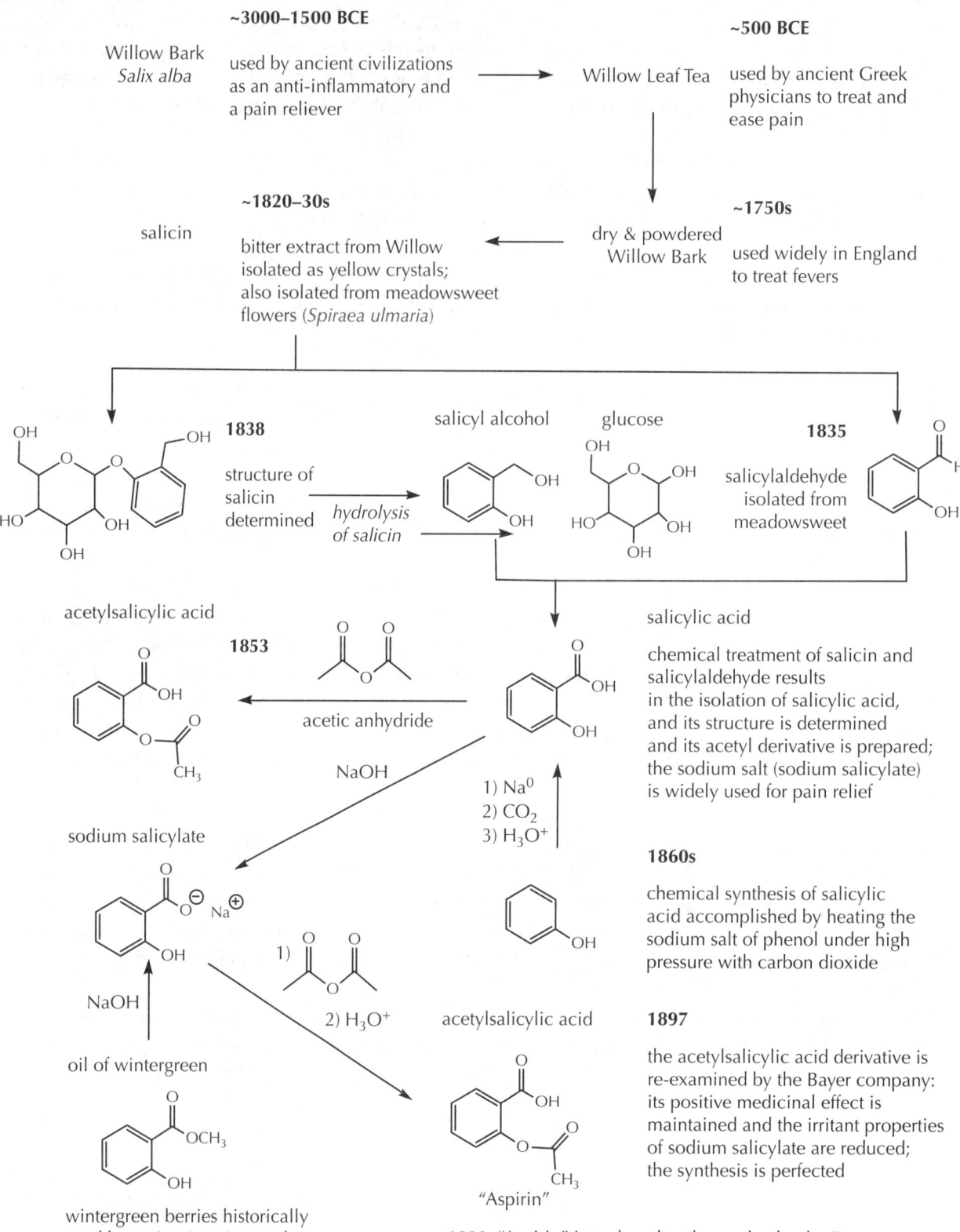

Nature only grows willow bark, and the willow plant only manufactures salicin. Synthetic chemistry has allowed for both the identification of aspirin, a better-tolerated alternative, and the efficient manufacture of that molecule from nonplant sources.

One version or another of this story repeats itself over and over in the history of drug discoveries. Nature, through certain molds and fungi, provides a small handful of naturally occurring antibiotic compounds known as penicillin, cephalosporin, tetracycline, and others. Because bacteria can evolve rapidly, the bacterial mutants that are not affected by these natural antibiotics can proliferate, as the ones that were sensitive to the natural antibiotics were killed off. The resistant strains could then thrive.

Through synthetic chemistry, new, unnatural antibiotics based on the natural structures have been prepared; these penicillin derivatives have familiar names to you, perhaps: ampicillin and amoxicillin. The cephalosporins were discovered in a fungus, and a wide array of structural variations on the natural compound has been prepared over time.

Naturally occurring compounds (called "natural products") have often inspired chemists to explore thousands and thousands of structural variations as they seek to improve the effectiveness, or diminish the side effects, of potential medicinal substances. Being able to plan and then having multiple options for carrying out these changes in molecular architecture is ultimately a net benefit to humanity.

Summary

In Chapter 4, you were introduced to the addition reactions of strong and weak Brønsted acids to carbon-carbon pi bonds. The electrophilic addition reaction mechanism of strong acids allowed you to understand the Markovnikov regioselectivity of these reactions: Protonation of the pi bond to give the more stable carbocation intermediate is faster than the protonation to give the less stable carbocation. Because the two bonds in the addition reaction form in different mechanistic steps, the stereochemical consequence is a mixture of syn and anti stereoisomers, experimentally detectable when two stereocenters have formed in the reaction.

The experimental conditions for these reactions are intuitively appealing. The addition reactions of the strong acid "HCl" use HCl as a reagent, while the addition reactions of the weak acid "H_2O" use water, and both show the Markovnikov regioselectivity.

Not all chemical reactions are so intuitive.

Reactions that proceed through less stable carbocation intermediates undergo fast rearrangement reactions of their hydrocarbon structures, eventually ending up in a structure of a more stable carbocation before further productive reaction takes place. With small molecules, and for carbocations with adjacent heteroatoms, some of these rearrangements are predictable. For the most part, however, the outcome is a matter of experimental observation.

In the mid-1950s, during research studies that were completely unrelated to the addition reactions of water, some chemists observed that a two-reaction sequence (hydroboration-oxidation), neither of which involving water as a reagent, resulted in a product in which the element of water ("H" and "OH") ended up added to pi bonds. The addition reaction also ended up with the opposite regioselectivity to the typical Markovnikov outcome, and the stereochemical outcome was syn addition. The result was unexpected, but the two-reaction hydroboration-oxidation procedure remains today as an exceptionally useful way to add water (H/OH) to an organic molecule with a "syn, anti-Markovnikov" outcome.

CHAPTER QUESTIONS

8.16

A. The following series of equilibria were proposed for the C_5H_7 cation.

(a) Provide the curved arrow mechanism required to explain each of these steps (*J Org Chem,* **1999**, *64,* 7768). Include the arrows for both the forward and reverse reaction.

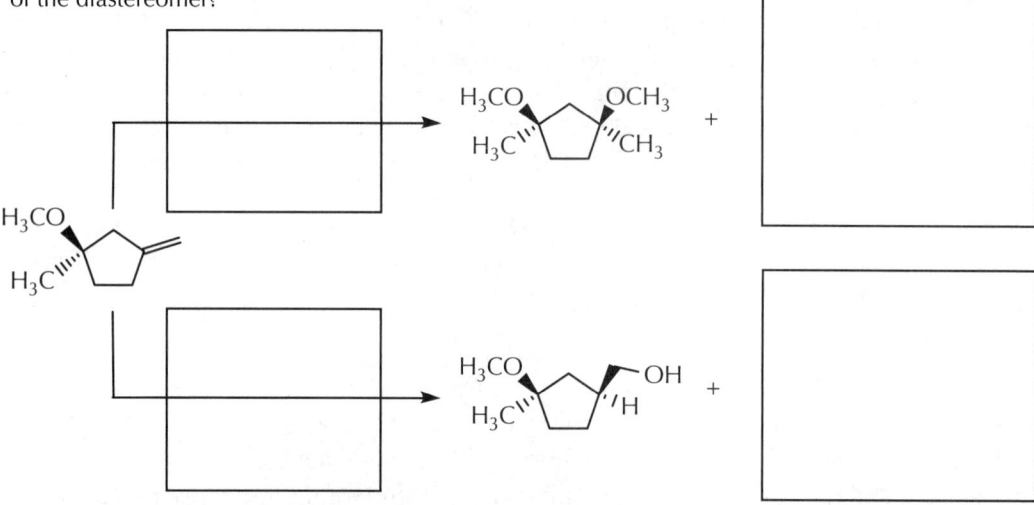

(b) Provide a starting material and the conditions under which it could result in the first of these carbocations.

B. The starting material, shown below (left) was used to carry out a pair of reactions that each gave a pair of diastereomers. What are the experimental conditions needed for each transformation and what is the structure of the diastereomer?

C. The hydroboration-oxidation of optically active alcohol D also gives two stereoisomeric (not regioisomeric) products E and F ($C_7H_{14}O_2$). Draw these two products and indicate whether or not they are optically active.

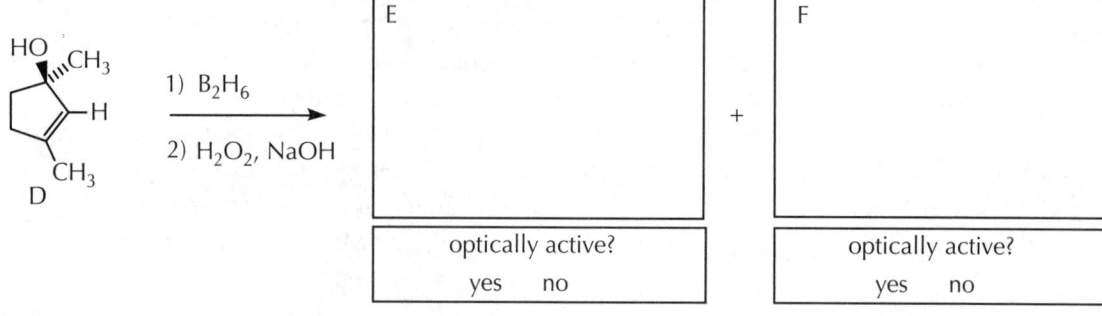

optically active?   yes   no

optically active?   yes   no

What is the stereochemical relationship between E and F?

8.17 Some Brønsted bases are called ambident bases because they have two different sites where they can be protonated. Often, these different sites are the atoms that share a delocalized negative charge, as depicted by a set of resonance contributors. For instance, thioacetate anion is an ambident base. It can be protonated to give a mixture of two different conjugate acids, as shown below.

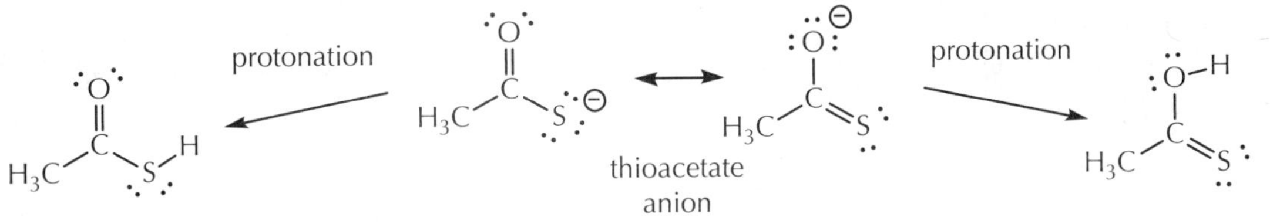

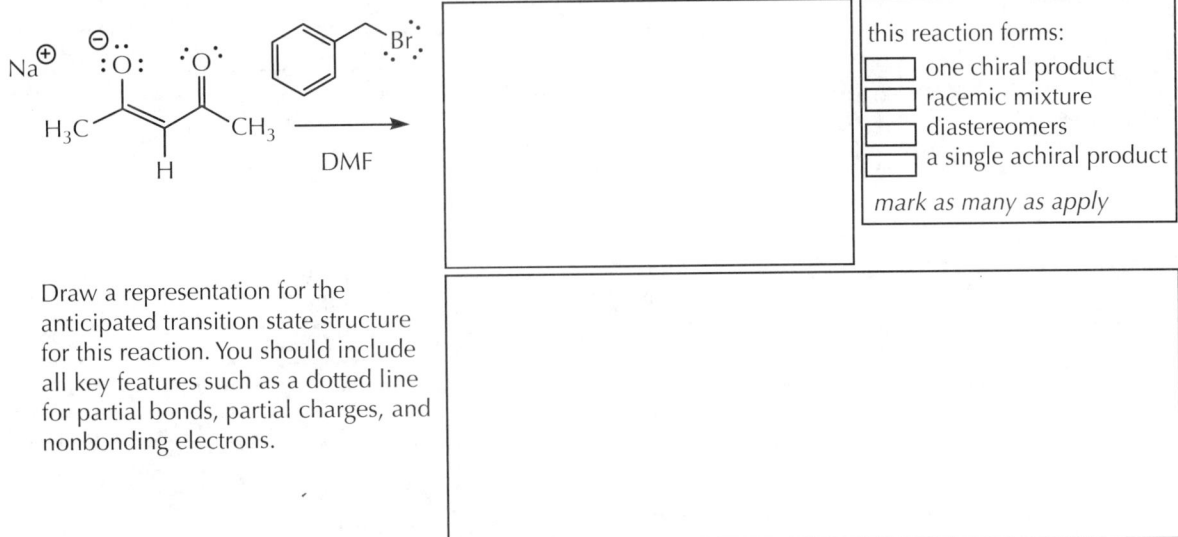

Some Lewis Bases are called ambident nucleophiles (*Angew Chem Internat Ed Eng*, **2011**, *50*, 6470). This means that there are different nucleophilic sites on the anion because of the contribution from different resonance contributors. The anions in the following two examples are ambident nucleophiles (only one of the resonance contributors is shown). In each part, a statement is made that will help guide you in understanding which of the resonance contributors is more significant in explaining the outcome from the reaction. Draw the structure of the anticipated product from these substitution reactions. If more than one stereoisomer results, then just draw one of them. Draw the most significant resonance contributor, if there is a choice.

(a) STATEMENT: "the rate of C-C bond formation is faster than C-O bond formation"

this reaction forms:
☐ one chiral product
☐ racemic mixture
☐ diastereomers
☐ a single achiral product

mark as many as apply

Draw a representation for the anticipated transition state structure for this reaction. You should include all key features such as a dotted line for partial bonds, partial charges, and nonbonding electrons.

(b) STATEMENT: "the atom that is a better hydrogen bond acceptor reacts with the electrophilic carbon atom"

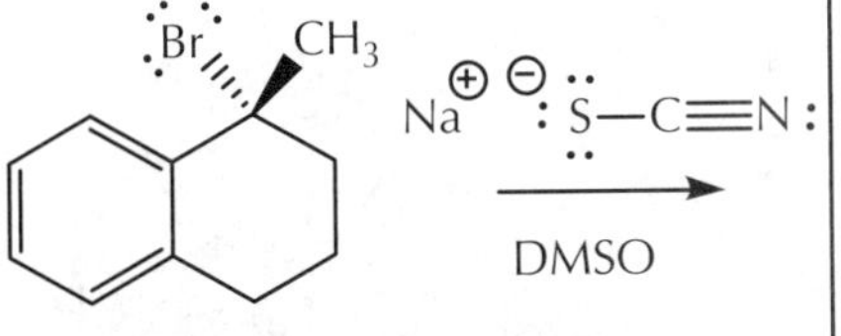

this reaction forms:
☐ one chiral product
☐ racemic mixture
☐ diastereomers
☐ a single achiral product

mark as many as apply

If the molecule with an iodine atom in place of the bromine atom was used, under the exact same conditions, answer the following:

(i) The reaction rate would (*mark one*):
☐ increase ☐ decrease ☐ remain the same

(ii) The distribution of organic reaction products would (*mark one*):
☐ be different ☐ remain the same

664 CHAPTER 8 Electrophilic Addition I: Brønsted Acids

8.18 Complete the following reactions; include stereochemistry where it is needed. Separate experimental steps should be numbered.

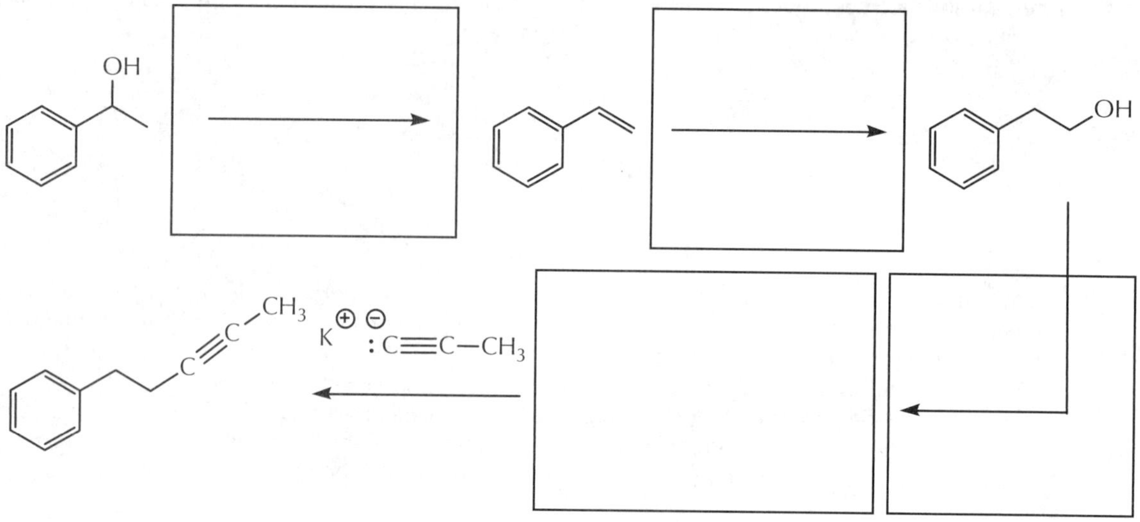

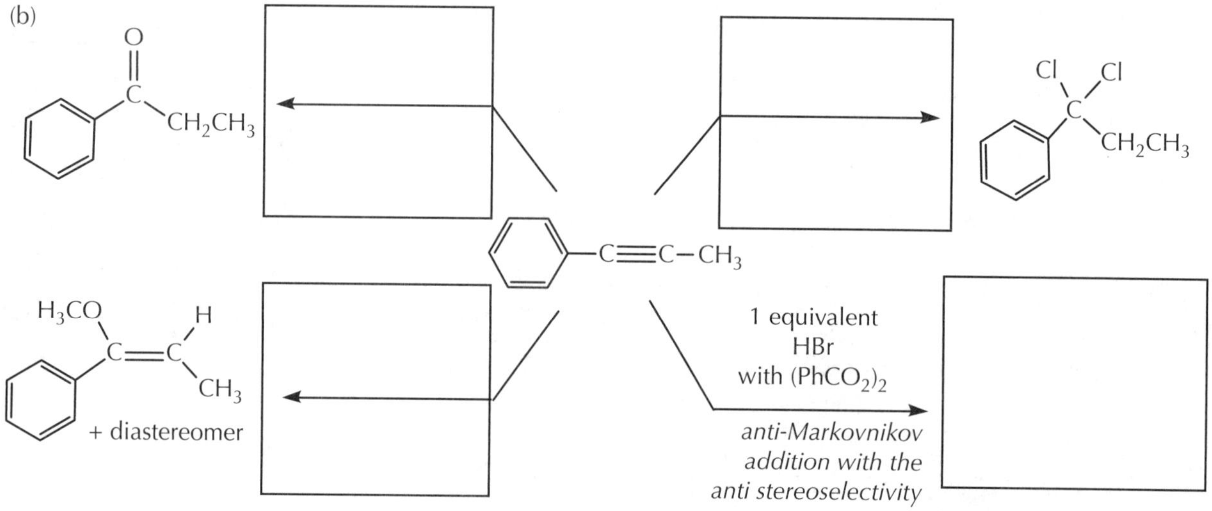

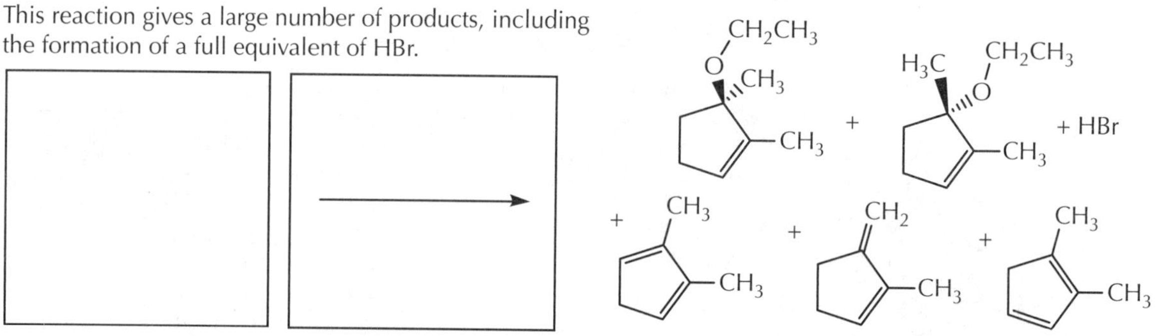

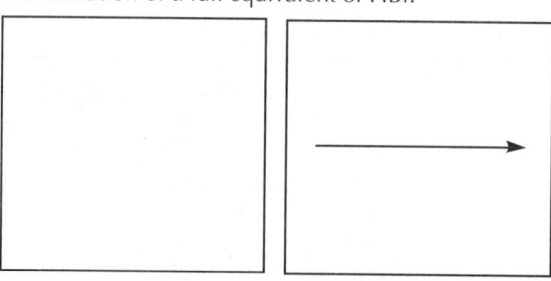

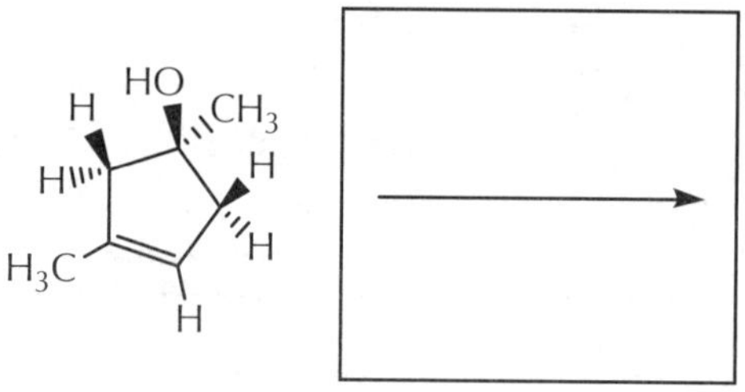

8.19

A. A new reaction was reported that results in an anti-Markovnikov addition of a hydrogen atom and a PPh$_3$ group to an alkene. This reaction provides an alkylphosphonium salt as the product. Given this information, draw the structure of the cationic product formed in the reaction shown below.

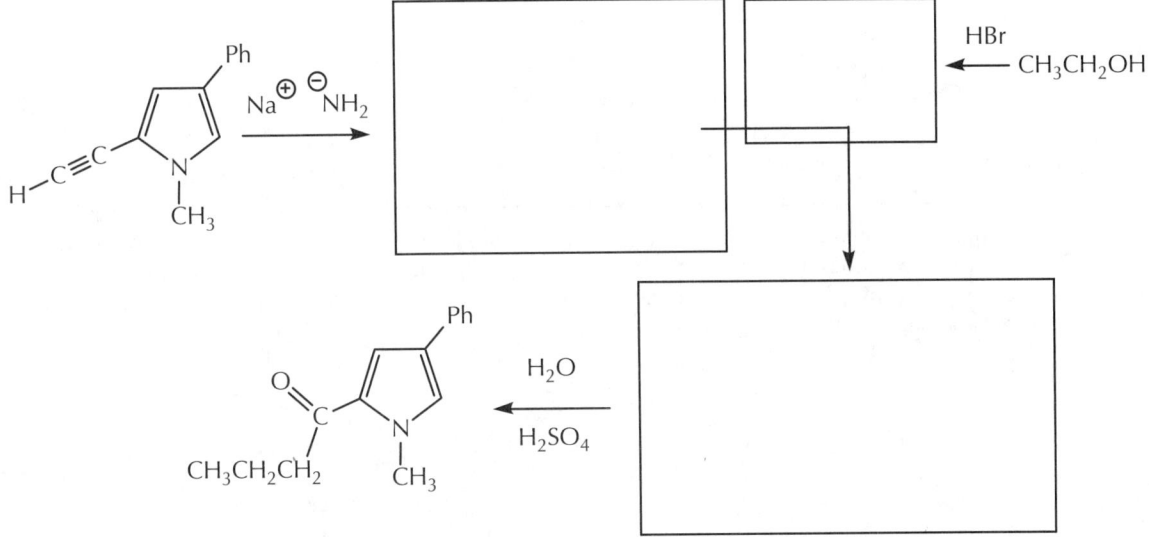

B. A seemingly simple acid-catalyzed addition of methanol to compound P was carried out in an attempt to make product Q. However, molecule R was formed as the major product instead of Q. Molecule R results from a single favorable rearrangement step. Draw the structure of product R.

C. Provide the reagents or structures as necessary to complete the following reaction scheme. If more than one synthetic step is required, number them sequentially.

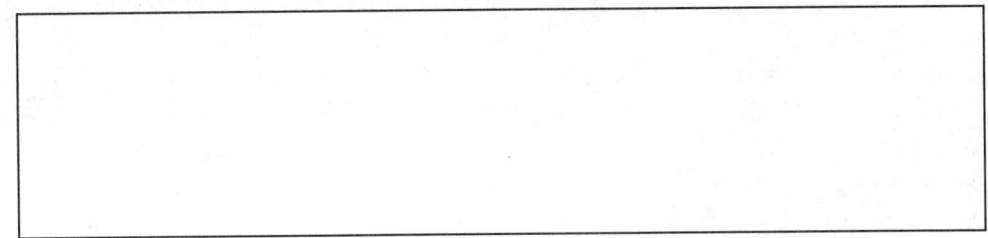

D. Draw the structure of (2Z,4E)-3-bromoocta-2,4-dien-6-yne.

8.20 (a) Provide the complete, stepwise mechanism for the following observed reaction, which is proposed to occur through a carbocation intermediate that also undergoes a rearrangement.

Note 1: Experiments using isotopic labels show that the oxygen atom in the product is derived from the alcohol group directly attached to the cyclopentyl ring in the starting material. You may use H-B and B⊖ for any Brønsted acid and base that you need.

Note 2: In acid catalyzed mechanisms, there should not be any strongly basic intermediates.

(b) In thinking about how to prepare the starting material for this reaction, the following plan was proposed. The Markovnikov addition of water to this double bond looks quite reasonable, but there are at least two good reasons why this plan will fail to create the desired product. Give one explanation why the following transformation will not go as planned.

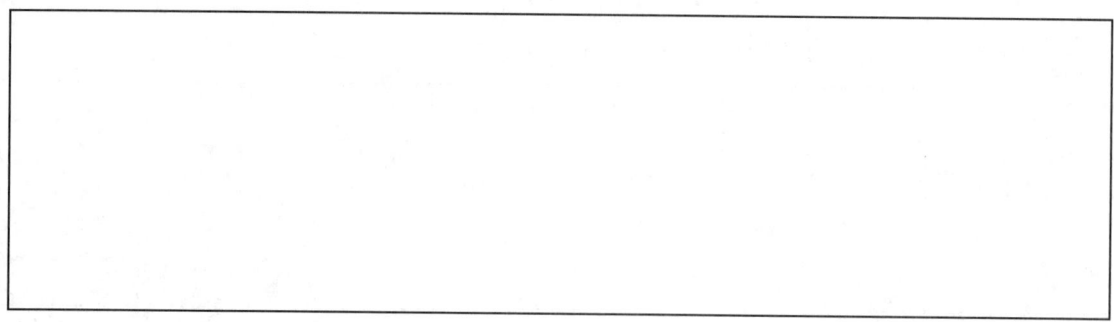

8.21

(a) The mechanism for the following reaction is proposed to begin with the formation of a carbocation derived from the loss of the hydroxyl group in compound A. The resulting carbocation intermediate then acts as the electrophile in an intramolecular (within the same molecule) electrophilic addition reaction with the double bond, which is the first step in the mechanism to form compound B. Not including compound B itself, how many other stereoisomers could be formed in this reaction?

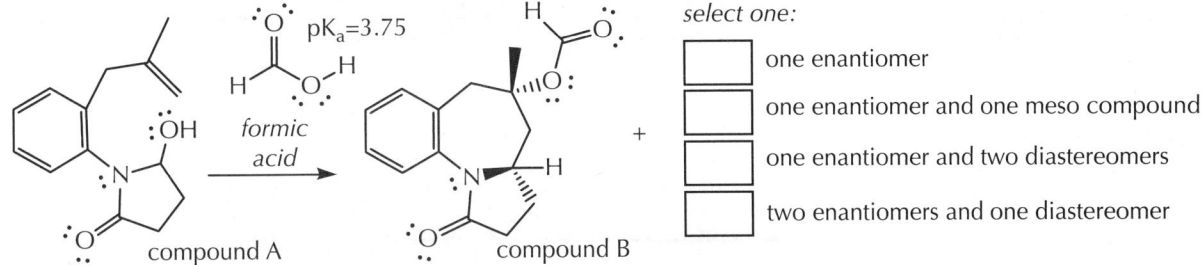

select one:
- [ ] one enantiomer
- [ ] one enantiomer and one meso compound
- [ ] one enantiomer and two diastereomers
- [ ] two enantiomers and one diastereomer

(b) Protonation of an alcohol by a formic acid (see above) can lead to the loss of a water molecule as a leaving group and the formation of a carbocation. If this happens with compound A, draw the resulting carbocation in box (i). In box (ii), explain why this carbocation is more stable than an ordinary secondary carbocation.

(i) carbocation formed	(ii) explanation of stability

(c) The mechanism of the formation of compound B from compound A starts with forming the carbocation you drew in part (b) followed by an intramolecular electrophilic addition reaction. Using this information, and the formic acid, draw the curved arrow mechanism for the conversion of A to B. Ignore stereochemistry in your answer.

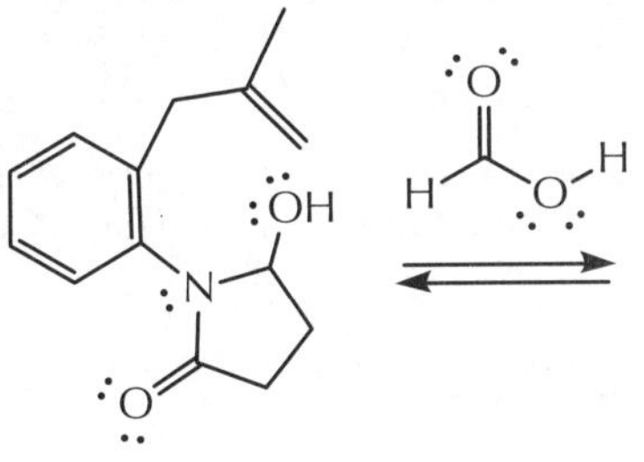

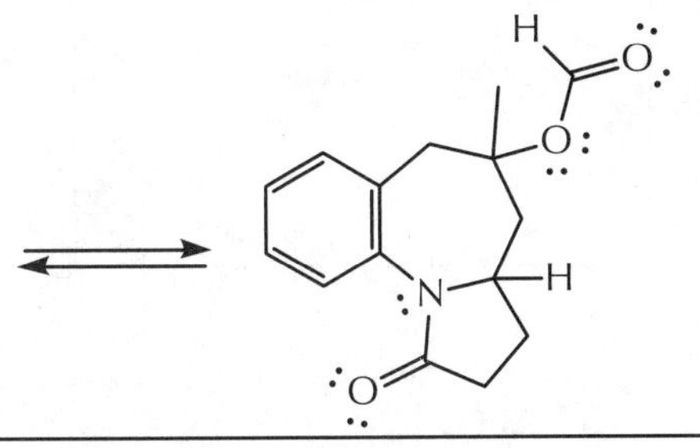

668 CHAPTER 8 Electrophilic Addition I: Brønsted Acids

8.22

A. When treated with a Brønsted acid catalyst, compound P was transformed into compound Q (isocumene) through a mechanism involving two reactive intermediates (*JACS* **1981**, *103*, 82). Provide the structure of the first intermediate (resulting from protonation), the mechanism for its transformation to the second intermediate, and the structure of the second intermediate. The second intermediate is deprotonated to give compound Q). The protonation and deprotonation mechanisms do not need to be shown.

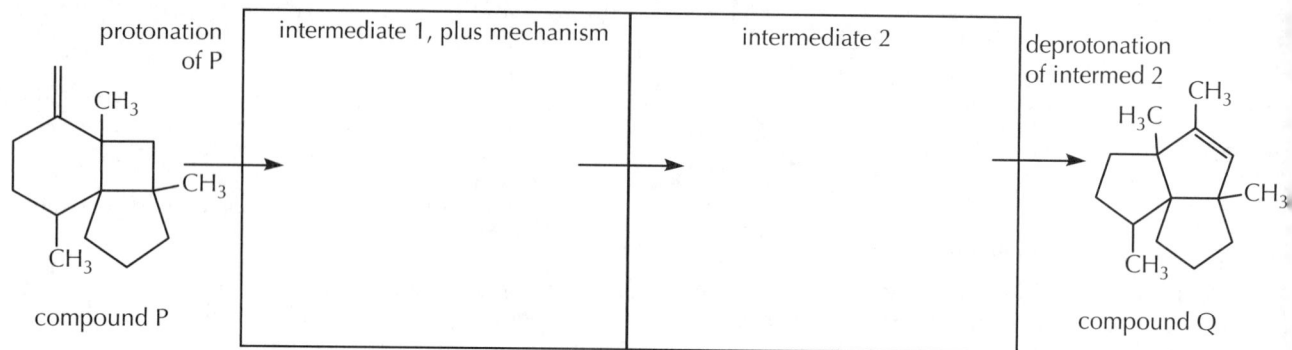

B. When dimethyl malonate (compound J; pK_a 13) is deprotonated, its conjugate base has a number of different resonance contributors. These resonance contributors are used to explain how the conjugate base of compound J can show nucleophilic reactivity from both its oxygen and carbon atoms.

When the conjugate base of compound J reacts with methyl triflate (CH_3-OSO_2CF_3), the different closed shell atom resonance contributors are used to explain how different S_N2 reactions can occur. Using the resonance contributors as nucleophiles, provide the structures for compounds M and N (important experimental data are provided).

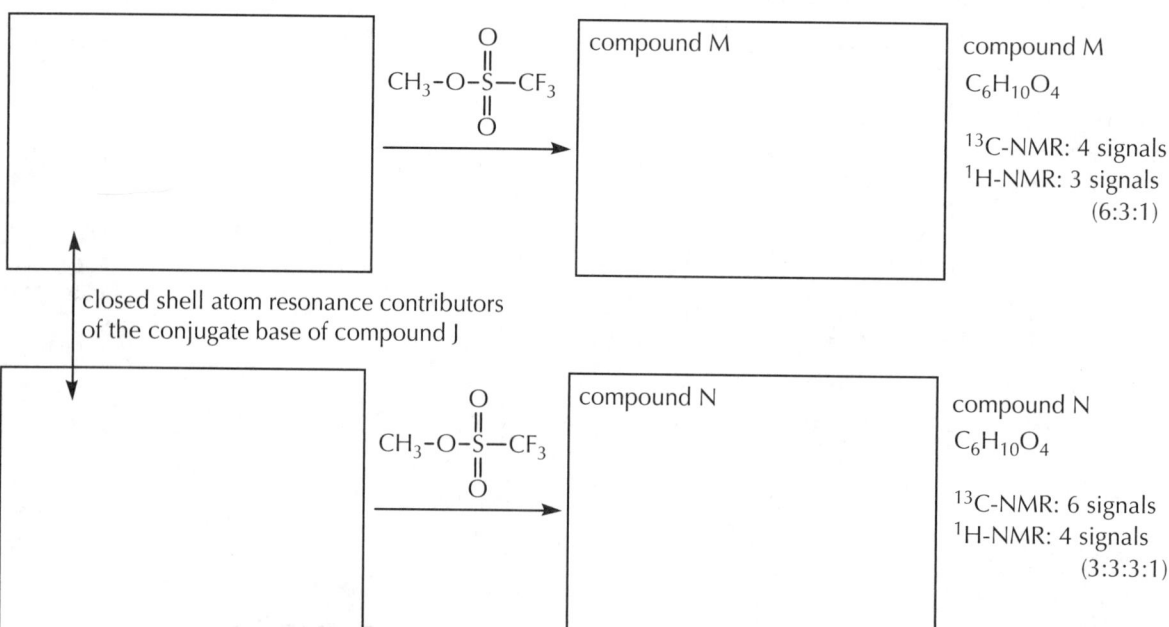

8.23

(a) In a highly regioselective addition reaction, **only** compound P forms when compound O undergoes a acid-catalyzed addition of water (hydration) reaction. Over time, compound Q begins to appear in the acidic reaction mixture. Compound Q is a structural isomer of compound O, and it forms **without** evidence of rearrangement. Ignoring stereochemistry, draw the connectivites of compounds O and Q.

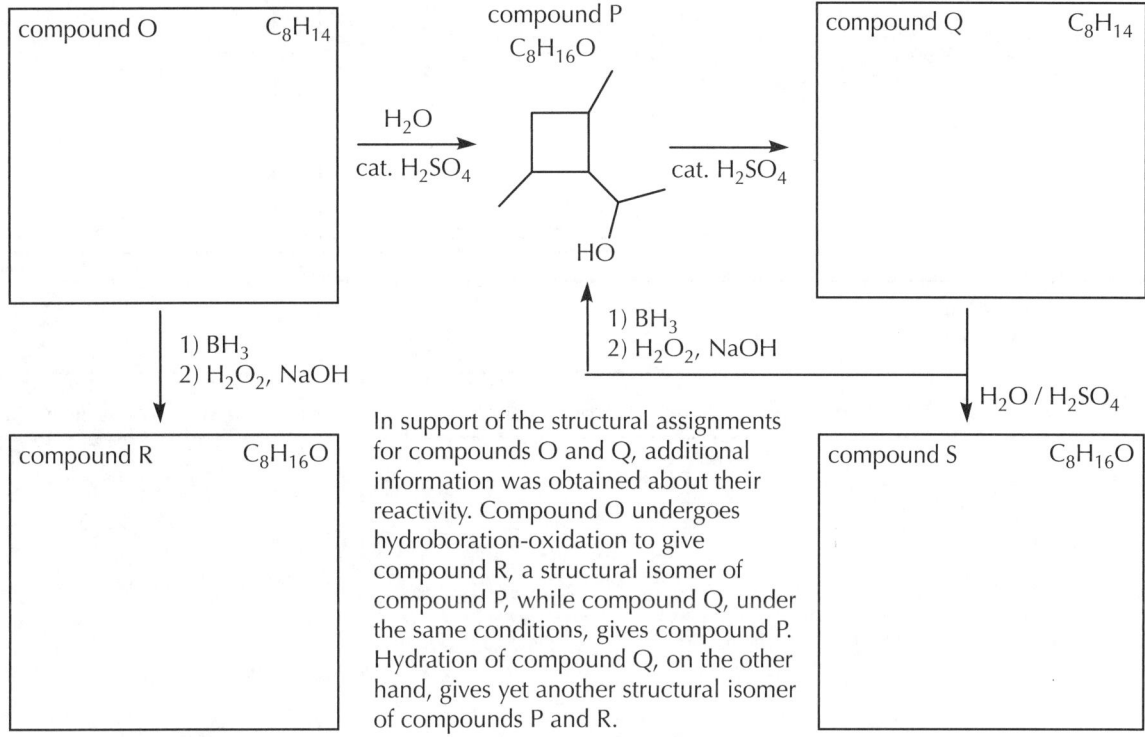

In support of the structural assignments for compounds O and Q, additional information was obtained about their reactivity. Compound O undergoes hydroboration-oxidation to give compound R, a structural isomer of compound P, while compound Q, under the same conditions, gives compound P. Hydration of compound Q, on the other hand, gives yet another structural isomer of compounds P and R.

(b) When compound T (below), which is the methylated version of compound P, is treated with dilute acid, two dehydration products with molecular formula C_9H_{16} form in high yield. Both of them result from a single rearrangement step followed by a deprotonation reaction. Provide the missing carbocation intermediates and the final dehydration products.

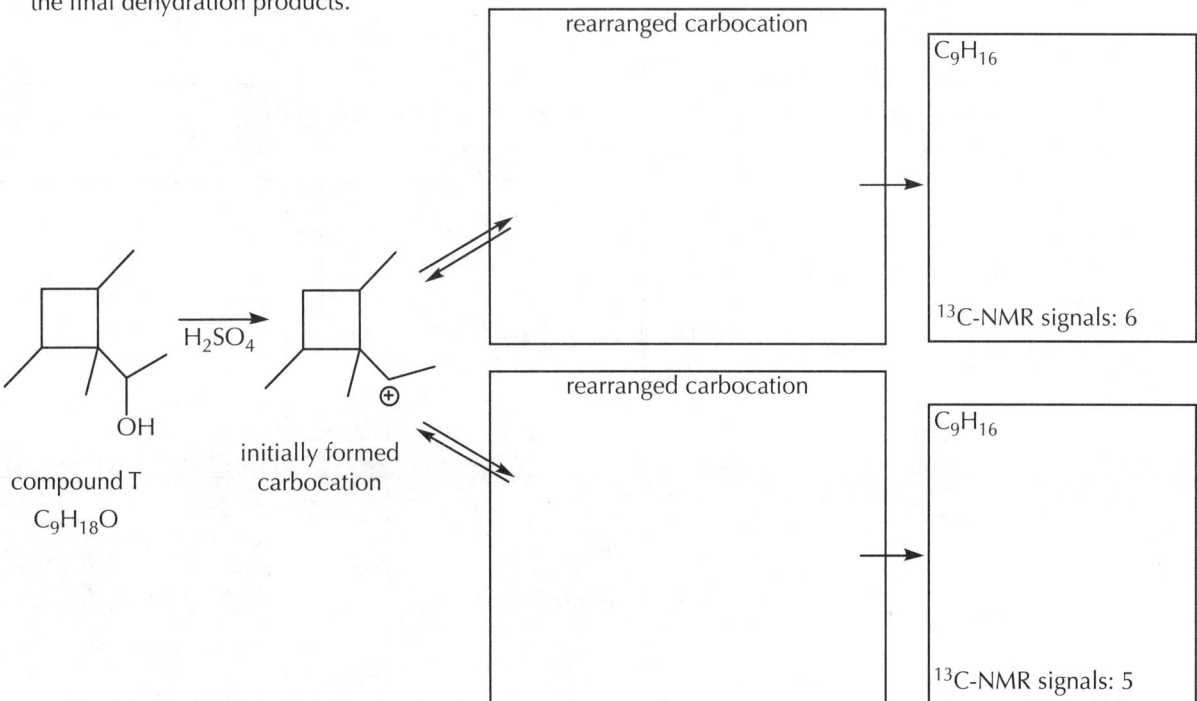

8.24

A. When compound A is treated with water in the presence of an acid catalyst, two achiral stereoisomeric products result. Draw these products, including stereochemistry.

(a)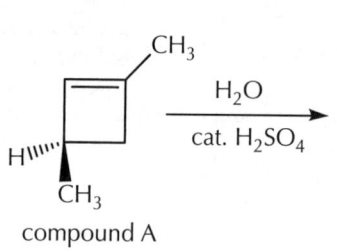

(b) Provide the complete IUPAC name, including stereochemistry, for compound A.

(c) Provide the complete, stepwise, curved arrow mechanism for formation of either one of your products.

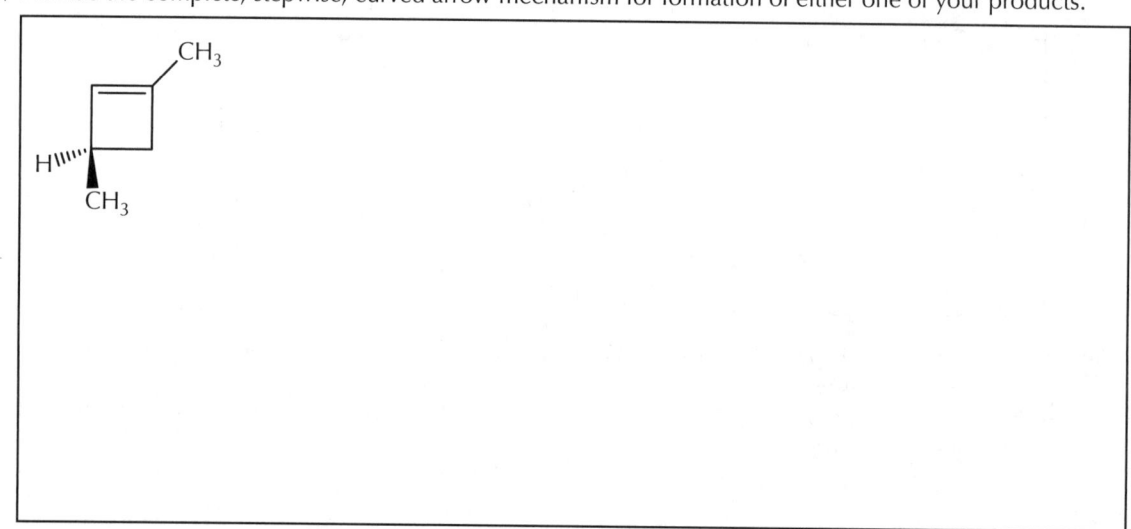

B. The following reaction yields a pair of stereoisomeric products. Draw the two products and answer the questions about them.

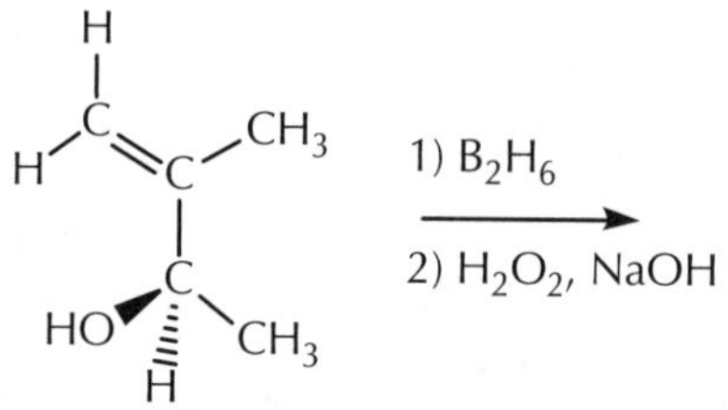

These two products are: (select all that apply)

- [ ] a diastereomeric pair
- [ ] a racemic mixture
- [ ] both are optically active
- [ ] both are optically inactive

8.25

A. For each of the following reactions, tell how many stereoisomeric products are expected to form. Draw one of them, and indicate if that particular molecule is chiral or achiral.

(a)

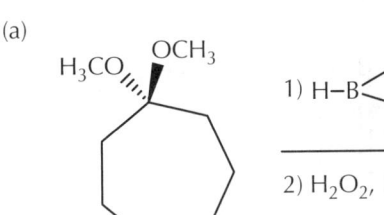

OCH3
H3CO
1) H–B
2) H2O2, NaOH

total number of stereoisomers formed in the given reaction?

is the stereoisomer that you drew a chiral molecule?

(b)

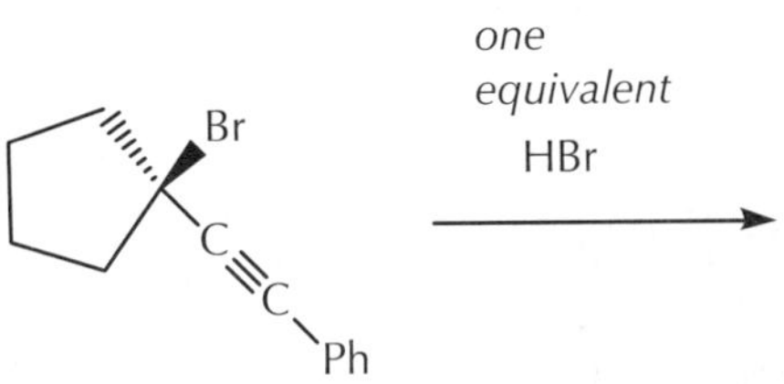

Br
C
C
Ph
one equivalent
HBr

total number of stereoisomers formed in the given reaction?

is the stereoisomer that you drew a chiral molecule?

(c)

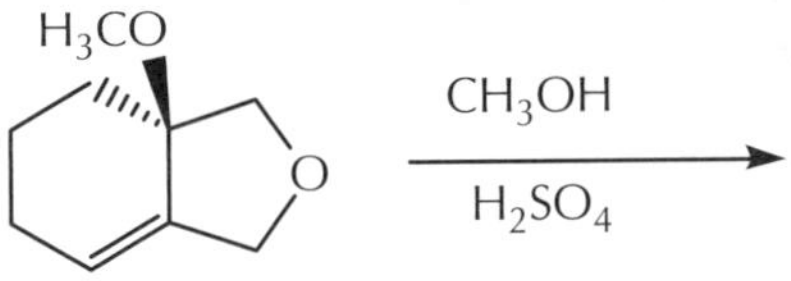

H3CO
O
CH3OH
H2SO4

total number of stereoisomers formed in the given reaction?

is the stereoisomer that you drew a chiral molecule?

B. When the natural terpene α-pinene is treated with hydrochloric acid (HCl), a new compound is formed called bornyl chloride. Provide the stepwise, curved arrow mechanism for this transformation.

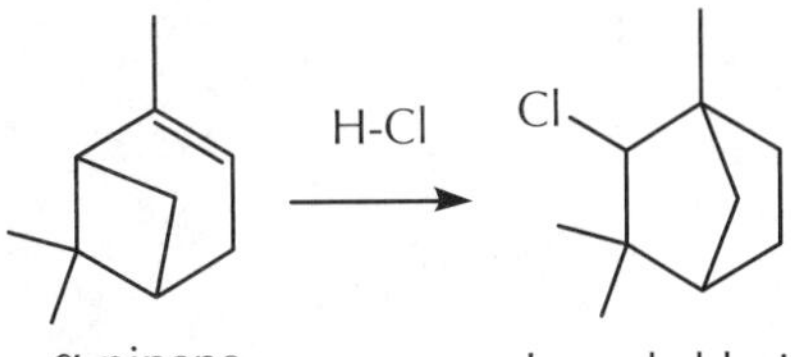

H-Cl
Cl

α-pinene bornyl chloride

H—Cl:

α-pinene

:Cl

bornyl chloride

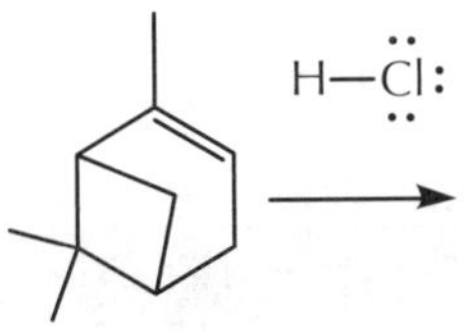

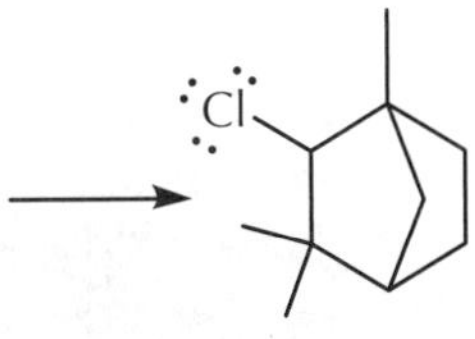

8.26 Complete the following reaction sequences. Be sure to number any sequential experimental steps.

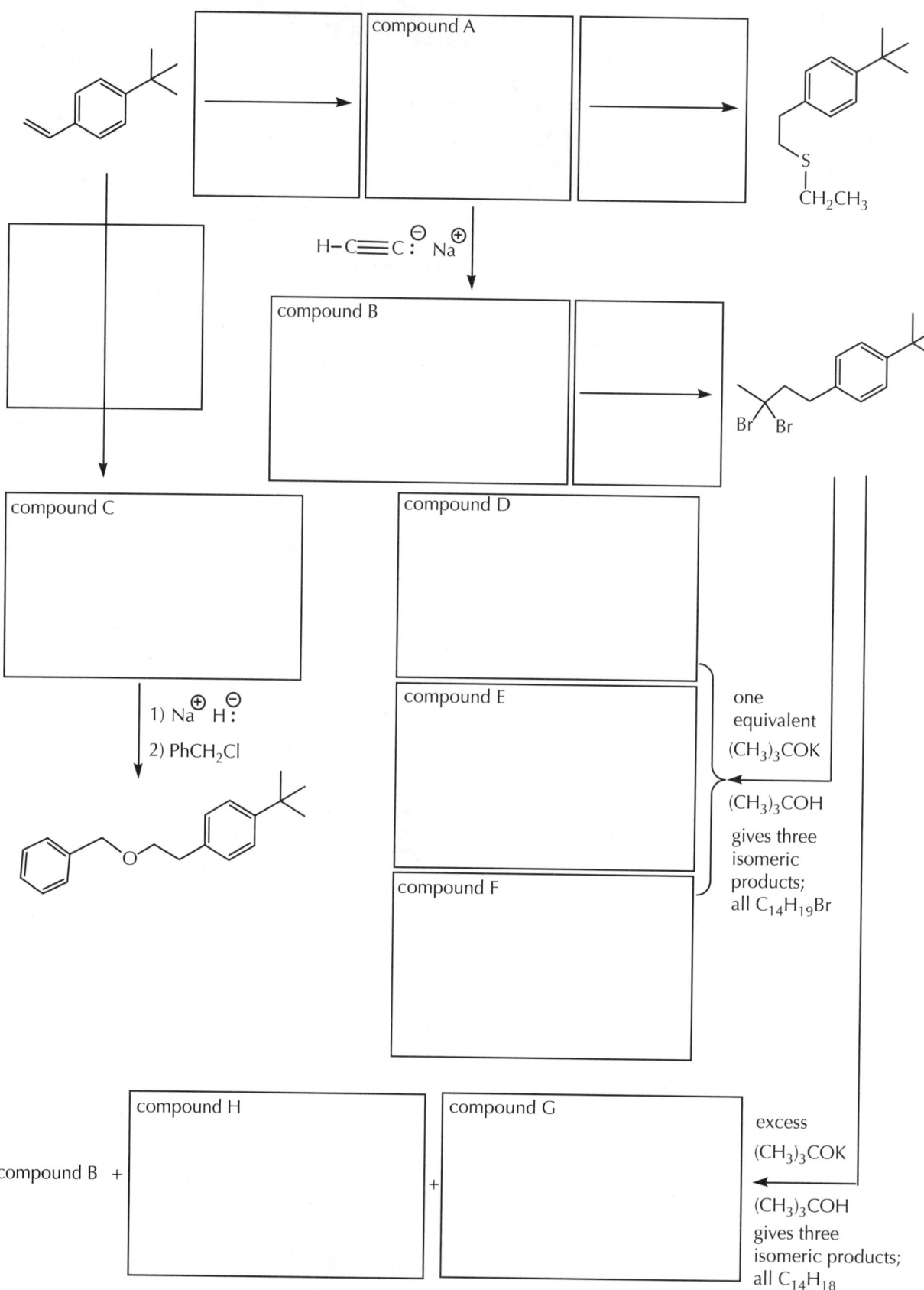

8.27

A. In 2007, Frontier and co-workers reported a new method to access spirocyclic compounds via an interrupted Nazarov cyclization. The formation of spirocyclic carbocation M from dienone N proceeds through two carbocation intermediates. Draw a curved arrow mechanism showing how spirocyclic carbocation M can be formed from dienone N.

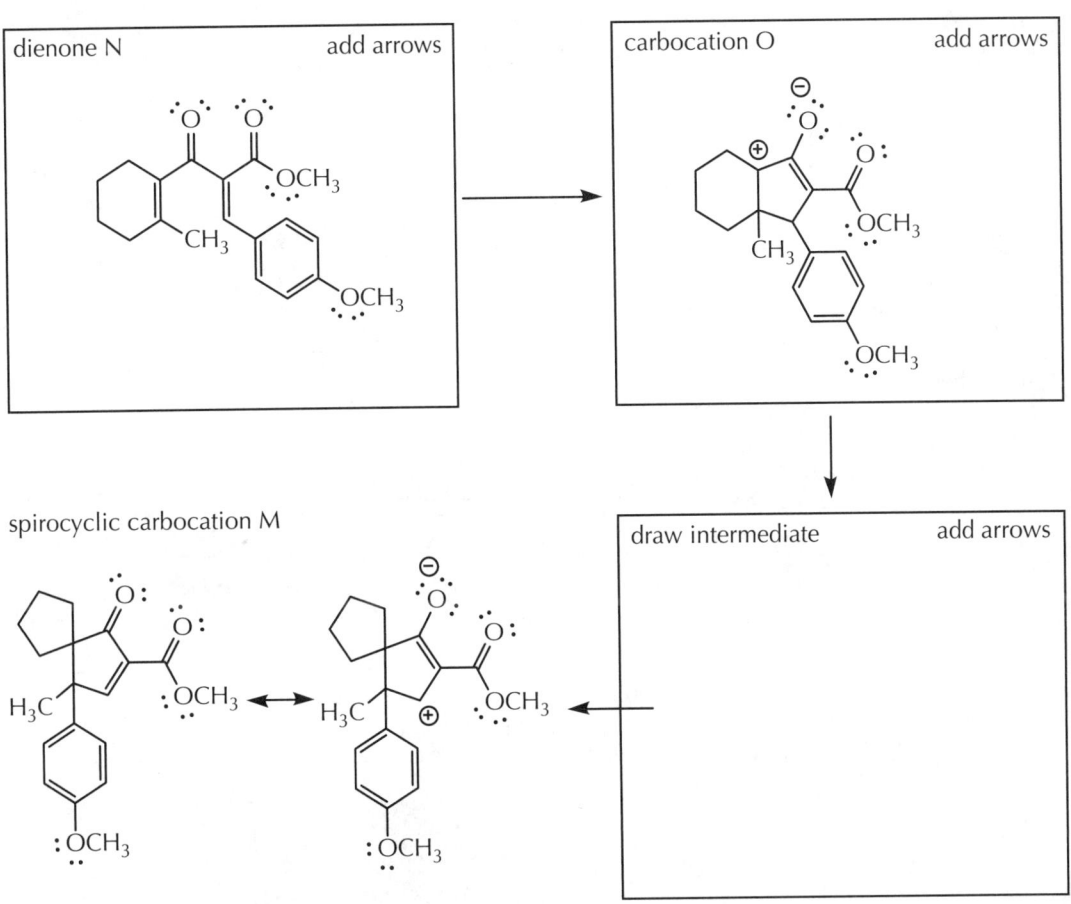

B. As described in the chapter, the Lewis acidic borane (BH_3) monomer is so reactive that is forms the diborane dimer (B_2H_6). Commercially available BH_3 is usually sold as a solution in diethyl ether or tetrahydrofuran (THF), the structures for which are shown below. Based on this information, suggest why these solvents are used instead of hydrocarbons such as pentane or cyclohexane.

8.28

A. The synthesis of 9-borabicyclo[3.3.1]nonane, also known as 9-BBN, is accomplished by the reaction of 1,5-cyclooctadiene with one equivalent of borane (BH_3).

(a) Draw the intermediate in the formation of 9-BBN-D_3 when BD_3 is used (D = deuterium). Be sure to indicate the stereochemistry of the reaction intermediate using lines, dashes, and wedges.

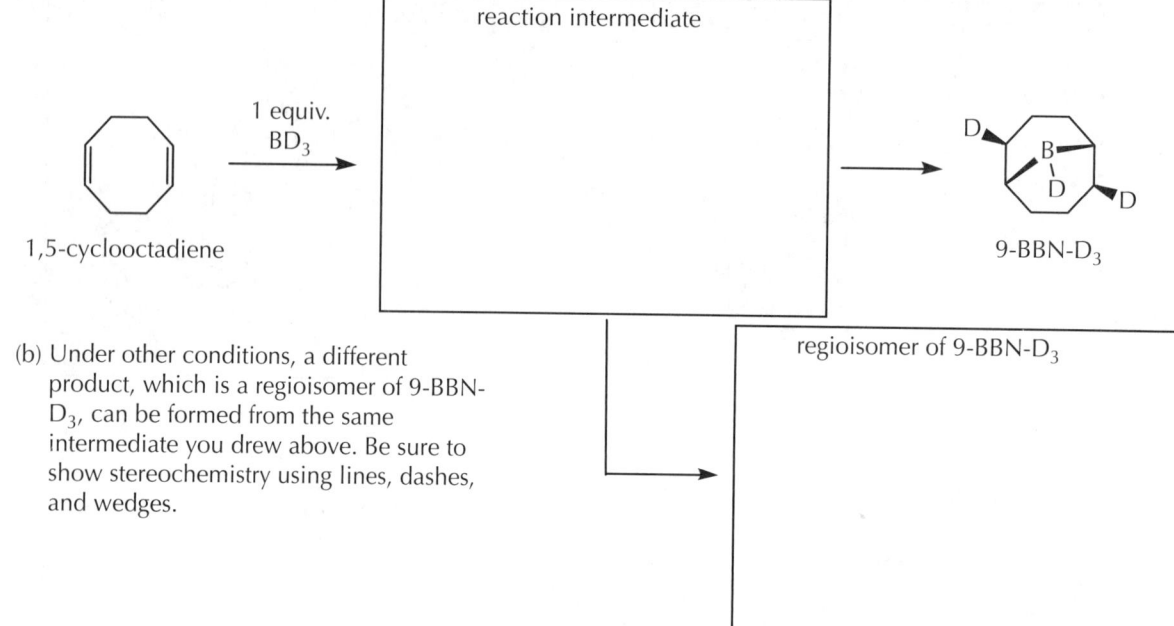

(b) Under other conditions, a different product, which is a regioisomer of 9-BBN-D_3, can be formed from the same intermediate you drew above. Be sure to show stereochemistry using lines, dashes, and wedges.

B. Draw the starting alkene that would yield compound J as its only significant product after treatment with water in the presence of sulfuric acid. If not isolated, compound J will continue to react in the presence of acid to yield (as a minor product) a new alkene that is a structural isomer of the starting alkene - draw this, too.

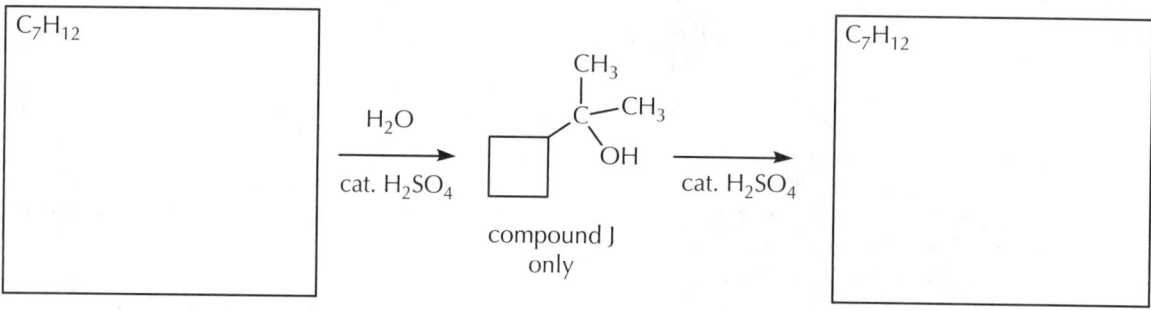

C. The following reactions were carried out during research into the stereochemistry of a hydroboration-oxidation reaction.

Draw the more stable conformation of compound A.

Compounds A and B are stereoisomers. They are formed in roughly equal amounts. Compound A is thermodynamically more stable than compound B.

8.29

A. The following reaction is observed when the starting material is heated in dilute acid. Using H_3O^+ as your Brønsted acid and H_2O as your Brønsted base, provide the curved arrow mechanism for this transformation.

B. The following hydroboration-oxidation reaction produces a pair of diastereomeric products in a nearly 99.5:0.5 ratio.

(a) What are the two products?

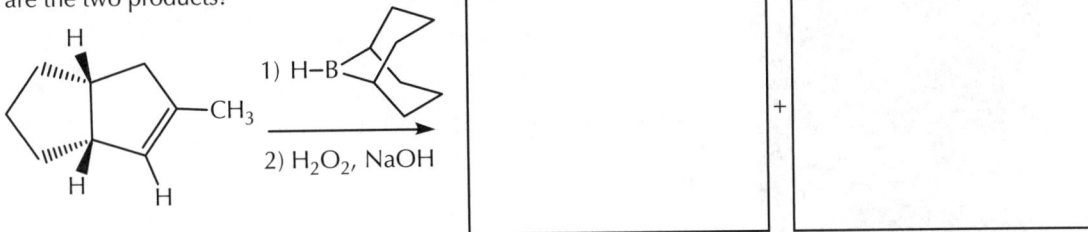

(b) A 3D perspective drawing for the folded shape made by these two rings is shown below. Think about the geometry of the transition states leading to the two products and use that to predict and explain which of the products is the major isomer and which is the minor isomer.

CHAPTER 9

Electrophilic Addition II: Halogenation, Oxidation, and Reduction Reactions

Setting the Stage: Saturated, Partially Hydrogenated, Brominated, and Polyunsaturated …

9.1 Halogenation and Other Reactions of the Halonium Ion Intermediate
  A. Transformation: Anti Addition of Halogens
  B. Stereoselectivity via the Halonium Ion Intermediate
  C. Halohydrins and Related Regioselective Reactions of the Halonium Ion Intermediate
  D. Intramolecular and Other Mixed-Atom Reactions

PRACTICE QUESTIONS

9.2 Oxidation Reactions
  A. Peroxyacid Reactions
  B. Syn Dihydroxylation Reactions
  C. Ozonolysis Reactions
  D. Atom-Swap Analogies

PRACTICE QUESTIONS

9.3 Reduction Reactions
  A. Catalytic Hydrogenation with Transition Metal Catalysts
  B. (Z)-Alkenes from Catalytic Hydrogenation with Poisoned Catalysts
  C. (E)-Alkenes from Dissolving Metal Reduction
  D. Multistep Transformations and Synthesis

PRACTICE QUESTIONS

Reflections About Science: The Power of Mechanistic Organic Chemistry

Summary

CHAPTER QUESTIONS

EXAM QUESTIONS (CHAPTERS 7–9)

CHAPTER 9

Electrophilic Addition II: Halogenation, Oxidation, and Reduction Reactions

Setting the Stage: Saturated, Partially Hydrogenated, Brominated, and Polyunsaturated...

Understanding organic chemistry automatically provides you with a new view of your familiar world. Things that you have known as only words become molecular structures—with connectivity, stereochemistry, and conformations—combined with a sense of chemical reactivity. Words related to nutrition, for instance, surround you, appearing on every food label or in discussions about diet: proteins, added sugars, carbohydrates, saturated fats, cholesterol, caffeine, sodium benzoate, monosodium glutamate, and so on.

Addition reactions provide one entrée into the connection between organic chemistry and some everyday topics. You have previously seen addition reactions between carbon-carbon pi bonds with both strong and weak Brønsted acids, and also the hydroboration-oxidation reaction as an addition of water (Chapters 4 and 8). This chapter introduces a broader range of addition reactions, which still (by definition) convert unsaturated compounds into more saturated ones.

Bromination (addition of bromine) is an example of an addition reaction included in this chapter. In the US, about 10% of all citrus-flavored beverages contain brominated vegetable oil (BVO). Introduced in the 1930s, BVO allows the dense and water-insoluble citrus oils to become suspended in water, and keeps your beverage from separating into layers, as oils and water tend to do. Consuming excessive amounts of bromine-containing compounds can lead to a condition called bromism, a psychological and neurological disorder that is far less common today than it was years ago. Knowing what it means, chemically, to be a vegetable oil, then how and why bromination is used, helps you to build a more chemistry-literate understanding about foods and nutrition.

Hydrogenation (addition of hydrogen) is another reaction included in this chapter. And another commonly encountered material in foods is called "partially hydrogenated vegetable oil." This oil is found in a wide variety of food products, particularly baked goods. Cakes, cookies, and crackers that contain "shortening" are made, according to their labels, with "fully or partially hydrogenated vegetable oil."

During the hydrogenation reaction, some of the double bonds in natural oils, which mainly have the (Z)- configuration, are isomerized to the more stable (E)- configuration. These oils, with double bonds in the (E)- or *trans* configuration, are called trans fats. Trans fats typically occur naturally, but in low concentrations, so unnaturally higher concentrations of trans fats result as byproducts under the hydrogenation reaction conditions. The trans fats can affect cardiovascular health by increasing the concentration of low-density lipoprotein (LDL) and lowering the concentration of high-density lipoprotein (HDL).

Food manufacturers are in business to sell food, and they use partially hydrogenated oils for lots of reasons that do not relate to the health of the consumer. The partially hydrogenated oils save money, they can extend the shelf life of food products, and they add a pleasing texture to baked goods. While you can easily check online for the full list of ingredients in any food you buy, partially hydrogenated oils are commonly used in packaged snacks, premade baked goods, coffee creamers, flavored and microwave popcorn, ready-made dough, including pizza crusts, and in vegetable shortening, margarine, and foods that have been deep-fried.

9.1 Halogenation and Other Reactions of the Halonium Ion Intermediate

A. Transformation: Anti Addition of Halogens

Molecular bromine was discovered in the 1820s, and its large-scale production began in 1858. By the mid-1890s, chemists had discovered the rapid addition reaction between unsaturated organic molecules and molecular bromine. Experimental methods for direct structural analysis were not yet known, but the defining feature of "addition" reactions was observed: The molecular formula of the product was a combination of the sum of the elements from the starting materials (e.g., $C_6H_{10} + Br_2$ gives $C_6H_{10}Br_2$). The reaction is characterized by an immediate color change, and so it became a popular chemical test for the presence of double and triple bonds in an unknown organic compound. Molecular bromine is a deep red-brown color. The addition reaction consumes the molecular bromine, and it generally does so rapidly, which means that when a red-brown bromine solution is dripped into a solution with an unknown organic compound, a fast discoloration of the bromine solution is potential evidence for the presence of a double or triple bond in the structure. Persistence of the bromine color is indicative of a saturated structure with which no addition can take place, or having reached the point in an addition reaction where no further addition can take place (Figure 0901).

Figure 0901

An early history of the bromination of organic compounds.

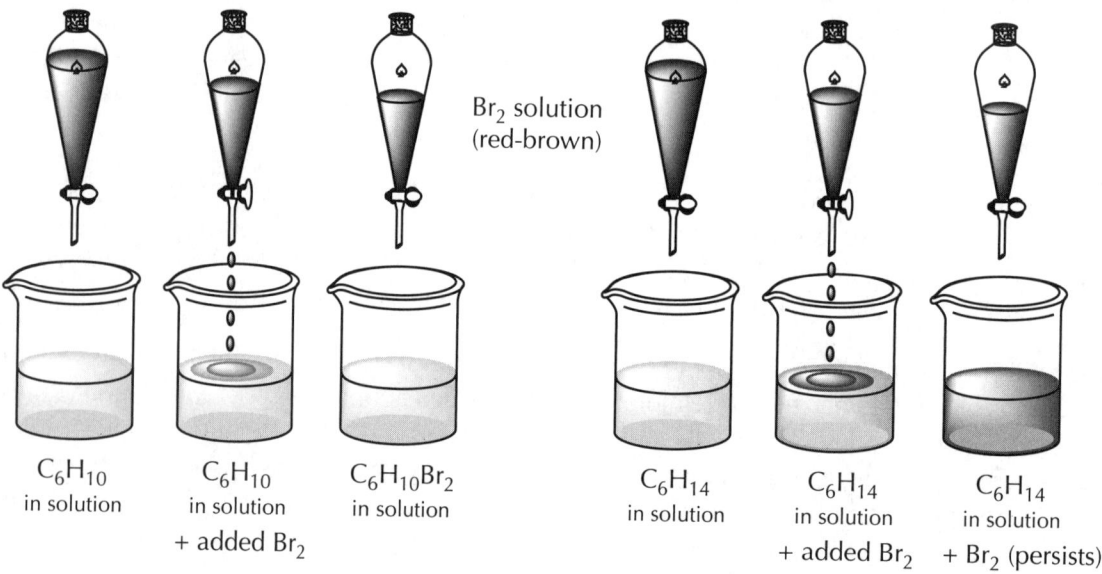

9.1 Halogenation and Other Reactions of the Halonium Ion Intermediate

By the 1920s, the bromination reaction had been determined to give the anti addition product and not the syn product. Comparable anti addition reactions were also observed for molecular chlorine and molecular iodine. At this point, historically, chemists still believed that all atoms were tetrahedral, and so the anti addition mechanism was a bit of a puzzle. It was easy to imagine a syn addition, because the edge between the two tetrahedral atoms could be imagined to split open. No one could really understand how the two carbon atoms could slide along their shared edge, opening up the anti bonding sites (Figure 0902).

Figure 0902

Anti stereochemistry of halogenation created a mechanistic puzzle.

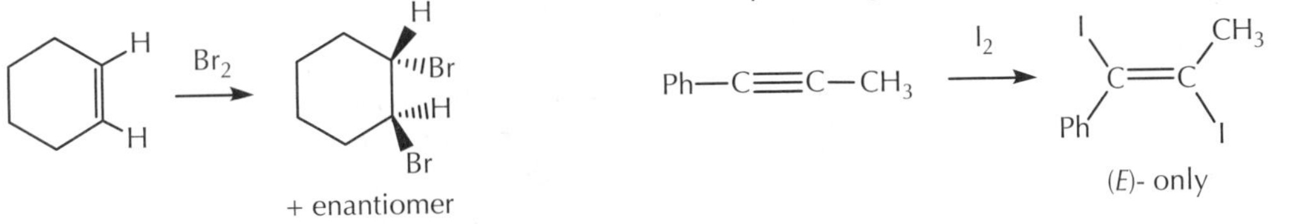

syn addition not observed
but a mechanism easily imagined in the 1920s: two tetrahedral atoms snap open and grab Cl_2

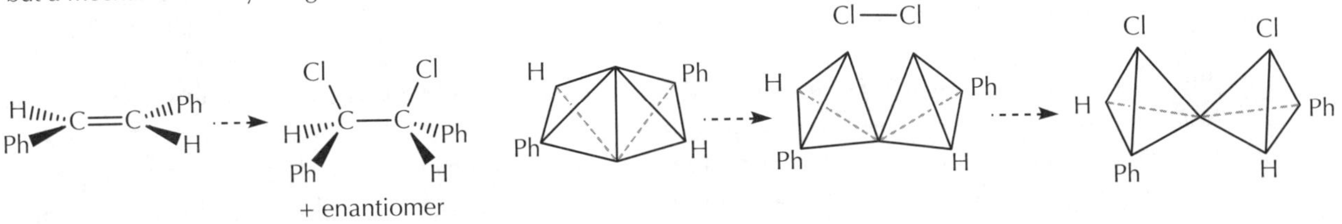

anti addition observed
but a mechanistic mystery in the 1920s: how to slide along an edge and grab Cl_2 from opposite sides?

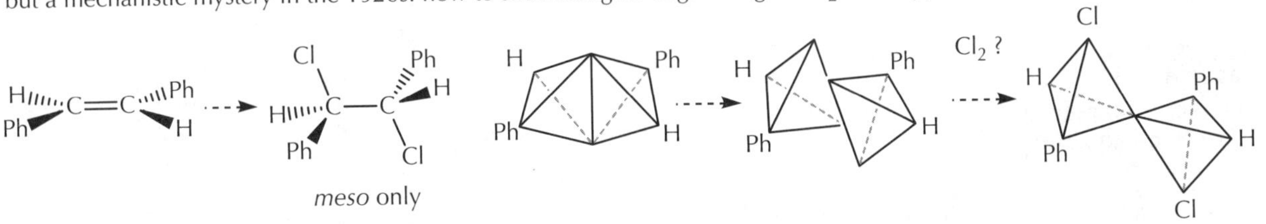

Ten years later, in the mid-1930s, the differences in sigma and pi bonding were being worked out, as was the general acceptance of the electron-based model of bonding. The curved arrow formalism was also introduced in the 1930s, and mechanisms for the electrophilic addition of Brønsted acids to pi bonds that you would recognize were among the first curved arrow mechanisms drawn (Figure 0903).

Figure 0903

Addition of HCl and Cl_2 to a double bond (1934).

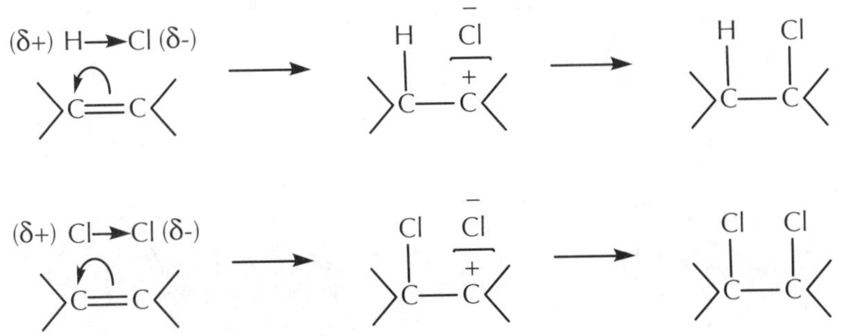

Adapted from p 272, C. K. Ingold, "Principles of an Electronic Theory of Organic Reactions" *Chem. Rev.* **1934**, *15*(2), 225.

In 1934, the mechanism of the reaction with molecular chlorine was proposed to be a direct analogy to the pi bond protonation mechanism using HCl. However, the open shell carbocation intermediate, which gives a mixture of syn and anti addition products in the HCl addition, did not explain the observation of the exclusive anti stereoselectivity in the chlorination reaction.

680 CHAPTER 9 Electrophilic Addition II: Halogenation, Oxidation, and Reduction Reactions

B. Stereoselectivity via the Halonium Ion Intermediate

In 1937, thanks in great part to the new electronic model of atoms and molecules, the halogenation reaction was proposed to occur through a cationic, three-membered ring intermediate (Figure 0904).

Figure 0904

General formation and reaction of halonium ion intermediates in halogenation reactions (1937).

"Another possible structure of the ion is one in which the positive charge is on the halogen."

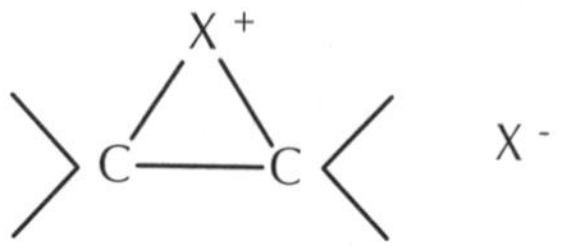

"The new atom will approach one of the carbon atoms from the side opposite to the X atom already present. A bond to this carbon will be formed while the bond from the original X to the carbon is broken, with simultaneous neutralization of the charge of the ion."

"This process will always lead to *trans* addition..."

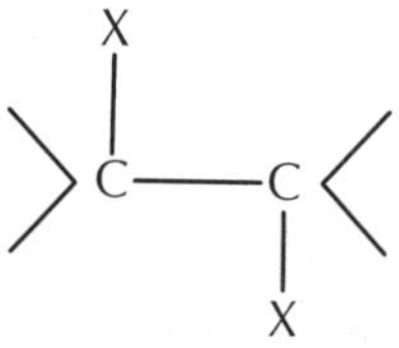

Adapted from p 948, I. Roberts & G. E. Kimball, "The Halogenation of Ethylenes" *J. Am. Chem. Soc.*, **1937**, *59*, 947.

The cationic intermediate, which is called a halonium ion (or, more specifically: chloronium, bromonium, or iodonium ions), solved a few problems. First, the reaction did not need a high energy, open shell carbocation intermediate, and so the relatively fast reaction rate could be explained. Second, the ring-opening reaction of the halonium ion has features of the S_N2 reaction, including a collision trajectory in which the released halide ion collides with one of the carbon atoms of the halonium ion to give inversion of configuration at that carbon. The ring opening reaction of the intermediate halonium ion explains the anti stereoselectivity because the syn addition pathway is inaccessible because of steric hindrance (Figure 0905).

Figure 0905

Halogenation reactions and the mechanisms to give anti addition products.

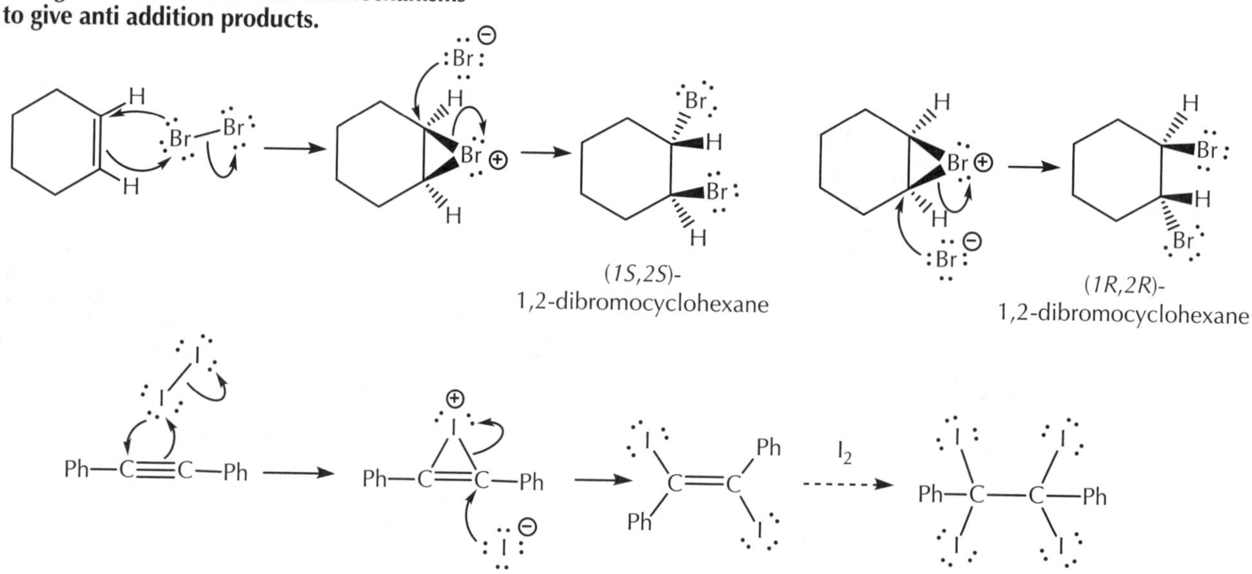

(1S,2S)-1,2-dibromocyclohexane

(1R,2R)-1,2-dibromocyclohexane

Mechanistically, there are a number of features to notice.

In the first step of the reaction mechanism, all of the bonding changes happen at the same time: Two new carbon-halogen bonds form and the halogen-halogen bond breaks, resulting in a positively-charged 3-membered ring. The stereochemical results are predictable. Because the two new carbon-halogen bonds are formed at the same time to give the 3-membered halonium ion ring: (a) the carbon-halogen

9.1 Halogenation and Other Reactions of the Halonium Ion Intermediate

bonds are formed with the syn stereochemistry, and (b) the stereochemical relationships of the groups on the double bond are preserved in the product.

In the second step of the reaction mechanism, the halide ion that was released in the first step comes around as a nucleophile and attacks one or the other of the carbon atoms in the halonium ion. There are two noteworthy characteristics of this second step: (a) the reaction at the carbon of the halonium ion ring is an S$_N$2 mechanism, so a new carbon-halogen bond forms as one of the carbon-halogen bonds in the ring breaks, with the positively charged halogen behaving as a leaving group (note that only one of the bonds breaks, on account of the collision occurring at only one of the carbon atoms); and (b) inversion of configuration at the ring carbon accounts for the overall anti addition outcome.

In all of these examples, the pi bonds are symmetrical, and so the two carbon atoms in the halonium ion are attacked at the same rate. In these symmetrical cases, there is no regioselectivity. This topic will be taken up in the next section. And as seen previously, in the addition reactions of Brønsted acids, triple bonds can add two equivalents of the electrophile.

A complete stereochemical analysis for a halogenation reaction depends on the specific structure of the alkene or alkyne. In Figure 0906, the stereochemical outcome from the bromination reactions of (Z)- and (E)-3-hexene are shown.

Figure 0906

Stereochemical analyses for the bromination reactions of (Z)- and (E)-3-hexene.

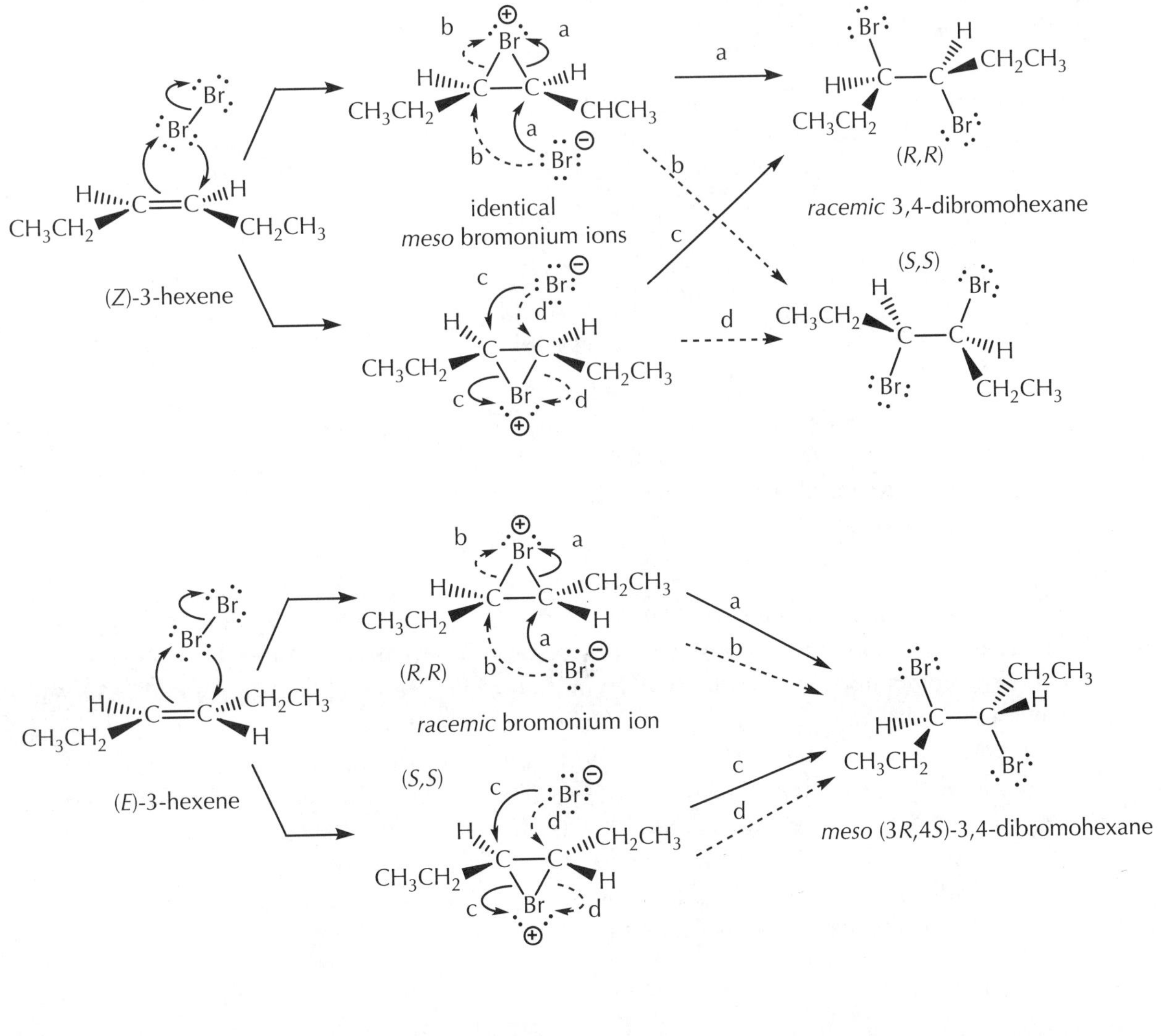

As in the earlier reactions of Brønsted acids, it is useful to start by visualizing the molecule with the alkene substituents perpendicular to the plane of the page (or writing surface), so that you can more easily draw the outcomes from reactions taking place both above and below the planar sp² atoms.

In the case of (Z)-3-hexene, the resulting bromonium ion is a meso compound, and the same intermediate is produced regardless of whether the bromonium ion forms from above or below the original reactant. The two carbon atoms of the bromonium ions are equivalent, and they are attacked at the same rate, resulting in a racemic mixture of the (3R, 4R)- and (3S, 4S)-3,4-dibromohexane enantiomers.

In contrast, a pair of enantiomeric bromonium ions results from adding bromine to the double bond of (E)-3-hexene. All four of the possible ring-opening reactions give the same *meso*-(3R, 4S)-3,4-dibromohexane isomer.

If the alkene is unsymmetrical, the procedure for the stereochemical analysis is the same, but some of the outcomes may be less predictable (Figure 0907).

Figure 0907

Stereochemical analysis for the bromination reactions of (E)-3-heptene.

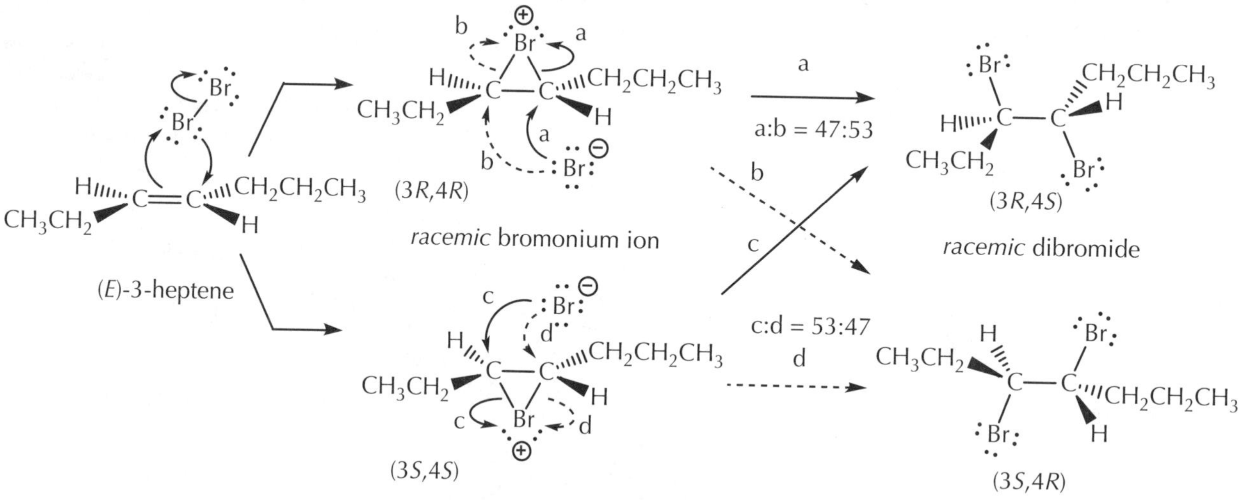

The bromination of (E)-3-heptene is analyzed here, to contrast with the bromination of (E)-3-hexene (from the previous example).

A pair of enantiomeric bromonium ions forms. Pathways a and c (solid lines in Figure 0907) give the same anti relationship (left-up, right-down) and the same product, (3R, 4S)-3,4-dibromoheptane. Pathways b and d (dashed lines) give the alternative anti relationship (left-down, right-up), which is the enantiomer, (3S, 4R)-3,4-dibromoheptane.

The starting material is achiral, and so the product must be formed as an achiral racemic mixture. Understanding that both of the bromonium ions are forming is key to understanding how a 50:50 mixture results, because the reaction rates at the two reactive carbons of both these bromonium ions are not equal. In the top bromonium ion in Figure 0907, pathways a and b are slightly different because in one case (pathway a), the bromide ion collides with a carbon atom that bears a propyl group, while in the other case (pathway b), the bromide ion collides with a carbon atom that bears an ethyl group. The reaction rates are close, but the experimental a:b ratio is 47:53 (propyl:ethyl), which may be due to a slight difference in steric hindrance (the propyl group is a little bulkier than the ethyl group). In the lower bromonium ion, the propyl:ethyl ratio (pathway d:c) is also 47:53. The sum of pathways a and c (47+53), which give one of the enantiomers, is the same as sum of pathways b and d (53+43), and so the end result is a racemic mixture, where the two enantiomers are formed in equal amounts.

9.1 Halogenation and Other Reactions of the Halonium Ion Intermediate

You may encounter other text or online resources in which the curved arrow mechanism for forming the halonium ion from the pi bond is shown with 2 arrows rather than 3 arrows. The logic of the 2-arrow mechanism is that the pi bond is a single electron pair and should only call for one arrow; the logic of the 3-arrow mechanism is that it accounts for the three changes taking place (2 bonds form and 1 bond breaks). In this text, we will always use the 3-arrow version because it better represents the changes in connectivity that happen in forming halonium ions (Figure 0908).

Figure 0908

Comparison of the 2- and 3-arrow versions of the mechanism for forming halonium ions.

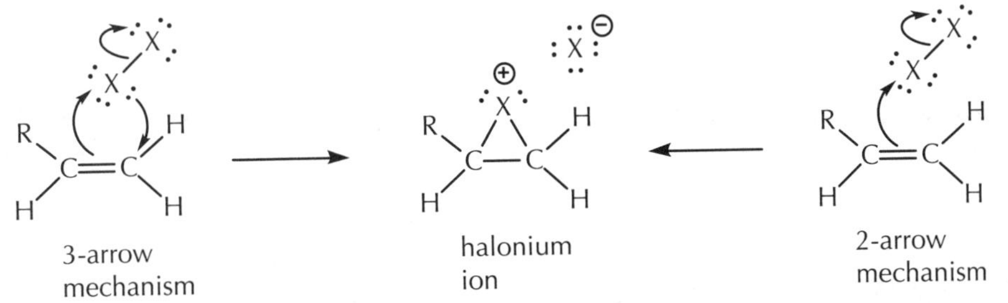

3-arrow mechanism → halonium ion ← 2-arrow mechanism

Historically, halonium ions were first proposed as intermediates in 1937, to account for the anti stereoselectivity in these addition reactions. No direct structural evidence for halonium ions existed until the early 1990s, when examples of chloronium, bromonium, and iodonium ions were isolated and their molecular structures determined by x-ray crystallography.

C. Halohydrins and Related Regioselective Reactions of the Halonium Ion Intermediate

In the addition reaction of molecular halogens, the formation of a halonium ion is accompanied by the formation of a halide ion. In the second mechanistic step, the resulting halide ion opens the halonium ion ring in a nucleophilic attack via an S_N2 mechanism.

Nucleophiles other than the halide ion, which is generated during the formation of the halonium ion, are also able to attack and open the halonium ion ring. If reaction conditions are meant to imply a nucleophile other than the halide, it is usually shown along with the molecular halogen in the line of reagents. For example, bromination reactions can be carried out with an excess of an added anionic reagent, or with water or alcohol, also used as the solvent (Figure 0909).

Figure 0909

A halonium ion can be opened by nucleophiles other than the original halide.

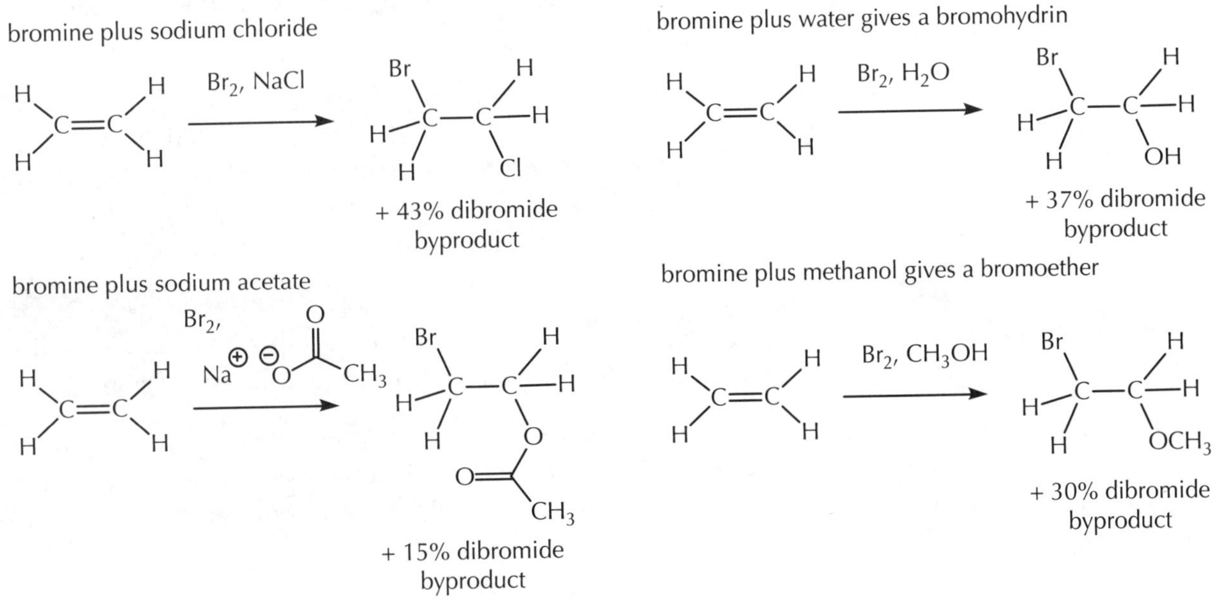

bromine plus sodium chloride: Br_2, NaCl
+ 43% dibromide byproduct

bromine plus water gives a bromohydrin: Br_2, H_2O
+ 37% dibromide byproduct

bromine plus sodium acetate: Br_2, Na⁺ ⁻O-C(=O)CH₃
+ 15% dibromide byproduct

bromine plus methanol gives a bromoether: Br_2, CH_3OH
+ 30% dibromide byproduct

When excess chloride ion was added as the nucleophile (top left case in Figure 0909), the product is a mixed alkyl halide. Other good anionic S_N2 nucleophiles, such as the conjugate base of an organic acid (e.g., sodium acetate), can also be used. If water is the nucleophile, used in excess as the solvent, the product has a hydroxyl group and a halogen atom on adjacent carbon atoms. Such compounds are known as halohydrins. And when an alcohol is used as the solvent, the corresponding product is known as a haloether. In all of these reactions, bromide ion, released when the bromonium ion forms, is present in the reaction mixture and the usual dibromide bromination product is formed as an undesired byproduct.

In all of these cases, the mechanisms are comparable (Figure 0910).

Figure 0910

Mechanisms for halonium ion ring-opening by nucleophiles other than the original halide.

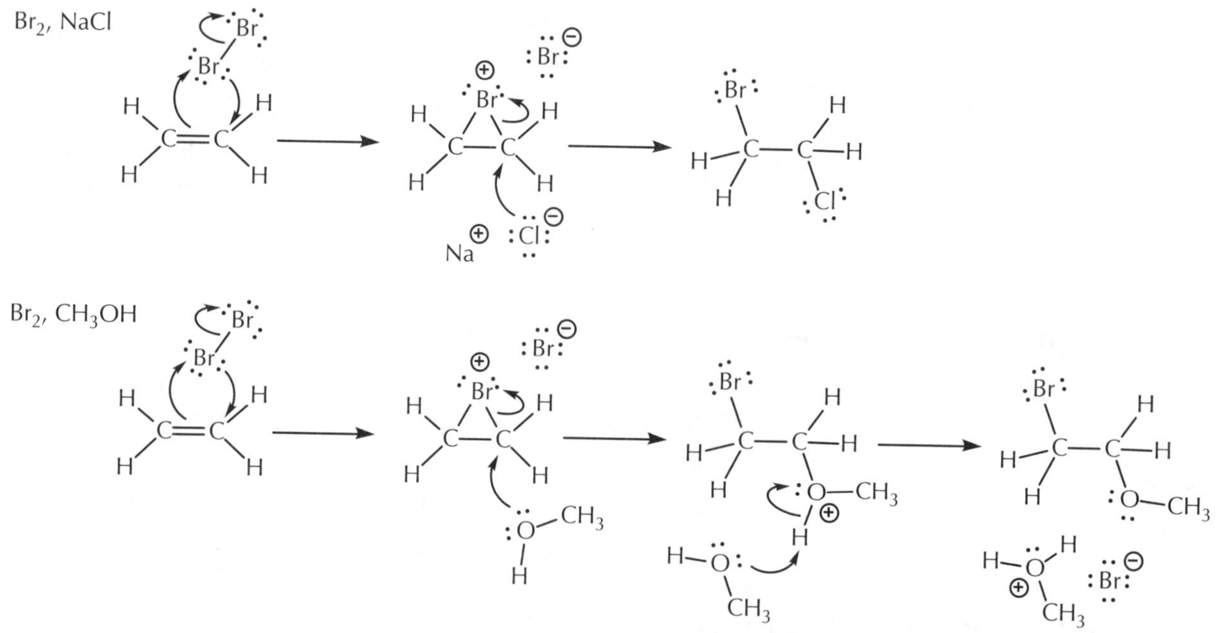

First, the intermediate bromonium ion forms, then it is opened by a nucleophile. If the nucleophile is an anion, the one-step opening completes the reaction. If the nucleophile is an uncharged molecule, such as water or methanol, then a deprotonation step needs to follow the bromonium ion ring-opening reaction, giving the uncharged organic product (FIGURE 0910).

In cases where obvious mixtures of products might result, a writer will often do something to signal to the reader which outcome is being implied. One way to signal to a reader that the mixed addition product is desired, rather than the dibromide, is by writing the reagents above the arrow, separated by a comma. The specific experimental conditions to optimize the yield of the desired product might take a long time to work out, and these conditions are rarely-to-never included in the depiction of the reaction—so some sort of visual clue about the intended reaction is useful to have.

In the mixed addition reactions, the regioselectivity of the ring opening becomes a factor to consider when the pi bond is unsymmetrical. Some examples of experimental results are shown in Figure 0911.

The unsymmetrical pi bond gives an unsymmetrical halonium ion, and the rate of ring opening at one of the carbons of the halonium ion is faster than the other. The experimental observations are consistent with the faster rate of nucleophilic attack into the halonium ion at the carbon atom that is better capable of stabilizing positive charge.

9.1 Halogenation and Other Reactions of the Halonium Ion Intermediate

Figure 0911

Examples of regioselectivity in halonium ion ring-opening reactions.

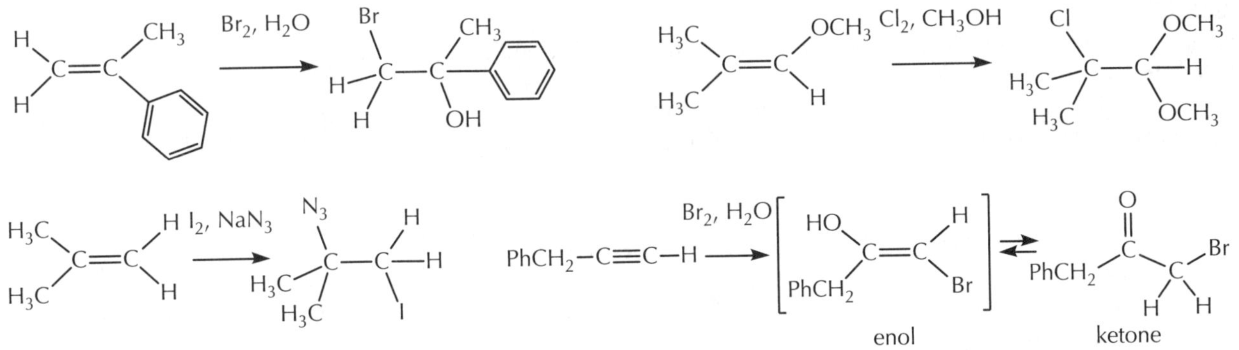

Chemists believe that these halonium ion rings are structurally unsymmetrical. As a way to stabilize the positively charged halogen atom in the three-membered ring, the carbon-halogen bond associated with the carbon atom that can better stabilize positive charge is longer and weaker (Figure 0912).

Figure 0912

Unsymmetrical halonium ions are used to explain regioselectivity.

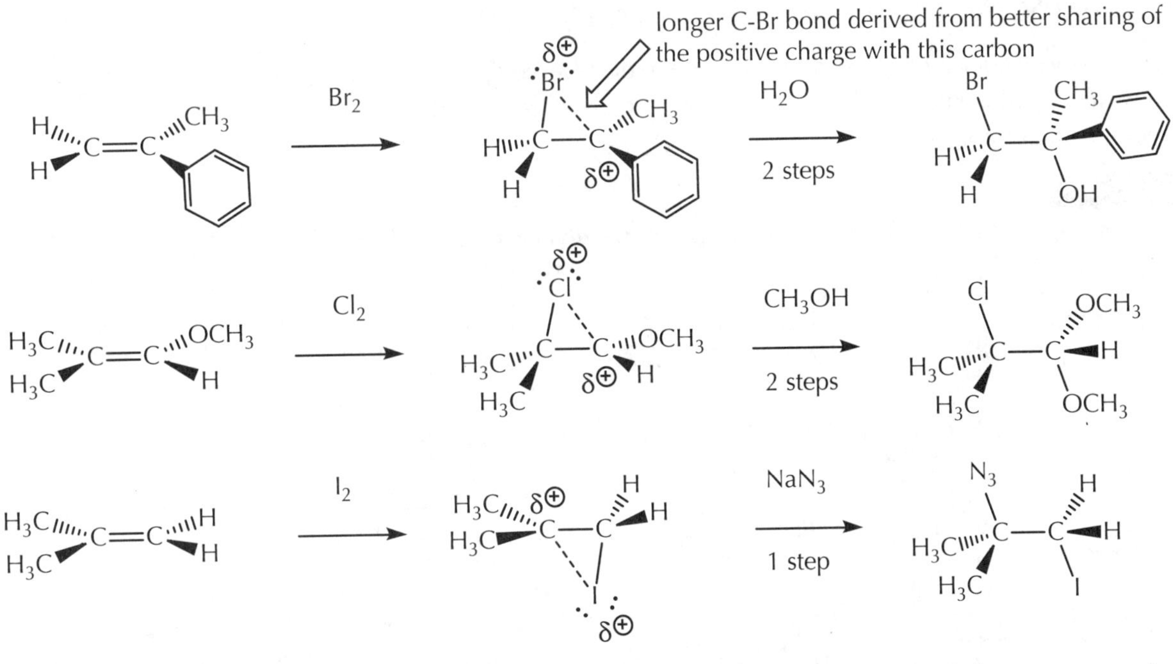

The longer bond derives from better delocalization of the positive charge from the halogen atom to the carbon atom that can better stabilize that charge. The unsymmetrical structure of the halonium ion is therefore more electrophilic at the carbon atom carrying more of the positive charge. The longer bond also changes the geometry around that more electrophilic carbon atom because its substituents can be more flattened out, making the carbon atom more accessible for attack even though it is generally more highly substituted.

The unsymmetrical halonium ions are exhibiting features of both S_N1 reactions (reaction is faster at the carbon atom that can better stabilize positive charge) and S_N2 reactions (the ring opening reaction occurs with inversion of configuration). The overall regioselectivity is also consistent with the Markonikov Rule, in that the electrophilic atom (the halogen, in this case) ends up on the end of the double bond that is least capable of stabilizing positive charge. These parallels are illustrated in Figure 0913.

Figure 0913

Parallels in halonium ion chemistry: features that derive from S_N1 and S_N2 pathways, and from Markovnikov's Rule.

Markovnikov hydrobromination (addition of HBr)

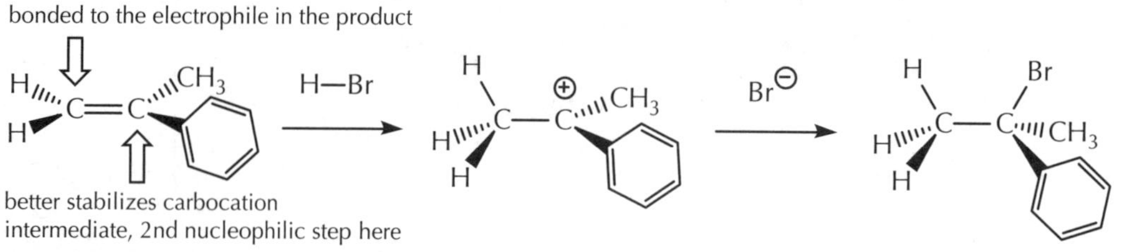
bonded to the electrophile in the product
H—Br
Br⊖
better stabilizes carbocation intermediate, 2nd nucleophilic step here

Bromohydrin formation (addition of Br_2/H_2O)

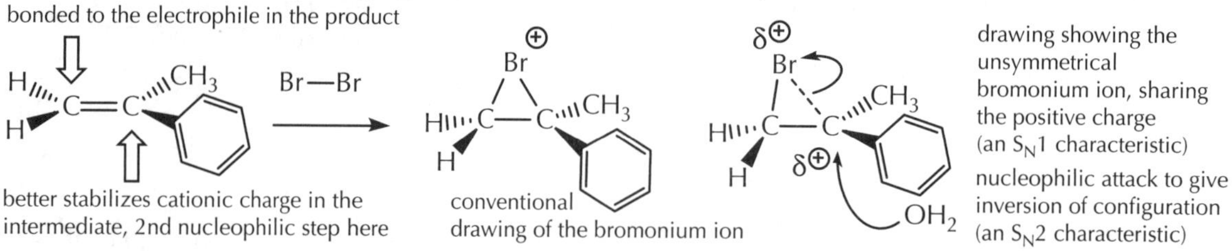
bonded to the electrophile in the product
Br—Br
better stabilizes cationic charge in the intermediate, 2nd nucleophilic step here
conventional drawing of the bromonium ion
drawing showing the unsymmetrical bromonium ion, sharing the positive charge (an SN1 characteristic)
nucleophilic attack to give inversion of configuration (an SN2 characteristic)

Over the years, a variety of reagents have been discovered that can carry out the halonium ion reactions in which an added nucleophile opens the ring to give a mixed addition product. Some of the most common examples are shown in Figure 0914.

Figure 0914

Common reagents that form halonium ions which react with added nucleophiles.

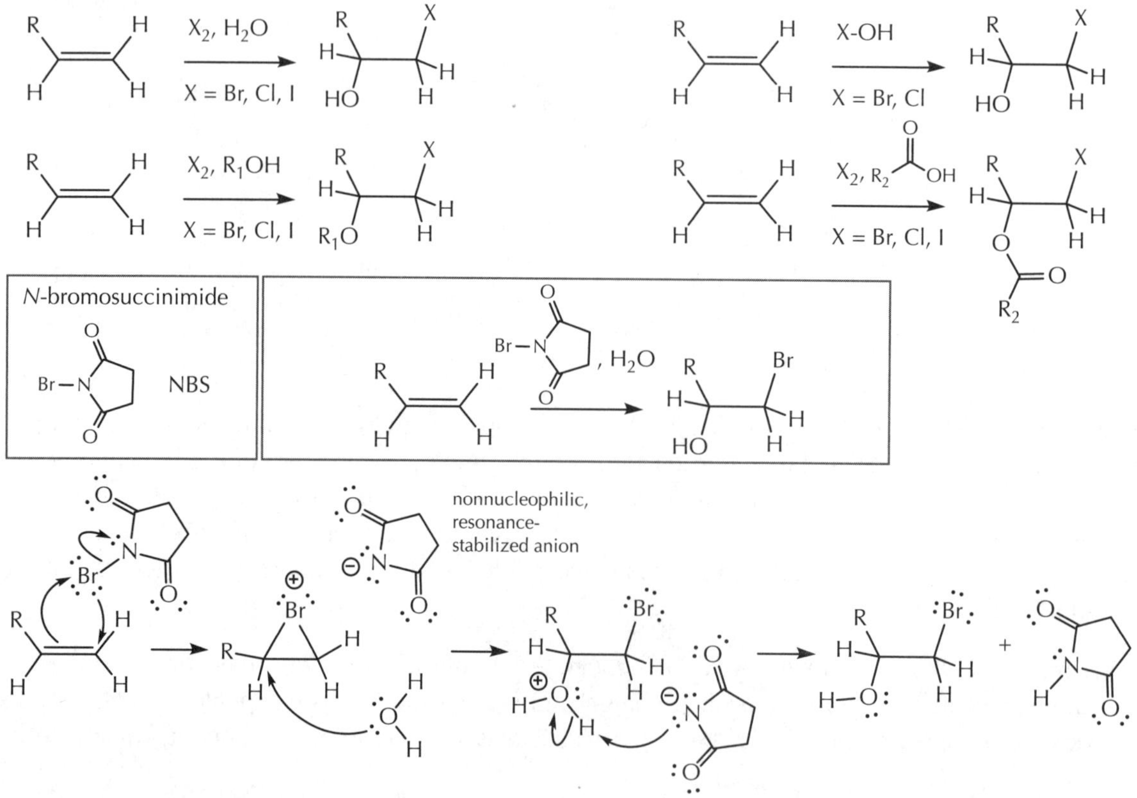
X2, H2O
X = Br, Cl, I
X-OH
X = Br, Cl
X2, R1OH
X = Br, Cl, I
X2, R2COOH
X = Br, Cl, I
N-bromosuccinimide
NBS
, H2O
nonnucleophilic, resonance-stabilized anion

9.1 Halogenation and Other Reactions of the Halonium Ion Intermediate

Molecular chlorine, bromine and iodine, in combination with water, alcohols, and carboxylic acids (or their conjugate bases), all of which give the anti addition of a halogen atom and the oxygen atom nucleophile.

Hypochlorous acid (HOCl) and hypobromous acid (HOBr) also give their corresponding chlorohydrins and bromohydrins, respectively, via a direct analogy to the reaction of molecular halogens. The X-OH reagents, as their X-X counterparts do, give the formation of a halonium ion while releasing hydroxide ion. The hydroxide ion opens the halonium ion, giving the same overall reaction products as the "X_2, H_2O" reaction conditions.

N-bromosuccinimide (NBS) is an experimentally stable and convenient source of electrophilic bromine atoms that form bromonium ions and can be combined with nucleophilic reagents, including water, alcohols, and carboxylic acids. These X-N reagents form halonium ions, releasing a nonnucleophilic, resonance-stabilized anion that often participates in the reaction as a weak Brønsted base.

D. Intramolecular and Other Mixed-Atom Reactions

Intramolecular (within the same molecule) reactions are generally more favorable than their intermolecular (between different molecules) counterparts. If the reacting partners are held close to one another within the same molecule, the rate of reaction between them is faster than if they separate reagents that need to find one another in solution. Intramolecular reactions that form unstrained ring sizes, that is, five- and six-membered rings, are particularly fast.

Every intermolecular reaction that you have seen, so far, can also be performed intramolecularly if the reactive partners are compatible with being in the same molecule and they can physically reach one another. A few examples of pairwise comparisons between familiar intermolecular reactions and their intramolecular counterparts are shown in Figure 0915.

Figure 0915

Familiar intermolecular reactions and their intramolecular counterparts.

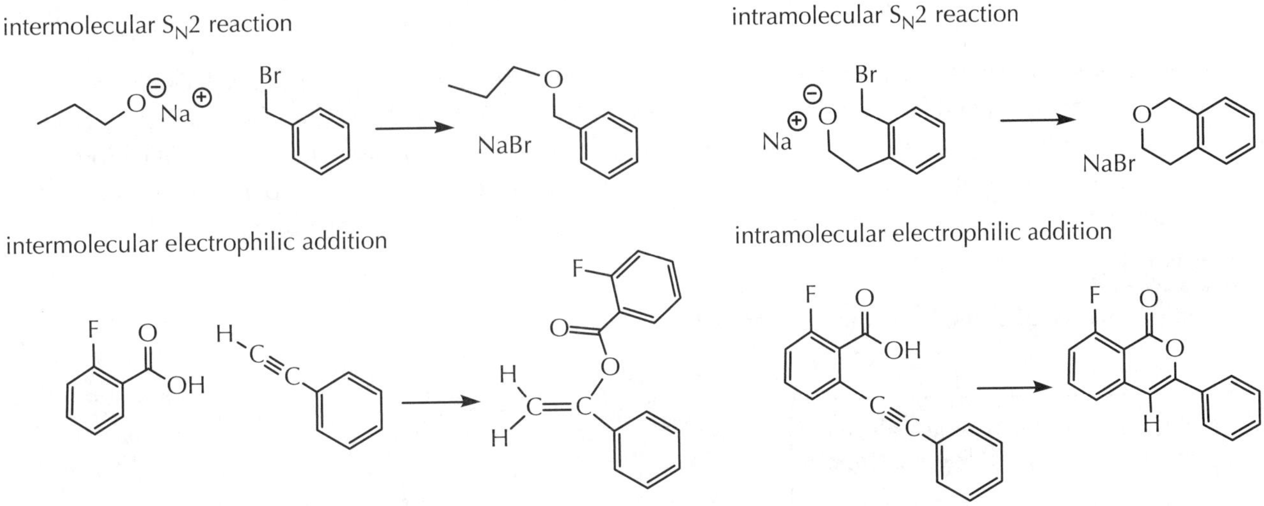

Intramolecular versions of capturing a halonium ion intermediate are well known. A nucleophile that can attack a halonium ion can be a part of the same molecule as the halonium ion. As in the intermolecular reactions, the nucleophiles that attack the halonium ion can be anions, resulting in a simple, one-step mechanism for the opening reaction, or the nucleophile can be neutral, in which case a deprotonation step is needed after the opening takes place. Examples of both types are illustrated in Figure 0916.

The most common molecular halogens are familiar to you. They are the homo-diatomic compounds you know as their elemental forms (F_2, Cl_2, Br_2, I_2).

Figure 0916

Halonium ion opening: intramolecular counterparts of the intermolecular reactions.

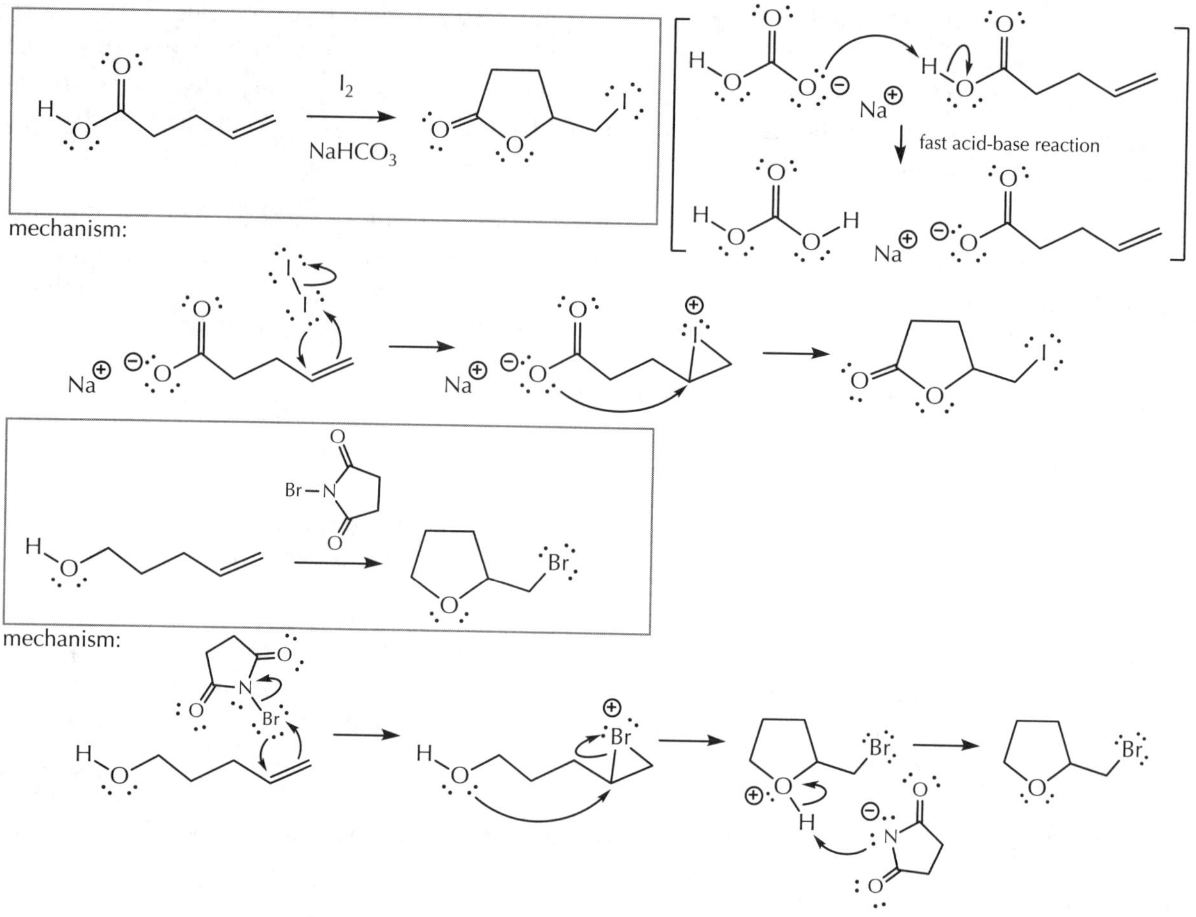

Nine different heterodiatomic compounds, called the interhalogens, are known. These molecules are ones in which there are two different halogen atoms, such as BrF and ICl. Of these nine, three of them are observed to participate in addition reactions with organic compounds: chlorine monofluoride (ClF), bromine monochloride (BrCl), and iodine monochloride (ICl). The others are too reactive and/or unstable to be used (Figure 0917).

Figure 0917

Addition reactions of diatomic interhalogen reagents.

Nine known diatomic interhalogens (three are used in addition reactions)

F	Cl	Br	I	At	
	Cl-F	Br-F	I-F		F
		Br-Cl	I-Cl	At-Cl	Cl
			I-Br	At-Br	Br
				At-I	I
					At

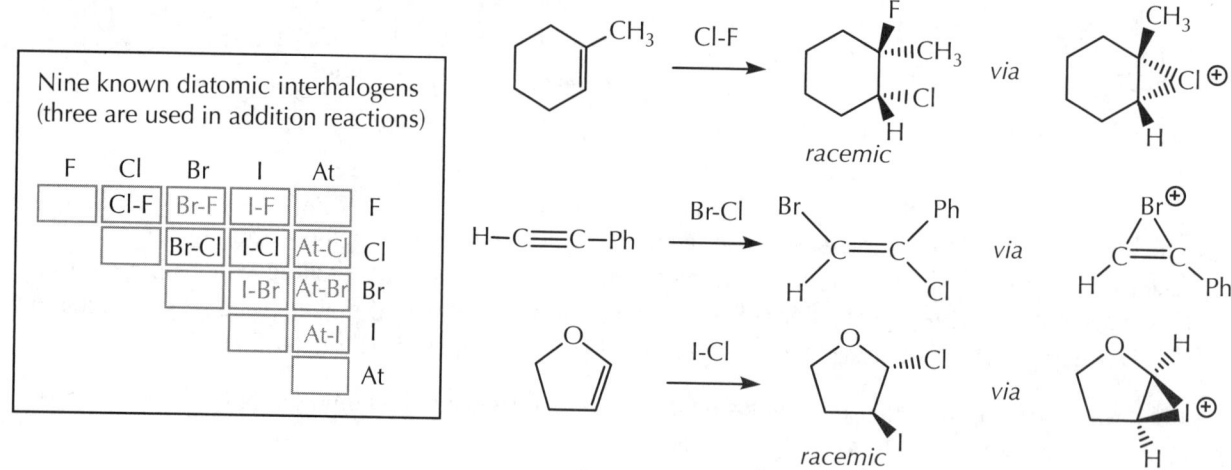

9.1 Halogenation and Other Reactions of the Halonium Ion Intermediate

When an interhalogen compound reacts with a pi bond, the halonium ion forms using the most electrophilic of the two halogen atoms (i.e., the less electronegative atom). Thus, chlorine monofluoride (ClF) reacts with pi bonds to give a chloronium ion that is then captured by the released fluoride ion; bromine monochloride (BrCl) gives a bromonium ion, and iodine monochloride (ICl) gives an iodonium ion.

The regioselectivity in the addition reactions of diatomic interhalogens follows the same Markovnikov-like generalization discussed throughout this section: the halonium ion intermediate is structurally unsymmetrical and the positive charge is shared between the halogen atom and the carbon atom that can better stabilize positive charge. Finally, the reactivity of halogen atoms, to give halonium ion intermediates, is not unique and it is mirrored by other atoms (Figure 0918).

Figure 0918

Addition reactions with benzenesulfenyl chloride and benzeneselenenyl chloride.

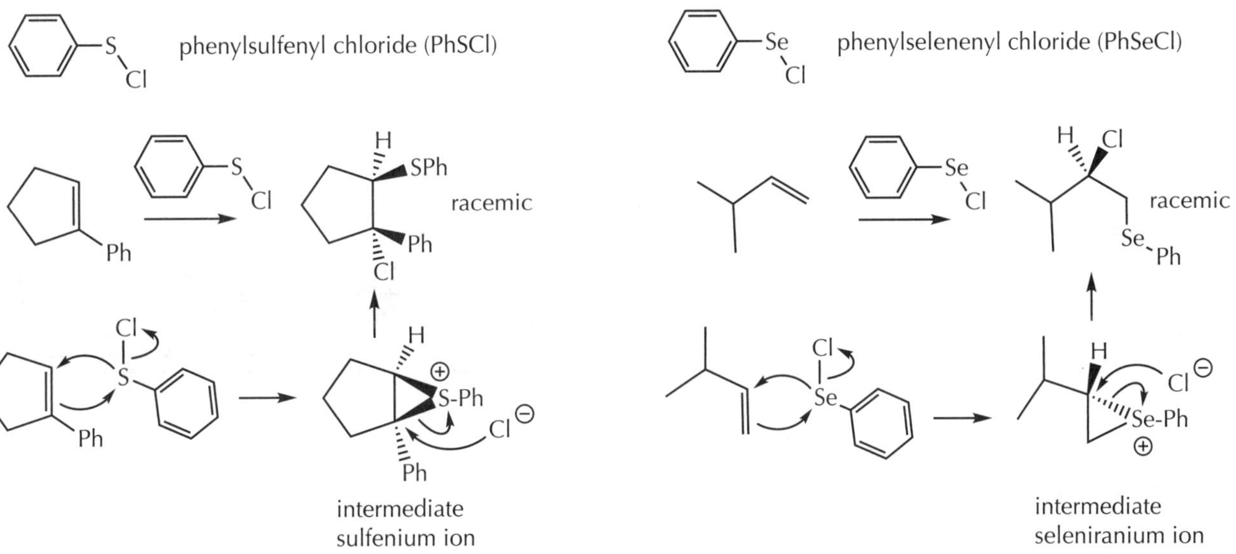

In organic chemistry, two atoms that are commonly observed to give intermediates analogous to halonium ions are sulfur and selenium atoms. Reagents of the general form RSCl and RSeCl give addition reaction products that are consistent with cationic three-membered rings involving sulfur and selenium, respectively, followed by the nucleophilic opening of the positively charged ring according to the Markovnikov-like regioselectivity.

As with the diatomic interhalogen compounds, chlorine is more electronegative than either sulfur or selenium, making sulfur and selenium the electrophilic atoms. There is no fundamental mechanistic difference in how these reagents react with pi bonds to give anti addition products compared with any of the other examples in this chapter.

The reactions and mechanisms for two frequently used reagents, PhSCl (benzenesulfenyl chloride) and PhSeCl (benzeneselenenyl chloride), both begin with the 3-arrow mechanism to give a cationic, three-membered ring intermediate. As with the halonium ions, the nucleophile that opens the positively charged ring can be the released chloride ion, other anions, or neutral nucleophiles. These ring opening reactions can also be carried out intermolecularly or intramolecularly.

An important feature about the role of analogies in organic chemistry is compellingly exemplified in this first section of Chapter 9: Understanding the details of the anti addition reaction of molecular bromine serves as a powerful mechanistic concept that allows for the understanding of, and prediction for, the reactions of many other chemical reagents.

690 CHAPTER 9 Electrophilic Addition II: Halogenation, Oxidation, and Reduction Reactions

PRACTICE QUESTIONS

9.01 Complete the following equations. Provide the structures of stereoisomeric products, if anticipated, and identify the stereoisomeric relationship (enantiomers, diastereomers) or stereochemical identity (an achiral molecule with no stereocenters, an achiral *meso* isomer)

(a)
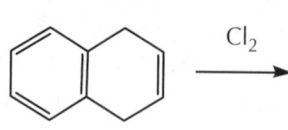

stereochemical relationship or identity:

(b)
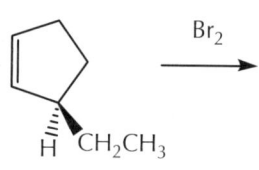

stereochemical relationship or identity:

(c)
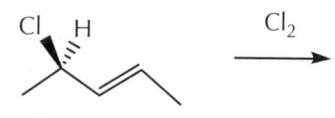

stereochemical relationship or identity:

(d)

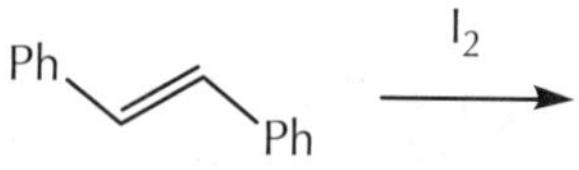

Ph⧸═⧹Ph $\xrightarrow{I_2}$

stereochemical relationship or identity:

(e)

$Ph-C{\equiv}C-CH_3 \xrightarrow[]{\text{1 equivalent} \atop Br_2}$

stereochemical relationship or identity:

9.02 Provide a complete stereochemical analysis for the following halogenation reaction mechanisms: (1) Draw the halonium ions resulting from each face of the pi bond. (2) Describe the resulting halonium ions as identical, enantiomeric, or diastereomeric. (3) Draw all of the products that can result from the halonium ions and demonstrate that they completely explain the observed outcome.

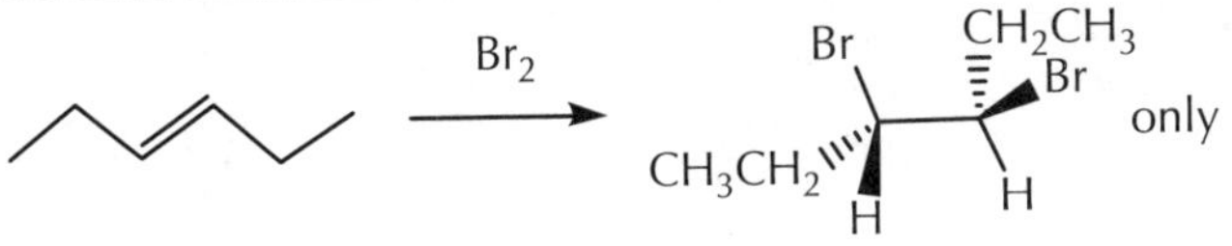

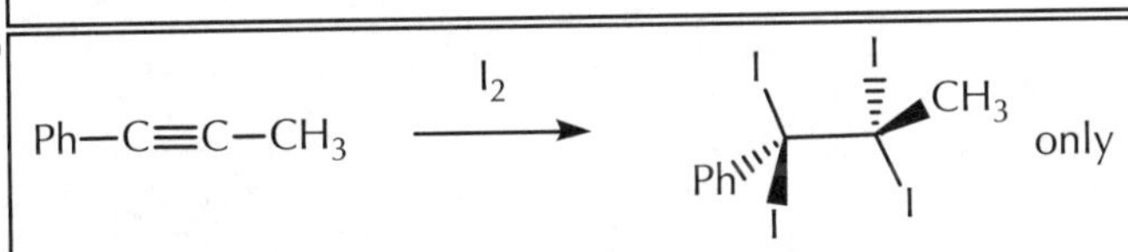

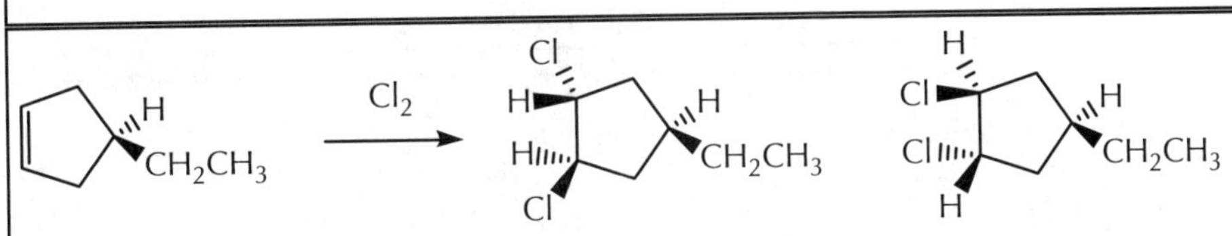

9.03 Complete the following equations. Predict the major regioisomer and draw the complete stereochemical outcome. If there are two stereoisomers, also describe their relationship.

(a) 1-phenylcyclohex-1-ene + BrCl →

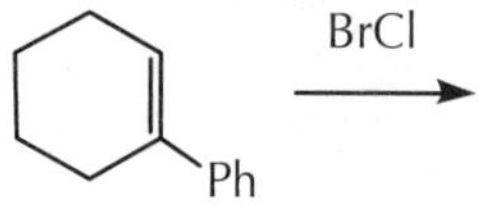

(b) 3-(2-phenylethyl)-2,5-dihydrofuran + Cl$_2$, CH$_3$OH →

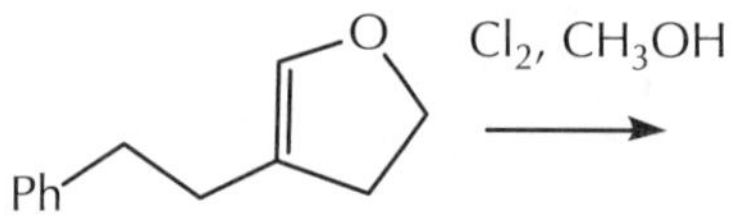

(c) 2-methyl-5-methylhex-2-ene (2,5-dimethylhex-2-ene) + HOBr →

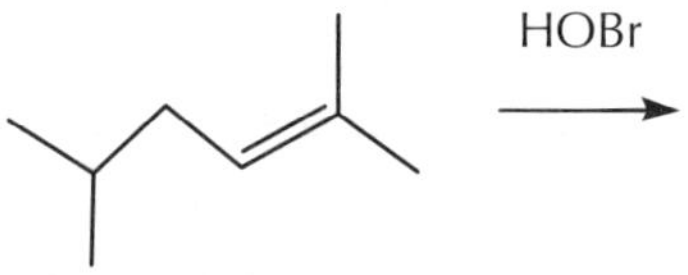

(d) H–C≡C–Ph + N-bromosuccinimide (NBS), H$_2$O →

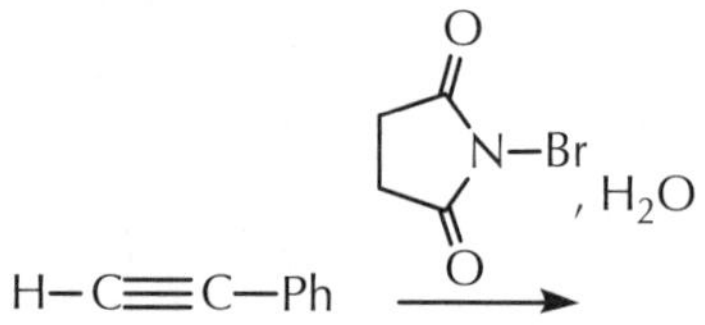

(e) 6-methoxy-3,4-dihydronaphthalene (H$_3$CO-substituted) + PhSeCl →

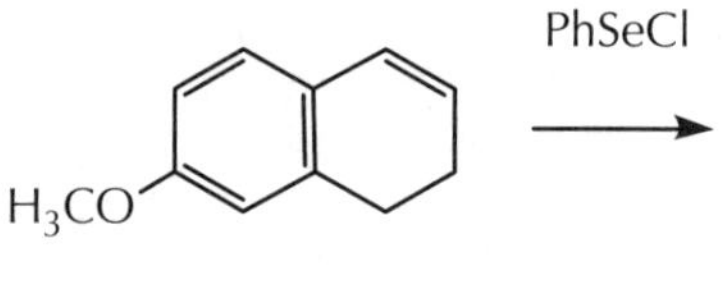

(f) 2,3-diphenylprop-1-ene (CH$_2$=C(CH$_2$Ph)(Ph)) + Br$_2$, NaCl →

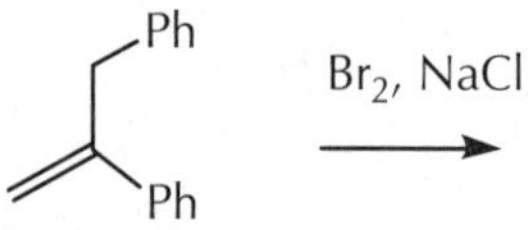

(g) 1-methoxy-5-phenylcyclopent-1-ene (with H and Ph shown stereochemically at C5) + NBS, CH$_3$OH →

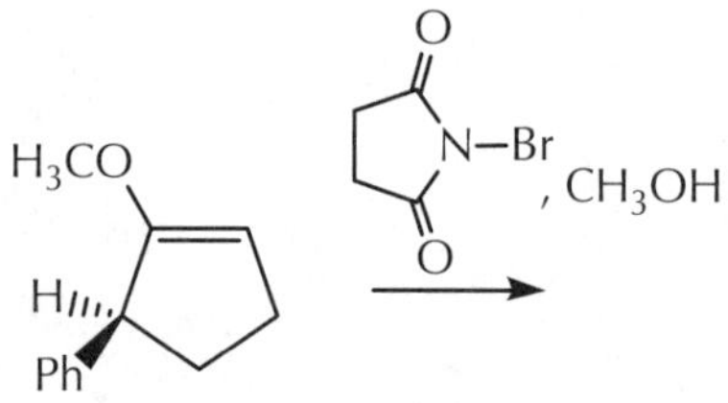

9.1 Halogenation and Other Reactions of the Halonium Ion Intermediate

9.04 Provide the complete, curved arrow mechanism for the following reactions.

(a)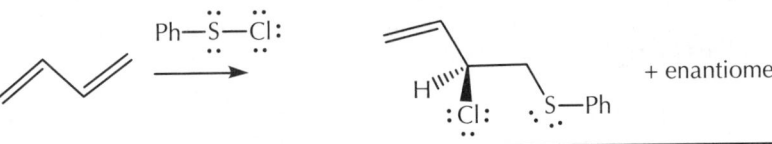

(b)

9.05 Complete the following equations.

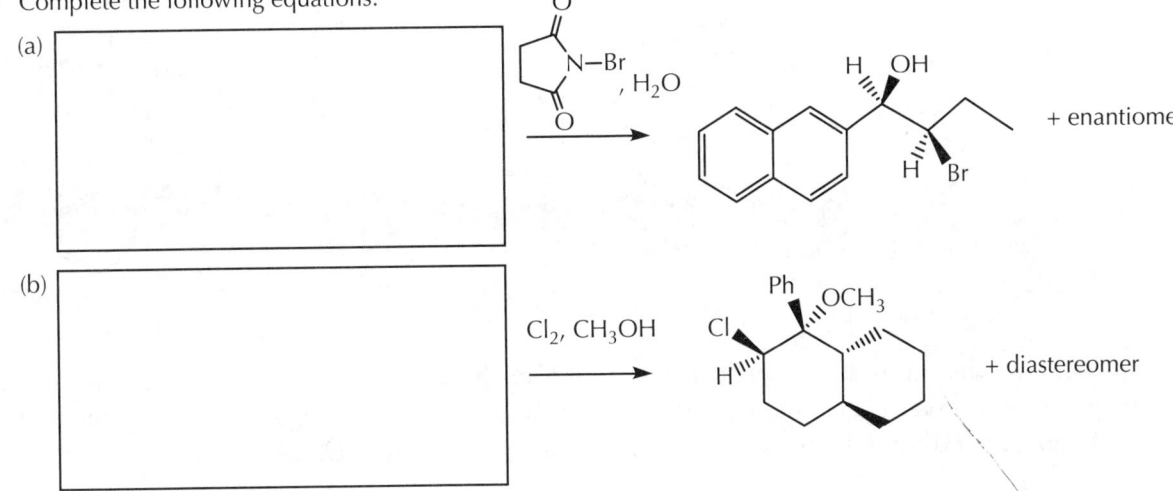

9.2 Oxidation Reactions

A. Peroxyacid Reactions

In chemistry, the terms oxidation and reduction date back to the 1860s. As our theories of chemistry have evolved, so have the definitions of these terms. Among the simplest meanings for oxidation is a reaction in which the percentage of oxygen in an organic molecule increases, while reduction, a reaction that achieves the opposite, decreases the percentage of oxygen.

In 1909, Nikolai Alexandrovich Prilezhaev reported the following oxidation reaction of alkenes: Double bonds plus peroxybenzoic acid yield epoxides, the common name for oxirane, which is a three-membered ring containing an oxygen atom (Figure 0919), and the overall transformation is called an epoxidation reaction.

Figure 0919

The Prilezhaev reaction, 1909: epoxidation of double bonds.

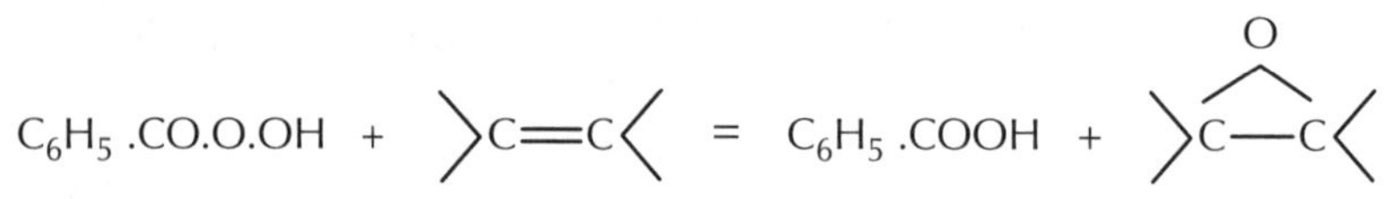

$$C_6H_5 \cdot CO.O.OH + \;\; \rangle C=C\langle \;\; = \;\; C_6H_5 \cdot COOH + \;\; \rangle\overset{O}{\overset{\triangle}{C-C}}\langle$$

Adapted from: Prileschajew, Nikolaj (1909). *Berichte der deutschen chemischen Gesellschaft* (in German). 42 (4): 4811–4815

A great deal is known about the epoxidation reaction.

The reaction is general for organic peroxy acids. Mainly for experimental convenience, one of three peroxy acids is typically used: peroxyformic acid, peroxyacetic acid, or *meta*-chloroperoxybenzoic acid. The structures for these three molecules, and their abbreviations, are shown in Figure 0920.

Figure 0920

Three common peroxy acids used for epoxidation reactions.

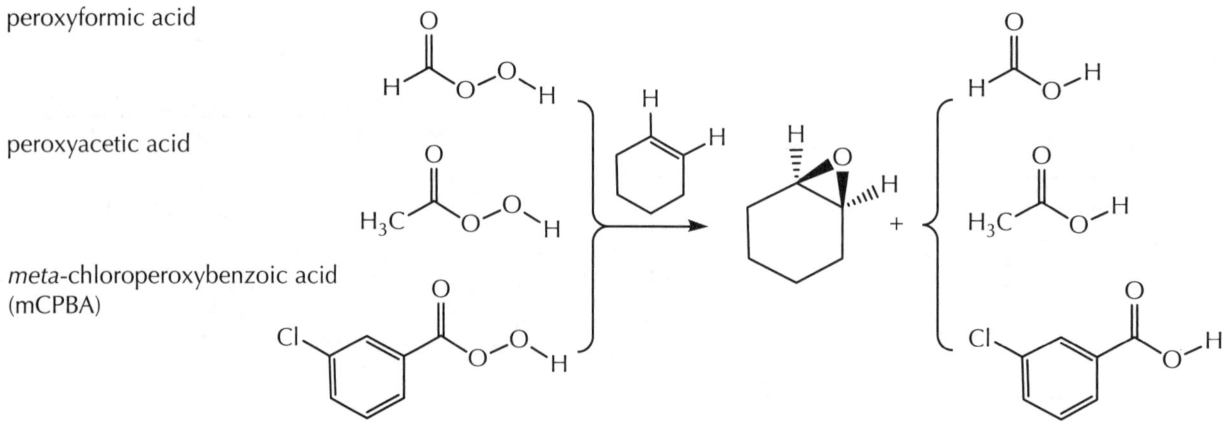

peroxyformic acid

peroxyacetic acid

meta-chloroperoxybenzoic acid (mCPBA)

The reaction mechanism is only one step, and so all of the bonding changes happen at the same time. In general, chemists prefer the 5-curved-arrow version of the mechanism (Figure 0921) over the 4-arrow version, but both versions formally account for the observed bonding changes.

9.2 Oxidation Reactions

Figure 0921

Curved arrow mechanism for the epoxidation reaction.

5-arrow version (preferred)

4-arrow version

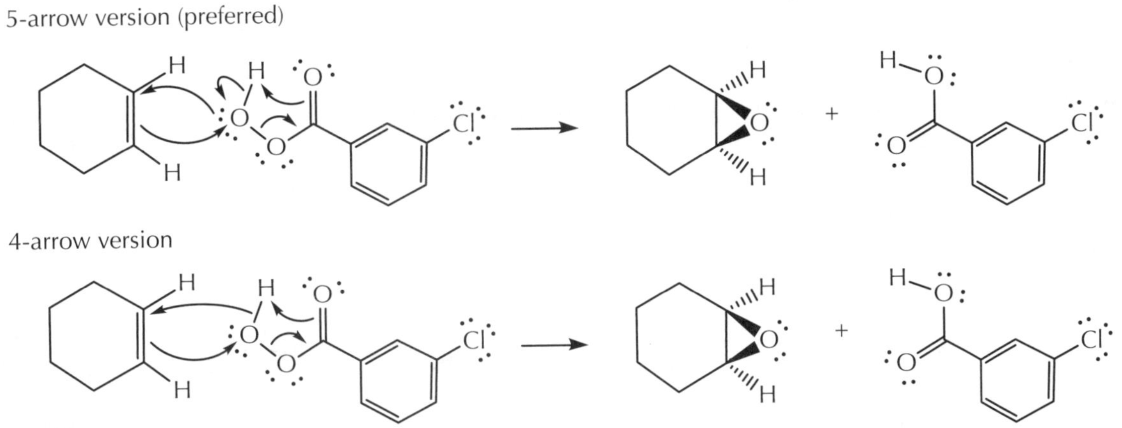

As in the first step of the halogenation reactions (see Figure 0905), the stereochemical results are predictable. In the epoxidation reaction, the two new carbon-oxygen bonds are formed at the same time to form the small, 3-membered ring (Figure 0922).

Figure 0922

Stereochemical features of the epoxidation reaction.

identical structures of the single *meso* compound

enantiomers (1:1 racemic mixture)

diastereomers (unequal amounts)

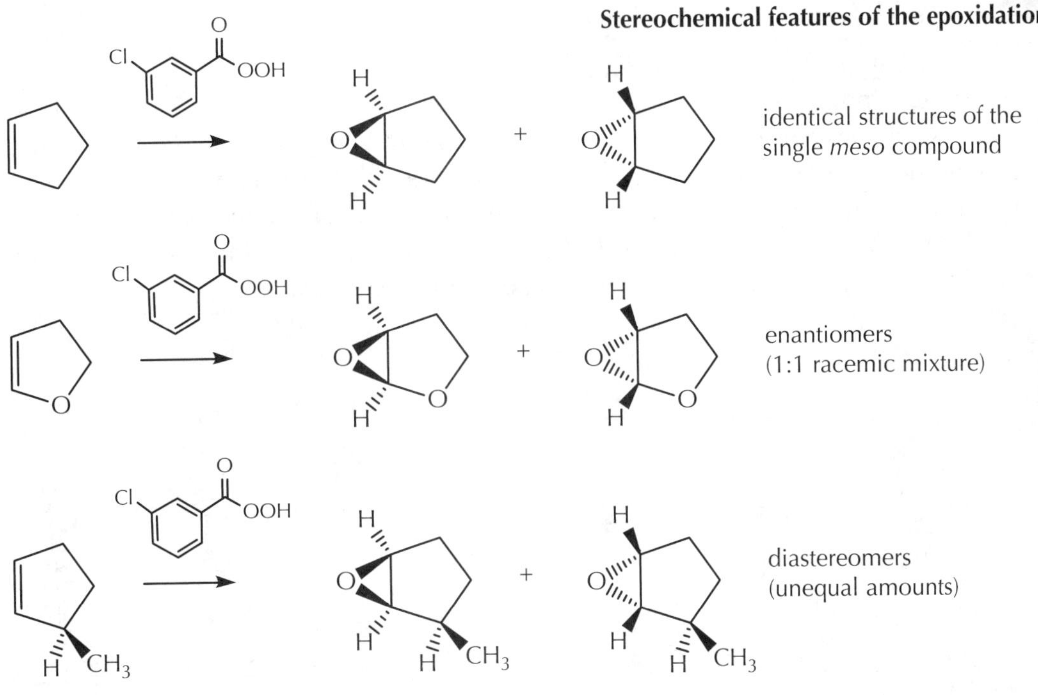

Unlike the positively charge halonium ions, the uncharged epoxide ring is stable under the reaction conditions. The same two important structural features are observed in this reaction as in the formation of the halonium ions: (a) the carbon-oxygen bonds add with the syn stereochemistry; and (b) because both of the carbon-oxygen bonds are formed at the same time, the stereochemical relationships of the groups on the double bond are preserved in the product. The epoxidation reaction can happen at either face of the pi bond, and the stereoselectivity of the reaction depends on whether the products formed from the two faces are identical, enantiomers, or diastereomers.

When there are two or more double bonds in the structure, the general observation is that the more substituted one reacts faster (Figure 0923). The more substituted double bond (analogous to a more substituted carbocation) is more electron-rich, and so more nucleophilic.

Figure 0923

Selectivity in epoxidation reactions for molecules with more than one double bond.

major product
minor product
OOH
Cl
O
H
>
+ enantiomer
+ enantiomer
Ph
HO
C
diastereomers

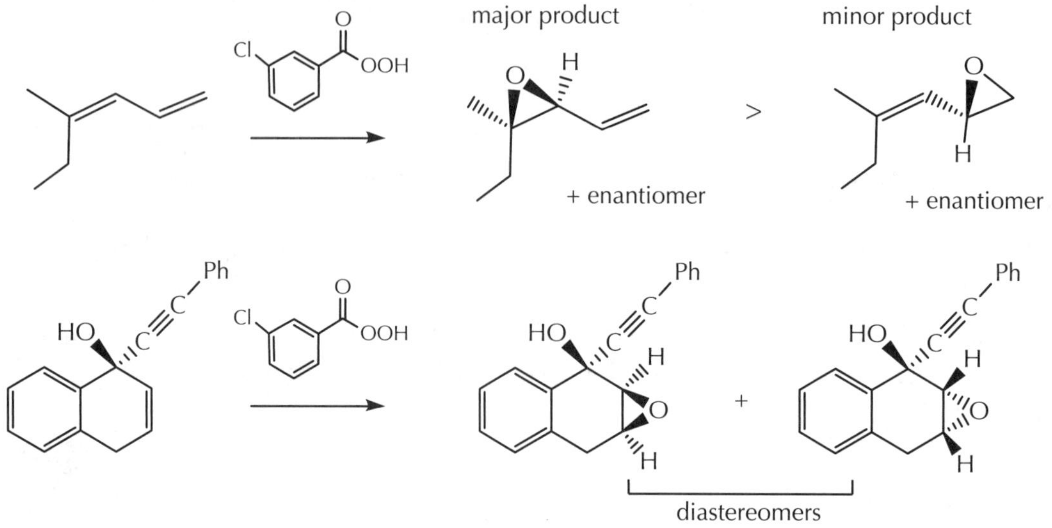

The second example in Figure 0923 is important to remember. As has been true throughout the text, the double bonds in benzene rings are unreactive towards the common electrophilic reagents that non-benzene double bonds react with quite readily. Triple bonds are also inert to epoxidation, and while this might be explained by the unfavorable ring strain of keeping a double bond in a 3-membered ring, the actual reason is a much larger topic that we will take up in Chapter 10 and will not even summarize here. At this point, the observation stands: Triple bonds do not form stable epoxides.

Figure 0924

Baeyer Test for unsaturation.

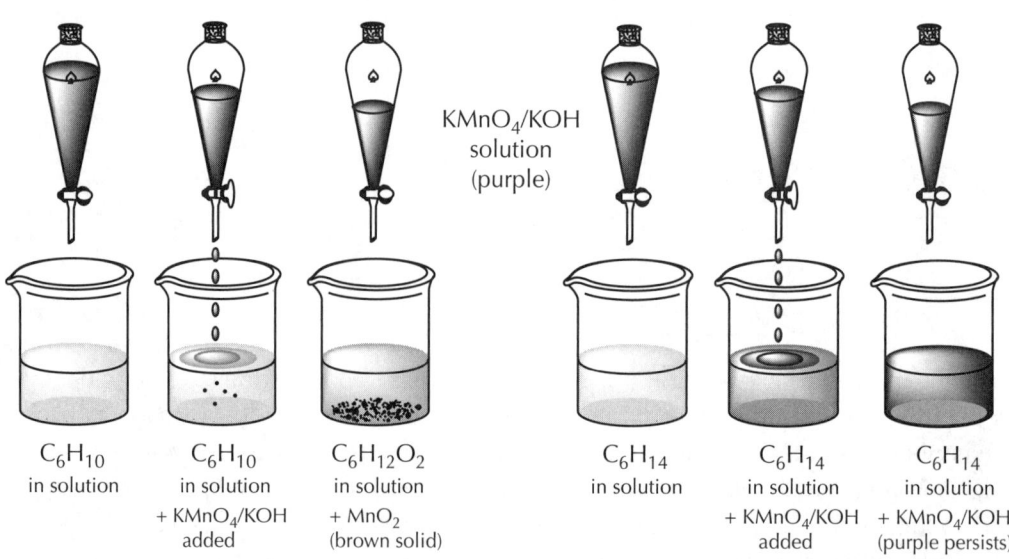

KMnO4/KOH
solution
(purple)
C6H10
in solution
C6H10
in solution
+ KMnO4/KOH
added
C6H12O2
in solution
+ MnO2
(brown solid)
C6H14
in solution
C6H14
in solution
+ KMnO4/KOH
added
C6H14
in solution
+ KMnO4/KOH
(purple persists)
Unsaturated compounds react with purple permanganate; the color disappears and a brown precipitate (MnO2) appears
Saturated compounds do not react with permanganate, and the deep purple color persists with no accompanying precipitate

B. Syn Dihydroxylation Reactions

Around the time bromination reactions were being reported in the 1890s, the reaction between double bonds and potassium permanganate ($KMnO_4$) was also being studied. As in the bromination reaction, the presence of a double bond could be detected by a fast color change. Permanganate solutions are a deep purple color. When permanganate solution is added to alkenes, the organic molecule is oxidized and the permanganate is reduced to manganese dioxide (MnO_2), which is a dull, brown precipitate (Figure 0924).

Historically, this color change from a purple solution ($KMnO_4$) to a brown solid (MnO_2) was called the Baeyer Test for unsaturation. In the days before spectroscopy, this result, when combined with the discoloration of a bromine solution, was used as strong evidence for the presence of a double bond in a molecular structure.

The chemical reaction of the Baeyer Test is a syn dihydroxylation of a double bond. The observed outcome, as the name implies, is shown in Figure 0925.

Figure 0925

Syn dihydroxylation reactions using basic potassium permanganate.

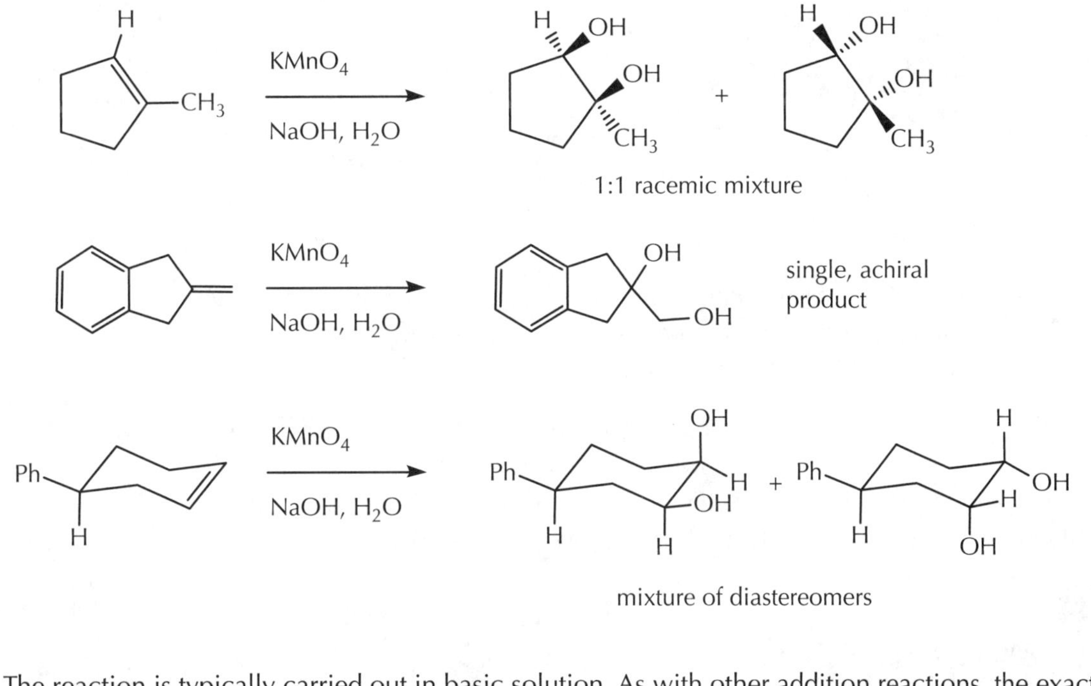

The reaction is typically carried out in basic solution. As with other addition reactions, the exact stereoisomeric outcome depends on how many stereocenters are formed during the reaction, in combination with how many stereocenters were present in the starting material. The reaction can be represented by simply writing "$KMnO_4$" or it can be represented as "$KMnO_4$/NaOH, H_2O." Be prepared to recognize either of these... and be sure to find out your instructor's preference when you are responsible for indicating the reaction conditions for this reaction.

The mechanism for the syn dihydroxylation reaction of permanganate begins with a 1-step addition reaction in which both of the new carbon-oxygen bonds are formed at the same time. As in the epoxidation reaction, earlier, the simultaneous construction of the new bonds means that the stereochemical information in the starting alkene is preserved in the product. The addition step involving permanganate forms a 5-membered ring intermediate (Figure 0926 on the next page).

Figure 0926

Syn dihydroxylation: addition mechanism using basic potassium permanganate.

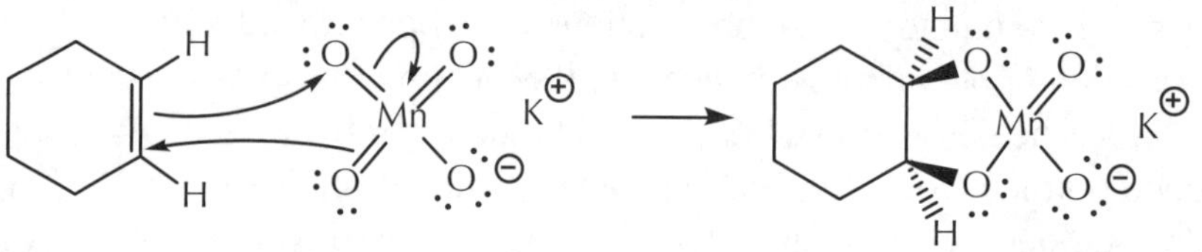

In basic solution, the 5-membered ring intermediate undergoes hydrolysis ("hydrolysis" means "to break apart, or loosen, with water") to break the oxygen-manganese bonds and form the hydroxyl groups observed in the product. The hydrolysis mechanism can be drawn in different ways, but the simplest version is shown in Figure 0927.

Figure 0927

Syn dihydroxylation: hydrolysis of the permanganate addition intermediate.

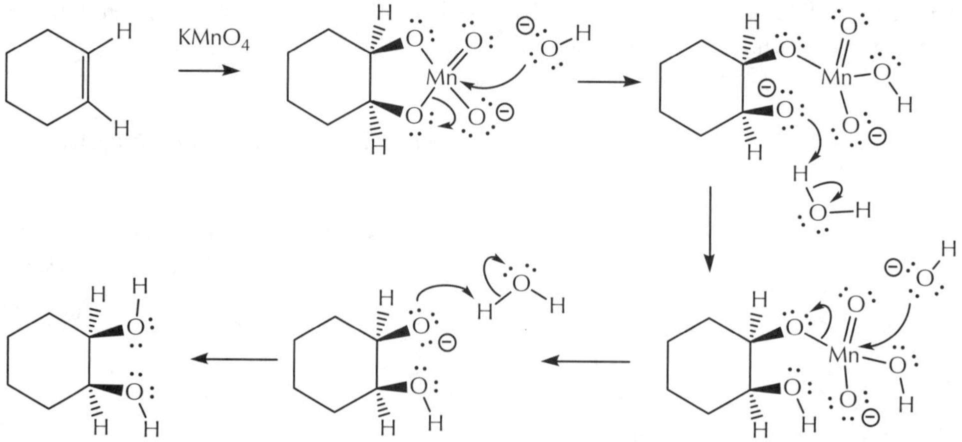
$KMnO_4$

Hydroxide ion, which is in excess, participates in a substitution reaction at the manganese atom, releasing the anionic oxygen that, in turn, is protonated by the water in which the reaction is carried out. This process happens twice, resulting in the diol product.

Over the years, chemists have discovered new reagents for syn dihydroxylation. In the early 1900s, osmium tetroxide, a compound which had been known for about a hundred years, was observed to be an extremely effective reagent for this reaction. The reaction can be represented by simply writing "OsO_4" or it can be represented in two steps, as "(1) OsO_4; (2) Na_2SO_3, H_2O" (Figure 0928).

Figure 0928

Syn dihydroxylation reactions using osmium tetroxide.

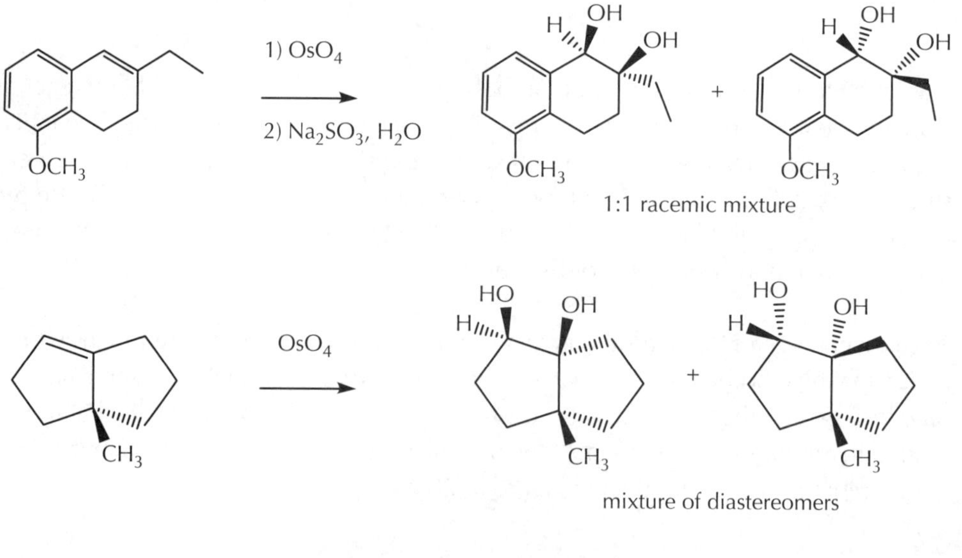
1) OsO_4
2) Na_2SO_3, H_2O
OCH_3
1:1 racemic mixture
OsO_4
CH_3
mixture of diastereomers

The reaction mechanism is comparable to that in the permanganate reaction. In a 1-step addition reaction, the two new carbon-oxygen bonds are formed along with an intermediate 5-membered ring that includes the osmium atom (Figure 0929).

Figure 0929

Syn dihydroxylation mechanism using osmium tetroxide.

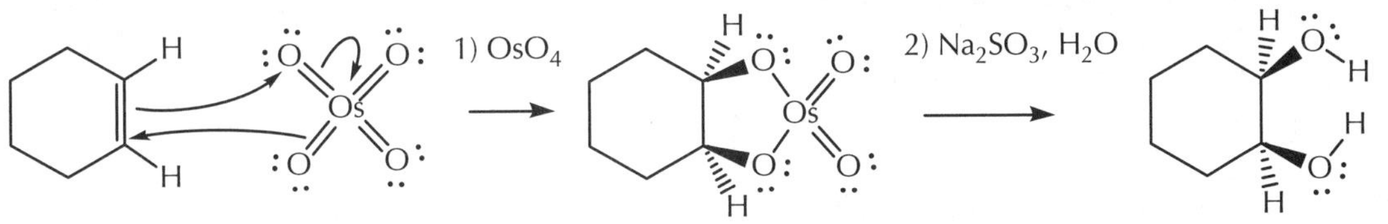

All of the stereochemical features are the same as in the permanganate reaction. The 5-membered ring intermediate is hydrolyzed in a second experimental step, breaking the oxygen-osmium bonds and forming the hydroxyl groups.

Over time, you are likely to see many different reaction conditions that include osmium tetroxide. Osmium tetroxide is quite toxic, expensive, and foul smelling (the element, osmium, was named using the Greek word for odor, *osme*), and researchers have explored many ways to maximize its benefits while minimizing the exposure and environmental hazards.

Numerous methods for using catalytic amounts of osmium tetroxide have been developed. In all of these cases, the OsO_4 reaction, as shown above, is still the central process. During a typical oxidation using only OsO_4, the osmium atom goes from a +8 oxidation state to a +6 state, giving the 5-membered ring intermediate. The reduced form of osmium (+6) is useless for additional oxidation reactions.

In the 1930s, a chemist discovered that by using hydrogen peroxide as the oxidant, only a catalytic amount (about 1%) of osmium tetroxide was needed. The hydrogen peroxide is too weak to oxidize the alkene, but it can reoxidize the osmium (+6) back to the original osmium (+8) under the reaction conditions, thus allowing the use of a small amount of osmium that keeps getting internally recycled during the course of the reaction. A depiction of this catalytic cycle is shown in Figure 0930.

Figure 0930

Syn dihydroxylation: catalytic cycle using osmium tetroxide and another oxidant.

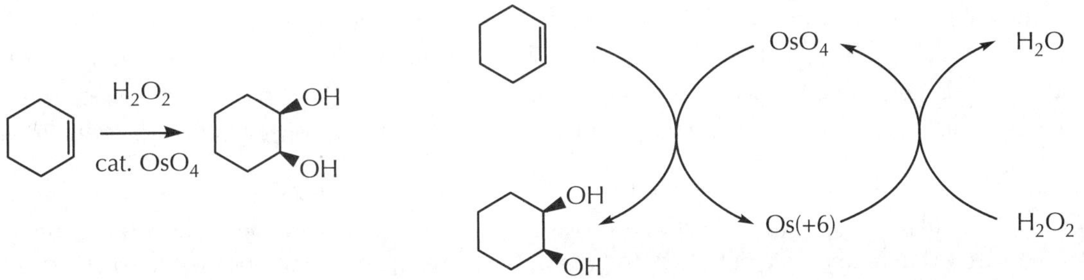

Over the last century, other oxidants have been discovered in addition to hydrogen peroxide, as well as different conditions for the hydrolysis step. In all of these modifications, osmium tetroxide is still the unique, central reagent and allows the reader to identify these conditions as a syn dihydroxylation. A summary of some different oxidants used in the catalytic reaction, as well as some differences in the hydrolysis conditions, is shown in Figure 0931 (on the next page).

Figure 0931

Syn dihydroxylation: combinations of conditions used in catalytic osmium tetroxide reactions.

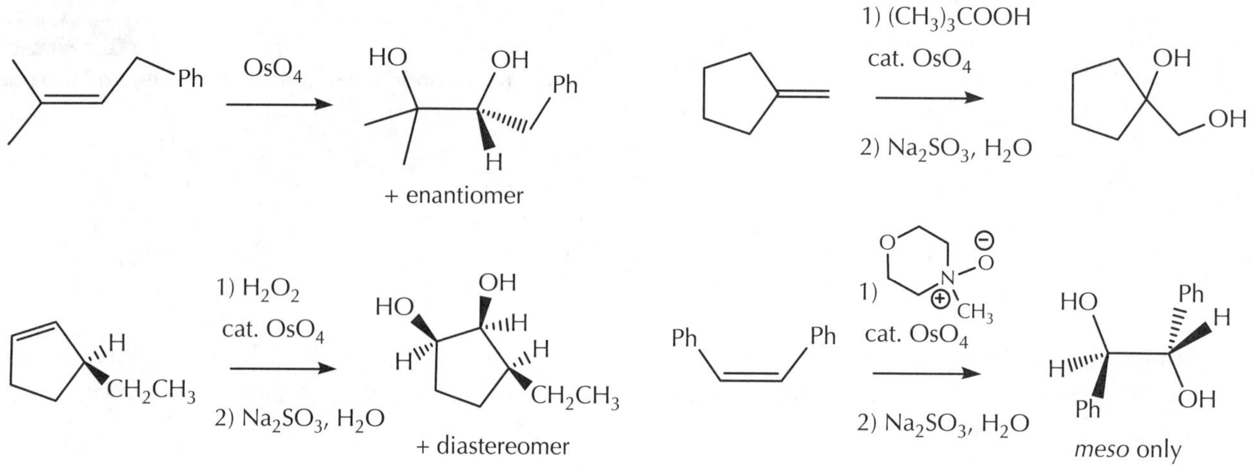

Please pay careful attention to the example on the bottom right corner of Figure 0931. Although the molecule is drawn in a conformation where the two hydroxyl groups are anti, this does not mean that the two hydroxyls added with an anti stereoselectivity. Once the addition takes place, the molecule is free to rotate around the central C-C bond, and any one of many conformations might be drawn. Take a moment to prove to yourself that this product is, in fact, the result from the syn dihydroxylation reaction.

As mentioned earlier, many chemists simply write "OsO$_4$" when referring, collectively, to all of these reactions. You should recognize all of these alternatives as representing the same reaction outcome, and your instructor should provide guidance for how you should represent the reaction when required.

As in the case of the epoxidation reactions, the syn dihydroxylation reactions do not work well with triple bonds, and so both of the oxidation reactions in these first two sections are limited to double bonds.

C. Ozonolysis Reactions

A distinctive odor associated with lightning storms has been reported for thousands of years. In the late 1700s, scientists studying ways to generate electricity noted "the smell of electricity." Fifty years later, a German-Swiss chemist, Christian Schönbein, doing experiments on the electrolysis of water, noted the same odor and attributed it to a substance he named ozone (Greek: *ozein*, meaning "to smell"). Schönbein isolated the substance and carried out its first reactions with organic compounds, noting that the odor disappeared (implying that a reaction had occurred). The structure of ozone (O$_3$) was determined in 1865. Methods for its large-scale production were developed in the 1880s, and procedures for carrying out reactions with alkenes and alkynes were refined to the point of usefulness over a period of about 20 years.

As the name implies, the ozonolysis reaction uses ozone to lyse (to "break" or "loosen") organic compounds. In particular, an ozonolysis reaction literally breaks apart, or separates, the two carbon atoms that are bonded to each other in a double or triple bond. The overall observations for ozonolysis reactions of alkenes and alkynes are shown in Figure 0932.

The ozonolysis experiment is carried out in two steps. The first step of the experimental procedure is the addition of ozone to the organic molecule. The second step is carried out as either an "oxidative workup" (hydrogen peroxide, H$_2$O$_2$) or a "reductive workup" [metallic zinc, Zn; or dimethyl sulfide, (CH$_3$)$_2$S]. The term "workup" typically refers to experimental procedures used to isolate the desired product from the reaction mixture.

9.2 Oxidation Reactions

Figure 0932

Ozonolysis reaction outcomes for alkenes and alkynes under oxidative and reductive workup conditions.

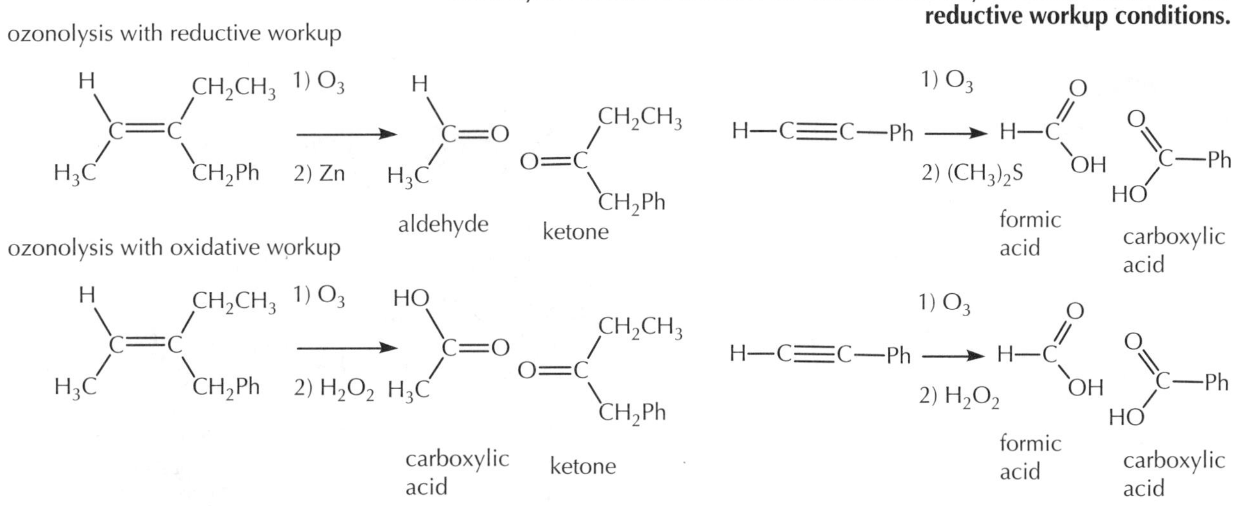

ozonolysis with reductive workup

[Alkene with H, CH₃ on one carbon and CH₂CH₃, CH₂Ph on other] 1) O₃, 2) Zn → aldehyde (H, H₃C, C=O) + ketone (CH₂CH₃, CH₂Ph, C=O)

H—C≡C—Ph 1) O₃, 2) (CH₃)₂S → formic acid (H—C(=O)OH) + carboxylic acid (HO—C(=O)—Ph)

ozonolysis with oxidative workup

[Alkene with H, CH₃ on one carbon and CH₂CH₃, CH₂Ph on other] 1) O₃, 2) H₂O₂ → carboxylic acid (HO, H₃C, C=O) + ketone (CH₂CH₃, CH₂Ph, C=O)

H—C≡C—Ph 1) O₃, 2) H₂O₂ → formic acid (H—C(=O)OH) + carboxylic acid (HO—C(=O)—Ph)

All of the molecules that form in ozonolysis reactions possess a carbonyl (C=O) group. The functional group classification for carbonyl-containing compounds depends on the two groups attached to the carbon atom. Some of these functional groups were introduced in the hydration reaction of triple bonds, and their formal introduction is presented here, along with their common abbreviations, because they are integral to describing the outcome from the ozonolysis reaction.

Ketone: a carbonyl carbon attached to two alkyl groups (R_2CO).
Aldehyde: a carbonyl carbon attached to an alkyl group and a hydrogen atom (RCHO).
Carboxylic acid: a carbonyl carbon attached to an alkyl group and a hydroxyl group (RCO_2H).

There are two unique compounds in these categories that do not contain alkyl groups:

Formaldehyde: a carbonyl carbon attached to two hydrogen atoms (H_2CO).
Formic acid: a carbonyl carbon attached to a hydrogen atom and a hydroxyl group (HCO_2H).

The ozonolysis reaction mechanism took many years to figure out, and the first full proposal was not made until 1953. The most familiar and generally applicable part of the mechanism is the first step, the reaction of the pi bonds with ozone. This step, called ozonation (addition of ozone), looks comparable to all of the other oxidation reactions in this section: Two new carbon–oxygen bonds form at the same time, resulting in a cyclic intermediate. The initially formed, 5-membered ring intermediate is called a molozonide (Figure 0933).

Figure 0933

Ozonolysis mechanism: ozonation of alkenes and alkynes gives molozonides.

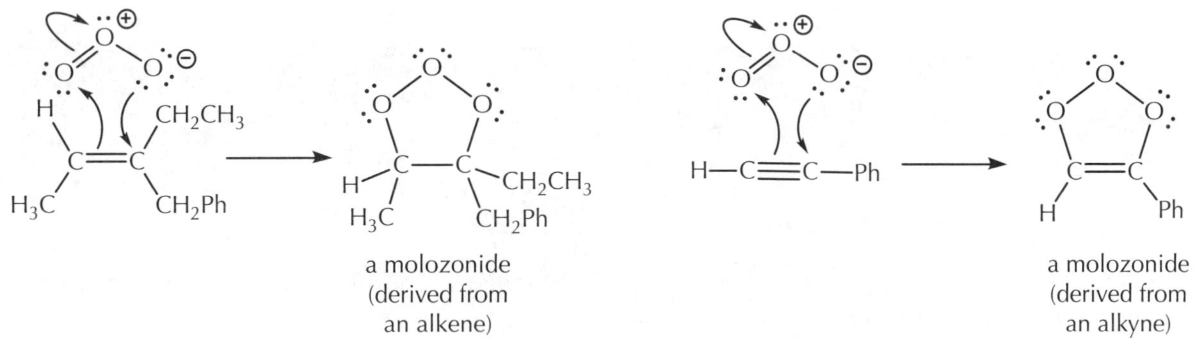

a molozonide (derived from an alkene)

a molozonide (derived from an alkyne)

The remaining details of the ozonolysis mechanism are unique to this reaction. The fate of the molozonides derived from alkenes and alkynes is a complex isomerization process resulting in new intermediates called ozonides (Figure 0934).

Figure 0934

Ozonolysis mechanism: isomerization of molozonides to ozonides.

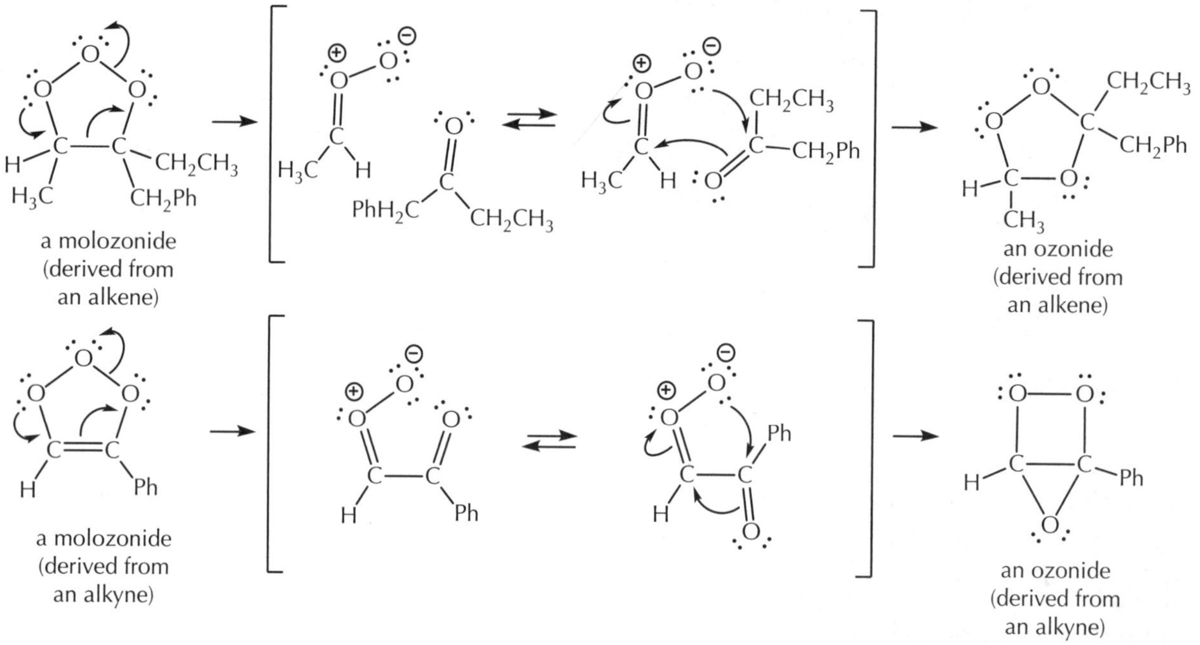

Under the reaction conditions that form the molozonide from an alkene, the molecule fragments into two pieces, and those two pieces recombine to give a new and more stable intermediate called an ozonide. The molozonide derived from the alkyne ozonation is believed to undergo an analogous reaction, although it does not fall apart into separate fragments. These ozonide intermediates are the actual results from the first experimental step (reaction with ozone).

In the second, workup step, the ozonides are decomposed to give the observed products. Here is where a broader view of the outcomes can help you remember what happens, rather than the smallest details. The result from alkynes is least complex: Under all reaction conditions, the two carbon atoms involved in a triple bond are separated from one another, and each of the carbon atoms ends up as a carboxylic acid (Figure 0935).

Figure 0935

Ozonolysis of alkynes under both reductive and oxidative workup conditions.

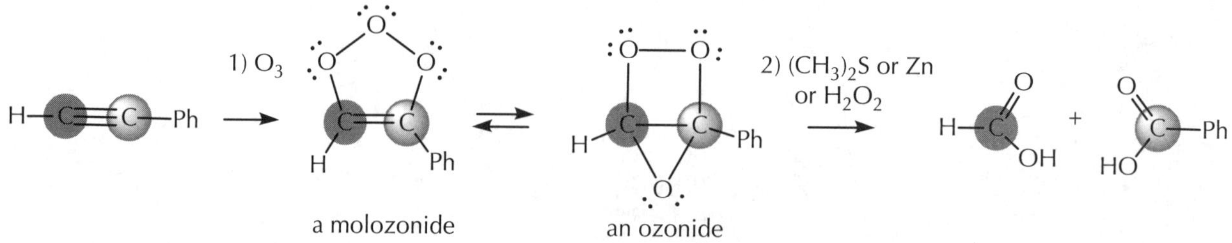

The result from alkenes under the reductive workup conditions is a little more complex because the molecules that form have different classifications (Figure 0936).

9.2 Oxidation Reactions

Figure 0936

Ozonolysis of alkenes under reductive workup conditions.

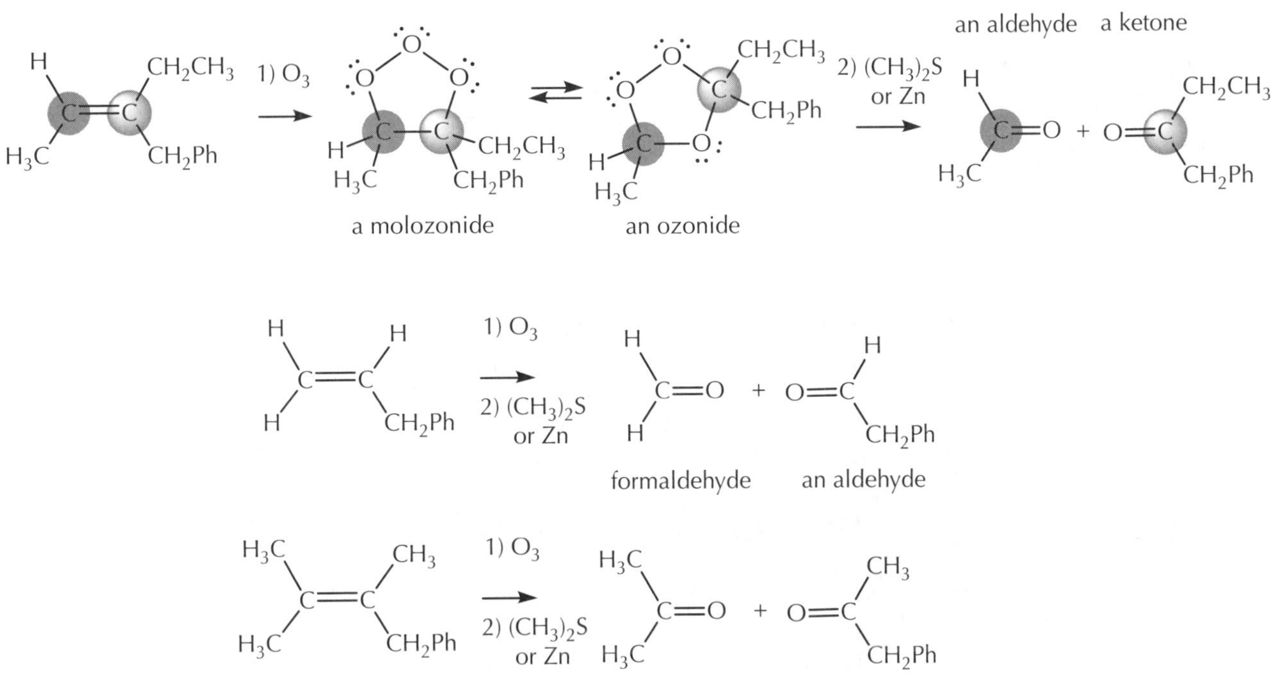

As in the case of the triple bond, the two carbon atoms involved in a double bond are separated from one another, and each of them (under reductive conditions) ends up as a carbon-oxygen double bond (carbonyl group), and the specific functional group classification depends on what else is attached to the carbon atom.

The mechanisms for the decomposition of the ozonides into the carbonyl-containing products depend on the reaction conditions. The proposed curved arrow mechanism for the decomposition of the ozonide derived from an alkene, using dimethyl sulfide, is shown in Figure 0937.

Figure 0937

Ozonolysis of alkenes under reductive work-up conditions: ozonide decomposition mechanism with dimethyl sulfide.

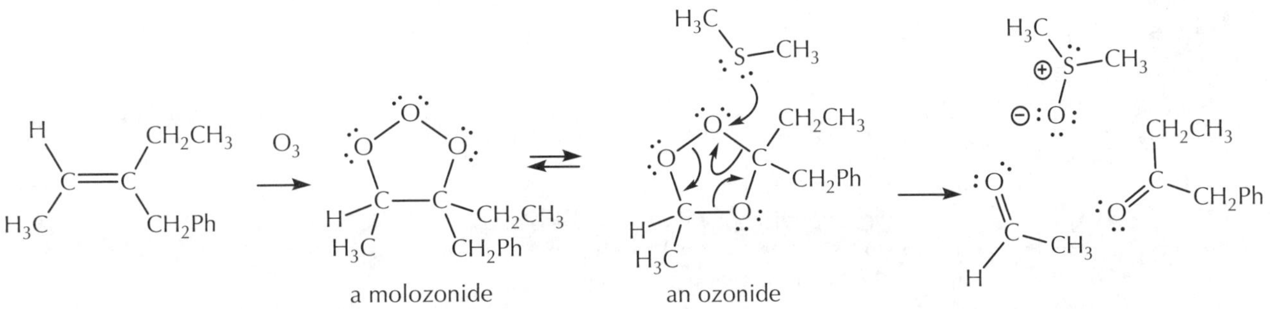

The outcome from the ozonolysis of alkenes under oxidative workup conditions is initially exactly the same as under the reductive conditions. However, one of the chemical properties of aldehyde functional groups is that they are easily oxidized to carboxylic acids (Figure 0938 on the next page).

Figure 0938
Ozonolysis of alkenes under oxidative workup conditions.

[Reaction 1: (E/Z)-2-methyl-1-phenyl-... type alkene (H₃C, H on one carbon; CH₂CH₃, CH₂Ph on other) → 1) O₃ 2) H₂O₂ → [an aldehyde + a ketone] → H₂O₂ → a carboxylic acid + a ketone]

[Reaction 2: CH₂=CH–CH₂Ph type alkene → 1) O₃ 2) H₂O₂ → [formaldehyde + an aldehyde] → H₂O₂ → formic acid + a carboxylic acid]

[Reaction 3: (CH₃)₂C=C(CH₃)(CH₂Ph) → 1) O₃ 2) H₂O₂ → [a ketone + a ketone] → H₂O₂ → a ketone + a ketone]

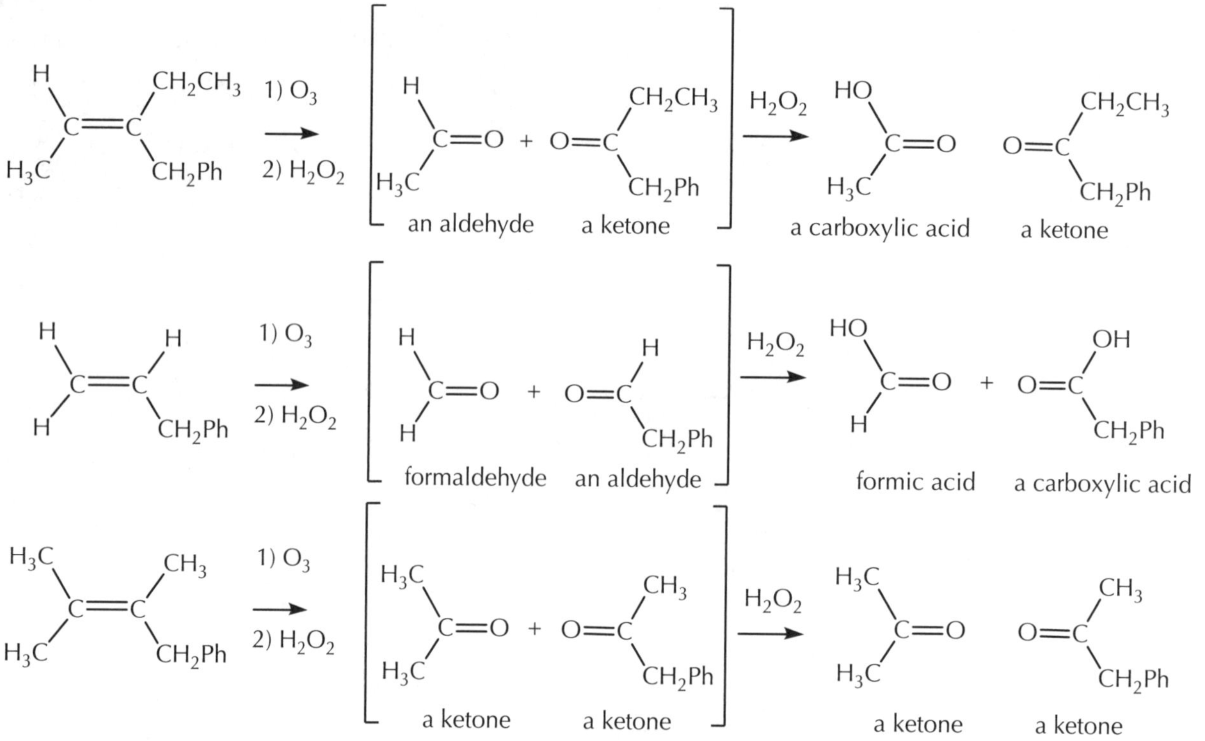

Consequently, in any ozonolysis reactions in which aldehydes are proposed to form, the aldehydes will be further oxidized to their corresponding carboxylic acids when the oxidative workup conditions are used. Ketones remain unoxidized, as ketones; aldehydes are oxidized to carboxylic acids; formaldehyde is oxidized to formic acid.

Historically, the reactions from this chapter, including the ozonolysis reaction, were important tools for helping chemists determine the structures of new, complex molecules (Figure 0939). Not that long ago,

Figure 0939
A structure determination example, circa 1950: experimental observations.

procedure/test	observation/result	structural conclusion
analysis	$C_{12}H_{22}$	two units of unsaturation = two double bonds; two rings; a double bond and a ring; a triple bond
$C_{12}H_{22} \xrightarrow{KMnO_4, NaOH, H_2O}$	purple color disappears, brown precipitate forms	presence of at least one double bond or a triple bond
$C_{12}H_{22} \xrightarrow{Br_2}$	red-brown color disappears, reaction is colorless; titration with Br_2 indicates a 1:1 reaction; analysis of the bromination product gives $C_{12}H_{22}Br_2$	one double bond
$C_{12}H_{22} \xrightarrow{1) O_3, 2) Zn}$	$C_6H_{10}O$ + $C_6H_{12}O$ → ?	an unknown compound; no prior reports of a substance with its properties
$C_{12}H_{22} \xrightarrow{1) O_3, 2) H_2O_2}$	$C_6H_{10}O$ + $C_6H_{12}O_2$ → cyclohexanone, hexanoic acid	known compounds; based on comparison of reported properties

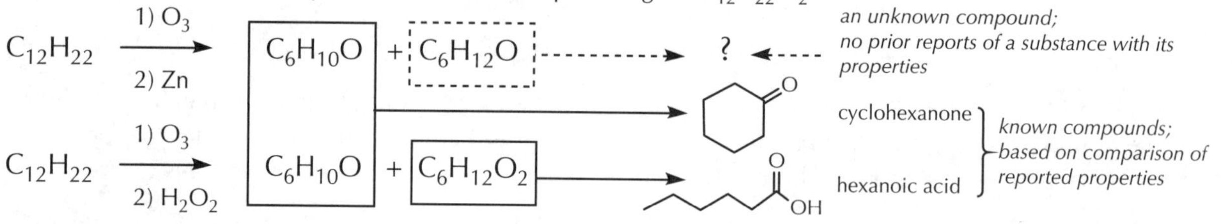

9.2 Oxidation Reactions 705

observing color changes, precipitation, and carrying out predictable decomposition reactions with ozone were standard experimental methods used for structure determination.

It is 1950 again, and a molecule with the molecular formula $C_{12}H_{22}$ has been isolated from a natural source. After comparing its properties to as many other $C_{12}H_{22}$ compounds you can locate in the chemistry journals, you conclude that this must be a new compound.

You know from the formula that there are two units of unsaturation, but you do not know if they are two rings, two double bonds, a triple bond, or one ring and one double bond. One of the first things you might do is add a few drops of bromine solution to a small sample of your unknown compound, and a few drops of potassium permanganate to another sample. Multiple tests for any given structural property were common, to provide affirming and independent corroboration of a structural conclusion. If the bromine hits your sample and goes colorless, and the purple permanganate color disappears and is replaced by a brown precipitate, you conclude that there is at least one double bond in the structure.

The bromination reaction can be carried out quantitatively, as a titration, and you can also conclude that there is a 1:1 reaction occurring between your molecule and bromine, which means there is only one double bond, and that the other unit of unsaturation is a ring.

Given that the molecule appears to have a double bond in it, ozonolysis is carried out. The molecule is predictably split into two pieces, which you have to recover and purify, and then you can check their properties.

After carrying out an ozonolysis reaction with a reductive workup step, you isolate two products: $C_6H_{10}O$ and $C_6H_{12}O$. Carrying out the ozonolysis and using the oxidative workup conditions, you isolate $C_6H_{10}O$, which is identical to the earlier $C_6H_{10}O$ product, and a new compound, $C_6H_{12}O_2$.

You determine some physical properties for these new molecules and head back to the library, and you find reported compounds whose properties exactly match the properties of some of your ozonolysis products. The $C_6H_{10}O$ compound was known (cyclohexanone), as was the $C_6H_{12}O2$ compound (hexanoic acid). The $C_6H_{12}O$ compound, however, appears to be previously unreported.

By understanding the structural relationships in an ozonolysis reaction, this information can be used to create a backwards assembly of the original structure and propose a structure for the unknown $C_6H_{12}O$ compound (Figure 0940).

Figure 0940

A structural determination example, circa 1950: conclusion.

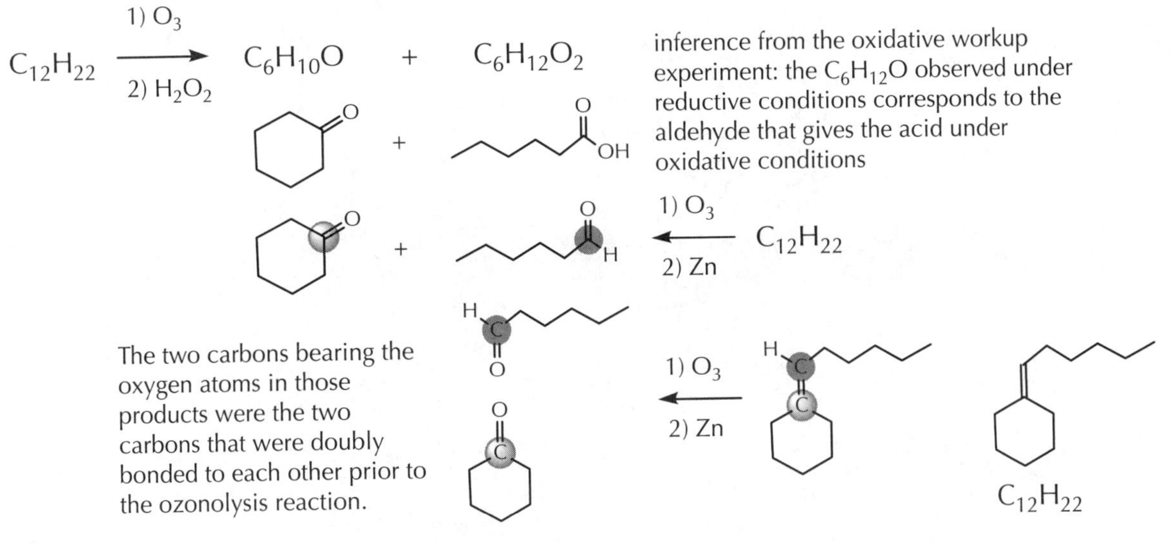

706 CHAPTER 9 Electrophilic Addition II: Halogenation, Oxidation, and Reduction Reactions

In the example shown in Figure 0940, you need to think about the reaction products from ozonolysis and then imagine what the structure of the starting material might be. This ozonolysis reaction (oxidative conditions) produced a $C_6H_{10}O$ ketone and a $C_6H_{12}O_2$ carboxylic acid, both of whose identities were known.

The exact same ketone was observed in the ozonolysis using the reductive workup. And although the properties of the $C_6H_{12}O$ product from the reductive workup not found in handbooks, its identity can be inferred from the $C_6H_{12}O_2$ acid, which was observed under the oxidative workup conditions. In an oxidative workup, any initially formed aldehydes are oxidized to their corresponding carboxylic acids, and the change in the molecular formula was consistent with that functional group relationship (reductive conditions: $C_6H_{12}O$; oxidative conditions: $C_6H_{12}O_2$). The inference is that the $C_6H_{12}O$ compound is the aldehyde that got oxidized to the observed carboxylic acid.

With the $C_6H_{10}O$ ketone and the $C_6H_{12}O$ aldehyde, all of the atoms of the original $C_{12}H_{22}$ unknown are accounted for except for the two oxygen atoms. The two carbons bearing the oxygen atoms in those products were the two carbons that were doubly bonded to each other prior to the ozonolysis reaction. Based upon the chemistry of the ozonolysis reaction, deleting the two oxygen atoms and reconnecting the two carbons gives the double bond of the original $C_{12}H_{22}$ starting material.

Prior to the development of fast, information-filled spectroscopic methods, these are the data that people would gather to determine a new molecular structure. Ozonolysis was quite useful because it chopped molecules apart in a highly predictable way.

When a molecule has two or more double and/or triple bonds, all of them will undergo ozonolysis. In questions where the products need to be predicted, you need to pay attention to whether or not the reaction is under oxidative or reductive workup conditions, because the results from the ozonolysis of some double bonds will vary, accordingly. Double bonds that are part of a ring can undergo ozonolysis as easy as any other, and so the rings will be broken open (Figure 0941).

Figure 0941

Ozonolysis reactions on molecules with multiple double and/or triple bonds.

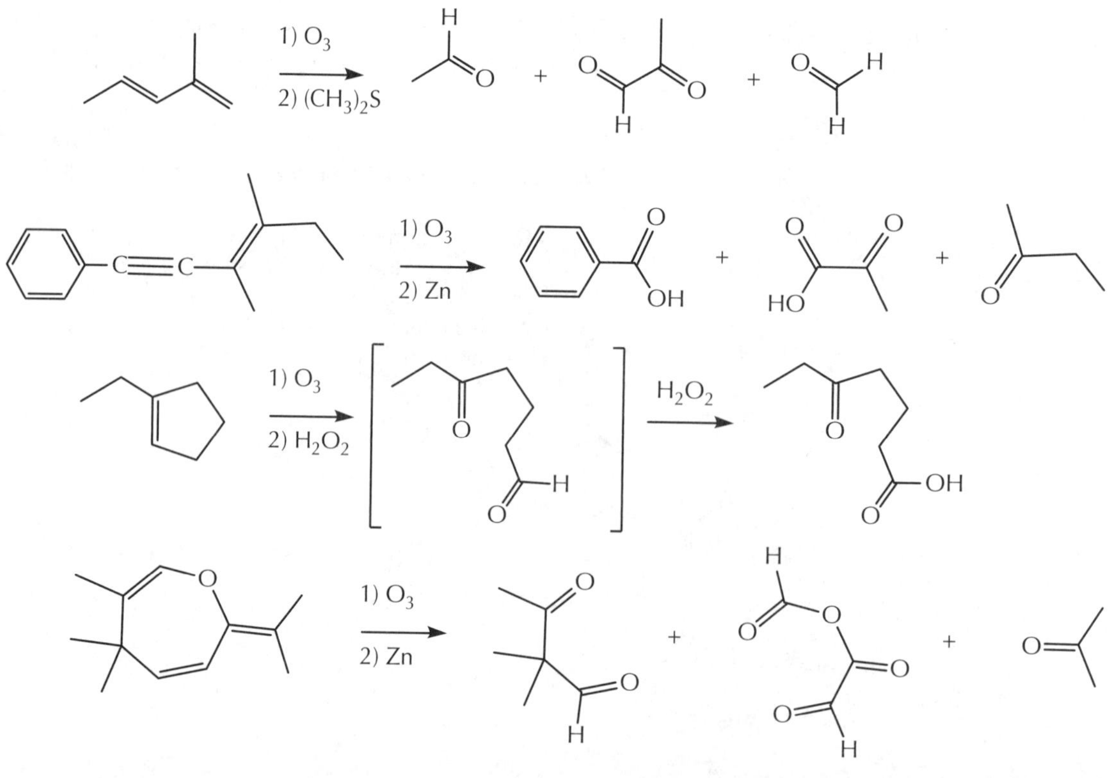

9.2 Oxidation Reactions

As has been true throughout the text, none of the addition reactions from this chapter are predicted to take place on the double bonds of a benzene ring. Although some oxidation reactions of benzene rings have been accomplished under strenuous conditions, you may always assume, unless otherwise specified, that the benzene ring is inert to these additions. There are other unsaturated rings that are comparable to benzene, and that topic will be taken up in Chapter 10.

Even historically, structural determination problems in which there is more than one double or triple bond, or in which the double bond has an (E)- or (Z)-configuration, were too complex to be solved with the information from the ozonolysis reaction alone. The ozonolysis products narrow down the options considerably, often creating a handful of possibilities rather than hundreds. Examples of possible structures derived from the chemical tests on some $C_{14}H_{22}$ unknowns are shown in Figure 0942.

Figure 0942

Ozonolysis outcomes narrow down the possible structures for unknowns.

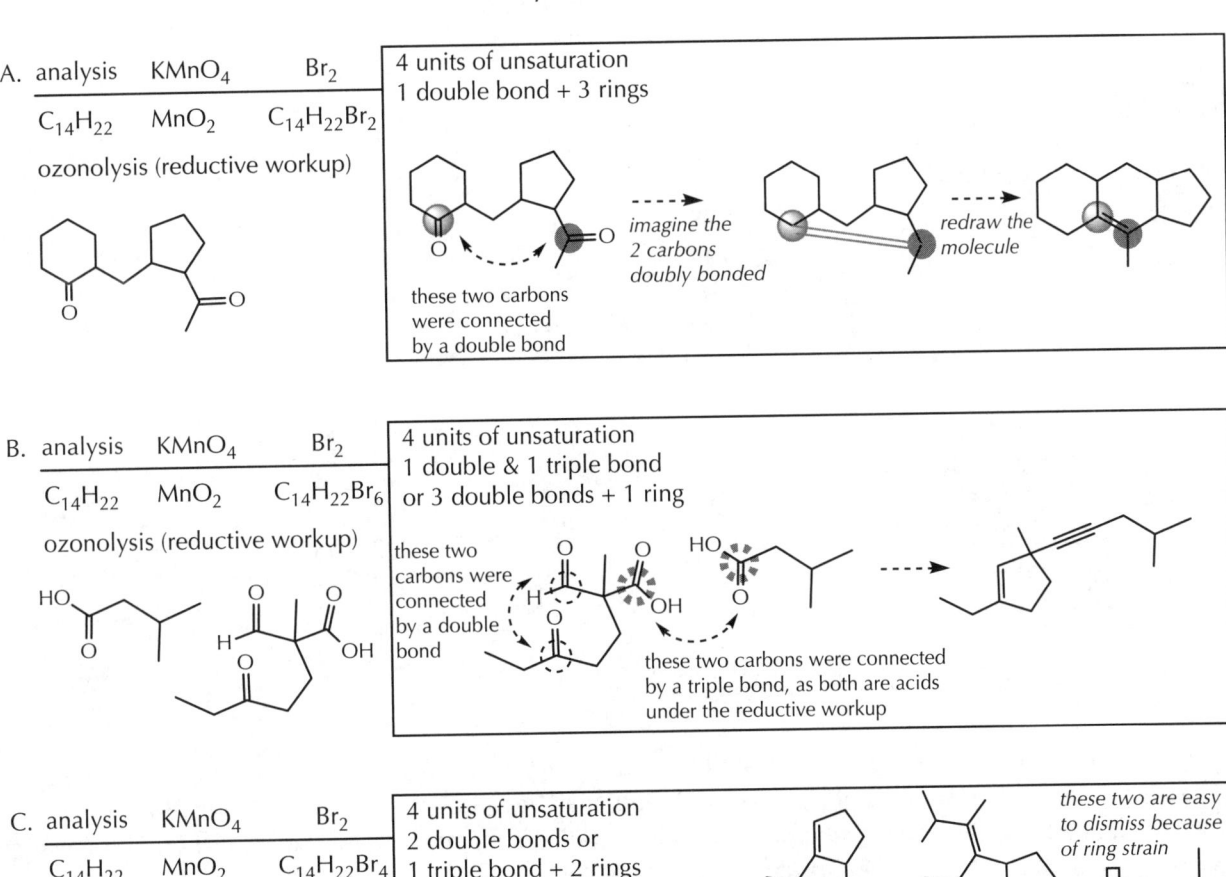

Although these chemical methods for structure determination have been replaced by the faster and more information-rich spectroscopic methods, the reactions themselves are still quite commonly used to carry out chemical transformations.

708 CHAPTER 9 Electrophilic Addition II: Halogenation, Oxidation, and Reduction Reactions

D. Atom-Swap Analogies

As in the earlier section on halogenation, the oxidation reactions in this section provide a baseline of reactivity for understanding a larger number of chemical reagents.

Review, in Figures 0917 and 0918, how the basic halogenation mechanism could be used to understand the reactions of interhalogen reagents in addition to the sulfur- and selenium-based reagents.

Here, we will look at reagents where, although atoms have been swapped out from some of the oxidizing agents, the general reactivity for addition reactions is analogous to the original reagents.

Reagents have been created in which nitrogen atom groups replace one or more of the oxygen atoms in osmium tetroxide. These reagents give the syn addition of two nitrogen atom groups, or, depending on the reagent, one nitrogen atom group and one hydroxyl group (Figure 0943). Applying the concepts from the osmium tetroxide reaction, the mechanism would be proposed to be the 1-step addition reaction to give a 5-membered ring intermediate, followed by hydrolysis.

Figure 0943

Analogies in addition reactions based on osmium tetroxide.

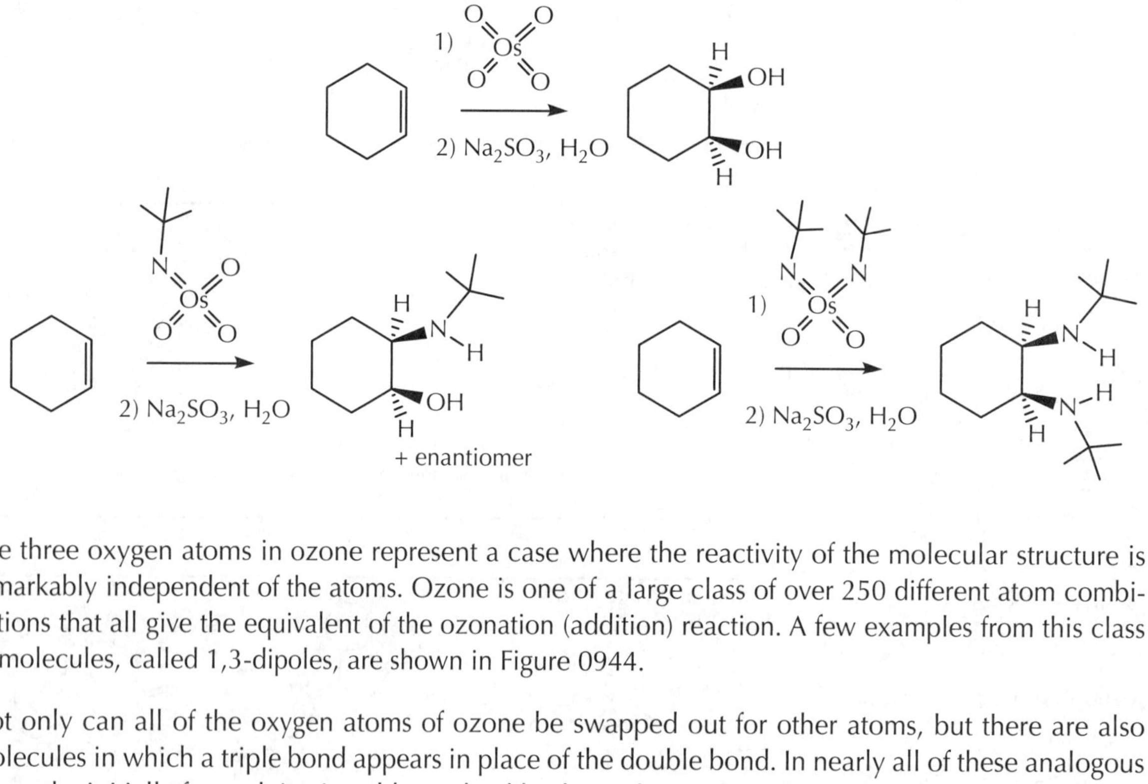

The three oxygen atoms in ozone represent a case where the reactivity of the molecular structure is remarkably independent of the atoms. Ozone is one of a large class of over 250 different atom combinations that all give the equivalent of the ozonation (addition) reaction. A few examples from this class of molecules, called 1,3-dipoles, are shown in Figure 0944.

Not only can all of the oxygen atoms of ozone be swapped out for other atoms, but there are also molecules in which a triple bond appears in place of the double bond. In nearly all of these analogous cases, the initially formed ring is stable, and unlike the molozonide, it does not rearrange and go on to break the original sigma bond. Interestingly enough, the overall reaction with ozone is the exception rather than the rule.

9.2 Oxidation Reactions

Figure 0944

Analogies in addition reactions based on ozone.

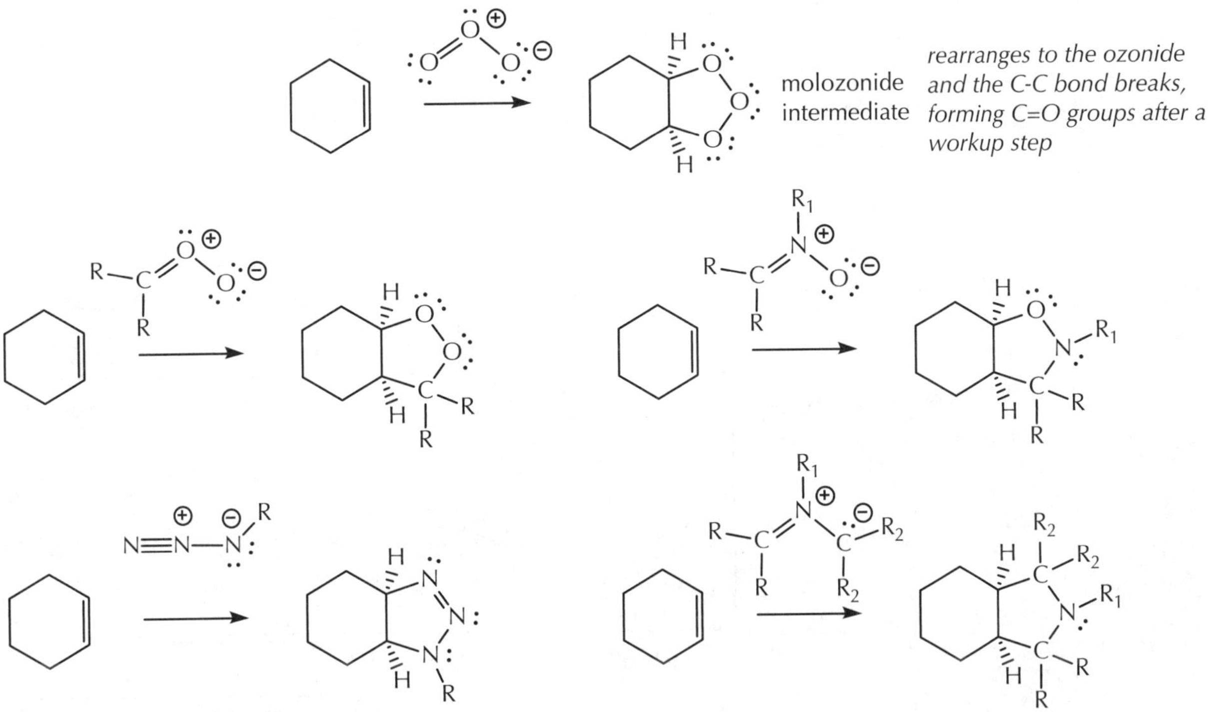

rearranges to the ozonide and the C-C bond breaks, forming C=O groups after a workup step

molozonide intermediate

PRACTICE QUESTIONS

9.06 Complete the following equations. Provide the structures of stereoisomeric products, if anticipated, and identify the stereoisomeric relationship (enantiomers, diastereomers) or stereochemical identity (an achiral molecule with no stereocenters, an achiral *meso* isomer)

(a) 1,2-dihydronaphthalene + KMnO₄ / NaOH, H₂O ⟶

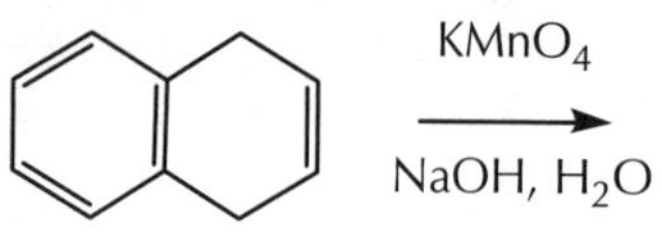

stereochemical relationship or identity:

(b) (R)-3-ethylcyclopentene + m-chloroperoxybenzoic acid ⟶

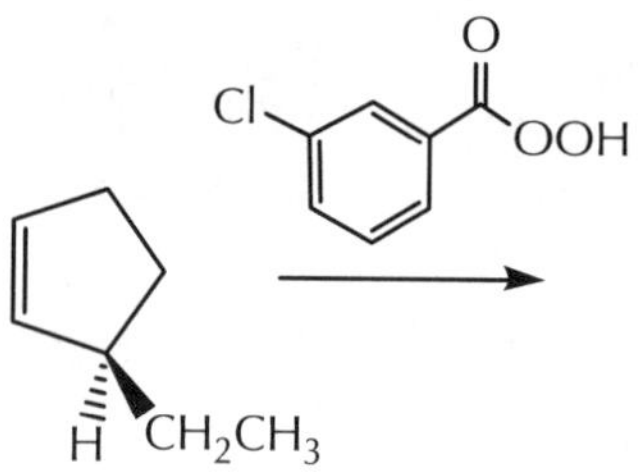

stereochemical relationship or identity:

(c) (S,E)-3-chloro-pent-2-ene (Cl on wedge, H on dash)
1) H₂O₂, cat. OsO₄
2) Na₂SO₃

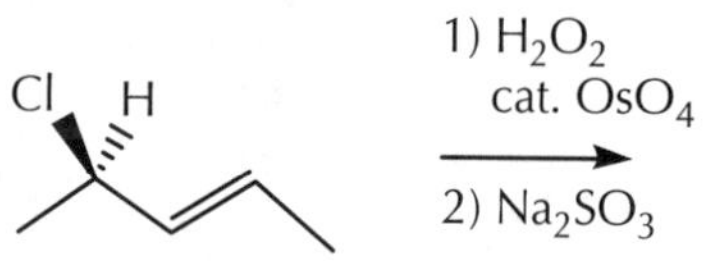

stereochemical relationship or identity:

(d) Ph-CH=CH-Ph (trans-stilbene)
1) O₃
2) (CH₃)₂S

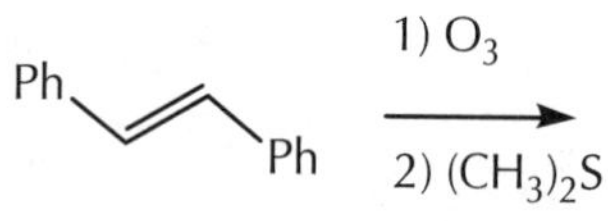

stereochemical relationship or identity:

(e) Ph—C≡C—CH₃
1) O₃
2) Zn

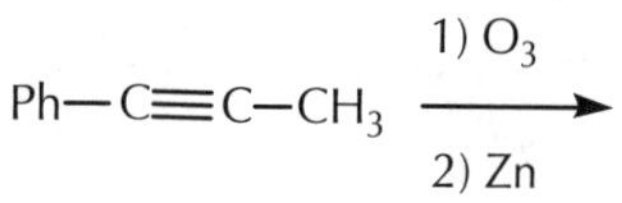

stereochemical relationship or identity:

9.07 Provide the complete, curved arrow mechanism for the following reactions.

(a)

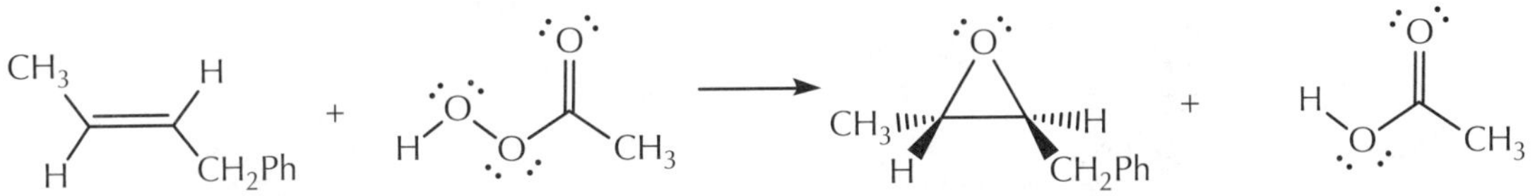

(b)

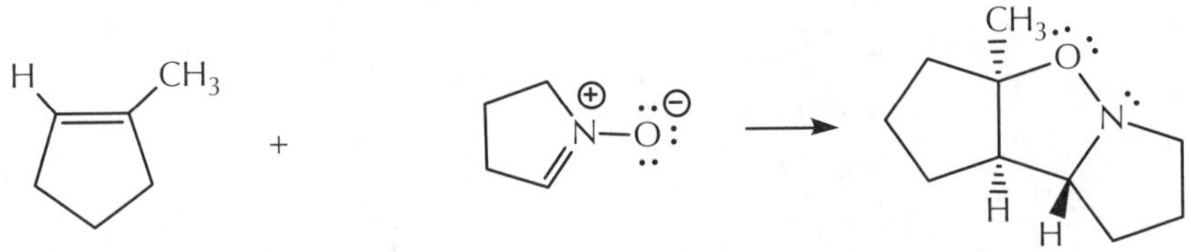

9.08 Complete the following equations.

(a)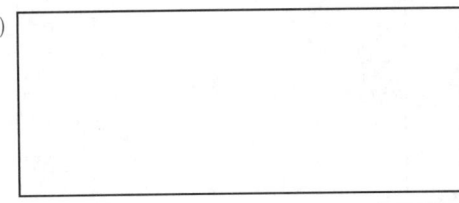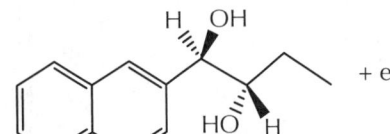

(b)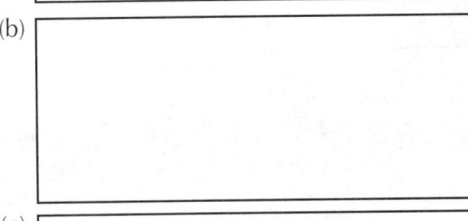

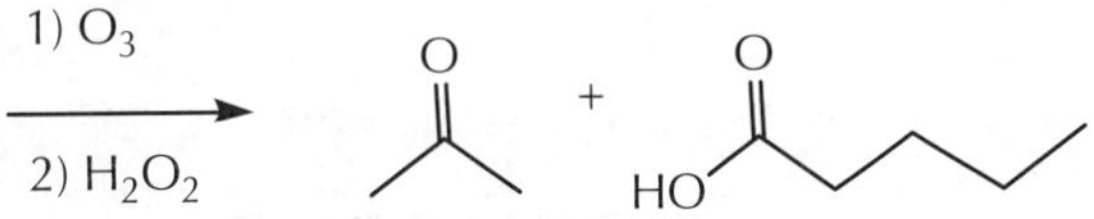

(c)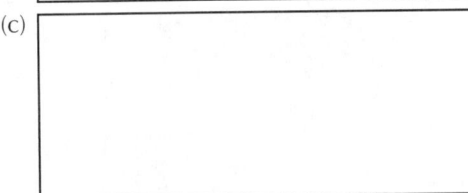

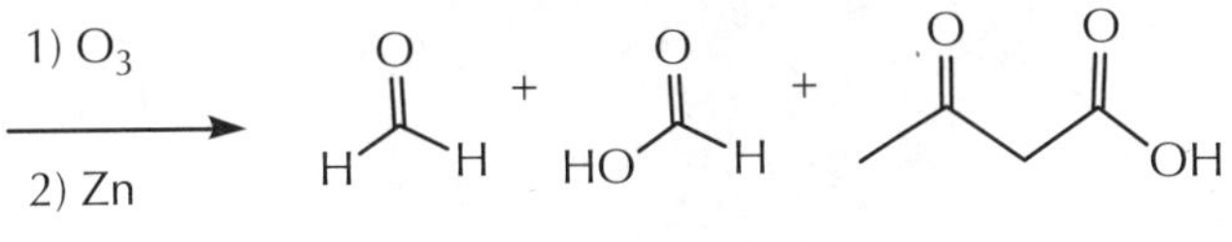

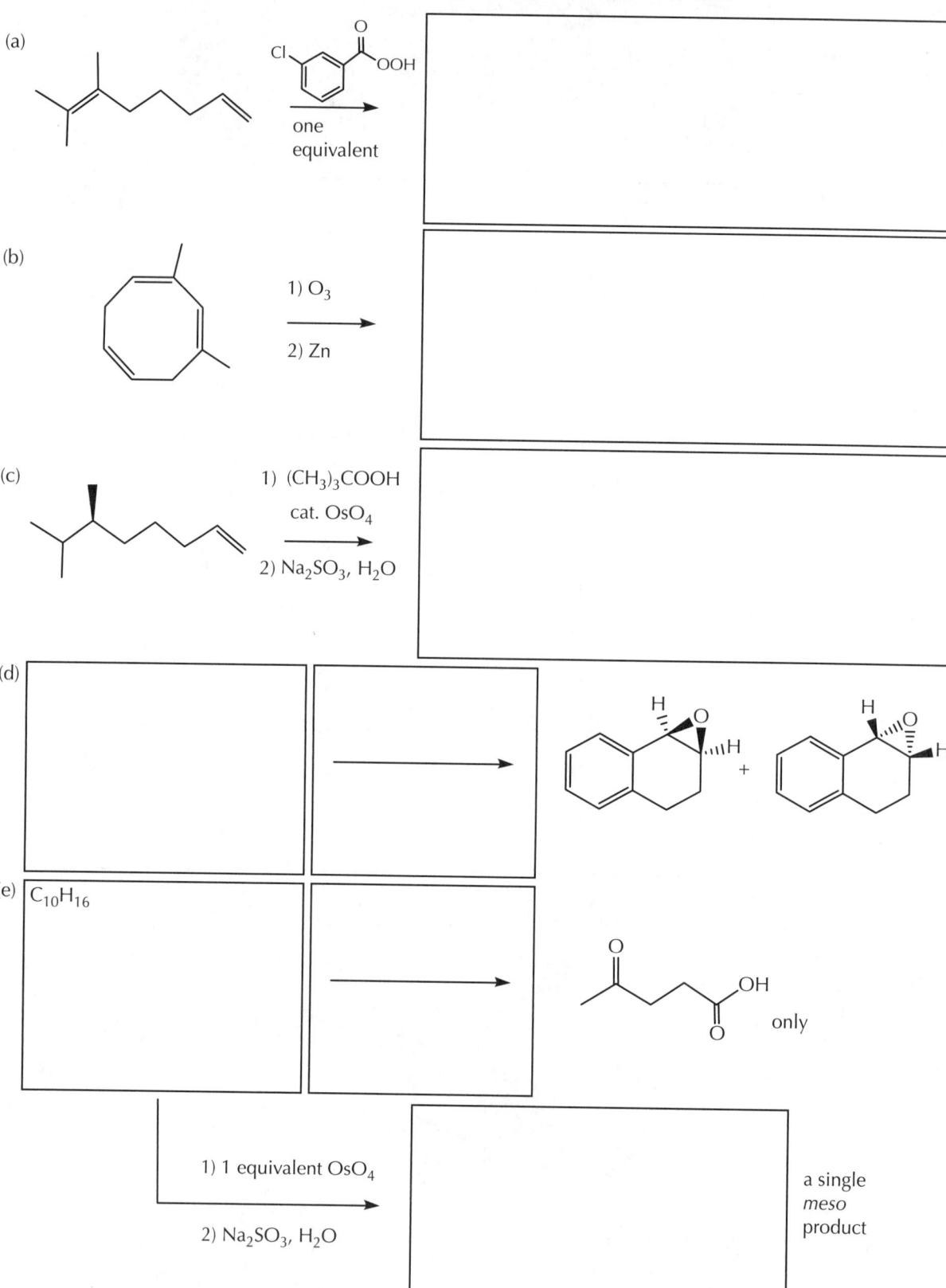

9.3 Reduction Reactions

A. Catalytic Hydrogenation with Transition Metal Catalysts

Reduction is a term with a long history. In modern chemistry, the dual referencing of reduction as the counterpart of oxidation ("redox") is the most widespread use of the terms. Inasmuch as oxidation increases the percentage of oxygen in an organic molecule, reduction is any process that decreases that percentage.

Historically, defining oxidation and reduction based upon the relative amounts of hydrogen has also been used to define these terms:

Reduction: increasing the percentage of hydrogen in a molecule, including addition reactions of molecular hydrogen (hydrogenation)

Oxidation: decreasing the percentage of hydrogen in a molecule (in earlier times, in fact, the oxidation reactions were called dehydrogenations, because the structural change was the elimination of a molecule of hydrogen)

Many biological oxidations are still called dehydrogenation (loss of hydrogen) reactions, and the associated enzymes (biological catalysts) are called dehydrogenases.

In 1897, the addition of hydrogen to carbon dioxide gas (CO_2), to give methane (CH_4), was discovered to take place in the presence of nickel metal. The discoverer of this reaction, Paul Sabatier, thought that the process could only be carried out on gaseous substances. Soon after, however, in 1901, a German chemist (Wilhelm Norman) demonstrated that the double bonds in nongaseous organic molecules could add hydrogen, too. Norman transformed oleic acid (derived from olive oil) into stearic acid (derived from animal fat) (Figure 0945) using hydrogen gas in the presence of powdered nickel.

Figure 0945

Catalytic hydrogenation of oleic acid to stearic acid.

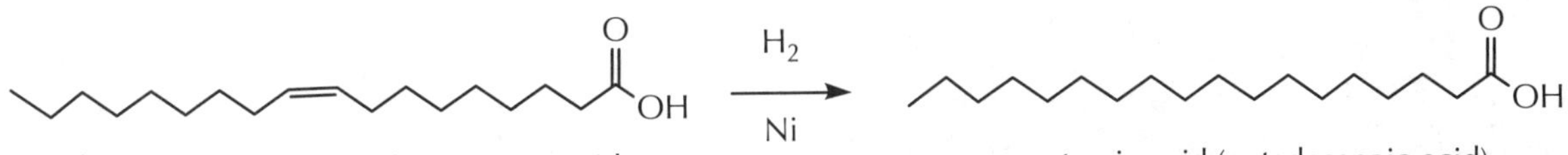

oleic acid ((Z)-octadec-9-enoic acid) → stearic acid (octadecanoic acid)

The molecules that Norman used are called fatty acids because they are the principal components in the molecular structure of the substances that we call fats and oils. Fats and oils have the same basic molecular structures, except that fats are generally solid at room temperature, whereas oils are liquids. Oils, which mainly derive from fish and plants, have a greater fraction of unsaturated fatty acids in their molecular compositions (such as oleic acid), while fats, which mainly derive from animal meat and dairy products, have a greater fraction of saturated fatty acids (such as stearic acid). Norman discovered that you could start with a liquid vegetable oil and only hydrogenate some of its double bonds (Figure 0946).

Figure 0946

Catalytic hydrogenation ("hardening") of vegetable oil

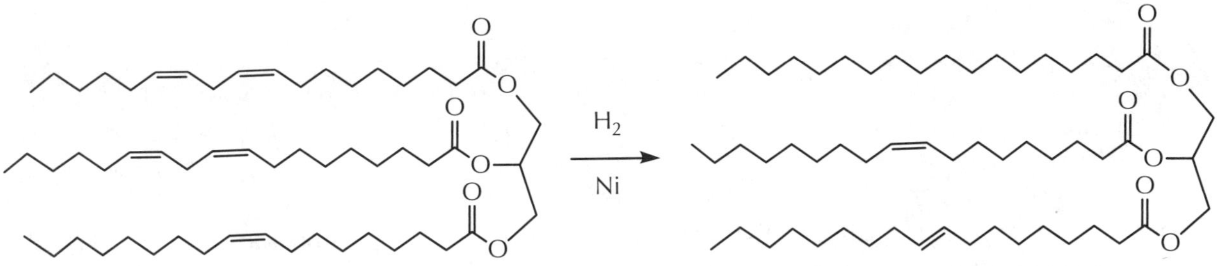

a vegetable oil (liquid) → a partially hydrogenated vegetable oil (semisolid; some *cis*-to-*trans* isomerization)

714 CHAPTER 9 Electrophilic Addition II: Halogenation, Oxidation, and Reduction Reactions

A higher degree of saturation in the hydrocarbon chain raises the melting point of the substance and turns the liquid oil into a soft (and spreadable) solid. These so-called hardened oils, or partially hydrogenated vegetable oils, are also known as margarine and shortening. Carrying out the hydrogenation reaction with excess molecular hydrogen over extended periods of time would, in principle, completely add hydrogen to any and all of the double bonds in a structure and give the fully saturated compounds (saturated fats) as products.

The overall reaction, hydrogenation, has been observed to be catalyzed by many different transition metals and their compounds, including nickel (Ni), platinum oxide (PtO_2), rhodium (Rh), ruthenium (Ru), and the most frequently used one, finely divided palladium metal (Pd) adsorbed onto the surface of powered charcoal (carbon), which is symbolized as Pd-C and read as "palladium on carbon." The experiment itself is not complex. A solution of the compound to be reduced is combined with a small amount of the catalyst, and the mixture is stirred under a pressurized atmosphere of hydrogen gas. The most significant safety caution is that air (oxygen) needs to be rigorously excluded to prevent catalyzing the explosive reaction between hydrogen and oxygen gases.

The catalytic hydrogenation reaction of pi bonds gives a syn addition of hydrogen. The reactions of both double and triple bonds result in the corresponding fully saturated structure (Figure 0947).

Figure 0947

Catalytic hydrogenation of double and triple bonds.

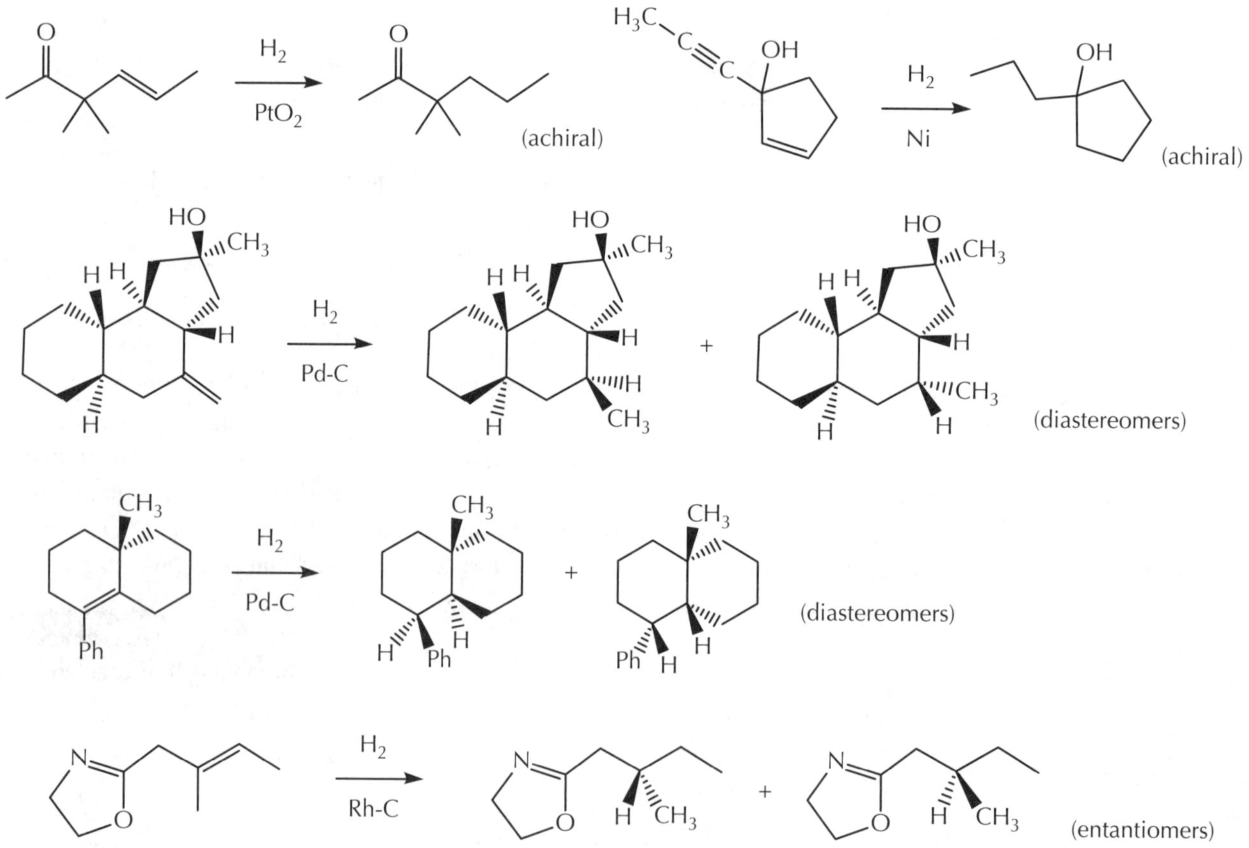

Under the typical catalytic hydrogenation conditions, the reactions of triple bonds are generally difficult to stop after one addition when using the common catalysts because both addition reactions are quite fast, and the first reduction (triple bond to double bond) is difficult to intercept before the second reduction begins. However, the rate of hydrogenation for carbon-oxygen and carbon-nitrogen pi bonds is significantly slower than for carbon-carbon pi bonds, and so functional groups with the carbon-oxygen and carbon-nitrogen double bonds are not reduced under conditions where carbon-carbon double and triple bonds react.

The exact details of the catalytic hydrogenation mechanism are difficult to study. The reaction involves dissolving a gas (hydrogen) in a solution with the reaction substrate (the molecule being reduced) and what then happens occurs on the surface of an insoluble particle of charcoal embedded with clusters of metal atoms (Pd-C). The steps are not easily represented using curved arrows, either. A picture of the catalytic process is illustrated in Figure 0948.

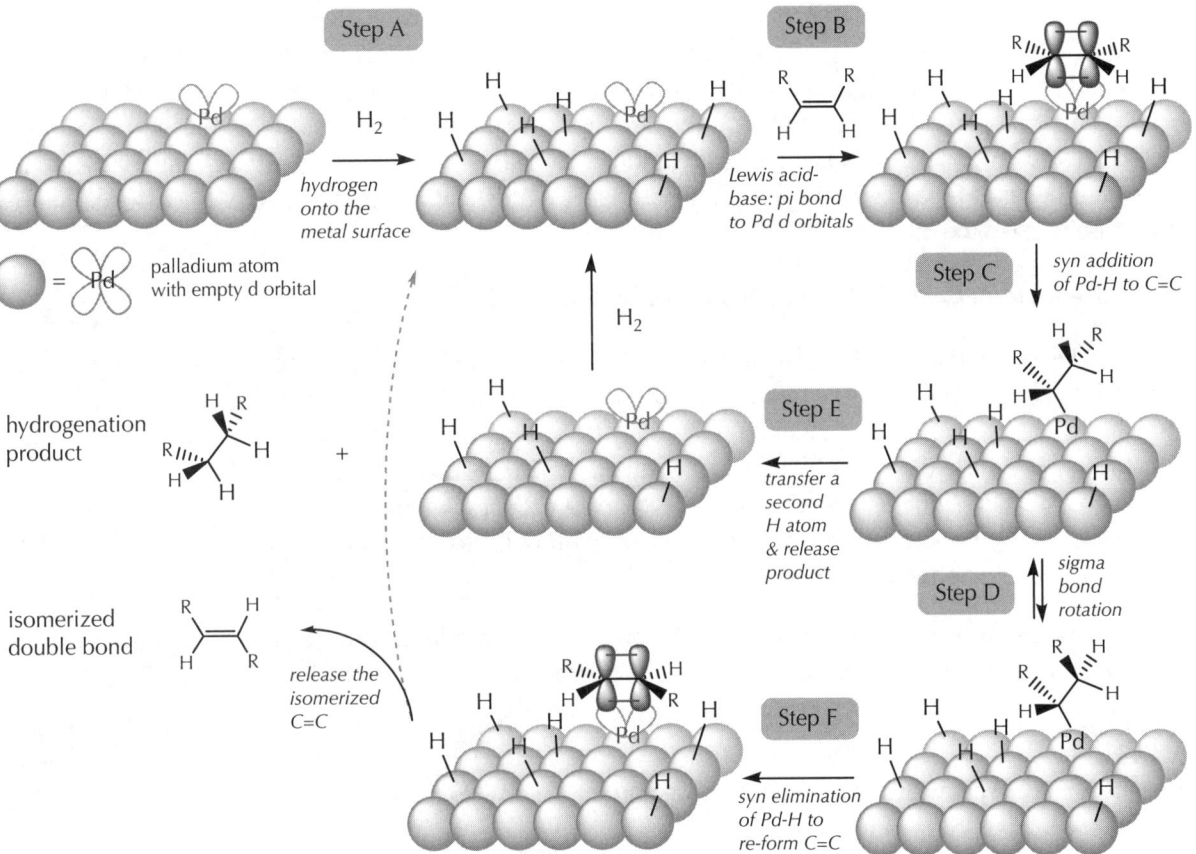

Figure 0948

Mechanism for catalytic hydrogenation of double and triple bonds.

The significant steps in the catalytic hydrogenation mechanism are:

Step A: the Pd-C surface adsorbs molecular hydrogen, and some of the hydrogen molecules are separated into hydrogen atoms that are associated with the palladium atoms

Step B: palladium atoms have open shell structures with empty d orbitals that are capable of forming a Lewis acid-base complex with the pi bond of the substrate

Step C: the equivalent of a Pd-H syn addition reaction takes place, where a hydrogen atom is transferred to one end of the pi bond and the other end is bonded to a palladium atom on the surface

Step D: after the pi bond has been broken, the sigma bond can rotate freely, and two reaction pathways are possible:

Step E: a second hydrogen atom can be transferred from the surface, completing the syn addition of hydrogen, and the catalyst returns for another cycle; or

Step F: the original Pd-H addition can be reversed (elimination of Pd-H), reforming the pi bond in a mixture of stereochemical configurations, resulting in a pathway in which a cis double bond might be changed to a more stable trans double bond (which can still return to the surface for reduction)

Steps D and F acknowledge an experimental reality: Reduction is not the only process that can happen on the surface of the metal catalyst. Stereochemical and structural isomerization can happen, also. These outcomes are not easily predicted, and so experimental evidence is needed to know they have happened. Also, the isomerized structures are still capable of being reduced, and so the fact that the isomerization happened might never be detected because the reduced products are all the same compound.

The cis-trans isomerization can have important consequences, and it is likely you have heard about it without realizing it. Returning to Wilhelm Norman and his partially hydrogenated vegetable oils: According to Steps D and F in the description of the catalytic hydrogenation mechanism, some of the *cis* double bonds present in the naturally occurring vegetable oils are reduced, some of them do not undergo reaction at all, and some of them are isomerized to the more stable *trans* configuration. This isomerization, which takes place during the partial hydrogenation process, is one origin of the so-called "trans fats" found in foods that are prepared with these compounds (many natural animal fats also have trans fats in them). The concern with dietary trans fats is a correlation with heart disease, first detected in the 1950s. Our bodies are better at breaking down fats with cis double bonds, and so the trans compounds persist and can accumulate in the arteries.

B. (Z)-Alkenes from Catalytic Hydrogenation with Poisoned Catalysts

In 1952, Herbert Lindlar developed a useful way to carry out addition of one equivalent of hydrogen to a triple bond, where the first syn addition reaction of hydrogen takes place, giving the (Z)- double bond, and where the second addition reaction is extremely slow or does not occur at all. The hydrogenation catalysts listed previously (Pd-C, PtO_2, Rh, Ru, Ni, etc.) are actually too effective to do this, as both the first and second hydrogenation reactions of the triple bond can occur competitively. Lindlar reasoned that what was needed was a poorer catalyst that would slow down both hydrogenation reactions and disfavor the addition to a double bond more than a triple bond.

Lindlar's catalyst, known as a "poisoned catalyst," involves two modifications: (1) the palladium atoms are supported on an oxygenated surface, usually calcium carbonate ($CaCO_3$) or barium sulfate ($BaSO_4$), which

Figure 0949

Hydrogenation of alkynes with poisoned catalysts stops at the (Z)-alkenes.

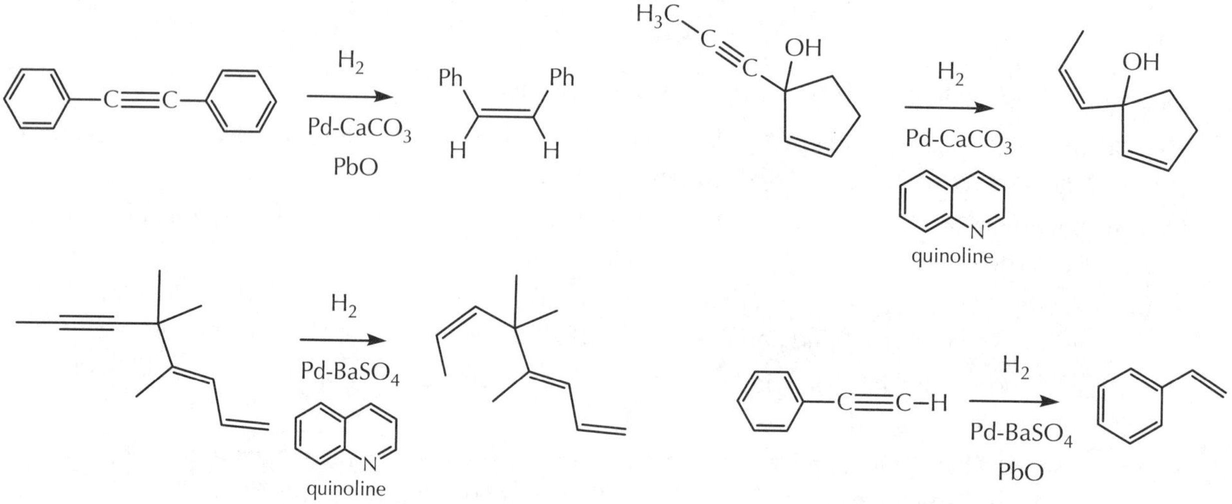

is a less absorptive surface than carbon and likely reduces the reactivity of the palladium atoms; and (2) an additive is used (lead compounds such as lead oxide, PbO, or organic amine bases, such as quinoline) that are also likely to change the reactivity of the metal atoms as well as clutter the surface of the catalyst, making it harder for the substrate to reach and react. Using this information, four different but equivalent "poisoned catalyst" reaction conditions can be written (Figure 0949).

An imagined representation for the surface for one of the poisoned catalyst conditions is shown in Figure 0950.

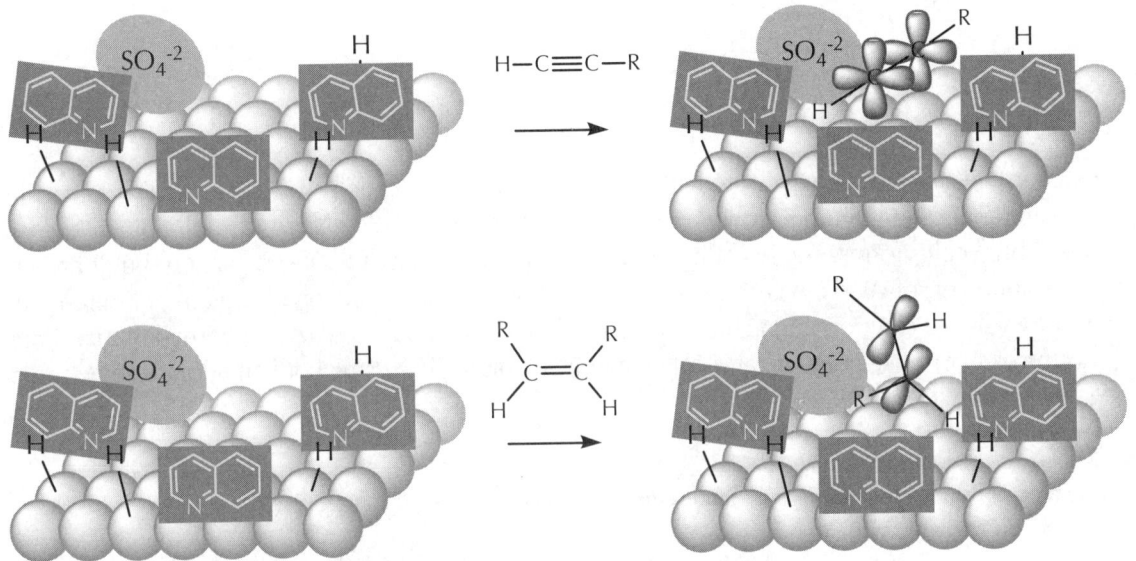

Figure 0950

A possible difference for the rate of hydrogenation between alkynes and alkenes with poisoned catalysts.

To begin with, the reactions are slower: The concentration of hydrogen atoms on the surface is lower because of the additives, and the polar ions of the supports ($BaSO_4$ or $CaCO_3$) will block some of the catalytic sites.

The faster hydrogenation rate for an alkyne is also due to its ability to squeeze into spots more easily, on that crowded surface, than an alkene can do. The linear ends of the alkyne, which have less exposed molecular surface area compared with trigonal planar ends of the alkene, allows for its easier complexation to the cluttered surface than the greater surface area demands of the corresponding alkene.

C. (E)-Alkenes from Dissolving Metal Reduction

Alkynes are also reduced by sodium or lithium metal in liquid ammonia. In contrast to catalytic hydrogenation, this reaction produces *trans* alkenes. For example, 3-octyne is reduced to (E)-3-octene with sodium metal in liquid ammonia. Figure 0951 (on the next page) shows a few examples of this reaction, which, similarly with the poisoned catalyst hydrogenation, is observed to work effectively for triple bonds and not for double bonds, and so double bonds are not reduced under these conditions.

Figure 0951

Reduction of alkynes to (E)-alkenes using the dissolving metal reduction.

$$CH_3CH_2-C{\equiv}C-CH_2CH_2CH_2CH_3 \xrightarrow[NH_3 \text{ (liq)}]{Na} \text{(E)-alkene}$$

$$Ph-C{\equiv}C-Ph \xrightarrow[NH_3 \text{ (liq)}]{Li} \begin{array}{c} PhH \\ C{=}C \\ HPh \end{array}$$

(alkyne with two OCH₃ stereocenters) $\xrightarrow[NH_3 \text{ (liq)}]{Na}$ (E)-alkene with two OCH₃ stereocenters

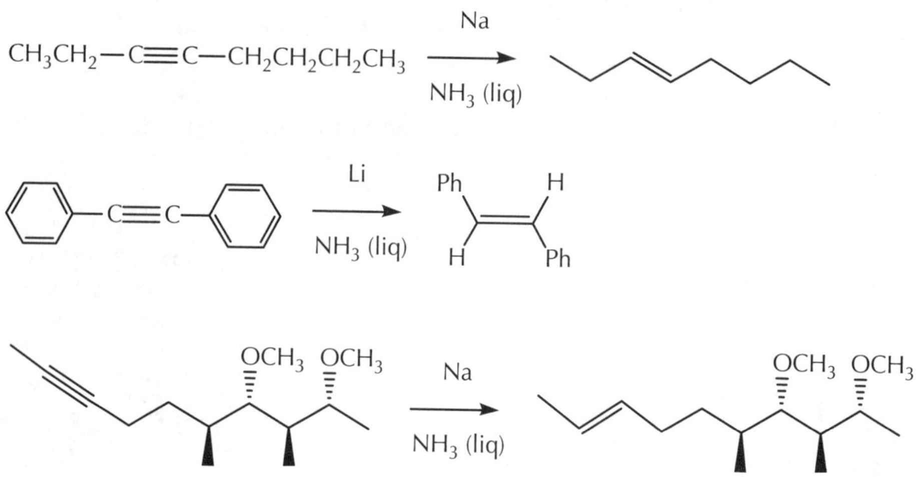

The reaction mechanism is complicated. In liquid ammonia (boiling point -33°C), a reaction takes place between ammonia and lithium or sodium metal, in which the metal dissolves and a solution with a beautiful, deep blue color forms. The exact nature of this solution is a matter of speculation, but it has been called a solution of solvated electrons. Formally, we think the usual reaction of sodium or lithium metal, to transfer a valence electron and achieve a closed shell configuration, takes place. The liquid ammonia accepts the electron to become what is called a radical anion, possessing both an unpaired electron and a negative charge (Figure 0952).

Figure 0952

Schematic mechanism for the dissolving metal reduction of alkynes to (E)-alkenes.

$$Na \text{ (metal)} + NH_3 \text{ (liquid)} \xrightarrow{-33°C} Na^{\oplus} [NH_3]^{\cdot\ominus} \text{ (deep blue color)}$$

$$Na^{\oplus}[NH_3]^{\cdot\ominus} + R-C{\equiv}C-R \xrightarrow{\text{electron transfer}} R-\overset{\ominus}{\underset{\cdot}{C}}{=}\overset{\cdot}{C}-R \xrightarrow[\text{proton transfer}]{NH_3} \begin{array}{c} R \\ C{=}\overset{\cdot}{C}-R \\ H \end{array}$$

$$\xrightarrow[\text{electron transfer}]{Na^{\oplus}[NH_3]^{\cdot\ominus}} \begin{array}{c} RR \\ C{=}C{:}^{\ominus} \\ H \end{array} \underset{\text{cis-trans equilibrium}}{\rightleftharpoons} \begin{array}{c} R\ominus \\ C{=}C \\ HR \end{array} \xrightarrow[\text{proton transfer}]{NH_3} \begin{array}{c} RH \\ C{=}C \\ HR \end{array}$$

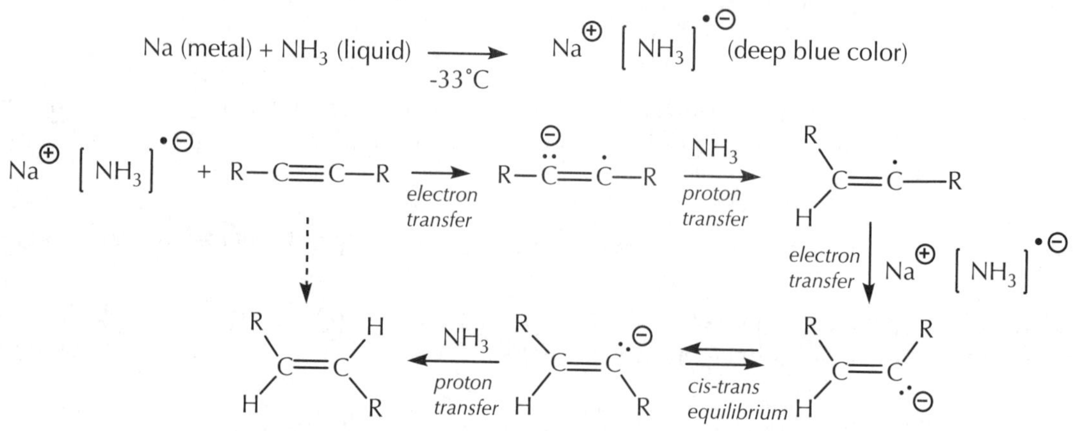

When the alkyne is added to this blue solution, a series of two separate, single-electron transfers and two proton transfers takes place, resulting in the alkene with more thermodynamically stable (E)- configuration. The cis-trans equilibration of the anion that gets protonated in the last step of the mechanism establishes the stereochemical outcome.

Reproducing the mechanisms for all of these reduction reactions are not typical tasks that you are asked to do on exams. The generalizable features of the mechanisms are not being described here and are usually part of a more advanced or intermediate level course. However, it is still useful for you to see that there are rationales for these reactions, and they can help you to picture and remember the outcomes.

The reduction reactions in this section demonstrate that alkynes and alkenes can be converted to their saturated alkane counterparts, or, by adjusting the reaction conditions, that alkynes can be transformed selectively into either the corresponding (E)- or (Z)-alkene, with no further reaction taking place to those or any other double bonds in the molecule.

D. Multistep Transformations and Synthesis

The concepts of multistep transformations and synthesis were introduced in the last section in Chapter 8. You have undoubtedly noted the high diversity of reactions appearing in this chapter, which increases substantially the number of longer and more complex transformation sequences that becomes possible.

Organic chemists are constantly developing new methods for synthesizing compounds having industrial, medicinal, or biological importance. For example, modifications of natural hormones and antibiotics are synthesized in attempts to understand how the natural substances function and to create more effective ones. Nature only provided one molecule, penicillin G, from mold, that started to be used as an antibiotic by the military in 1942. Only organic synthesis gives us ampicillin, amoxicillin, and over 100 next generation antibiotics based on the penicillin structure.

Individual reactions are comparable to puzzle pieces that can be assembled in many different ways to create completely different pictures.

The characteristic features of reactions need to be considered together:

(1) the overall bonding changes—complexation/decomplexation (forming or breaking a sigma bond), substitution (sigma to sigma exchange), addition (breaking a pi bond to give two new sigma bonds that are adjacent), and elimination (breaking two adjacent sigma bonds to form a new pi bond)

(2) mechanism and its effect on regioselectivity and stereoselectivity (e.g., the hydroboration-oxidation reaction of double bonds gives a syn, anti-Markovnikov addition of water; primary, sp^3 carbon-leaving group bonds react with anions whose conjugate acid pK_a values are less than 30 via an S_N2 mechanism to give the product of substitution with inversion of configuration, etc.)

(3) functional group relationships in transformations and the reagents for accomplishing them (e.g., a double bond with a (Z)- configuration can be prepared from a triple bond using hydrogenation with a poisoned catalyst (Pd-$CaCO_3$, PbO); the carbon-carbon bond that connects a triple bond to a primary carbon can be prepared by the S_N2 reaction between the molecule in which the primary carbon has a leaving group and the sp carbon of the triple bond is an anion)

The cumulative nature of the organic chemistry subject matter is critical to appreciate. At this point in the text, nearly every fundamental concept you need to predict products and explain mechanisms has been introduced. From here on out, the explanations will be built upon the content in the first 8 chapters, comparable to the way you use the same words to construct many different sentences that can convey many different meanings.

Figure 0953 (on the next page) provides one summary of the reactions and functional group relationships that have been introduced so far. This type of table is exactly the type of summary that you should, in principle, be constructing for yourself after reading and developing an understanding of the chemistry. Figure 0953 is not a dictionary of answers. Each item points to a concept that can be generalized and applied to new and unfamiliar examples in nearly unlimited different combinations. The map that you create and then use to help answer question is more important than this one because you will have optimized and internalized it through practice. Authors and instructors hesitate to provide these summaries for students precisely because providing them might seem like a convenient shortcut for learning. *Caveat emptor!*

720 CHAPTER 9 Electrophilic Addition II: Halogenation, Oxidation, and Reduction Reactions

Figure 0953
One of many summaries of reactions and functional group relationships.

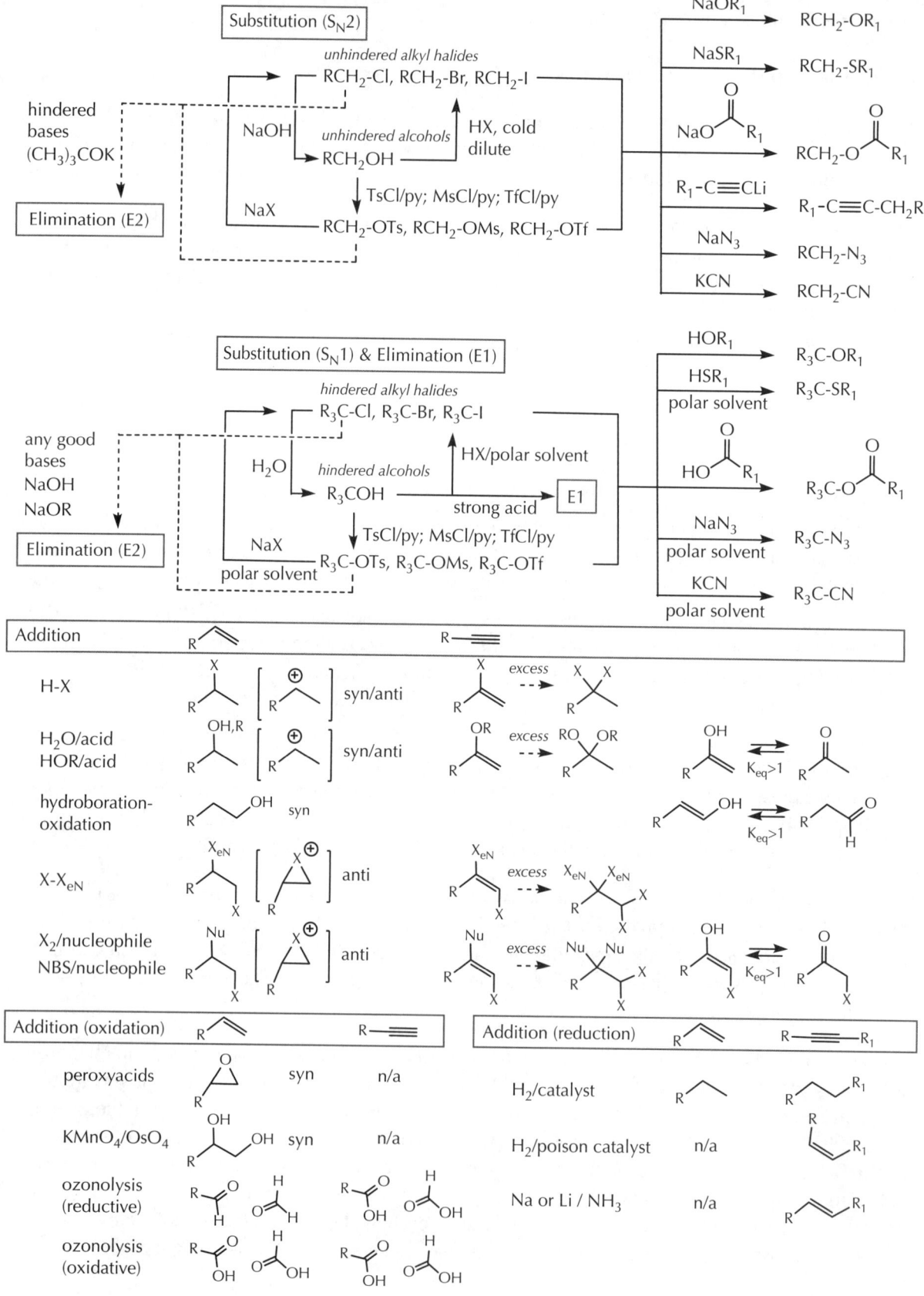

The next example demonstrates how a complex transformation question can be answered.

As stated at the end of Chapter 8, although you are being introduced to the thought process for answering this type of question, you still need some coaching to do them on your own. Here is the transformation (Figure 0954):

Figure 0954

An organic synthesis challenge: the transformation of acetylene to disparlure.

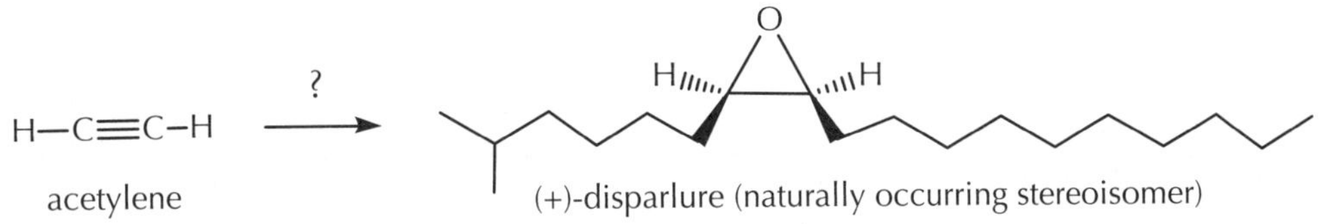

H—C≡C—H   →[?]   (+)-disparlure (naturally occurring stereoisomer) + enantiomer

acetylene

Disparlure is the compound used by the female gypsy moth to attract the male moths for mating. The species name, from 1758, is *Lymantria dispar dispar*, and naming the sex attractant "disparlure" is both self-evident and a little amusing.

Insects communicate with each other, and with their environment, by means of organic chemicals secreted in minute amounts. These substances, known as pheromones, have a wide range of structures and include a variety of functional groups.

The gypsy moth is a devastating pest in hardwood forests. The sex pheromone for the gypsy moth is a chiral oxirane. The dextrorotatory compound, (+)-disparlure, is a potent attractant for male gypsy moths, but the enantiomer, the levorotatory compound, has no such activity. Once its structure was determined, large amounts of the racemic disparlure were easily prepared starting from acetylene (ethyne), hence the transformation problem, above.

The chemists who isolate natural products and determine their structures face huge experimental challenges. How to start with an organism, such as a moth, and find the miniscule amount of an unknown compound that is responsible for luring another moth? Even after isolating the suspected target molecule(s), the structural determinations need to be carried out on only a few milligrams or less of the isolated material—often too little to be seen with the naked eye. In the modern era, advances in methods for separation and isolation, as well as spectroscopic techniques, have made these searches much more efficient (and possible) than they were even 50 years ago.

A typical hunt might look this way. When disparlure was isolated in 1970, the extraction of the abdominal tips of 500,000 virgin gypsy moth females yielded only 75 mg of the pheromone that attracts males. This quantity of active material had to be separated from approximately 250 g of fatty acids and their esters and 70 g of steroids, mostly cholesterol. The purification process, which is like looking for the proverbial needle in a haystack, was aided by biological assay methods in which male gypsy moths were exposed to the fractions obtained in various separation steps. The fractions that did not excite the male moths were discarded and those that did were purified further until finally a single component that was highly attractive to the males was isolated.

Although the question posed in Figure 0954 is formatted as a left-to-right transformation, knowing that the analysis needs to begin with the structure of the product is a critical insight. The connection between acetylene and the disparlure structure is not obvious. If you simply start to write down some reactions where a triple bond is the starting material, you are not likely to end up at disparlure.

722 CHAPTER 9 Electrophilic Addition II: Halogenation, Oxidation, and Reduction Reactions

When faced with this kind of situation, working backwards from the product is the best strategy: imagining what some possible last steps might have been and worrying about the acetylene, later. Working backwards from a desired final product will usually create diverse options, as branches, that all converge on the product, providing the experimentalist with multiple pathways rather than only one. The term used for the "working backwards" strategy for solving synthesis problems is "retrosynthetic analysis."

Figure 0955

The retrosynthetic analysis of disparlure: epoxidation of a (Z)-alkene.

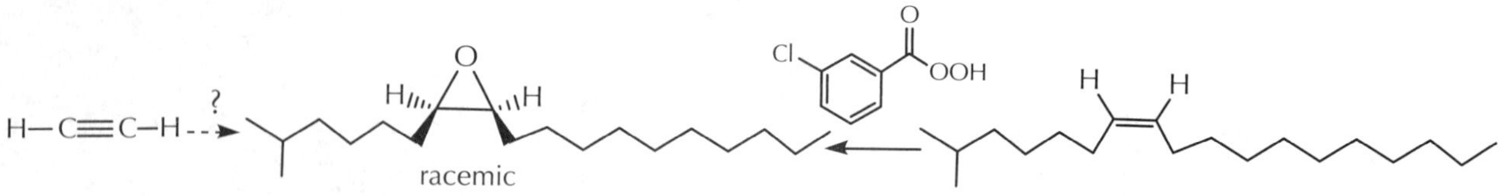

The only functional group in disparlure is an epoxide. Fortunately, thanks to the information in this chapter, you have a reaction that results in the formation of epoxides: the epoxidation reaction of double bonds using peroxy acids. The alkyl groups on disparlure's oxirane ring are shown to be cis. Because the epoxidation reaction is a syn addition, the cis stereochemistry in disparlure must have resulted from capturing the cis stereochemistry of the double bond. Thus, epoxidation of the corresponding (Z)-alkene will give the racemic disparlure.

The connection between acetylene and the (Z)-alkene is no more obvious than the epoxide was, but you can connect the (Z)-alkene to its corresponding triple bond because (Z)-alkenes result from the hydrogenation of alkynes using the poisoned catalyst conditions (Figure 0956).

Figure 0956

The retrosynthetic analysis of disparlure: forming a (Z)-alkene.

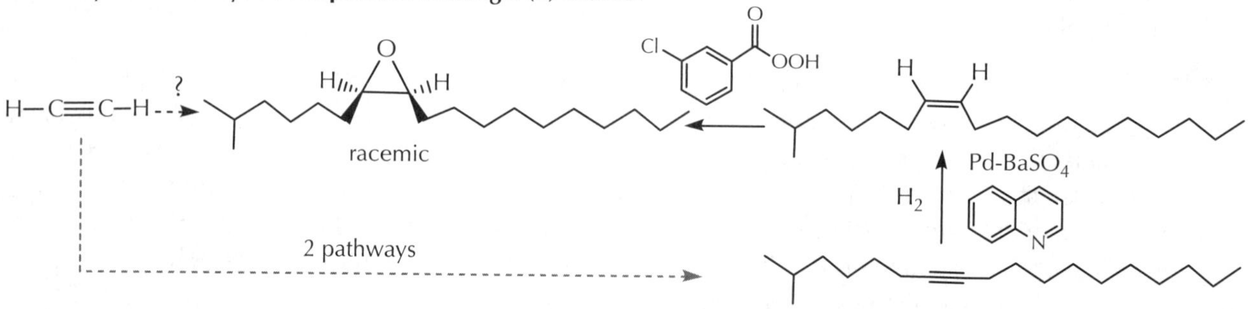

Now there are two direct connections between acetylene and the alkyne (Figure 0957).

Figure 0957

The retrosynthetic analysis of disparlure.

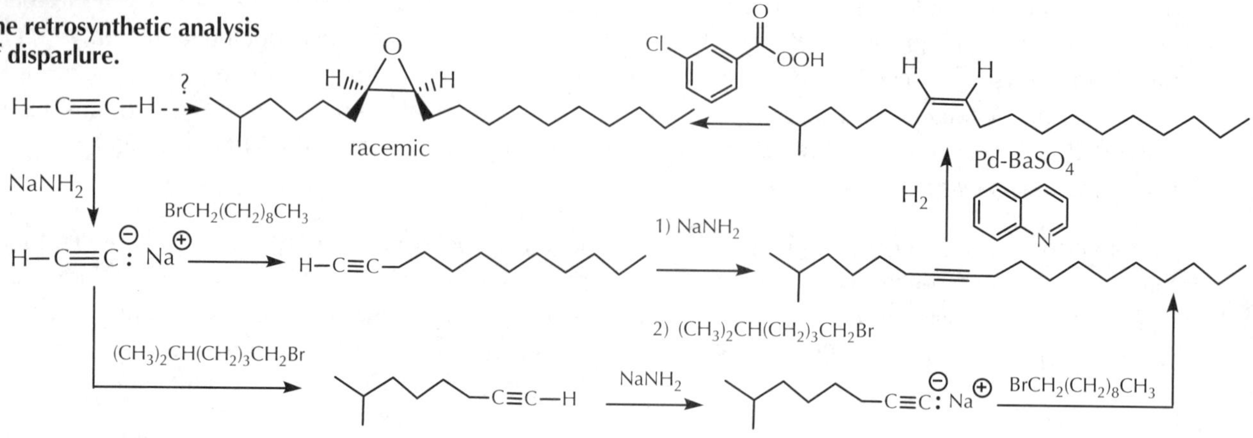

Both of the alkyl groups attached to the sp carbons or the alkyne are primary, so the carbon-carbon bonds between both of the sp carbons and the primary carbons to which they are attached could have resulted from S_N2 reactions. The sp C-H bonds of acetylene are acidic enough (pK_a 26) to be deprotonated by a strong base, for example, $NaNH_2$, and used as a nucleophile with primary carbon-leaving group bonds. The two substitution reactions must be done in sequence, not simultaneously. Either of them could be done first (both sequences are shown), and that completes the analysis.

Once the miniscule amount of disparlure was isolated from its natural source, and had its structure determined, kilograms of the compound could be prepared using exactly the methods proposed by this retrosynthetic analysis. Pest control using traps that are baited with sex pheromones are more environmentally friendly compared with widespread use of pesticides.

Just as in the stories about antibiotics, however, pheromone-based traps have contributed to the evolution of the moth species. Those few male moths who did not respond to the disparlure because of a mutation would not usually have been successful in reproducing. However, as the baited traps reduce the population of those moths that do respond to disparlure, the mutant moths have greater success in mating than they would otherwise, and their "resistance" to disparlure becomes a new trait, leading scientists on another hunt to find the new pheromones being used.

Compare the disparlure transformation with the ones shown in Figure 0958.

Figure 0958

Transforming an alkyne to the corresponding *meso*-dibromide and *meso*-diol.

(1*R*,2*S*)-1,2-dibromo-1,2-diphenylethane

(1*R*,2*S*)-1,2-diphenylethane-1,2-diol

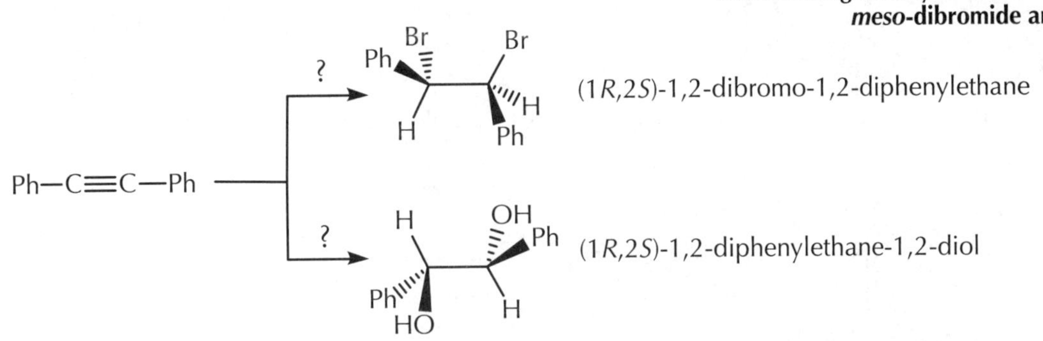

These questions ask for how to transform the single alkyne into the *meso* stereoisomers of a pair of structurally similar products: a dibromide and a diol.

The direct transformation of the alkyne into either of these products does not match any of the known reactions, and so a multistep transformation is implied. As before, retrosynthetic analysis, working backwards from the final product, is the recommended strategy (Figure 0959).

Figure 0959

Stereochemical analysis for transforming of an alkyne to the corresponding *meso*-dibromide and *meso*-diol.

redraw via bond rotation to place the 2 bromines *anti* (staggered)

redraw via bond rotation to place the 2 hydroxyls *syn* (eclipsed)

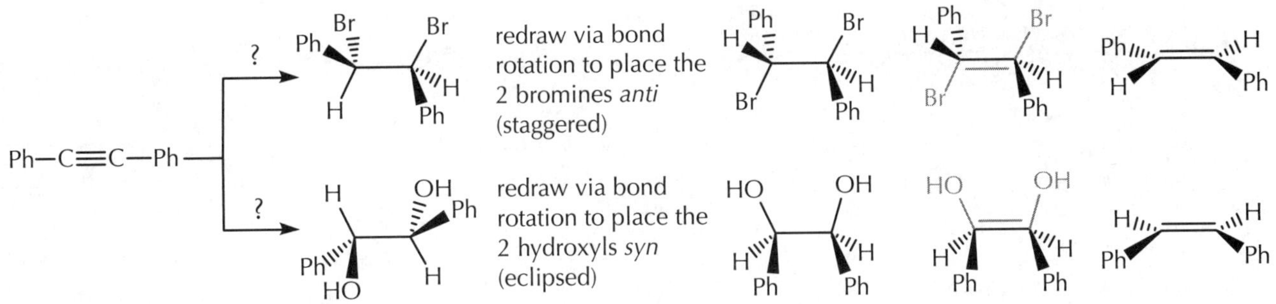

724 CHAPTER 9 Electrophilic Addition II: Halogenation, Oxidation, and Reduction Reactions

From this chapter, you have seen that the anti addition reaction of molecular bromine to a double bond can give a dibromide product, while the syn addition of two hydroxyl groups is possible using reagents such as basic potassium permanganate or osmium tetroxide. In both of these examples, the stereochemistry of the alkene that is needed must be determined by a stereochemical analysis of the structure of the product.

To determine whether the (E)- or (Z)-alkene is needed for the bromination reaction, the best strategy is to redraw the product molecule into a conformation where the anti relationship of the bromine atoms is clearly present. Placing the two bromine atoms opposite from one another, drawn in the plane of the page, gives the clearest picture of the relationship between the other four groups attached to those carbons. By imagining that the two bromine atoms are faded away from the structure, the *trans* relationship of the two phenyl rings is evident, and so the (E)-alkene would undergo anti addition of bromine to give the observed *meso*-dibromide product.

When performing the stereochemical analysis on the diol, a conformational drawing is needed in which the two hydroxyl groups look as they would after a syn addition. In this case, fading them away leaves the two phenyl rings *cis*, and so the (Z)-alkene would undergo a syn dihydroxylation reaction to give the observed *meso*-diol product.

Under the proper conditions, the alkyne can be transformed stereoselectively to either the (E)-alkene, using the dissolving metal reduction, or the (Z)-alkene, via hydrogenation with a poisoned catalyst (Figure 0960).

Figure 0960

The completed transformation of an alkyne to the corresponding *meso*-dibromide and *meso*-diol.

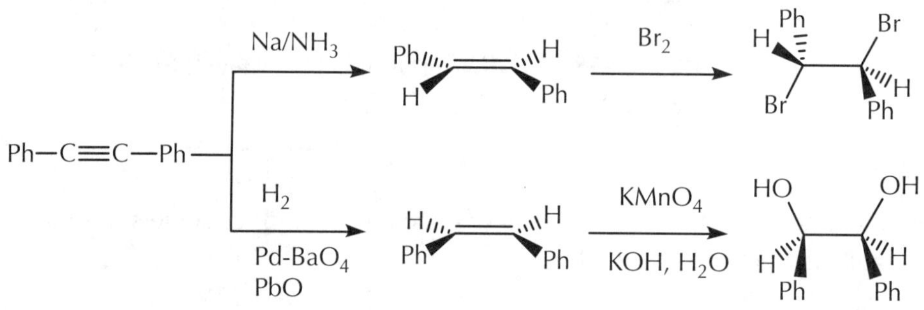

9.3 Reduction Reactions

PRACTICE QUESTIONS

9.10 Complete the following equations. Provide the structures of stereoisomeric products, if anticipated, and identify the stereoisomeric relationship (enantiomers, diastereomers) or stereochemical identity (an achiral molecule with no stereocenters, an achiral *meso* isomer). If stereoisomers are not formed, simply draw the one product.

(a)

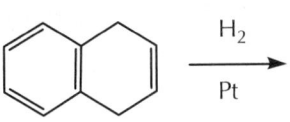

stereochemical relationship or identity:

(b)

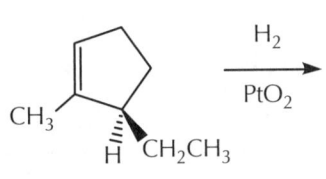

stereochemical relationship or identity:

(c)

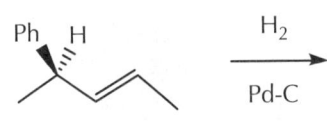

stereochemical relationship or identity:

(d)

Ph₂C=CPh₂ (Ph, Ph / Ph, Ph) → H₂, Pd-C

stereochemical relationship or identity:

(e)

Ph—C≡C—CH₃ → H₂, PtO₂

stereochemical relationship or identity:

9.11 Complete the following equations. Provide the structures of stereoisomeric products, if anticipated, and identify the stereoisomeric relationship (enantiomers, diastereomers) or stereochemical identity (an achiral molecule with no stereocenters, an achiral *meso* isomer). If stereoisomers are not formed, simply draw the one product.

(a) Ph—C≡C—CH₃ $\xrightarrow{\text{H}_2,\ \text{Pd-CaCO}_3,\ \text{PbO}}$

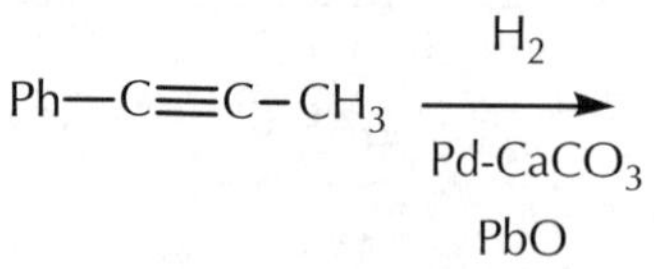

stereochemical relationship or identity:

(b) [cyclohexene with methyl and propargyl substituent] $\xrightarrow{\text{H}_2,\ \text{Pd-BaSO}_4,\ \text{quinoline}}$

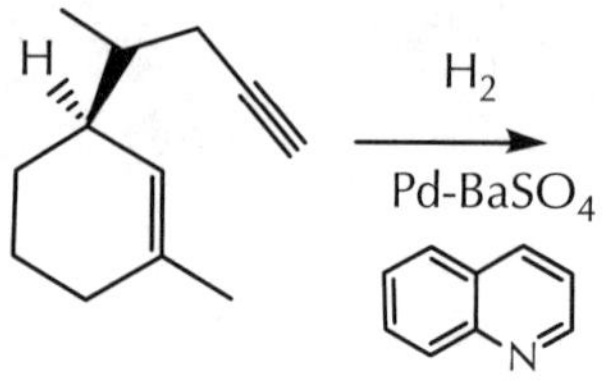

stereochemical relationship or identity:

(c) [cyclopentane with CH₃ and isopropyl-alkyne substituents] $\xrightarrow{\text{Li, NH}_3\ (\text{liq})}$

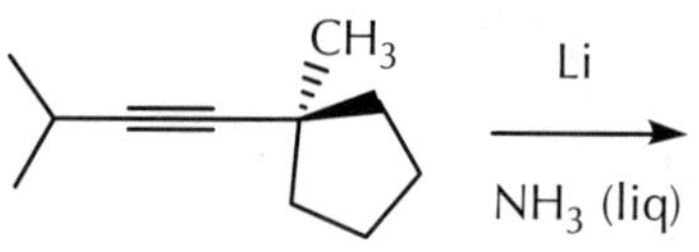

stereochemical relationship or identity:

(d) Ph—C≡C—C≡N $\xrightarrow[\text{30 atm, 85 °C}]{\text{H}_2,\ \text{Ni}}$ $C_9H_{13}N$

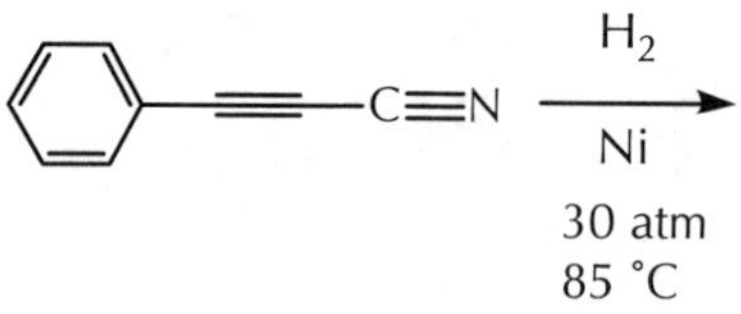

stereochemical relationship or identity:

(e) [cyclohexane with acetyl and propynyl substituents] $\xrightarrow{\text{H}_2,\ \text{Pd-BaSO}_4,\ \text{PbO}}$

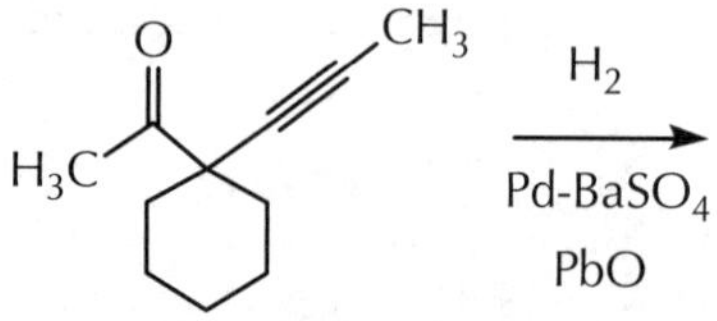

stereochemical relationship or identity:

9.12 The following alkyne can be selectively transformed into a *meso* epoxide or the stereoisomeric racemic mixture. Complete the following reaction sequences as required. A note about naming: The epoxidation of propene results in a racemic mixture of (*R*)- and (*S*)-2-methyloxirane.

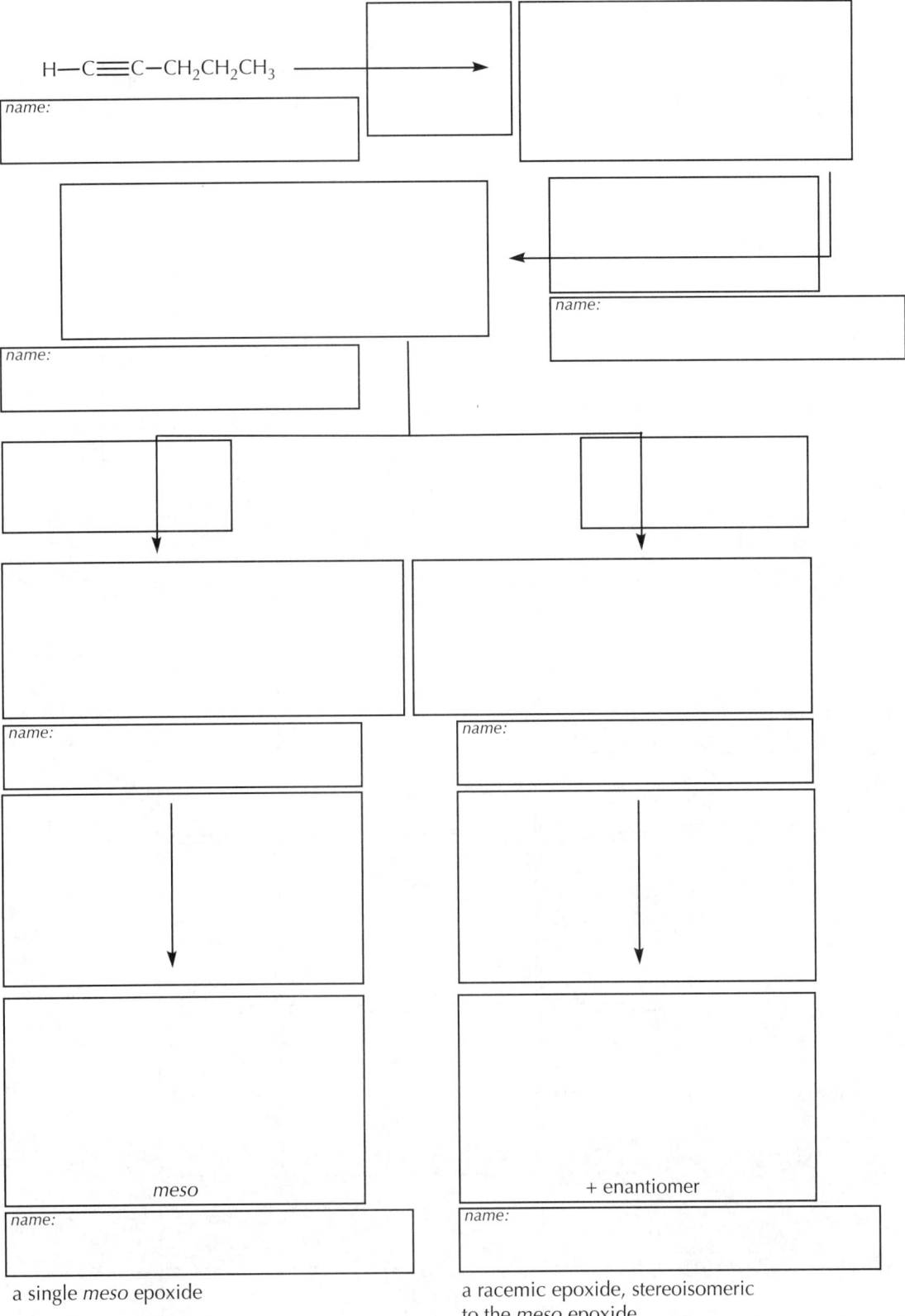

728 CHAPTER 9 Electrophilic Addition II: Halogenation, Oxidation, and Reduction Reactions

9.13 Carry out the following transformations. The suggested number of steps is indicated by the answer spaces. Note that there may be alternatives that are longer and equally valid solutions.

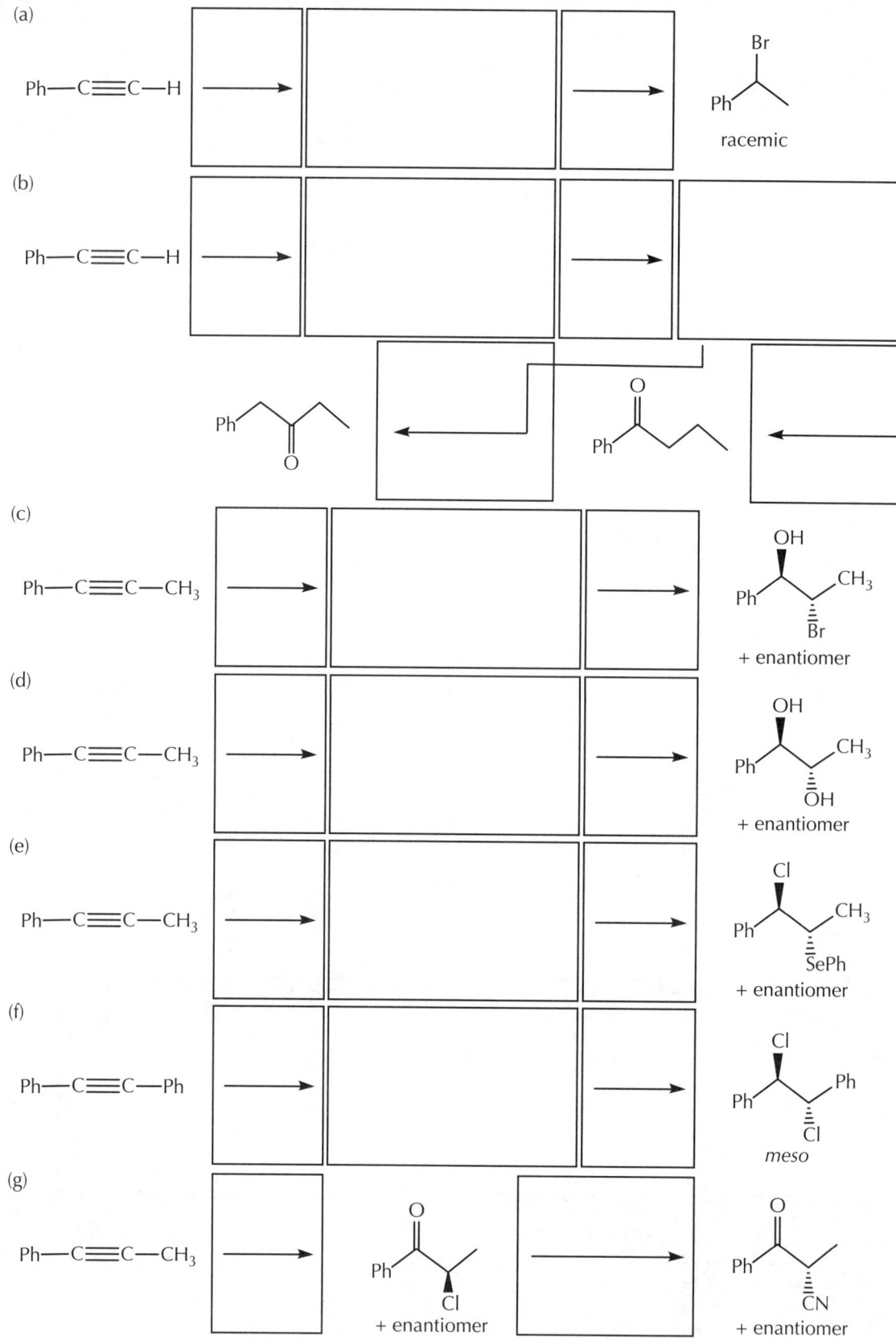

9.14 The two reactions in the following short transformation were both observed to occur with a high degree of selectivity.

(a) Complete the transformation.

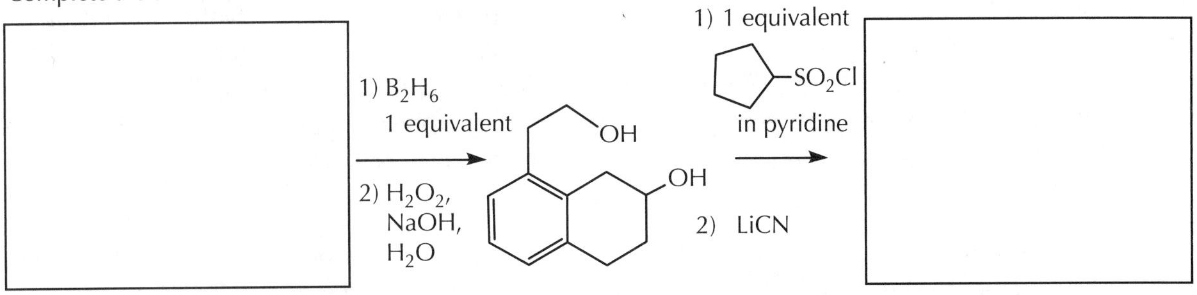

The compound shown as an intermediate in this transformation was observed to be sensitive to acidic conditions, under which an unusual dehydration product formed.

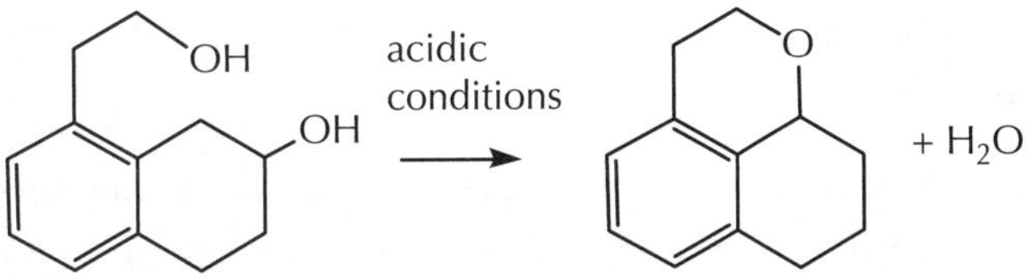

The reaction mechanism was proposed to involve a carbocation rearrangement reaction, followed by an intramolecular capture of that carbocation.

(b) Given that description, draw a complete, stepwise mechanism for the acid-catalyzed dehydration reaction. Use HB as your generic Brønsted acid and B⊖ as its conjugate base.

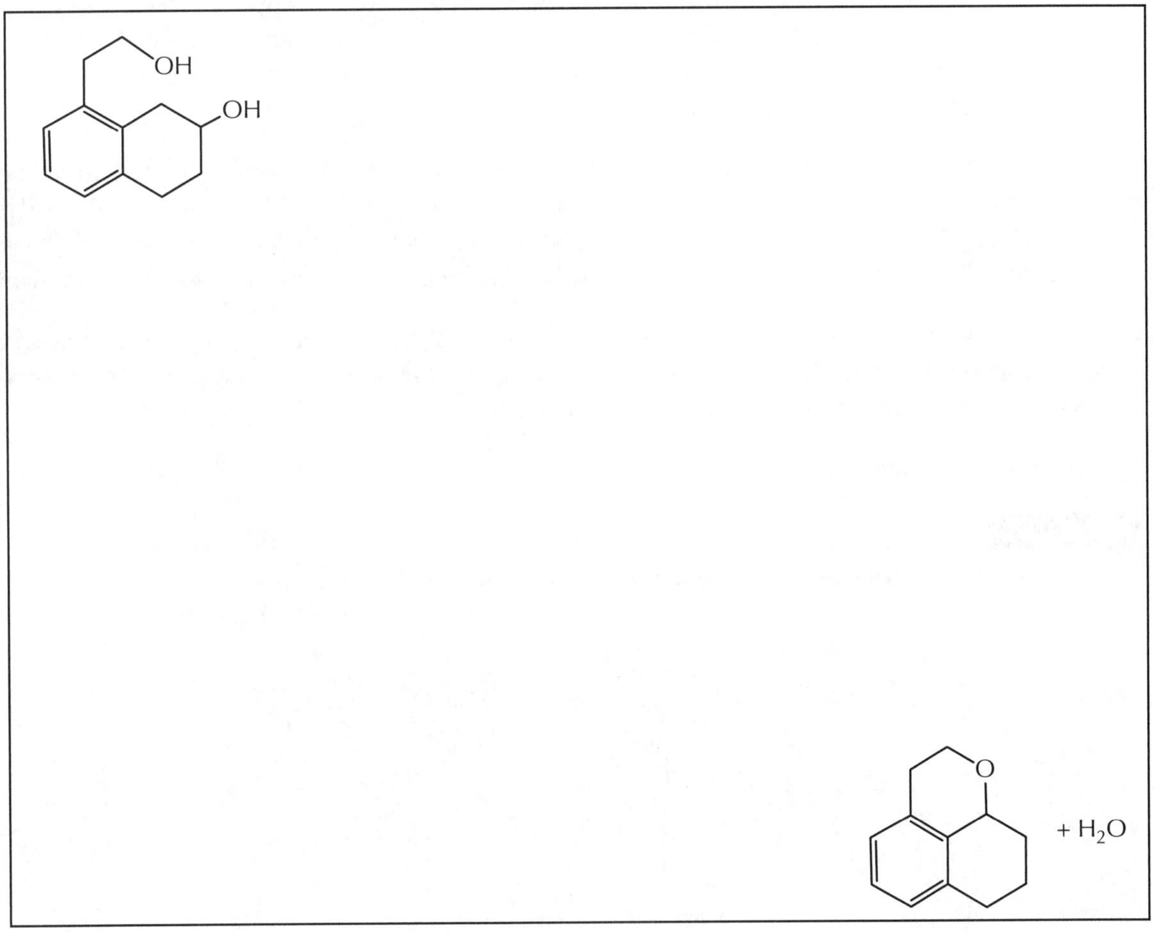

Reflections About Science: The Power of Mechanistic Organic Chemistry

For most of its first century, about 1850–1950, organic chemistry was organized according to functional group relationships. An alcohol or an alkene was defined more by its reactivity than its molecular structure, because detailed knowledge about structure was limited. Subatomic particles were not discovered until 1900–1920, and the electronic model of bonding, including sigma and pi bonds, did not emerge until the 1930s. And so, although "1-pentene plus bromine gave a dibromo compound" and "1-pentene plus a peroxy acid gave an epoxide," there was no underlying mechanistic model that linked these reactions because the functional groups were simply different.

Mechanism needed to evolve with the theory of chemistry. For a hundred years, the idea of all atoms being tetrahedral dominated our model for bonding and created dilemmas such as how to explain anti addition (through tetrahedra sliding on their edges) versus syn addition (the quite rational opening of a shared edge between two tetrahedra). It's tempting to look back at those tetrahedral atoms and see them as a bit silly, sort of the "flat earth model" of chemistry, based on our contemporary ideas, but that would be a grievous error of hubris.

If the past is any indication, which it certainly is in science, the people from our next century will look back on most of our models and see them as equally naïve. And then it will happen a hundred years after that, and a hundred more after that. The fact is that the models we have today will not only continue to improve as they evolve, but they probably all qualify as merely the next best version of flat earth relative to any absolutely true description of nature.

This idea does not take away from our progress, but it reminds us why we still pursue science. The phenomena do not change (an apple falls from a tree, bromine discolors when it reacts with an alkene), but our understanding and explanations are always (hopefully) improving and allow us to make better predictions about the things we have not yet done.

Through the 1950s and 1960s, the first models for a core mechanistic picture of organic chemistry became clear. There appeared to be strong underlying ideas that transcended functional group identity and related to bonding and changes in bonding. Today, if presented with a new chemical observation, the first question is not "what is this new thing?" but rather reasoning by analogy: "what familiar thing is it like?"

Reasoning by analogy is the name of the game in understanding contemporary organic chemistry. Recall how the atoms in ozone can be swapped out in hundreds of different combinations and yet all of these variations give exactly the same addition reaction reactivity (Figure 0943).

Look at the observation reported in Figure 0961.

Figure 0961

Reasoning by analogy in the oxymercuration reaction: structural comparison.

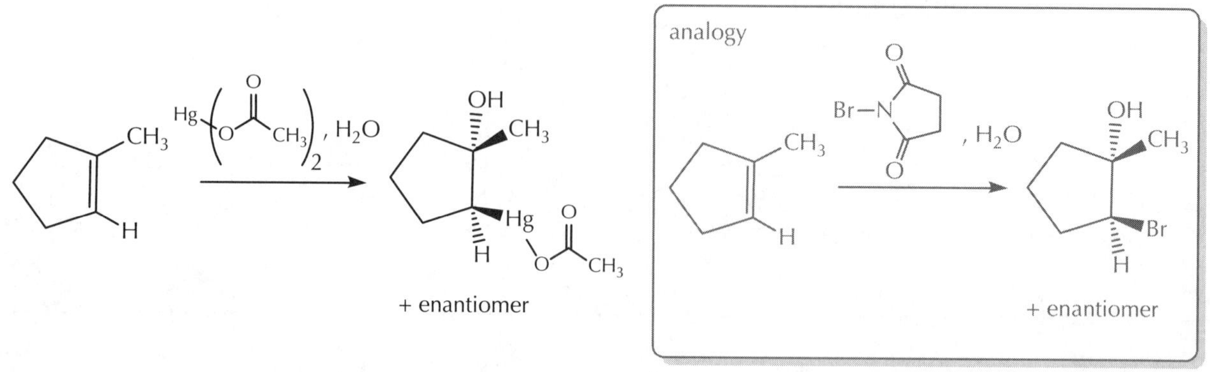

This transformation, called oxymercuration, is new and unfamiliar to the chemistry in this text. Either it is a totally new type of reaction, or it falls into one of the previous categories and can be understood by analogy.

The reaction is an anti addition of the mercury group and a hydroxyl group, where the hydroxyl group has ended up at the carbon that can better support positive charge. The structural result is comparable to the reaction of a bromonium ion with water (Figure 0911).

In the absence of additional information, the mechanism for this reaction would be imagined as exactly the same as the reaction of an alkene with *N*-bromosuccinimide and water (Figure 0962).

Figure 0962

Reasoning by analogy in the oxymercuration reaction: mechanistic comparison.

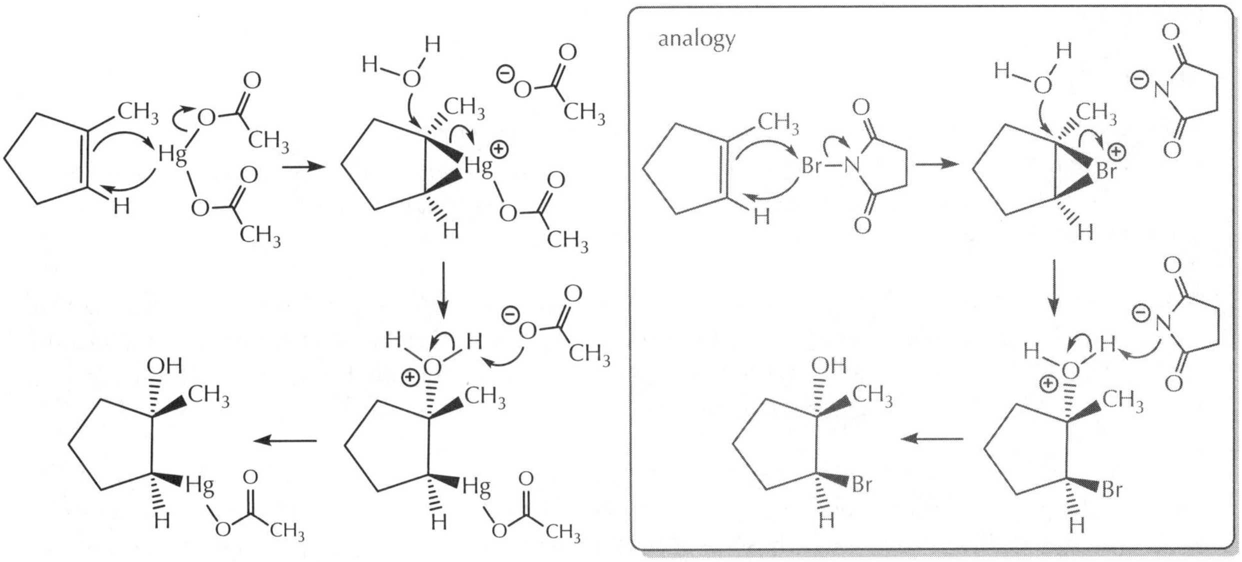

The addition reactivity of pi bonds is not at all restricted to alkenes and alkynes. Starting in Chapter 4, and then revisited in Chapter 8, the acid-catalyzed addition reactions of alcohols to carbon-carbon double and triple bonds are described (Figure 0963).

Figure 0963

Acid-catalyzed addition reactions of alcohols to alkenes and alkynes.

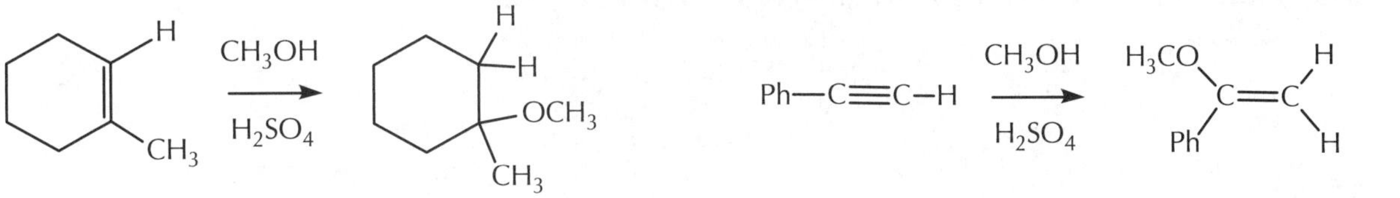

Many other atoms can form double and triple bonds, and the acid-catalyzed addition of alcohols can be carried out with these compounds, too (Figure 0964).

Figure 0964

Acid-catalyzed addition reactions of alcohols to double and triple bond functional groups beyond alkenes and alkynes.

Understanding these reactions as being fundamentally the same as the reactions of alkenes and alkynes, rather than as something totally new, begins to make learning more organic chemistry easier because it builds upon your prior knowledge.

The downside to that fact is that if you have not learned the prior information as well as you might, then this foundation you need for building upon is not as strong as it could be. As you move forward in learning organic chemistry, it is worth real time to go back often and revisit, even superficially, all of the previous chapters. Having a good set of notes in a course can be incomparably valuable as a review tool, particularly if you can begin to enhance your prior notetaking with insights gained from later on in your learning.

In exactly this way, your own more naïve ("flat earth") version of your own understanding about organic chemistry, captured as it was at the beginning of your study, can evolve based upon seeing a more complete picture.

Summary

Addition reactions were first introduced in Chapter 4 using the Markovnikov addition of strong and weak Brønsted acids to double and triple bonds. Carbocation intermediates are involved in these reactions. As seen in Chapter 8, these reactions give a mixture of syn and anti addition products, so there is no significant stereoselectivity.

A much broader picture of addition reactions is introduced in Chapters 8 and 9. A wide array of functional group relationships exist, and the reactions are nearly all stereoselective: the syn hydroboration-oxidation, epoxidation, and dihydroxylation reactions, the anti result in halonium ion reactions, and the syn or anti possibility in the reduction reactions of an alkyne.

All of these addition reactions, taken in combination with substitution and elimination reactions, can be sequenced in a wide variety of different ways. These multistep transformations allow complex molecular structures to be built up from relatively simpler starting materials, providing scientists with the ability to design and construct previously unknown or important molecular structures, ranging in application from new medicines to new materials.

This point in the study of organic chemistry represents a significant milestone because the fundamental basics of both structure (connectivity, stereochemistry, and conformations) and reactivity (complexation/decomplexation, substitution, addition, and elimination reactions) have all been introduced in some detail. The remainder of this introduction to organic chemistry builds upon these fundamentals.

CHAPTER QUESTIONS

9.15 A. Oxidation of optically active alcohol A with OsO$_4$ gives two stereoisomeric products B and C (C$_5$H$_{10}$O$_3$).

(a) Product B is optically active; product C is optically inactive. Draw the structures for products B and C.

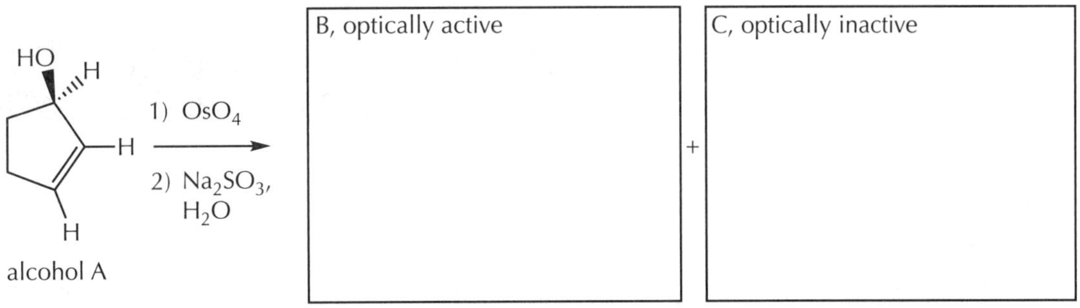

B, optically active	C, optically inactive

(b) What is the stereochemical relationship between B and C?

(c) Name alcohol A:

(d) Complete the following transformation of alcohol A.

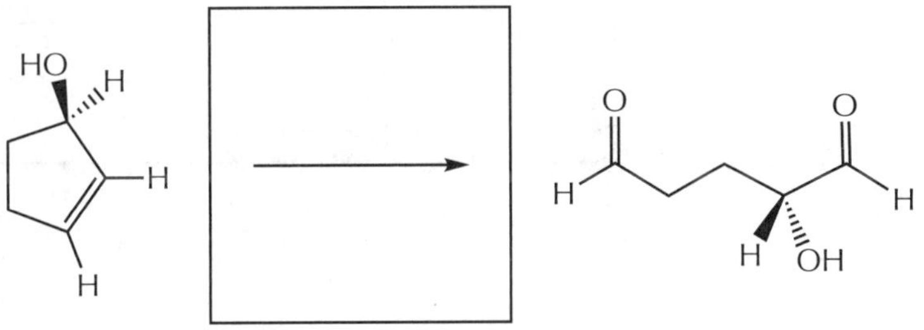

B. The following reaction is called an iodolactonization. Based on the information given below:

(a) draw the structure of the anticipated iodonium ion intermediate in this reaction, including its stereochemistry, that gives this particular stereoisomer of the product; and

(b) provide the curved arrow mechanism for the transformation of the intermediate to the product.

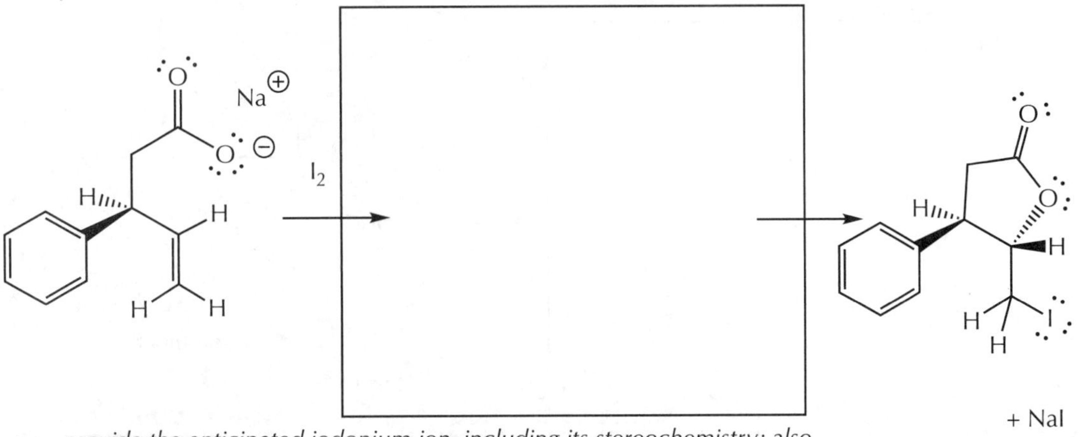

provide the anticipated iodonium ion, including its stereochemistry; also show the mechanism of its transformation to the observed product

(c) The iodonium ion is formed from an unsymmetrical double bond. Does the intramolecular attack onto the iodonium ion take place at the carbon atom that you anticipate for this unsymmetrical structure? Explain.

9.16 Dehydrohomoancepsenolide (compound E) is a naturally occurring substance that is used by gorgonians (a soft coral) to defend themselves from predators in their natural marine environment. Compound A was converted to compound E through a sequence of reactions.

(a) Complete this reaction sequence by providing the necessary reagents or chemical structures, as required.

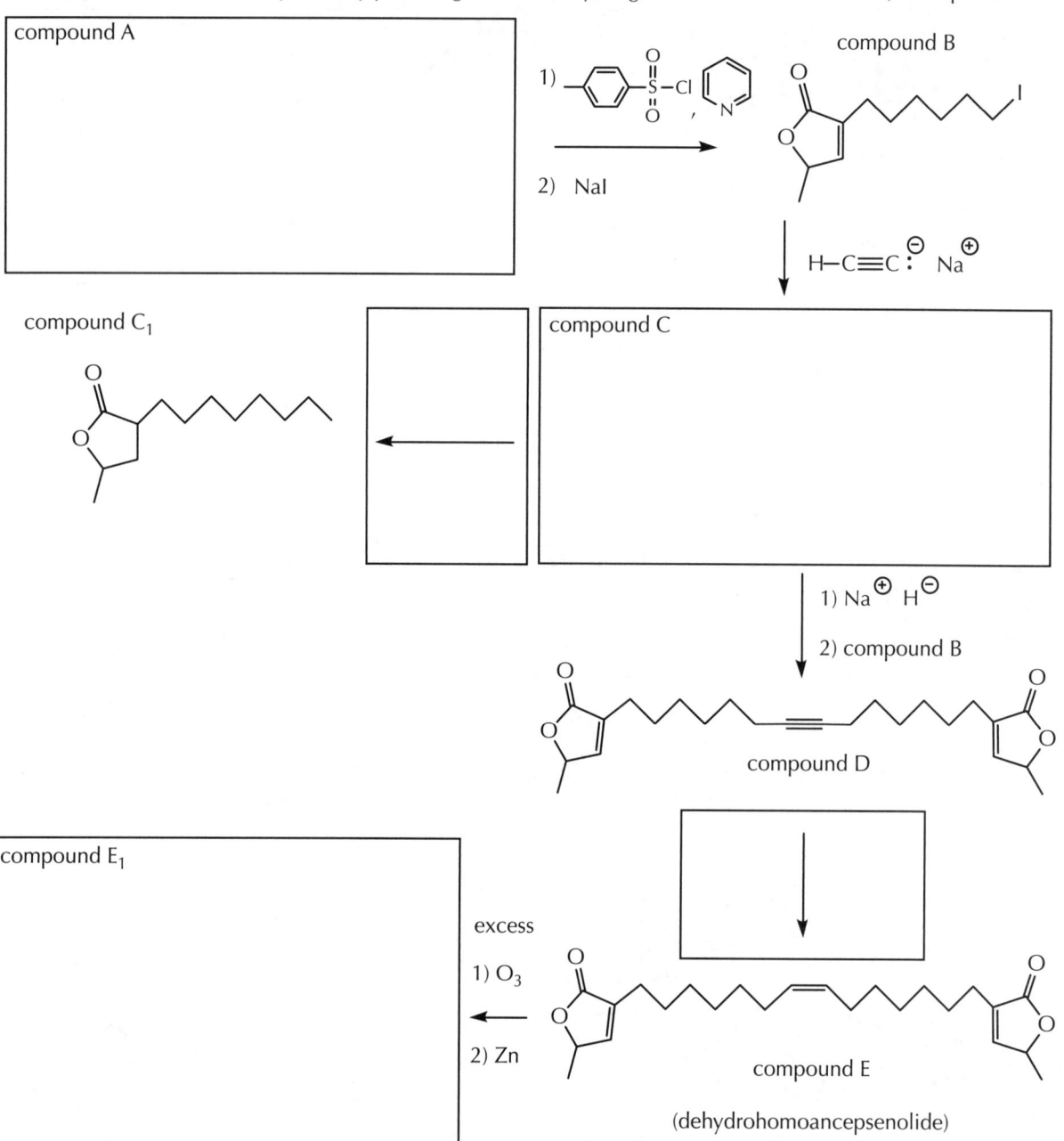

Two equivalents of compound E_1 are formed in this last step, so just draw one molecular structure here that corresponds to compound E_1.

(b) Which of these compounds would discolor a purple solution of potassium permanganate ($KMnO_4$) and show the brown precipitate of MnO_2?

compound:	A	B	C	C_1	D	E	E_1
$KMnO_4$ to MnO_2 (put an "X" in all that apply)							

9.17 Bromine in methanol is one of many experimental conditions that give electrophilic addition reactions.

(a) Complete the following by providing each of the intermediate structures along the way to the products. If a mixture of stereoiosmers forms for any of these steps, you only need to draw one of them. Curved arrow mechanisms are not required.

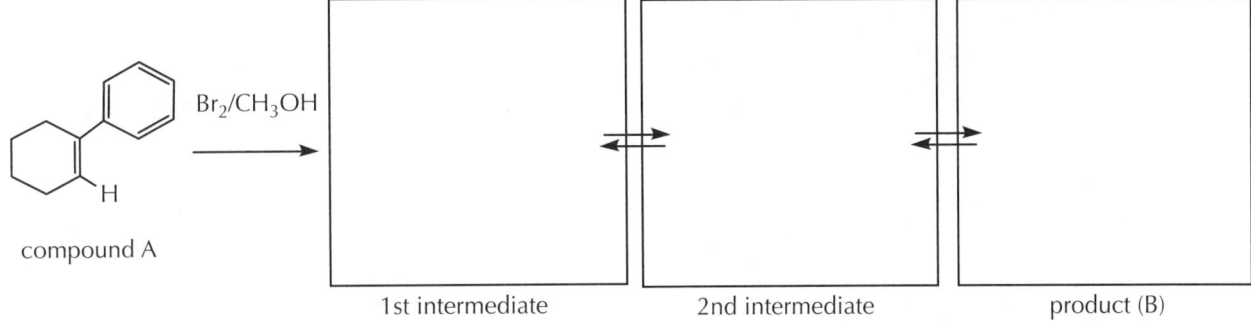

(b) Product (B) would be formed along with an equivalent amount of what byproduct?

(c) Product (B) would be formed as: (circle one)

    (a) a mixture of two stereoisomers   (b) a mixture of four stereoisomers

    (c) a single enantiomer   (d) a single diastereomer

(d) The IUPAC name of compound A is:

(e) When compound C is treated with molecular iodine, it undergoes a reaction that is analogous to the one in part (a). The difference is that the first intermediate (in the reaction of compound C) is captured intramolecularly by the alcohol in the molecule instead of an intermolecular capture by methanol. What is the structure of the product D? No stereochemistry is required to be shown.

product D, $C_6H_{11}IO$

(f) Product (D) would be formed as: (circle one)

    (a) a mixture of two stereoisomers   (b) a mixture of four stereoisomers

    (c) a single enantiomer   (d) a single diastereomer

(g) The IUPAC name of compound C is:

9.18 N-Bromosuccinimide (NBS) is a much more pleasant source of electrophilic bromine atoms than the noxious molecular bromine (Br₂), and it reacts to produce a bromonium ion intermediate just as easily as Br₂. Using the reaction with Br₂ as a mechanistic analogy, provide the complete, curved arrow mechanism for the following reaction. Provide the mechanism for the formation of the specific stereoisomer of the product that is shown. No other reagents are necessary to include.

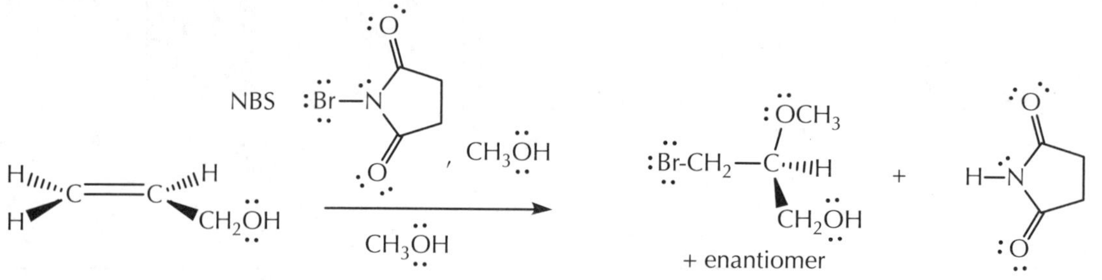

Provide the mechanism for the formation of <u>the specific stereoisomer</u> of the product that is shown.

9.19 Complete the following transformations as required. Separate experimental steps should be numbered.

(a)

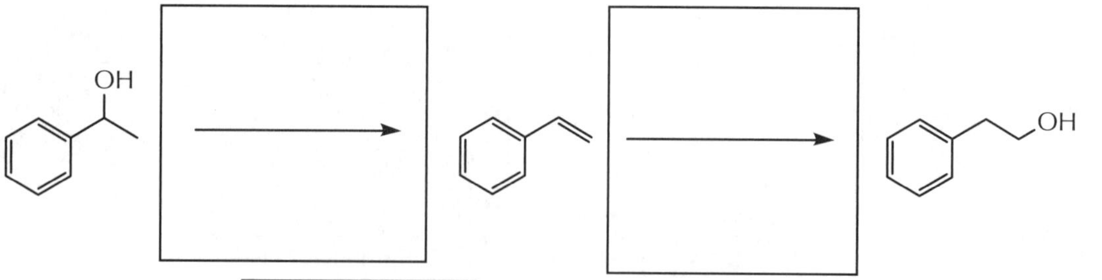

(b) 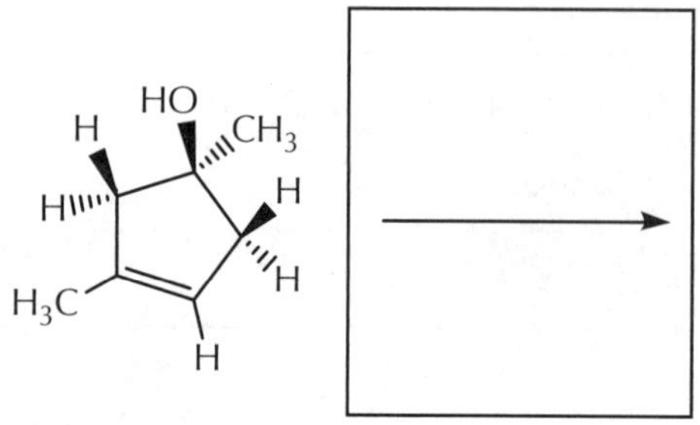 the products are a pair of diastereomers that includes a *meso* compound

9.20 Provide the reagents or structures as necessary to complete the following reaction scheme. If more than one synthetic step is required, number them sequentially.

9.21 Using lines, dashes, and wedges to indicate stereochemistry if/when appropriate, draw the structure of (2Z,4E)-3-bromoocta-2,4-dien-6-yne.

9.22 Complete the following reaction schemes, both of which give specific stereochemical outcomes.

9.23 This reaction gives a large number of products, including the formation of a full equivalent of HBr.

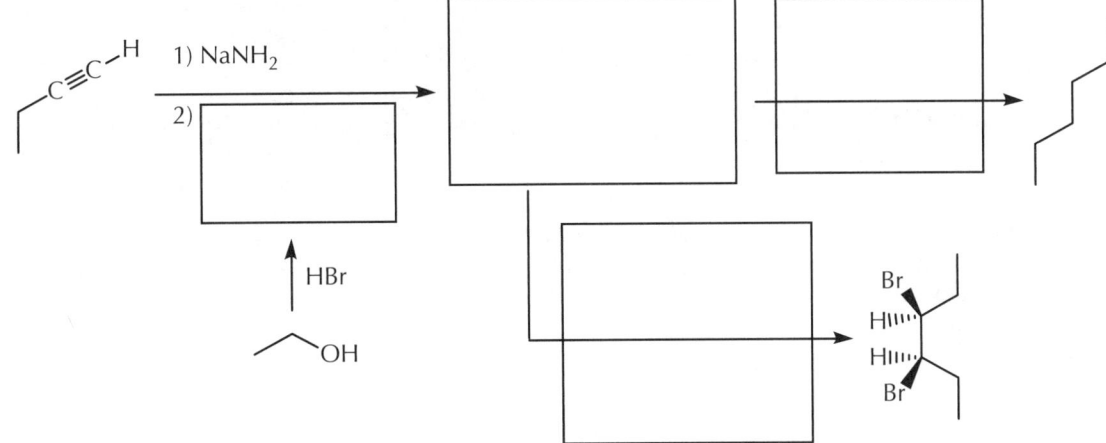

9.24 When compound C was treated with molecular bromine (Br$_2$) at 0°C in CCl$_4$, the rapid formation of the unexpected product, compound D, was observed, and not the expected dibromination reaction product.

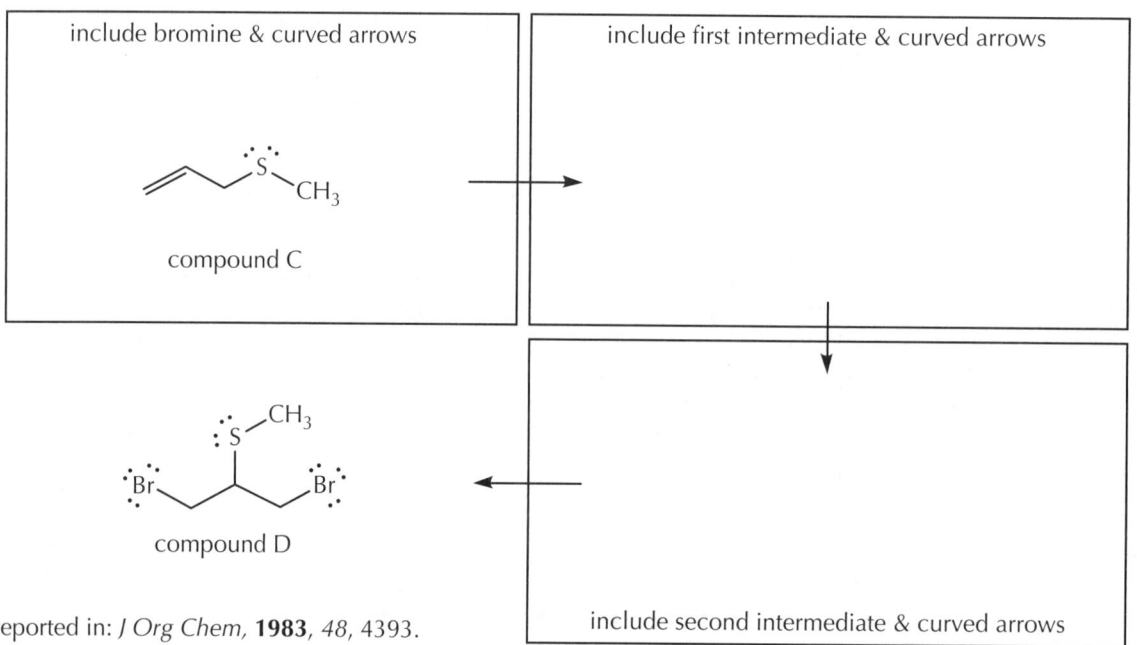

(a) Based on what you know about the regular dibromination reaction mechanism, and given the additional information [Experimental Information (A)–(C), on the right], propose a complete, stepwise, curved arrow mechanism for this reaction.

reported in: *J Org Chem,* **1983**, *48*, 4393.

The authors report that it was "immediately clear" in the NMR spectra that compound D had formed and not the expected dibromination reaction product.

(b) How many signals are in the ^{13}C-NMR spectrum of compound D?

(c) How many signals would have been expected in the ^{13}C-NMR spectrum for the expected (but unobserved) dibromination product of compound C?

(d) Briefly, why doesn't compound E (see box at top of page) give the same unusual reaction that compound C does? Please limit your words to a concise explanation.

9.25 Complete the following as required. Sequential experimental steps should be numbered.

(a) *J Am Chem Soc,* **1979**, *101*, 259.

(b) *J Am Chem Soc,* **1980**, *102*, 2117.

(c) *J Chem Soc,* **1965**, 384.

(d) *J Am Chem Soc,* **1992**, *114*, 10082.

(e) *J Am Chem Soc,* **1998**, *120*, 6425.

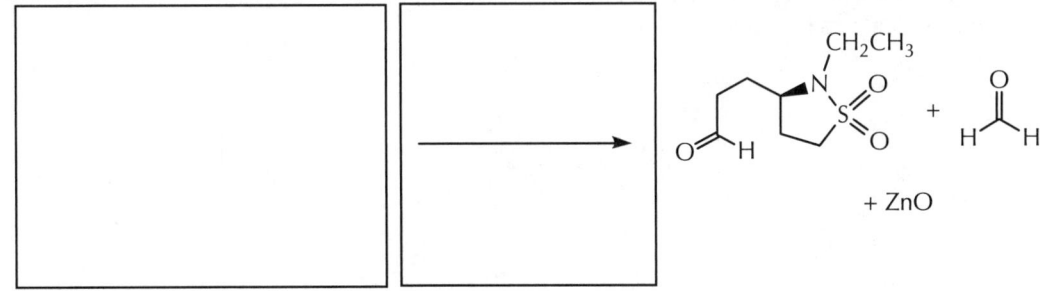

9.26 Fill in the missing intermediates and reagents in this synthesis of one component of the sex attractant pheromone of the Tea Tortrix Moth (*App Ent Zoo,* **1979**, *14*, 1).

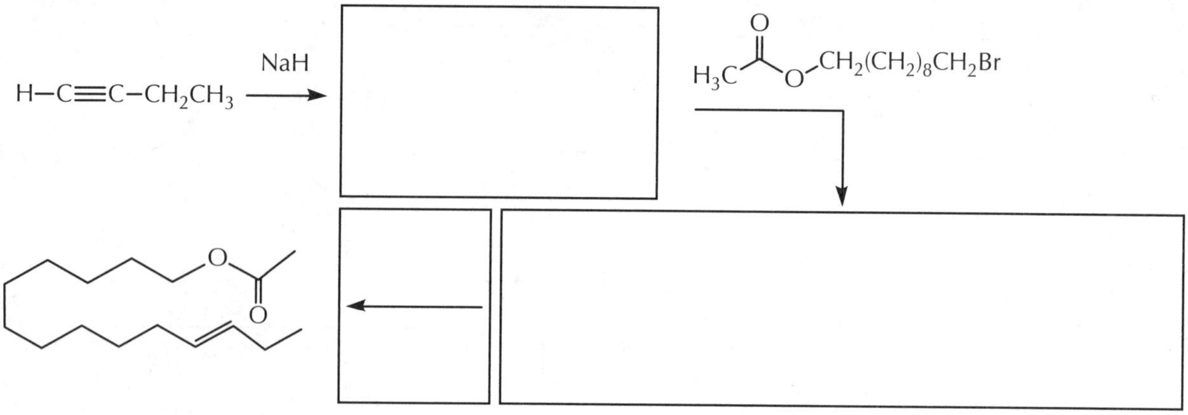

9.27 The following reactions, from the synthesis of an antitumor agent, give two diastereomeric products.
  (a) The major diastereoisomer from the hydrogenation reaction has the (R)-configuration at the new stereocenter bearing a methyl group (adapted from *ACS Med Chem Lett,* **2017**, *8*, 746).
  (b) The reaction with mCPBA shows the same stereoselectivity as the hydrogenation reaction (i.e., both occur on the same side/face of the alkene). Draw the complete structure of the major diastereoisomer in each case.

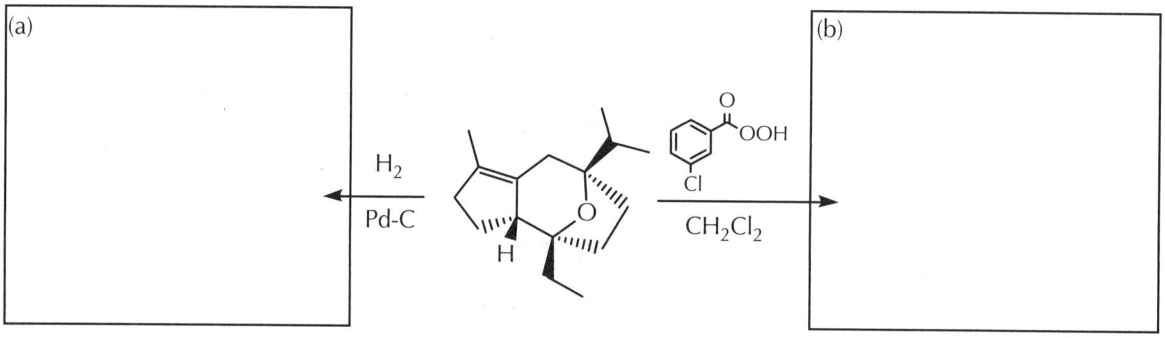

9.28 The following reaction produces two isomeric products, both of which add two equivalents of molecular bromine to give two different tetrabrominated products (*Org Lett,* **2017**, *19*, 5446). What are the structures of the original two products?

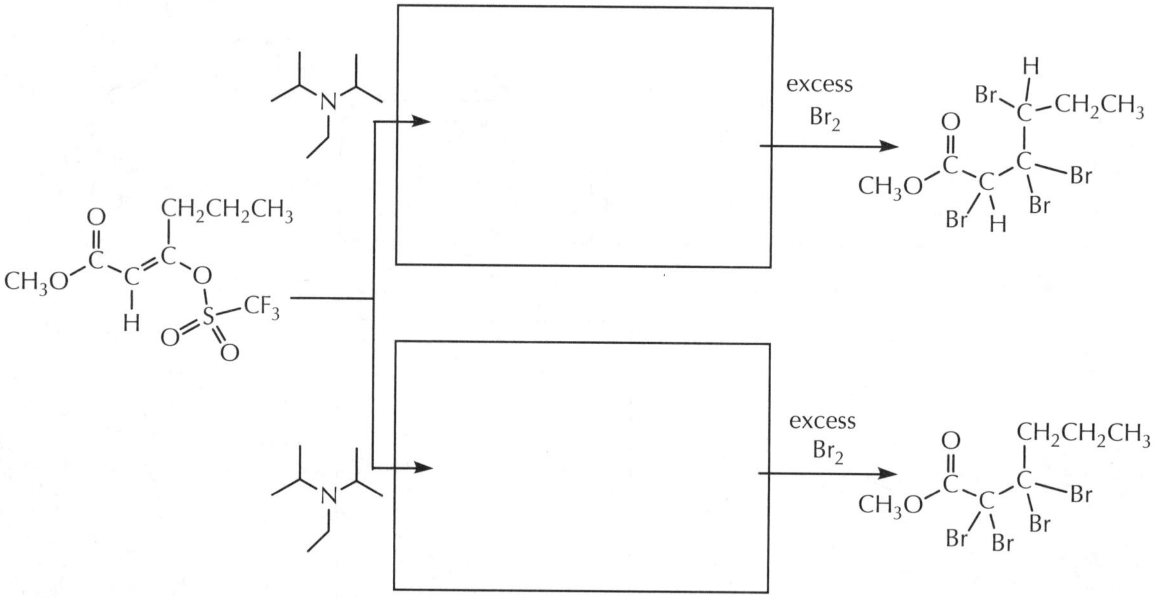

9.29 Provide the reagents or structures as necessary to complete the following sequences of reactions. If more than one synthetic step is required, number them sequentially.

If any reaction produces a mixture of stereoisomers, follow the specific instructions to draw the single indicated stereoisomer and write "+ enantiomer" or "+ diastereomer" in the box.

(a) Draw the product that does not have an (S) stereocenter.

(b) Draw the product that has only one (R) stereocenter.

(c) Provide the reaction conditions for this highly stereoselective reaction.

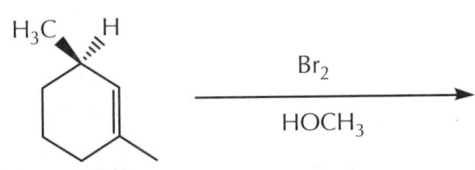

(d) This reaction produces a mixture of diastereomers.

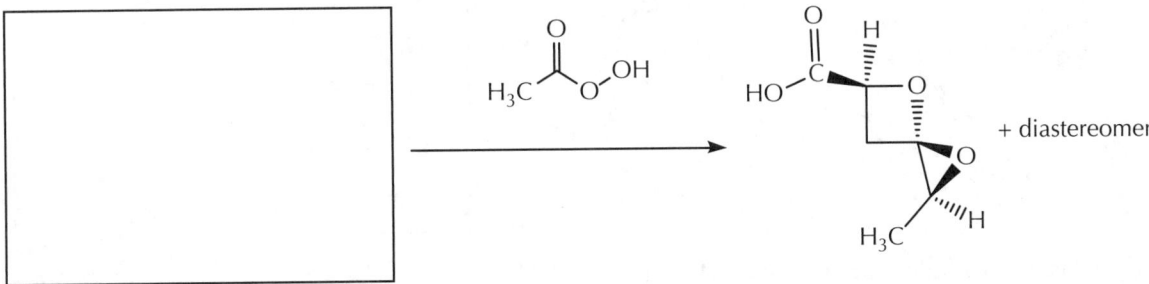

+ diastereomer

(e) This reaction produces a mixture of diastereomers.

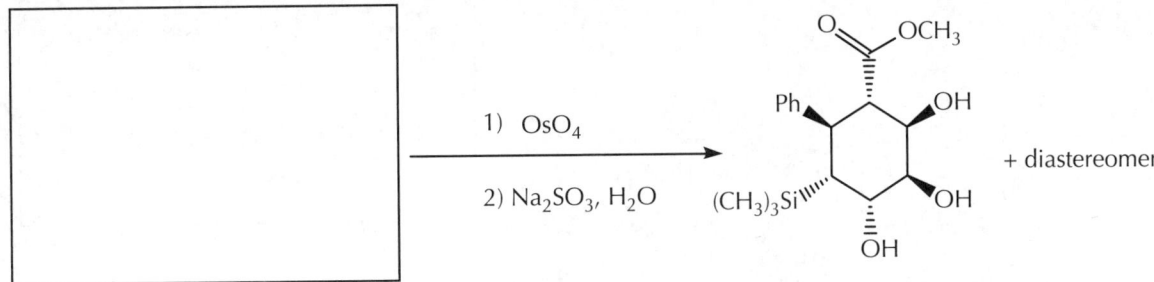

+ diastereomer

9.30 The synthesis of iodohydrins using N-iodosuccinimide (NIS) as a source of electrophilic iodine was studied on a large series of compounds (*Internat J Cancer*, **2014**, *53B*, 1425). Provide the complete, stepwise curved arrow mechanism for this reaction. Include stereochemistry where appropriate, leading to the stereoisomer explicitly shown here.

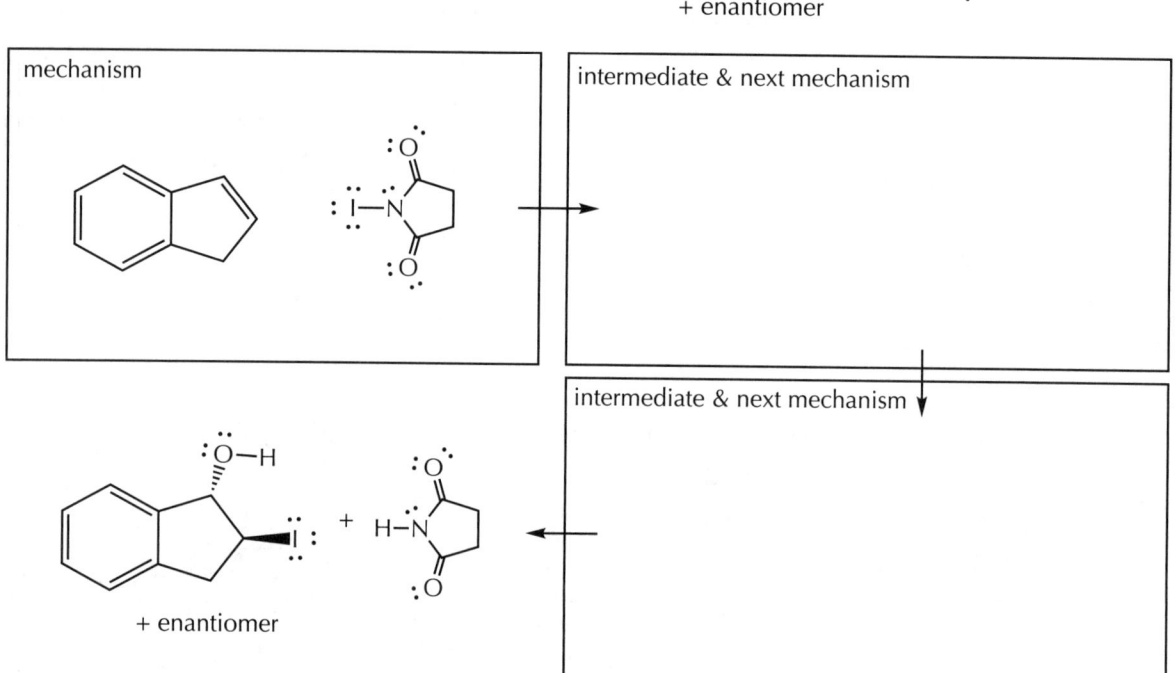

9.31 N-Bromosuccinimide (NBS) and methanol were used to carry out an addition reaction to the following alkyne. Under the reaction conditions, the product is a single, achiral compound ($C_{10}H_9BrO_3$), formed purely as one stereoisomer.

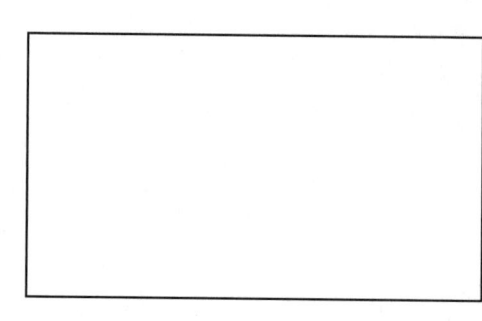

9.32 Draw: (R)-1,3-dimethylcyclohexa-2,4-dien-1-ol

9.33 Compound A, which is a degradation product of the antibiotic vermiculine, has the following structure:

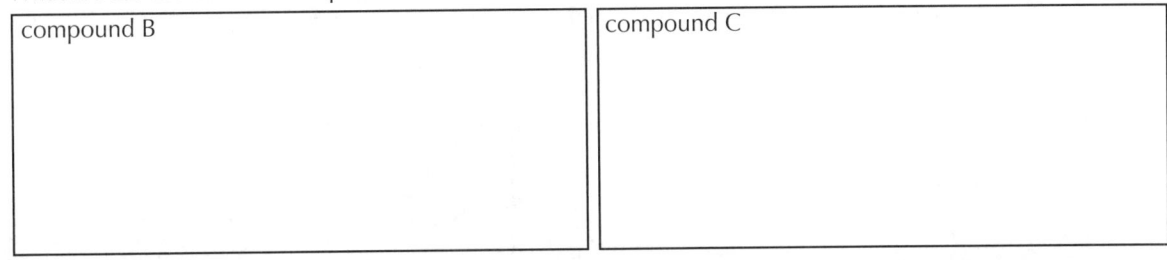

The structure was confirmed by converting compound A to compound B, $C_{11}H_{18}O_4$, which was also prepared by ozonolysis of compound C, $C_{11}H_{18}O_2$.

compound A $\xrightarrow{\text{H}_2, \text{Pd-C}}$ compound B $\xleftarrow[\text{2) (CH}_3)_2\text{S}]{\text{1) O}_3}$ compound C

$C_{11}H_{14}O_4$ \hspace{1cm} $C_{11}H_{18}O_4$ \hspace{1cm} $C_{11}H_{18}O_2$

What are the structures of compounds B and C?

compound B	compound C

9.34 A report in *Nature* contains the astonishing news that female moths and female elephants use the same ester as a sex pheromone (a chemical used in communication) to signal their readiness to mate. The ester has the structure shown below.

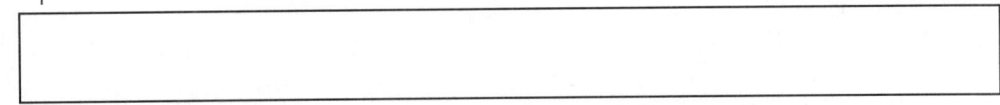

The alcohol corresponding from which this ester can be prepared has the structure shown below.

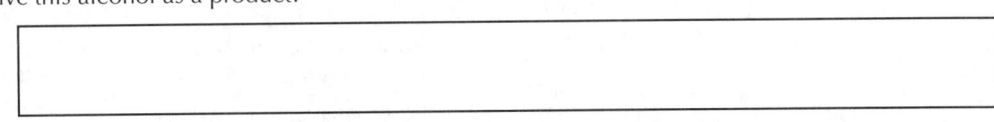

(a) What is the complete IUPAC name for this alcohol?

(b) What is the name of the starting material that can be reduced, using hydrogenation with a poisoned catalyst, to give this alcohol as a product?

9.35 When 1-methylcyclohexene reacts with N-bromosuccinimide, a source of electrophilic bromine atoms, in the presence of a source of fluoride ions, an anti addition of bromine and fluorine atoms occurs to give 2-bromo-1-fluoro-1-methylcyclohexane. Draw an accurate three-dimensional representation of the two chair conformations for one of the enantiomers of the product of the reaction described here.

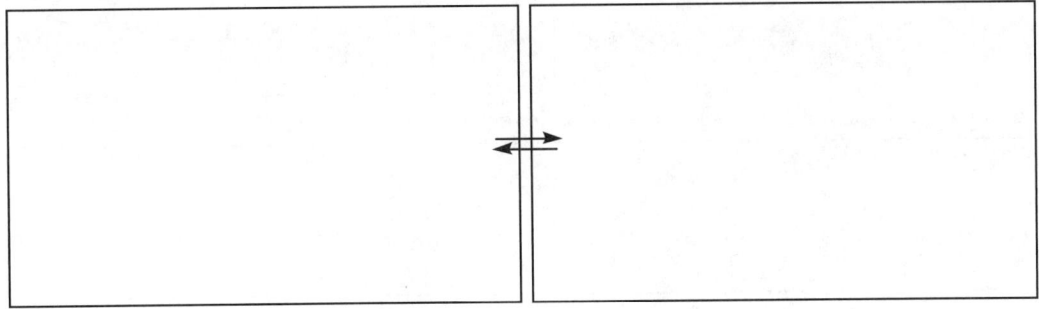

9.36 Aziridinium ions contain a positively charged nitrogen in a three-membered ring. The reactivity of these cations is analogous to halonium ions. For example, the following reactions were observed for this aziridinium ion.

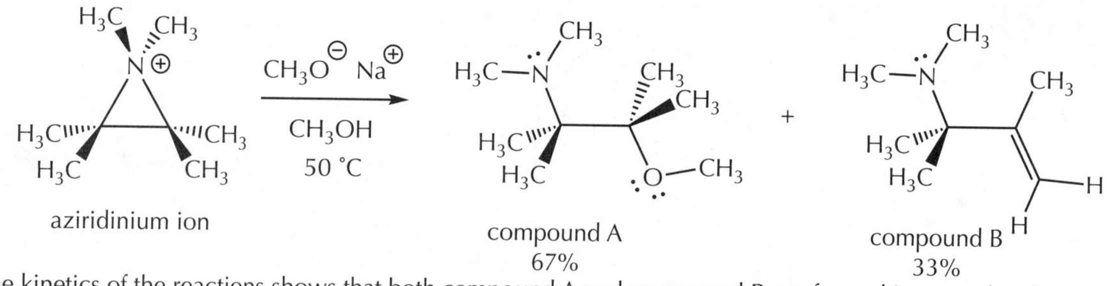

[Reaction scheme: aziridinium ion + CH₃O⁻ Na⁺ in CH₃OH at 50 °C → compound A (67%) + compound B (33%)]

The kinetics of the reactions shows that both compound A and compound B are formed in second-order reactions. If the aziridinium ion is dissolved in methanol without sodium methoxide, only the first product, compound A, is formed.

(a) Draw curved arrow mechanisms for the two reactions of methoxide ion with the aziridinium ion shown above.

mechanism of formation for compound A	mechanism of formation for compound B

(b) Why is compound A the only product formed when the aziridinium ion is put in methanol, while both compounds A and B form when sodium methoxide is the reagent?

9.37 Aziridinium ions contain a positively charged nitrogen in a three-membered ring, and undergo ring-opening reactions with nucleophiles (Nu:) such as water, alcohols, and halide ions, comparable to halonium ions.

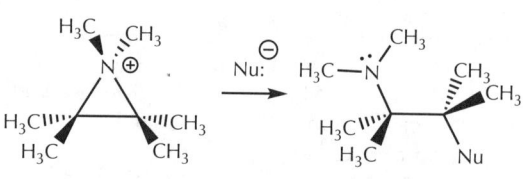

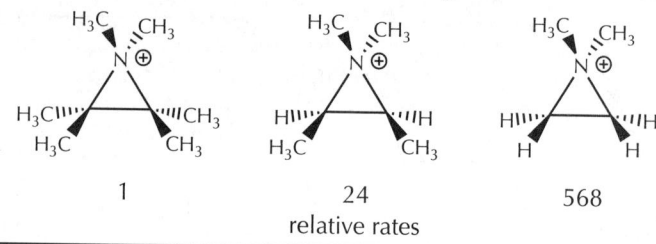

The effect of structure on the rate of the ring-opening reaction of aziridinium ions was studied. The following results were found:

relative rates: 1, 24, 568

How would you explain this trend in the rates of the reaction?

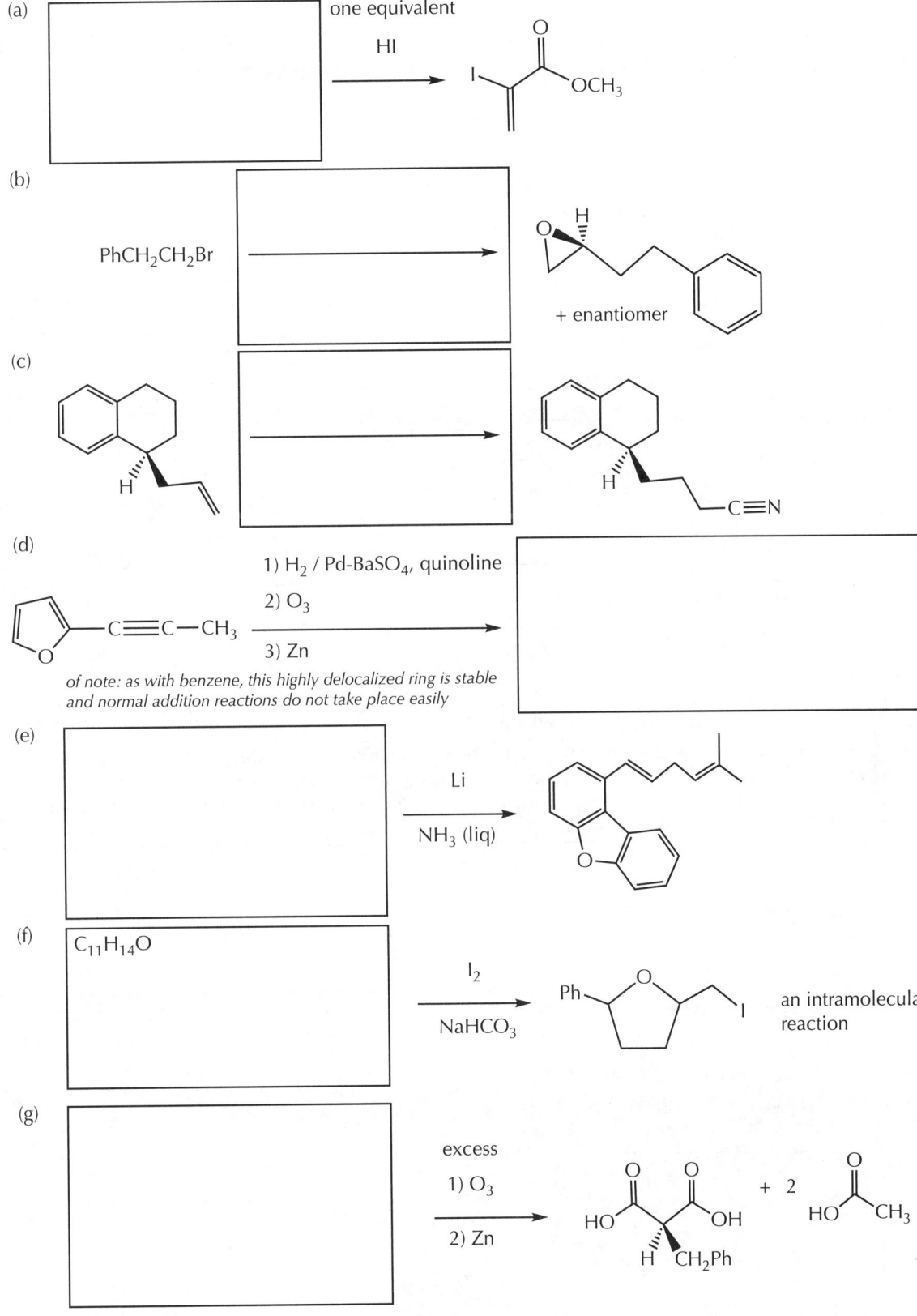

9.39 The sex attractant of the fall armyworm moth is produced by the female. This material, which attracts the male moth at very low concentrations, is called spodoptol and is a primary alcohol with the molecular formula $C_{14}H_{28}O$. Hydrogenation of this alcohol gives 1-tetradecanol, $C_{14}H_{30}O$. Ozonolysis of spodoptol is known to give two aldehydes, aldehyde A, $C_5H_{10}O$, and aldehyde B, $C_9H_{18}O_2$, indicating the position of the double bond in spodoptol.

At this point the structure of the sex attractant was known, except for the stereochemistry of the double bond. So little of the material is isolated from the female moths that the compound had to be synthesized before the stereochemistry could be determined.

(a) Draw the structures for compounds (or intermediates) D, E, and F, formed in the synthesis of spodoptol.

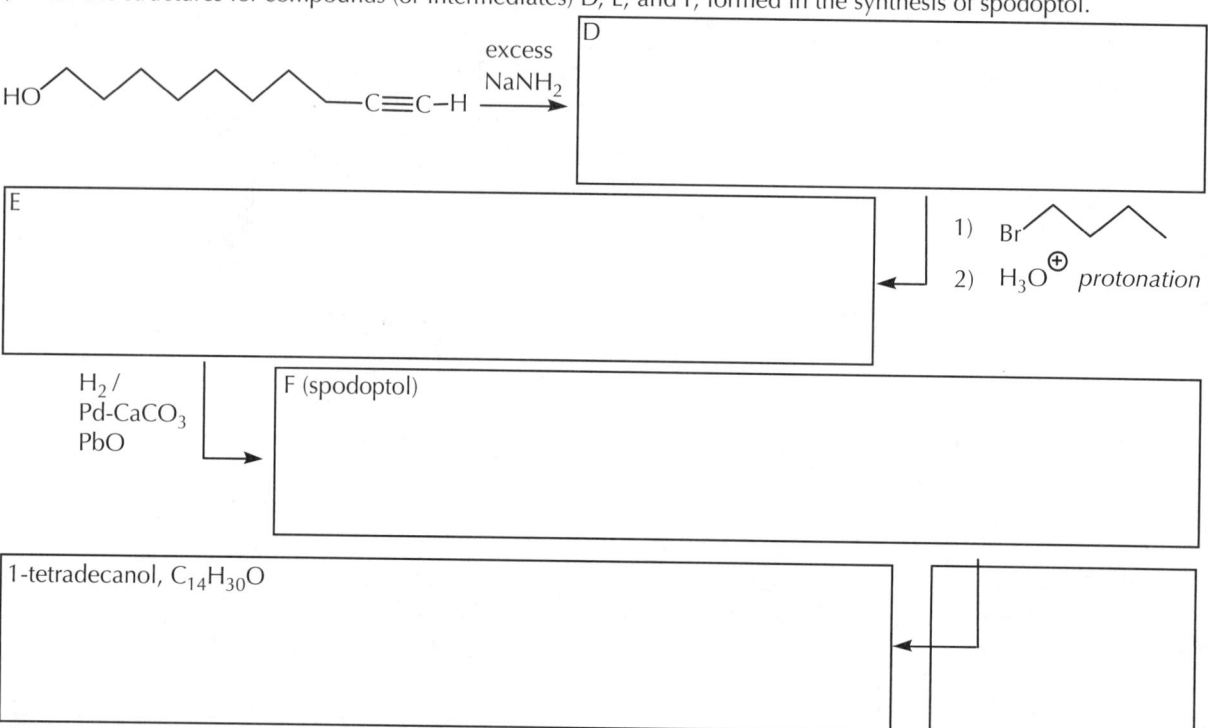

(b) As a check on the stereochemistry of the double bond, compound E, prepared above, was also reduced with sodium metal in liquid ammonia. The alkene that resulted attracted no male moths, indicating that this was not the stereochemistry found in the natural product. What is the structure of the alkene formed in the sodium metal reduction of compound E?

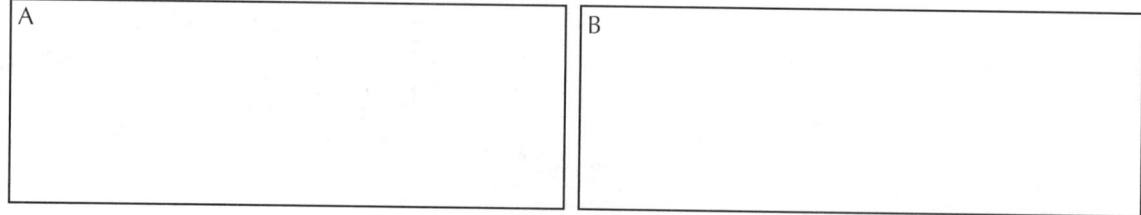

(c) Assign structures to aldehyde A and aldehyde B.

F (spodoptol) $\xrightarrow[\text{2) }(CH_3)_2S]{\text{1) }O_3}$ aldehyde A + aldehyde B
$\qquad\qquad\qquad\qquad\qquad\qquad C_5H_{10}O \qquad C_9H_{18}O_2$

A

B

9.40 2-Methylheptadecane is a sex pheromone for certain species of insects. It has been synthesized using 1-undecyne as the starting material. How would you carry out this transformation?

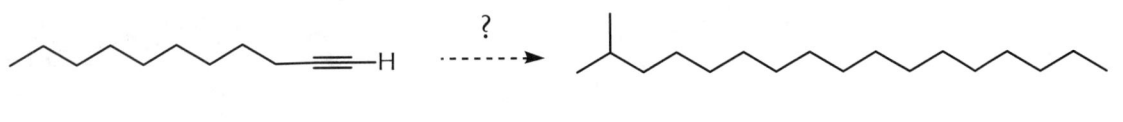

9.41 Synthesis of a compound isolated from a marine sponge involved the following set of reactions. Provide the complete, curved arrow mechanisms for each step in this transformation.

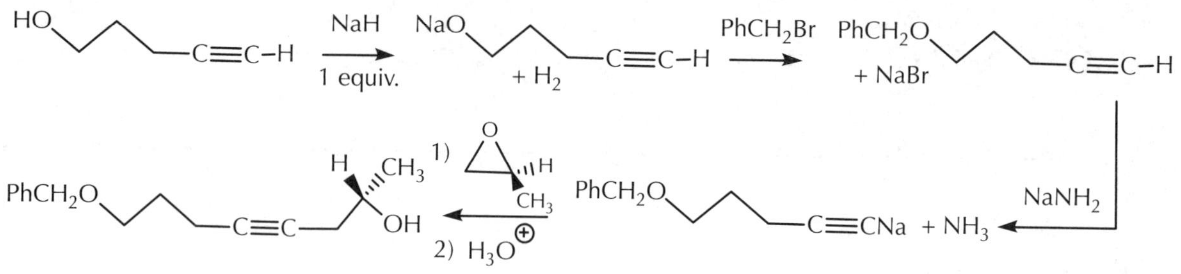

9.42 Phthaloyl peroxide reacts with alkenes and provides reaction products the same as would be obtained from the reaction of alkenes with OsO$_4$ and KMnO$_4$. However, toxic ions such as osmium and manganese are not used, so reagents such as phthaloyl peroxide are more environmentally friendly. The workup step in this case is mild, aqueous base.

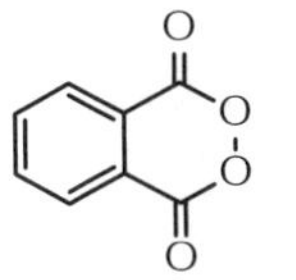

phthaloyl peroxide

Each of the following reaction sequences produces a specific diastereomer (step 1) followed by a specific pair of enantiomers (steps 2 & 3). In the case of the final products, draw either one of the anticipated enantiomers (using lines, dashes, and wedges).

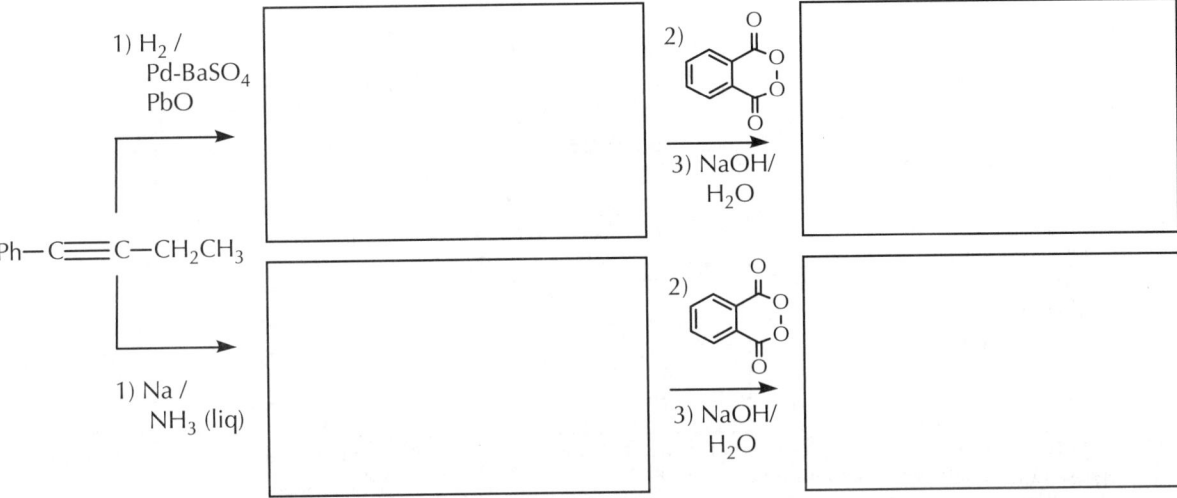

9.43 When 1-phenyl-1-butyne reacts with bromine and water, the product that forms (compound M) can undergo a variety of S$_N$2 reactions (e.g., with NaCN), whereas the reaction between 1-phenyl-1-butyne with bromine and benzyl alcohol (PhCH$_2$OH) does not give a product that can undergo S$_N$2 reactions (compound N). When compound N is treated under mild hydrogenation conditions, compound M forms (note: this is not a familiar reaction).

(a) What are the structures of compounds M and N?

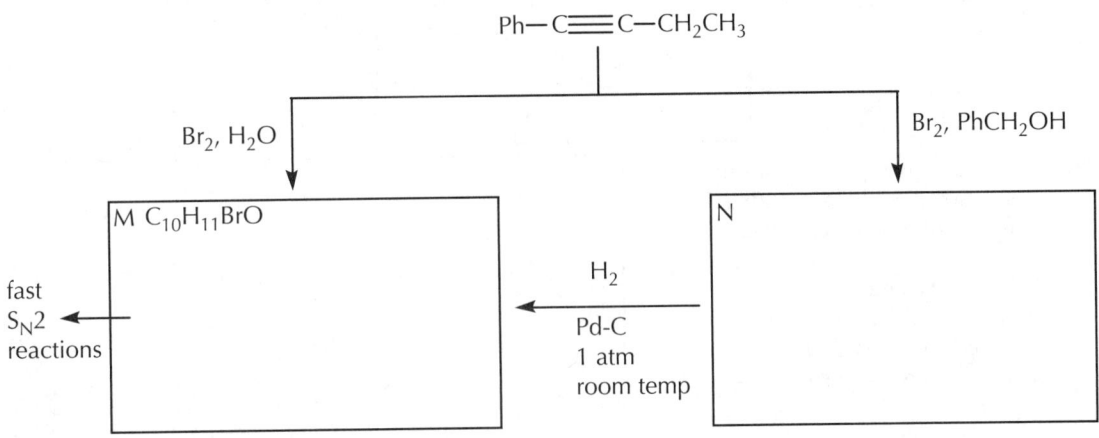

(b) The reaction of hydrogen with compound N is not a reaction you have seen before. Given the result (the formation of compound M), the reaction of compound N with hydrogen first forms an intermediate structure that then undergoes a spontaneous and rapid equilibration to compound. What is the most likely intermediate structure?

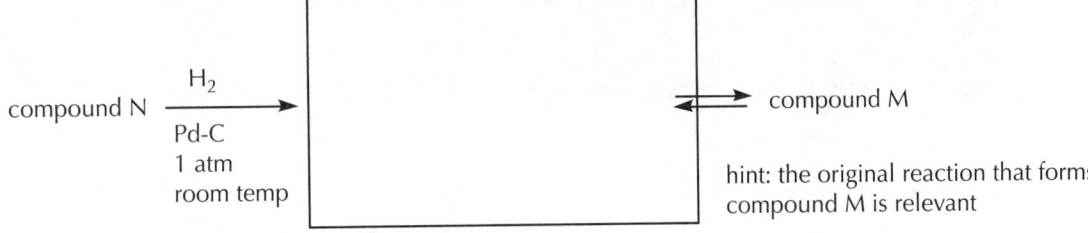

9.44 A study of electrophilic addition reactions reported on the reactivities of cyclopropanated compounds (*J Org Chem*, **2002**, *67*, 4100). In the following reaction, two different non-isomeric products were observed. One of the products (A) was a single *meso* compound, while the other (B) was a racemic mixture.

a) Draw the structures for these products, including a careful representation of the stereochemistry.

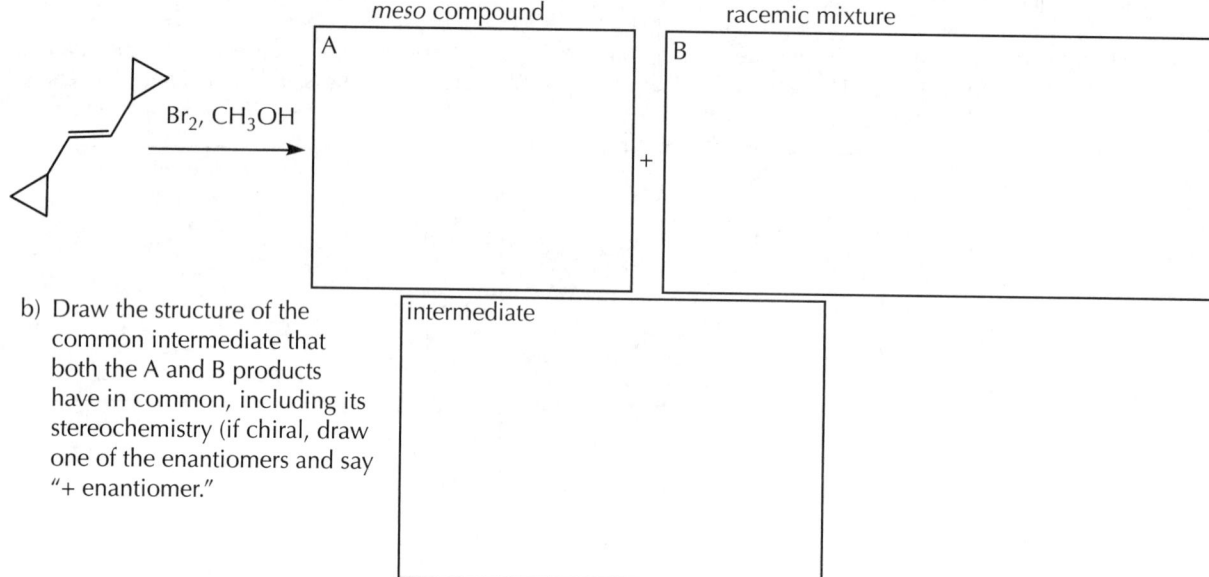

b) Draw the structure of the common intermediate that both the A and B products have in common, including its stereochemistry (if chiral, draw one of the enantiomers and say "+ enantiomer."

9.45 In the reaction below, HOBr was used and two achiral, regioisomeric products (C and D) resulted. The regioselectivity was relatively high (~ 85:15) in the normally anticipated direction.

(a) Provide drawings of the two regioisomeric products.

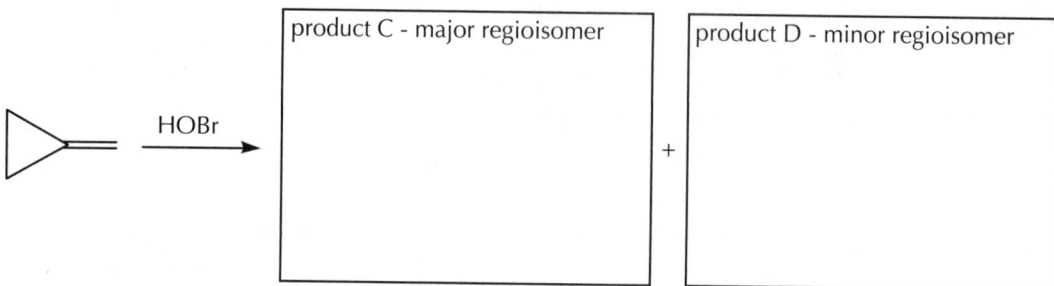

(b) Provide a drawing of the transition state which directly leads to the formation of product C above.

The drawing should be:
(i) 3D (lines, dashes, wedges)
(ii) include any partial bonds
(iii) include any partial charges

(c) The regioselectivity for the molecule below is higher than that of the reaction reported, above. Suggest a reason why the transition state energy for forming the major regioisomer with this molecule may be lower than the one you drew in part (b).

9.46 Complete the following reactions as directed. Transformations requiring sequential experimental steps should be numbered appropriately. Show the major organic product(s) unless otherwise specified. If a product forms as a stereoisomeric mixture, draw one and write "+ enantiomer" or "+ diastereomer."

(a)

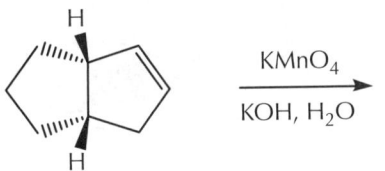

(b)

| | any product(s) with three ^{13}C-NMR signals | any product(s) with two ^{13}C-NMR signals |

(c) Draw the optically active compound that is produced in each of these reactions.

9.47 When compound Y is treated with either of the following reagents, a stereoisomeric mixture of products results.

(a) Draw all of the stereoisomeric products for both reactions.

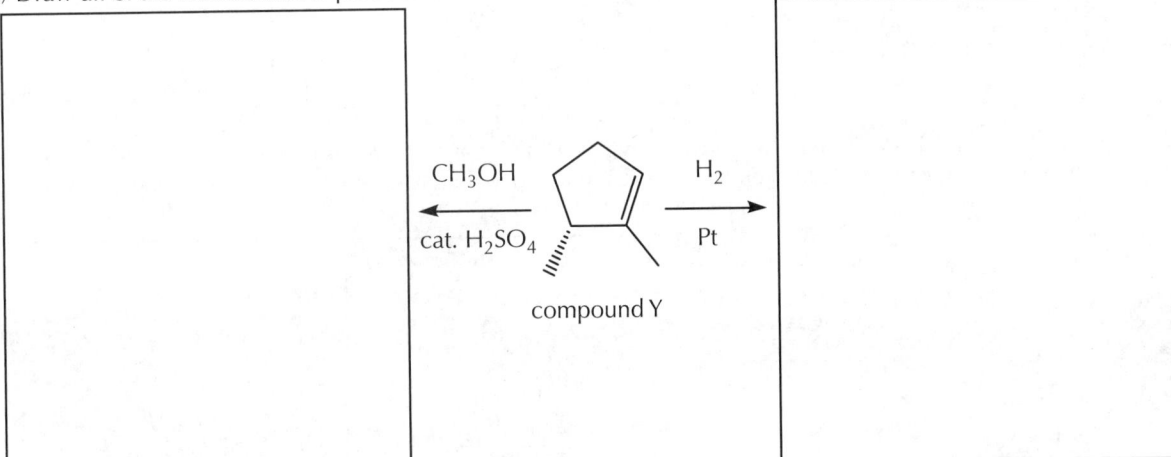

(b) Provide the IUPAC name, including stereochemistry, for compound Y.

9.48 Cyanogen bromide (BrCN) is a good source of electrophilic bromine. When used in addition reactions, it displays the same regioselectivity and stereoselectivity predicted for reagents like HOCl or HOBr. For the following reaction, one product is drawn for you. Draw the other products predicted to form.

(a)

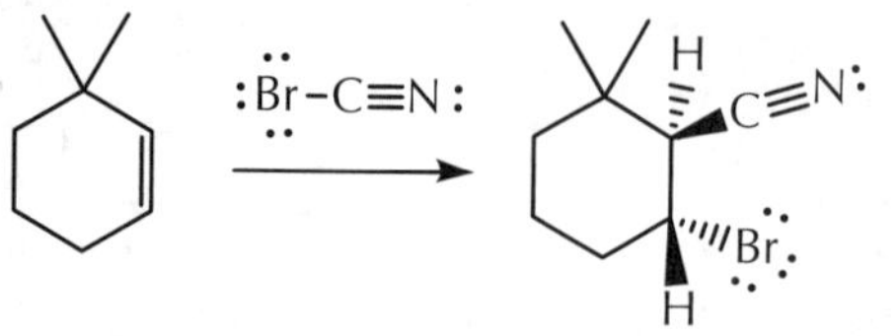

(b) Draw the complete, two-step mechanism of formation for the specific stereoisomer shown above. Include the stereochemistry of the intermediate.

(c) Treating the product (shown above) with a strong, bulky base leads to a single elimination reaction product. Draw the complete curved arrow mechanism for this E2 reaction, using the appropriate chair conformation. Also: draw the elimination product.

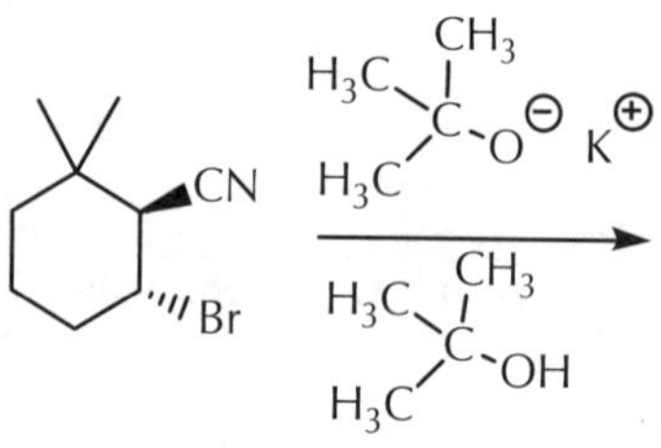

reactive chair form & mechanism 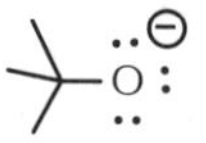 | product

(d) Given the anticipated stereoselectivity and regioselectivity of cyanogen bromide, draw the structures for the two stereoisomeric products predicted to form in the following reaction.

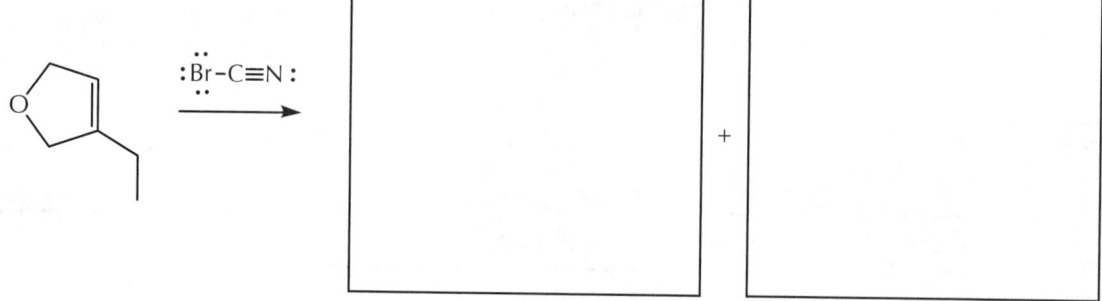

9.49 In a 2013 paper (*Org Lett,* **2013**, *15*, 1306), the authors report the selective reduction of alkynes in the presence of a nitro group. Complete these reactions.

(a)

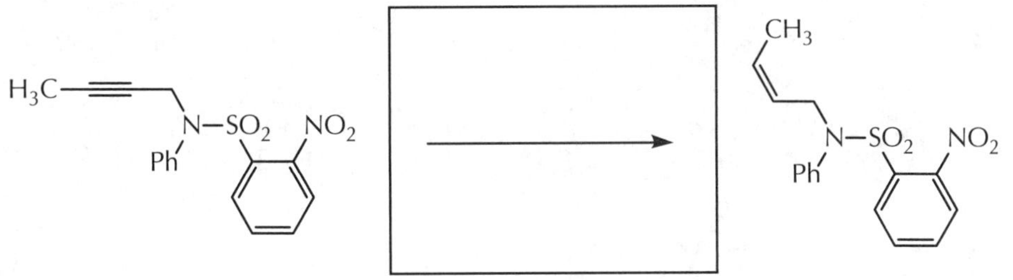

(b)

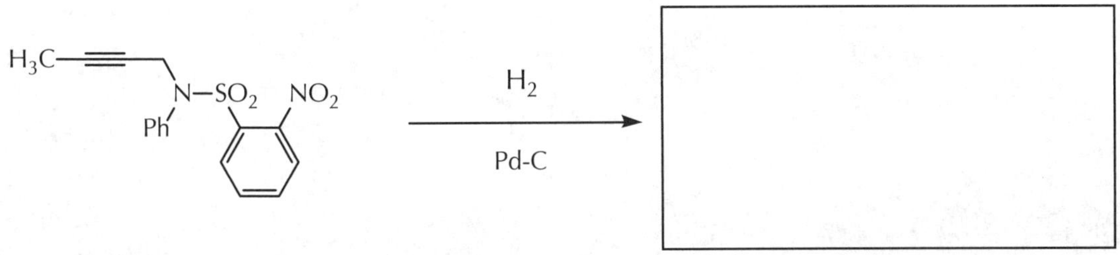

9.50 The haloamidation of an alkene has been reported (*Org Lett,* **2013**, *15*, 1310). The reaction involves an electrophilic chlorine source (Ph₂SeCl⁺) that is analogous to, but more reactive than, HOCl. The halonium intermediate then reacts with acetonitrile (CH₃CN) to form a product, which undergoes further reaction with water to generate the haloamide. Include stereochemistry in your drawings.

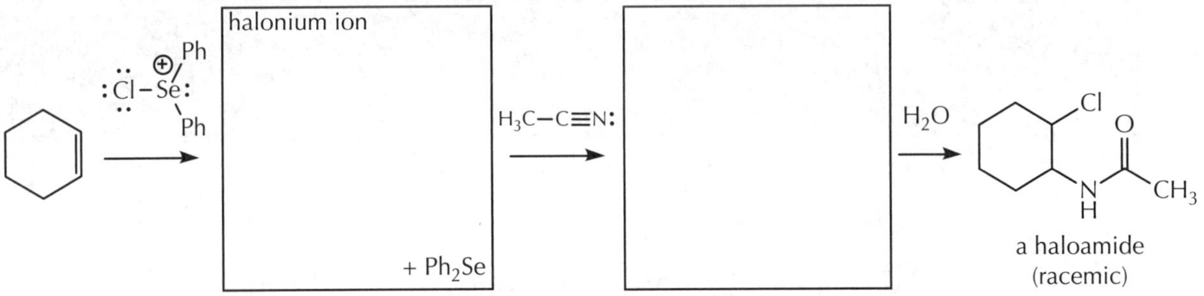

9.51 Complete the following reaction scheme as required. Number any sequential experimental steps.

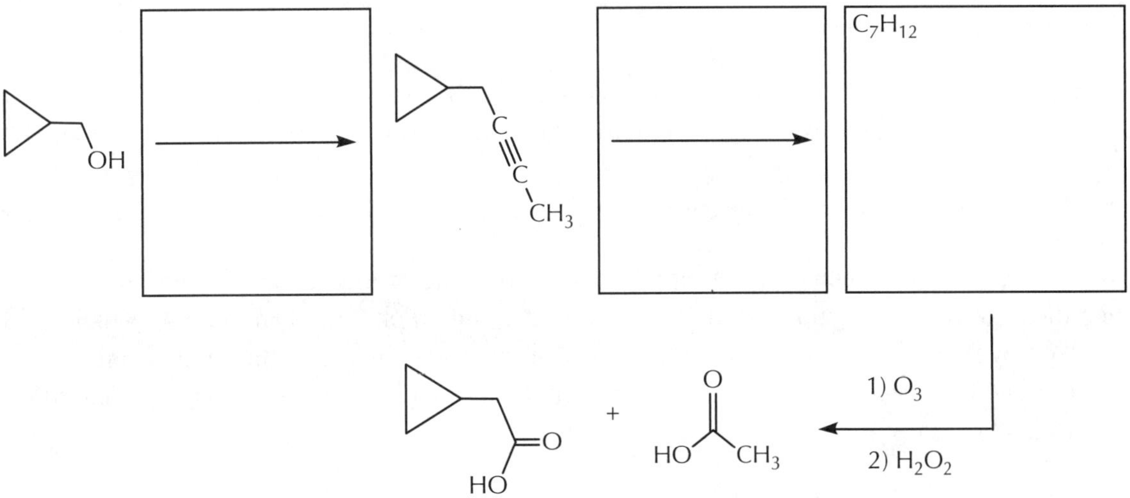

Organic Chemistry Lecture, 1893.

Chemical History at the University of Michigan

The original chemistry building, then called the Chemical Laboratory of the University of Michigan, is thought to be the first structure on the North American continent that was designed, erected, and equipped solely for instruction in chemistry.

Although there were multiple additions made to the original 1856 building, the number of students who needed chemistry was increasing at a faster pace. The natural growth of the University, and particularly the development of professional training in dentistry, engineering, medicine, and pharmacy, together with enhanced interest in chemistry for teacher training, led to crowded lecture halls with every seat taken.

Just after the turn of the century, amphitheater space became available in the old Medical Building when the new one was completed in 1903. Soon after, the amphitheater in the old Dental Building was taken over for lectures to the larger classes in elementary chemistry. But these rooms had been built for other purposes and were not provided with facilities adequate for setting up lecture demonstrations in chemistry.

Public domain image from the University of Michigan Collection, kindly provided by the Bentley Historical Library (File BL002006).

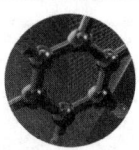

CHAPTERS 7–9
Exam Questions

The third mid-term examination in the first-term organic chemistry course at the University of Michigan is typically made up of five pages of questions that integrate the topics from Chapters 7–9.

The following pages of "Exam Questions" (EQ 03.01–EQ 03.30) were used in examinations between 1998–2019.

The 5-page examinations are written for a 60-minute time period, and students have 90 minutes to take the tests.

Any set of five pages from this group of 30 questions might be used to create a practice exam.

Select any five of these pages and time yourself for 90 minutes, without accessing any resources, to self-test your mastery of the main topics from Chapters 7–9.

Chapter 7
- Possible Outcomes in Substitution and Elimination Reactions
- Possible Mechanistic Pathways for Substitution and Elimination Reactions
- Application of Substitution and Elimination Mechanisms
  - Unified View of S_N2, E2, S_N1, and E1 Pathways
  - Extended Definition of Leaving Groups
- A Predictive Model for S_N/E Reactions
- Rate and Reactivity in S_N/E Reactions
  - Leaving Group Ability
  - Solvent Effects
  - Nucleophilicity
  - Heteroatom Electrophiles

Chapter 8
- Regioselectivity and Stereoselectivity in Electrophilic Addition of Brønsted Acids
- Carbocation Rearrangement Reactions
- Hydration via Hydroboration-Oxidation
- Introduction to Synthesis

Chapter 9
- Halogenation and Other Reactions of the Halonium Ion Intermediate
- Oxidation Reactions
  - Peroxyacid Reactions
  - Syn Dihydroxylation Reactions
  - Ozonolysis Reactions
- Reduction Reactions
  - Catalytic Hydrogenation with Transition Metal Catalysts
  - (Z)-Alkenes from Catalytic Hydrogenation with Poisoned Catalysts
  - (E)-Alkenes from Dissolving Metal Reduction

EQ 03.01

Complete the following reaction schemes. Include stereochemistry where appropriate. Multiple steps in answers should be numbered.

(a) *Tet Lett,* **1999**, *40*, 7485.

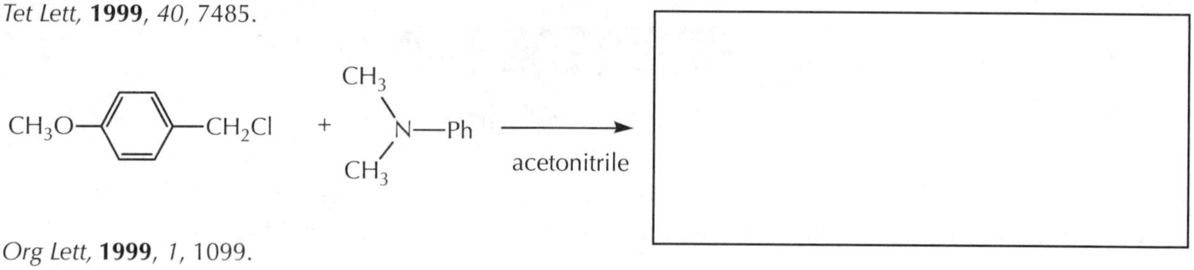

(b) *Org Lett,* **1999**, *1*, 1099.

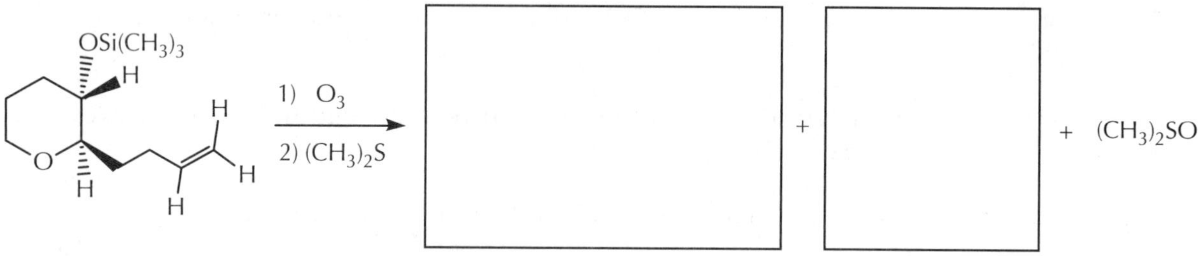

(c) *J Org Chem,* **1999**, *64*, 8399.

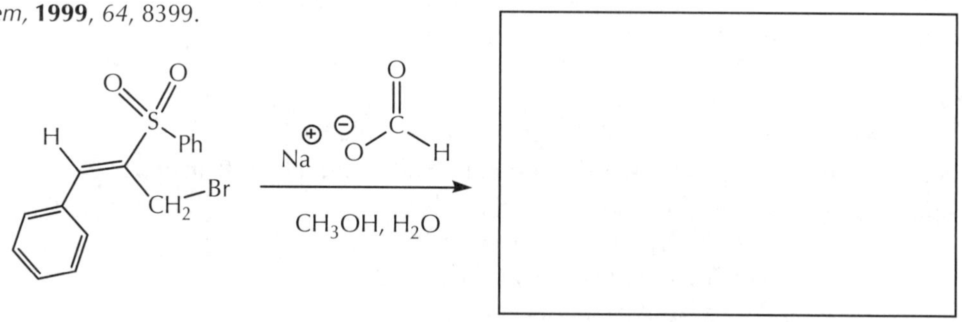

(d) *J Org Chem,* **1999**, *64*, 8311.

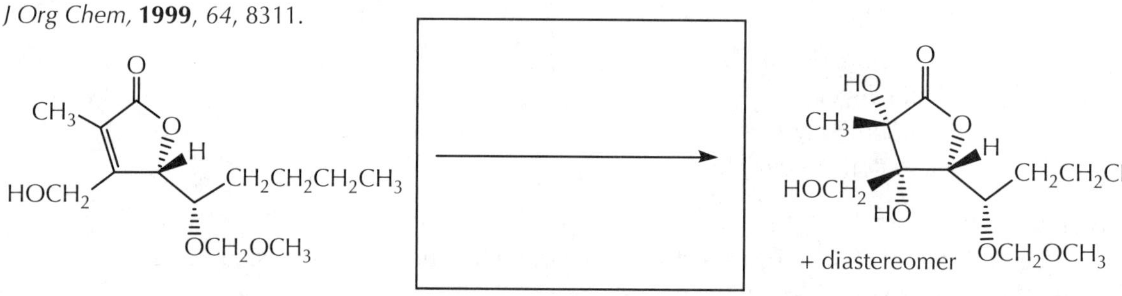

(e) *Org Lett,* **1999**, *1*, 1439.

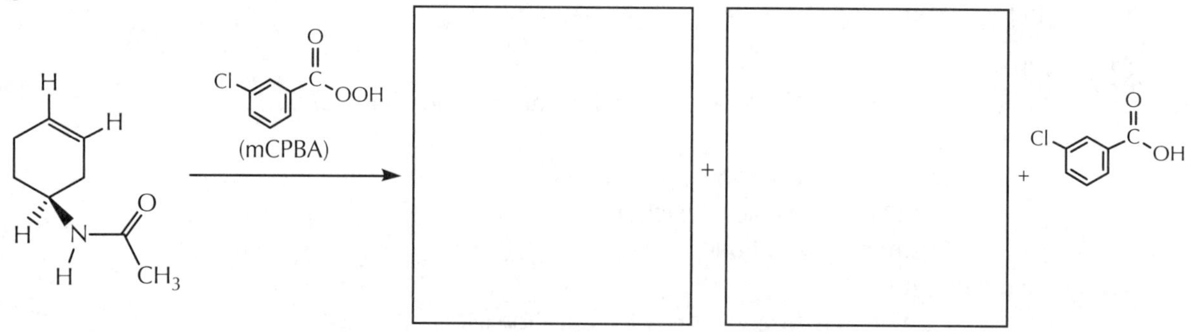

these are stereoisomeric products

EQ 03.02

Provide the reagents or structures as necessary to complete the following. Do not use acronym abbreviations for structures (e.g., "OTs").

(a) *J Org Chem*, **2000**, *65*, 7399.

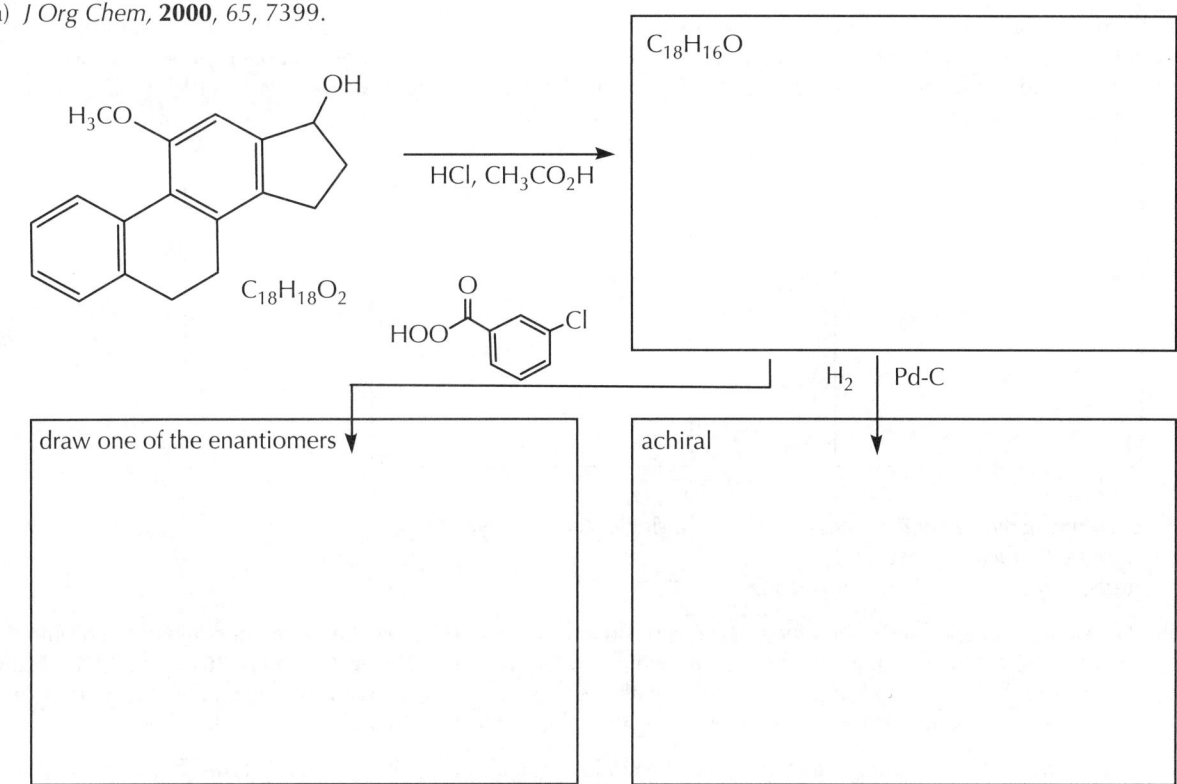

(b) Complete the following equation. Also: provide the 3D transition state structure for the formation of the final product. Use dotted lines for partial bonds and indicate any needed partial charges. Include necessary non-bonding electron pairs. Pay close attention to the stereochemistry.

758 CHAPTER 9 Electrophilic Addition II: Halogenation, Oxidation, and Reduction Reactions

EQ 03.03

One of the reasons why the reactions of epoxides are of interest is because they have been implicated in some of the mechanisms for how cancers get started. The following problem relates to some epoxide chemistry.

(a) Complete the following reaction sequence by providing the necessary reagents, starting materials, or products, as needed. Abbreviations may be used for reagents (e.g., Ph, TsCl) if they are used completely correctly; however, acronyms (e.g., mCPBA) may not be used. HINT: part (a) might be easier if worked backwards! Pay attention to the information written below the boxes.

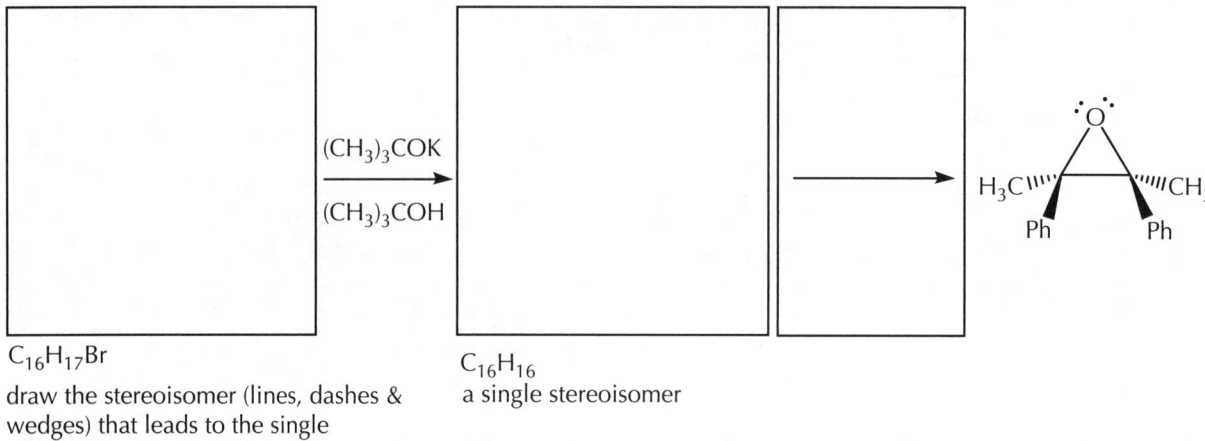

$C_{16}H_{17}Br$
draw the stereoisomer (lines, dashes & wedges) that leads to the single stereoisomer implied in the next step

$C_{16}H_{16}$
a single stereoisomer

(b) The reactivity of epoxides, when the oxygen is positively charged, parallels that of the opening of the bromonium ion intermediate in bromination reactions. In particular, Ollevier et al. (*Chem Commun,* **2012**, *48*, 3806) studied how a Lewis acidic iron (+2) ion can combine with the Lewis basic oxygen atoms of epoxides to give a 1:1 Lewis acid-base complex that, in turn, can undergo ring opening reactions with a variety of nucleophiles.

(i) Following this description, draw the pair of enantiomeric products, all uncharged atoms, from this nucleophilic ring-opening reaction.

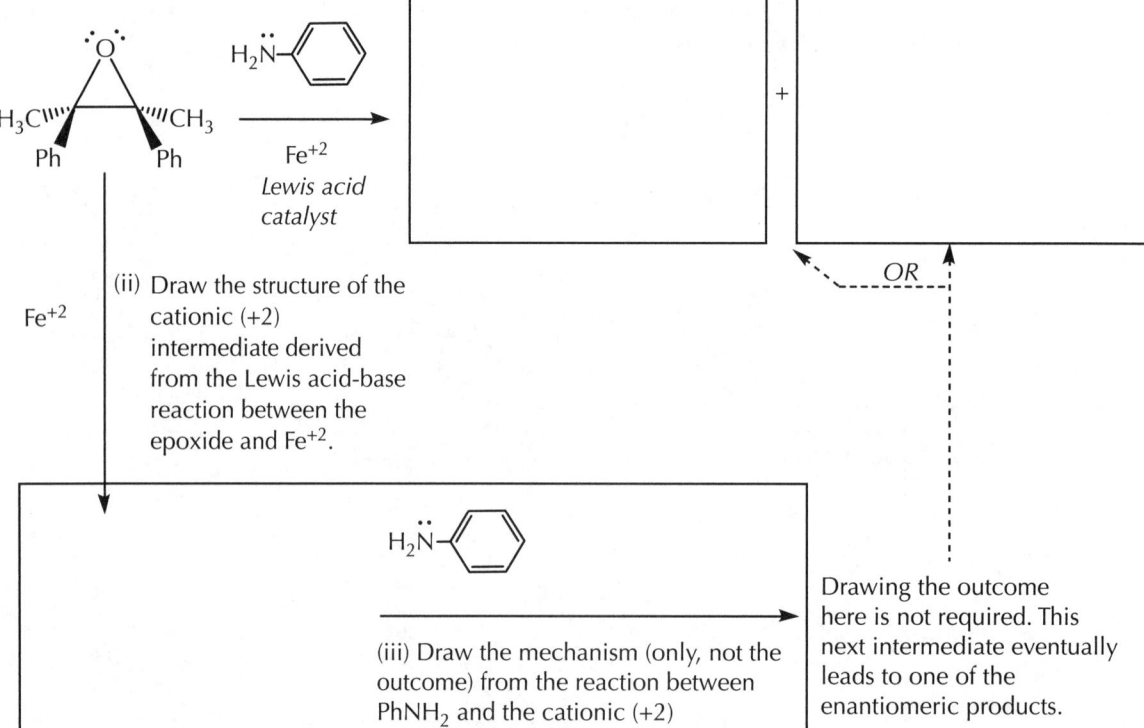

(ii) Draw the structure of the cationic (+2) intermediate derived from the Lewis acid-base reaction between the epoxide and Fe^{+2}.

(iii) Draw the mechanism (only, not the outcome) from the reaction between $PhNH_2$ and the cationic (+2) intermediate you have drawn to the left.

Drawing the outcome here is not required. This next intermediate eventually leads to one of the enantiomeric products.

Exam Questions (Chapters 7–9)

EQ 03.04

Provide the reagents or structures needed to complete the following. Do not use acronym abbreviations for structures (e.g., "OTs"). Sequential experimental steps should be numbered; only the major organic product is needed. If stereochemistry is indicated, it should be used consistently in your answer. Take care to represent accurate geometries for atoms according to their hybridizations.

(a) *J Org Chem,* **2000**, *65*, 7475.

(b) *J Am Chem Soc,* **2002**, *124*, 4628.

(c) *Org Lett,* **2000**, *2*, 3473.

(d) *J Org Chem,* **2009**, *74*, 8826.

(e) *J Org Chem,* **2000**, *65*, 7051.

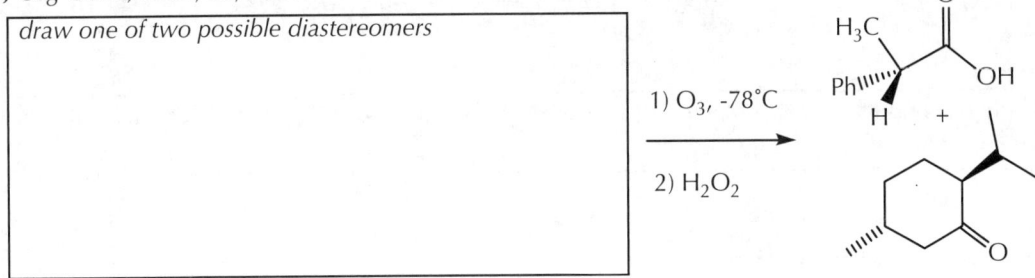

EQ 03.05

This question contains some turn-of-the-century chemistry from the year 2000 (*J Org Chem*, **2000**, *65*, 7158). When 1-phenyl-1-cyclopentanol (compound A) is combined with butyl azide in the presence of trifluroacetic acid, product B is formed.

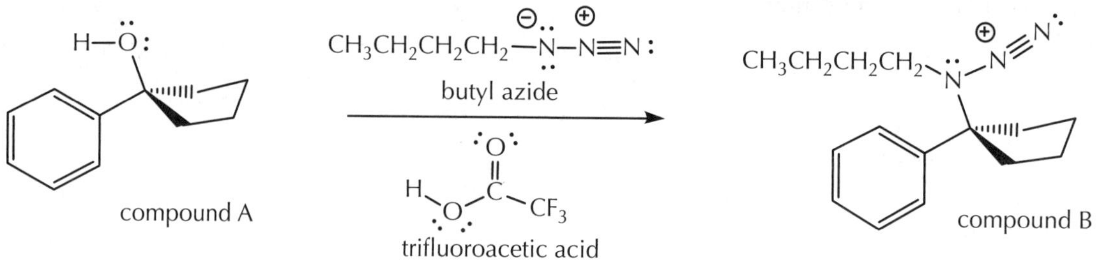

(a) What is the specific mechanistic classification for this reaction?

(b) Provide a complete, stepwise mechanism for this reaction using curved arrow notation. Use trifluroacetic acid as a general Brønsted acid source, and its conjugate base as a general Brønsted base, as needed.

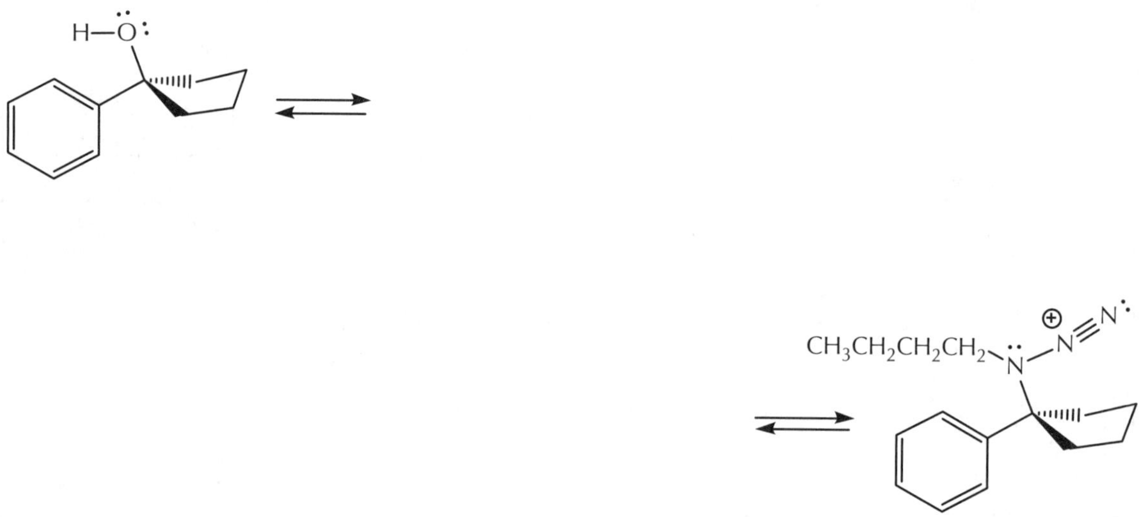

(c) Increasing the concentration of butyl azide will cause the rate of formation of compound B to (circle one):
(i) increase.   (ii) decrease.   (iii) remain the same.
Briefly explain why:

(d) Increasing the concentration of compound A will cause the rate of formation of compound B to (circle one):
(i) increase.   (ii) decrease.   (iii) remain the same.
Briefly explain why:

(e) Starting from compound A ($C_{11}H_{14}O$), the major byproduct is compound C ($C_{11}H_{12}$). What is the name of compound C?

Exam Questions (Chapters 7–9)

EQ 03.06

Complete the following reactions as indicated; include stereochemistry as needed. Separate experimental steps should be numbered.

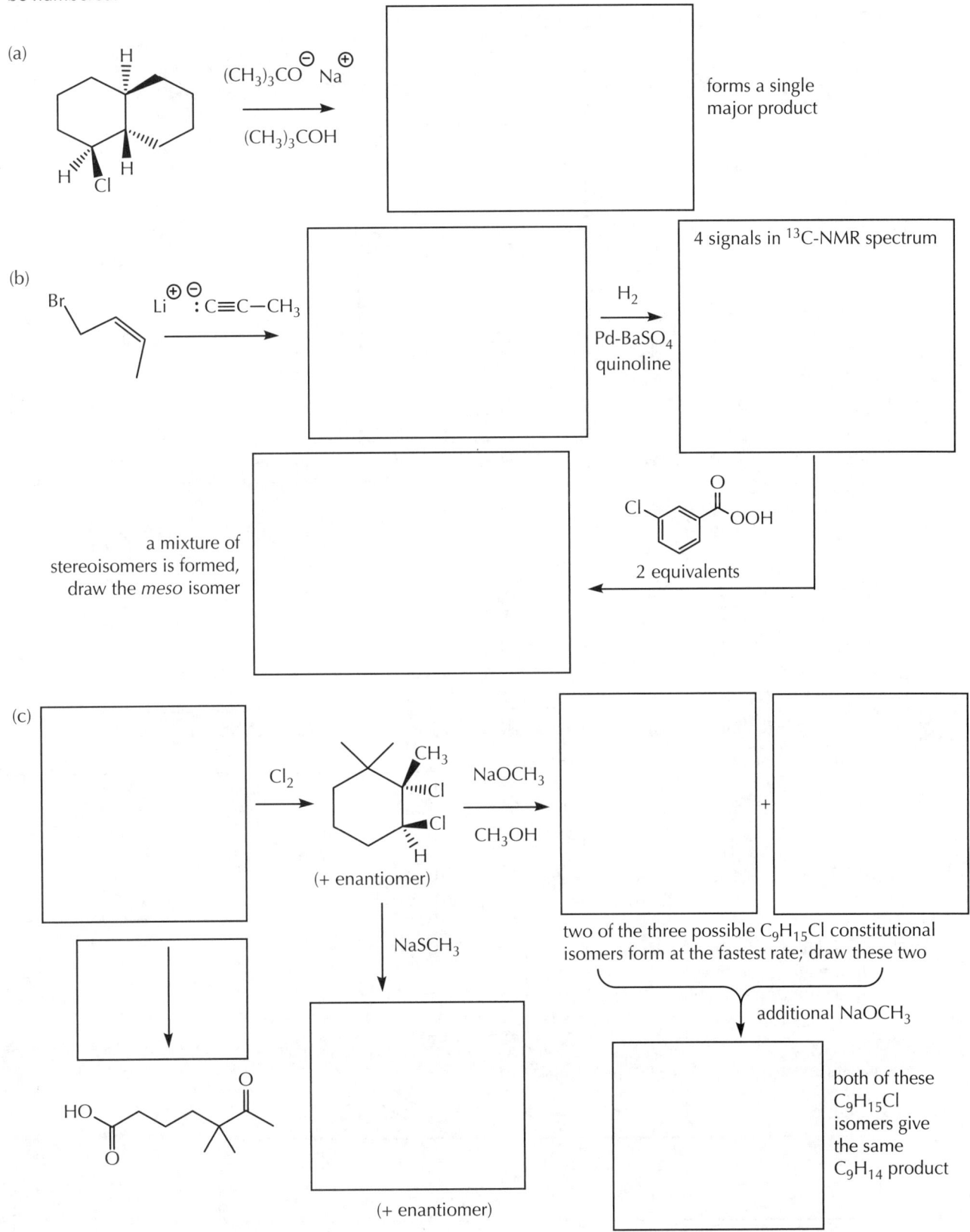

762 CHAPTER 9 Electrophilic Addition II: Halogenation, Oxidation, and Reduction Reactions

EQ 03.07

A. These reactions were carried out in the preparation of some glycolysis inhibitors being used to test anticancer treatments (*ACS Med Chem Lett,* **2016**, *7*, 217). Complete the following as directed.

(a) the following S$_N$1 reaction shows high selectivity

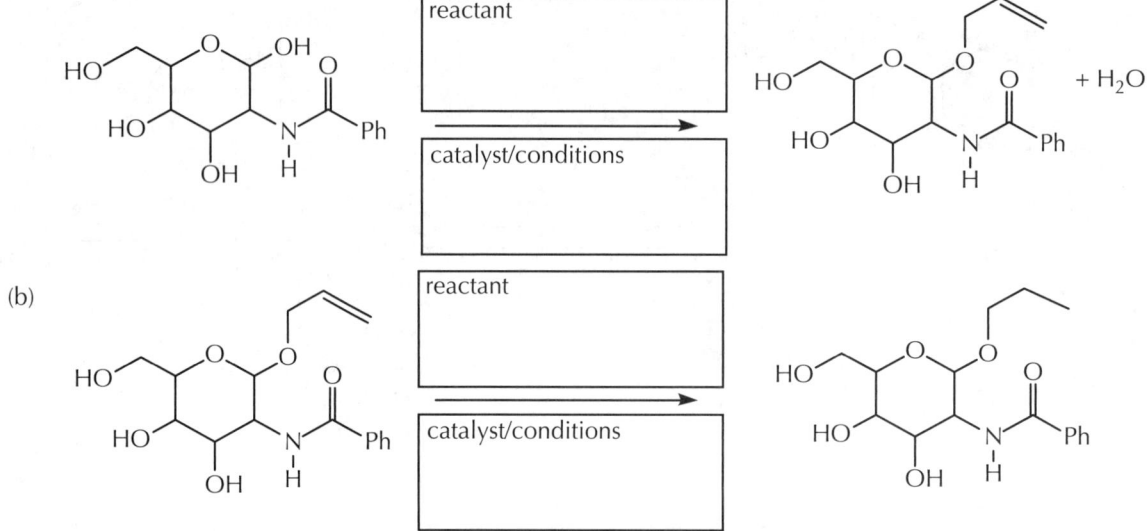

(c) p-toluenesulfonyl chloride shows high selectivity for less hindered nucleophiles

B. The following hydroborating reagent gives complete regioselectivity in its reactions with this diene.

(a) Draw one of the anticipated stereoisomeric products.

(b) In addition to the one you have drawn, how many other stereoisomeric products result from this reaction?

(c) What is the IUPAC name for this molecule?

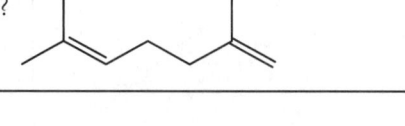

EQ 03.08

A. *N*-Bromosuccinimide (NBS) is a laboratory reagent that provides a safer and more convenient source of electrophilic bromine (compared with using Br_2). As with bromine, NBS also reacts to give bromonium ion intermediates, releasing an anion whose conjugate acid has a pK_a of 9.5. Provide the complete, stepwise, curved arrow mechanism for this reaction.

B. Some molecules, such as compound A, when heated up in water, give highly predictable products because there is a single, clearly best carbocation that can form after rearrangements take place. What is the structure of the predictably most stable carbocation that forms after rearrangements occur, and what are the structures of the two products (which have no stereoisomers) that result?

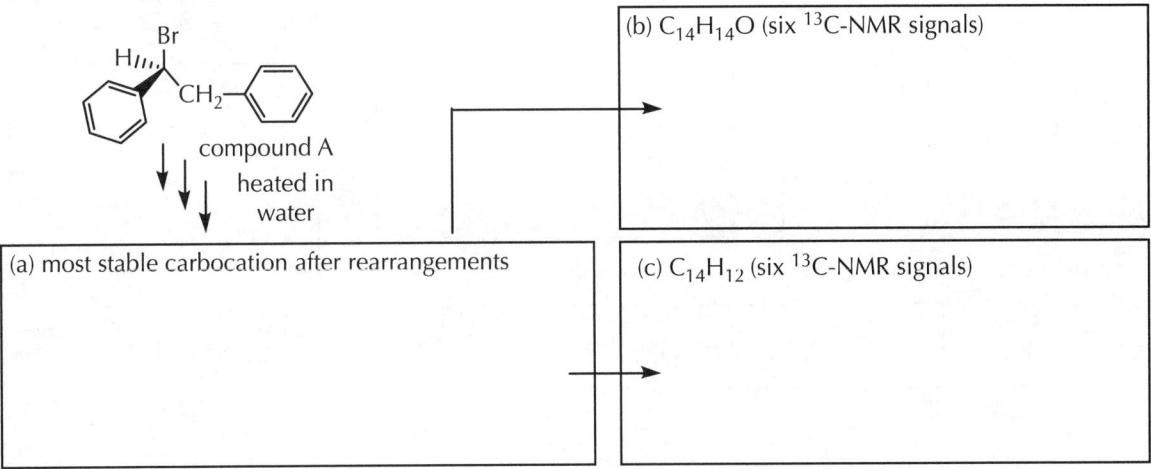

(a) most stable carbocation after rearrangements

(b) $C_{14}H_{14}O$ (six ^{13}C-NMR signals)

(c) $C_{14}H_{12}$ (six ^{13}C-NMR signals)

764 CHAPTER 9 Electrophilic Addition II: Halogenation, Oxidation, and Reduction Reactions

EQ 03.09

A. In 2007, Frontier and co-workers reported a new method to access spirocyclic compounds via an interrupted Nazarov cyclization. The formation of spirocyclic carbocation J from dienone H proceeds through <u>two</u> carbocation intermediates. Draw a curved arrow mechanism showing how spirocyclic carbocation J can be formed from dienone H.

dienone H   carbocation I   spirocyclic carbocation J

draw curved arrows for each step and the missing intermediate between I and J

provide the intermediate between I and J

B. For the reaction shown below, answer the following questions.

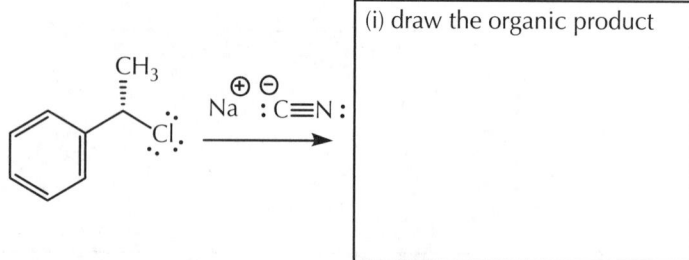

(i) draw the organic product

(ii) The reaction rate will depend on the concentration(s) of:

(circle one)

electrophile   nucleophile   both

(iii) Complete the energy diagram, including any intermediates.

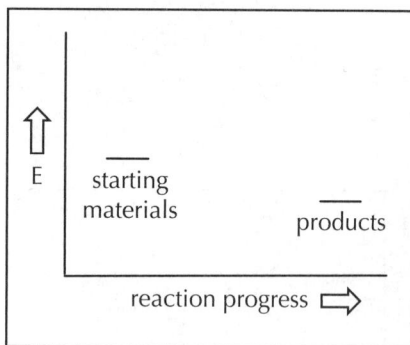

EQ 03.10

A. Heating a variety of different C_8H_{16} isomers with H_3PO_4 (phosphoric acid) gives a new alkene (Q), which also has the C_8H_{16} formula. Ozonolysis of Q gives two equivalents of (only) 2-butanone as the observable product. Hydrogenation of Q gives a racemic mixture of an alkane (C_8H_{18}). And bromination of Q gives a *meso*-compound with the molecular formula $C_8H_{16}Br_2$. What is the structure and stereochemistry of compound Q?

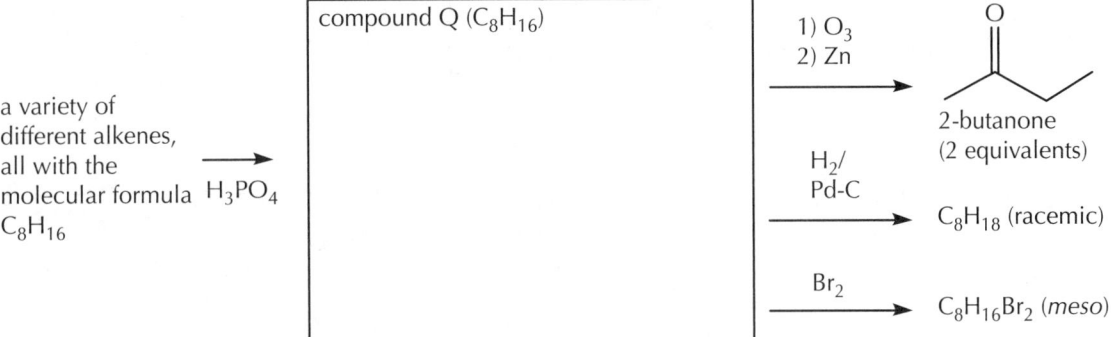

B. When compound K is treated with bromine and a base, an intermediate L is proposed to form. Intermediate L then undergoes an equilibirum isomerization process that results in a second intermediate (M). Pay attention to the accuracy of the geometries of your sp^3, sp^2, and sp atoms in your drawings.

(a) Provide the structure for cationic intermediate L (containing a bromonium ion), including stereochemistry, its mechanism of formation, and then the mechanism for its subsequent transformation to intermediate M (*J Org Chem*, **2000**, 65, 7253).

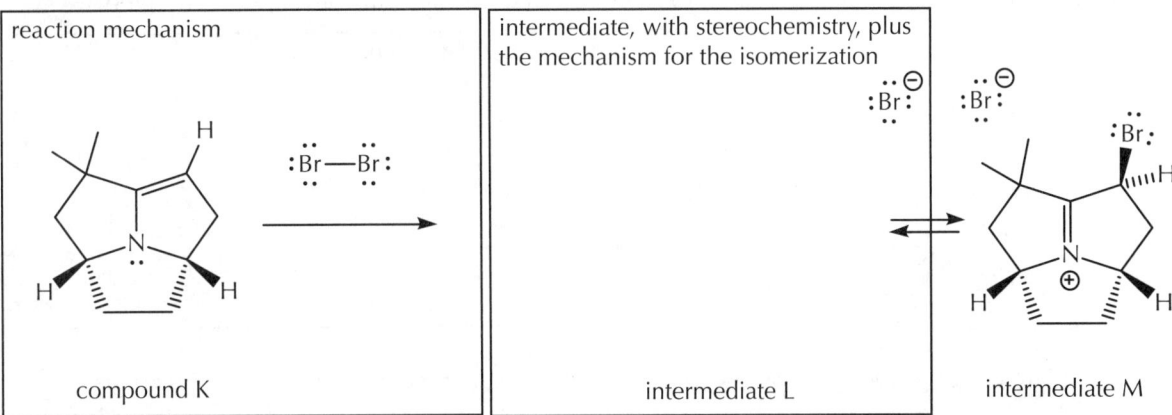

(b) The reaction is performed in the presence of triethylamine [$(CH_3CH_2)_3N$], which is a base. Intermediate M is deprotonated to give compound N ($C_{11}H_{16}BrN$), a molecule with all atoms closed shell and uncharged. What is the structure of compound N?

CHAPTER 9 Electrophilic Addition II: Halogenation, Oxidation, and Reduction Reactions

EQ 03.11

Complete the following reaction schemes as directed. Do not use abbreviations for reagents; use structures.

(a) *Org Lett,* **2017**, *19*, 3422; the reaction is regioselective and stereoselective (via iodonium ion)

(b) *Org Lett,* **2017**, *19*, 4231; a nonacidic, two-step process

(c) *Org Lett,* **2017**, *19*, 4275; the reaction is regioselective and stereoselective, giving a racemic product

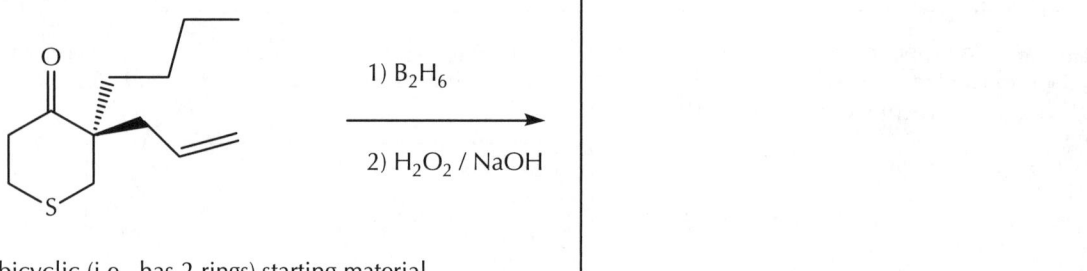

(d) *Org Lett,* **2017**, *19*, 5007; the reaction is regioselective

(e) a bicyclic (i.e., has 2 rings) starting material

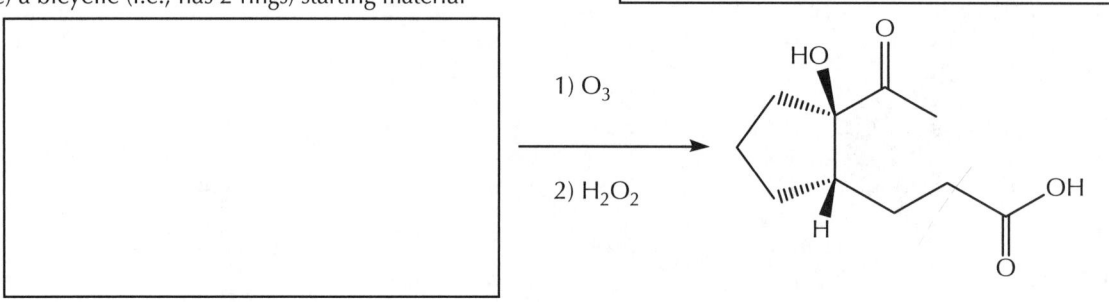

EQ 03.12

The following chemical reactions were reported over the course of about 100 years. Because some of the reaction conditions are atypical, pay attention to the information provided about the observed outcomes.

(a) *ACS Med Chem Lett,* **2015**, *6*, 266 (conditions promote an S$_N$2 reaction with a usually poor nucleophile)

(b) *Accts Chem Res,* **1971**, *4*, 9 (conditions promote azido-iodination reaction via an iodonium ion)

racemic; draw one of the enantiomers

(c) *J Chem Soc,* **1952**, 1208 (conditions give 3 E2 reaction products, where nitrate NO$_3^\ominus$ is a leaving group)

(d) *Annalen,* **1929**, *468*, 98 (conditions promote an S$_N$1 reaction at the primary C-Br bond)

provide the single most significant resonance contributor for the first anticipated intermediate

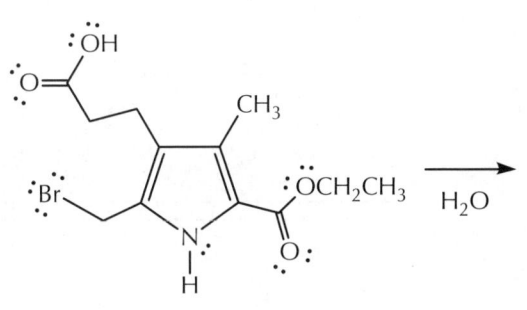

768 CHAPTER 9 Electrophilic Addition II: Halogenation, Oxidation, and Reduction Reactions

EQ 03.13

A. Cylindricine A and B are natural products isolated from a marine sea squirt (*Chem Rev*, **2006**, *106*, 2531). These two natural products are observed to rapidly interconvert.

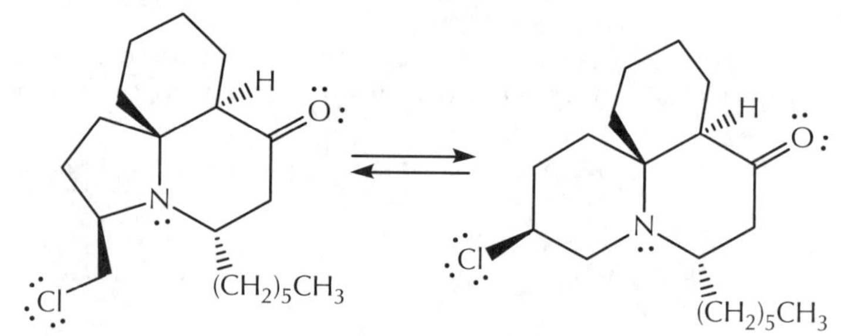

cylindricine A ⇌ cylindricine B

(a) Based on the following description, provide the curved arrow mechanism for this two-step interconversion process, both of which are S_N2 reactions. In the first S_N2 rection, cylindricine A forms a new ring and generates a positively charged intermediate Z, with all closed shell atoms and with chloride as its counterion. The second S_N2 reaction opens the newly formed ring in intermediate Z. No additional reagents are required for these reactions.

mechanism for A to Z, structure of Z, & the mechanism for Z to B

cylindricine A → intermediate Z → cylindricine B

(b) The rates of reaction of cylindricine A and B with NaSH are significantly different enough that only one product is observed when NaSH is added to this interconverting mixture. Draw the structure of the expected single product.

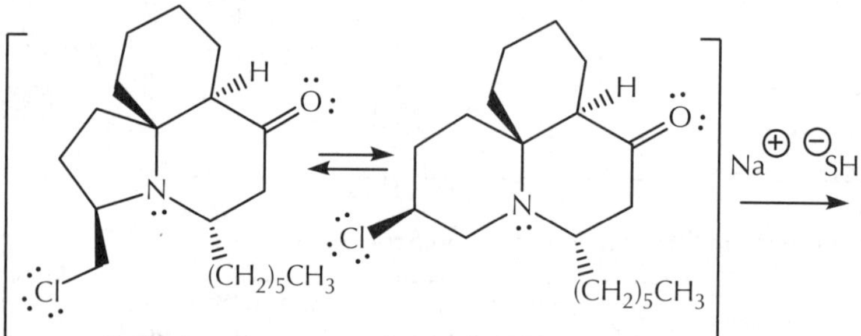

[cylindricine A ⇌ cylindricine B] + Na⊕ ⊖SH →

B. (a) Provide the name for the following:

(structure: HC≡C–CH₂–CH₂–CH₂–OH)

(b) Draw the following:

(R,E)-3-chloro-4-methylhexa-1,4-diene

EQ 03.14

(R)-(-)-α-Phellendrene is a naturally occurring compound found in the oil of peppermint leaves. It has the molecular formula $C_{10}H_{16}$. From the chemical and spectroscopic information given below, deduce the structure of (R)-(-)-α-phellendrene, as well as the structures for compounds D, E, F, G, I and J. NOTE: these types of problems can rarely be solved linearly! Although (R)-(-)-α-phellendrene is required in the first answer space, it is not at all likely to be the first answer you will write. Working backwards from known product(s) is perhaps the best and most common strategy. Check and cross-check the data.

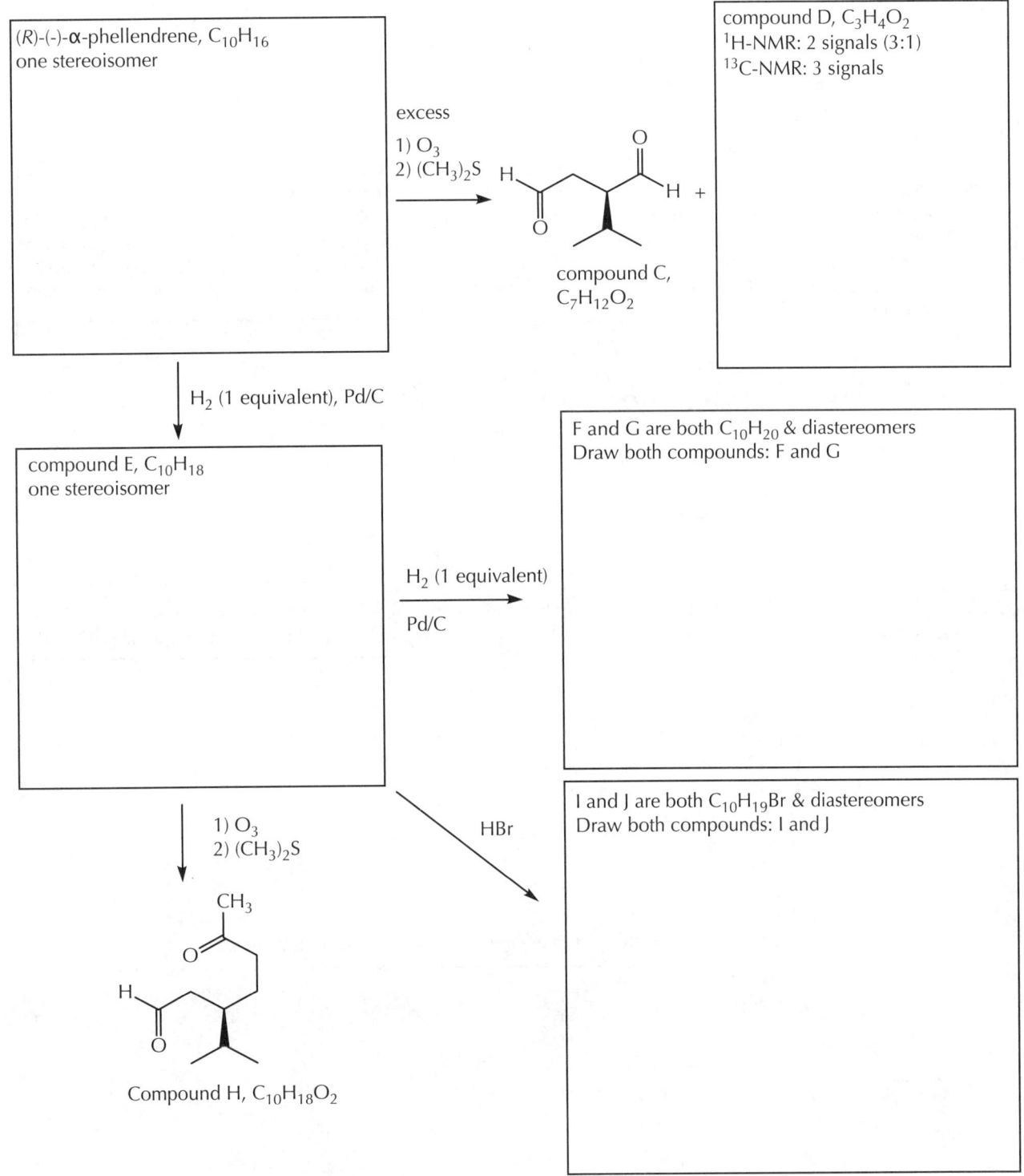

CHAPTER 9 Electrophilic Addition II: Halogenation, Oxidation, and Reduction Reactions

EQ 03.15

A. Aluminum is directly below boron on the periodic table. Not surprisingly, aluminum hydride reagents (containing an Al-H bond) give many reactions that are analogous to boron hydride reagents (containing the B-H bond). With that in mind, predict the structure of the product formed (including the proper configuration) when diisobutylaluminum hydride reacts with the following compound.

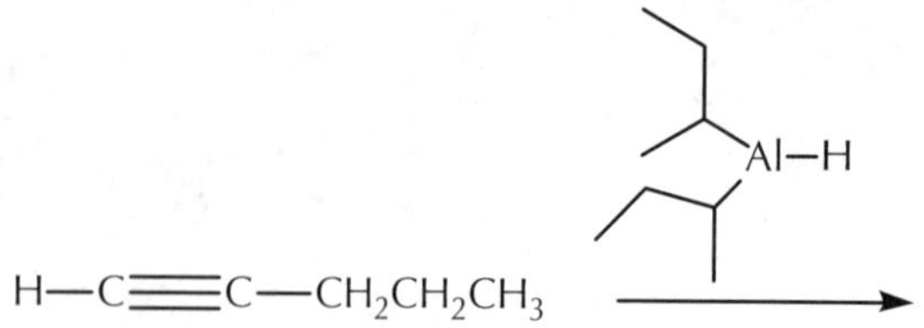

$C_{13}H_{27}Al$

H—C≡C—CH$_2$CH$_2$CH$_3$ + (iBu)$_2$Al—H →

B. Oxymercuration reactions of alkenes with Hg(OAc)$_2$ (mercuric acetate; see below) in the presence of alcohols such as CH$_3$OH (methanol) are used to carry out the electrophilic addition of "-HgOAc" and "-OR" (e.g., in this case, "-OCH$_3$") to alkenes. The reactions proceed with Markovnikov selectivity such that the Hg(OAc)$_2$ behaves as the electrophilic component and the alcohol as the nucleophilic component. These reactions provide products resulting from anti-addition. Draw the complete structure for one of the enantiomers of the major product from the following reaction.

mercuric acetate Hg(OAc)$_2$

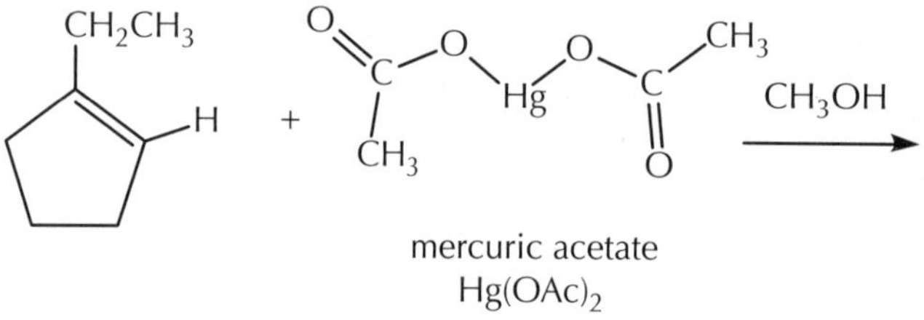

C. Carbonyl oxides (connectivity: R$_2$COO) give addition reactions that are analogous to those of ozone. Predict the structure of the product from the following reaction (note: the regioselectivity of the reaction can be predicted on the basis of steric considerations).

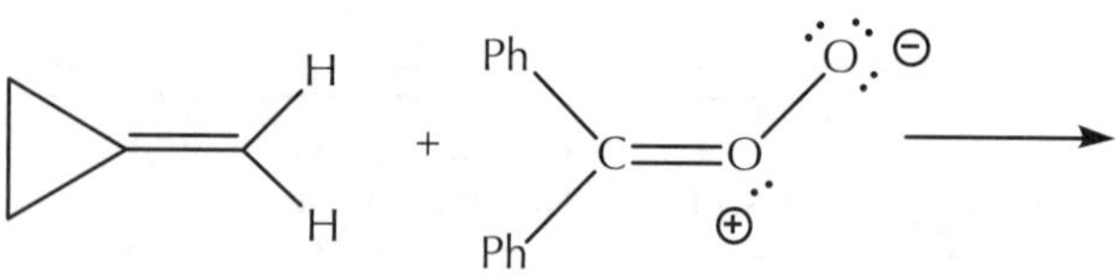

D. Speaking of ozone... provide the complete set of organic products from the following reaction.

1) O$_3$ excess
2) H$_2$O$_2$

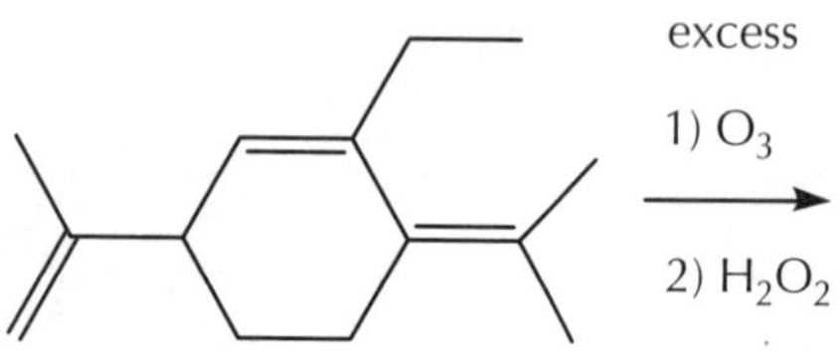

EQ 03.16

From a 2013 report on the concise synthesis of Lubeluzole (*Org Lett,* **2013**, *15*, 1158). This compound exhibits promising therapeutic effects for the treatment of acute ischemic stroke, and is also associated with neuroprotective activities.

(a) Epoxides (three-membered rings containing oxygen atoms) react with nucleophiles with a mechanism similar to the nucleophilic ring-opening of halonium ions, but with different regioselectivity. Based on this information and the product shown below (right), fill in the missing neutral product and the reagent for the second reaction in the appropriate boxes.

(b) Subsequently, molecule P reacts with thiazole Q to generate product R. The mechanism proceeds in three steps. The first step involves the most basic group in P reacting with the most electrophilic position in Q, without the loss of the leaving group. The second step involves the departure of the chloride ion leaving group. The final step is a deprotonation reaction using water as the base.

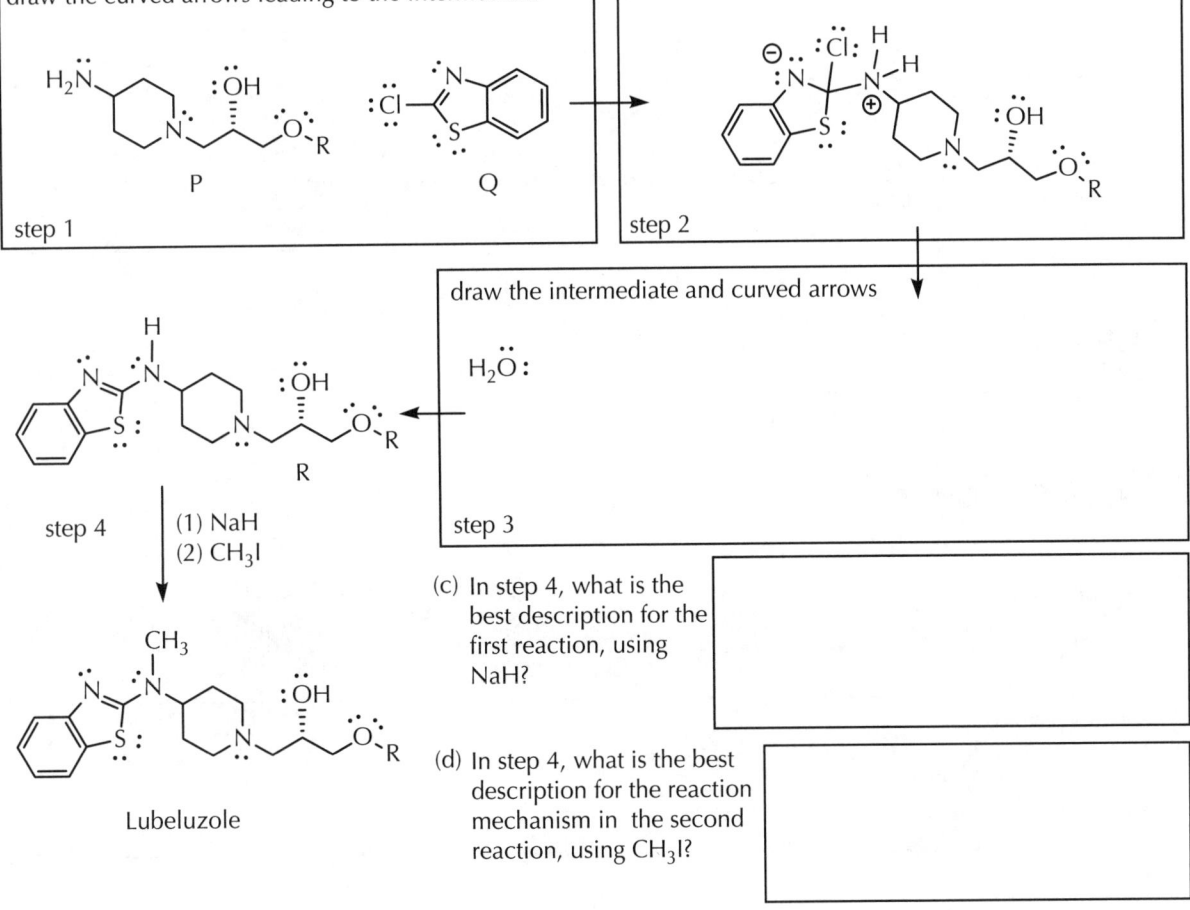

(c) In step 4, what is the best description for the first reaction, using NaH?

(d) In step 4, what is the best description for the reaction mechanism in the second reaction, using CH$_3$I?

EQ 03.17

A. Compound W reacts with potassium *tert*-butoxide to give a mixture of diastereomeric (*E*)- and (*Z*)-alkenes (*J Org Chem*, **2000**, *65*, 7248).

(a) Complete the Newman projection for the conformation of W that gives the fastest rate of formation for the (*E*)-isomer of the product (C_a in front, C_b in back).

(b) Show the curved arrow mechanism for the formation of the (*E*)-isomer.

compound W (*Z*)-isomer (*E*)-isomer

conformation of W whose reaction gives the (E)-isomer and the corresponding curved arrow reaction mechanism

(*E*)-isomer

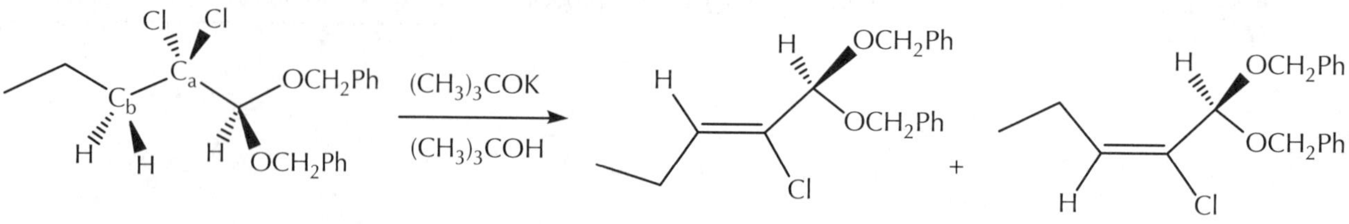

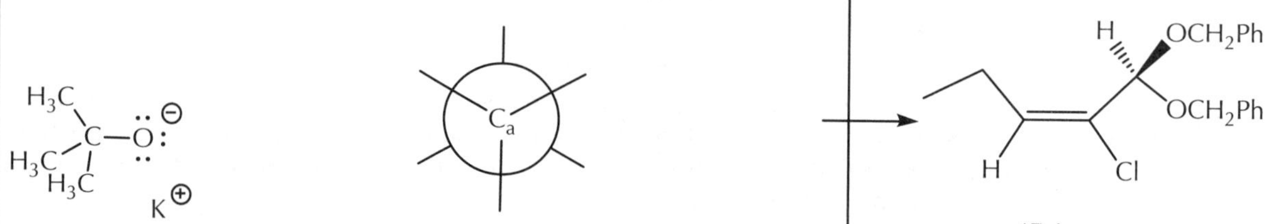

B. Alkenes react with I_2 in the presence of good nucleophiles, such as the azide ion, $N_3^{\ominus}$, to give azido iodides. Starting with (*E*)-3-methyl-2-pentene (below), provide the product from this highly regioselective reaction (draw clearly the 3-dimensional representation using the line, dash, and wedge notation). Show one of the enantiomers from this racemic product.

(*E*)-3-methyl-2-pentene

+ enantiomer

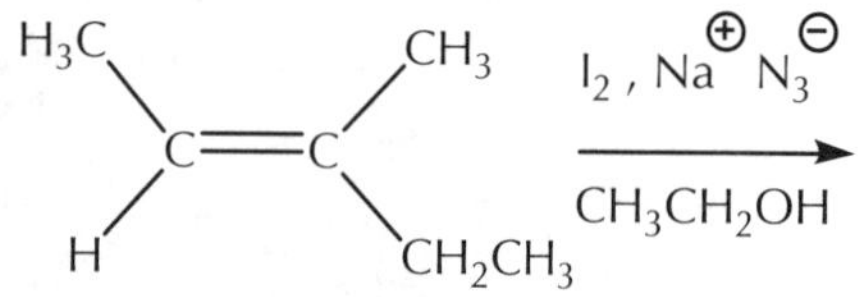

C. Provide the complete IUPAC name for this compound:

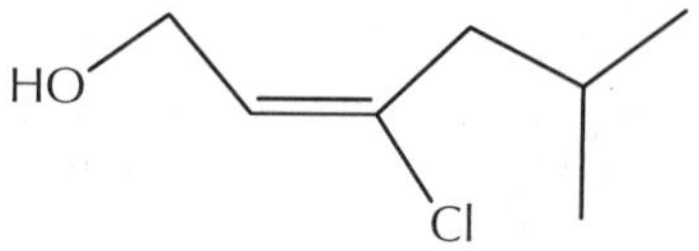

EQ 03.18

The following chemical reactions were first reported in 1969 (*Helv Chim Acta*, **1969**, *52*, 1102).

(a) Based on the following data, what is the structure of compound A, which is a constitutional isomer of compound D?

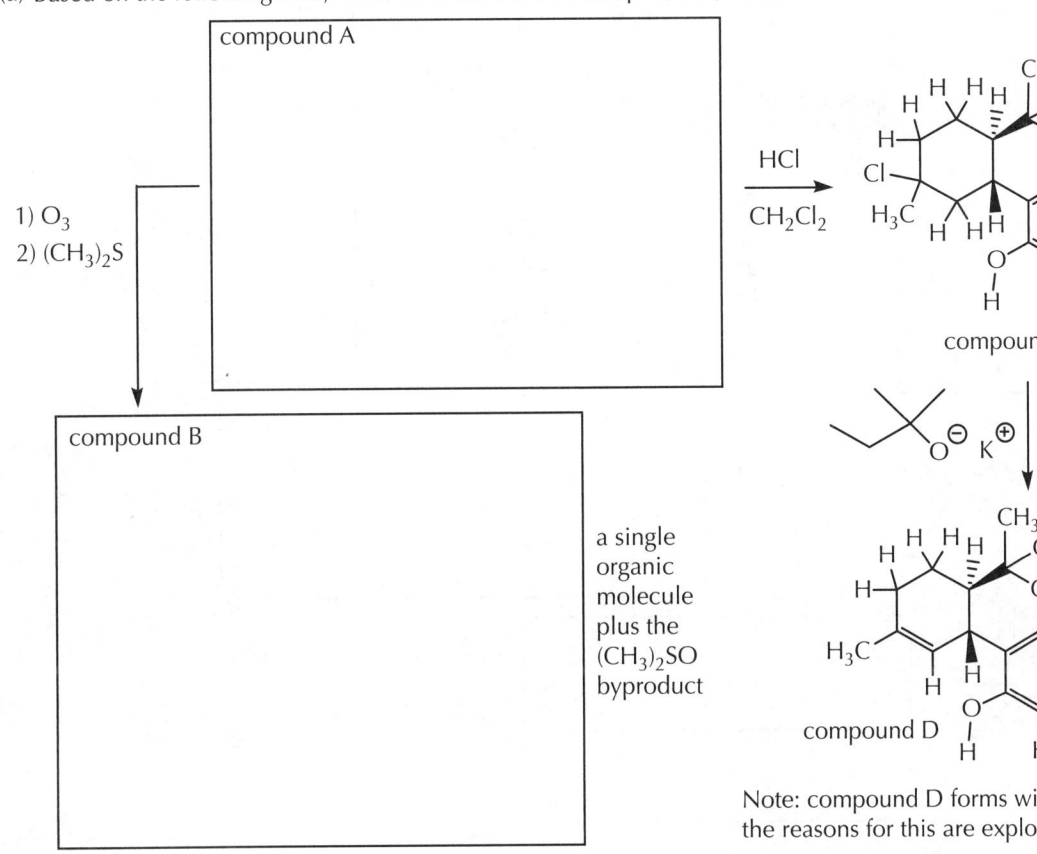

(b) The mechanism for the reaction of compound C to compound D is actually more interesting than it first appears.

Compound C has an acidic proton (pK_a ~ 10) that undergoes a fast deprotonation reaction. The conjugate base of compound C then forms compound D via an intramolecular (within the molecule) reaction, with no other reagents required.

Provide the appropriate structure for the ion generated by deprotonating compound C, and then the curved arrow mechanism for the reaction to give the observed product, compound D.

CHAPTER 9 Electrophilic Addition II: Halogenation, Oxidation, and Reduction Reactions

EQ 03.19

Complete the following as required. Complete structures should be shown, not abbreviations or acronyms. Sequential experimental steps should be numbered.

(a) [blank box] → (mCPBA: 3-chloroperoxybenzoic acid) → [intermediate epoxide shown] → a few steps → [alkene intermediate shown] → [blank box] → [final product shown]

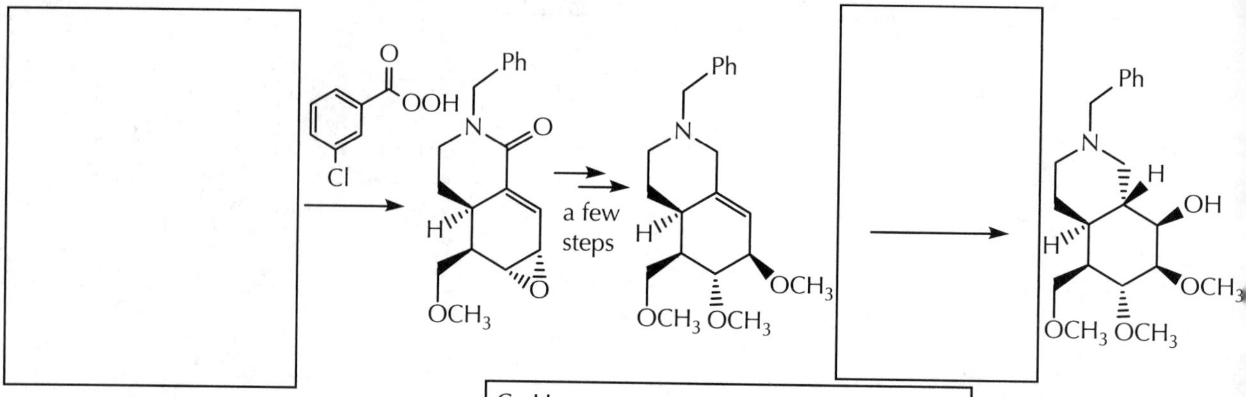

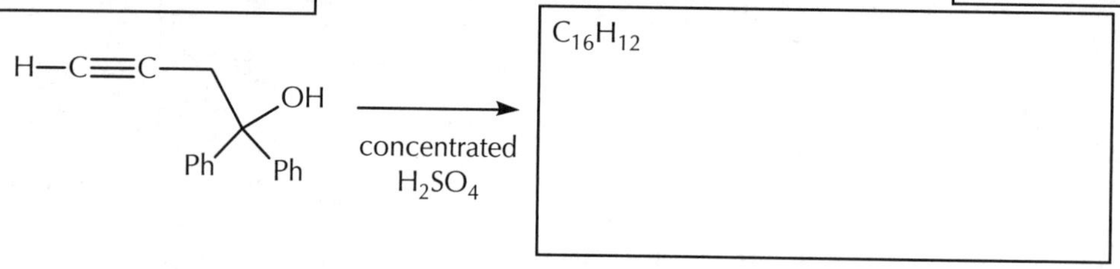

(b) H−C≡C−C(OH)(Ph)(Ph) + concentrated H₂SO₄ → [blank box] $C_{16}H_{12}$

(c) CH₃(CH₂)₅CH₂—Br + [blank box] → CH₃(CH₂)₅CH₂—C≡C—(CH₂)₅CH₃ + NaBr

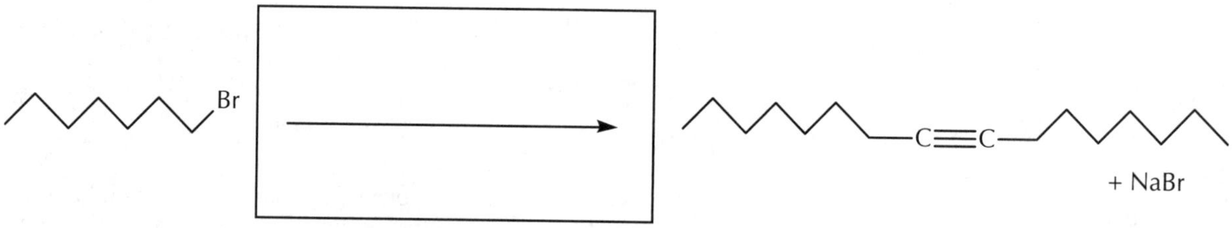

(d) CH₃(CH₂)₅—C≡C—(CH₂)₅CH₃ + H₂, Pd-CaCO₃, quinoline → [blank box]

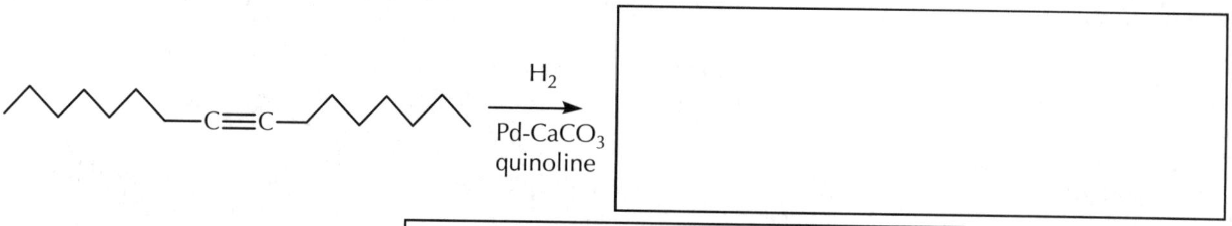

(e) [bicyclic lactone with NH] + Br−CH₂−C(=O)−OCH₃, Na₂CO₃ (mild base) → [blank box] $C_9H_{13}NO_4$

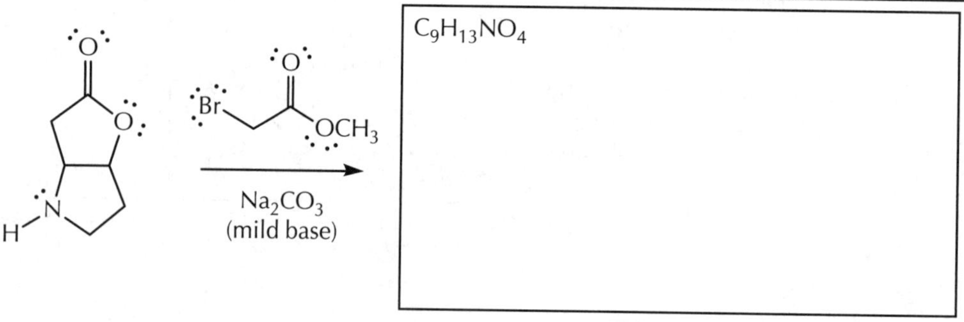

(f) [5-chloro-1-methyl-benzimidazol-2(3H)-one with N⁻ Li⁺] + (CH₃)₃C−I → [blank box]

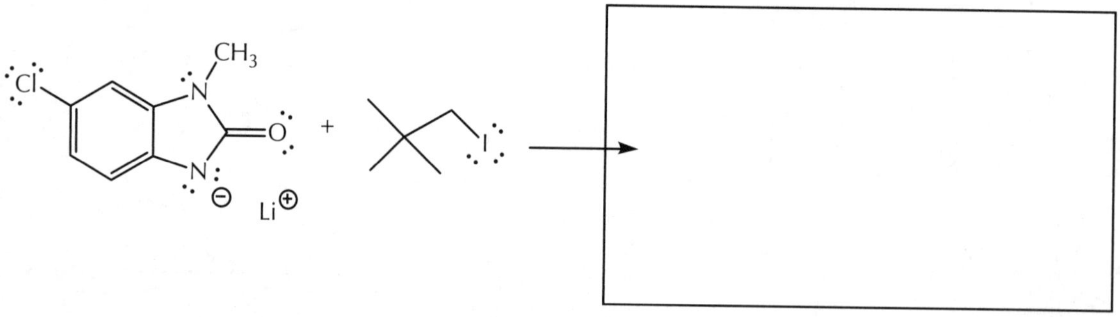

Exam Questions (Chapters 7–9)

EQ 03.20

Complete the following reaction schemes.

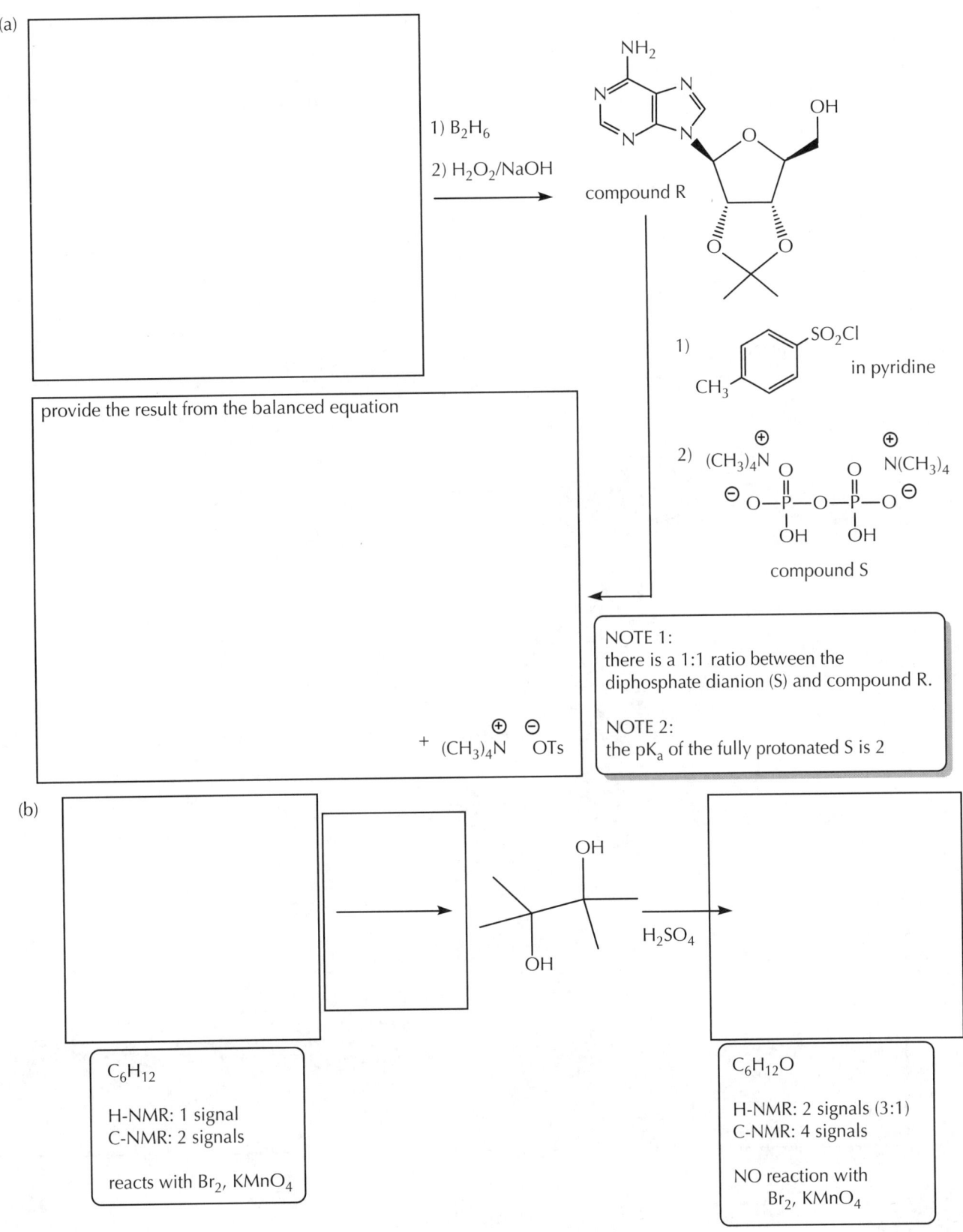

EQ 03.21

Upon treatment of compound G with fluorosulfonic acid (FSO$_3$H; a strong Brønsted acid), a rearrangement occurs to provide compound H. Given the following information, provide a complete curved arrow mechanism for the conversion of G to H. You may use H–B as a generic Brønsted acid and B:⊖ as a generic Brønsted base as necessary in your mechanism.

The reaction mechanism is proposed to occur in 4 steps:

(a) The alkene reacts with the strong acid to form a 3° carbocation that is delocalized.
(b) 1,2-Alkyl shift to form a 2° carbocation that is delocalized.
(c) 1,2-Alkyl shift to form a 3° carbocation that is not delocalized.
(d) Deprotonation to form the product.

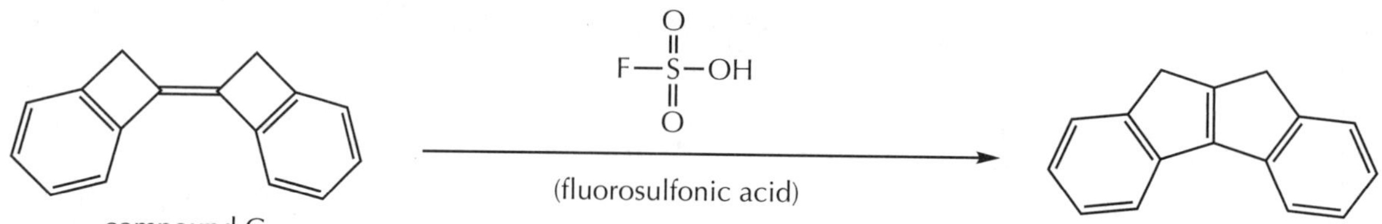

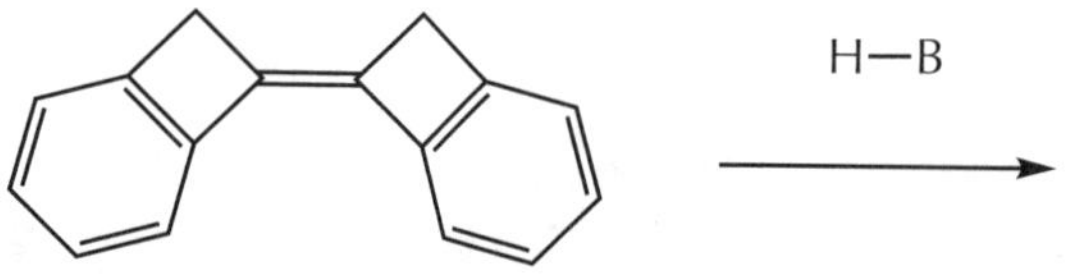

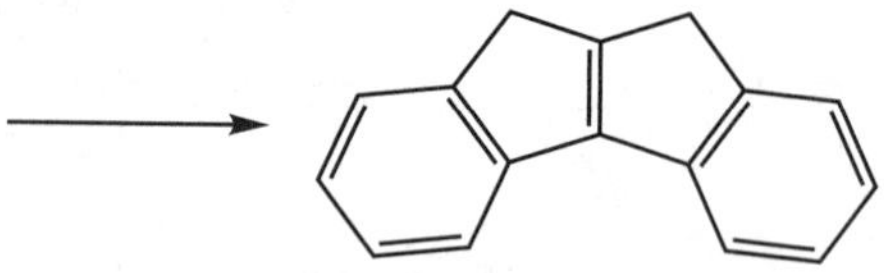

EQ 03.22

A. The major E2 reaction product from the following reaction is an oxygen-substituted alkene (*Tet Lett*, **2006**, *48*, 8353).
   (a) Predict the structure of the major alkene product, and clearly indicate its stereochemistry.

   (b) Provide the single Newman projection for the conformation of the starting material that leads to the fastest E2 reaction, and therefore to the predicted product.

   The carbon atom with the leaving group should be in the front, and the carbon atom with the β-H should be in the back.

B. The mechanism for the following transformation takes place in four steps. First, compound U is protonated to give intermediate V. Next, intermediate V undergoes a fast intramolecular substitution reaction where the nucleophilic sulfur atom forms a positively charged, three-membered ring called an episulfonium ion (intermediate W), and releases water as a leaving group. Third, the episulfonium ion (W) undergoes a second intramolecular substitution reaction, comparable to those in halonium ions, to give intermediate X. And finally, in the last step, X is deprotonated by water to give the observed product, compound Y. Draw structures for intermediates V, W, and X. No curved arrows are required.

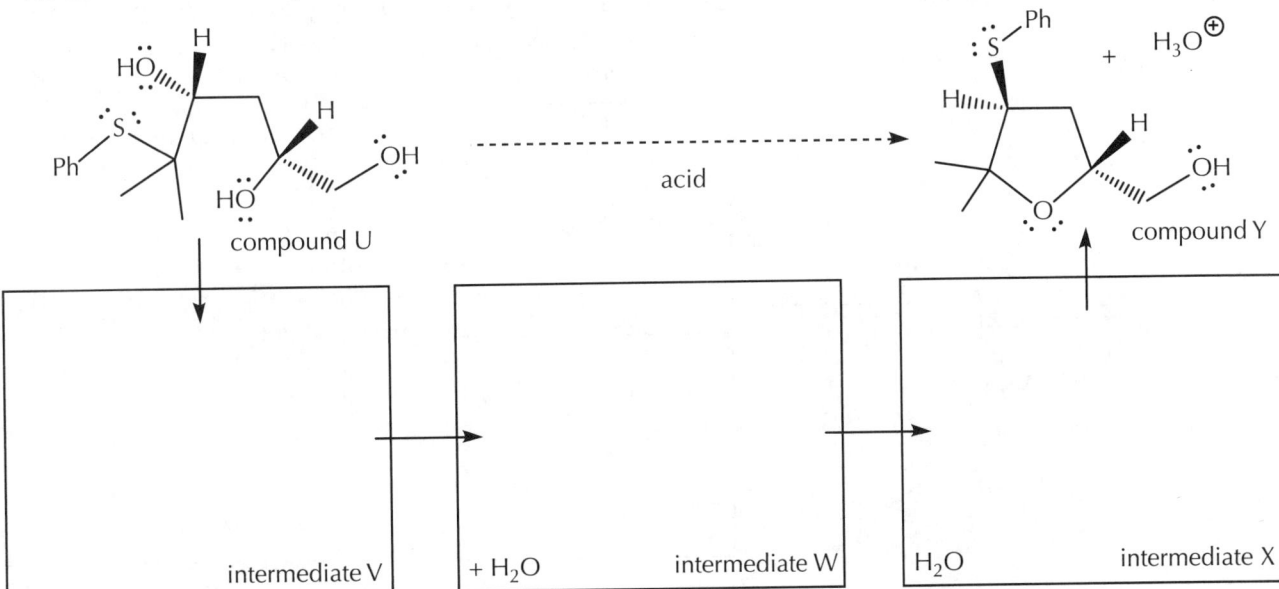

EQ 03.23

A. Provide the missing information in the following reaction.

(a)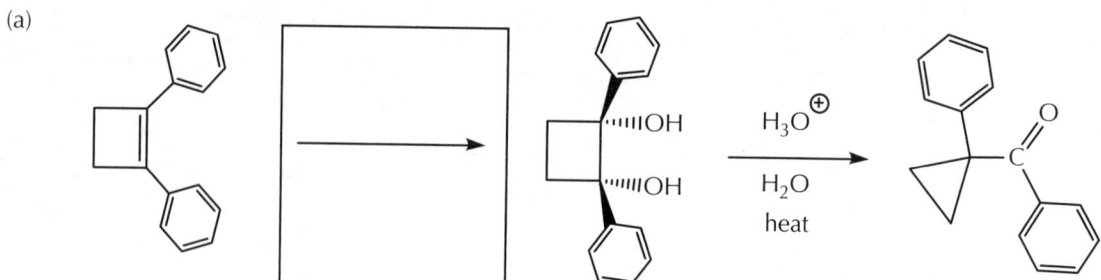

(b) In the second step, above, the diol undergoes an acid-catalyzed pinacol rearrangement reaction. Provide the complete, stepwise, curved arrow mechanism for this transformation. For intermediates where you have a choice of resonance contributors, draw the most significant one. Ignore stereochemistry.

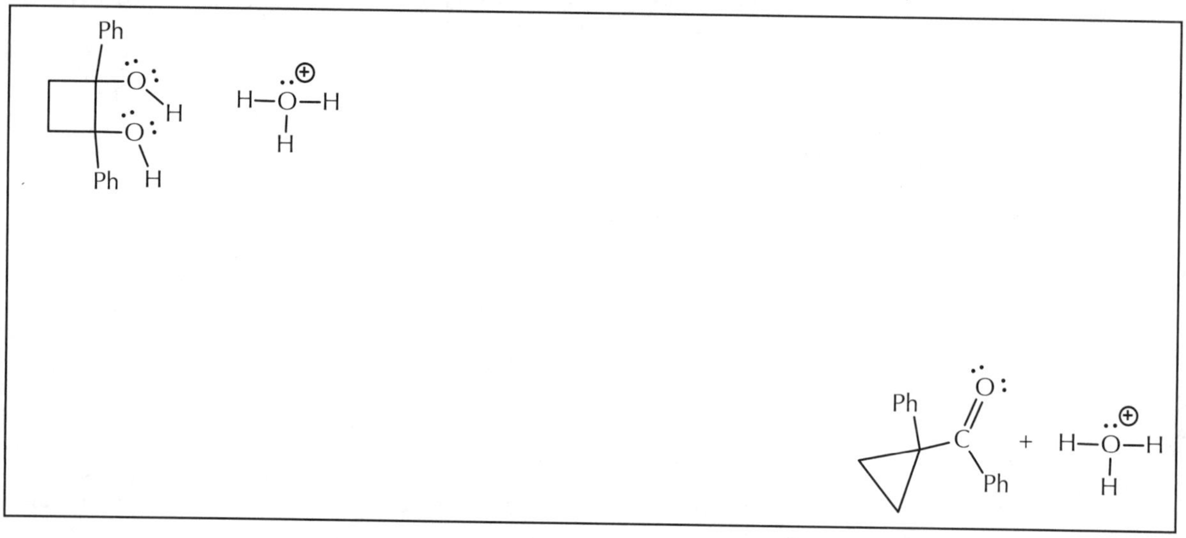

(c) In the reaction above, what is the stereoisomeric outcome? (circle one)

| a single chiral molecule | a single achiral molecule | a pair of enantiomers | a pair of diastereomers |

B. In the reaction below, a single, swift bimolecular elimination leads to the formation of two stereoisomeric alkene products, the major product is kinetically favored because the conformation leading to it is in greater abundance.

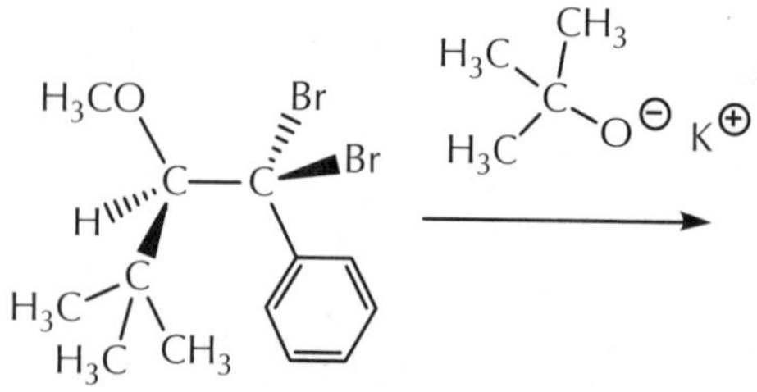

(a) Draw the two stereoisomeric alkene products

major product

minor product

(b) Draw a Newman projection of the conformation that leads to the major product; the carbon with the bromine atoms should be in the front.

EQ 03.24

A. (a) Provide the missing information for the following transformation.

(b) In this reaction, compound A forms: (check all that apply):

☐ alone, as a single product
☐ with a diastereomer
☐ with an enantiomer
☐ with >1 other stereoisomer

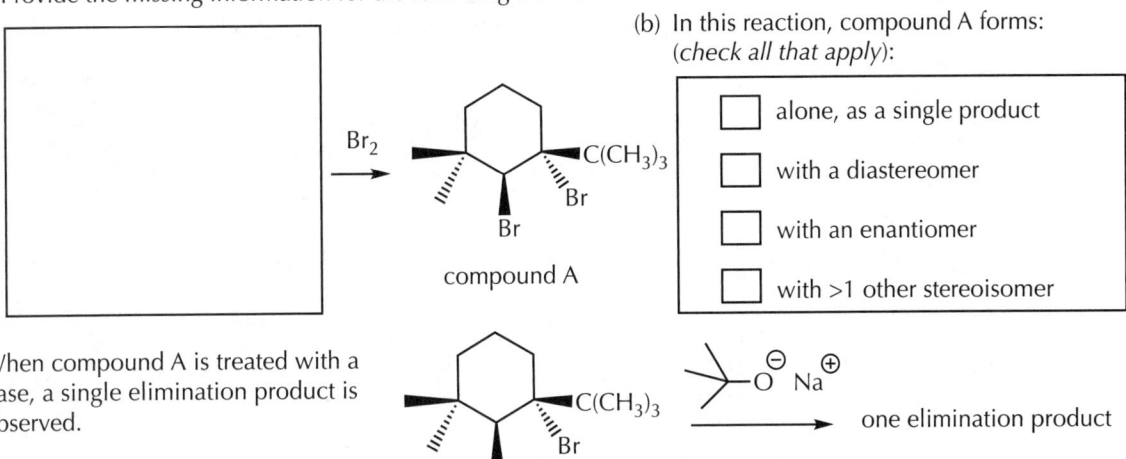

When compound A is treated with a base, a single elimination product is observed.

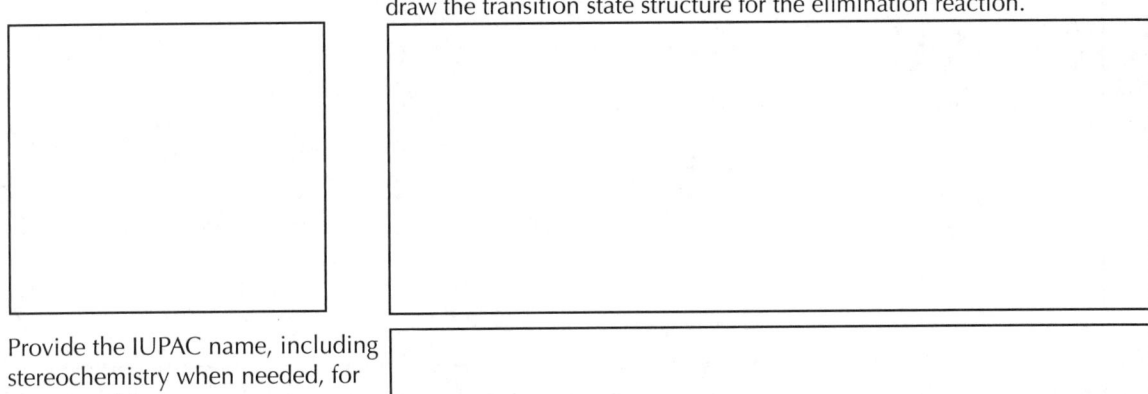

(c) Draw the elimination product.

(d) Starting with the chair conformation that best explains the E2 mechanism, draw the transition state structure for the elimination reaction.

(e) Provide the IUPAC name, including stereochemistry when needed, for compound A.

B. Both of the following transformations results in two distinct products: one that is optically active, and one that is optically inactive. Provide the structures of these products.

(a) optically inactive product

optically active product

(b) optically inactive product

optically active product

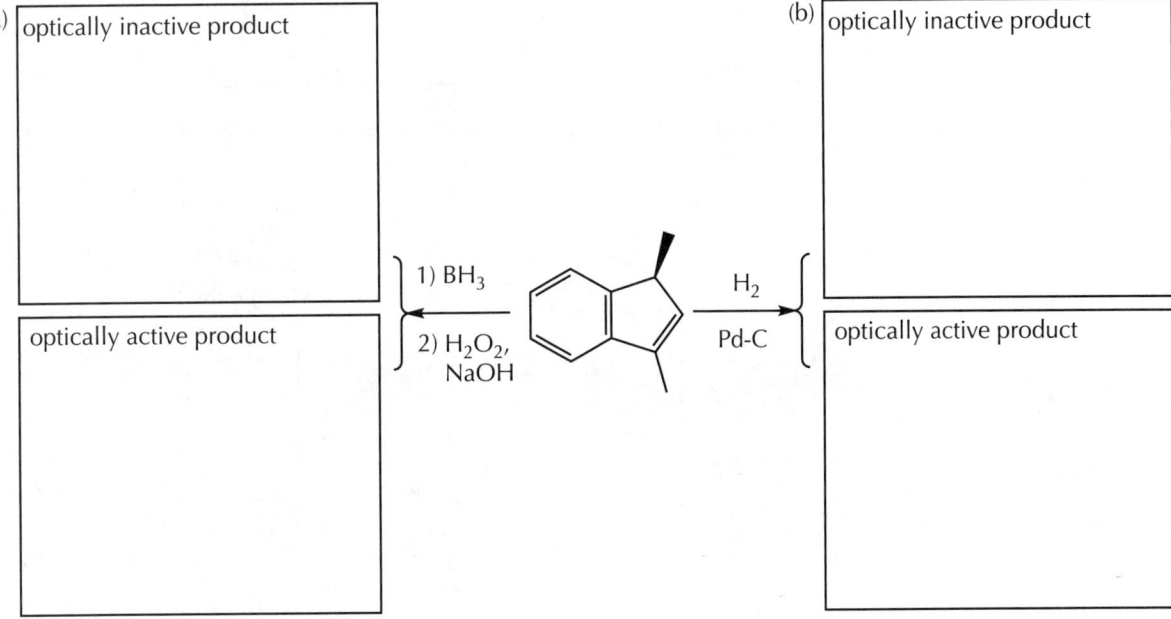

EQ 03.25

A. In the following reaction, there are two different rearrangement mechanisms that could give the same product. By building the molecule with a ^{13}C isotope to strictly identify and track the atom bearing the hydroxyl group, it is observed that only one of the mechanisms is operating.

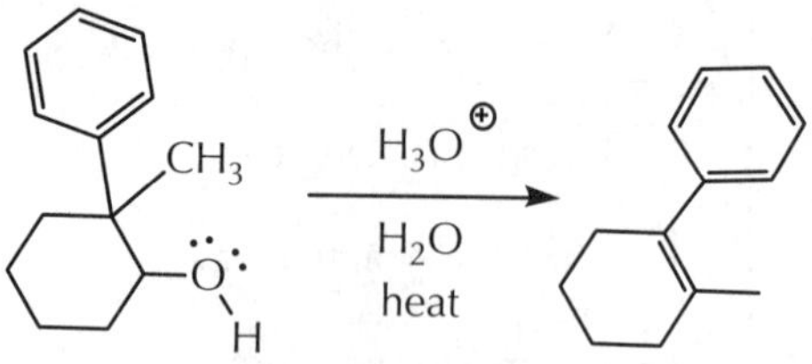

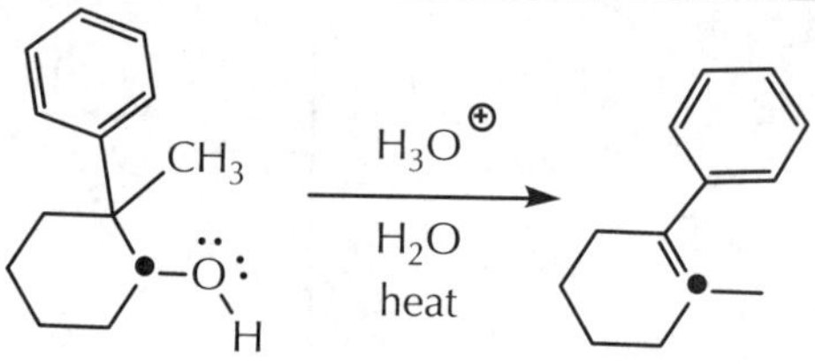

(a) Based on these results, and using H_3O^+ and H_2O as needed, draw the curved arrow mechanism for the reaction.

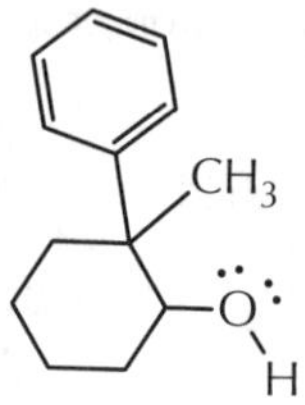

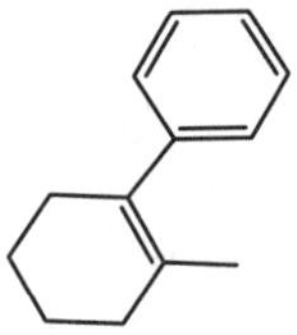

(b) Briefly explain what the results from the ^{13}C experiment allow you to conclude about the mechanism.

B. In the reaction below, a single, fast bimolecular elimination leads to the loss of nitrogen gas, N_2, and the stereoselective production of a single alkene product.

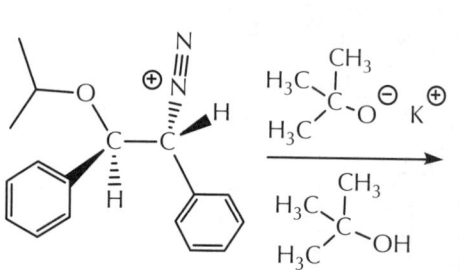

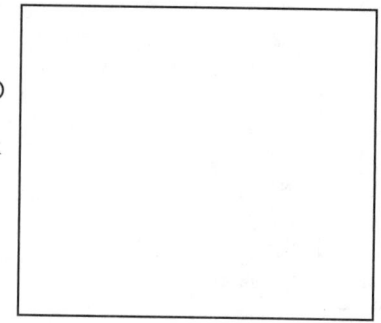

(a) alkene product

(b) Use a Newman projection for the conformation that leads to the kinetically favored E2 product; the carbon with the leaving group should be in front.

Exam Questions (Chapters 7–9)

EQ 03.26

Complete the following reactions, as directed. Transformations that require sequential experimental steps should be numbered appropriately. Abbreviations for reagents should not be used.

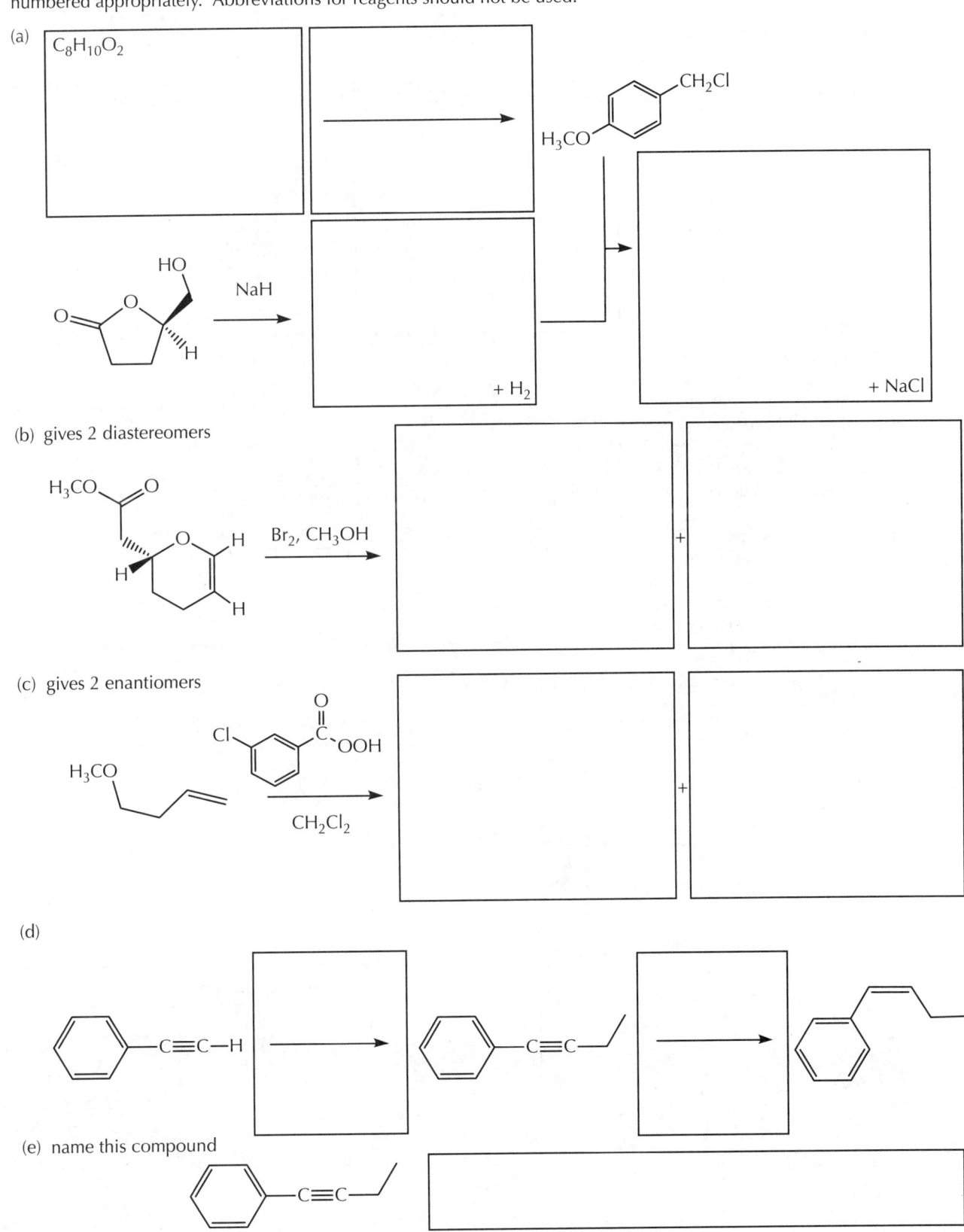

782 CHAPTER 9 Electrophilic Addition II: Halogenation, Oxidation, and Reduction Reactions

EQ 03.27

The following reaction sequence was proposed for carrying out an S_N2 reaction on the secondary alcohol groups in the following starting material. In the top tier equation ("predicted"), draw the structures for anticipated results (including stereochemistry). In the second tier equation ("observed"), some experimental information about what happened is included. Draw the structures (including stereochemistry) for the observed results, based on this information.

(a) predicted

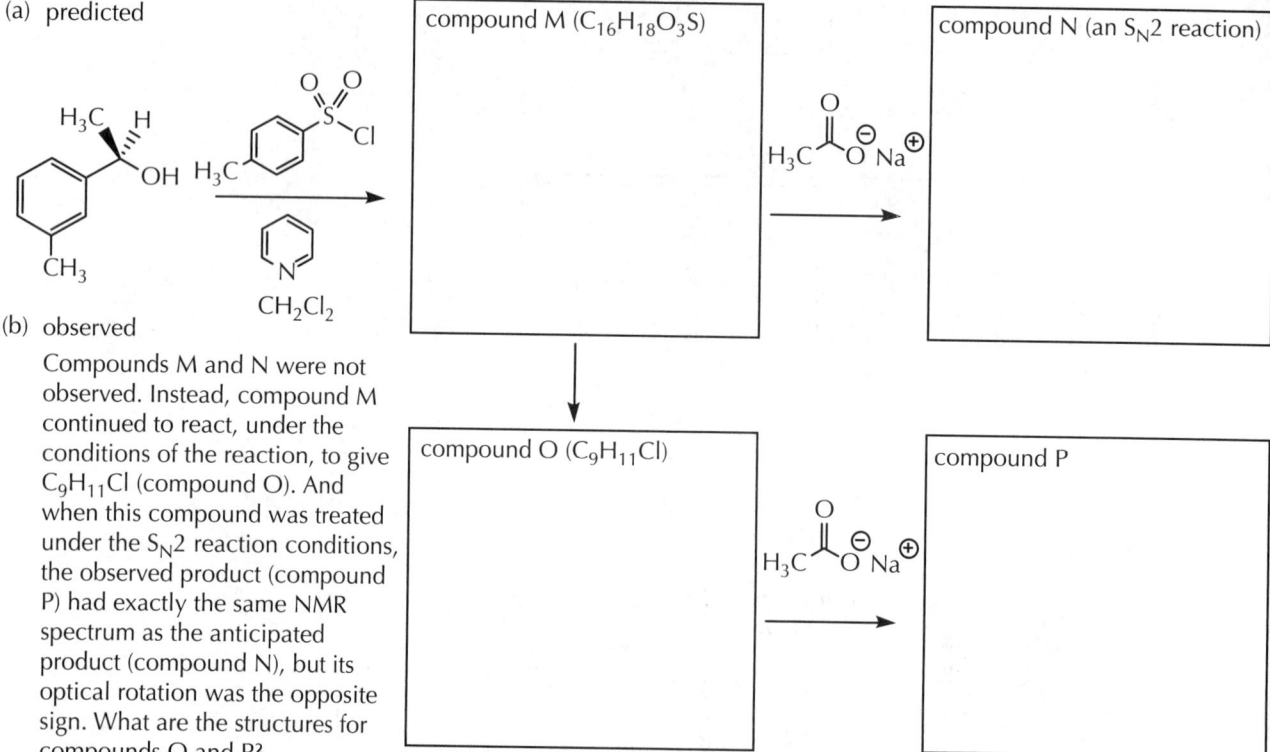

(b) observed

Compounds M and N were not observed. Instead, compound M continued to react, under the conditions of the reaction, to give $C_9H_{11}Cl$ (compound O). And when this compound was treated under the S_N2 reaction conditions, the observed product (compound P) had exactly the same NMR spectrum as the anticipated product (compound N), but its optical rotation was the opposite sign. What are the structures for compounds O and P?

(c) This starting material was used in some additional reactions. Complete the following. Be sure to number any sequential experimental steps.

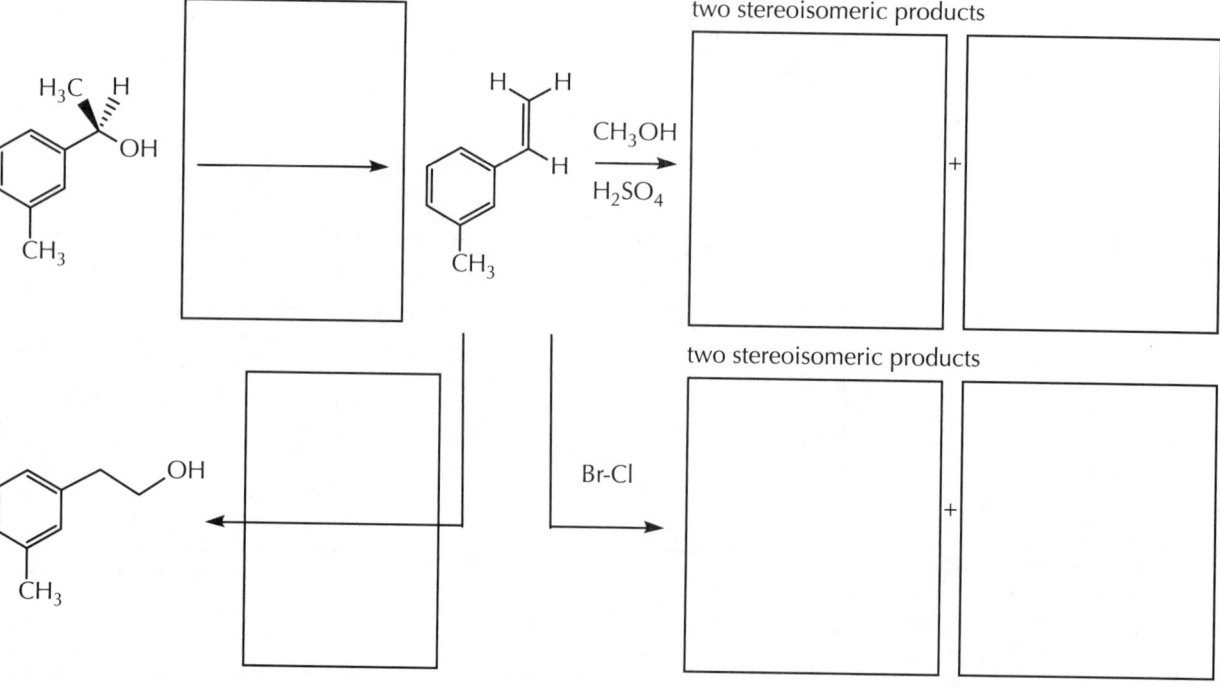

Exam Questions (Chapters 7–9)

EQ 03.28

A. The structure of aspicilin, first isolated from a lichen in 1900, was determined in 1985, and synthesized in 1991 (*Tetrahedron: Asymmetry*, **1991**, *32*, 801). An osmium tetroxide oxidation was used in the 1991 synthesis. What is the structure of the starting material for this (draw the stereochemistry properly and clearly)?

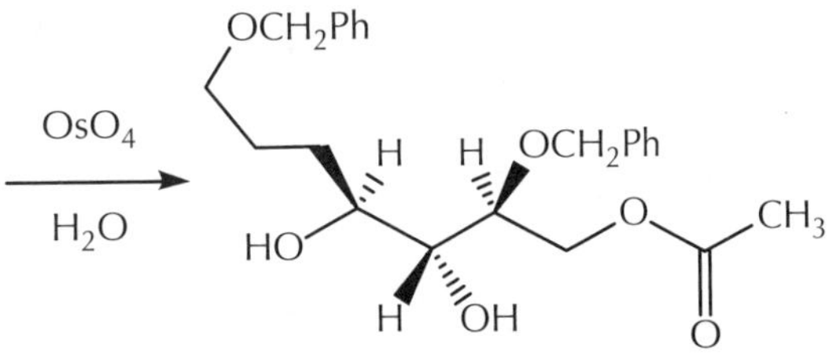

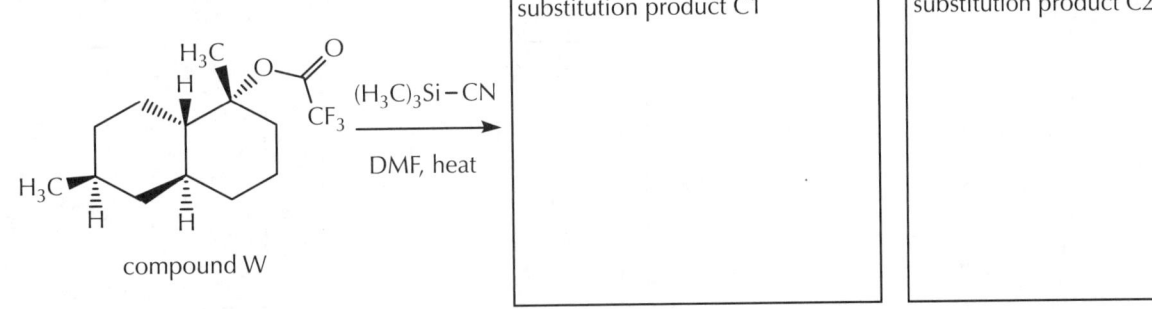

plus a diastereomer

B. Draw the structure for (*S,E*)-nona-2,8-dien-5-yne-1,7-diol

C. Trimethylsilyl cyanide [(CH$_3$)$_3$Si-CN] is a source of nucleophilic cyanide ion that, unlike sodium cyanide (NaCN), can dissolve readily in organic solvents. There are five different substitution and elimination products observed in the following reaction with compound W. What are they?

substitution product C1	substitution product C2

elimination product C3	elimination product C4	elimination product C5

When compound W reacts with potassium tert-butoxide [(CH$_3$)$_3$COK], only three of these five products are observed. Of those three products, they can be ranked from major to minor based solely on the steric accessibility of the β-H that is being removed (more accessible = faster reaction). Using the C1 through C5 labels above, according to the products as you have entered them, rank the three anticipated outcomes from the reaction of compound W with potassium tert-butoxide as major to minor.

major > intermediate > minor

☐ > ☐ > ☐

enter the C1-C5 label

EQ 03.29

A. The following C₈H₁₄ compound (3,3-dimethylcyclohexene) can be protonated to form two possible carbocations. One of them can undergo a single, most favorable rearrangement that, after deprotonation, gives a different C₈H₁₄ compound exhibiting 4 ^{13}C-NMR signals. Provide the intermediates in this mechanism, as specified, in addition to the final product.

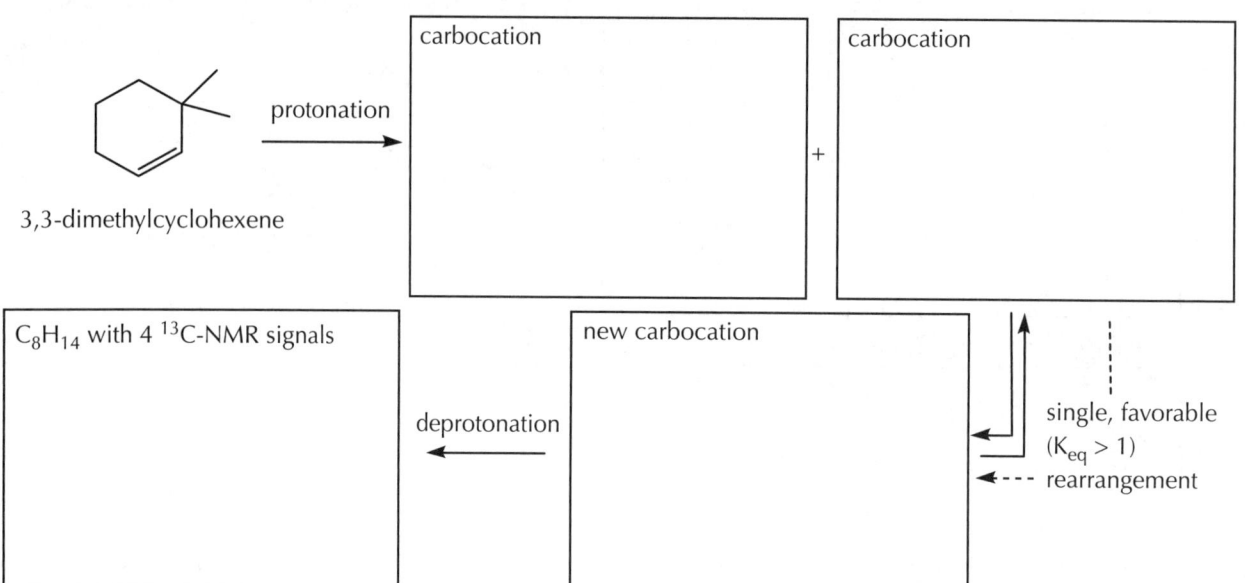

B. Complete the following reactions as directed. Show the major organic product(s) as specified.

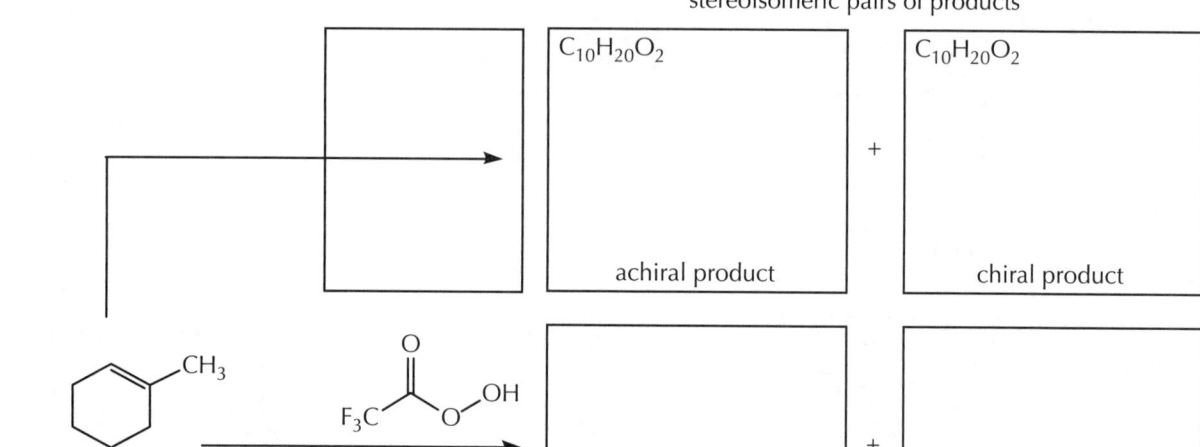

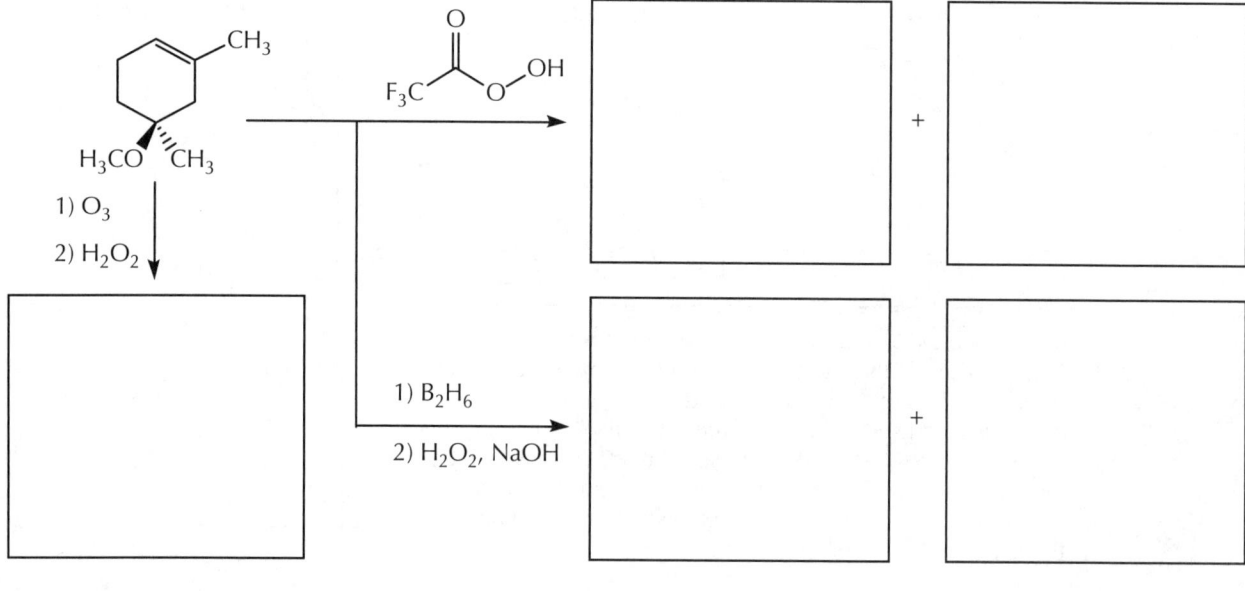

EQ 03.30
Reaction analysis: Provide the requested information in each of the following chemical transformations.

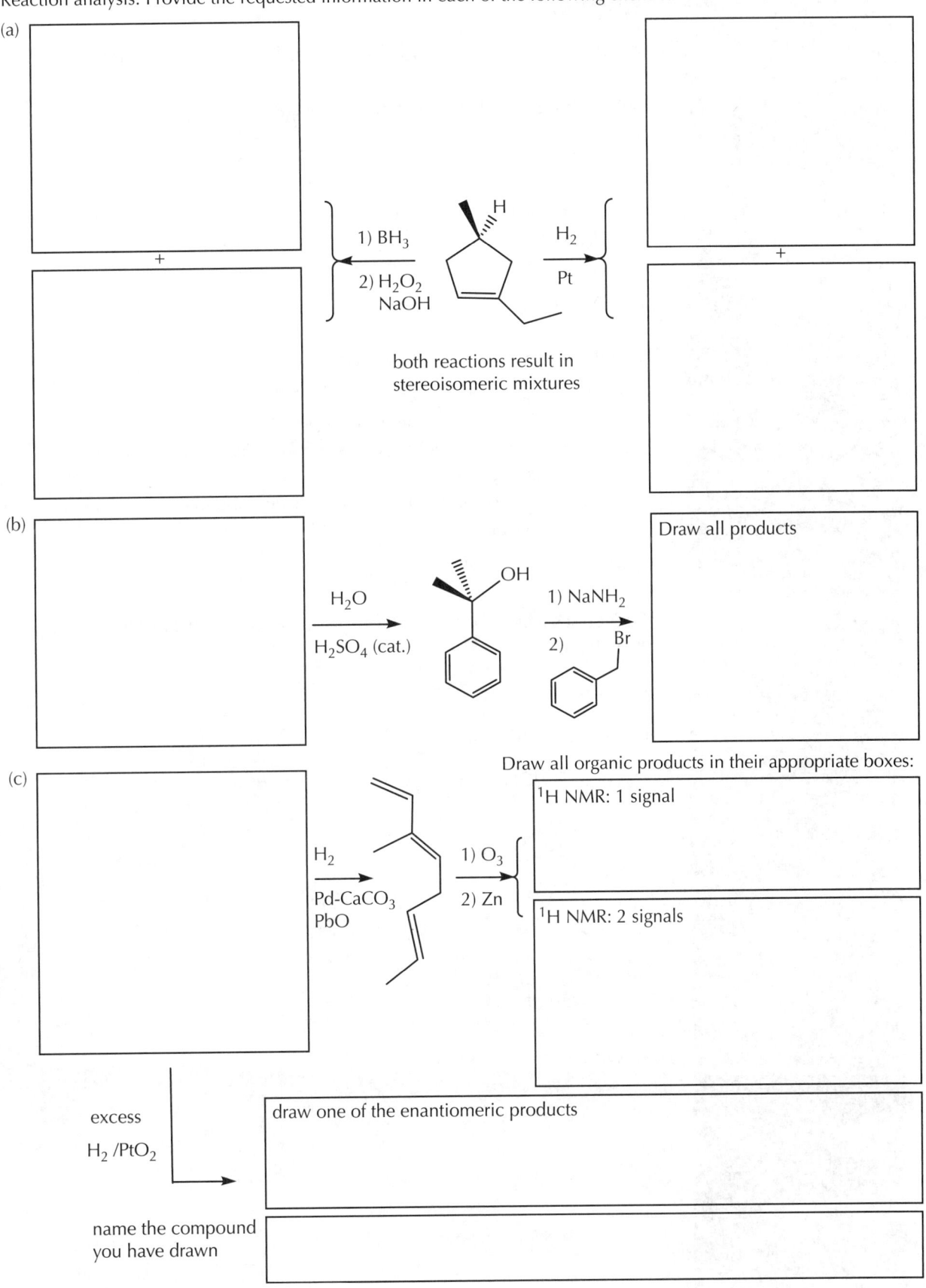

CHAPTER 10

Aromaticity and Electrophilic Aromatic Substitution Reactions

Setting the Stage: Word Meanings Change

10.1 Aromaticity

    A. Benzene and the Aromatic Sextet
    B. Hückel Aromaticity (4N+2 Rule) and "Antiaromatic" Compounds
    C. Cations, Anions, and Heterocyclic Compounds
    D. Chemical and Physical Properties of Aromatic Compounds

PRACTICE QUESTIONS

10.2 Electrophilic Aromatic Substitution

    A. Halogenation, Sulfonation, and Nitration Reactions
    B. Friedel-Crafts Alkylation and Acylation Reactions
    C. Regioselectivity of Electrophilic Aromatic Substitution Reactions (Directing Effect)
    D. Relative Rates of Electrophilic Aromatic Substitution Reactions (Activation)

PRACTICE QUESTIONS

Reflections About Science: Occam's Razor (Don't Invoke More Than You Need)

Summary

CHAPTER QUESTIONS

EXAM QUESTIONS (CHAPTERS 1–10)

CHAPTER 10

Aromaticity and Electrophilic Aromatic Substitution Reactions

Setting the Stage: Word Meanings Change

Words are created to capture the meaning of an idea, and ideas in science are based upon whatever the current theories are. As theories change, some words disappear while others get redefined according to the new theories.

This chapter begins with the word "aromaticity." This is not the first word you have encountered that has been coopted for a specific meaning in organic chemistry (e.g., "addition"), but it is one where the original use of the term in chemistry was for its evocative and familiar meaning (a pleasant odor) and its meaning evolved to represent an exceptionally significant resonance stabilization.

Before starting in on the history of how the term "aromatic" developed and changed, we can visit the important idea that word meaning is never permanent, in general, and that it is a common practice in science to evolve some word meanings as theories evolve. The terms "acid" and "oxygen," whose meanings intertwine, provide a nice lesson.

A Brief History of the Terms "Acid" and "Oxygen"

When the word "acid" was invented in the mid-1600s, about 50 years before even atoms were a marginally accepted idea, it represented anything that had a sharp, sour, tart, or metallic taste. The taste of vinegar (a word that means "sharp wine") was the common point of reference. In chemistry, heating elements in humid air was a reliable way to produce acids. Sulfur gave "the acid of sulfur" (sulfuric acid) and phosphorus gave phosphoric acid. Consequently, chemists looked to the air for the presence of an active chemical agent.

Words using the prefix "oxy" were already common in botany. *Lepidium oxycarpum* is a plant with thorny projections that protect its flowers, also known as the sharp-fruited pepperweed. The adjective "oxycanthous" is used to describe plants with sharp thorns or prickles, such as roses.

Antoine Lavoisier was one of the many chemists who was studying the chemical properties of air. In 1777, Lavoisier, who knew the existing meaning of "oxy," proposed the name *principe oxygine* for the gaseous material that was the essential principal (substance) required to form acids—substances known for their sharp, thorny taste. Within 10 years, the term changed to *principe oxygène* and then settled in as *oxygène*. Breaking that word apart reveals a chemical process: "oxy" (the sharpness of acids) and "gen," meaning "bringer of" (as in "genesis of"). Oxygen was a word invented to represent a substance in air that was "the bringer of acids." Lavoisier could then write and think about the sour and metallic-tasting compounds formed from sulfur, mercury, and silver as the oxyds (or oxides) of sulfur, mercury, and silver. And by 1790, Lavoisier was writing about the gaseous material, oxygen, as a fractional component of the normal atmosphere.

We understand combustion and the role of oxygen better today than we did 250 years ago. A prevailing model for combustion in the 1700s held that there was a substance (called phlogiston) that was released from materials when they burned, in part to account for the apparent loss of mass as a log is reduced to ashes. When Joseph Priestly inadvertently discovered oxygen in 1774, he called it "dephlogisticated air" (a gas that was so deprived of phlogiston that it could easily draw phlogiston out of other materials, hence explaining how the gas supported combustion so readily).

The word "phlogiston" is barely familiar today because the word was discarded when the theory was discarded. Some words persist because the underlying theories get more useful and the words are repurposed; other words disappear along with their appended, rejected theories.

The words oxygen, oxides, and oxidation all persisted, evolving beyond their original association with the word acid. Plenty of oxides were discovered that were not acidic, and the term "oxygen" (and eventually the dioxygen molecule) carried the broader meaning of the word, as a fundamental element, forward. Christian Arrhenius and Johannes Brønsted took the general meaning of acid in its new direction, built upon the idea of hydrogen ions.

"Oxidation," as the combination of any chemical body with oxygen, was a meaningful term, emerging in the early 1800s. And as mentioned in the previous chapter, increasing and decreasing the percentage of oxygen in a substance became the most common use of the two terms "oxidation" and "reduction" for many years. The most contemporary definition of these two terms (loss and gain of electrons) did not happen until the late 1920s and early 1930s, after the discovery and acceptance of subatomic particles (protons, electrons, and neutrons), as well as the models for the electronic model of atoms and bonding.

This history of aromatic compounds and the development of the chemical concept called "aromaticity" begins with incense.

Trade routes developed along with the ancient civilizations of Eurasia and North Africa. From around 3000 BCE through 500 CE, major east-west routes moved the goods that were plentiful in one location to those where they were scarce. The Amber Road, the Incense Route, the Silk Road, the Spice Route, the Salt Route and the Tea Route are all names that tell of commodities that spread, along with language and culture.

Incense has been used for perfumes and for religious or other ceremonial reasons for millennia. The popularity of the dried sap of specific trees found on the Arabian Peninsula (frankincense and myrrh) began to grow in the broader Mediterranean region as the domestication of camels, in about 1000 BCE, proved a reliable form of transportation from the Middle East. The spice trade, providing Europe with exotic flavorings such as cinnamon, turmeric, nutmeg, clove, cardamom, and ginger, originated in SE Asia, particularly from Sumatra, Java, and the Moluccas (or "Spice Islands"), all of which are a part of present-day Indonesia.

A popular Indonesian incense, called *lubān jāwiyy* (لبَان جَاوِيّ) by the Arabians, was brought in through the maritime trade routes and then exported to Europe. Its name means "Javanese frankincense" and was roughly pronounced "luban javi." Language evolves rapidly, and according to local oral customs as spoken by individuals, particularly in an era without methods for mass communication. Speakers of the Romance languages of Europe heard the sounds of "luban javi" as "lu banjavi" where the "lu" sound was interpreted as a definite article ("le" in French, "lo" in Italian, "la" in Spanish). Over the years, the name of this incense, with its definite article long discarded, was known as *benzoi* (Italian), *beijoim* (Portuguese), *benjuí* (Spanish), and *benjoin* (French).

By the 1400s and 1500s, new pronunciations led to new words. The raw, hardest sap of this incense, which started off as "luban javi," was widely known as either "gum benjamin" or "gum benzoin"—and can be purchased from Indonesia by those names, even today. The Sumatran tree from which this sap derives was officially designated as *styrax benzoin*.

The notable fragrances of these incenses, spices, and balms (sap from balsam trees) were all classified as aromatic for their strong and pleasant odors. Alchemists, and then chemists, were interested in isolating the "essential oils" or "essences" of these highly aromatic substances, and other natural products, as a way to concentrate and potentially intensify their effects, including for medical purposes.

The history of aromaticity overlaps here with the history of acids. In 1556, Nostradamus isolated an acidic substance by heating samples of gum benzoin in the air and called the product benzoic acid. Almost three hundred years later, in 1832, in Germany, Justus von Liebig and Friedrich Wöhler determined that the hydrocarbon portion (called a Radical) of benzoic acid was highly unsaturated. The benzoic acid molecule, stripped of its acid, was the unsaturated but relatively unreactive compound that was named benzene (C_6H_6).

A few years earlier, in 1825, the British chemist, Michael Faraday, had also isolated a highly unsaturated yet unreactive compound from the oil-gas used to fuel streetlamps and named it "phene," meaning "glowing." Before too long, it was understood that Faraday's phene and Liebig's benzene were the same compound. The German word ended up persisting as the noun, benzene, while you may have even recognized that we use Faraday's version to represent the attached benzene group, "phenyl."

As more and more substances were isolated from incenses and spices, the vast majority of them contained benzoic acid or various compounds that included benzene derivatives. These and many related compounds were also isolated from coal tar and petroleum during an active period in the early 1840s, led by a chemist named August Hofmann, who was an assistant to Liebig. The chemical properties of these benzoic acid and benzene derivatives was distinctive, particularly their unusual lack of reactivity as highly unsaturated substances. Sometimes new ideas get new words; sometimes an old word gets used and picks up a new meaning. Hofmann, in 1855, introduced the term "aromatic" as a scientific classification for these compounds, using the word "aromaticity" to define a chemical property and not as something with an aroma.

And that brings you to Chapter 10, where we will deal with the definition and use of the term "aromatic" as both a physical and chemical property.

The general concept of asking words to take on meaning and doing lots of heavy work for us was delightfully taken up by Lewis Carroll in the 1871 sequel to *Alice's Adventures in Wonderland*.

When Alice meets Humpty Dumpty in *Through the Looking Glass*, she objects to the way he has used the word "glory":

> "But 'glory' doesn't mean 'a nice knock-down argument,'" Alice objected.
>
> "When I use a word," Humpty Dumpty said, in rather scornful tone, "it means just what I choose it to mean—neither more nor less."
>
> "The question is," said Alice, "whether you can make words mean so many different things."
>
> "The question is," said Humpty Dumpty, "which is to be the master—that's all."
>
> Alice was much too puzzled to say anything; so after a minute Humpty Dumpty began again. "They've a temper, some of them—particularly verbs: they're the proudest—adjectives you can do anything with, but not verbs—however, I can manage the whole lot of them! Impenetrability! That's what I say!"
>
> "Would you tell me, please," said Alice, "what that means?"
>
> "Now you talk like a reasonable child," said Humpty Dumpty, looking very much pleased. "I meant by 'impenetrability' that we've had enough of that subject, and it would be just as well if you'd mention what you mean to do next, as I suppose you don't mean to stop here all the rest of your life."
>
> "That's a great deal to make one word mean," Alice said in a thoughtful tone.
>
> "When I make a word do a lot of work like that," said Humpty Dumpty, "I always pay it extra."
>
> Lewis Carroll, *Through the Looking-Glass*, 1871

10.1 Aromaticity

A. Benzene and the Aromatic Sextet

The structure for benzene (C_6H_6) as a six-membered ring with alternating double bonds was first proposed in 1865 by August Kekulé, using structural drawings that are quite recognizable today, and who wrote about how these two drawings were suggestive of a shared equivalence in the affinity of all six carbon atoms for each other (Figure 1001). Recall that the discovery of electrons was still over 30 years away at this point, and both the acceptance of electrons as chemical bonds and the concept of electron delocalization (resonance contributors) were over 50 years away. Whatever Kekulé was drawing in 1865, the exact meaning was quite different than any modern conception of bonding.

Figure 1001

Kekulé's drawings for benzene (1865) and the existence of only three disubstituted benzene derivatives.

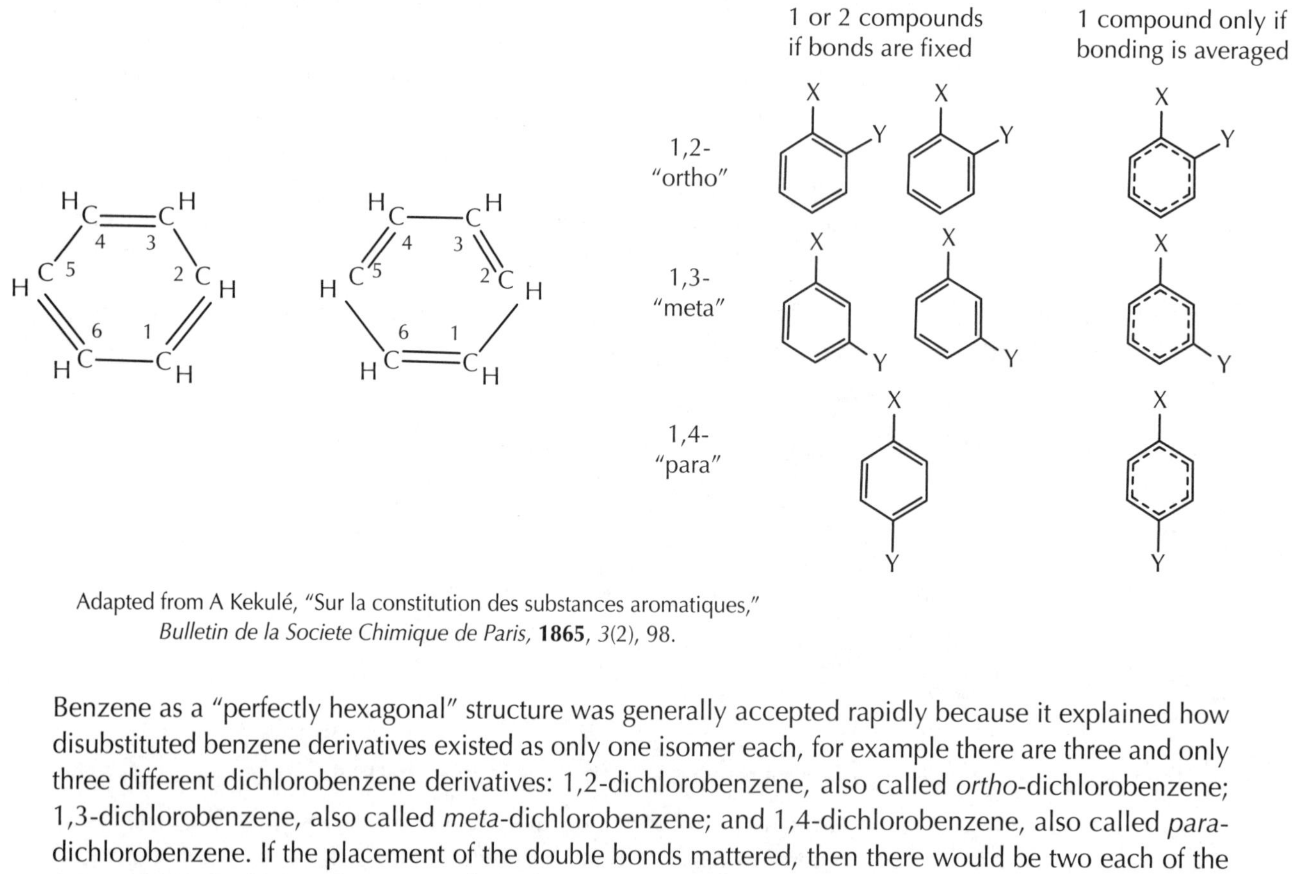

Adapted from A Kekulé, "Sur la constitution des substances aromatiques," *Bulletin de la Societe Chimique de Paris,* **1865,** *3*(2), 98.

Benzene as a "perfectly hexagonal" structure was generally accepted rapidly because it explained how disubstituted benzene derivatives existed as only one isomer each, for example there are three and only three different dichlorobenzene derivatives: 1,2-dichlorobenzene, also called *ortho*-dichlorobenzene; 1,3-dichlorobenzene, also called *meta*-dichlorobenzene; and 1,4-dichlorobenzene, also called *para*-dichlorobenzene. If the placement of the double bonds mattered, then there would be two each of the 1,2- and 1,3-disubstituted compounds.

Following from the early history, more and more structures were determined for the essential oils of spices and incenses, and they all contained benzene rings (Figure 1002).

The chemical concept called "aromaticity" took about 80 years to work out. J. J. Thompson, who discovered electrons, suggested that the six carbon-carbon bonds in benzene were equivalent and each carried three electrons (ca. 1905). Sir Robert Robinson (1925) introduced the idea that there was something special about the closed ring of six electrons beyond those needed to make the six carbon-carbon bonds and called these electrons the "aromatic sextet."

Figure 1002

Some of the isolated essential oils: the aromatic compounds that derive from aromatic substances.

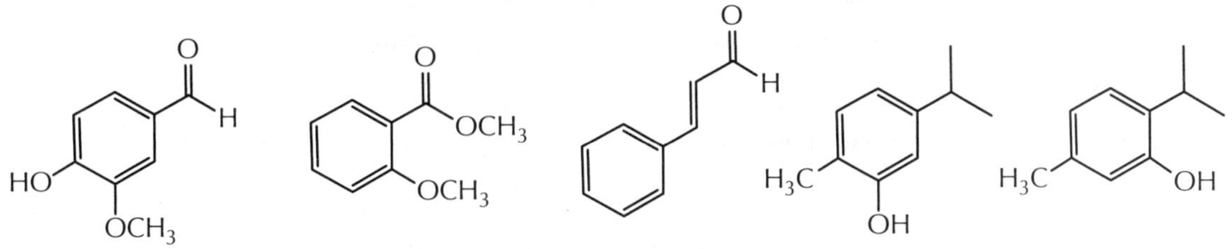

vanillin (in vanilla) methyl salicylate (in wintergreen) cinnamaldehyde (in cinnamon) carvacrol (in oregano) thymol (in thyme)

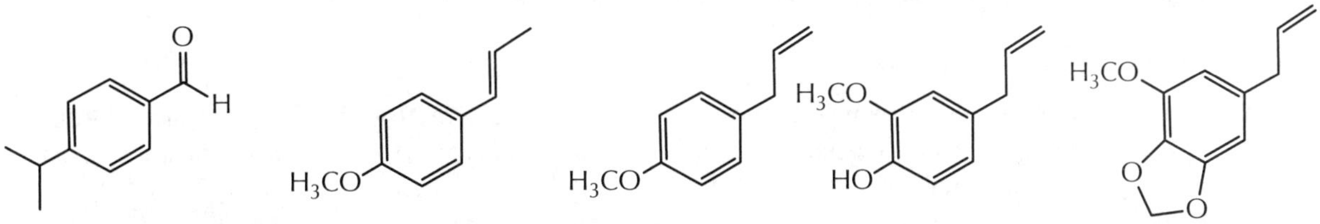

cuminaldehyde (in cumin) anethole (in star anise) estragole (in basil) eugenol (in clove) myristicin (in nutmeg)

Then, in 1931, following the development of the electronic model of bonding and the new and powerful theory of quantum mechanics, German chemist Erich Hückel revolutionized bonding and molecular structural theory. Hückel separated out bonding electrons into sigma and pi bonds and changed completely the way we picture double and triple bonds. His defining model for the chemical property of aromaticity (Hückel aromaticity) is the one we use today.

B. Hückel Aromaticity (4N+2 Rule) and "Antiaromatic" Compounds

In modern terms, the property called aromaticity is an exceptional delocalization stability (resonance) associated with a molecular structure, or any part of a molecular structure, that meets precisely the following criteria. Erich Hückel proposed these criteria in 1931, and so these are called the rules for Hückel aromaticity.

(1) A set of atoms is covalently bonded in a monocyclic ring. If there is more than one ring in the structure, then each one is considered separately.

(2) Every atom in the ring has, or has the possibility for having, a p orbital (filled, shared, or empty).

(3) The atoms contributing to the ring can achieve coplanarity, resulting in the appended p orbitals being mutually parallel around the perimeter of the ring (one at each vertex).

(4) (a) When the number of delocalizable electrons (i.e., the electrons in that set of parallel p orbitals) is an even number but not a multiple of 4 (i.e., 2, 6, 10, 14…), then the molecule will exhibit an exceptional delocalization (resonance) stability and is called an aromatic ring.

(b) When the number of delocalizable electrons (i.e., the electrons in that set of parallel p orbitals) is an even number and a multiple of 4 (i.e., 4, 8, 12, 16…), then the molecule will exhibit an exceptional delocalization (resonance) instability and is called an antiaromatic ring.

Hückel demonstrated that if the number of delocalized electrons in the p orbitals was a number from the series "2, 6, 10, 14…" (abbreviated as the "4N+2" series in mathematics, where N is 0 or any integer), then this situation was equivalent to a closed shell configuration for these cyclic pi systems covering greater than two atoms.

In contrast, if a molecule fit all the structural requirements but had an electron count in the second series (4, 8, 12, 16...; the "4N" series of even numbers), then it would possess an unstable form with two unpaired electrons.

Seeing how these criteria apply to different molecules is key to understanding how they are used.

Figure 1003 shows six molecular structures. The figure begins with a useful order for stepping through the criteria for defining an aromatic ring, as they are laid out above but restated as questions that you can use.

Figure 1003

Is this an aromatic ring? Applying the aromaticity criteria to six hydrocarbons.

(1) Is there at least one ring of atoms?
Yes. If there is only one, proceed; if there is more than one, proceed and treat each ring separately from the other(s).
No. If there is no ring, then the molecule, or any part of it, cannot be defined as aromatic or antiaromatic, so the rest of the analysis is irrelevant, and you are done.

(2) If there is more than one ring, consider each separately. For each ring, is there, or is there the possibility for, a p orbital at each vertex (atom in the ring)? If yes, then proceed to counting; if no, then the molecule, or that part of it, cannot be defined as aromatic or antiaromatic, so the rest of the analysis is irrelevant, and you are done.

(3) Look only at the electrons that are being contributed by the p orbitals and count them (a p orbital may be filled, carrying 2 electrons in a nbe pair; or it may be in a shared pi bond, with each atom carrying 1 electron; or it may be empty and carrying 0 electrons).

(4) (a) If the pi-electron count is in the 4N+2 series (2, 6, 10, 14...), then if the ring can achieve planarity, it is an aromatic ring and exhibit, among other properties, exceptional stability due to delocalization of the pi electrons. If the ring cannot achieve planarity, then it is considered a nonaromatic ring because the delocalization is being prevented.

(b) If the pi-electron count is in the 4 series (4, 8, 12, 16...), then the ring is called an antiaromatic ring, and in its planar form the delocalization will lead to an unstable electronic structure with two unpaired electrons called a ground state diradical.

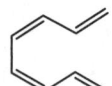

compound A
(1) ring of atoms? no

compound B (benzene)
(1) ring of atoms? yes
(2) p orbital at each vertex? yes
(3) number of electrons in pi system? 6
(4) aromatic, non-, or antiaromatic?

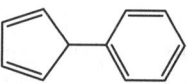

compound C
(1) ring of atoms?
   yes (left); yes (right)
(2) p orbital at each vertex?
   no (left); yes (right; continue below)
(3) number of electrons in pi system? 6
(4) aromatic, non-, or antiaromatic?

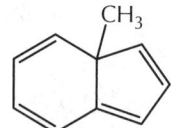

compound D
(1) ring of atoms?
   yes (left); yes (right)
(2) p orbital at each vertex?
   no (left); no (right)

compound E
(1) ring of atoms? yes
(2) p orbital at each vertex? no

compound F (1,3-cyclobutadiene)
(1) ring of atoms? yes
(2) p orbital at each vertex? yes
(3) number of electrons in pi system? 4
(4) aromatic, non-, or antiaromatic?

In Figure 1003, compound A fails at question (1) because there is no ring of atoms, and as a ring is a basic requirement, there is no reason to count electrons or continue asking other questions about potential aromaticity. Compounds B, E, and F each have one ring of atoms, and compounds C and D have two, so we can proceed to question (2). Compound E, both rings in compound D, and the left-hand ring in compound C fail at question (2) because there is a saturated, sp³ carbon in each ring, with no p orbital or the possibility of a p orbital. Compound B (benzene), the right-hand ring in compound C, and compound F (1,3-cyclobutadiene) all have sp² atoms at every vertex of the ring, which means we can proceed to step (3): counting the electrons in the p orbitals. These molecules are relatively easy to analyze because every atom in the ring is associated with a double bond. Each double bond has one pi bond with two electrons. In the case of compound B and the right-hand ring of compound C, there are a total of 6 electrons in the p orbitals (usually stated as "6 pi electrons"), and so these two rings are classified as aromatic. In compound F, there are 4 pi electrons, and this is an antiaromatic ring.

An important analogy to consider at this point is that defining an aromatic or antiaromatic ring is similar to defining a functional group. When you identify an sp³ carbon with a hydroxyl group on it, then its label as an alcohol is correlated with a set of properties, such as having a pK_a value of about 16 or being able to be transformed into a sulfonate ester leaving group. Within a molecular structure, recognizing an aromatic or antiaromatic ring is also correlated with a set of properties, and just as with a hydroxyl or any other functional group, a molecule can include more than one of them, and the same is true for aromatic rings. Some of the properties of these rings are summarized in Figure 1004 (on the next page), using benzene and 1,3-cyclobutadiene as representative examples.

Figure 1004: Benzene has a perfectly planar ring structure with six carbon-carbon bonds of equal length. Each bond is 1.40 Å, approximately halfway between the length of standard single bond (1.54 Å) and an isolated double bond (1.34 Å). Its two significant resonance forms are equal contributors. Although, by convention, we draw one of the significant resonance contributors, the pi bond in benzene is imagined to be a cyclic overlap that involves all six p orbitals. For a short period of time, drawing benzene was done using a hexagon with a full or dotted circle in it, as a way to express the aromatic delocalization; this drawing has been largely abandoned, likely because drawing curved-arrow mechanisms using these pi electrons is impossible using the established rules.

Using 1,3-cyclohexadiene and 1,4-cyclohexadiene as reference points, you can see a few things. First, the degree of delocalization (resonance) stabilization gained in going from the 1,4- to the 1,3- isomer is significant yet quantitatively low, about 2–3 kcal/mol, and both these two molecules undergo all of the reactions that are typical of alkenes. Benzene, on the other hand, has been experimentally determined to have about 40 kcal/mol of resonance stability, a quantitatively large value that means disrupting this pi system requires strong conditions. For instance, whereas the two cyclohexadiene isomers undergo hydrogenation in a matter of minutes at a slightly positive pressure and at room temperature, the hydrogenation of benzene requires using a container pressurized to over 65 atmospheres of hydrogen and a temperature of 100°C.

Antiaromatic compounds are molecules in which delocalization leads to an exceptionally unstable diradical form. These compounds undergo structure and/or reactivity equilibria to disrupt the unstable delocalized structure. These equilibria serve to negate one or more of the conditions for assessing the aromaticity of the molecule, such as creating new bonds at one or more of the ring atoms (removing p orbitals), and/or undergoing reactions that change the electron count (oxidation-reduction), and/or changing the geometry from the idealized planar conformation, which is needed for the best delocalization.

794 CHAPTER 10 Aromaticity and Electrophilic Aromatic Substitution Reactions

Figure 1004

Summary of chemical and physical properties for benzene and 1,3-cyclobutadiene due to their aromatic and antiaromatic character.

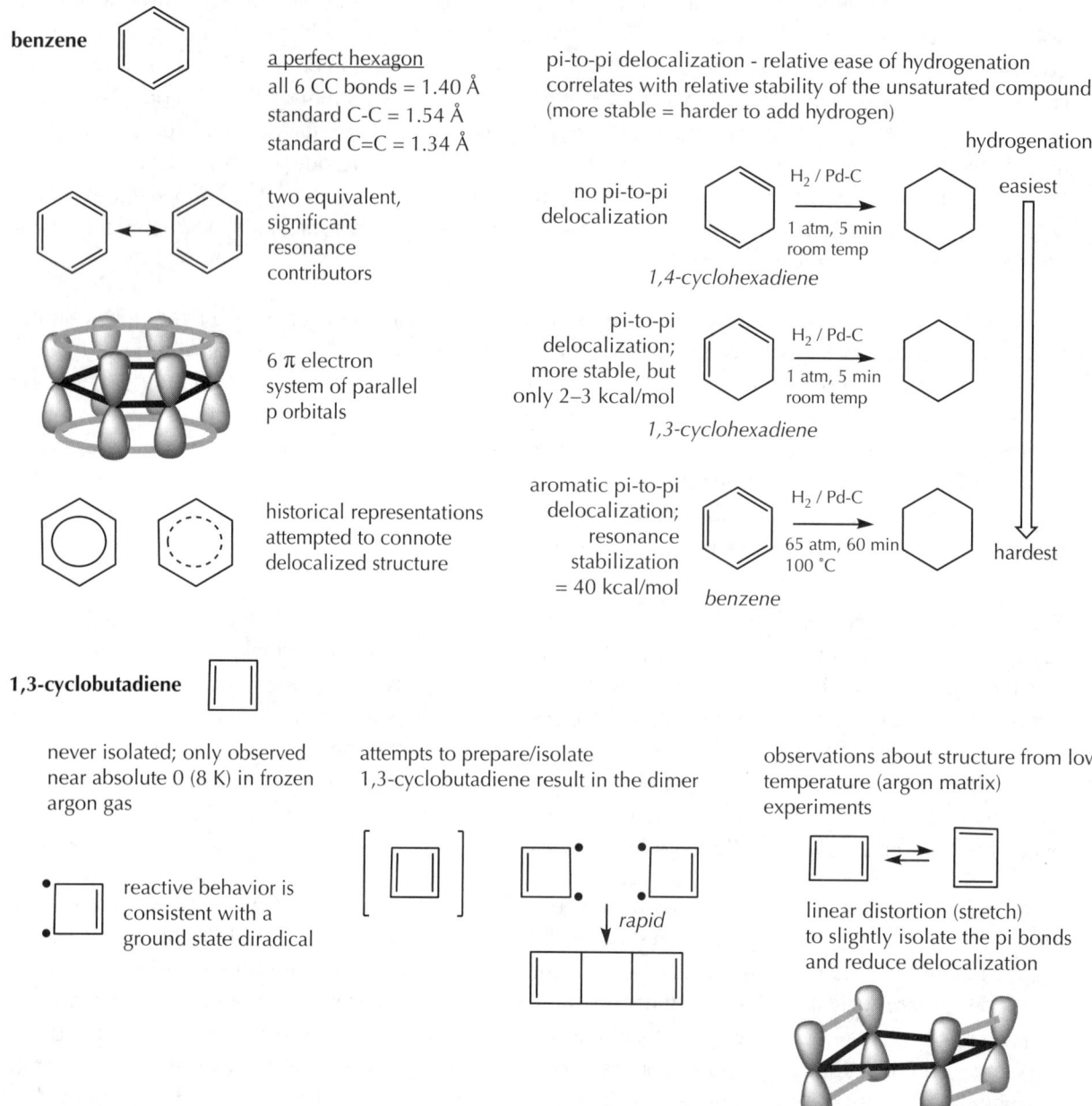

Although conformational equilibria (e.g., bending the ring) are probably the easiest way to disrupt the delocalization, the small ring of 1,3-cyclobutadiene is inflexible and can do little to distort the geometry away from planarity.

Although never isolated, the 1,3-cyclobutadiene molecule has been trapped and studied in a solid matrix of frozen argon gas (8 Kelvin), where the molecules cannot move around and react with one another or with the inert argon atoms. Some data about the structure of the molecule could be determined by spectroscopic studies. Its structure is not a perfect square: Elongating into a rectangle allows it to avoid a little bit of the perfectly delocalized geometry; the rectangular structure may also allow the two ends to twist a little bit, moving away from perfect coplanarity. Still, all attempts to prepare the molecule under normal laboratory conditions results in the same dimerization outcome shown in Figure 1004. As anticipated by the Hückel criteria, the antiaromatic ring behaves as a ground state diradical, and the most viable equilibrium that can disrupt the antiaromatic delocalization is for two of the molecules to snap together and give the nonaromatic cyclobutadiene dimer.

Summary: Classifying a ring as aromatic, nonaromatic, or antiaromatic is comparable to a functional group identification because it can be used to explain and predict both the structure and the reactivity of that molecular entity. So far, from this chapter, an aromatic ring has exceptional delocalization stability and so it is unreactive towards addition reactions that are easily performed with other unsaturated compounds. The exceptional delocalization stability also favors a strongly held planar geometry in which all of the contributing p orbitals are parallel. An antiaromatic ring shows exceptional instability upon delocalization, and so it reacts and/or adopts geometries that disrupt the delocalization whenever possible.

Six more hydrocarbons are shown in Figure 1005, along with the criteria for identifying whether any of the rings are aromatic, nonaromatic, or antiaromatic.

Figure 1005

Applying the aromaticity criteria to large and polycyclic ring systems.

1,3,5,7-cyclooctatetraene
(1) ring of atoms?
(2) p orbital at each vertex?
(3) number of electrons in pi system?
(4) aromatic, non-, or antiaromatic?

(1Z,3Z,5Z,7Z,9Z)-   (1Z,3Z,5E,7Z,9E)-
[10]annulene
(1) ring of atoms?
(2) p orbital at each vertex?
(3) number of electrons in pi system?
(4) aromatic, non-, or antiaromatic?

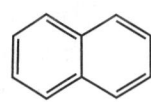

naphthalene
(1) ring of atoms?
(2) p orbital at each vertex?
(3) number of electrons in pi system?
(4) aromatic, non-, or antiaromatic?

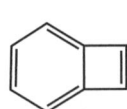

benzocyclobutadiene
(1) ring of atoms?
(2) p orbital at each vertex?
(3) number of electrons in pi system?
(4) aromatic, non-, or antiaromatic?

[18]annulene
(1) ring of atoms?
(2) p orbital at each vertex?
(3) number of electrons in pi system?
(4) aromatic, non-, or antiaromatic?

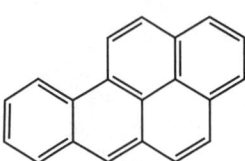

benzopyrene
(1) ring of atoms?
(2) p orbital at each vertex?
(3) number of electrons in pi system?
(4) aromatic, non-, or antiaromatic?

Before reading on, you should try using the Hückel criteria and classify the rings in these compounds.

An immediate and important topic that comes up in three of these examples is how to classify rings when the molecule has more than one ring in it and when, unlike benzene, the rings from different resonance contributors end up with different classifications.

Think about how that might be handled and recognize that you have been doing this all along when evaluating the relative significance of resonance contributors. A quick reminder lesson about this concept is worth thinking about before analyzing these examples.

Reminder lesson: the relative significance of resonance contributors in explaining reactivity.

The regioselective electrophilic addition reaction of HBr with 2-methylpropene is faster than that of propene, which are both in turn faster than that of ethene.

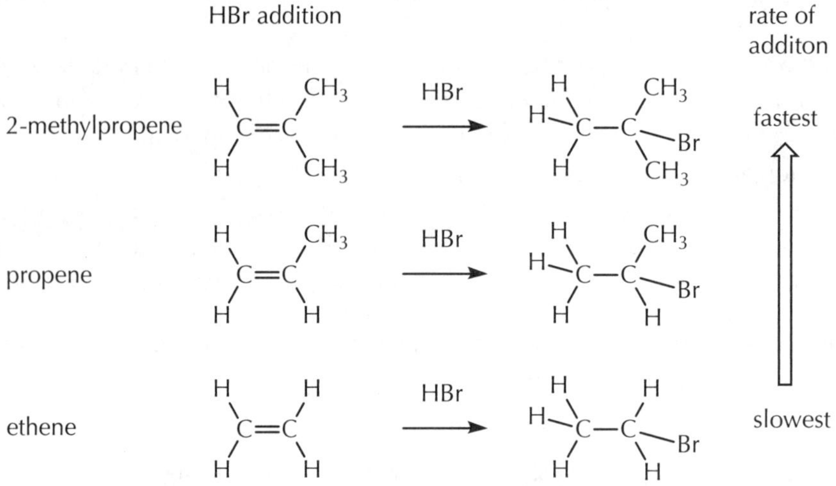

All three double bonds have three analogous sets of resonance contributors: one in which the atoms have all closed configurations, and two in which the pair of electrons from the pi bond is localized on one of the carbon atoms from the pi bond or the other (leaving its partner with an open shell). According to the fundamental concepts of structure, none of these resonance contributors represents the actual structure; the actual structure is a weighted average of them, with individual contributions (hence the term "contributions") being higher when the resonance form displays more stabilizing features.

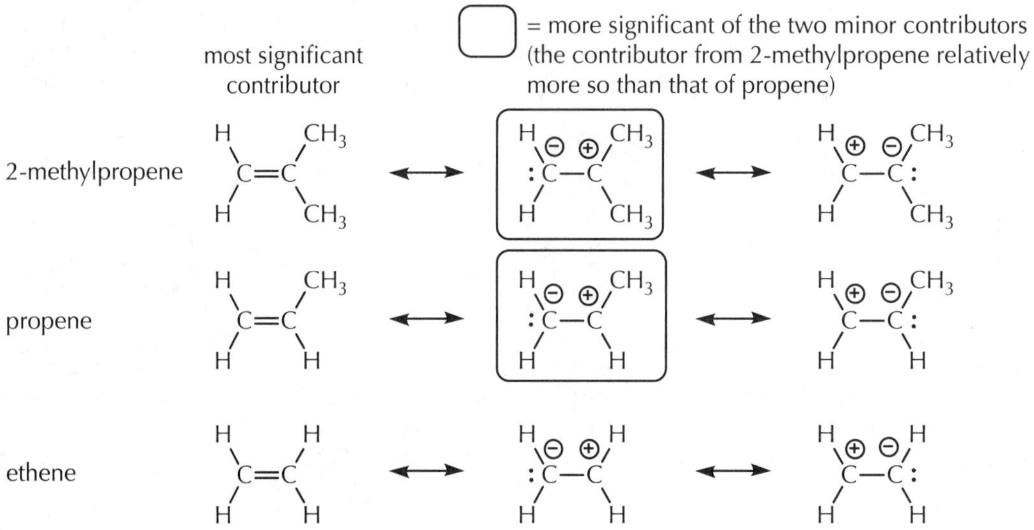

For these three alkenes, the resonance form with all closed shell atoms is certainly the most significant. Among the others, the ones where the open shell atoms are stabilized by an increased degree of substitution means that we think those forms are contributing more to the actual structure and reactions of that compound. In the case of 2-methylpropene, the contribution from the resonance form with a tertiary (3°) carbocation is higher than that of the analogous form in propene, in which the contributor exhibits a secondary (2°) carbocation. And both of these contributors are more significant than the comparable resonance form of ethene.

We explain that the addition reaction for 2-methylpropene is faster because of the differentially higher contribution from the more significant one of its minor resonance forms relative to the comparable contributions from those of the other two compounds.

Going back to the experimental observation: The regioselective electrophilic addition of HBr with 2-methylpropene is faster than the others. We imagine that the terminal carbon is more reactive towards protonation because the partial negative charge is relatively higher than in the other alkenes, and that the transition state is lower in energy than the others because the developing positive charge is more stable.

The term "resonance contributor" and its relationship to an actual single structure is an important concept to keep in mind for both the structure and reactions of aromatic compounds, and we will be revisiting the fundamental concepts about resonance throughout the rest of this chapter.

Here we return to the six molecules from Figure 1005. The analyses of these molecules (Figures 1006–1011) contain important lessons, based on experimental observations, that go beyond the information provided from making the predictions about aromaticity. Chemists often use these examples as reference points or analogies when trying to understand the aromatic character of a new compound about whose structure and/or reactions they are explaining.

Figure 1006

Applying the aromaticity criteria to 1,3,5,7-cyclooctatetraene.

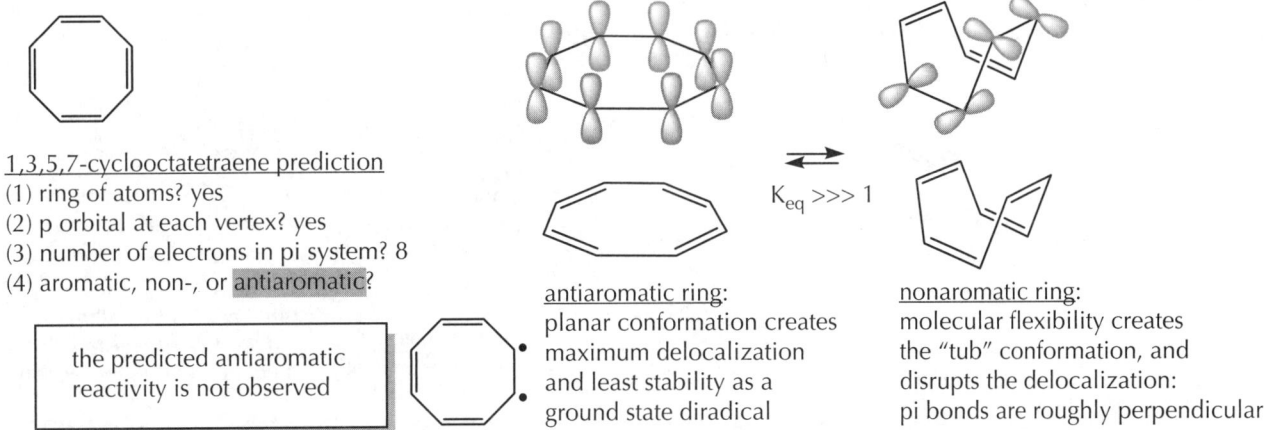

1,3,5,7-cyclooctatetraene prediction
(1) ring of atoms? yes
(2) p orbital at each vertex? yes
(3) number of electrons in pi system? 8
(4) aromatic, non-, or antiaromatic?

the predicted antiaromatic reactivity is not observed

antiaromatic ring:
planar conformation creates maximum delocalization and least stability as a ground state diradical

nonaromatic ring:
molecular flexibility creates the "tub" conformation, and disrupts the delocalization: pi bonds are roughly perpendicular

An antiaromatic compound exhibits structure and/or reactivity equilibria that favorably disrupts the delocalization associated with the unstable form on the compound.

Figure 1006: 1,3,5,7-cyclooctatetraene. Using the Hückel criteria, the 1,3,5,7-cyclooctatetraene ring qualifies as antiaromatic, which means that meeting all of the structural criteria creates an unstable ring in the delocalized form. Unlike the earlier example of cyclobutadiene, however, cyclooctatetraene is not restricted to be planar, and the delocalization can be favorably disrupted by a conformational equilibrium.

As seen in Figure 1006, the 1,3,5,7-cyclooctatetraene molecule folds up on itself, creating what is called a "tub" conformation. In this shape, all of the adjacent double bonds are nearly perpendicular to each other, preventing the unfavorable delocalization from occurring.

So, is 1,3,5,7-cyclooctatetraene actually an antiaromatic ring or is it a nonaromatic ring?

Defining antiaromatic rings is a place where you will find considerable discrepancy and debate between different texts, online sources, and chemists in general. This 8-membered ring does not exist in its unstable form because it folds into the nonaromatic tub conformation, but it folds that way in the first place because the planar, delocalized structure would be antiaromatic and unstable. The debate is quite a genuine dilemma for how to define a word (recall the deliberations between Alice and Humpty Dumpty in the opening essay for this chapter). Should 1,3,5,7-cyclooctatetraene simply be called nonaromatic, as you might label cyclohexene, because its folding disrupts the Hückel criteria? Or should it be called antiaromatic because that is the reason it folds to begin with?

Consider again 1,3-cyclobutadiene, which has a four-membered ring that cannot undergo a conformational change and is more or less trapped in its planar, delocalized structure. Should it be called antiaromatic because it cannot fold? Yet, quite like its 8-membered ring counterpart, it also participates in an equilibrium process that shifts it away from its unstable form: It reacts with itself (unless it happens to be captured at 8K in solid argon and cannot actually move anywhere).

There is no consensus on the answers to these questions or regarding the definition of whether a ring should ever be called antiaromatic.

In this text, we will classify a ring as being aromatic or antiaromatic if the planar conformation is available to it, so that a structural (e.g., folding) or a reactivity (e.g., dimerization) equilibrium, towards its favorable form, is as available to it as its unfavorable form. If the planar form is unavailable, and the aromatic or antiaromatic ring is not a part of the equilibrium, then the ring is classified as nonaromatic.

This definition is put to the test in the next example, [10]annulene (Figure 1007).

Figure 1007

Applying the aromaticity criteria to [10]annulene.

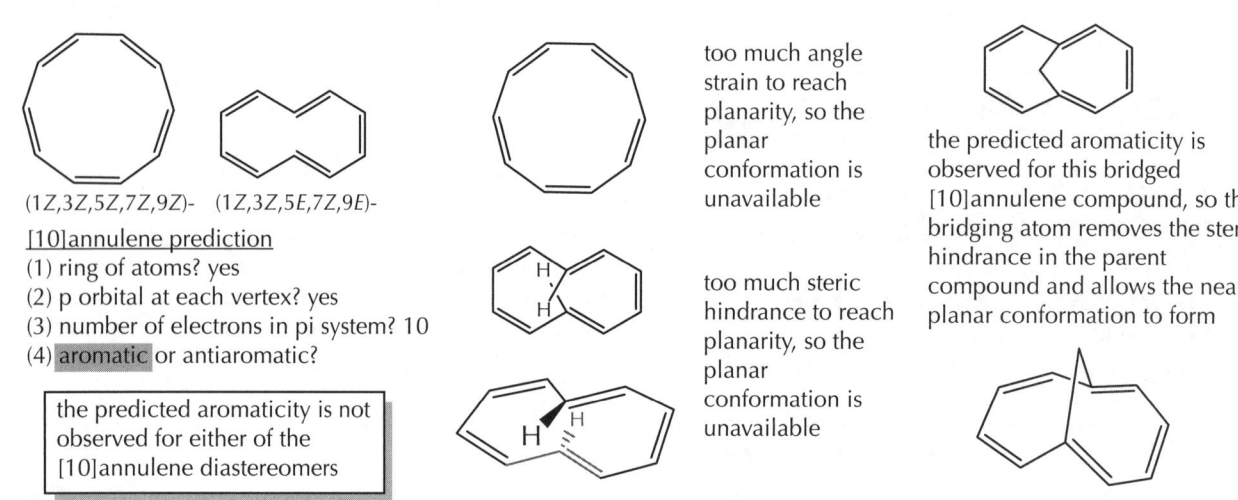

Figure 1007: [10]annulene. By all the Hückel criteria, both of these [10]annulenes are classified as aromatic rings; it's a reasonable conclusion without the experimental results, so rest assured that it is a good prediction. However, the experimental observations on these [10]annulenes do not show any aromatic stability (e.g., addition reactions happen readily), so one of the criteria for aromaticity must be getting disrupted in a way that is not obvious from the original drawings for the structures.

Aromaticity is a delocalization phenomenon, so the first hypothesis was that these molecules cannot achieve the planarity they need for the fully delocalized state. The structure for the all-*cis* (1Z,3Z,5Z,7Z,9Z)-[10]annulene is now known to be nonplanar, and chemists think the planer structure puts too much angle strain on the double bonds (even given the potential stabilization from the aromatic ring). In the (1Z,3Z,5E,7Z,9E)-[10]annulene, we now know that there is too much steric crowding from those internally-pointed carbon-hydrogen bonds, and the planar conformation is once again disrupted. The result is surprising. You would not predict it based on the Hückel criteria, but due to the experimental results both of these [10]annulenes are classified as nonaromatic compounds.

To demonstrate that the CH bonds were the problem, the (1Z,3Z,5E,7Z,9E)-[10]annulene was prepared with those two hydrogen atoms removed and replaced by a CH_2 group bridging across the ring. This compound exhibited aromatic properties (about 26 kcal/mol of aromatic stabilization, compared with 40 kcal/mol for benzene). And so, by removing the steric conflict, the molecule can now adopt the more planar conformation.

This example raises an important question for students: When will you know if there is a situation such as in the [10]annulenes, where the Hückel criteria appear to predict aromaticity but the compound is actually nonaromatic? The answer is exactly the one that was true for chemists, originally. You will need to be presented with the conflicting experimental data; without that, both of these compounds are quite confidently predicted to be aromatic rings.

Figure 1008

Applying the aromaticity criteria to naphthalene.

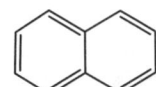

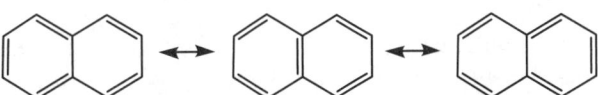

naphthalene has three significant resonance contributors

naphthalene prediction
(1) ring of atoms? yes: 2 rings (left & right)
(2) p orbital at each vertex?
   yes (left); yes (right)
(3) number of electrons in pi system?
   6 (left); and 6 (right) - *see analysis*
(4) aromatic, non- or antiaromatic?
   aromatic (left); aromatic (right)

naphthalene is not treated as a 10 electron ring for the analysis of its aromatic character; the two rings are considered separately for their individual contributions to the overall structure - aromatic ring plus non-aromatic ring components in the resonance contributors = aromaticity, but less so than if all contributors were aromatic

naphthalene is observed to be aromatic, but somewhat less so than two separate benzene rings

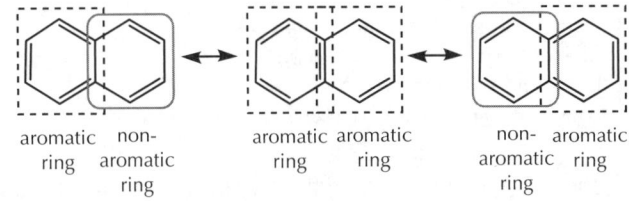

aromatic   non-   aromatic aromatic   non-   aromatic
ring   aromatic   ring   ring   aromatic ring
           ring   ring

Figure 1008: The Hückel criteria apply to monocyclic rings containing p orbitals, and so the two six-membered rings in naphthalene are analyzed independently. For nearly all of the polycyclic, fully unsaturated molecules, such as naphthalene, you need to consider the contribution of different resonance contributors when you are counting the electrons in the pi system.

Examine the individual rings in each of the resonance contributors for their potential classification as aromatic rings in that form. As always, the actual structure is a weighted average of all the resonance contributors. And also, as always, you are looking for whether or not the resonance contributor possesses a source of stability or instability as you assess the properties of the compound. The ring on the left-hand side of naphthalene is an aromatic ring in two of the three resonance contributors, and a nonaromatic ring in the third. The same is true for the right-hand ring. In general, if a ring has a combination of aromatic ring and nonaromatic ring contributors, the rings are classified as aromatic, but it might be predicted to show less delocalization stability than a compound where every resonance contributor displayed an aromatic ring.

The electron-count follows from the analyses of the individual rings. Thus, naphthalene has two 6 pi electron aromatic rings. This is not a contradiction to the fact that there are only a total of 10 pi electrons, but rather an artifact of the Hückel rules for how an aromatic ring is defined. And so we say naphthalene is made up of two aromatic rings rather than being considered as a single, 10 pi electron compound, as in the previous analysis of the [10]annulenes. Note also that the bridged [10]annulene in Figure 1007 is not equivalent to naphthalene, either, because the bridging atom is saturated and the [10]annulene is still a single, continuous ring.

Experimentally, the two rings in naphthalene do not end up creating a molecule that has twice the stability as benzene. The contributions from nonaromatic rings diminish the overall stabilization slightly (naphthalene's resonance stability is about 60 kcal/mol compared with benene's 40 kcal/mol).

It is tempting to count the 10 pi electrons in naphthalene and call it an aromatic molecule, and you can find many (perhaps too many) sources that repeat this error. The next example illustrates the wisdom of the Hückel rules and why the rings need to be analyzed separately (Figure 1009).

Figure 1009
Applying the aromaticity criteria to benzocyclobutadiene.

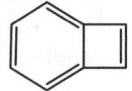

benzocyclobutadiene
(1) ring of atoms? yes: 2 rings (left & right)
(2) p orbital at each vertex?
    yes (left); yes (right)
(3) number of electrons in pi system?
    6 (left); and 4 (right) - *see analysis*
(4) aromatic, non- or antiaromatic?
    aromatic (left); antiaromatic (right)

> benzocyclobutadiene is observed to behave as the combination of aromatic and antiaromatic rings; it does not behave as a single 8 electron antiaromatic pi system

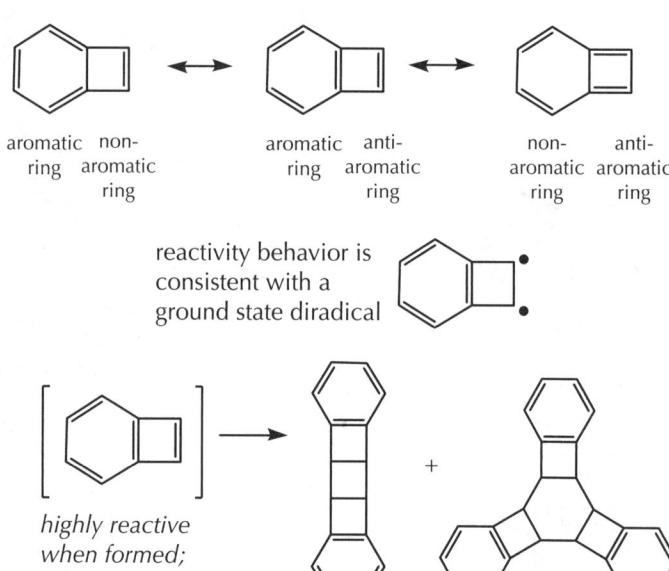

Benzocyclobutadiene is structurally similar to naphthalene, and it illustrates why keeping the analysis of the rings separate from one another is important. If the entire molecule were being counted, benzocyclobutadiene might be classified as an 8 pi electron antiaromatic compound. It cannot fold on itself the way cyclooctatetrene does and would be predicted to be quite unstable.

The molecule exists, and although it is reactive, the reactivity is quite specific for only one of the rings. There are two rings in benzocyclobutadiene, a 6-membered ring that is aromatic and a 4-membered ring that is antiaromatic. The aromaticity provided by the 6-membered ring gives the overall structure stability relative to a wholly antiaromatic compound. Unsurprisingly, the double bond in the 4-membered ring shows the same type of reaction products that cyclobutadiene does. Its reactivity is consistent with a ground state diradical that dimerizes and, in this case, trimerizes.

10.1 Aromaticity

Figure 1010
Applying the aromaticity criteria to [18]annulene.

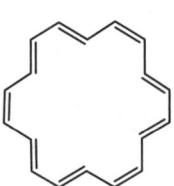

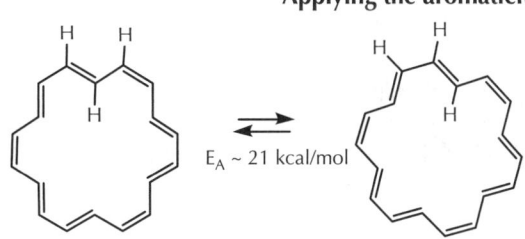

[18]annulene
(1) ring of atoms? yes
(2) p orbital at each vertex? yes
(3) number of electrons in pi system? 18
(4) aromatic, non- or antiaromatic?

experimentally, ~ 37 kcal/mol of resonance stabilization (about the same amount as benzene, but spread out over 18 atoms rather than 6), so the overall molecule is more weakly aromatic than benzene

conformational flexibility between identical aromatic contributors; high E_A because the aromaticity is broken as the molecule folds and bonds rotate

first prepared in 1962

less overall aromatic character than benzene; adds bromine and hydrogen readily

its partial bond lengths (1.42 Å and 1.38 Å) are comparable to benzene's 1.40 Å, but two values indicate that the delocalization is not uniformly even around the ring; the bonds alternate as slightly longer and slightly shorter

Figure 1010: Chemists have prepared large annulenes, such as 10[annulene] (Figure 1007) and this one, [18]annulene, to test the Hückel predictions about aromaticity. In the larger [18]annulene rings, the cross-ring steric interference does not happen as it did in [10]annulene, and these large annulenes show aromatic properties. However, the degree of aromatic stability is not as significant as in benzene. This [18]annulene, for example, adds bromine and hydrogen more readily than benzene. The larger rings are more flexible, existing in different conformations that, while all aromatic rings, also have shapes that are not perfectly planar thanks to some steric hindrance and/or eclipsing interactions that get relieved upon slightly twisting the structure.

Figure 1011
Applying the aromaticity criteria to benzopyrene.

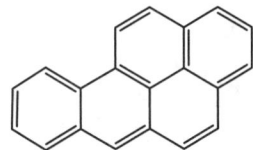

benzopyrene
(1) ring of atoms?
    yes, there are 5
(2) p orbital at each vertex?
    yes, in all 5 rings
(3) number of electrons in pi system?
    6 electrons each
(4) aromatic or antiaromatic?
    all 5 are aromatic rings

experimentally: aromatic properties, but not as unreactive as benzene; the site of highest reactivity depends on the reagents and reaction conditions

representative resonance contributors (there are others) show how the rings are sometimes aromatic in the contributor and sometimes nonaromatic; net result = all 5 rings are 6 electron aromatic rings, all are predicted to be more weakly aromatic than benzene, and understanding differences in reactivity depend on experimental observations rather than prediction

carcinogenic action: bonds to DNA

benzopyrene is an example of a polycyclic aromatic hydrocarbon (PCAH), found in tobacco smoke, soot, coal tar and charbroiled foods; PCAH compounds are universally considered cancer-suspect agents

802 CHAPTER 10 Aromaticity and Electrophilic Aromatic Substitution Reactions

As with naphthalene, the five rings in benzopyrene need to be considered independently. Benzopyrene has five 6 pi electron aromatic rings that have resonance contributions from nonaromatic forms as well. The rings are all more reactive than benzene, but an assessment of relative reactivity across the five rings is difficult to predict from an analysis of resonance contributors. Explaining experimental results is more common than predicting new results.

Benzopyrene is an example of a polycyclic aromatic hydrocarbon (PCAH). These compounds are found in charred or barbequed meats and also in tobacco smoke. If you consume these molecules, you cannot use them for nutrition, and your body uses oxidation reactions to try and increase their water solubility for excretion. Unfortunately, if these PCAH epoxides get to the wrong place at the wrong time, they can slide into your DNA and end up covalently bonded to it. If this happens in a way that your body cannot correct or discard, your own biochemistry will begin to create useless tissue and copies of the damaged DNA, otherwise known as cancer. And so PCAHs are universally considered to be cancer-causing agents (carcinogens).

C. Cations, Anions, and Heterocyclic Compounds

Many examples of unsaturated, cyclic hydrocarbons are known in which one or more of the ring positions are a carbocation and/or a carbanion center. These structures can be analyzed for their aromatic character using the Hückel criteria. Eight examples of ions, along with the questions derived from Hückel criteria, are shown in Figure 1012.

Figure 1012
Applying the aromaticity criteria to hydrocarbon-derived ions.

cycloproprenyl cation
(1) ring of atoms?
(2) p orbital at each vertex?
(3) number of electrons in pi system?
(4) aromatic, non- or antiaromatic?

cycloproprenyl anion
(1) ring of atoms?
(2) p orbital at each vertex?
(3) number of electrons in pi system?
(4) aromatic, non- or antiaromatic?

cyclobutenyl cation
(1) ring of atoms?
(2) p orbital at each vertex?
(3) number of electrons in pi system?
(4) aromatic, non- or antiaromatic?

cyclobutadienyl dianion
(1) ring of atoms?
(2) p orbital at each vertex?
(3) number of electrons in pi system?
(4) aromatic, non- or antiaromatic?

cyclopentadienyl anion
(1) ring of atoms?
(2) p orbital at each vertex?
(3) number of electrons in pi system?
(4) aromatic, non- or antiaromatic?

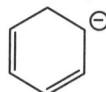

cyclohexadienyl anion
(1) ring of atoms?
(2) p orbital at each vertex?
(3) number of electrons in pi system?
(4) aromatic, non- or antiaromatic?

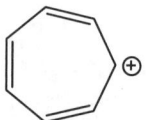

cycloheptatrienyl cation
(1) ring of atoms?
(2) p orbital at each vertex?
(3) number of electrons in pi system?
(4) aromatic, non- or antiaromatic?

cyclooctatetraenyl dication
(1) ring of atoms?
(2) p orbital at each vertex?
(3) number of electrons in pi system?
(4) aromatic, non- or antiaromatic?

As you think about these examples, recall the rules for assigning hybridization to atoms based upon delocalizable electrons and the possibility for resonance contributors that was introduced in Chapter 2.

One of the key steps in using the Hückel criteria is deciding whether a p orbital is present or possible at each atom in the ring, and the details of how to anticipate delocalization from the earliest lessons in Chapter 2 are going to be important in all of these examples. Each of the ions from Figure 1012 is analyzed separately in Figures 1013–1020.

Figure 1013
Applying the aromaticity criteria to cyclopropenyl cation.

cyclopropenyl cation
(1) ring of atoms? yes
(2) p orbital at each vertex? yes
(3) number of electrons in pi system? 2
(4) aromatic, non- or antiaromatic?

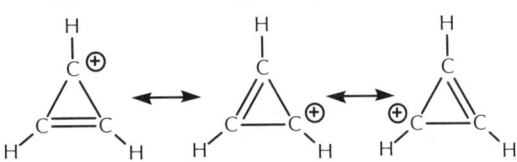
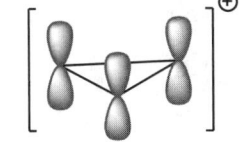

typical analysis for delocalization = all 3 carbon atoms are sp²
and there are 2 electrons in these 3 p orbitals

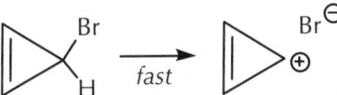

the S_N1 rate for the ionization of 3-bromocyclopropene is much greater than for bromocyclopropane because the resulting carbocation is not only stabilized by typical delocalization, but by the aromaticity of the cation

Figure 1013: The cyclopropenyl cation is planar, with an sp² atom at each vertex. In the typical analysis for resonance contributors carried out in earlier chapters, delocalization is predicted. As there are 2 pi electrons in these 3 p orbitals, the ring is classified as aromatic and the planar, delocalized form is predicted to be exceptionally stable.

Figure 1014
Applying the aromaticity criteria to cyclopropenyl anion.

cyclopropenyl anion
(1) ring of atoms? yes
(2) p orbital at each vertex? yes
(3) number of electrons in pi system? 4
(4) aromatic, non-, or antiaromatic?

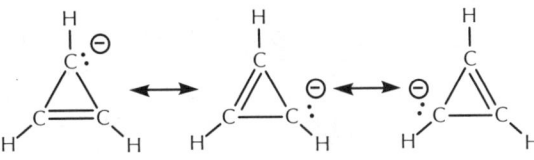

typical analysis for delocalization = all 3 carbon atoms are sp²
and there are 4 electrons in these 3 p orbitals

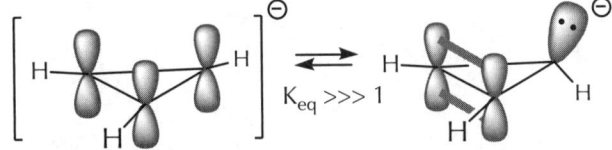

all planar, delocalized
structure is least stable

localized sp³ anion structure
disrupts some of the delocalization

Figure 1014: In the analysis for resonance contributors, delocalization is predicted. Based on the information in Chapter 2, all three of the carbon atoms are predicted to be sp². The ring meets all of the criteria for analysis of its aromaticity, and as there are 4 pi electrons in these 3 p orbitals, the ring is classified as antiaromatic if it remains in the planar, delocalized form. This anion will be unstable if it is fully delocalized, and so there will be some kind of equilibrium that localizes the electrons to the greatest possible degree. Conformational changes have been illustrated already (cyclooctatetraene) as have reactions (cyclobutadiene, benzocyclobutadiene). The cyclopropenyl anion uses yet another strategy: It adopts a more localized hybridization at one of the atoms.

As a 3-membered ring, there is no option for the ring to bend as a way to disrupt the delocalization, but partial localization by adopting an sp³ hybridization for a nbe pair diminishes the delocalization by taking one of the purely p orbitals out of the ring. If the carbon bearing the nonbonding electron pair is sp³ (a more localized structure) rather than sp² (delocalized structure), the effect of the antiaromatic ring is disrupted, and gives a more stable form of the ring as a pi bond using two of the atoms with the third atom as an sp³ carbanion. The sp³ atom still has some p character, so the delocalization is not completely eradicated. This anion would have a higher-than-expected reactivity (e.g., as a Brønsted base, a topic that will be taken up in the next section).

Figure 1015

Applying the aromaticity criteria to cyclobutenyl cation.

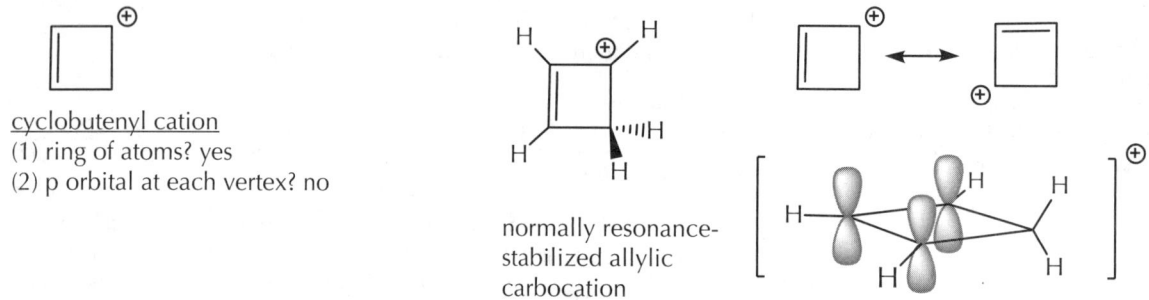

cyclobutenyl cation
(1) ring of atoms? yes
(2) p orbital at each vertex? no

normally resonance-stabilized allylic carbocation

Figure 1015: In the fundamental analysis for resonance contributors, delocalization is predicted. The ring fails the test for considering its aromaticity, however, because one of the atoms is saturated. This molecule is simply a resonance-stabilized carbocation.

Figure 1016

Applying the aromaticity criteria to cyclobutadienyl dianion.

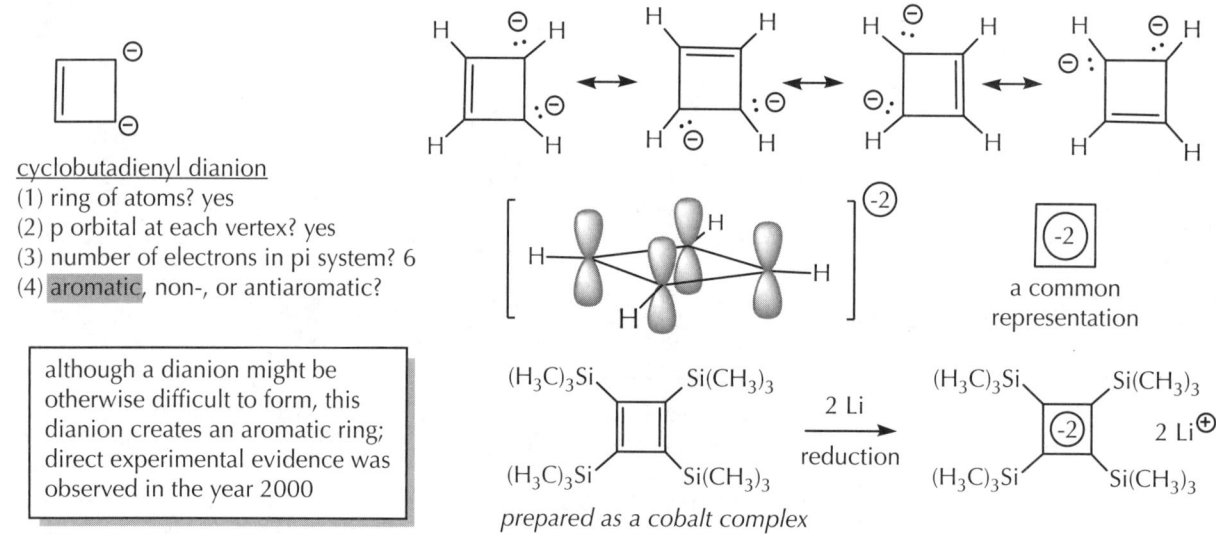

cyclobutadienyl dianion
(1) ring of atoms? yes
(2) p orbital at each vertex? yes
(3) number of electrons in pi system? 6
(4) aromatic, non-, or antiaromatic?

a common representation

although a dianion might be otherwise difficult to form, this dianion creates an aromatic ring; direct experimental evidence was observed in the year 2000

prepared as a cobalt complex

Figure 1016: In the fundamental analysis for resonance contributors in this dianion, delocalization is predicted. All the atoms of the ring are sp² and the ring meets the criteria of analysis for aromaticity. Counting the electrons, there are 6 pi electrons in these 4 p orbitals, and the ring is classified as aromatic. Even though two negative charges shared over such a small number of atoms would be hard to otherwise achieve, this planar, delocalized form is relatively stable as an aromatic ring.

Figure 1017
Applying the aromaticity criteria to cyclopentadienyl anion.

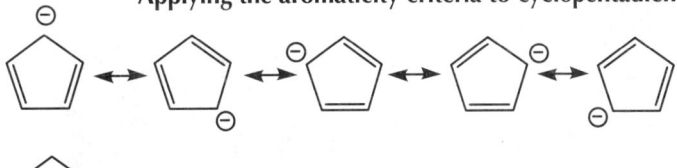

cyclopentadienyl anion
(1) ring of atoms? yes
(2) p orbital at each vertex? yes
(3) number of electrons in pi system? 6
(4) aromatic, non-, or antiaromatic?

"Cp" is also a commonly used abbreviation for the aromatic cyclopentadienyl anion

Figure 1017: Fundamental delocalization is predicted. All the atoms of the ring are sp² and the ring meets the criteria of analysis for aromaticity. There are 6 pi electrons in these 5 p orbitals, and the ring is classified as aromatic. The planar, delocalized form is exceptionally stable.

Figure 1018
Applying the aromaticity criteria to cyclohexadienyl anion.

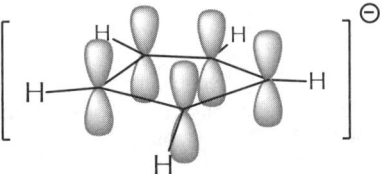
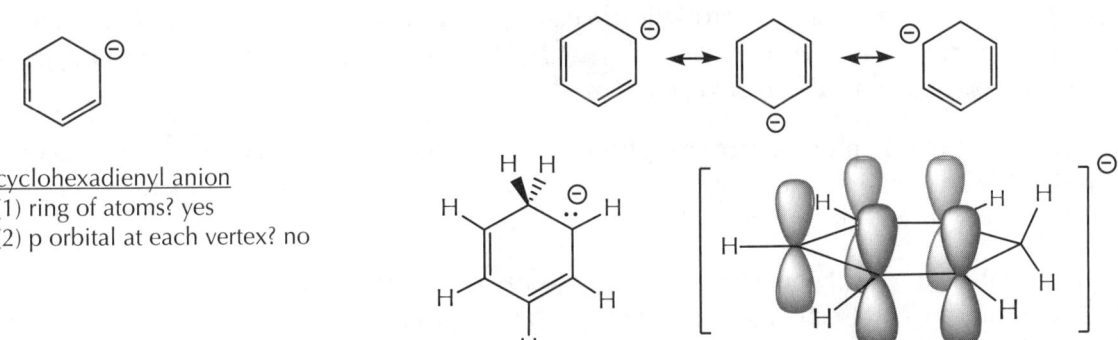

cyclohexadienyl anion
(1) ring of atoms? yes
(2) p orbital at each vertex? no

Figure 1018: In the simple analysis for resonance contributors, delocalization is predicted. The ring fails the test for considering its aromaticity, however, because one of the atoms is saturated. This molecule is simply a resonance-stabilized carboanion.

Figure 1019
Applying the aromaticity criteria to cycloheptatrienyl cation.

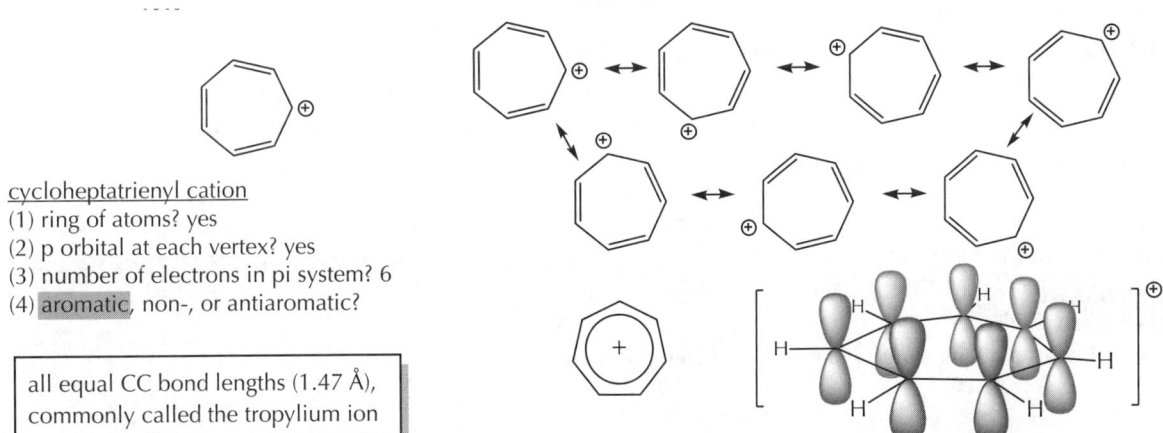

cycloheptatrienyl cation
(1) ring of atoms? yes
(2) p orbital at each vertex? yes
(3) number of electrons in pi system? 6
(4) aromatic, non-, or antiaromatic?

all equal CC bond lengths (1.47 Å), commonly called the tropylium ion

Figure 1019: Fundamental delocalization is predicted for this cation. All of the atoms of the ring are sp² and the ring meets the criteria of analysis for aromaticity. As there are 6 pi electrons in these 7 p orbitals, the ring is classified as aromatic and the planar, delocalized form is predicted to be exceptionally stable.

Figure 1020

Applying the aromaticity criteria to cyclooctatetraenyl dication.

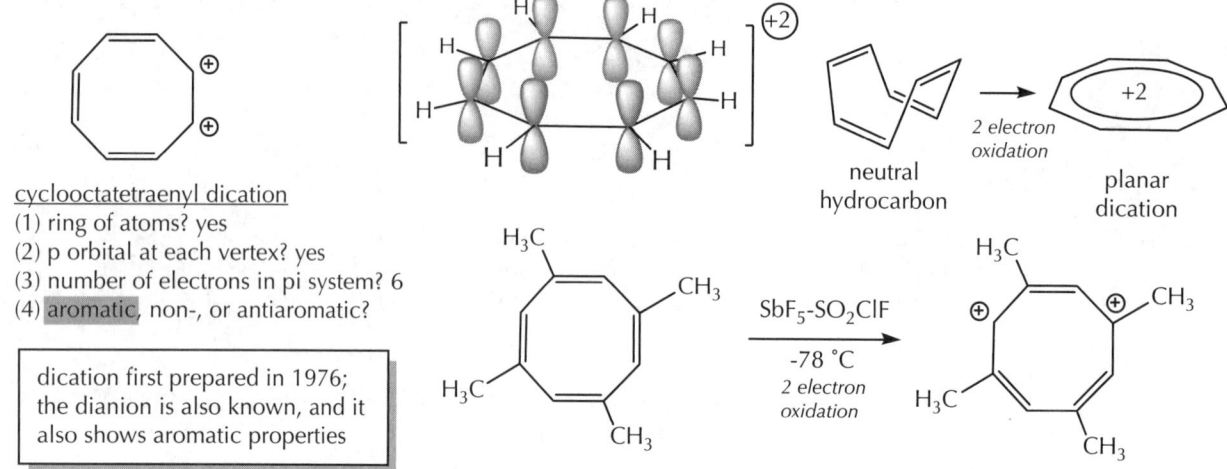

cyclooctatetraenyl dication
(1) ring of atoms? yes
(2) p orbital at each vertex? yes
(3) number of electrons in pi system? 6
(4) aromatic, non-, or antiaromatic?

dication first prepared in 1976; the dianion is also known, and it also shows aromatic properties

Figure 1020: In a simple analysis for resonance contributors, delocalization is predicted. All the atoms of the ring are sp2 and the ring meets the criteria of analysis for aromaticity. There are 6 pi electrons in these 8 p orbitals, and the ring is classified as aromatic. Even though a dication shared over such a small number of atoms might be hard to achieve, this planar, delocalized form is stable. The dianion, which is a planar, 10 pi electron system, has also been prepared.

Hydrocarbons are not the only source of aromatic compounds. Millions of different molecular structures exist in which heteroatoms have replaced carbon atoms, and the Hückel criteria can apply equally well for those structures as they do for hydrocarbons.

The proper analysis to classify heteroaromatic rings relies significantly on the early fundamentals of delocalization and the construction of resonance contributors: Which electrons are delocalizable versus localized, what is the assignment of hybridization, and what is the 3D meaning for the location of non-bonding electrons?

To this latter point: The assignment of antiaromaticity in a ring affects hybridization in a way that is counter to past generalizations. Until this chapter, delocalization was always a stabilizing effect, and so the hybridization of an atom was the one that allows for delocalization to take place. Antiaromatic character is the only situation where that reasoning is reversed: Delocalization leads to instability, and so for systems that risk being antiaromatic, adopting strategies to localize electrons is favorable. In cyclopropenyl anion (Figure 1014), what you would have concluded in Chapter 2 to be a delocalized system is more stable in its localized sp³ hybridization than what would be initially assumed because the delocalized form, with all sp² hybridized atoms, is antiaromatic.

Eight heterocyclic structures are shown in Figure 1021. The analyses of these examples for aromaticity are gathered in Figure 1022. As before, you should try these analyses yourself, before reading further.

Figure 1021

Applying the aromaticity criteria to heterocyclic compounds.

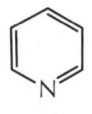  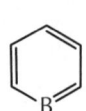 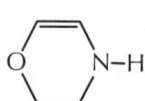 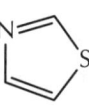 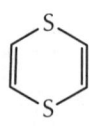

pyridine　　furan　　oxirene　　borabenzene　　3,4-dihydro-2H-1,4-oxazine　　thiazole　　1,4-dithiine

10.1 Aromaticity **807**

Figure 1022

Applying the aromaticity criteria to heterocyclic compounds: analyses.

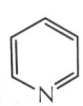

pyridine
(1) ring of atoms? yes
(2) p orbital at each vertex? yes
(3) number of electrons in pi system? 6
(4) aromatic, non-, or antiaromatic?

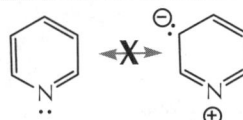

sp^2 N atom, nbe in an sp^2 orbital; not delocalizable

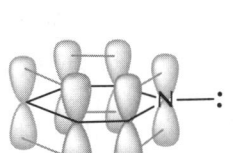

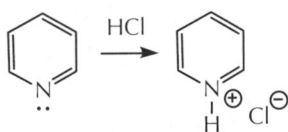

nbe are available for usual reactions, e.g., acid-base; protonation does not disrupt the aromatic ring

---

furan
(1) ring of atoms? yes
(2) p orbital at each vertex? yes
(3) number of electrons in pi system? 6
(4) aromatic, non-, or antiaromatic?

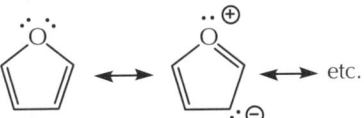 etc.

one delocalizable nbe pair by the usual guidelines for resonance contributors

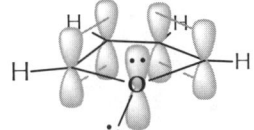

sp^2 O atom, 1 delocalizable nbe pair and 1 nbe pair in an sp^2 orbital

---

oxirene
(1) ring of atoms? yes
(2) p orbital at each vertex? yes
(3) number of electrons in pi system? 4
(4) aromatic, non-, or antiaromatic?

one delocalizable nbe pair by usual guidelines for resonance contributors (sp^2 O); but the delocalized structure is unstable, resulting in a favorable equilibrium towards the localized structure (sp^3 O); unobserved compound

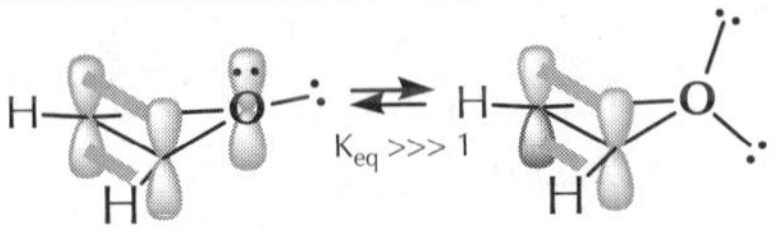

this is why epoxides do not form from alkynes

---

borabenzene
(1) ring of atoms? yes
(2) p orbital at each vertex? yes
(3) number of electrons in pi system? 6
(4) aromatic, non-, or antiaromatic?

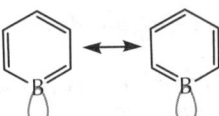

sp^2 B atom with an empty sp^2 orbital; not delocalizable to the empty orbital

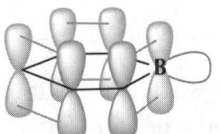

super-strong Lewis acid that has never been isolated in an uncomplexed form - it even forms a complex with N_2

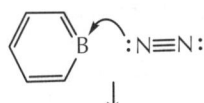

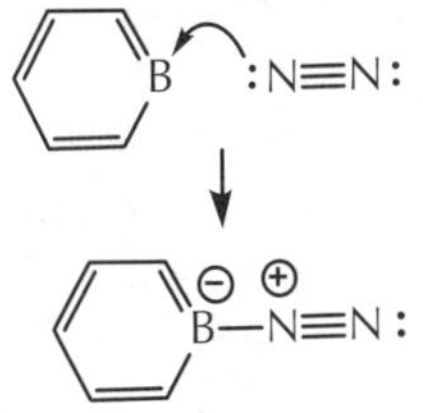

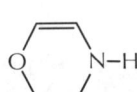

3,4-dihydro-2H-1,4-oxazine
(1) ring of atoms? yes
(2) p orbital at each vertex? no

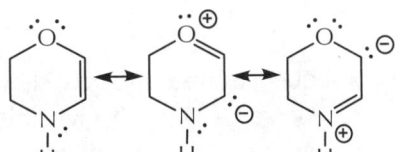

sp^2 O and sp^2 N under the usual guidelines for resonance contributors

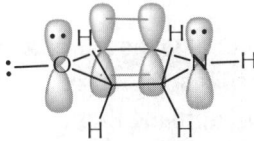

simple nonaromatic compound with resonance contributors

Analyses continue on next page . . .

Figure 1022 continued . . .

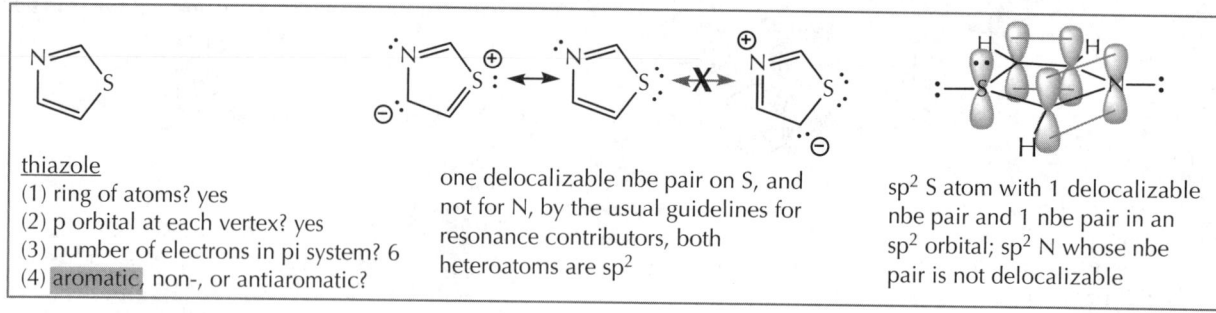

thiazole
(1) ring of atoms? yes
(2) p orbital at each vertex? yes
(3) number of electrons in pi system? 6
(4) aromatic, non-, or antiaromatic?

one delocalizable nbe pair on S, and not for N, by the usual guidelines for resonance contributors, both heteroatoms are sp²

sp² S atom with 1 delocalizable nbe pair and 1 nbe pair in an sp² orbital; sp² N whose nbe pair is not delocalizable

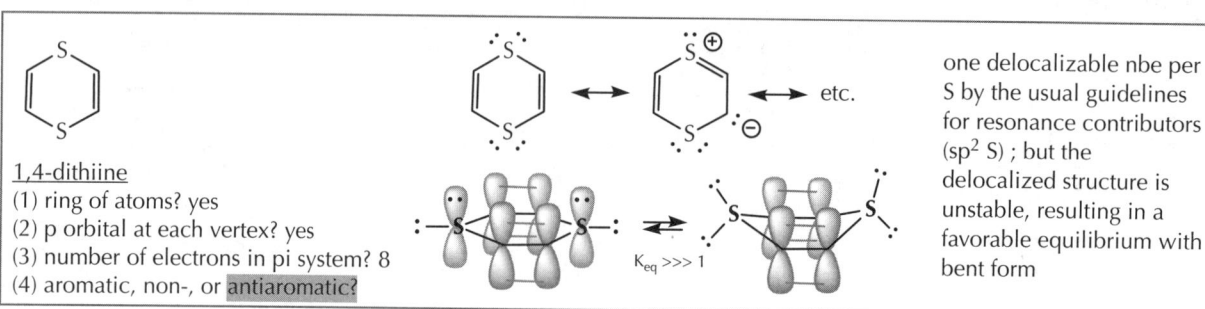

1,4-dithiine
(1) ring of atoms? yes
(2) p orbital at each vertex? yes
(3) number of electrons in pi system? 8
(4) aromatic, non-, or antiaromatic?

one delocalizable nbe per S by the usual guidelines for resonance contributors (sp² S) ; but the delocalized structure is unstable, resulting in a favorable equilibrium with bent form

In summary: The nonbonding electron pair of any heteroatom that is already participating in a formal pi bond (e.g., pyridine, the N in thiazole) will be localized, while one (1) nonbonding electron pair on any heteroatom that is not participating in a formal pi bond (e.g., furan, oxirene, the N and O in the oxazine, the S in thiazole, the S atoms in 1,4-dithiine) is delocalizable by the usual rules for anticipating resonance contributors. The relative stability of the delocalized structures is evaluated by assessing whether they are aromatic rings (pyridine, furan, borabenzene, thiazole), antiaromatic rings (oxirene, 1,4-dithiine), or simply nonaromatic compounds with their usual resonance contributors (3,4-dihydro-2H-1,4-oxazine). The antiaromatic compounds will be reactive and/or adopt conformations that disrupt the delocalized structure.

D. Chemical and Physical Properties of Aromatic Compounds

The analysis for aromaticity is an extension of the electron delocalization principles introduced in Chapter 2. In that introduction, the effect of delocalization on structure, in the hybridization of atoms, was presented: Delocalized electrons are in unhybridized p orbitals, while localized electrons are in sigma bonds or are hybridized nonbonding electron pairs. And only a general statement about reactivity was presented: Delocalized electrons are more stable than localized electrons.

The analysis for aromaticity adds an additional dimension to your understanding of delocalization. Aromatic delocalization of electrons creates exceptional stability, while antiaromatic delocalization creates exceptional instability.

Structurally, there are two main consequences resulting from the aromatic/antiaromatic analysis (Figure 1023). First, because aromaticity is an energetically highly favorable delocalization, a planar conformation of an aromatic ring keeps the associated p orbitals all mutually parallel and best suited for pi bonding. Aromatic rings are expected to be planar. Second, because pi bonding is how delocalization can occur, the hybridization of the aromatic ring atoms will strongly favor hybridizations that provide an unhybridized p orbital (generally sp²). In antiaromatic rings, delocalization leads to instability, so the molecule will undergo reactions, favor nonplanar conformations, and/or adopt localized hybridizations (generally sp² to sp³)—all of which disrupt delocalization.

Figure 1023
Summary of the effects of aromaticity on molecular structure.

aromatic compounds

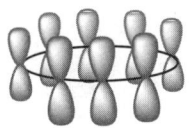

cyclic array of parallel p orbitals; one at each vertex

4N+2 pi (delocalized) electrons:
planar ring geometry
exceptionally high delocalization (resonance) stability
resist typical addition reactions that disrupt aromaticity

antiaromatic compounds

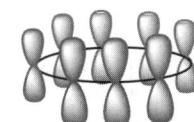

cyclic array of parallel p orbitals; one at each vertex $K_{eq} \ggg 1$

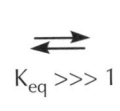

conformation(s) and/or reaction pathway(s) that disrupt the delocalized structure

4N pi (delocalized) electrons:
delocalization is unstable, resulting in reaction and/or conformational equilibria that disrupt the delocalized structure in favor of localized form(s)

Chemical reactivity is dramatically affected by the aromatic/antiaromatic property.

Although highly unsaturated, aromatic rings do not undergo addition reactions as readily as their nonaromatic unsaturated counterparts. Throughout the text, the exceptionally strong resonance stability in benzene rings excluded those double bonds from electrophile addition of acids, hydrogenation, hydroboration, and so on. Antiaromatic rings that cannot disrupt delocalization by changes in conformational shape and/or hybridization end up being highly reactive as the third option for losing the unstable, high-energy ring.

Addition reactions are particularly difficult to achieve on an aromatic ring because aromaticity is disrupted; but doing elimination reactions to produce an aromatic ring ought to be particularly easy. Shaking dihydronaphthalene with a palladium catalyst causes a dehydrogenation reaction that aromatizes the second ring and gives naphthalene. Similarly, both of the dihydronaphthalene alcohol derivatives undergo a fast dehydration reaction under mild reaction conditions (Figure 1024).

Figure 1024
The effects of aromaticity on reactivity: addition and elimination reactions.

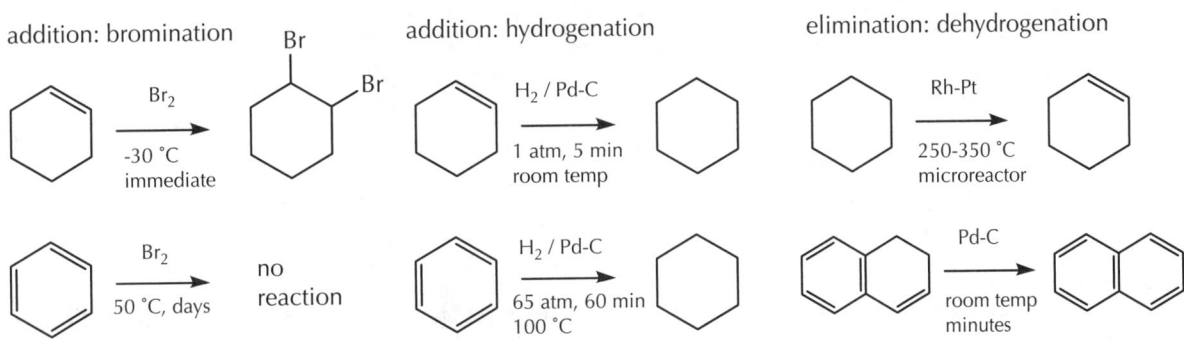

The availability and relative stability of electrons is a core concept in acid-base chemistry, and so aromaticity and antiaromaticity can have significant effects on pK_a values (Figure 1025). Deprotonation of a nonaromatic compound to give an aromatic one is highly favorable, and will give the acid form a low pK_a value. Cyclopentadiene has a pK_a value of 15, as deprotonation gives the aromatic cyclopentadienyl anion, while cyclopentene has a pK_a value of 45. The conjugate acid of pyrrole has a pK_a of 0.04, while the conjugate acid of the saturated version of that same ring, pyrrolidine, has a pK_a of 11.3. And similarly, a deprotonation reaction that gives an antiaromatic ring will be harder to do than its corresponding nonaromatic counterpart: The pK_a of cyclopropene is 61, while the pK_a of cyclopropane is 46. In the usual complementary manner, the basicity of electron pairs that are part of a stable aromatic ring will be significantly lower than their nonaromatic counterparts.

Figure 1025

The effects of aromaticity on reactivity: acid-base strength.

The availability and relative stability of electrons is also a core concept is most other organic reactivity, which means that aromaticity and antiaromaticity will have an effect on both reaction rates and the position of equilibria (Figure 1026).

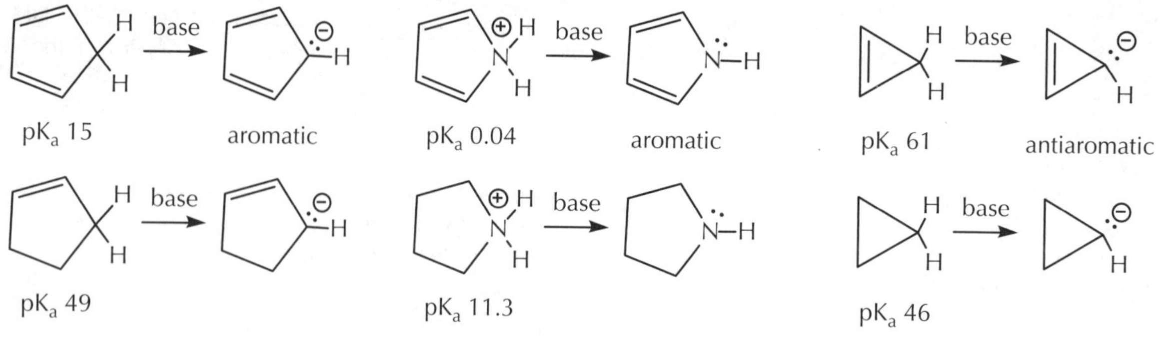

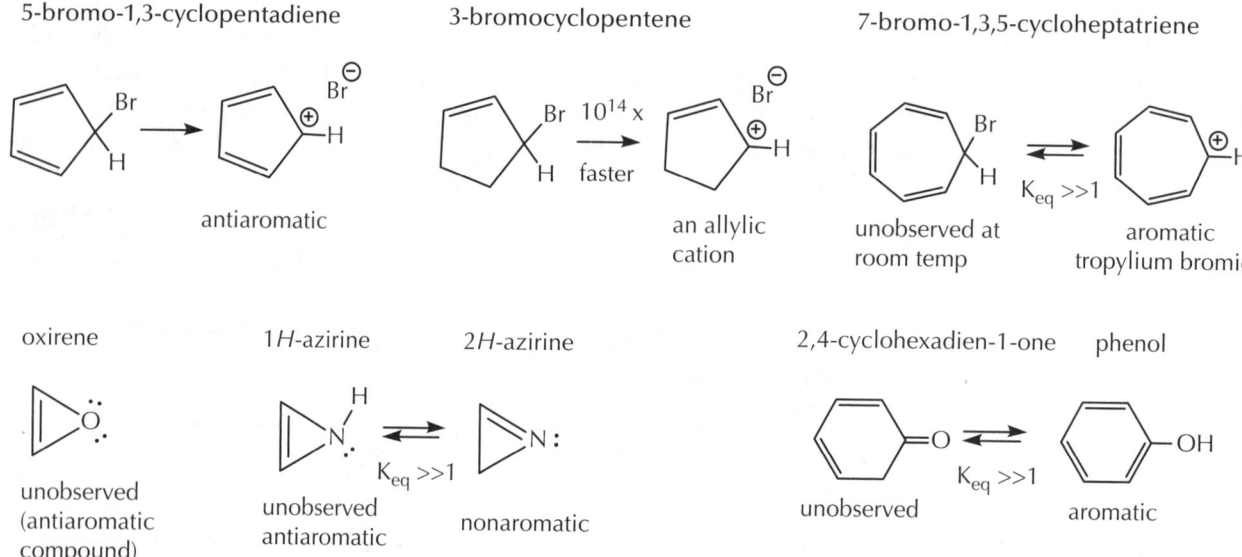

Figure 1026

The effects of aromaticity on reactivity: rates and equilibria.

The ionization rates for 3-bromocyclopentene, which gives a resonance-stabilized allylic carbocation, is 1014 times faster than 5-bromo-1,3-cyclopentadiene, which gives an antiaromatic cation. On the other hand, the rate of ionization of 7-bromo-1,3,5-cycloheptatriene is so favorable that the molecule is not only unobserved at room temperature, but the ionic compound is stable and can be purchased commercially.

Oxirene is only the hypothetical epoxidation reaction product derived from acetylene; it has never been observed, presumably due to the antiaromatic ring. The nitrogen-atom analogue, 1*H*-azirine, has also never been observed. Attempts to prepare 1*H*-azirine always end up giving the nonaromatic isomer, 2*H*-azirine thanks to a highly favorable equilibrium.

Similarly, attempts to prepare 2,4-cyclohexadien-1-one always result in forming phenol, as the comparable equilibrium places the third double bond into an aromatic ring.

10.1 Aromaticity

PRACTICE QUESTIONS

10.01 The following molecules contain many different rings. One resonance contributor for each of these compounds is shown (as usual), but the entire set of significant resonance contributors is implied. Which rings are aromatic (AR), which rings are antiaromatic (AA), and which are nonaromatic (NA)? Decide whether the various nonbonding electron pairs are delocalized (DE), localized (LE), or one of each (OE; if more than one).

(a) 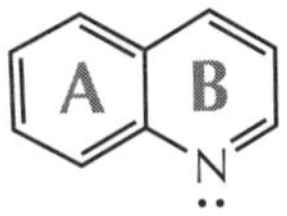 quinoline structure with rings A and B, N with lone pair
- ring A:
- ring B:
- N nbe:

(b) 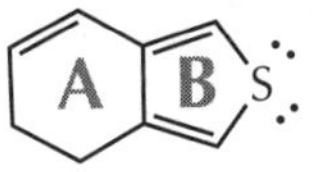 benzo-fused thiophene-like structure with rings A and B, S
- ring A:
- ring B:
- S nbe:

(c)  2,3-dihydrobenzothiophene structure with rings A and B, S
- ring A:
- ring B:
- S nbe:

(d) 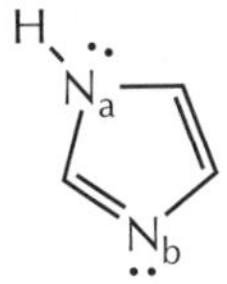 imidazole with H on N$_a$, N$_b$
- ring:
- N$_a$ nbe:
- N$_b$ nbe:

(e) 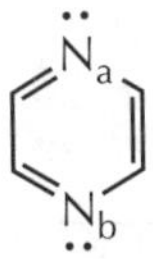 pyrazine with N$_a$ and N$_b$
- ring:
- N$_a$ nbe:
- N$_b$ nbe:

(f) 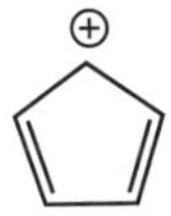 cyclopentadienyl cation
- ring:

(g) 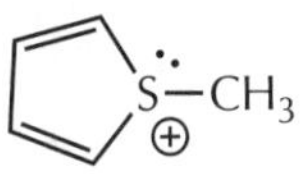 S-methyl thiophenium cation
- ring:
- S nbe:

(h) 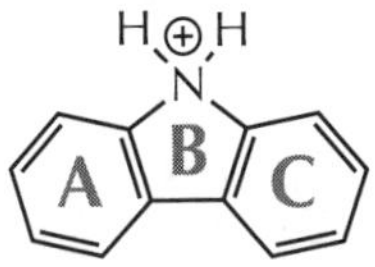 carbazole with NH$_2^+$, rings A, B, C
- ring A:
- ring B:
- ring C:

(i)  oxepine/oxacycloheptatriene
- ring:
- O nbe:

(j) 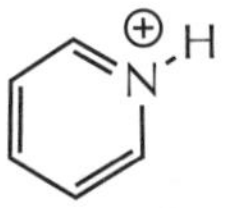 pyridinium (N–H$^+$)
- ring:
- O nbe:
- N nbe:

(k) 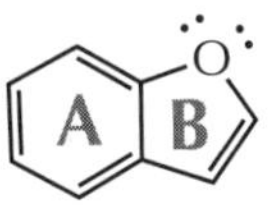 benzofuran rings A and B with O
- ring A:
- ring B:
- O nbe:

(l)  anthracene rings A, B, C
- ring A:
- ring B:
- ring C:

(m) 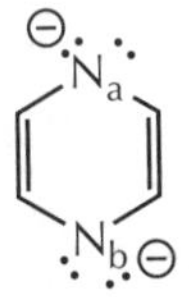 piperazine-like ring with N$_a^-$ and N$_b^-$
- ring:
- N$_a$ nbe:
- N$_b$ nbe:

(n) 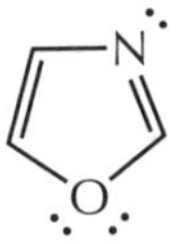 oxazole
- ring:
- O nbe:
- N nbe:

(o) 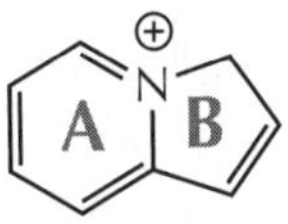 indolizinium cation, rings A and B
- ring A:
- ring B:

(p) 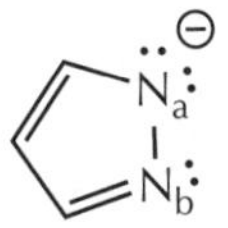 pyrazole anion with N$_a^-$, N$_b$
- ring:
- N$_a$ nbe:
- N$_b$ nbe:

(q) 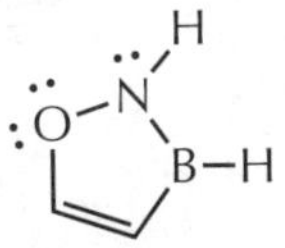 isoxazoline with B–H
- ring:
- O nbe:
- N nbe:

(r)  phenalenyl cation, rings A, B, C
- ring A:
- ring B:
- ring C:

(s) 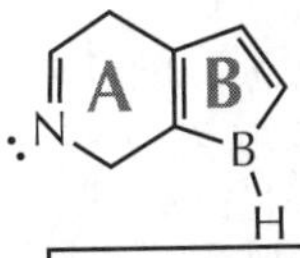 fused bicyclic with N, B–H, rings A and B
- ring A:
- ring B:
- N nbe:

(t) 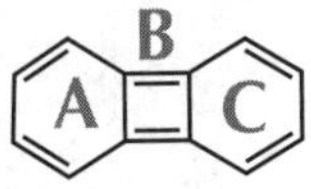 biphenylene-like structure rings A, B, C
- ring A:
- ring B:
- ring C:

812 CHAPTER 10 Aromaticity and Electrophilic Aromatic Substitution Reactions

10.02 Nucleic acids (RNA and DNA) are among the most important molecules on our planet. Each of them includes four nucleobases, heterocyclic molecules that comprise the so-called "base pairs" in the double helix structure. The four nucleobases from DNA are shown below. For each one, answer the following questions:

(i) Is the resonance contributor of the 6-membered ring, as drawn, an aromatic ring? Yes or No?

(ii) Is there at least one other resonance contributor that has an aromatic ring? Yes or No? If yes, draw it. If no, say "N/A"

(a) cytosine

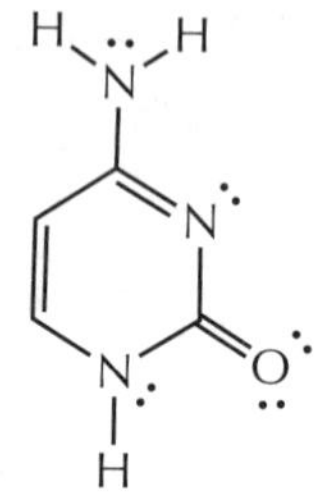

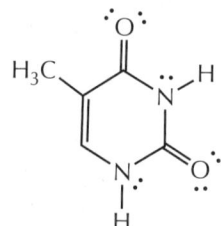

Is the resonance contributor of the 6-membered ring, as drawn, aromatic?
Yes   No

Is there another resonance contributor that is aromatic? If yes, draw it. If no, say "N/A"

(b) thymine

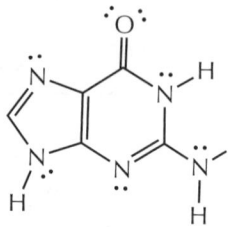

Is the resonance contributor of the 6-membered ring, as drawn, aromatic?
Yes   No

Is there another resonance contributor that is aromatic? If yes, draw it. If no, say "N/A"

(c) guanine

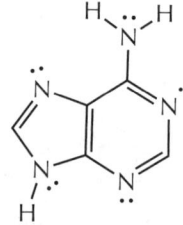

Is the resonance contributor of the 6-membered ring, as drawn, aromatic?
Yes   No

Is there another resonance contributor that is aromatic? If yes, draw it. If no, say "N/A"

(d) adenine

Is the resonance contributor of the 6-membered ring, as drawn, aromatic?
Yes   No

Is there another resonance contributor that is aromatic? If yes, draw it. If no, say "N/A"

10.03 A typical ketone (acetone) has a dipole moment of 2.91 Debye, and ketones with resonance contributors have slightly higher values (benzophenone is 2.97 Debye). Cyclopropanone has a dipole moment of 2.67 Debye, while that for cyclopropenone, first isolated in 1972, is 5.08 Debye.

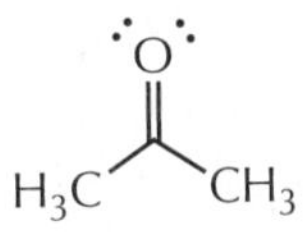

acetone	benzophenone	cyclopropanone	cyclopropenone
2.91 D	2.97 D	2.67 D	5.08 D

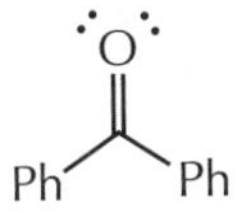

(a) Using words and drawings, explain the difference in dipole moment between acetone and benzophenone.

(b) Using words and drawings, explain the difference in dipole moment between cyclopropanone and cyclopropenone.

(c) Based on these two explanations, what must be true about the relative significance of the resonance contributors in acetone with respect to the resonance contributors in cyclopropanone?

(d) The conjugate acids of cyclopropanone and cyclopropenone (protonated at the oxygen atoms) have pK$_a$ values of either -12 and -5, or -5 and -12, respectively. Which value goes with which compound, and why?

10.04 The following molecules each have two heteroatoms that could be protonated. In all of these cases, one of the heteroatoms is observed to be significantly more basic than the other. Predict which of the heteroatoms, in each case, is more easily protonated and explain why, using words and pictures.

(a)

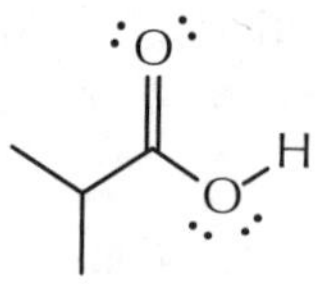

(b)

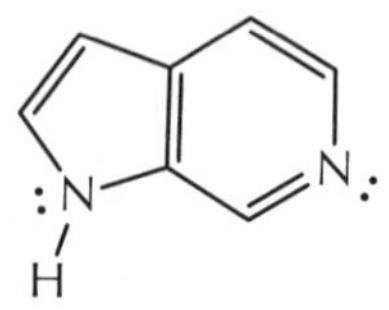

(c)

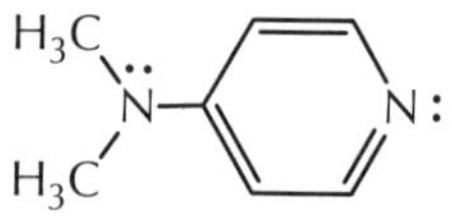

(d)

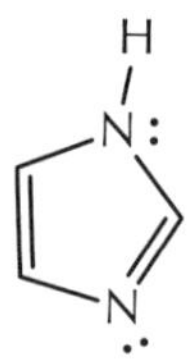

10.05 The S$_N$1 reaction rates were compared for these three molecules under identical reaction conditions. One of them reacted within seconds; one of them took about an hour; and one of them did not react at all. Which one is which, and why?

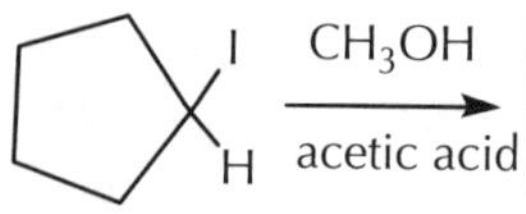

☐ reacted within seconds ☐ took an hour ☐ did not react at all
Explain:

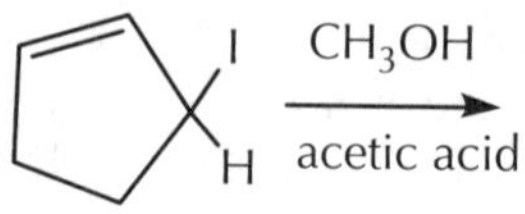

☐ reacted within seconds ☐ took an hour ☐ did not react at all
Explain:

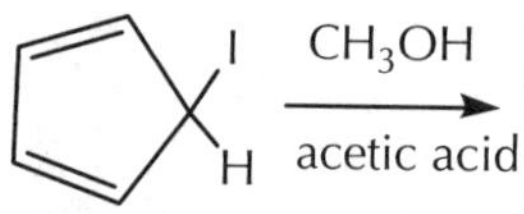

☐ reacted within seconds ☐ took an hour ☐ did not react at all
Explain:

10.2 Electrophilic Aromatic Substitution

A. Halogenation, Sulfonation, and Nitration Reactions

The historical development of the reactivity of aromatic compounds is difficult to sort out because the historical observations were so inconsistent. Some unsaturated compounds would react rapidly and completely with reagents such as molecular bromine and potassium permanganate, and others, particularly in the class of "aromatic compounds," would sometimes react completely, sometimes partially, and sometimes not at all. Doing organic chemistry in the era before a good structural theory would often have apparent inconsistencies (Figure 1027).

Figure 1027

Historical observations of reactivity related to unsaturated compounds.

A. $C_{10}H_{12}$ $\xrightarrow[\text{KOH, H}_2\text{O}]{\text{KMnO}_4}$ no discoloration

$\xrightarrow{\text{Br}_2}$ no discoloration

$\xrightarrow[\text{Pd-C}]{\text{H}_2}$ no change

B. $C_{10}H_{12}$ $\xrightarrow[\text{KOH, H}_2\text{O}]{\text{KMnO}_4}$ discoloration

$\xrightarrow{\text{Br}_2}$ discoloration briefly, then persists

$\xrightarrow[\text{Pd-C}]{\text{H}_2}$ $C_{10}H_{14}$

C. $C_{10}H_{12}$ $\xrightarrow[\text{KOH, H}_2\text{O}]{\text{KMnO}_4}$ discoloration

$\xrightarrow{\text{Br}_2}$ discoloration with excess bromine

$\xrightarrow[\text{Pd-C}]{\text{H}_2}$ $C_{10}H_{22}$

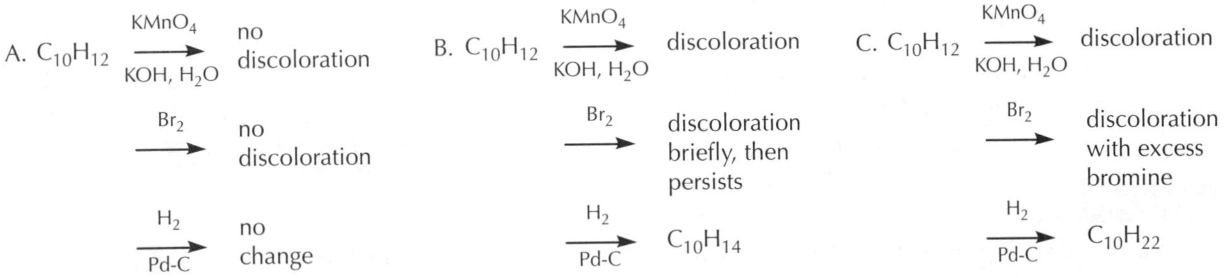

With structural theory, the lack of reactivity of benzene rings (and other aromatic compounds) was consistent, even if it was not understood. In point of fact, this is the way you have learned it so far in this text. When you look at the reactions in Figure 1028, which are the same ones as in Figure 1027, the observations are consistent with what you have seen previously.

Figure 1028

Structure-based reactivity of compounds with aromatic rings.

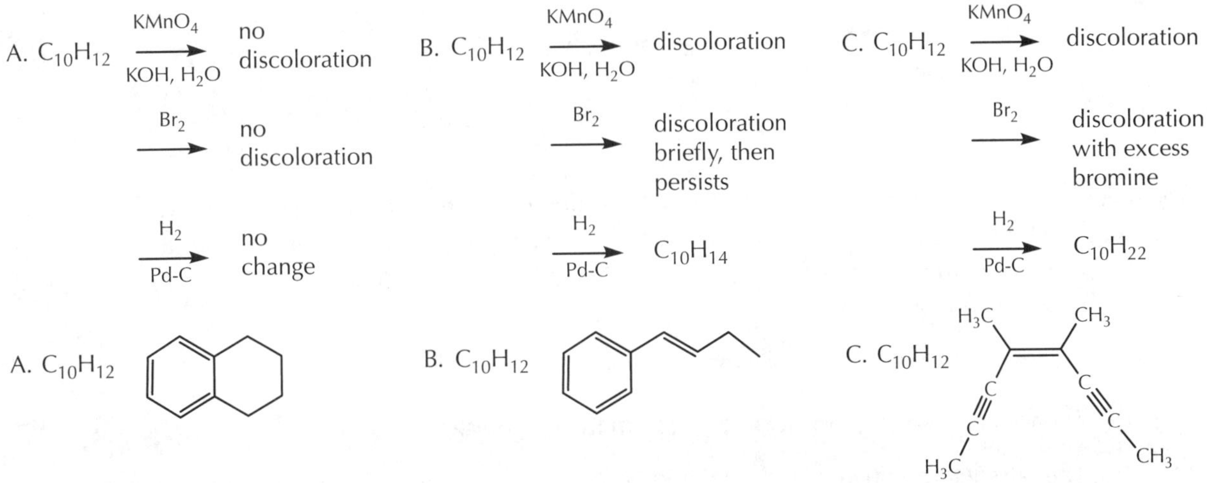

A. $C_{10}H_{12}$ $\xrightarrow[\text{KOH, H}_2\text{O}]{\text{KMnO}_4}$ no discoloration

$\xrightarrow{\text{Br}_2}$ no discoloration

$\xrightarrow[\text{Pd-C}]{\text{H}_2}$ no change

B. $C_{10}H_{12}$ $\xrightarrow[\text{KOH, H}_2\text{O}]{\text{KMnO}_4}$ discoloration

$\xrightarrow{\text{Br}_2}$ discoloration briefly, then persists

$\xrightarrow[\text{Pd-C}]{\text{H}_2}$ $C_{10}H_{14}$

C. $C_{10}H_{12}$ $\xrightarrow[\text{KOH, H}_2\text{O}]{\text{KMnO}_4}$ discoloration

$\xrightarrow{\text{Br}_2}$ discoloration with excess bromine

$\xrightarrow[\text{Pd-C}]{\text{H}_2}$ $C_{10}H_{22}$

A. $C_{10}H_{12}$ (tetralin structure)

B. $C_{10}H_{12}$ (phenyl-substituted alkene structure)

C. $C_{10}H_{12}$ (dimethyl-substituted diyne structure)

As aromatic rings were observed to be generally less reactive than other unsaturated structures, one of the reasonable questions was whether or not changing the reaction conditions might cause addition reactions to occur. Experimentally, the common strategies for overcoming low reactivity include heating the reaction, changing the solvent and/or concentration of the reagents, increasing the reaction time, and trying to catalyze the reaction.

816 CHAPTER 10 Aromaticity and Electrophilic Aromatic Substitution Reactions

History does not record all of the failed attempts to get benzene and molecular bromine to react, but it does record that iron metal (Fe) and iron (III) salts such as ferric bromide (FeBr$_3$) would readily cause the red-brown color to disappear when molecular bromine and benzene were combined. The observed reaction was not the expected addition product, however. The reaction produced a substitution product, bromobenzene, and an equivalent of HBr (Figure 1029). This overall reaction was reproducible and quite general for aromatic rings and became known as the Electrophilic Aromatic Substitution (EAS) reaction.

Figure 1029

Electrophilic aromatic substitution: bromination of benzene.

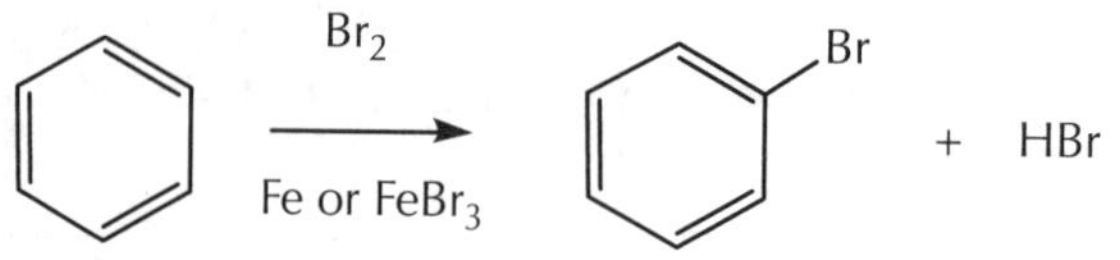

benzene + Br$_2$ →(Fe or FeBr$_3$) bromobenzene + HBr

The EAS reaction of benzene using molecular chlorine with iron metal or ferric chloride (FeCl$_3$) was also observed, as was the nitration reaction (using nitric acid, or a mixture of nitric and sulfuric acids) and the sulfonation reaction (using sulfuric acid, or a mixture of sulfur trioxide with sulfuric acid) (Figure 1030).

Figure 1030

Electrophilic aromatic substitution: chlorination, nitration, and sulfonation of benzene.

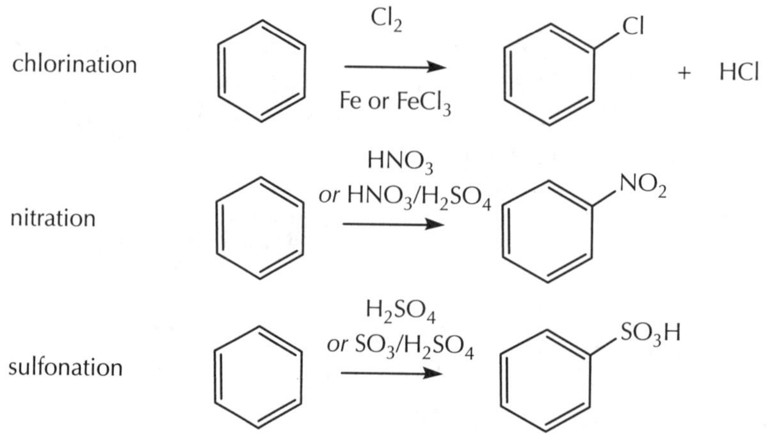

chlorination: benzene + Cl$_2$ →(Fe or FeCl$_3$) chlorobenzene + HCl

nitration: benzene + HNO$_3$ or HNO$_3$/H$_2$SO$_4$ → nitrobenzene

sulfonation: benzene + H$_2$SO$_4$ or SO$_3$/H$_2$SO$_4$ → benzenesulfonic acid (SO$_3$H)

The mechanism for the electrophilic aromatic substitution reaction is not another new, fundamental type of reaction mechanism. Instead, it is the first of many examples, illustrated throughout the remaining chapters, where a combination of fundamental mechanisms (S$_N$1, E2, electrophilic addition, etc.) can be used to explain a new observation.

Every EAS reaction mechanism exhibits three familiar steps (Figure 1031).

Step 1: The formation of a good electrophile from the reagents.

Step 2: The rate-determining electrophilic addition of the electrophile to the aromatic ring to give a resonance stabilized carbocation intermediate, breaking the aromaticity of the ring.

Step 3: The deprotonation (elimination reaction) of the carbocation intermediate, from the site of the electrophilic addition, restoring the aromatic ring.

10.2 Electrophilic Aromatic Substitution

Figure 1031
Electrophilic aromatic substitution: general features of the mechanism.

The formation of a good electrophile from the reagents.

reagents ⟶ $E^{\oplus}$

The rate-determining electrophilic addition of the electrophile to the aromatic ring to give a resonance stabilized carbocation intermediate, breaking the aromaticity of the ring.

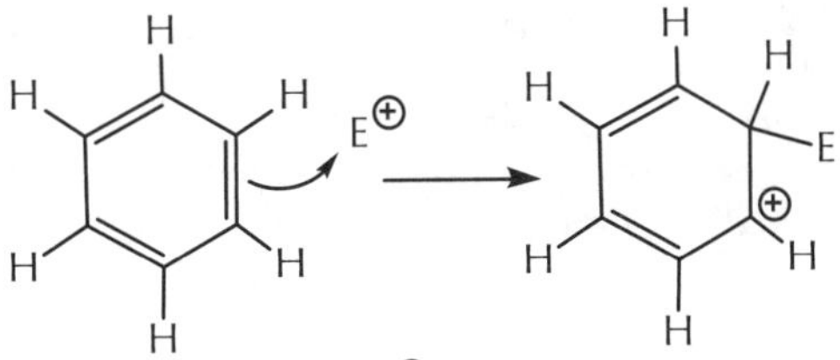

The deprotonation (elimination reaction) of the carbocation intermediate, from the site of the electrophilic addition, restoring the aromatic ring.

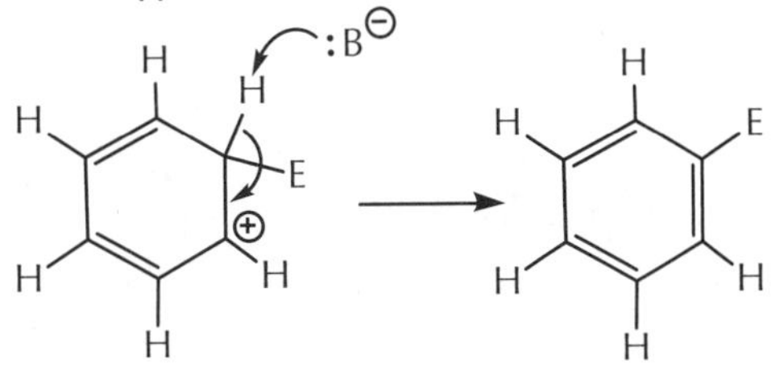

These general features are applied, below, to the bromination (Figure 1032), nitration (Figure 1033), and sulfonation (Figure 1034) EAS reactions. Note that the only significant variations in these three reactions are the details for generating the electrophile. After that, the mechanisms are exactly the same steps.

Figure 1032
Electrophilic aromatic substitution: bromination mechanism.

Br_2/Fe makes $FeBr_3$ (3 Br_2 + 2 Fe ⟶ 2 $FeBr_3$) so all of the reactions are Br_2 plus the Lewis Acid, $FeBr_3$

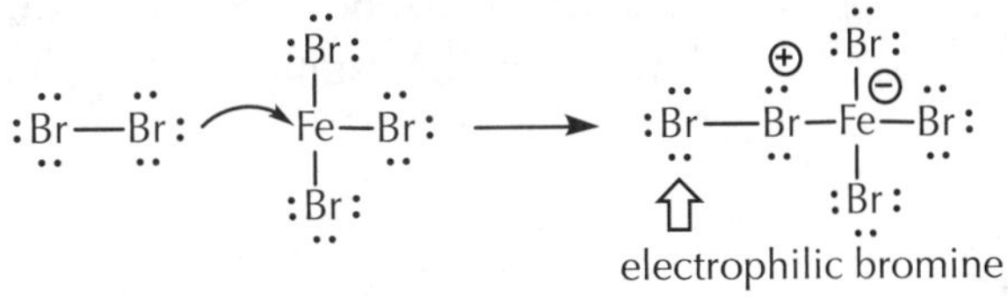

electrophilic bromine

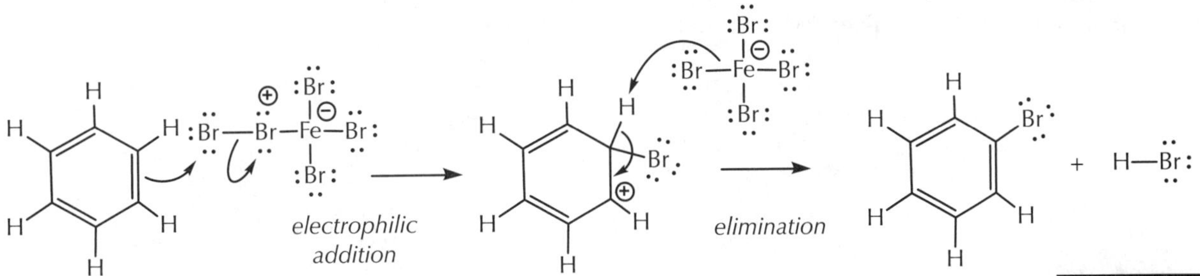

electrophilic addition

elimination

Figure 1033
Electrophilic aromatic substitution: nitration mechanism.

Protonation of the hydroxyl group in nitric acid by either sulfuric or nitric acid ("H-B"), and the subsequent loss of water, creates the NO_2^+ electrophile

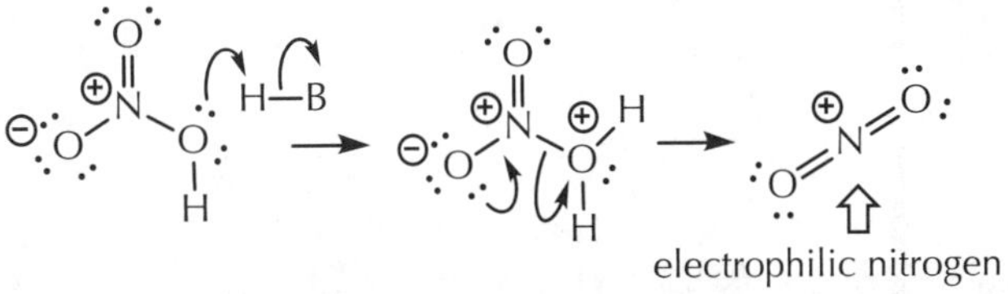

electrophilic nitrogen

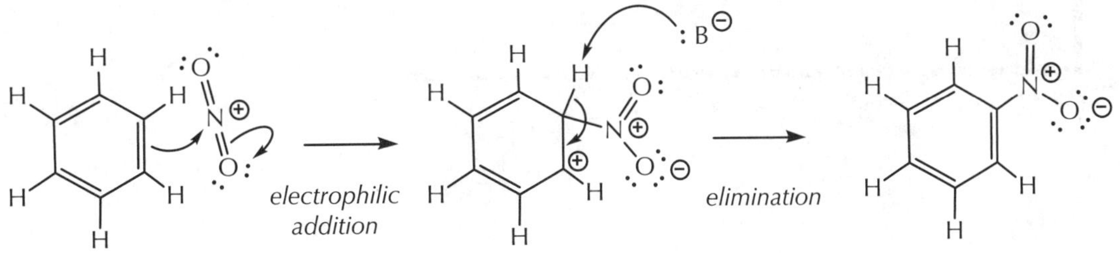

electrophilic addition

elimination

Figure 10.34

Electrophilic aromatic substitution: sulfonation mechanism.

Protonation of a hydroxyl group in sulfuric acid or SO₃ by another sulfuric acid molecule ("H-B") and the subsequent loss of water, creates the SO₃H⁺ electrophile

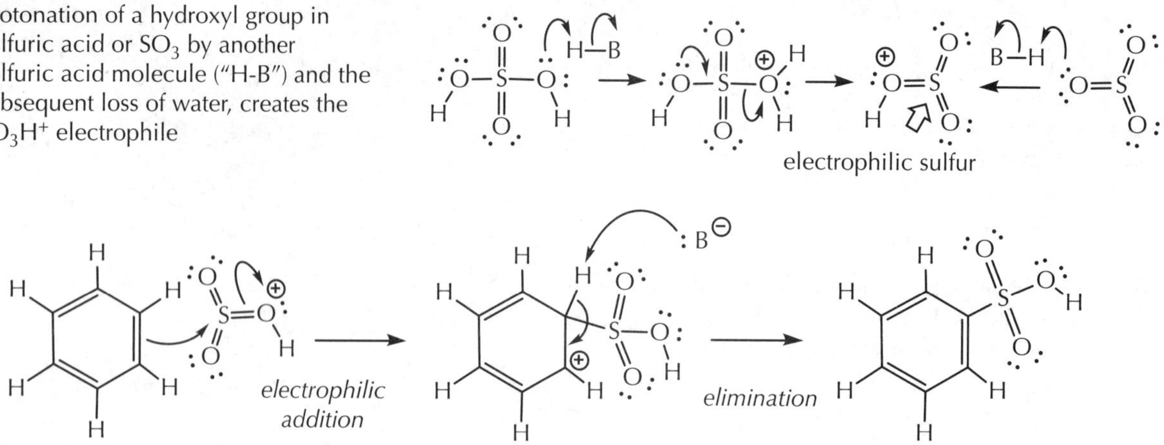

electrophilic sulfur

electrophilic addition → *elimination*

In all three of these reactions, the carbocation intermediate should, in principle, be able to be captured by a Lewis base and give the product from addition. The deprotonation reaction of these carbocation intermediates is the exclusive outcome, however, because the loss of the proton restores the exceptional stability of the aromatic ring, while the capture of the carbocation would result in a nonaromatic compound (Figure 10.35). The carbocation intermediate, in other words, is the conjugate acid of a base that is an aromatic ring (as in Figure 10.25), and so the pK$_a$ values for these carbocations are estimated to be about -25, significantly lower than for traditional carbocations, which are already considerably strong acids, in the –11 range.

Figure 10.35

Electrophilic aromatic substitution: energy diagram.

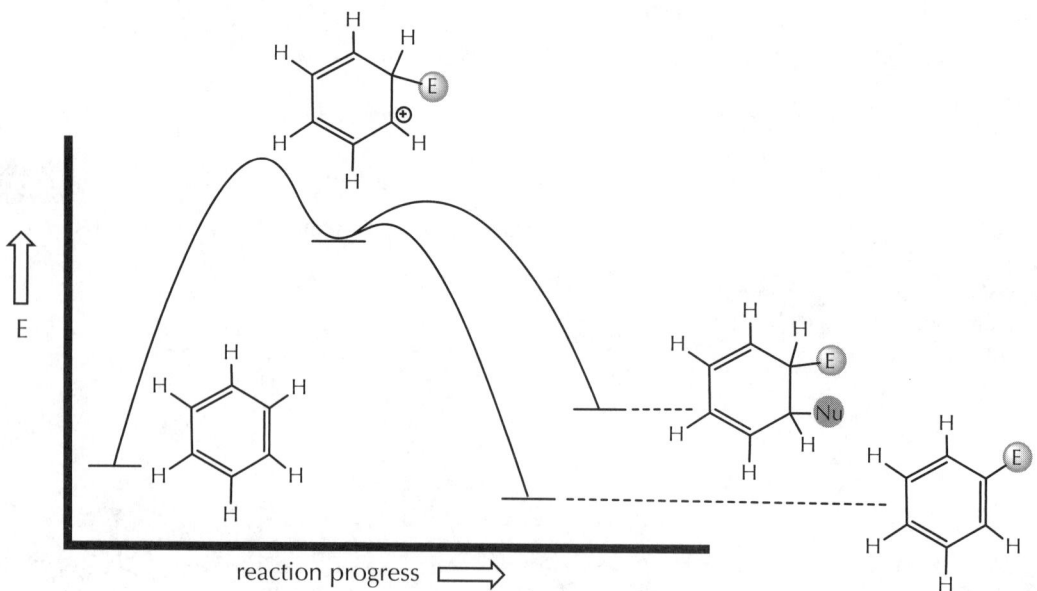

B. Friedel-Crafts Alkylation and Acylation Reactions

Many different electrophiles have been used in EAS reactions. The simplest reaction is protonation, which takes place in aqueous solutions of sulfuric acid under mild reaction conditions. The process is undetectable under typical conditions, though, because the protonation of benzene to give the carbocation intermediate, followed by the deprotonation of intermediate, gives benzene back again (Figure 1036). This would be classified as a nonproductive reaction because the product is the same as the starting material.

Figure 1036

Electrophilic aromatic substitution: nonproductive protonation reaction.

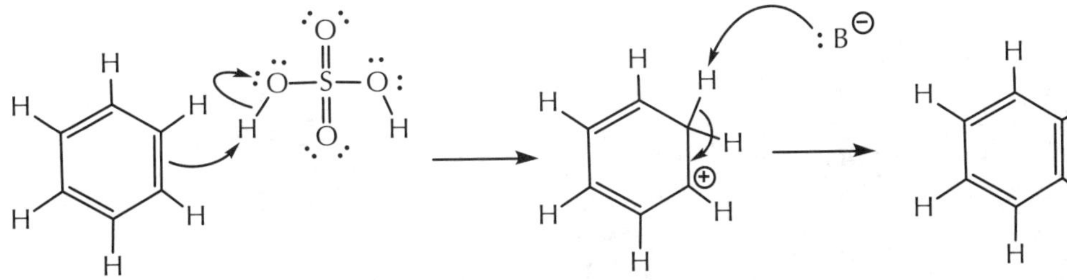

How can we determine, experimentally, that a reaction is actually taking place?

The protonation reaction can be demonstrated by using the deuterated version of sulfuric acid (D_2SO_4) in deuterated water (D_2O), which ends up making deuterated benzene. Or we can start with the deuterated benzene and see the C-D to C-H substitution taking place in regular sulfuric acid (Figure 1037).

Figure 1037

Electrophilic aromatic substitution: deuteration and protonation reactions.

benzene + D_2SO_4

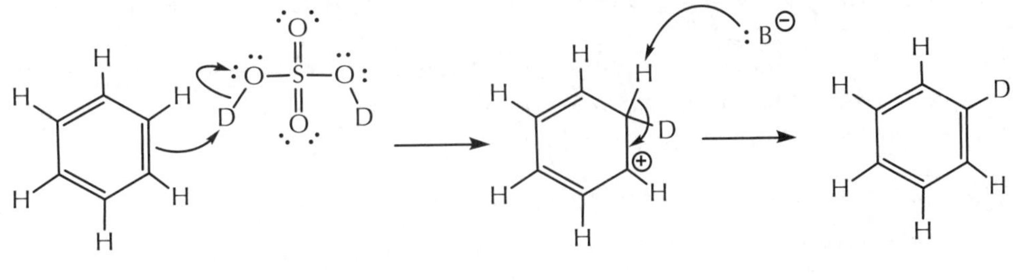

repeat, isolate, repeat... eventually forms C_6D_6

d_6-benzene + H_2SO_4

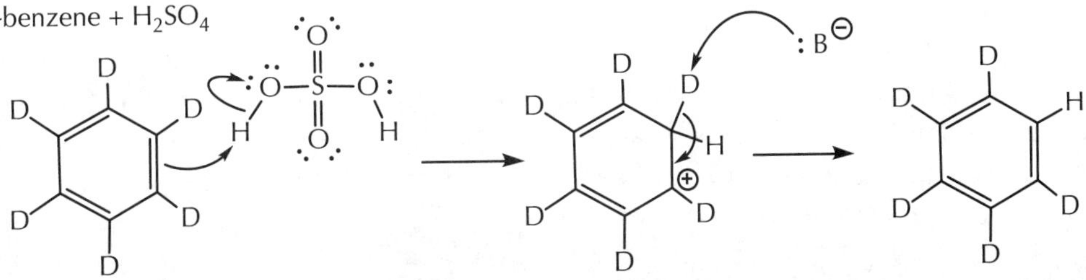

repeat, isolate, repeat... eventually forms C_6H_6

First, by analogy: With sufficient time and care to remove the proton-containing byproducts, benzene (C_6H_6) can be converted to its fully deuterated form (hexadeutereobenzene, d_6-benzene, C_6D_6). Second, the protonation reaction can be observed directly if d_6-benzene is used as the starting material. Both reactions demonstrate that the C-H bonds in benzene are undergoing electrophilic aromatic substitution in the presence of sulfuric acid, but that the reaction is nonproductive (i.e., undetectable) unless the isotope exchange experiments are performed.

820 CHAPTER 10 Aromaticity and Electrophilic Aromatic Substitution Reactions

Carbocations are an important source of electrophiles for EAS reactions because larger and more complex organic molecules can be prepared by using them. EAS reactions involving carbon atom electrophiles are called Friedel-Crafts reactions (Figure 1038).

Figure 1038

Electrophilic aromatic substitution: Friedel-Crafts alkylation.

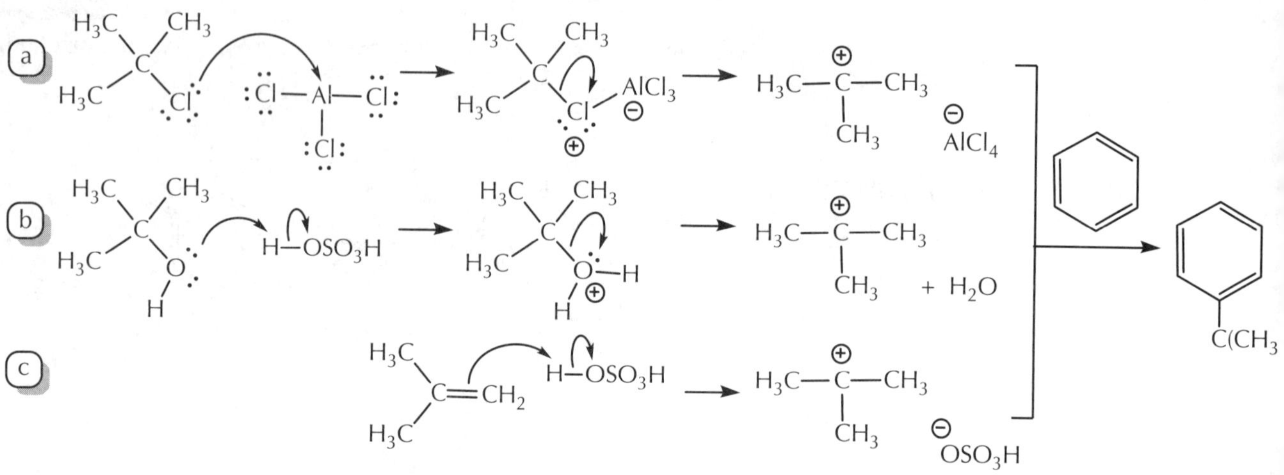

The two main methods for creating carbocations are used here: (1) protonation of a pi bond and (2) ionization of a carbon-leaving group bond. As seen previously (e.g., the summary in Chapter 8.2), the relatively stable *tert*-butyl carbocation can be formed by (a) ionization of the carbon-bromine bond in 2-bromo-2-methylpropane with aluminum trichloride, (b) ionization of the carbon-oxygen bond in 2-methyl-2-propanol with a strong acid, and (c) protonating 2-methylpropene with a strong acid.

In all three of these cases, the *tert*-butyl carbocation is the electrophile, and the *tert*-butyl alkyl group ends up attached to the benzene ring in the EAS reaction. When a Friedel-Crafts EAS reaction uses an alkyl-derived carbon electrophile, it is called a Friedel-Crafts alkylation reaction.

There are two things to keep in mind about the Friedel-Crafts alkylation reactions (Figure 1039): 1° and 2° carbons form unstabilized carbocations slowly, and when they do form, they can undergo rearrangement reactions.

The mechanism for the Friedel-Crafts methylation reaction is best shown as a direct reaction of the ring onto the unhindered methyl group possessing a cationic leaving group atom. Forming the completely unsubstituted methyl cation would be a slow process. As in the bromination reaction mechanism (Figure 1032), forming the free bromine cation would also be slow compared with the direct reaction of the ring onto the bromine-ferric tribromide complex.

In reactions that will form unstabilized 1° and 2° alkyl cations, the distribution of Friedel-Crafts alkylation products will depend on the reaction conditions: Higher temperatures will result in more products derived from carbocation rearrangement(s).

In contrast with an alkyl group, an acyl group is the name for the molecular unit comprising a carbon-oxygen double bond that is attached, at the carbon, to a hydrogen atom or an alkyl group (Figure 1040).

10.2 Electrophilic Aromatic Substitution

Figure 1039

Electrophilic aromatic substitution: methylation and propyl cation rearrangements in Friedel-Crafts alkylation reactions.

	at -6 °C	60%	40%
	at 35 °C	40%	60%

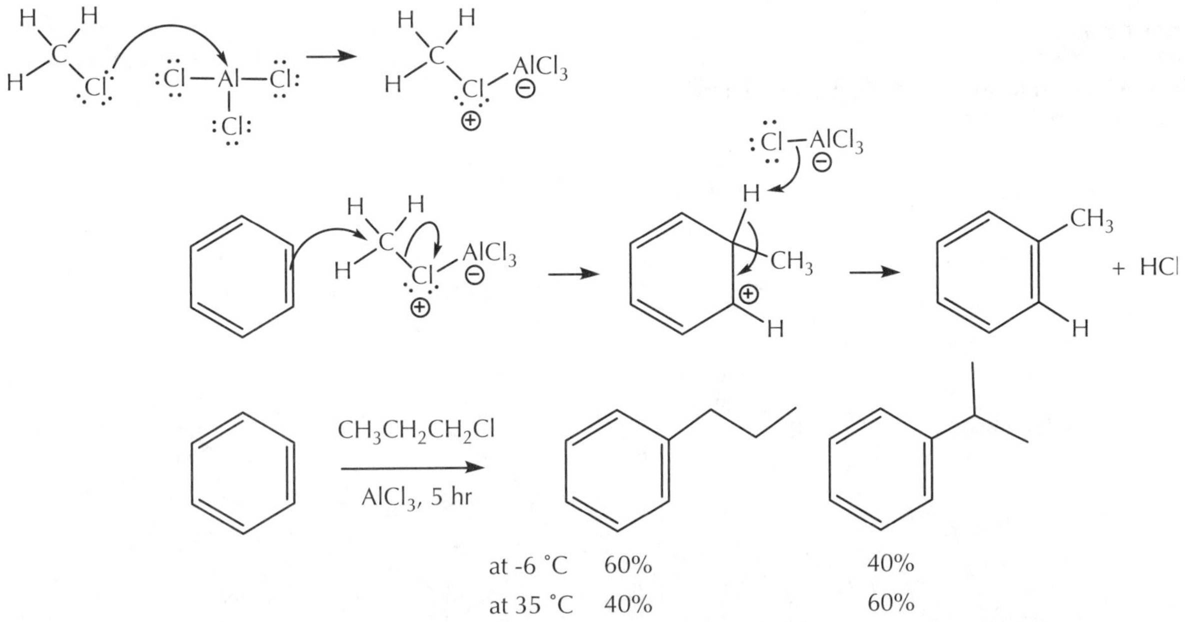

In direct analogy with Friedel-Crafts alkylation reactions, the carbocation derived from an acyl group can be formed by ionization of a carbon-chlorine bond in the presence of aluminum trichloride. The resulting carbocation, called an acylium ion, is resonance-stabilized, forms relatively easily, and does not rearrange.

The mechanism of the reaction, as in all of examples shown, is the same two steps (once the electrophile forms): electrophilic addition to the aromatic ring to give an intermediate, resonance-stabilized carbocation, and then an elimination reaction that restores the aromaticity of the ring.

Figure 1040

Electrophilic aromatic substitution: Friedel-Crafts acylation.

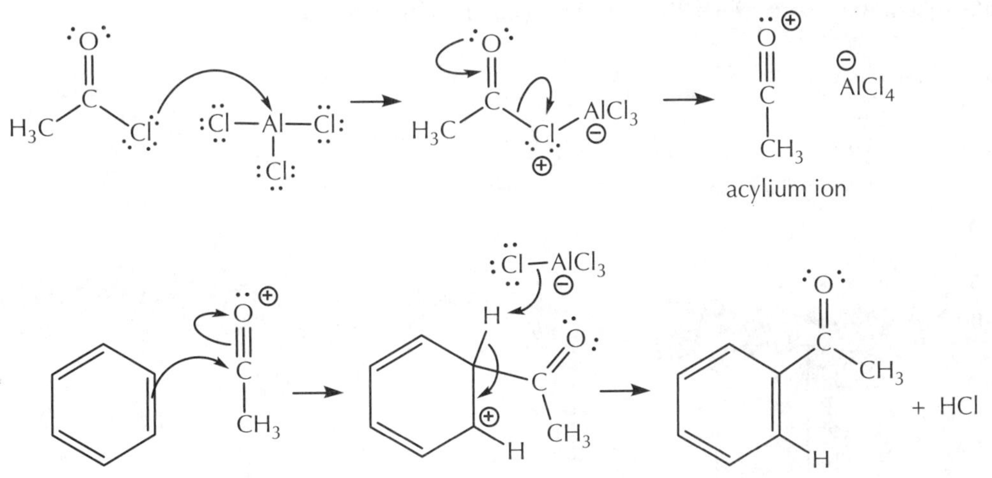

822 CHAPTER 10 Aromaticity and Electrophilic Aromatic Substitution Reactions

C. Regioselectivity of Electrophilic Aromatic Substitution Reactions (Directing Effect)

When an EAS reaction is carried out on an unsymmetrical aromatic ring, the different structural isomers that could form are an example of regioisomers (Figure 1041).

Figure 1041

Electrophilic aromatic substitution: regioselectivity in heteroaromatic compounds.

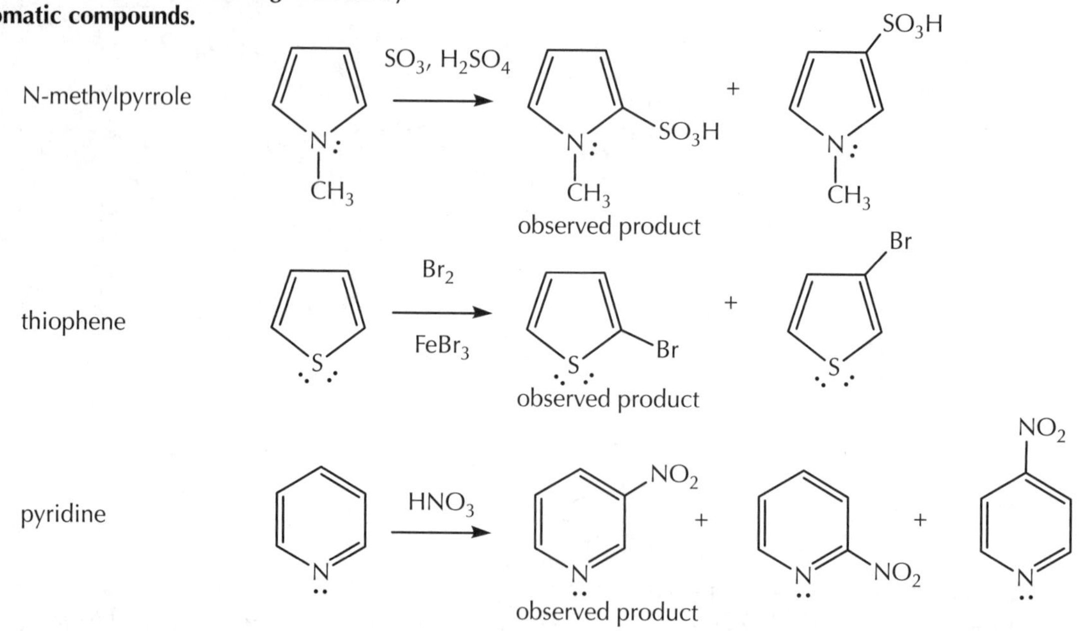

The sulfonation reaction of N-methylpyrrole and the bromination of thiophene, for example, could result in two different regioisomeric products, and the nitration of pyridine could result in three regioisomers. In all three cases, one major product is experimentally observed.

The mechanistic explanation for the regioselectivity includes a combination of relative rates and the relative stability of the proposed intermediates (Figure 1042).

Figure 1042

Electrophilic aromatic substitution: regioselectivity in the bromination of thiophene.

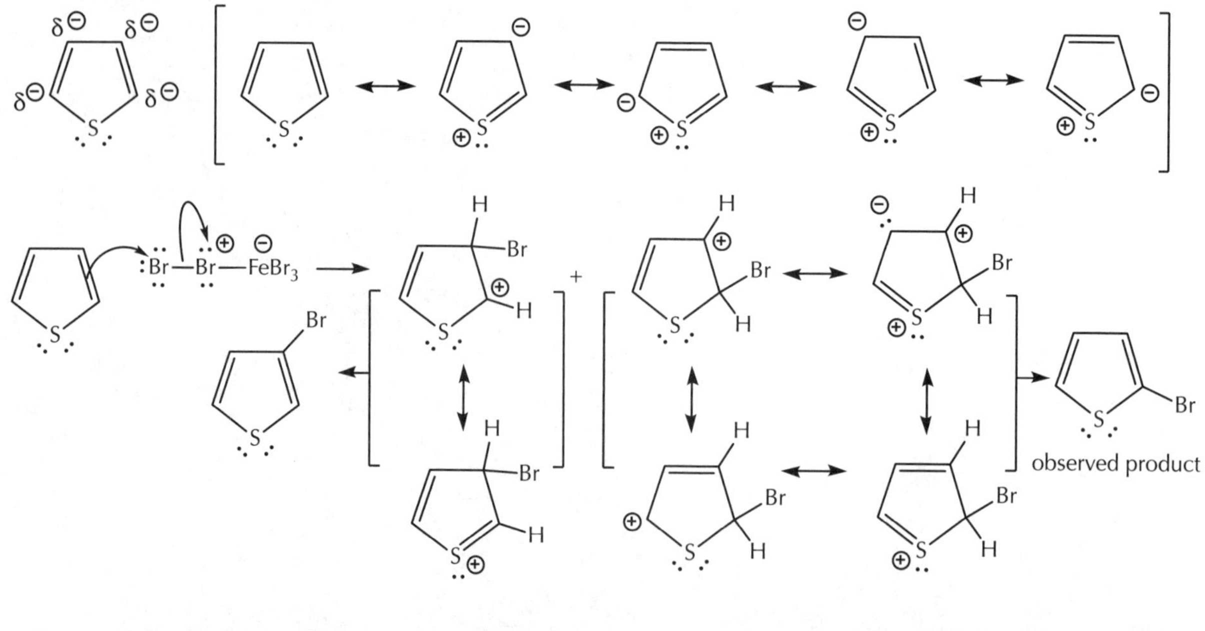

Usually, the experimental result is needed in these heteroaromatic compounds because there is no clear-cut prediction.

In this thiophene example, for instance, there are resonance contributors that place negative charge at both of the different carbon-atom positions on the ring, so predicting the more nucleophilic site is difficult.

Comparing the two carbocation intermediates is more revealing: There are more resonance contributors for the carbocation intermediate that gives the observed product, 2-bromothiophene, than for the unobserved regioisomer, 3-bromothiophene. The intermediate carbocation with more resonance contributors is more stable, and (as in S_N1 reactions) a more stable carbocation forms faster and yields the major product.

The EAS reactions of substituted benzene rings also create regioisomers. As seen in Figure 1043, the nitration reactions of monosubstituted benzenes result in mixtures of the three regioisomeric *ortho-*, *meta-*, and *para-*disubstituted products, usually with one of them occurring as the major isomer.

Figure 1043

Electrophilic aromatic substitution: regioisomeric outcomes from nitration reactions of monosubstituted benzenes.

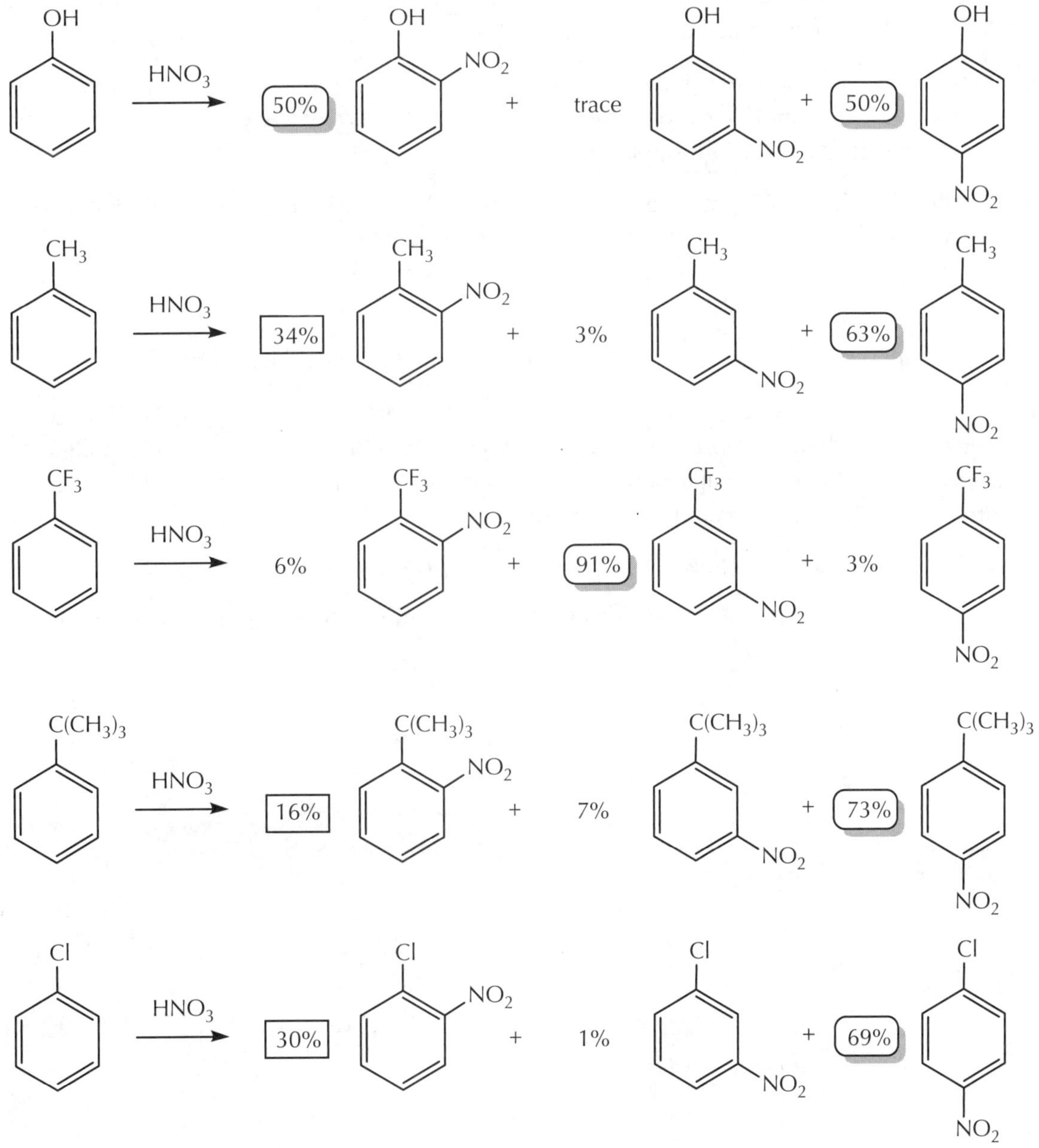

As in the thiophene example in the previous figure, the strategy for understanding the observed regioselectivities depends on using experimental observations to create a model for identifying the relative reaction rates at the three different reaction sites.

One of the early observations can be seen in the results in Figure 1043: The placement of the electrophile is not dictated by the electrophile, as it is the same in all five cases. Instead, the regioselectivity is correlated with the substituent that is already on the ring. The substituent that is already on the ring dictates, or "directs," the incoming electrophile in EAS reactions, so chemists called this the "directing effect" of the existing substituent group. Another pattern that you can see in that small collection of results is that the major product was either a mixture of the ortho and para products, with little to none of the meta product, or just the opposite, where the meta product was major and the ortho/para products were minor.

Over time, with enough experimental results, the ortho/meta/para pattern emerged for how the directing effect correlated with the structure of the existing substituent group. For monosubstituted benzenes, the two effects illustrated in Figure 1043 were the ones that were universally observed. The existing substituent on the ring was either an "ortho/para director" of the incoming electrophile, or it was a "meta director." Again, this relationship means that the group that was already on the ring is dictating the relative position of where the new electrophile ended up in an EAS reaction.

The result from those observations divides the effect of a substituent group on a ring into one of two categories: ortho/para director or meta director. A summary of these directing effects and a short list of substituents that behave this way follow, below:

ORTHO/PARA DIRECTORS: are simple alkyl groups and resonance-based electron-donating substituents, i.e., the atom attached to the ring has a nonbonding electron pair. These groups will direct an incoming electrophile to the positions *ortho* and *para* to itself, with a general preference for *para* > *ortho* due to steric effects.

Examples of ortho/para directors include: CH_3, Ph, Br, OH, OCH_3, $N(CH_3)_2$, SPh

META DIRECTORS: are alkyl groups bearing multiple electronegative atom groups and resonance-based electron-withdrawing substituents, i.e., the atom attached to the ring is part of a multiple bond, generally a carbon atom attached to an electronegative atom. These groups will direct an incoming electrophile to the position *meta* to itself.

Examples of meta directors include: CF_3, CN, and various C=O, N=O, S=O groups

The data from Figure 1043 are tabulated in Figure 1044, and the patterns are consistent with these generalizations.

Figure 1044

Electrophilic aromatic substitution: benzene substituents direct either ortho/para or meta regioselectivity.

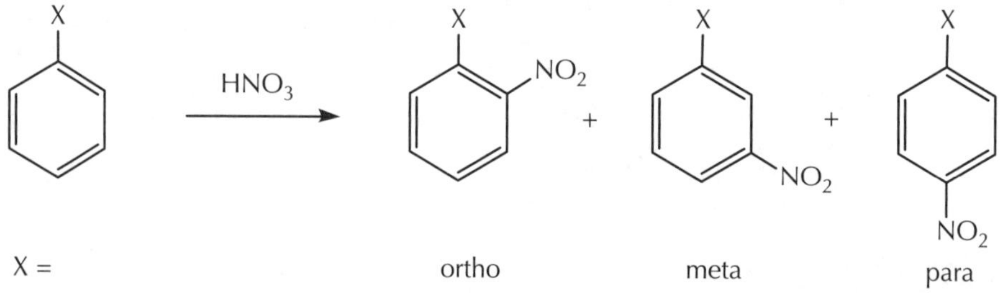

X =		ortho	meta	para
-OH	o/p director (p = o)	50%	trace	50%
$-CH_3$	o/p director (p > o)	34%	3%	63%
$-CF_3$	m director	6%	91%	3%
$-C(CH_3)_3$	o/p director (p > o)	16%	7%	73%
-Cl	o/p director (p > o)	30%	1%	69%

10.2 Electrophilic Aromatic Substitution

The relative significance of resonance contributors explains the directing effects, as illustrated with methoxybenzene in Figure 1045.

The resonance contributors of the starting material (the substituted benzene) can be used, indicating which of the ring positions bear more electron density (i.e., which would carry partial negative charge), or the resonance contributors of the intermediate carbocation in the EAS reaction can be used, indicating which carbocation is more stable and so forms faster. These sets of resonance contributors both tell the

Figure 1045

Electrophilic aromatic substitution: resonance contributors for the starting material or the intermediate carbocation are used to explain regioselectivity.

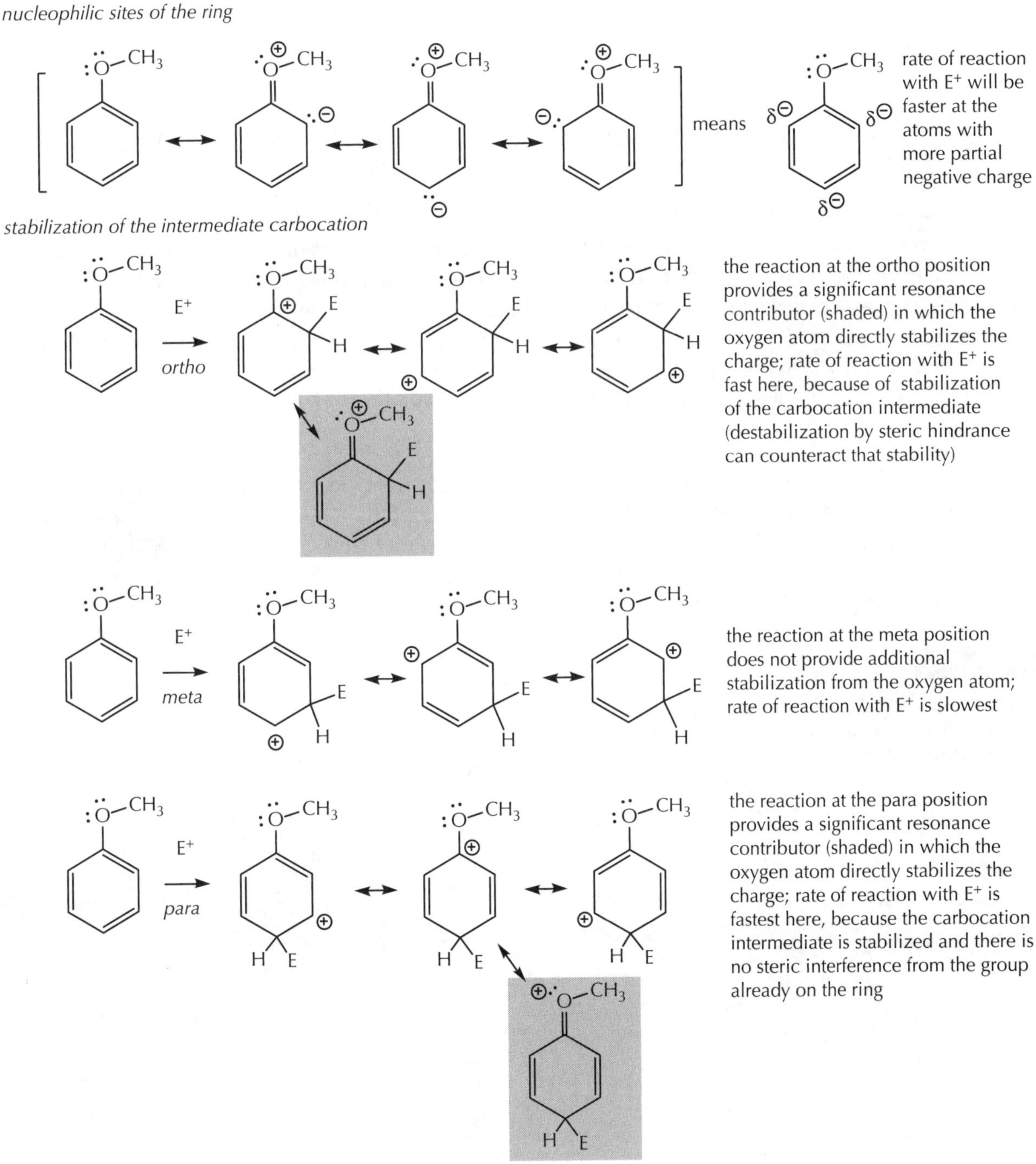

nucleophilic sites of the ring

rate of reaction with E$^+$ will be faster at the atoms with more partial negative charge

stabilization of the intermediate carbocation

ortho: the reaction at the ortho position provides a significant resonance contributor (shaded) in which the oxygen atom directly stabilizes the charge; rate of reaction with E$^+$ is fast here, because of stabilization of the carbocation intermediate (destabilization by steric hindrance can counteract that stability)

meta: the reaction at the meta position does not provide additional stabilization from the oxygen atom; rate of reaction with E$^+$ is slowest

para: the reaction at the para position provides a significant resonance contributor (shaded) in which the oxygen atom directly stabilizes the charge; rate of reaction with E$^+$ is fastest here, because the carbocation intermediate is stabilized and there is no steric interference from the group already on the ring

same story, so one or the other is sufficient for the explanation. In this example, the nitration of methoxybenzene (PhOCH$_3$), we will show both.

The aromatic ring is the nucleophile in EAS reactions, and any of the sites on the ring are prospective nucleophilic sites. An entire ring is not a type of nucleophile that you have seen previously, and so a strategy is needed that looks at the ortho/meta/para contributions of substituent groups on the nucleophilicity of the different sites on the ring. Resonance contributors are used to decide how the sites differ in their partial negative charges as a way to predict nucleophilic reactivity.

Methoxybenzene itself has four significant resonance contributors. Using the rules for predicting delocalization, a nonbonding electron pair on the oxygen can be delocalized onto the two positions that are ortho to the methoxy group, as well as onto the para position. The methoxy group is classified as an ortho/para director, then, because the rate of reaction for an incoming electrophile will be faster at these three positions, relative to the position meta to the methoxy group, because the methoxy group makes the para and two ortho positions more electron rich.

In the starting material, thanks to the delocalization, we imagine that the ortho and para ring positions, with respect to an ortho/para director, bear partial negative charges.

Alternatively, a completely consistent picture emerges when you look at the resonance contributors for the intermediate carbocation in the EAS reaction. Figure 1045 shows the full set of significant resonance contributors for the carbocation that derives from adding the electrophile to the ortho, meta, and para positions with respect to the methoxy group. When the electrophile ends up at the ortho or para positions, a resonance contributor can be drawn that places the positive charge on the same ring carbon as the oxygen atom of the methoxy group. When the substituent group can stabilize positive charge, such as this one, then the carbocation is more stable and, so, consistent with the Markovnikov Rule, that carbocation forms faster.

In summary, the methoxy group on a benzene ring makes the positions ortho and para with respect to itself more nucleophilic because of the electron-donating properties of the oxygen atom. Those electron-rich ortho and para sites will have faster reaction rates in the EAS reaction with incoming electrophiles. Once the electrophile has added to the ring and an intermediate carbocation forms, the carbocations that derive from adding to the ortho and para positions are more stable than that from adding at the meta

Figure 1046

Electrophilic aromatic substitution: regioselectivity varies based on the exact comparison between the substituent and the electrophile.

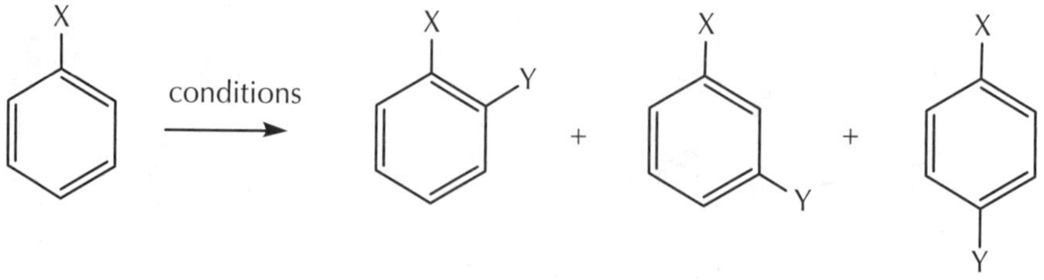

X =	conditions	Y =	ortho	meta	para
-OH	HNO$_3$	-NO$_2$	50%	trace	50%
-OH	Br$_2$/FeBr$_3$	-Br	12%	trace	88%
-OCH$_3$	HNO$_3$	-NO$_2$	35%	1%	64%
-OCH$_3$	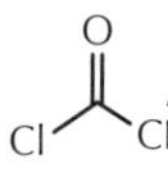 Cl-C(O)-CH$_3$ /AlCl$_3$	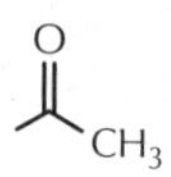 -C(O)-CH$_3$	8%	2%	90%

position, again due to resonance stabilization by the oxygen atom of the methoxy group. More stable carbocations form faster than less stable ones, so the EAS reaction is also faster at the ortho and para positions, relative to the methoxy group, when looking at the stability of the intermediates. These two factors result in the ortho/para directing effect of a methoxy group.

With ortho/para directors, the para product forms faster, and this is attributed to more favorable steric effect; that is, the reaction at the ortho position is more crowded (Figure 1046).

Figure 1047

Electrophilic aromatic substitution: simple alkyl groups are ortho/para directors.

nucleophilic sites of the ring

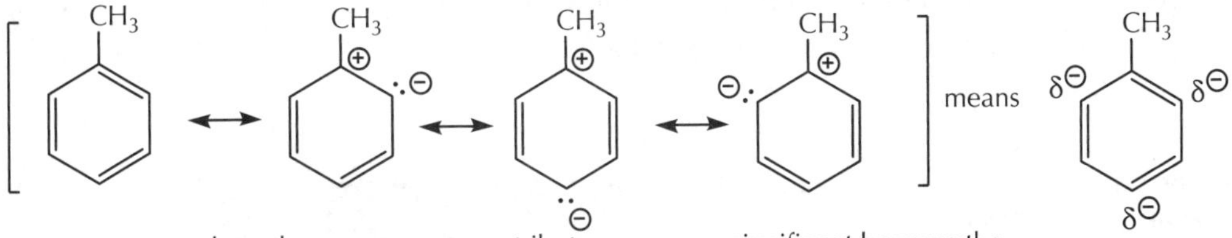

[resonance structures of toluene showing negative charge at ortho, para, ortho positions] means [toluene with δ⁻ at ortho and para positions] — rate of reaction with E⁺ will be faster at the atoms with more partial negative charge

these three resonance contributors are more significant because the open shell atom (carbocation) component is more highly substituted

stabilization of the intermediate carbocation

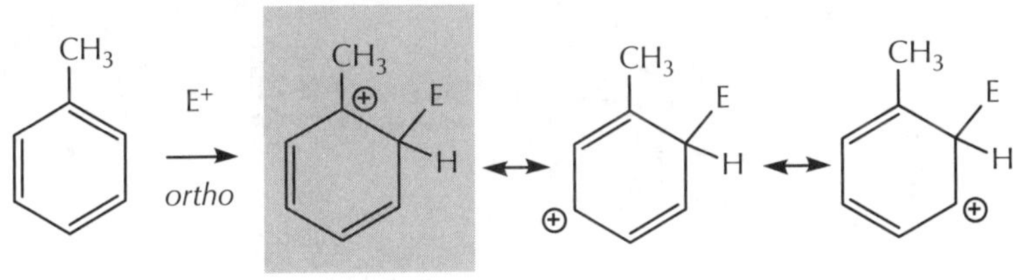

ortho: [toluene + E⁺ → three resonance structures of sigma complex, with the first one shaded] — the reaction at the ortho position provides a significant resonance contributor (shaded) in which the methyl group can directly stabilize the positive charge; rate of reaction with E⁺ is fast here, because of stabilization of the carbocation intermediate (destabilization by steric hindrance can counteract that stability)

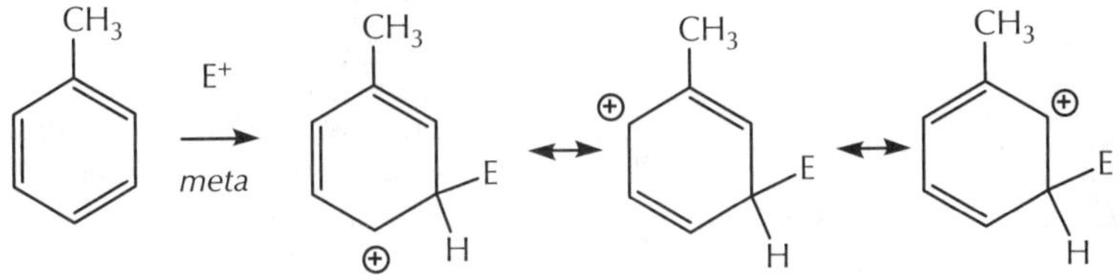

meta: [toluene + E⁺ → three resonance structures of sigma complex] — the reaction at the meta position does not provide additional stabilization from the oxygen atom; rate of reaction with E⁺ is slowest

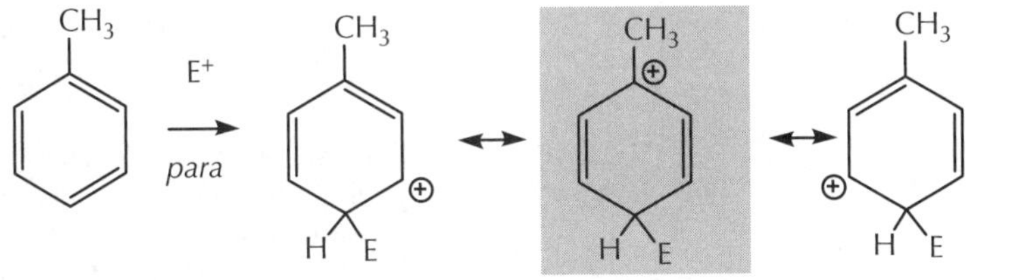

para: [toluene + E⁺ → three resonance structures of sigma complex, with the middle one shaded] — the reaction at the para position provides a significant resonance contributor (shaded) in which the methyl group directly stabilizes the charge; rate of reaction with E⁺ is fastest here, because the carbocation intermediate is stabilized and there is no steric interference from the group already on the ring

For small and electron-rich substituent groups, particularly the hydroxyl group, the ortho regioisomer is observed to be a significant product, in part due to a statistical advantage: There are two ortho positions and only one para position with respect to a substituent group.

Simple alkyl groups (methyl, ethyl, etc.) do not have nonbonding electron pairs, but they are also observed to be ortho/para directors, which is consistent with thinking of them as net electron-donating groups, as in assessing carbocation stability (Figure 1047).

Although additional closed shell atom resonance contributors are not possible for toluene (PhCH$_3$), the pi electrons of the ring are delocalizable. A resonance form in which a positive charge is on the same carbon as the methyl group represents a more significant contribution to the stability of the carbocation intermediate. In the starting material, this means that set of resonance contributors that were more significant for the methoxy group are the same ones that are significant for the methyl group, in which partial negative charge is anticipated at the positions ortho and para with respect to the substituent.

Comparable to the results from the methoxy group (Figure 1045), the intermediate carbocations derived from ortho and para electrophilic addition reactions, with respect to the methyl group, have resonance contributors with a more highly substituted open shell atom. The resulting carbocations are more stable and so form faster.

For the set of substituent groups observed to be meta directors, the meta position must end up being the more electron-rich site for attracting the reaction of an incoming electrophile. The nitration reaction of acetophenone (PhCOCH$_3$) is detailed as an example (Figure 1048).

Figure 1048

Electrophilic aromatic substitution: resonance contributors for the meta directors are electron-poor at the ortho/para positions.

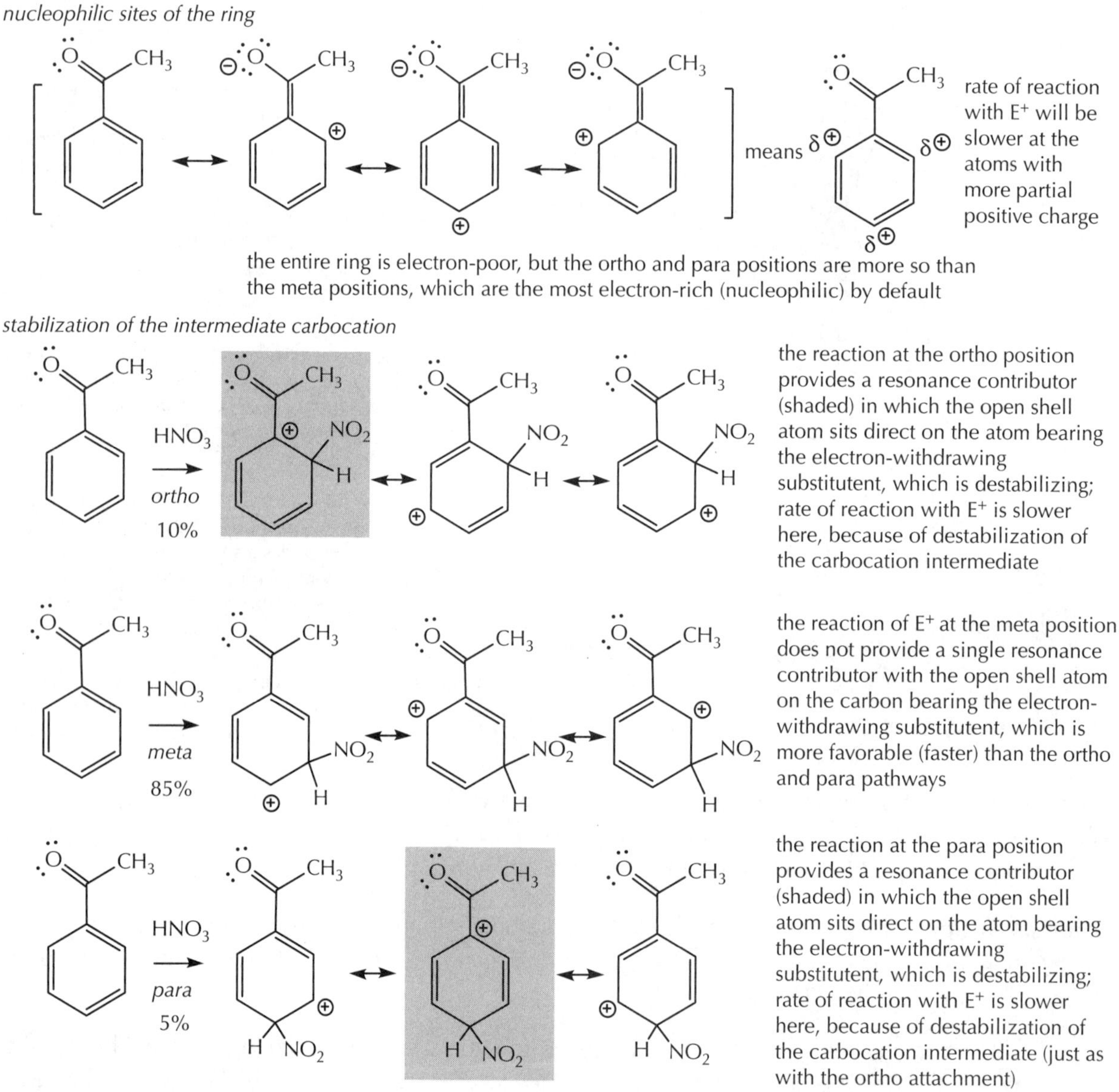

nucleophilic sites of the ring

the entire ring is electron-poor, but the ortho and para positions are more so than the meta positions, which are the most electron-rich (nucleophilic) by default

stabilization of the intermediate carbocation

ortho 10% — the reaction at the ortho position provides a resonance contributor (shaded) in which the open shell atom sits direct on the atom bearing the electron-withdrawing substitutent, which is destabilizing; rate of reaction with E$^+$ is slower here, because of destabilization of the carbocation intermediate

meta 85% — the reaction of E$^+$ at the meta position does not provide a single resonance contributor with the open shell atom on the carbon bearing the electron-withdrawing substituent, which is more favorable (faster) than the ortho and para pathways

para 5% — the reaction at the para position provides a resonance contributor (shaded) in which the open shell atom sits direct on the atom bearing the electron-withdrawing substitutent, which is destabilizing; rate of reaction with E$^+$ is slower here, because of destabilization of the carbocation intermediate (just as with the ortho attachment)

In the starting material, the delocalization of the ring electrons results in resonance contributors where the flow of electrons is not into the ring, as in the methoxy case, but where the electron flow is into the substituent group. Now, in the resonance contributors of starting material, the positions ortho and para with respect to the substituent group bear positive charges and the ring of acetophenone is represented with partial positive charge at those positions. An incoming electrophile, which is also positively charged, will react more slowly at the ortho and para positions thanks to charge/charge repulsion (or simply because those positions are less electron-rich). The meta position, with respect the meta-directing substituents, is the most electron-rich site for this set of groups, by default, because the ortho and para positions are demonstrably electron-poor.

The story from looking at the intermediate carbocations is the same: If the electrophile attacks ortho or para, there is a resonance contributor in which the positive charge is on the same carbon as the substituent, and a positive charge is destabilized, not stabilized, by the group. The resonance contributors derived from the addition to the meta site distribute the positive charge to either side of the substituent group and so represents the faster and more stable pathway.

Finally, if we compare the trifluromethyl group, observed to be a meta director, with the simple methyl group, the strong electron-withdrawing effect from the three fluorine atoms must be enough to destabilize an adjacent positive charge (Figure 1049).

Figure 1049

Electrophilic aromatic substitution: directing effects of trifluoromethyl (meta) versus methyl (ortho/para) substituent groups.

nucleophilic sites of the ring (methyl substituent)

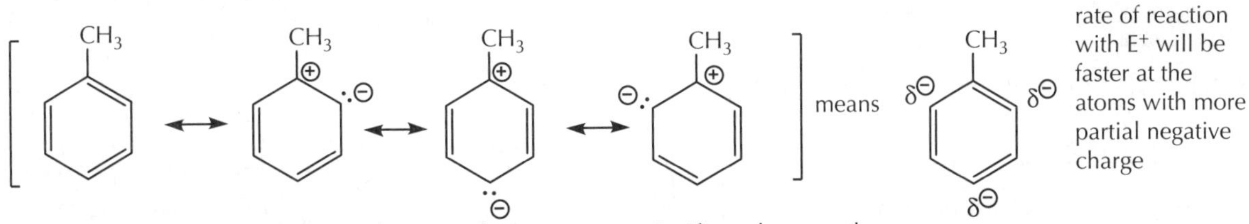

these resonance contributors are more significant because the open shell atom (carbocation) component is more highly substituted (ortho/para director)

means rate of reaction with E⁺ will be faster at the atoms with more partial negative charge

nucleophilic sites of the ring (trifluoromethyl substituent)

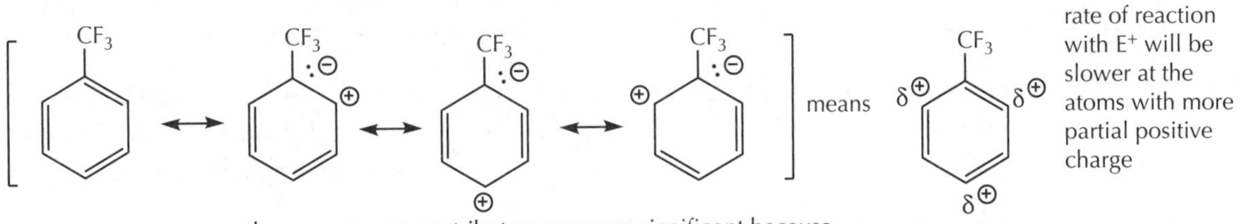

these resonance contributors are more significant because the anionic contribution is stabilized by the trifluoromethyl group (meta director)

means rate of reaction with E⁺ will be slower at the atoms with more partial positive charge

experimental results

toluene + HNO₃ → 34% ortho-nitrotoluene + 3% meta-nitrotoluene + 63% para-nitrotoluene

trifluoromethylbenzene + HNO₃ → 6% ortho + 91% meta + 3% para

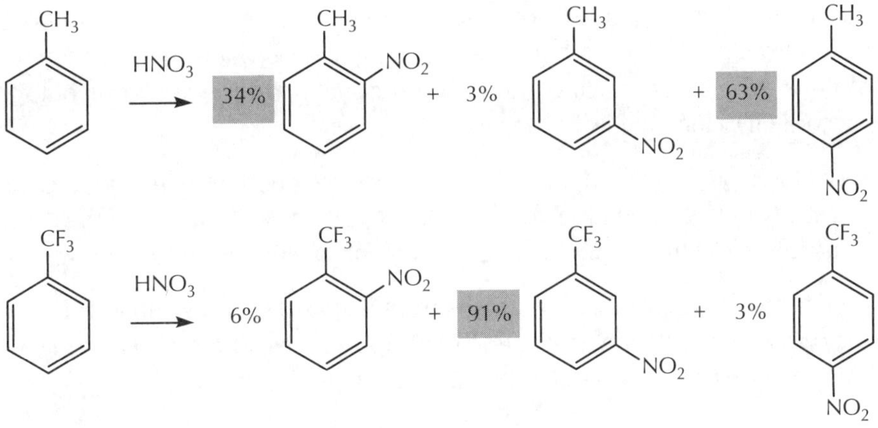

Looking at the experimental results, the trifluoromethyl group favors resonance contributors with the positive charge away from the ring carbon to which it is attached, which is a description of what is means to be a meta director.

D. Relative Rates of Electrophilic Aromatic Substitution Reactions (Activation)

In EAS reactions, the aromatic ring is the nucleophile with respect to the incoming electrophile.

A more electron-rich ring is a better nucleophile and predictably going to react faster with an electrophile than an electron-poor ring. Historically, dating back to the days before electrons had been discovered, aromatic rings that reacted faster were referred to as being more "active."

The rates at which different monosubstituted benzenes undergo EAS depend on the identity of the substituent groups. Unsubstituted benzene was used as the point of reference. If a monosubstituted benzene reacted faster in an EAS than benzene itself, then the ring was said to have been "activated" by the substituent group, and it was referred to as a ring-activating substituent, or more simply as a ring activator. And if monosubstituted benzene reacted slower than benzene in an EAS, then the ring was said to have been "deactivated" by the substituent group, and it was referred to as a ring-deactivating substituent, or more simply as a ring deactivator.

In today's language, think of "activation" as a synonym for nucleophilicity in EAS reactions. A substituent whose net effect is to donate electrons to a ring makes the ring more nucleophilic, and its reactions will be faster than a ring whose substituent has a net electron-withdrawing effect (Figure 1050).

Figure 1050

Electrophilic aromatic substitution: activating and deactivating groups.

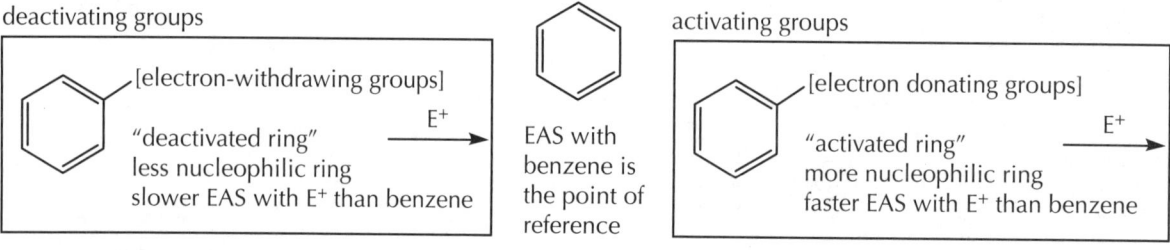

At the outset, it is important to say that predicting whether a substituent will be an electron donor, activating a ring to increase its nucleophilicity with respect to benzene, or an electron-withdrawing group, deactivating a ring to lower its nucleophilicity with respect to benzene, is an empirical question.

Look at methoxybenzene (PhOCH$_3$) again (Figure 1051).

An oxygen atom has both electron-withdrawing (electronegativity) and electron-donating (delocalization) properties. As do nitrogen atoms, halogen atoms, sulfur atoms, and any atom with nonbonding electron pairs that is more electronegative than carbon. The activating or deactivating effect of a group is whatever the net effect of the electron-withdrawing and electron-donating properties turns out to be, and this balance differs according to the identity of the atom.

If the reaction of methoxybenzene with an electrophile was slower than benzene, then we would say that the electronegativity of the oxygen was responsible for pulling electrons out of the ring, outweighing the resonance donation of electrons from the nbe pair that push electrons into the ring.

But if the reaction of methoxybenzene with an electrophile was faster than that of benzene, then we would say just the opposite: that the delocalization effect (pushing electrons into the ring) was overwhelming the intrinsic electronegativity of the oxygen atom (its inductive effect).

10.2 Electrophilic Aromatic Substitution

Figure 1051

Electrophilic aromatic substitution: is methoxy activating or deactivating?

Activating or Deactivating? - A balance between inductive effect and delocalization (resonance)

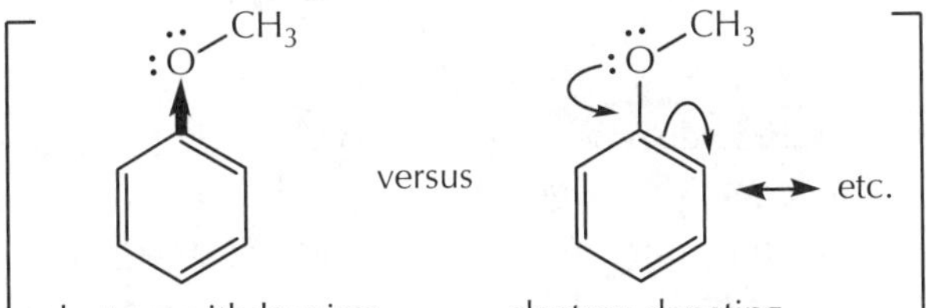

electron-withdrawing due to the electronegativity (inductive effect) of the oxygen atom

versus

electron-donating due to the delocalization (resonance effect) of the oxygen atom's nbe

The balance of electron-withdrawing and electron-donating effects of the OCH_3 group is only revealed through measuring the rate of EAS for $PhOCH_3$ with electrophiles (E^+) relative to the rate of EAS for benzene with the same E^+.

If $PhOCH_3$ is faster, the $-OCH_3$ is a "ring activator."
If $PhOCH_3$ is slower, the $-OCH_3$ is a "ring deactivator."

Ortho/Para or Meta Directing? - Primarily a delocalization (resonance) effect

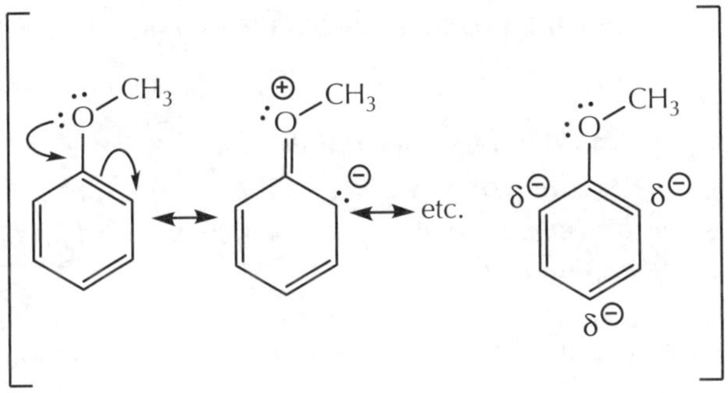

The directing effect is attributed to the delocalization (resonance) effect of the substituent group on the electron distribution of the ring.

The $-OCH_3$ is a resonance-donor, resulting in partial negative charge at the ortho and para positions with respect to itself, thus the reaction rate of electrophiles at those positions is faster than at the meta position.

Whether a substituent group makes an aromatic ring is more or less nucleophilic than benzene in an EAS (activated versus deactivated) has no bearing on the directing effect of that group. A methoxy group is an ortho/para director, regardless of this balance between delocalization and induction, because the directing effect is primarily a resonance effect.

The relative rate experiments to define whether substituent groups increased (activated) or decreased (deactivated) the nucleophilicity of a ring were performed long ago. The experimental order of reactivity of monosubstituted benzenes, based on the substituent group, is shown in Figure 1052.

Figure 1052

Electrophilic aromatic substitution: empirically-determined order of activation of monosubstituted benzenes.

Rate of electrophilic aromatic substitution reactions relative to unsubstituted benzene

slower EAS than benzene ⟶ *faster EAS than benzene*

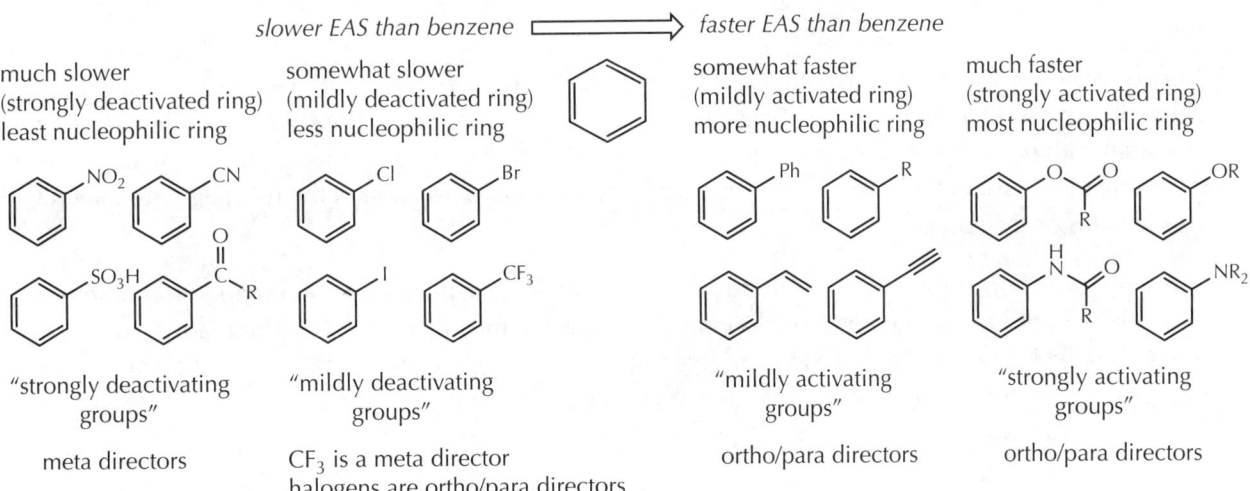

much slower (strongly deactivated ring) least nucleophilic ring

somewhat slower (mildly deactivated ring) less nucleophilic ring

somewhat faster (mildly activated ring) more nucleophilic ring

much faster (strongly activated ring) most nucleophilic ring

"strongly deactivating groups"
meta directors

"mildly deactivating groups"
CF_3 is a meta director
halogens are ortho/para directors

"mildly activating groups"
ortho/para directors

"strongly activating groups"
ortho/para directors

Oxygen and nitrogen atom groups with nonbonding electron pairs are strong activators. The net effect, being a strong electron donor, means that the delocalization (resonance) effect outwieghs the intrinsic inductive effect. Simple alkyl groups are classified as mild activators. Toluene (PhCH$_3$) is faster in EAS reactions than benzene, but not as fast as methoxybenzene (PhOCH$_3$). The simple alkyl group does not add electrons to a ring by delocalization.

Benzene is the point of reference. Halogen-atom substituted benzene rings are slower in EAS reactions than benzene, and halogen atoms are classified as mildly deactivating groups. Halogen-atom substituent groups, which are ortho/para directors because of the delocalization (resonance) effect, must end up with their intrinsic electronegativity outweighing the resonance effect. Alkyl groups filled with halogen atoms, such as the trifluoromethyl group, are also mild deactivators. The pi-bonded electronegative atom groups that withdraw electrons by delocalization are pulling electrons out of the ring by both resonance and induction, and so rings with these groups (C=O, N=O, S=O, CN) are the slowest, and the groups are called strong deactivators.

Activation relies on both resonance and inductive effects, while directing relies on resonance. As stated above:

> Whether a substituent group makes an aromatic ring more or less nucleophilic than benzene in an EAS (activated versus deactivated) has no bearing on the directing effect of that group. A methoxy group is an ortho/para director, regardless of this balance between delocalization and induction, because the directing effect is primarily a resonance effect.

Using the differences in activating effects becomes relevant when there are competitions to resolve (Figure 1053).

Figure 1053

Electrophilic aromatic substitution: intermolecular competition (monosubstituted rings).

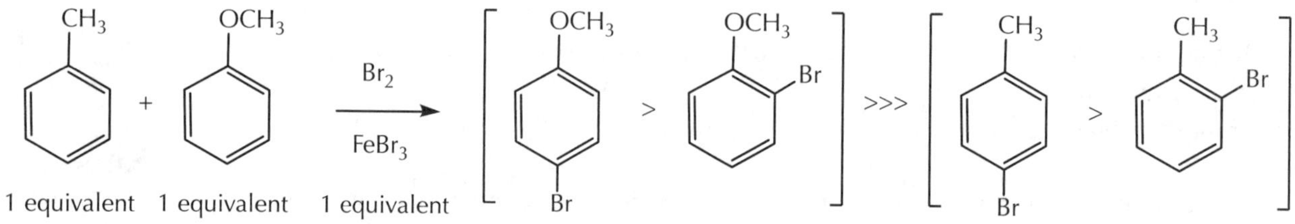

plus unreacted starting materials (PhCH$_3$ >>> PhOCH$_3$)

The simplest competition reactions to understand would be those that ask for the relative rate ranking of different monosubstituted benzenes. For instance, in a 1:1:1 mixture of the reagent (bromine plus ferric bromide) with methoxybenzene and toluene, the EAS of methoxybenzene is faster than toluene, and so the ratio of EAS products observed from methoxybenzene should overwhelm the reaction with toluene. This result is observed.

The same principle applies to disubstituted rings, where the substituents are in competition based on the relative rates (activating effect, nucleophilicity) at which they promote EAS. The reaction of 1-methoxy-4-methylbenzene with bromine and ferric bromide is, in effect, the same competition as the one in Figure 1053, but now it is an intramolecular competition (Figure 1054).

Figure 1054
Electrophilic aromatic substitution: intramolecular competition (directing effects not reinforcing).

1-methoxy-4-methylbenzene

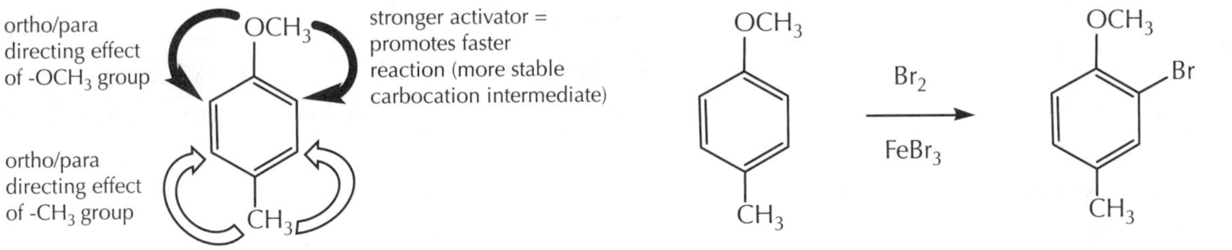

Both of these groups are ortho/para directors. In 1-methoxy-4-methylbenzene, the para position with respect to both of them is blocked, and so the competition is between which of the two possible ortho substitution products is predicted. In the methoxy versus methyl competition, the EAS reaction rates promoted by the methoxy group (a strong activator) are faster than methyl (a mild activator), and so the directing effect of the methoxy group wins out over that of the methyl group, and the 2-bromo product is predicted to be the major isomer over the 3-bromo regioisomer.

Sometimes, the directing effects of two different substituent groups are coincident, reinforcing one another, which makes predicting the outcome easier. The reaction of 1-(trifluoromethyl)-4-methoxy-benzene with bromine with ferric bromide is a nice contrast to the reaction with 1-methoxy-4-methylbenzene (Figure 1055).

Figure 1055
Electrophilic aromatic substitution: intramolecular competition (directing effects reinforcing).

1-(trifluoromethyl)-4-methoxybenzene

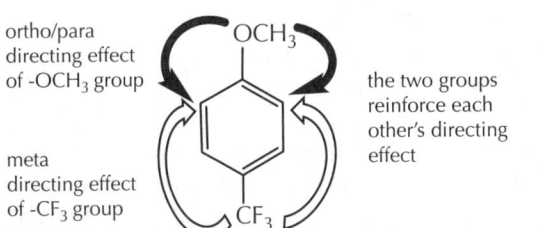

The methoxy group is still a strong activator and ortho/para director, and the trifluoromethyl group is a mild deactivator and a meta director. In this case, both of the substituent groups are directing the incoming electrophile to the same position on the ring, and the 2-bromo product is anticipated.

Another question for an intermolecular comparison might be: How do the relative rates of these two bromination reactions compare? Is the bromination of 1-(trifluoromethyl)-4-methoxy-benzene faster or slower than that of 1-methoxy-4-methylbenzene?

Both molecules carry the methoxy group, and so the difference is between the EAS reaction rate effect of a methyl group and a trifluoromethyl group (Figure 1056).

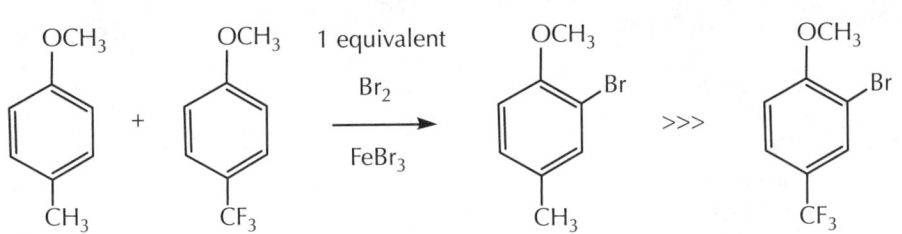

plus unreacted starting materials

Figure 1056
Electrophilic aromatic substitution: intermolecular competition (disubstituted rings).

834   CHAPTER 10   Aromaticity and Electrophilic Aromatic Substitution Reactions

The methyl group is a mild activator, so the combination of a strong activator and weak activator (1-methoxy-4-methylbenzene) is predicted to be a better nucleophile (faster EAS) than the combination of the same strong activator with a weak deactivator (1-(trifluoromethyl)-4-methoxy-benzene).

Armed with the combination of directing and activating effects, the outcome from more complex cases can be predicted and/or explained.

The nitration reaction of toluene gives 4-nitrotoluene. With excess nitric acid, a second nitration takes place. Three questions can be asked: (1) Where will the second nitro group go? (2) Are the directing effects of the two groups on the ring reinforcing each other or in competition? And (3), is the second nitration reaction faster or slower than the first one? (See Figure 1057.)

Figure 1057

Electrophilic aromatic substitution: nitration of toluene.

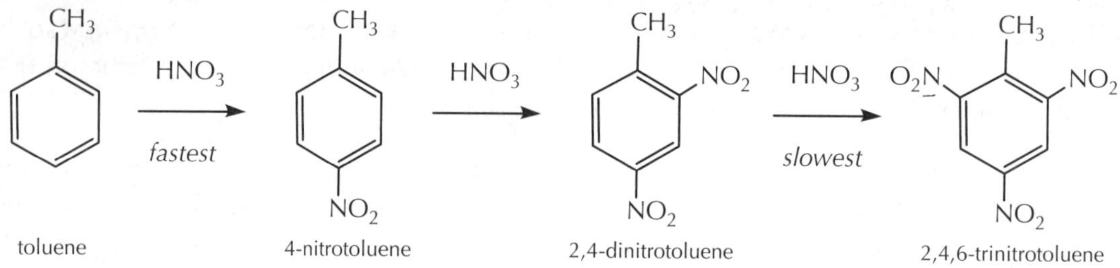

toluene → (HNO$_3$, fastest) → 4-nitrotoluene → (HNO$_3$) → 2,4-dinitrotoluene → (HNO$_3$, slowest) → 2,4,6-trinitrotoluene

In the first nitration, toluene is a benzene ring with a methyl group. The methyl group is a mild activator and an ortho/para director, and 4-nitrotoluene results as the major product. In 4-nitrotoluene, the methyl is still the same mild activator and ortho/para director, but the nitro group is a strong deactivator and it is a meta director. The second nitro group is directed to the same spot by both of the groups already on the ring, favoring the formation of the 2,4-dinitrotoluene. The 4-nitrotoluene is less nucleophilic (less activated) than toluene, so the second nitration should be slower.

What about the third nitration? The three groups in 2,4-dinitrotoluene are all coincident in their directing effects, and the 2,4,6-trinitrotoluene is expected. The third nitration should be the slowest of these nitration reactions, as the 2,4-dinitrotoluene contains two strong deactivators and the one mild activator (methyl) common to all of the rings in the sequence. The final product, 2,4,6-trinitrotoluene, is more familiarly known by its abbreviation, TNT.

Another type of competition is when two different rings are in the same molecule (Figure 1058).

Figure 1058

Electrophilic aromatic substitution: intramolecular competition (different rings).

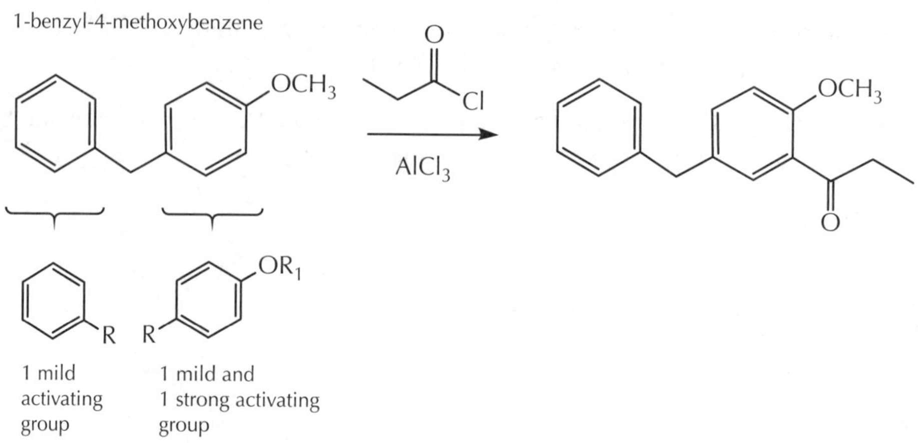

1-benzyl-4-methoxybenzene

1 mild activating group

1 mild and 1 strong activating group

10.2 Electrophilic Aromatic Substitution

In 1-benzyl-4-methoxybenzene, there are two rings. They can be analyzed separately and treated as you would the intermolecular competition (as in Figure 1053). Thus, 1-benzyl-4-methoxybenzene is made up of a disubstituted ring and a monosubstituted ring. The disubstituted ring has a strong activator and a mild activator, while the monosubstituted ring has only a mild activator that is comparable to the one in the disubstituted ring.

The disubstituted ring is more nucleophilic than the monosubstituted ring. Both the methoxy group and the benzyl group are ortho/para directors, but the reaction will favor the directing effect of the stronger activator, giving the product of Friedel-Crafts acylation ortho to the methoxy group.

Relative activation and directing effects also come to play in planning multistep transformations using EAS reactions.

Starting with benzene, designing ways to carry out the selective preparation of 1-cyclopentyl-3-nitrobenzene, 2-bromo-1-isopropyl-4-nitrobenzene, 2,4-diethyl-1-nitrobenzene, and 2-chloro-1,4-dimethoxybenzene creates a series of challenges (Figure 1059).

Figure 1059

Electrophilic aromatic substitution: multistep design challenges.

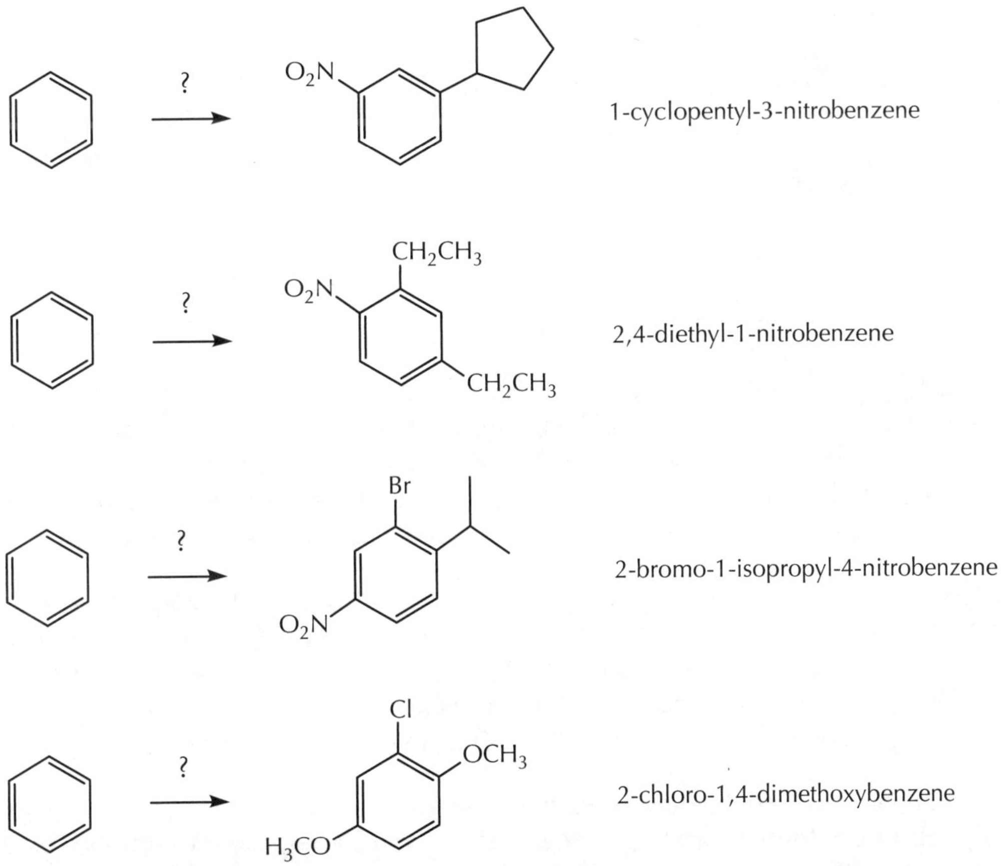

benzene —?→ 1-cyclopentyl-3-nitrobenzene

benzene —?→ 2,4-diethyl-1-nitrobenzene

benzene —?→ 2-bromo-1-isopropyl-4-nitrobenzene

benzene —?→ 2-chloro-1,4-dimethoxybenzene

Three principles underlie these four examples: (1) the order of the steps in multiple EAS can matter, (2) you are limited in EAS to the set of electrophiles that you know, and (3) there are plenty of molecules for which you might not be able to create reasonable designs (at least not yet).

The four examples are taken up separately in Figures 1060–1063 (which appear on the next pages).

Figure 1060

Preparation of 1-cyclopentyl-3-nitrobenzene.

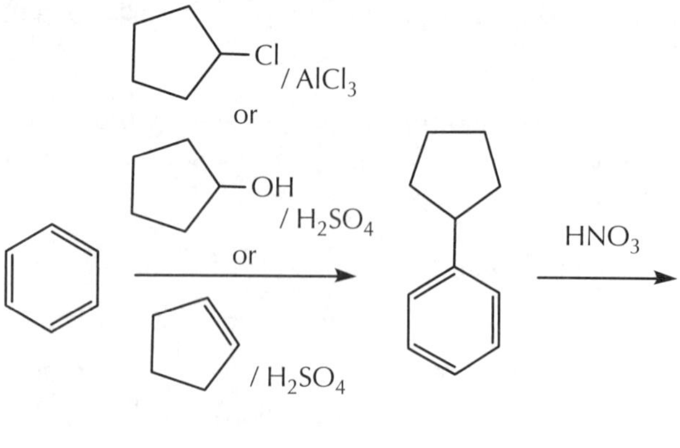

1-cyclopentyl-4-nitrobenzene

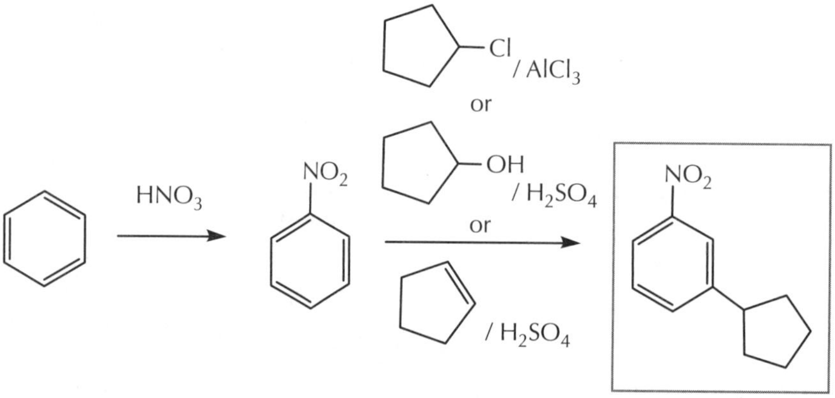

1-cyclopentyl-3-nitrobenzene

Figure 1060: 1-cyclopentyl-3-nitrobenzene. Both of these groups are in the set of your available electrophiles. The nitro group can be introduced onto the ring with either of the nitration reaction conditions (nitric acid or the combination of nitric acid and sulfuric acid). The cyclopentyl group is the result from doing a Friedel-Crafts alkylation reaction using any of three starting materials (chlorocyclopentane with aluminum trichloride, cyclopentanol with sulfuric acid, or cyclopentene with sulfuric acid).

The order matters: The group placed onto benzene first is the one whose directing effect will dictate the outcome from the second reaction. Looking at both options, benzene can undergo both the Friedel-Crafts reaction and the nitration reaction to give cyclopentylbenzene and nitrobenzene, respectively. If the nitration reaction is carried out on cyclopentylbenzene, the cyclopentyl group, an ortho/para director, directs the nitro group to the position corresponding to the undesired, regioisomeric product.

If the nitration was done first, then the nitro group in nitrobenzene is a meta director, and the Friedel-Crafts reaction would give the desired product. The meta relationship of the groups in the desired product means that you need to put the first group on the ring that will direct the second one to the proper position.

Worrying about the order of the reactions might seem like a limitation at first; it is not. Being able to control the regiochemical outcome from an EAS reaction provides a powerful example of how designing a new molecular structure according to your own interest can be carried out.

10.2 Electrophilic Aromatic Substitution

Figure 1061

Preparation of 2-bromo-1-isopropyl-4-nitrobenzene.

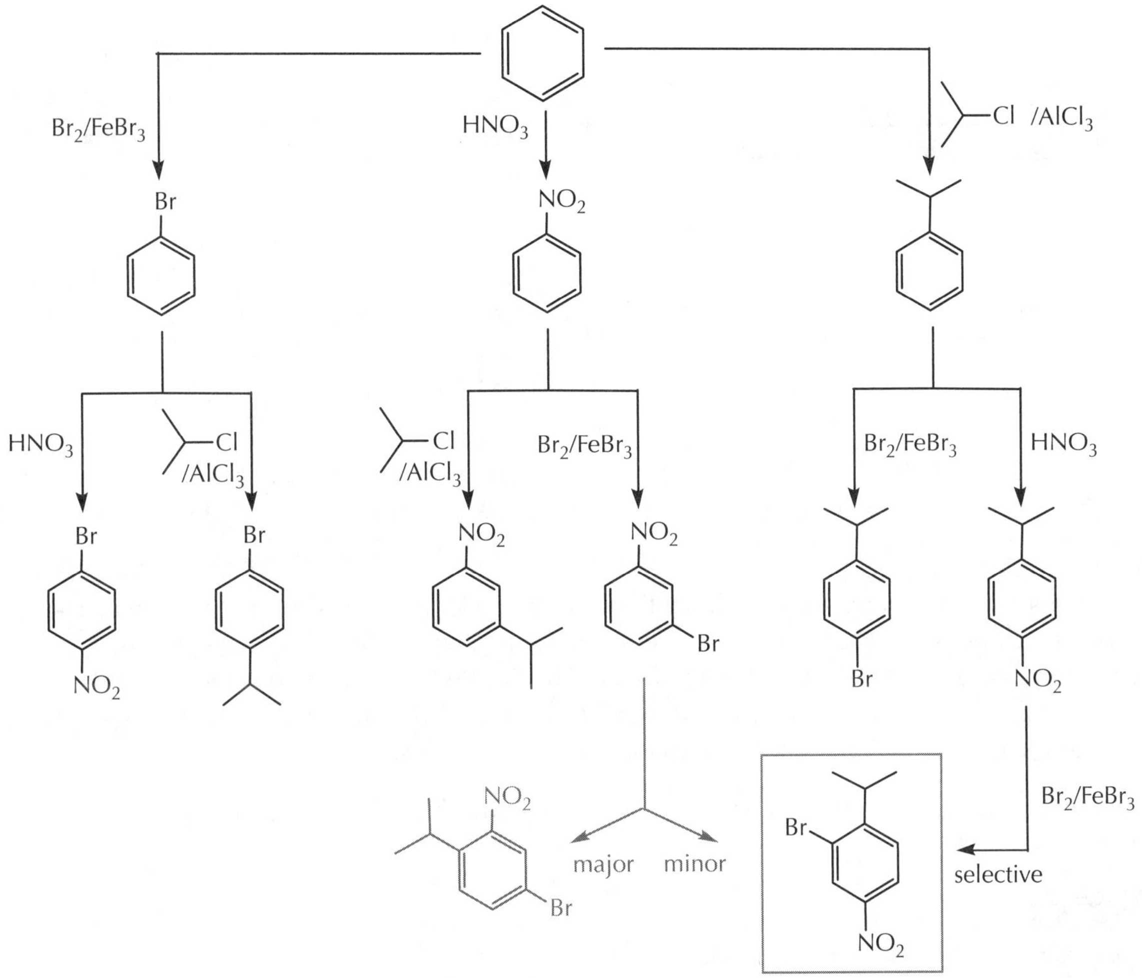

Figure 1061: 2-bromo-1-isopropyl-4-nitrobenzene. As the number of groups increases, so do the options. First, all three of the groups are in your set of electrophiles. One analytical approach is to play out the three starting options and then look at whether the expected second reaction places the group in a position that corresponds to the desired product.

Assuming you can count on sterics to favor the regioselection of para over ortho, attaching the bromo group first cannot lead to the product because there is nothing para to the bromine in the target molecule. Putting on the nitro group first can be used to establish the meta relationship with the bromine, but the outcome from the Friedel-Crafts reaction is not clear-cut. By putting on the isopropyl group first, however, the para relationship with the nitro group can be established, and these two groups are reinforcing their directing effects.

Thus, the most likely order is (1) the Friedel-Crafts alkylation of benzene to make isopropyl benzene, (2) nitration of isopropylbenzene to give the para-disubstituted 1-isopropyl-4-nitrobenzene, and then (3) bromination to give the product (where the directing effects of the isopropyl group and the nitro group are reinforcing each other).

838 CHAPTER 10 Aromaticity and Electrophilic Aromatic Substitution Reactions

Figure 1062

Preparation of 2,4-diethyl-1-nitrobenzene.

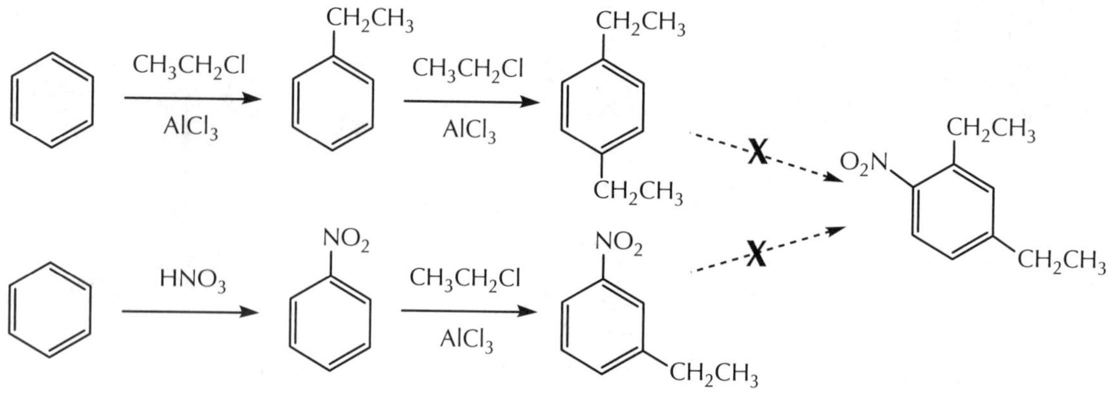

Figure 1062: 2,4-diethyl-1-nitrobenzene. Both of the groups are in your set of electrophiles. In this example, the position of the groups on the ring is problematic because the ethyl groups, which are ortho/para directors, are meta to each other.

It does not take long to run through the options. Starting with the ethylation to give ethylbenzene, the second ethylation would not go to the right location. The nitration of ethylbenzene puts the nitro group in the right place, but the second ethylation should be directed ortho to the other ethyl group and meta to the nitro group. If the nitration is carried out first, then the ethylation of nitro benzene does not put the ethyl group in either of the locations corresponding to the product.

The synthesis of this compound cannot be designed with much confidence based on the information in this chapter.

Figure 1063

Preparation of 2-chloro-1,4-dimethoxybenzene.

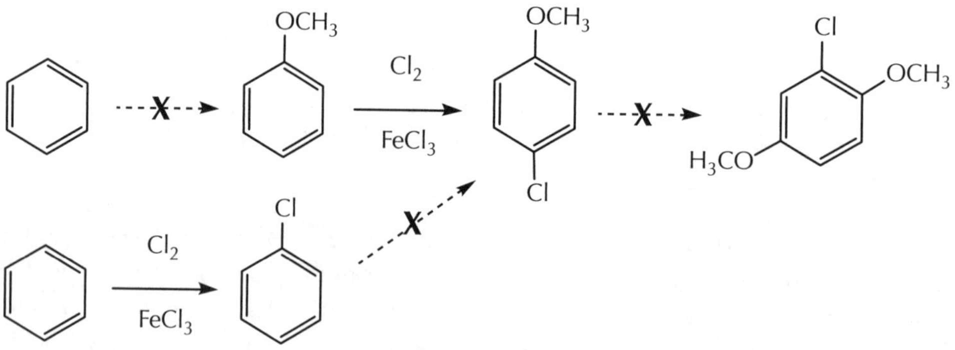

Figure 1063: 2-chloro-1,4-dimethoxybenzene. You have no source of electrophilic oxygen with which you can place a methoxy group on a ring. And even if you started with methoxybenzene (PhOCH3), the chlorination would give the para product, which does not have the regiochemistry of the target molecule. The synthesis of this compound cannot be designed with any confidence based on the information in this chapter, either.

PRACTICE QUESTIONS

10.06 Complete the following electrophilic aromatic substitution reactions, as required, and provide the complete, stepwise mechanisms (including the mechanism of formation for the electrophile).

(a)

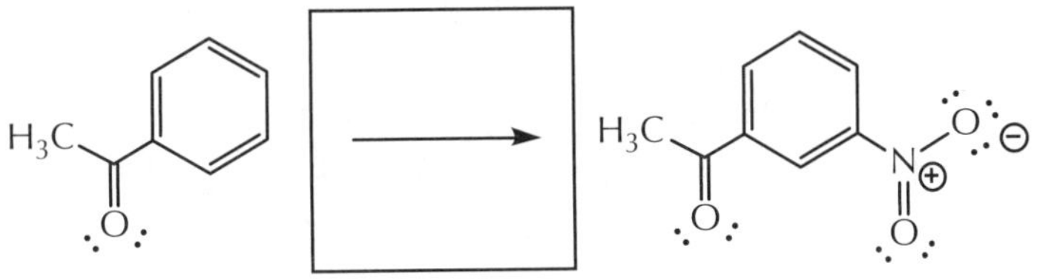

(b)

10.07 What is the major product of the following reactions?

(a)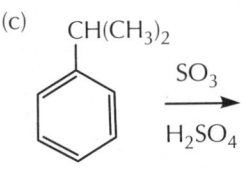

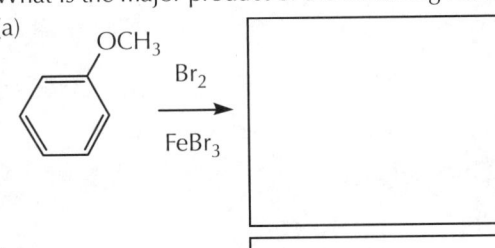

(b)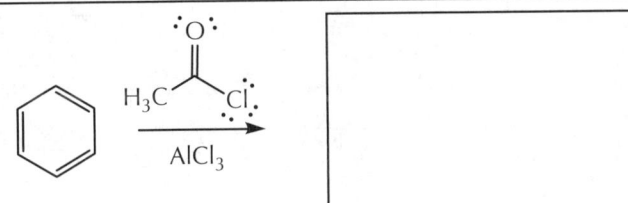

(c) (d)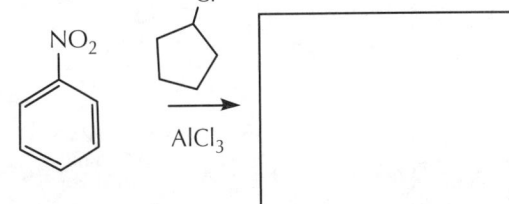

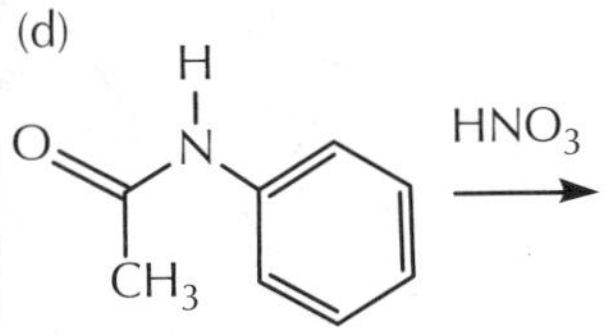

10.08 The nitrosyl (NO) group is observed to have the same effect as a halogen atom group in promoting electrophilic aromatic substitution reactions: It is a mild ring deactivator and it is an ortho/para director.

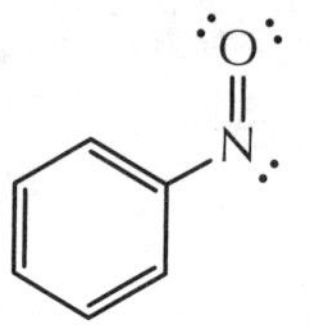

(a) Explain why NO is a mild ring deactivator.

(b) Use resonance contributors to explain the directing effect of NO, in terms of its effect on the structure of the starting material prior to the reaction with an electrophile.

(c) Use resonance contributors to explain the directing effect of NO, in terms of its effect on the structure of the intermediate, during the reaction with an electrophile (you may use any electrophile you would like).

(d) The origin of the nitrosyl group as an electrophile for EAS is nitrous acid (HNO$_2$; connectivity HO-N=O). Using the mechanism for nitration reactions as an analogy, propose the complete mechanism for formation of the NO$^+$ electrophile and its subsequent use in the EAS reaction of benzene.

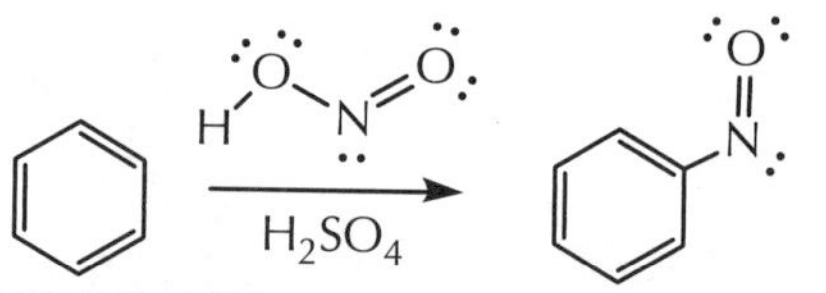

10.09 The 2-carbon substitutent with a double bond is called an ethenyl group. Historically, it was also called a vinyl group. The vinyl group is comparable to an alkyl group (e.g., ethyl) in EAS: It is a mild activator and an ortho/para director.

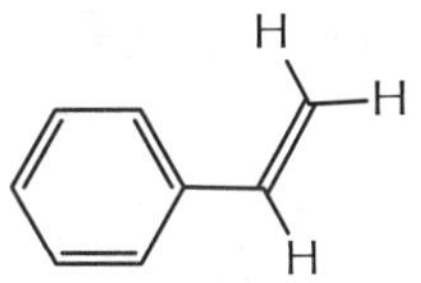

ethenylbenzene
vinyl benzene
1-phenylethene

(a) Use resonance contributors to explain the directing effect of the vinyl group, in terms of its effect on the structure of the starting material prior to the reaction with an electrophile.

(b) Use resonance contributors to explain the directing effect of the vinyl group, in terms of its effect on the structure of the intermediate, during the reaction with an electrophile (you may use any electrophile you would like).

10.10 Predict the major product anticipated to result from the following reaction sequences.

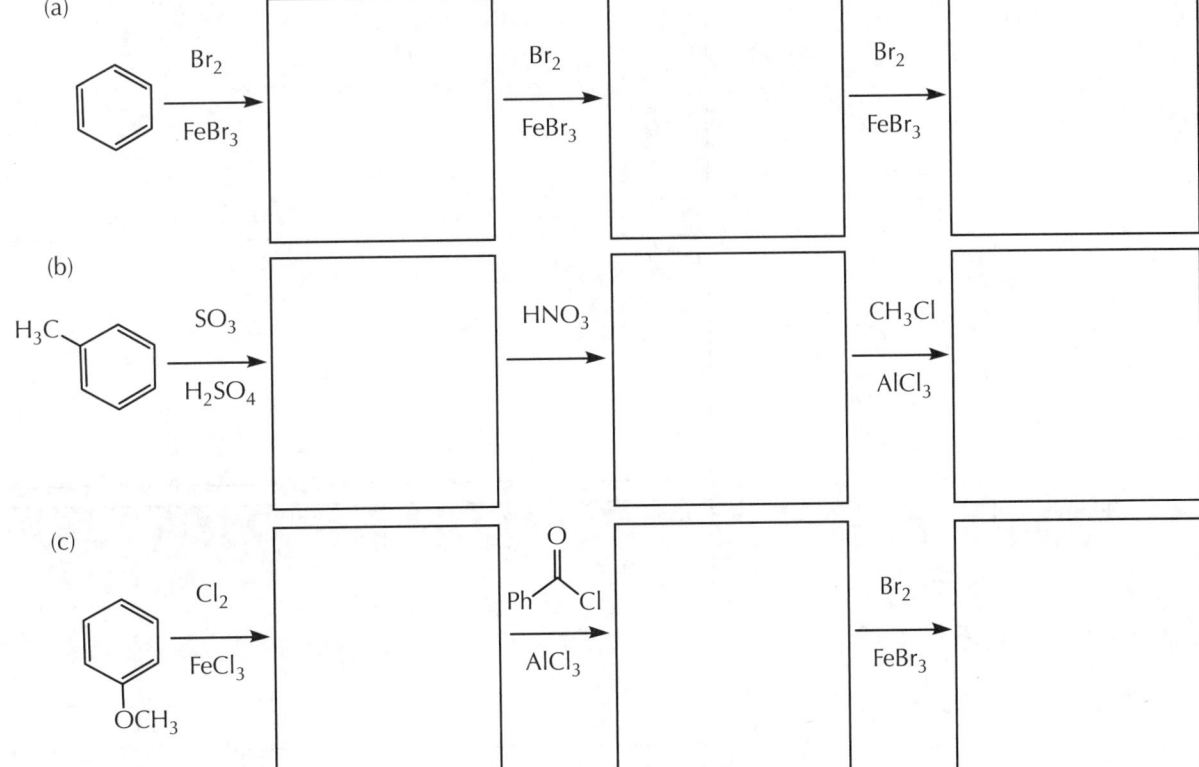

10.11 Design a way to transform benzene into each of the following. The desired product should be predicted as the major isomer.

(a) benzene → 3,5-dinitro(cyclohexyl)benzene

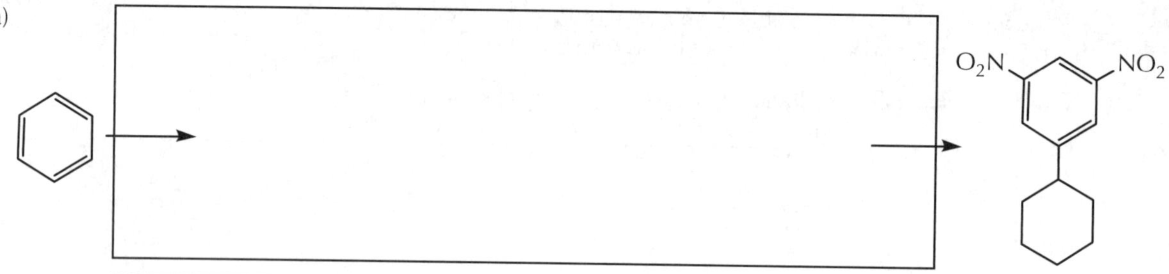

(b) benzene → 1-bromo-2-ethyl-4-acetylbenzene (Br and COCH₃ with CH₂CH₃)

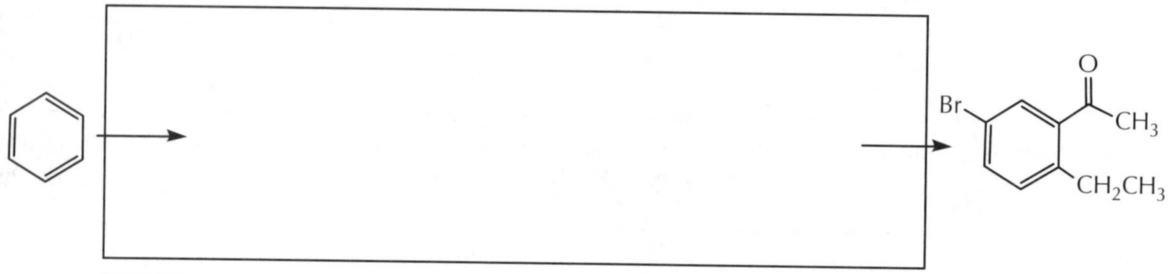

(c) benzene → 1-(1-methylcyclopentyl)-2-sulfo-4-nitrobenzene

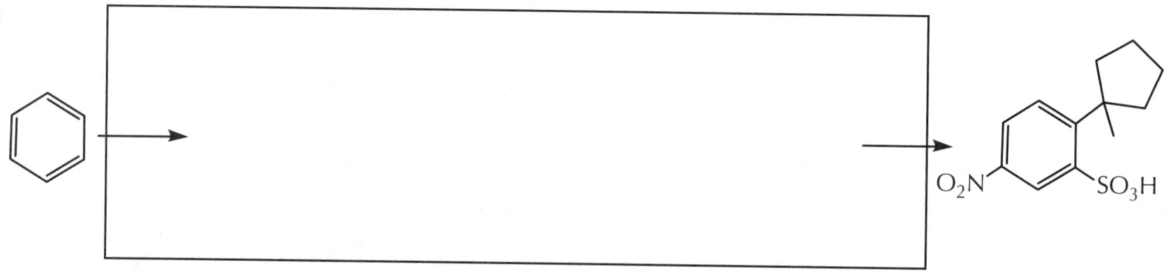

10.12 Predict the structure for the following mononitration reactions and provide your rationale.

(a) 4,5-dimethylxanthene + HNO₃ → product / rationale

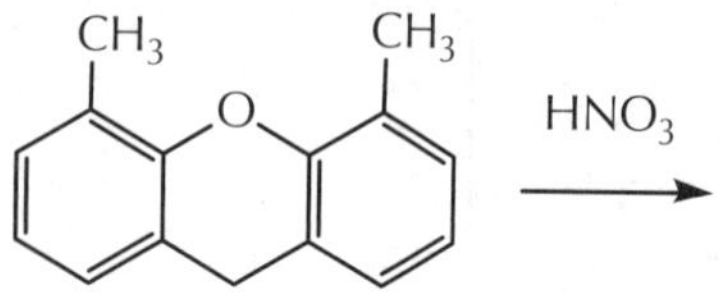

(b) phenyl phenylacetate + HNO₃ → product / rationale

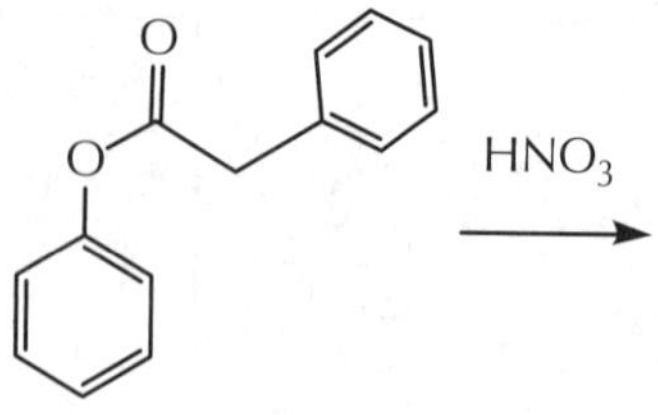

(c) 1,4-diacetyl-1,2,3,4-tetrahydroquinoxaline + HNO₃ → product / rationale

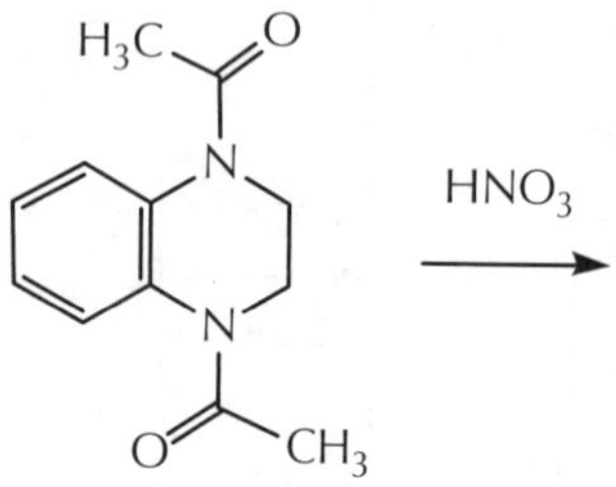

Reflections About Science: Occam's Razor (Don't Invoke More Than You Need)

Aromaticity, and the electrophilic aromatic substitution reactions, can be used to illustrate an important idea about how science operates. Whenever possible, do not create explanations that invoke more than you need to account for the observations, and stick with things that are known rather than inventing something new.

Viewed from a contemporary perspective, the EAS reaction mechanism starts with the observation of carrying out a substitution reaction at an sp^2 atom, and rather than creating a new type of fundamental reaction, every aspect of the explanation is built upon familiar ideas, starting with an electrophilic addition that gives the more stable carbocation intermediate at a faster rate than the less stable one. The individual steps are well established and consistent with the facts, and nothing new is needed except the stabilizing effect of an aromatic ring.

Philosophically, science attributes this core principle, of favoring the explanation that shaves away the greatest number of assumptions, to a 12th century scholar named William of Occam. The principle, known as Occam's Razor, was not named or popularized until the mid-1600s. A version of the saying attributed to Occam, *Entia non sunt multiplicanda praeter necessitatem* (Entities must not be multiplied beyond necessity), is interpreted as (a) invoke only what is needed to explain your evidence, (b) do not overclaim what your evidence implies, and (c) whenever possible, use as many well-established theories rather than unknown ones.

In popular culture, Occam's Razor is often misstated as "the simplest explanation is the best one," where the notion of "simple" means "not complex." The intended meaning of simple, here, is the explanation that requires the lowest level of unsupported assumptions, not that an explanation cannot be sophisticated.

Back when the model of the universe put the Earth at the center of everything, certain troublesome observations needed to be explained. Mercury and Venus, when viewed from Earth, do not appear to move in a smooth orbit around us. Instead, they move, stop, back up, stop and then start again (called retrograde motion). If Earth is the center of everything, and everything moves in perfect orbits, then how to explain this behavior? You can check out the history of astronomy for more about this, but the point is that assuming Earth to be the center of the universe was a problem, and this assumption ended up compounded by new assumptions about how to explain the inconsistent behavior of some of the other celestial bodies. The sun-centered universe required few assumptions, and by applying Occam's Razor, it is the better explanation; unfortunately, it was also religious heresy.

Both historical and contemporary science are littered with stories about ideas that were initially rejected—not because of the evidence, but because we did not like the story. Bacterial infection as the source of stomach ulcers was dismissed for 20 years, and it took about that same amount of time for the otherwise toxic nitric oxide (NO) to become accepted as a common neurotransmitter in mammals.

Sometimes the answer to the question "what is it, then?" is "I have no idea, but let's run with it and see what happens." This strategy is quite common. In the early days of organic chemistry, the fact that the unsaturated compounds derived from incense and spices showed a remarkably different reactivity than other unsaturated compounds was not explained by Hoffmann in 1855, when he named this class of molecules as 'aromatic.' Aromaticity was a placeholder assumption, unexplained but nonetheless useful for almost 100 years before the first explanation was proposed.

Using placeholder terms is a great compromise in science, a way to state an assumption but to not invest a lot of speculation on its explanation. My Physics colleagues might argue with me, but a term such as "Dark Matter" and "Dark Energy" seems to also fall into this category. The history of chemistry is replete with examples. In the early 1800s, when the arrangement of atoms was proposed to account for differences in properties, the imagined force holding the atoms together was called "chemical affinity." No one had any idea what chemical affinity was; the term was a useful placeholder with no explanation until electrons were discovered and they were integrated into the idea of "bonding." In organic chemistry, some comparable terms that you may eventually encounter are "the anomeric effect," "the cis effect," and "the gauche effect."

Placeholders can be terms invented to describe an unexplained effect, but our hypotheses and theories are also placeholders: holding the privileged spots as our current and best explanations for natural phenomena. Every idea, no matter how simple or complex, is a placeholder until new observations motivate new ideas to take their place.

Summary

Hückel aromaticity is the name we use to describe the exceptional delocalization (resonance) stabilization observed when a planar ring of atoms, each contributing a p orbital, contains the number of pi electrons, in those p orbitals, corresponding to a value in the "4N+2" series of integers (2, 6, 10, 14...). As a stabilizing effect, aromaticity affects both the structure and the reactivity of molecules that contain these rings, including conformation, acid-base properties, and the position of structural equilibria. Aromatic rings do not undergo addition reactions as readily as other unsaturated compounds, and instead, with strong electrophiles, undergo the electrophilic aromatic substitution (EAS) reaction.

Rings that meet the structural criteria for assessing aromaticity, but which have the number of pi electrons in the "4N" series of integers (4, 8, 12, 16...) are antiaromatic, which means that the ring, in the planar conformation needed for delocalization, is particularly reactive, and constitutes a ground state diradical. Such rings, when possible, will adopt conformations and/or undergo reactions that diminish or remove completely the delocalization, shifting the equilibrium away from the unstable planar, delocalized structure.

Specific sources of strong electrophiles are used in electrophilic aromatic substitution (EAS) reactions. Heteroatom electrophiles are known for nitration, halogenation, and sulfonation reactions. EAS reactions with carbocation electrophiles, known as Friedel-Crafts alkylation and acylation reactions, are important for creating structurally complex organic molecules.

The regioselectivity of EAS reactions is dictated by a combination of the resonance-based directing effect of the substituent (or substituents) that are already on the aromatic ring, with the relative rate-based effect (activation) that the substituent(s) have.

CHAPTER QUESTIONS

10.13 Provide the complete, stepwise mechanisms for the following reactions, including the formation of the electrophile.

(a)

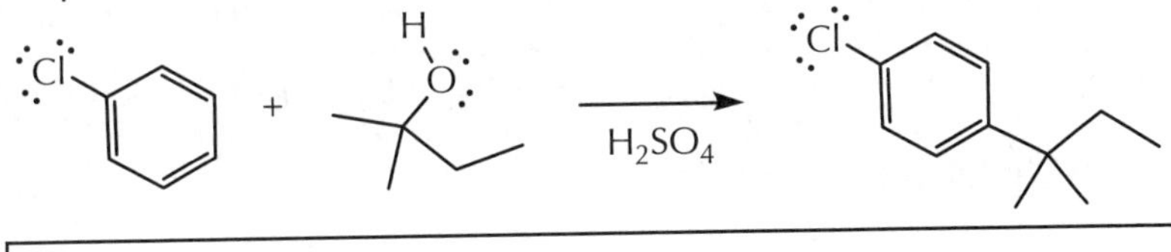

(b)

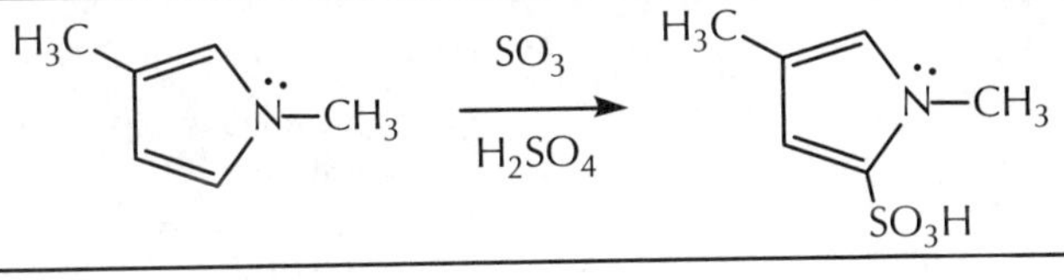

(c)

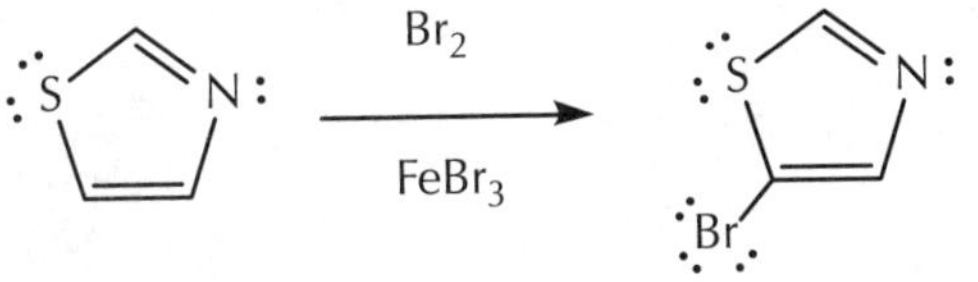

10.14 Azulene exhibits all of the properties of an aromatic compound, such as its resistence to undergo addition reactions and its ability to be nitrated and other EAS reactions. Despite the nonpolar looking structure (below), azulene has two rather unusual properties for an aromatic hydrocarbon: The molecule has a relatively large molecular dipole moment (1.09 Debye), and it is a deep blue color (hence the name, as azure is derived from the color of the blue mineral, *lapis lazuli*. For reference, the dipole moment of dimethyl ether (CH_3OCH_3) is 1.3 D and the dipole moment of BrCl is 0.52 D, and the dipole moment of HCl is 1.08 D.

Provide an analysis of the structure of azulene that argues for the aromaticity of its rings and explain the usually high dipole moment.

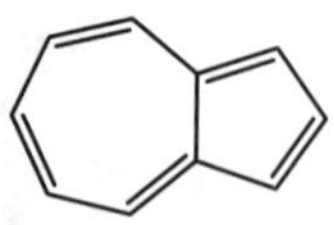

1.08 D
deep blue

10.15 Both of the following molecules, when treated with sulfuric acid, give the same $C_{14}H_{14}$ product. Some of the experimental data about that compound is shown below. Based on these data, propose a structure for the product in addition to a mechanism for its formation.

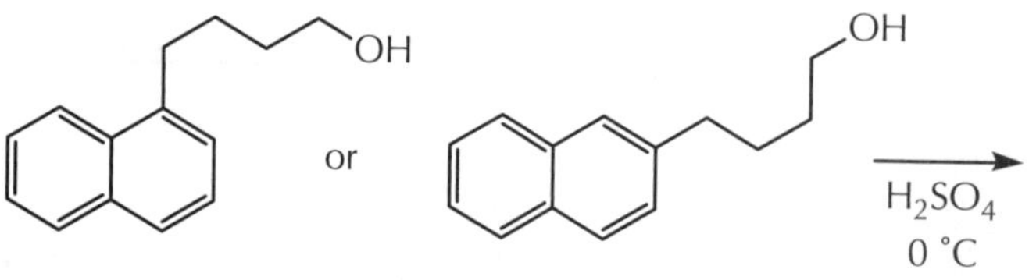

$\xrightarrow{H_2SO_4}_{0\,°C}$

both give the same product
molecular formula: $C_{14}H_{14}$
does not readily react with Br_2
does not readily react with $KMnO_4$
does not readily react with H_2/Pd-C
at higher temperatures, mixtures of products are observed, including this one and others that derive from rearrangements

10.16 Do you anticipate that the trimethylammonium group will be a ring activator or a ring deactivator? And do you anticipate that it will be an ortho/para or meta director? Explain.

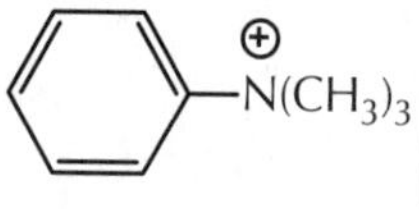

—$\overset{\oplus}{N}(CH_3)_3$

10.17 Complete the following equations as required.

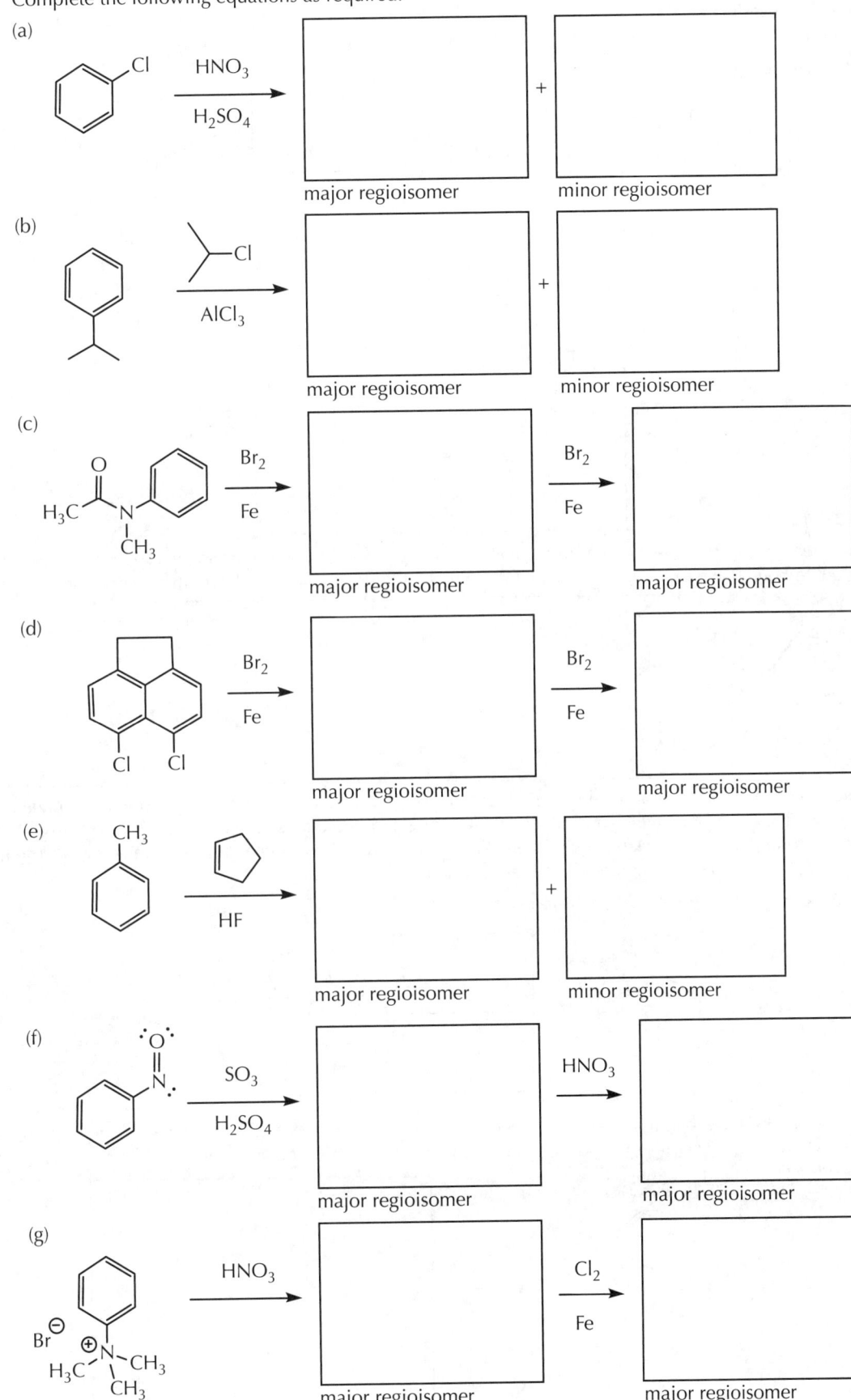

10.18 Some phenol derivatives have activity against hormones involved in blood clotting. During a study of some new drug candidates, the following two-step synthesis was performed. The first step, a Friedel-Crafts alkylation reaction, used an alcohol as the source of the electrophile. Complete the following.

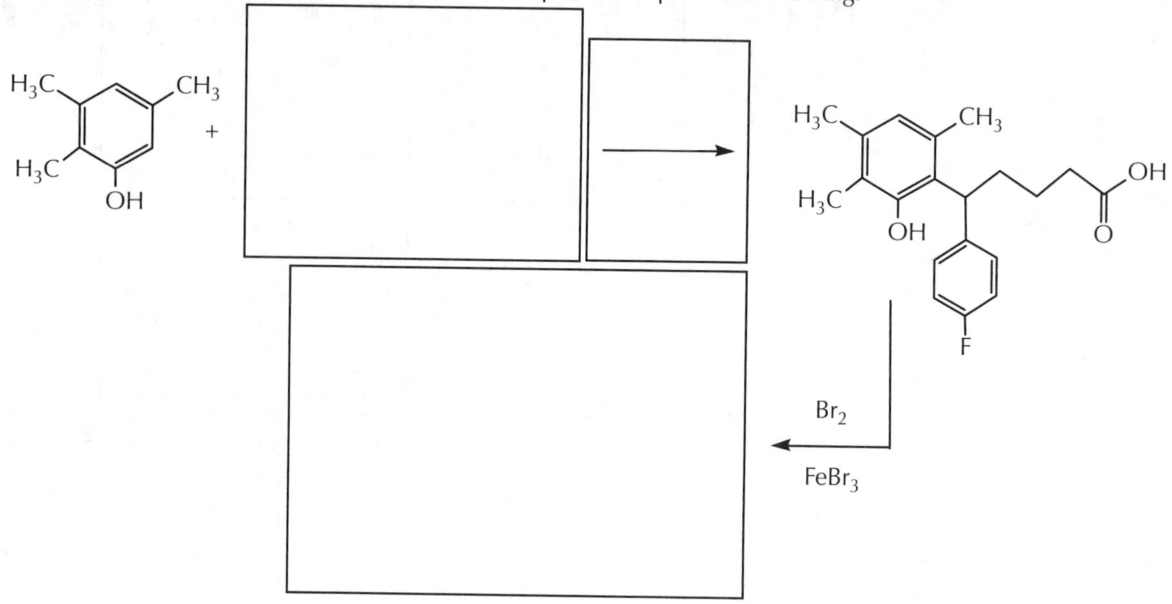

10.19 The enediol, fulvene-6,6-diol, is a strong acid, pK_a = 1.3. Usually, enols (compounds having a hydroxyl group on a double bond) have pK_a values of about 10. How can the unexpected acidity of this compound be explained?

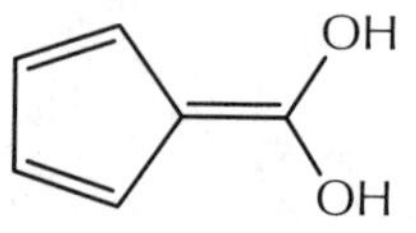

10.20 Corannulene was first synthesized in 1966 by Professor Richard Lawton, at the University of Michigan, as an interesting test of aromaticity. New interest in preparing corannulene emerged in the 1980s because it is a substructural portion of buckminsterfullerene (a C_{60} molecule with a soccerball-like connectivity). The last step in a 1999 synthesis of corannulene is shown here.

(a) What reagent is needed to carry out this transformation?

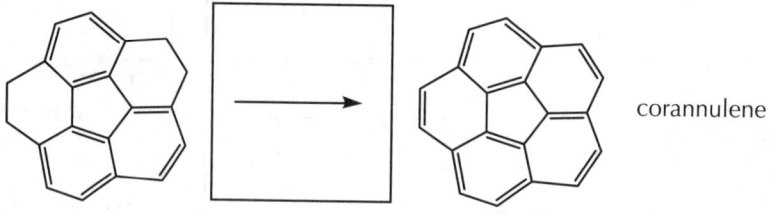

corannulene

(b) Lawton designed this structure (and named it) as a molecule that would have two aromatic rings: a larger outer one and an inner "core" one (hence, core-annulene). Draw a resonance contributor for corannulene that is consistent with this argument.

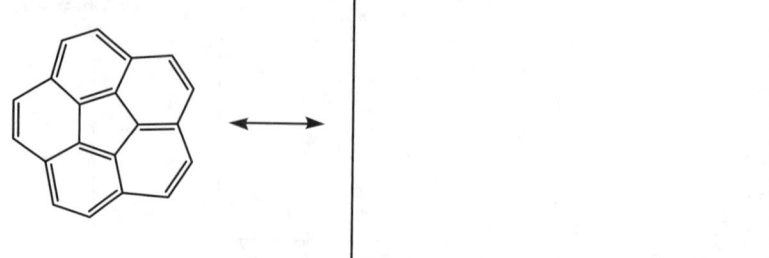

10.21 The typical regioselectivity for EAS reactions of naphthalene is illustrated by its bromination, where the ratio of 1-bromonaphthalene to 2-bromonaphthalene is 99:1.

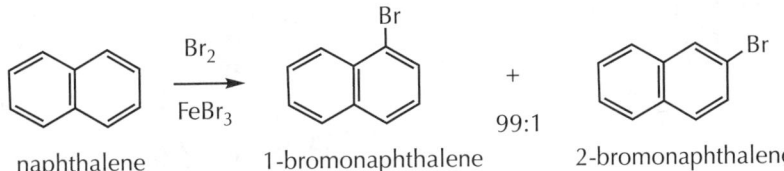

An analysis of the resonance contributors that are possible for the two carbocation intermediates in the EAS reaction leading to the two regioisomers can be used to explain the regioselectivity. Draw the carbocation intermediates that lead to each of the products and explain how they can be used to support an argument for the observed regioselectivity.

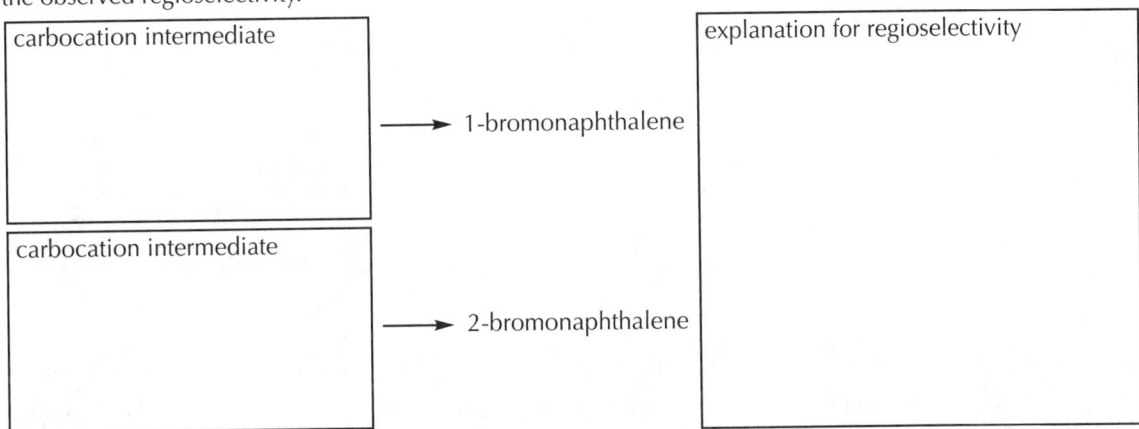

10.22 Comparable to all carbocations, the carbocation intermediates in EAS reactions can undergo rearrangements. In the following molecule, which is labelled with ^{13}C at the indicated atom, treatment with HCl causes an isomerization to take place, resulting in a 50:50 mixture of the two products. Molecules that only differ by the position of isotopic labels are called "isotopomers." What is the mechanism for this isomerization reaction?

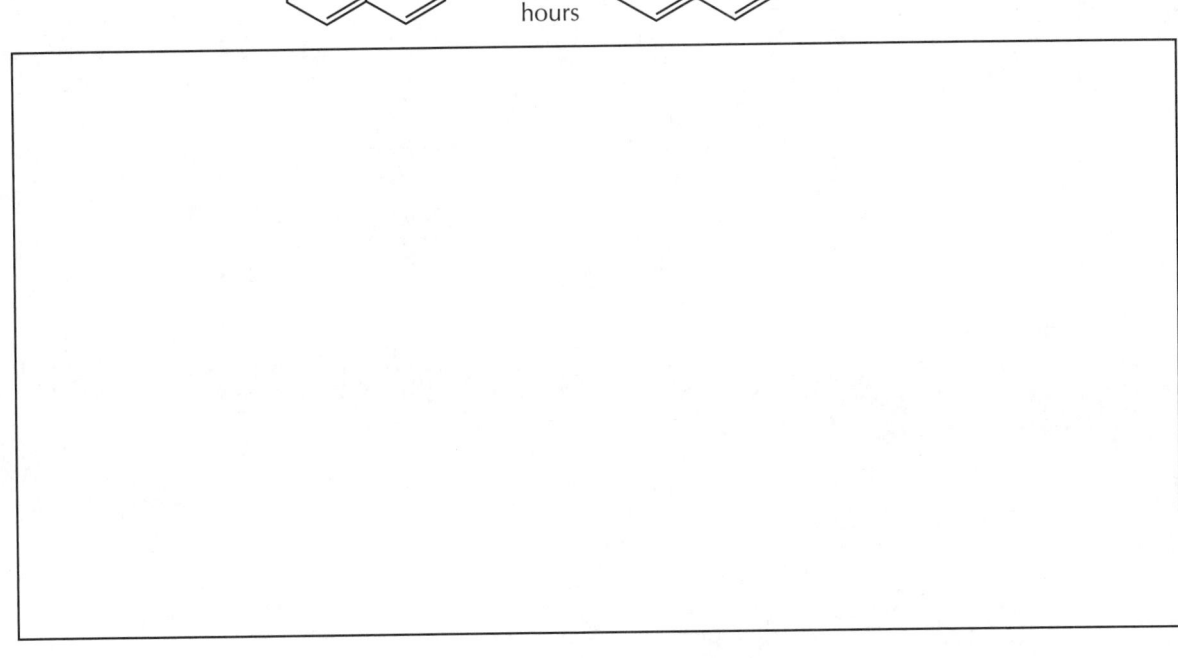

10.23 Heptafulvenes are compounds in which a carbon atom of the cycloheptatriene ring is doubly bonded to a carbon outside the ring. The parent compound, shown below, is highly unstable, but substitution of two cyano groups for the hydrogen atoms on the double bond outside the ring gives a stable compound. Explain this difference in stability (words and pictures).

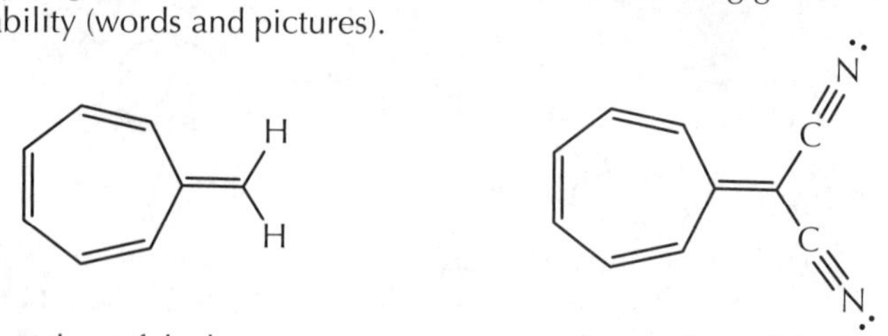

heptafulvalene
unstable

a dicyanoheptafulvalene
stable

10.24 When allyl alcohol is treated with hydrogen fluoride in the presence of benzene, two products are formed: 3-phenyl-1-propene and 1,2-diphenylpropane. What is the mechanism for the formation of these products?

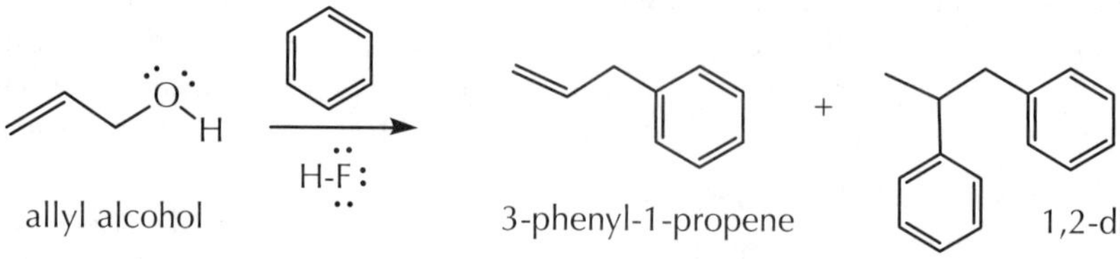

allyl alcohol   H-F: 3-phenyl-1-propene   +   1,2-diphenylpropane

10.25 1-Naphthol is a fluorescent organic molecule, which means it emits visible light when exposed to ultraviolet light (fluorescent inks are used in so-called "black light posters"). It is used in a variety of chemical tests to detect the presence of carbohydrates and proteins.

(a) What is the structure of the all closed shell atom intermediate in the following electrophilic aromatic substitution reaction of 1-naphthol?

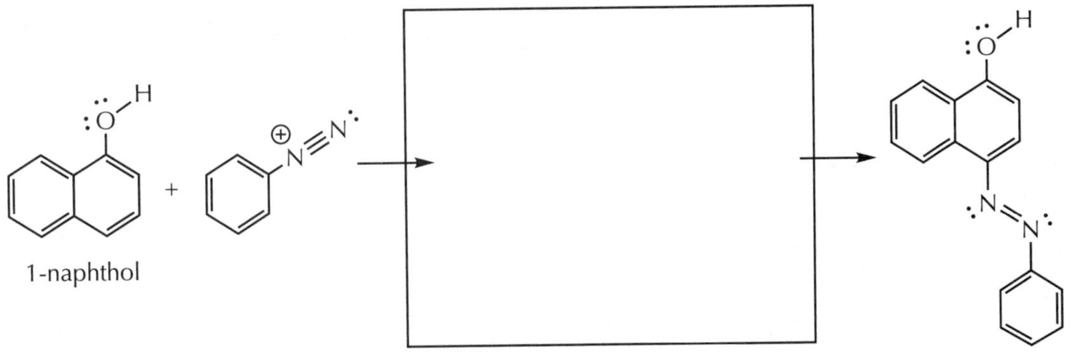

(b) Using what you learned about the regioselectivity of the EAS reactions with 1-naphthol in part (a), above, what 3-step sequence of reactions is required to accomplish the following transformation?

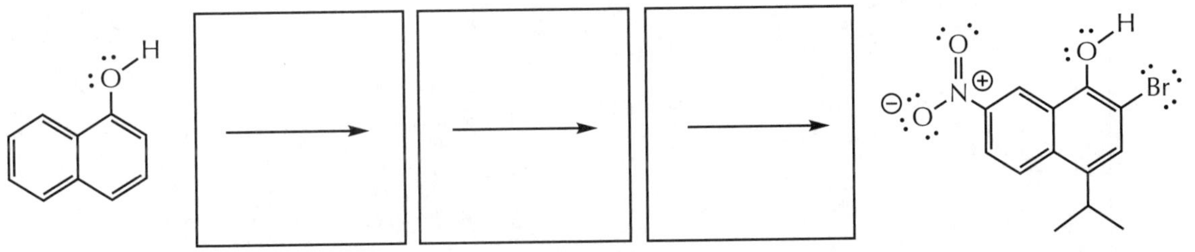

10.26 N,N-Dimethylaniline gives different regioselectivities in its EAS reactions, depending on the pH of the reaction conditions. This allows for the selective formation of Friedel-Crafts alkylation products. One of the results, when using strong Brønsted acid conditions, is shown here. What is the prediction for the other result?

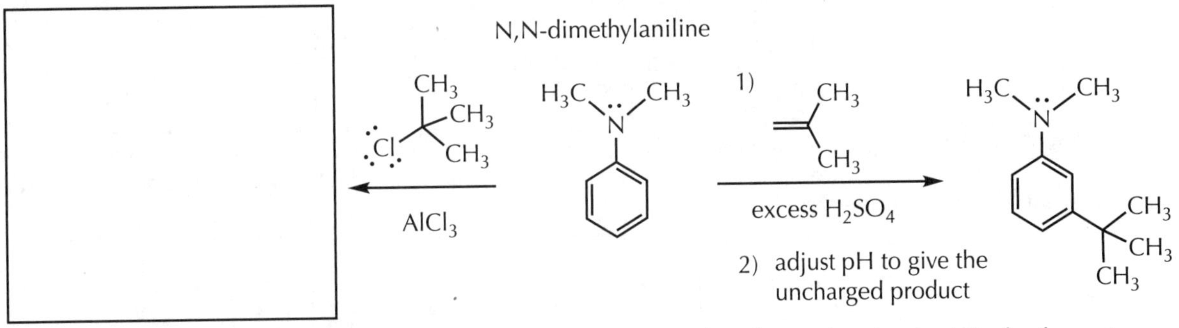

Using words and pictures, explain why N,N-dimethylaniline gives this observed regioselectivity for the *meta* regioisomer under these acidic conditions.

10.27 The following intramolecular (within the same molecule) electrophilic aromatic substitution was used to construct a synthetic steroid. Provide the complete, stepwise mechanism for this reaction. You may use "HB" for the Brønsted acid catalyst and "B⊖" for its conjugate base.

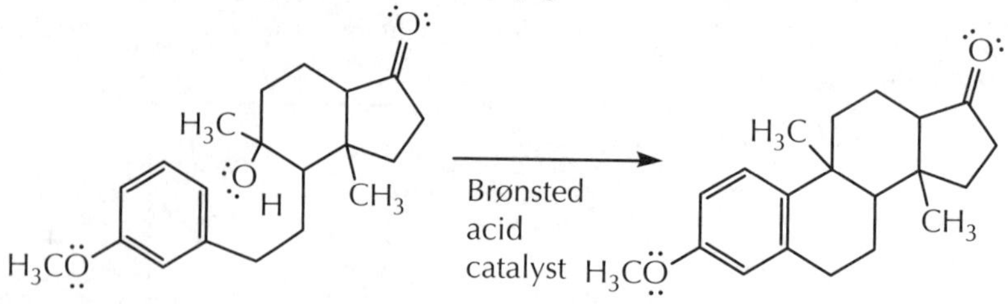

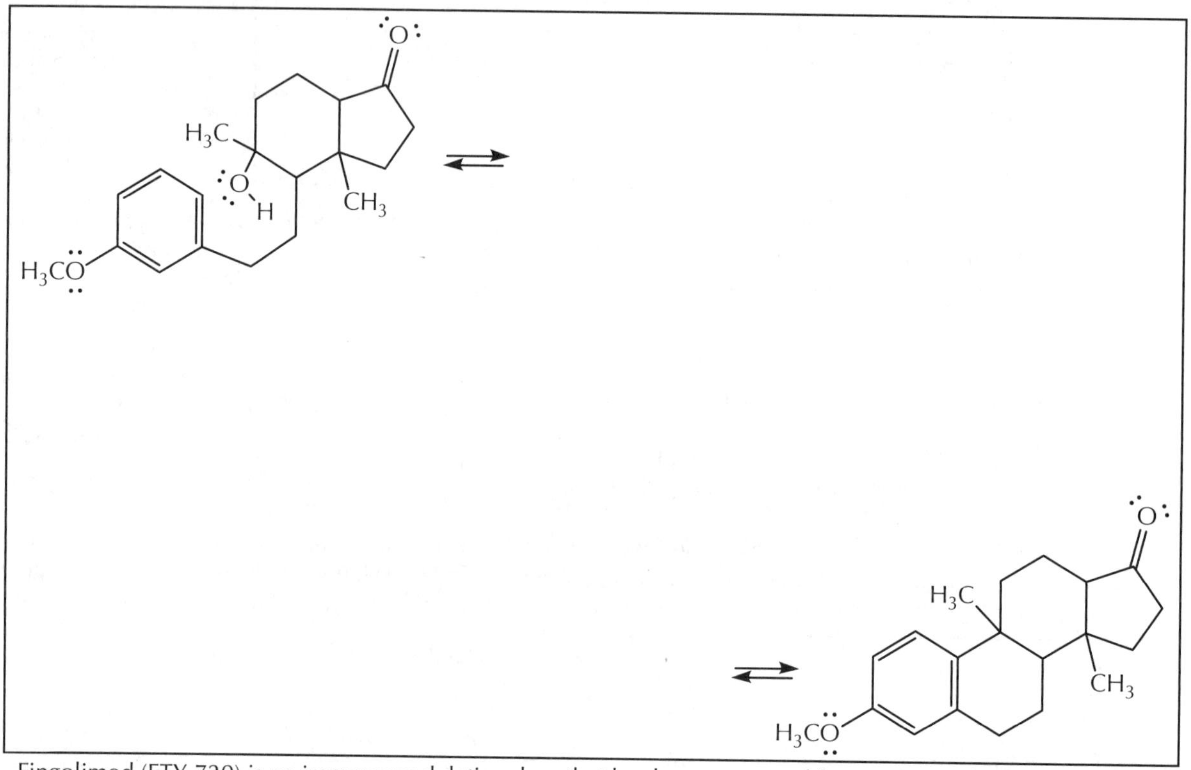

10.28 Fingolimod (FTY 720) is an immunomodulating drug that has been approved for treating multiple sclerosis. Complete the following reaction sequence (*Synthesis*, **2000**, *4*, 505), which describes the preparation of Fingolimod starting from octylbenzene.

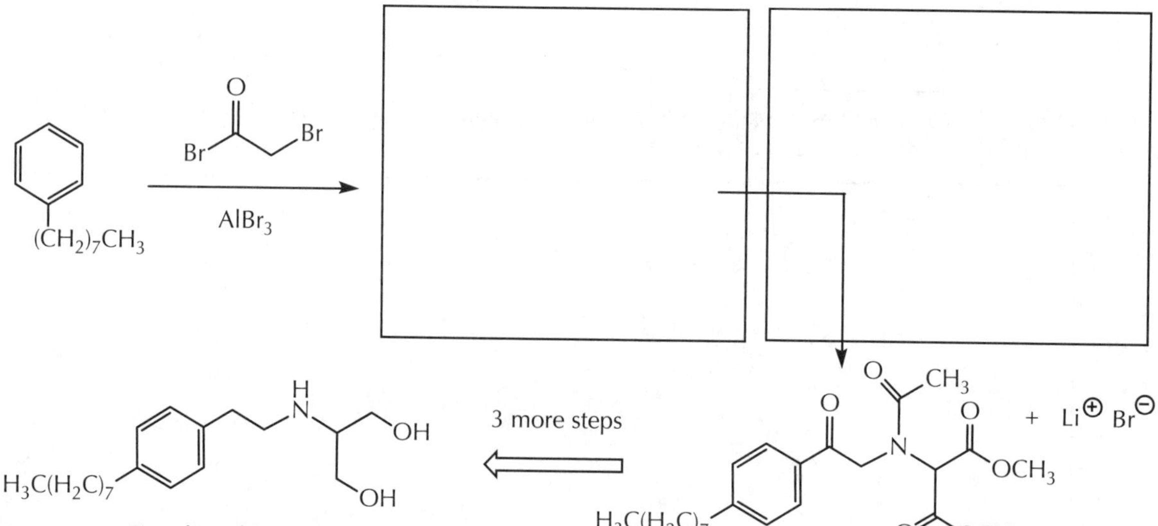

Fingolimod (FTY 720)

10.29 When a catalytic amount of a strong Brønsted acid (symbolized by "H-B") is added to compound Q, an intramolecular electrophilic aromatic substitution reaction is observed. Provide the complete, stepwise, curved arrow mechanism for this transformation. You may use "H-B" to represent the source of a strong Brønsted acid, and "B:⊖" as its conjugate base. Do not use or include any other reagents.

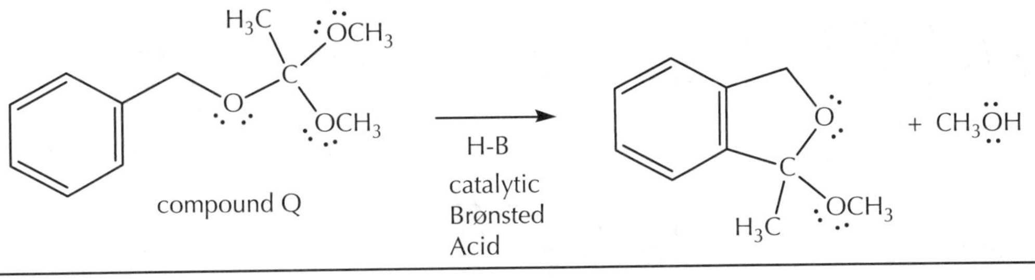

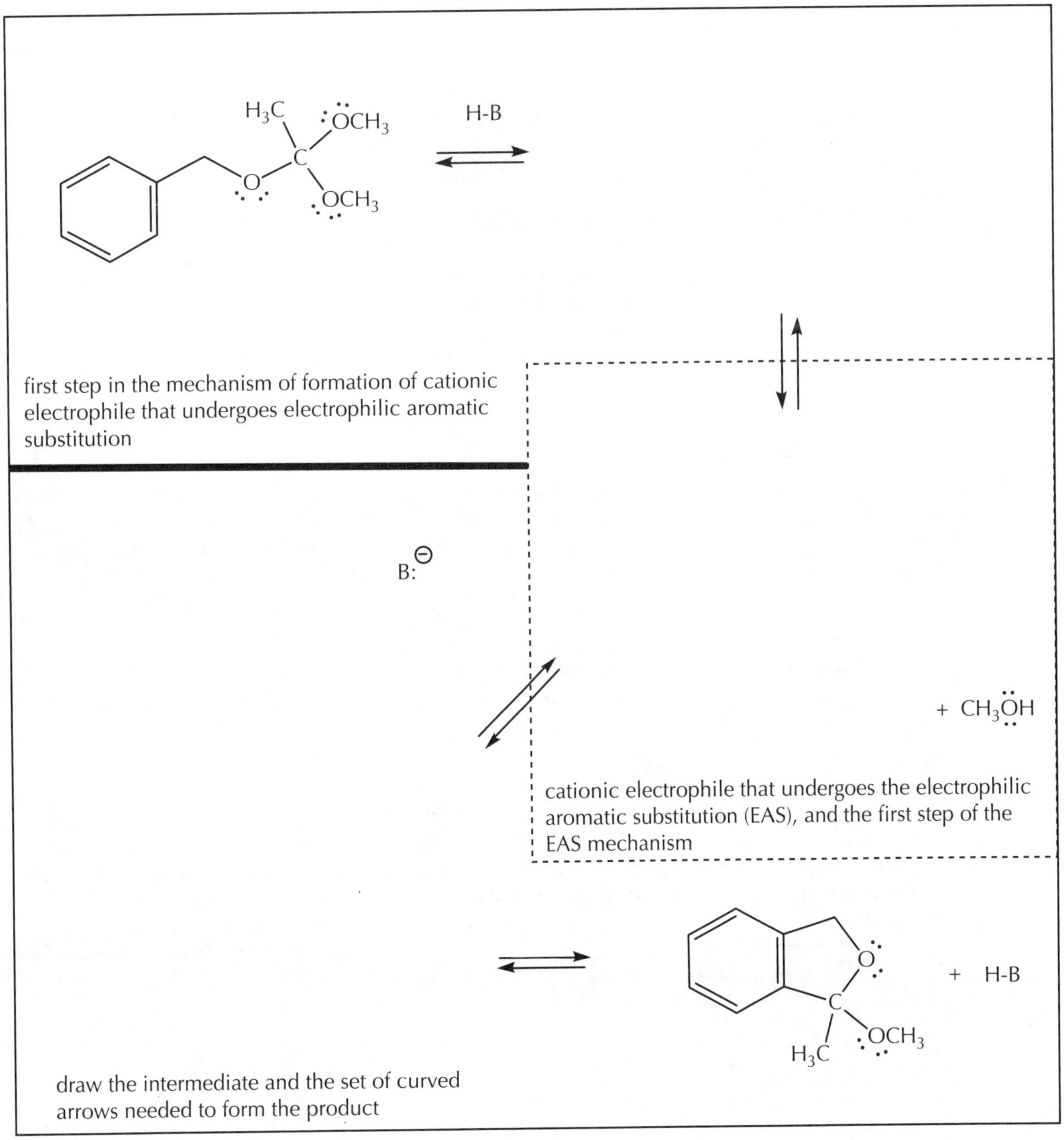

10.30 When treated with a strong base, both of these structural isomers undergo a fast elimination reaction. Compound L is observed to react according to the anticipated E2 mechanism, while compound M undergoes elimination by the E1cb (E1 conjugate base). Compound M is first deprotonated to give intermediate O, and then the leaving group leaves. The energy diagram for this competition is shown below.

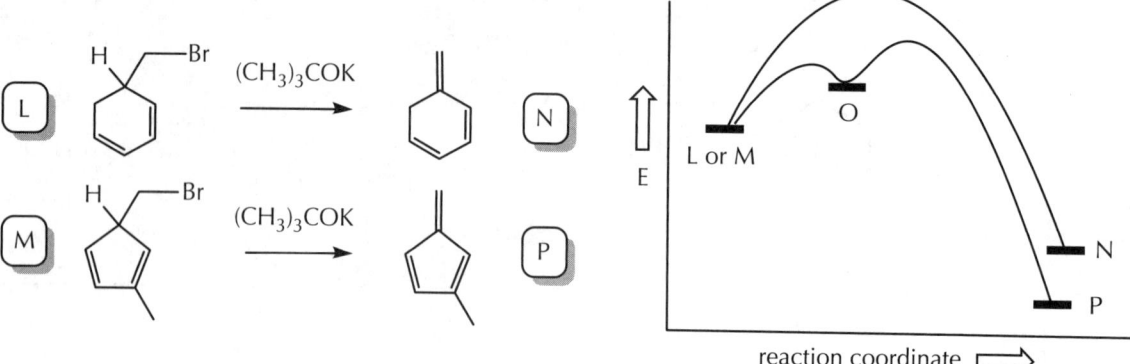

(a) Provide a drawing for the transition state for the L to N transformation using dotted lines and partial charges, as needed (3D is not required).

(b) What is the structure of intermediate O?

(c) Why does intermediate O form from compound M while it does not form from compound L?

10.31 Trifluoromethyl groups can be introduced onto aromatic rings using the following reaction. In this reaction sequence, there is only a minor amount of the para-substituted product (R) observed when the second equivalent of the reagent reacts with Q. The major intermediate product (S), then gives the final product (T) upon introduction of a third trifluoromethyl group.

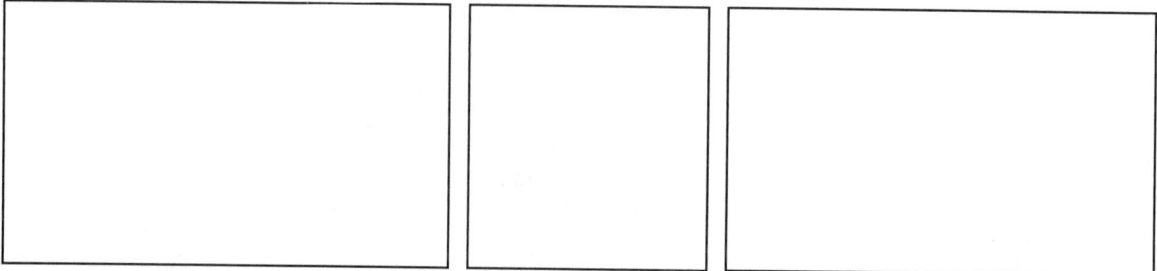

(a) Draw the single resonance contributor for the carbocation intermediate that forms in going from Q to R that best explains why compound R does not readily form.

(b) The ^{13}C-NMR spectrum for compound T shows 3 signals, and the ^{1}H-NMR spectrum shows 1 signal. What is the structure of T?

10.32 Draw a circle around the molecule that contains the most acidic hydrogen atom.

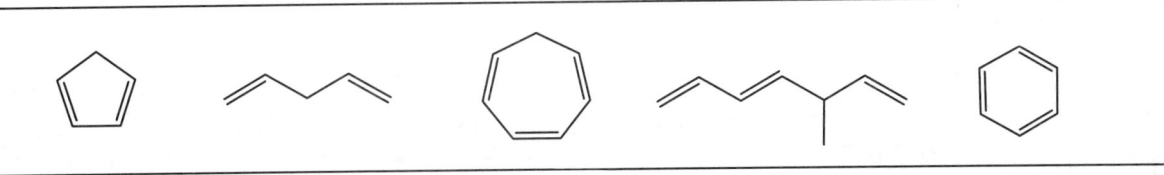

Explain (words and pictures):

10.33 Draw a circle around aromatic rings in the compounds below (assume that all can achieve planar geometries).

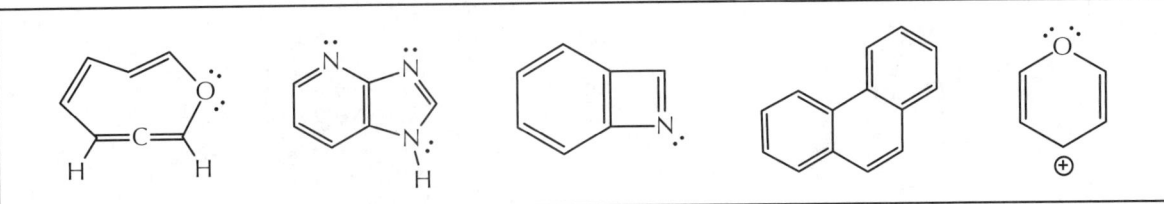

10.34 The structure drawn below is one possible resonance form for sesquifulvalene. One unusual property of this compound is that, unlike most hydrocarbons, this molecule is polar and has a dipole moment of about 1.85 Debye (where comparable hydrocarbons all have dipole moments of about 0 D).

(a) Draw the resonance contributor of sesquifulvalene that explains its unusually large dipole moment.

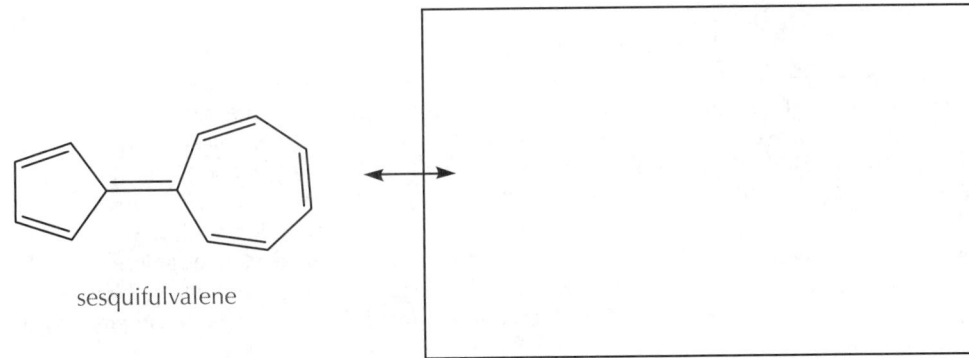

sesquifulvalene

(b) Add the symbol that is used to show the direction of the molecular dipole moment to your drawing in part (a). Just as a reminder, it looks like this: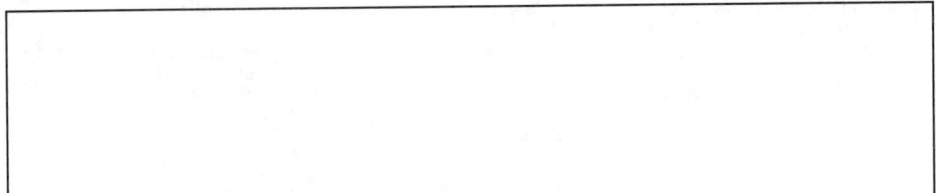

(c) As expected, the resonance contributor shown above is the major resonance contributor to the actual structure. It has all of the usual attributes: closed shell and uncharged atoms. Although it is significant, it only accounts for about 75% of the actual structure. Provide the structural explanation for why the other resonance contributor is so significant to the structure.

10.35 Use curved arrow notation to draw the complete mechanism for the following transformation. Use "H-B" as a Brønsted acid, and "B:⊖" as a base, as needed. Hint: This mechanism involves a single favorable carbocation rearrangement.

10.36 There are three $C_{10}H_8$ hydrocarbons that have two fused rings and an alternating array of five pi bonds around their perimeters. Their properties are quite different. Naphthalene is a stable, colorless compound (mp ~ 80 °C) with aromatic stabilization (61 kcal/mol) about 50% greater than that of benzene (40 kcal/mol). Azulene, a deep blue solid (mp ~ 100 °C) with a high dipole moment (1.08 D, about the same as HCl) also shows aromatic properties, although its aromatic stabilization is less than benzene (24 kcal/mol). Bicyclo[6.2.0]decapentaene, the third compound, is a deep red-orange oil, reacts with molecular oxygen, dimerizes at less than 100 °C, and has a calculated resonance destabilization of 4 kcal/mol. Although you can find many resources that want to claim that naphthalene is a 10 pi electron aromatic compound, how do these other two examples provide a compelling counterargument for that?

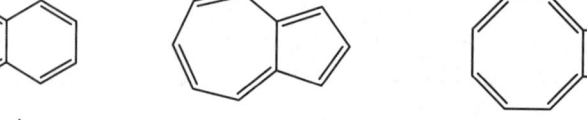

naphthalene   azulene   bicyclo[6.2.0]decapentaene

10.37 Electrophilic addition reactions of double bonds and aromatic rings is an extremely versatile bond-forming process. The following two reactions occur in three mechanistic steps. In the second step, both of the new carbon-carbon bonds are formed. Based on your experience with electrophilic addition and electrophilic aromatic substitution, propose complete, stepwise, curved arrow mechanisms for these two reactions.

(a)

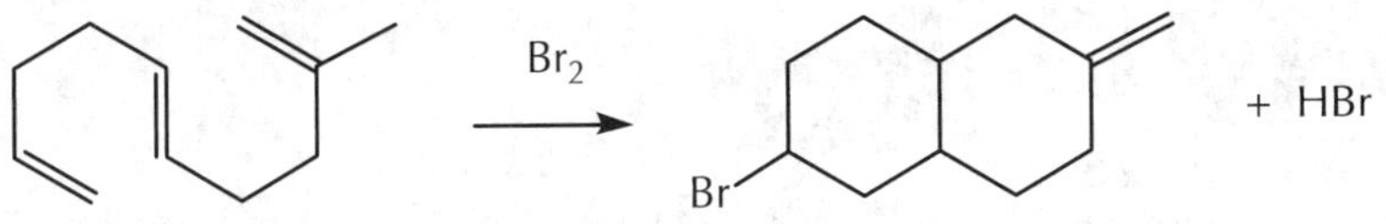

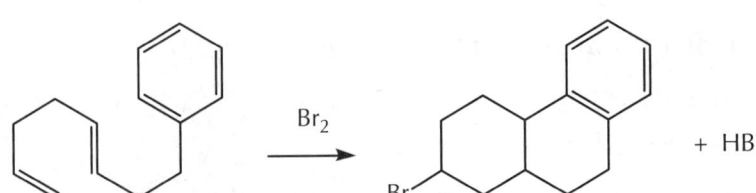

(b)

Work in an Analytical Chemistry Lab, February 1888.

Chemical History at the University of Michigan

Understanding reactions begins by knowing "what is in the flask?" and so integrating laboratory instruction on analysis techniques was central to the curriculum.

At the University of Michigan, Albert B. Prescott was a professor of organic and applied chemistry, dean of the school of pharmacy, and served as the director of the chemical laboratory (where this class was taking place). He was a strong supporter of the type of laboratory-based instruction that this location provided, and he was a staunch advocate for how important chemical analysis ("what is in the flask?") was to the advancement of chemical and pharmaceutical sciences.

In an August 22, 1893, address to the World's Congress of Chemists (*J Am Chem Soc*, **1893**, *15*, 376), Prescott advocated for the power of analysis in advancing chemistry.

> Chemical literature is thickly strewn with directions for analysis to the end of the identification of the integral molecule, representing matter in its living state, if such a figure of speech may be used. This is indeed the special task of analytical research, although the terms of analysis are also given to operations that accompany synthetic work, to wit: In studies of the molecular structure of bodies produced by nature, bodies mineral, vegetable, and animal, as well as those of artificial production. To classify analytical work strictly by definition, which, however, I have no desire to do, all the studies of molecular constitution come within the range of analytical inquiry. But even under customary classification of chemical labor, it will be observed that certain instruments of observation early used by analytical chemists have since been found most effective in studies of molecular structure even for what is termed the configuration of the molecule.

Public domain image from the University of Michigan Collection, kindly provided by the Bentley Historical Library (File HS3518).

CHAPTERS 1–10
Exam Questions

The final examination in the first-term organic chemistry course at the University of Michigan is typically made up of five or six pages (depending on the format, term, and/or instructor preference) of questions that integrate the topics from Chapters 1-10.

The following pages of "Exam Questions" (EQ 04.01–EQ 04.46) were used in examinations between 1998–2019.

The examinations are written for a 90-minute time period, and students have 120 minutes to take the tests.

Any set of five or six pages from this group of 46 questions might be used to create a practice exam.

Select a set of these pages and time yourself for 120 minutes, without accessing any resources, to self-test your mastery of the main topics from Chapters 1–10.

EQ 04.01

Synthesizing molecules where hydrogen atoms have been substituted by deuterium atoms is a common experimental strategy used for studying biochemical mechanisms.

(a) Draw the (R,R)-stereoisomer (compound H) for the deuterium derivative of 2-phosphoglycerate, shown here.

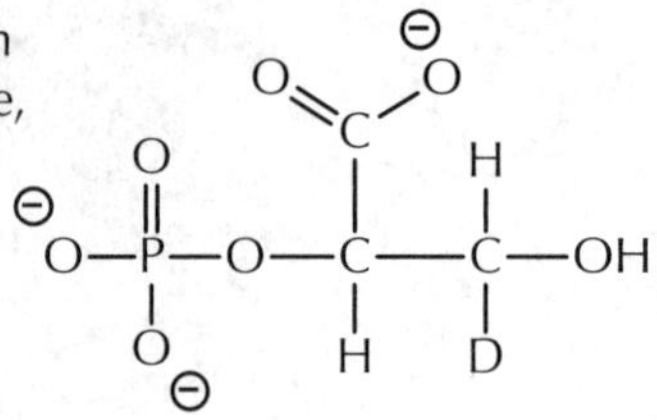

compound H

(b) Under natural enzymatic conditions, the (R,R)-stereoisomer (compound H) undergoes an elimination reaction of water in order to produce the specific stereosiomer (compound J) shown below.

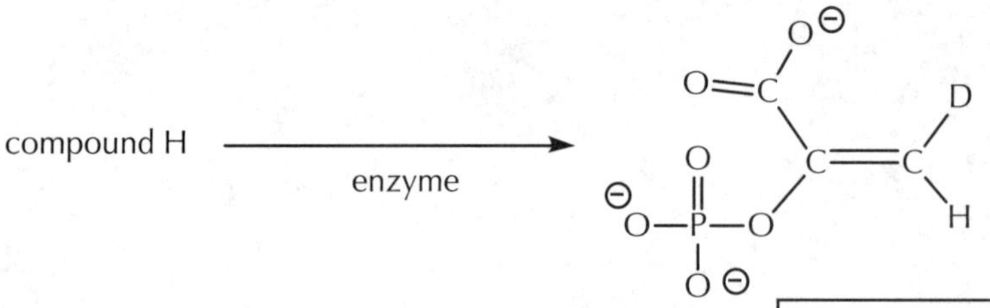

compound H $\xrightarrow{\text{enzyme}}$ compound J

(i) What is the stereochemical configurational label for compound J?

(ii) Based on the stereochemical information provided, does the enzyme promote a SYN or ANTI elimination reaction?

To answer this, you should provide a conformational drawing for the (R,R)-stereoisomer (compound H) that clearly demonstrates the stereochemical relationship between the groups that are eliminated and the stereochemistry of compound J.

SYN or ANTI elimination? (circle one:)   SYN   ANTI
(must be clear and consistent with compound H)

(c) The state of protonation/deprotonation of compound H depends on the pH of the environment in which it exists. In this molecule, the approximate pK_a values for the conjugate acids of the two phosphate bases are 2 and 7, while that of the conjugate acid of the carboxylate base is 5.

(i) Based on these values, what is the pH value at which at least 50% of the molecules in a solution of compound J would be in the trianionic form, as shown above?

(ii) Based on these values, what is the pH value above which the K_{EQ} for the trianionic form, as shown above, would be 10^4 with respect to its conjugate acid?

EQ 04.02

The following questions refer to this compound:

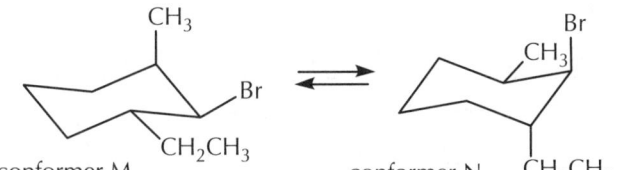

compound A

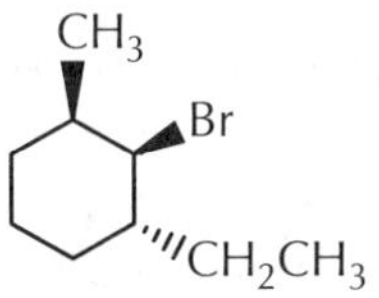

(a) A constitutional isomer of compound A is named (1*R*,2*R*,3*R*)-1-bromo-3-ethyl-2-methylcyclohexane. Draw this structure.

(b) The equilibrium between the two chair forms of compound A are shown here (conformers M and N), along with some experimental data.

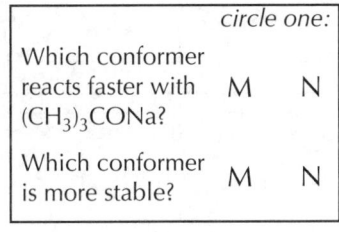

CH_3 (equatorial to axial) = +1.70 kcal /mol
CH_2CH_3 (equatorial to axial) = +1.75 kcal /mol
Br (equatorial to axial) = +0.50 kcal /mol

Br/CH_3 and Br/CH_2CH_3 gauche = 0.25 kcal /mol

(i) The rate of reaction from each of these conformers with $(CH_3)_3CONa/(CH_3)_3COH$ is anticipated to be significantly different. Which one is faster? What is the structure of the major product?

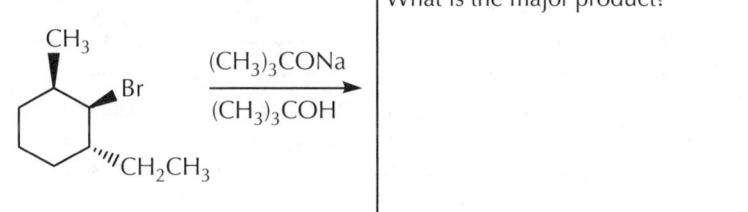

	circle one:	
Which conformer reacts faster with $(CH_3)_3CONa$?	M	N
Which conformer is more stable?	M	N

What is the major product?

(ii) A more detailed conformational analysis, using Newman Projections, of the o-p bond, is used to help explain this rate difference. Using the drawing below to define atoms o and p, and using conformational drawings M and N as references, complete the Newman projections for the o-p bond of both conformers M and N in which atom o is the front atom and atom p is the back atom.

conformation M (o front, p back)

conformation N (o front, p back)

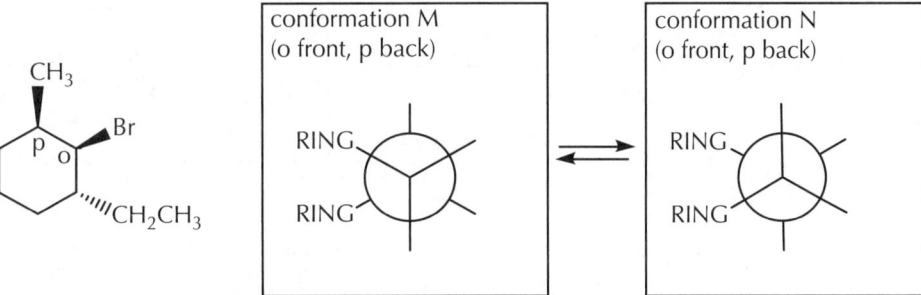

(iii) Finally, explain briefly how one of these two Newman Projections provides evidence to support the selection you made in part (ii) about which conformer reacts faster with $(CH_3)_3CONa/(CH_3)_3COH$.

(iv) As written (left to right, M to N), what is the value of the energy difference (i.e., $\Delta G°$, in kcal /mol)?

circle one:	-2.95	-0.55	-0.30	+0.30	+0.55	+2.95

862 CHAPTER 10 Aromaticity and Electrophilic Aromatic Substitution Reactions

EQ 04.03

A. The epoxidation reaction of (S)-3-phenyl-1-cyclopentene (SPC) results in two diastereomeric products, compounds H and J.

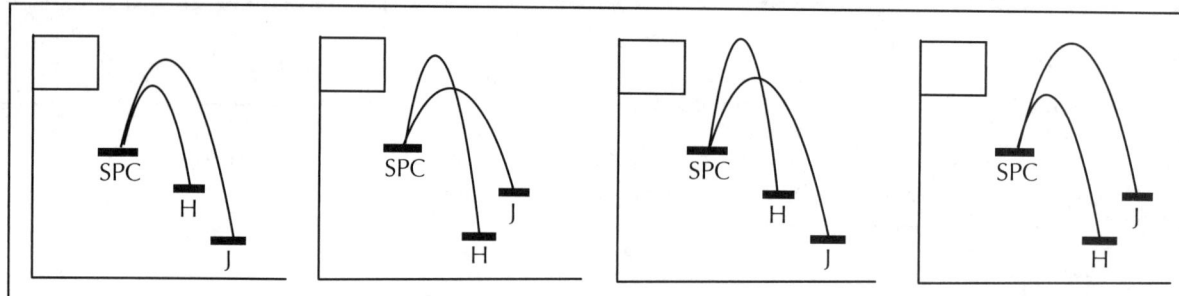

These two products differ in stability (thermodynamics), and they also differ in their rates of formation (kinetics). In both cases, steric arguments are used to explain the differences.

(a) Which one of the following energy diagrams best represents these reactions (mark with an "X")?

(b) Based on this information, explain the difference in transition state energies between leading to H and J. Be sure to point to specific structural differences and how those play a role (note that this answer can only make sense if the correct energy diagram has been selected).

B. Complete the following reactions as required.

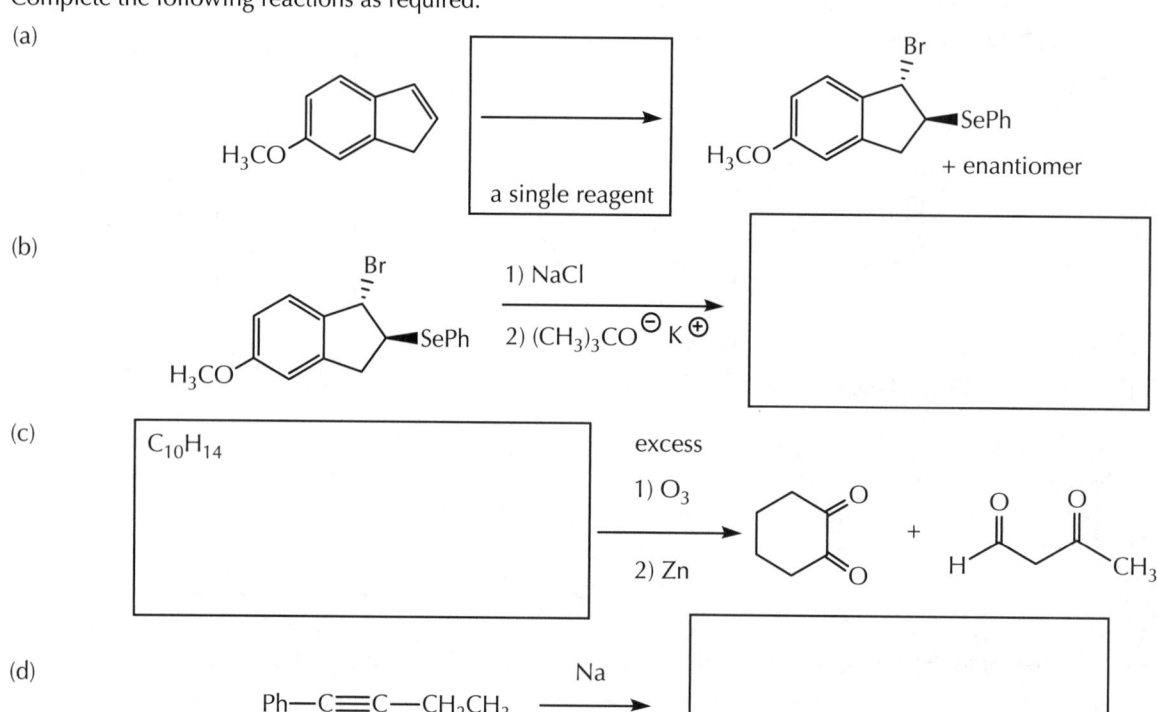

EQ 04.04

Complete the following reaction schemes as indicated.

(a) The reaction conditions reported here are the equivalent of nitric acid (*Angew Chem Int Ed Eng*, **2004**, *116*, 1259).

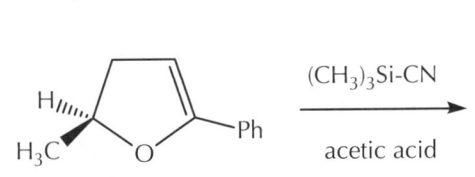

major product

This is an example of (*circle one*):

regioselectivity

stereoselectivity

(b) Note: [(CH$_3$)$_3$Si-CN] is a source of nucleophilic cyanide ion (*Org Lett*, **2017**, *19*, 6534).

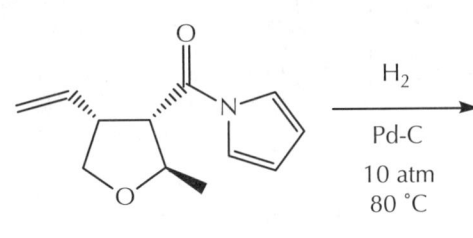

major product is the (*S,R*)-isomer

This is an example of (*circle one*):

regioselectivity

stereoselectivity

(c) *Org Lett*, **2017**, *19*, 6658.

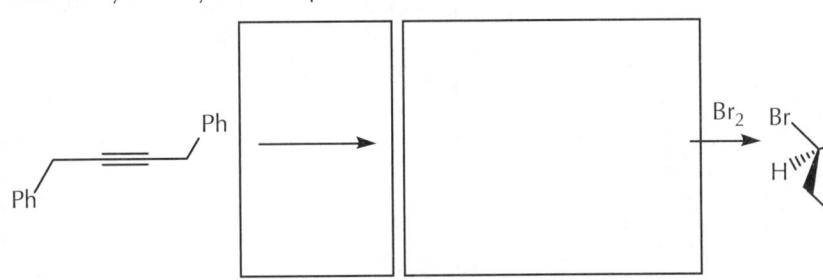

major product is C$_{12}$H$_{21}$NO$_2$

This product has how many units of unsaturation?

(d) How can you carry out this specific transformation?

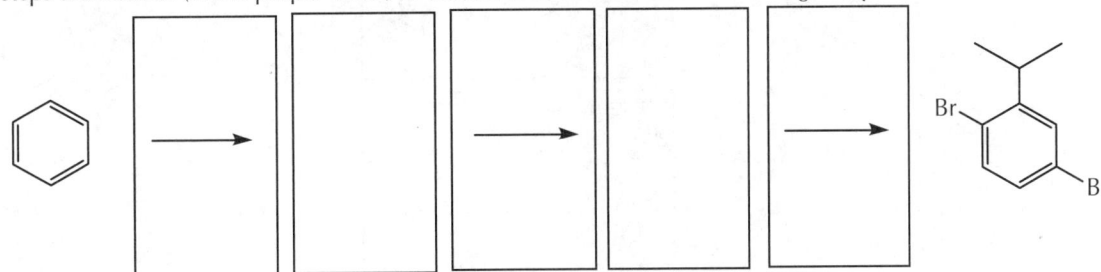

This product is (circle one):

optically inactive

optically active

(e) What is the name of the product in part (d)?

(f) What steps are needed (in the proper order) to transform benzene into the following compound?

EQ 04.05

A. The reaction shown here is observed when the organic molecule is placed in water. Provide the complete, stepwise, curved arrow mechanism according to the instructions.

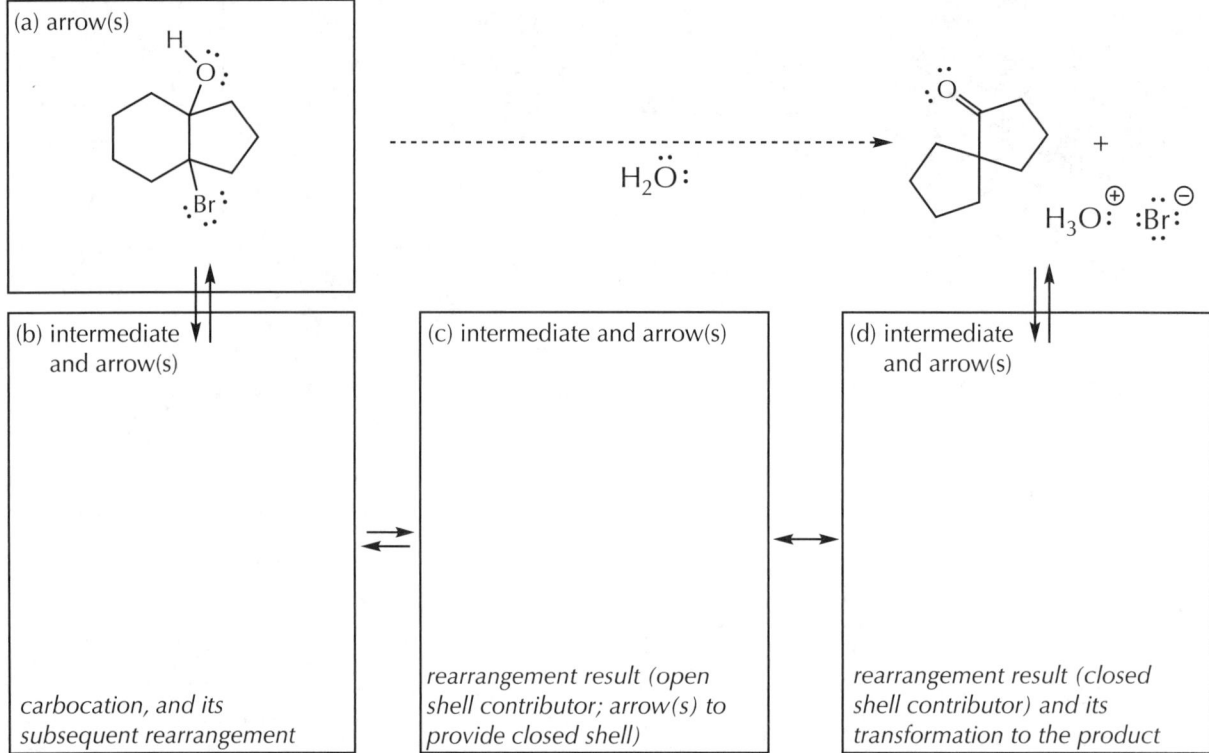

B. In the first step, an ionization gives a carbocation and an anion. In the second step, the anion then undergoes an electrophilic aromatic substitution with the carbocation. Step 2 includes an uncharged intermediate with all closed shell atoms. Draw the intermediate resulting from step 2.

C. Provide reagents to perform the following transformation (no abbreviations or acronyms).

EQ 04.06

A. Draw the products or provide reagents as necessary to complete the following reaction scheme. If more than one step is required to complete a transformation, clearly number steps.

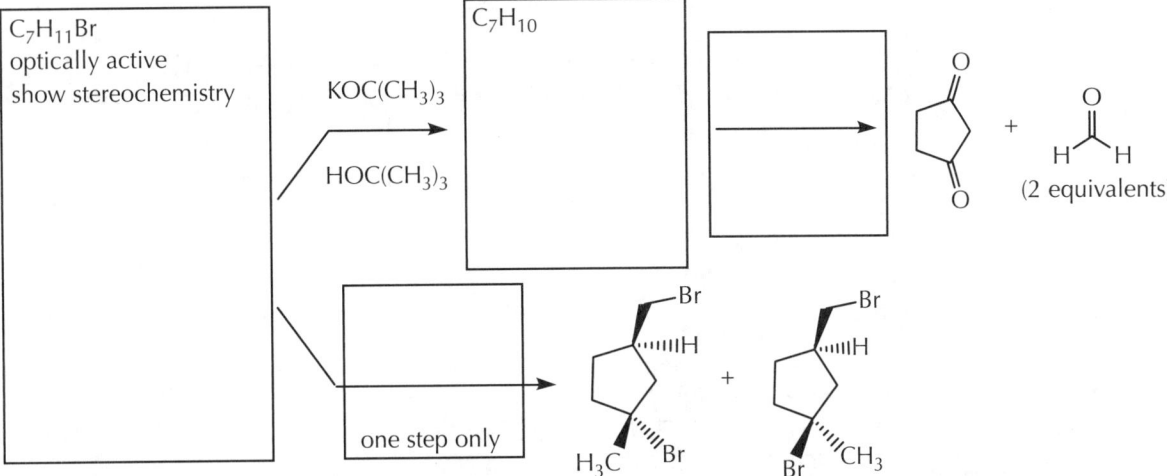

B. A recent study examined the reactivity of alkyl bromide (compound P) with lithium diethylamide (compound Q). Unexpectedly, the major product formed in the reaction is the amine (compound R), while the alkene (compound S) is the minor product.

(a) Given the information provided below, construct an energy diagram for the conversion of P + Q to give R and S. Clearly label transition states A and B on the diagram.

The formation of compound R proceeds through transition state A, and the activation energy for this process is + 12 kcal/mol.

The formation of compound S proceeds through transition state B, and the activation energy for this process is + 18 kcal/mol

(b) Provide reagents to carry out the synthesis of the original alkyl bromide (compound P) starting from the alkene (compound S). Number the separate experimental steps.

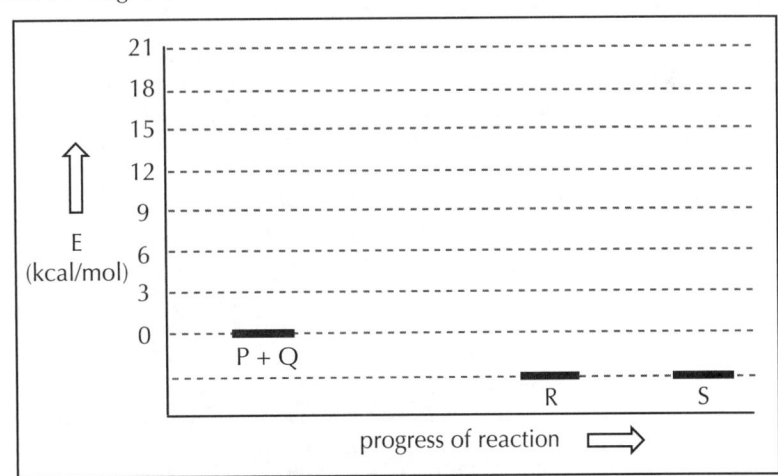

CHAPTER 10 Aromaticity and Electrophilic Aromatic Substitution Reactions

EQ 04.07

(a) Fill in the following reaction sequence. Where necessary be sure to draw the specific structures of the reagents involved in a given reaction, including stereochemistry as needed.

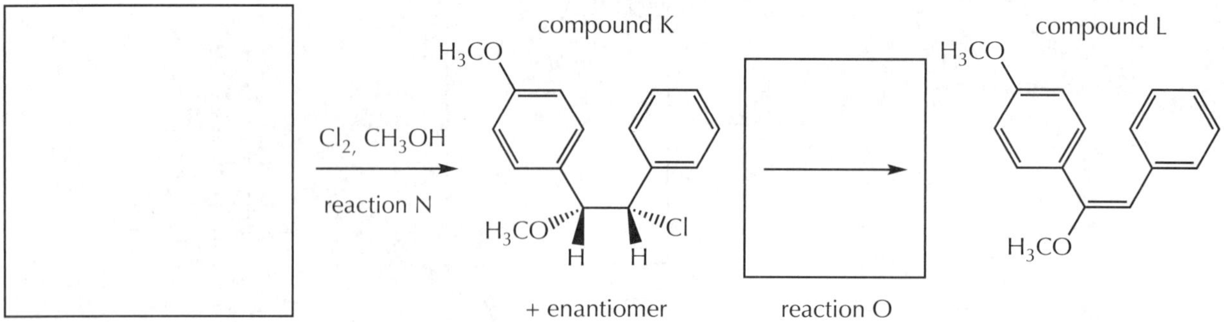

[Blank box] — Cl₂, CH₃OH, reaction N → compound K (structure shown: H₃CO–C₆H₄–CH(OCH₃)–CH(Cl)–C₆H₅ with H,H wedge/dash stereochemistry) + enantiomer → [Blank box] reaction O → compound L (structure shown: H₃CO–C₆H₄–C(OCH₃)=CH–C₆H₅ alkene)

(b) The mechanism of reaction N involves the formation of two cationic intermediates. In the boxes below, draw the structures of these two intermediates. Be sure to specify stereochemistry and provide the intermediates that form the specific stereoisomeric product from reaction N, shown above.

[Box: Intermediate 1 in reaction N]

[Box: Intermediate 2 in reaction N]

(c) Provide an explanation for the regioselectivity of this reaction. In particular, provide an explanation for the regioselectivity of the reaction of intermediate 1 to form intermediate 2. Use pictures and words.

[Blank box]

(d) Draw a Newman projection compound K, from reaction N, that leads to the stereochemical outcome shown for compound L.

Put the sp³-hybridized carbon bearing the OCH₃ group in the front.

[Blank box]

EQ 04.08

A. The pK$_a$ of the conjugate acids of compounds A, B, and C are –10, –8, and 0, respectively.

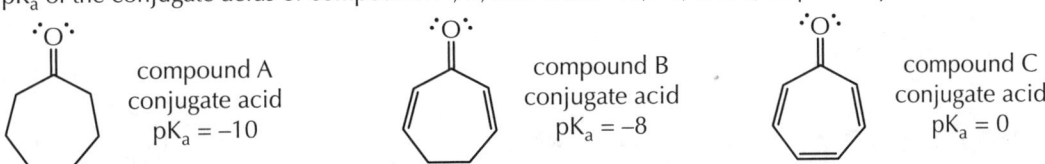

There are many resonance contributors for the conjugate acid of compound C, including one with all closed shell atoms and seven with a single open shell atom.

(a) Draw the all closed shell atom resonance contributor for the conjugate acid of compound C, as well as the most significant contributor from the seven other resonance forms with a single open shell atom.

all closed shell atoms	most significant contributor with one open shell atom
	↔

(b) Why are these resonance contributors, with open shell atoms, significant when, in general, they are not usually considered important?

B. The pK$_a$ of compound E is 11.

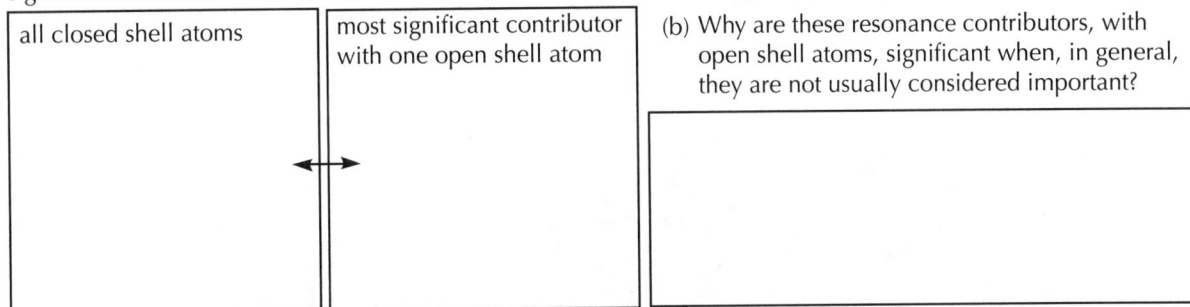

(a) Which of the following can completely deprotonate (K$_{eq}$ > 10^3) compound E? Mark all that apply with an "X" in the box:

☐ sodium azide (NaN$_3$)
☐ sodium hydride (NaH)
☐ sodium methanethiolate (NaSCH$_3$)
☐ sodium methoxide (NaOCH$_3$)
☐ sodium chloride (NaCl)

(b) The conjugate base of compound E has three significant resonance contributors with all closed shell atoms. Draw all three of these resonance structures.

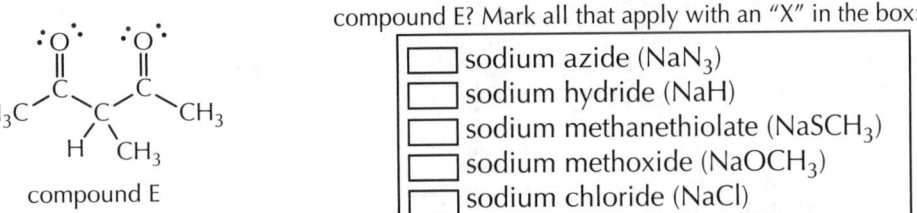

(c) The conjugate base of compound E reacts with bromomethane (CH$_3$Br) to give compound F. Interestingly, it reacts with methyl triflate (CH$_3$OSO$_2$CF$_3$) to give compound G, which is a structural isomer of compound F and a (Z)-stereoisomer. NMR data are provided. What are the structures of compounds F and G?

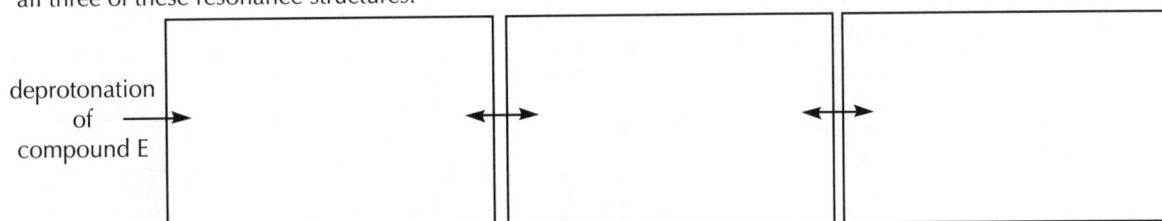

compound F
^{1}H-NMR:
  2 signals (1:1 ratio)
^{13}C-NMR:
  4 signals

compound G
^{1}H-NMR:
  4 signals (1:1:1:1)
^{13}C-NMR:
  7 signals

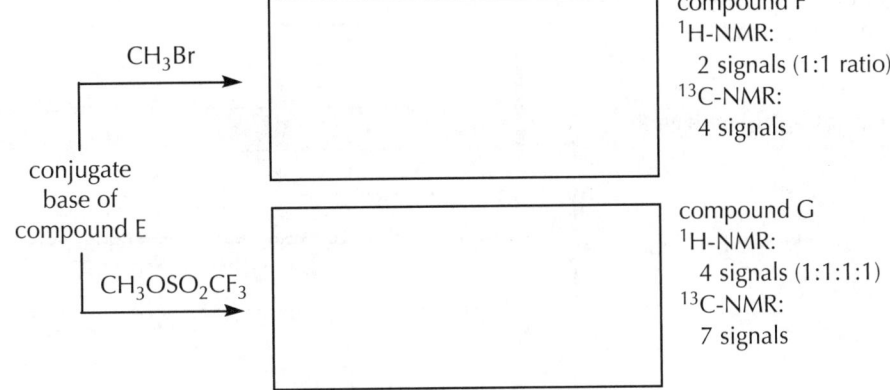

EQ 04.09

A. In a mixture of trifluoroacetic acid and water, compound F reacts to give a mixture of compounds G and H.

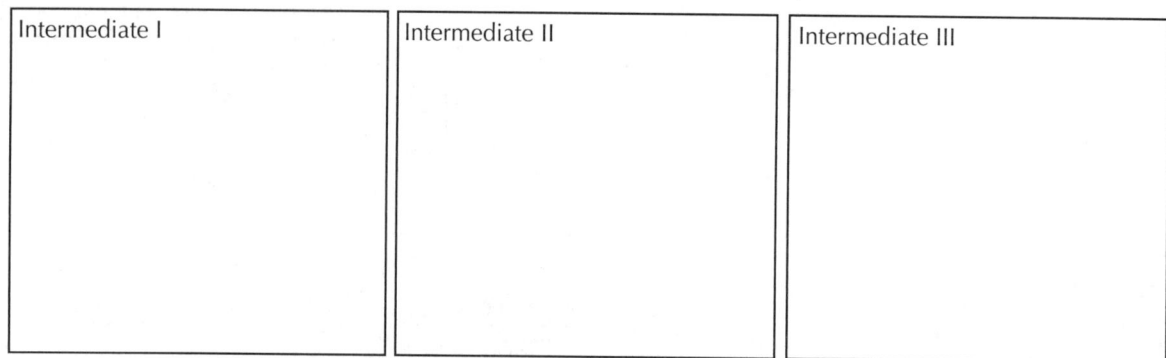

compound F   compound G   compound H

(a) There are three cationic intermediates in the reaction mechanism to form compound G from compound F. Draw all three (if more than one resonance structure is possible, then draw the most significant contributor). Note: Only the intermediates are called for here; mechanistic arrows are NOT required.

Intermediate I	Intermediate II	Intermediate III

(b) What is the stereochemical relationship between compounds G and H?

(c) If compound G is levorotatory [that is, it has a negative, or (−) rotation], then what can you predict about the optical rotation of compound H?

(d) Under the same reaction conditions, the enantiomer of compound F produces:

circle one: only G   only H   both G & H   neither G nor H

B. A chemical reaction starts with compound J, proceeds through a single, high energy intermediate (K), and gives two products (L and M).

Additional information: The overall reactions of both L and M are exothermic, starting from J. The formation of the intermediate is the rate-determining step for the forward reaction. Under kinetically controlled conditions, product L dominates. Under thermodynamic conditions, there is a 1:1 ratio of L and M. Draw an energy diagram consistent with these data.

Label K, L, and M; show the appropriate energy curves between them (J is provided).

EQ 04.10

A. Methyl azide (CH_3N_3) is a neutral molecule with the H_3CNNN connectivity.

(a) Draw the two most significant resonance contributors to the methyl azide structure.

[ ] ⟷ [ ]

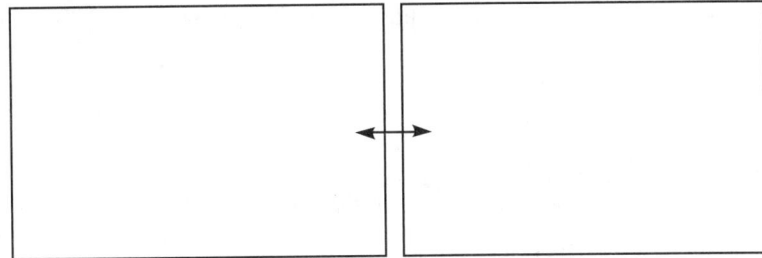

(b) The azido group (N_3) in methyl azide structure undergoes an addition reaction with triple bonds in a mechanism that is a direct analogy with the way ozone undergoes addition with triple bonds (the addition step, only). Thus the addition reaction between methyl azide and propyne creates two regioisomeric products, both with 5-membered rings and the molecular formula $C_4H_7N_3$ (all atoms closed shell and uncharged). In the presence of Cu^{+1}, the major product is the sterically less crowded regioisomer. Based on this information, draw the following:

methyl azide

CH_3N_3

\+ $\xrightarrow{Cu^{+1}}$

H–C≡C–CH_3

propyne

$C_4H_7N_3$ (major regioisomer) + $C_4H_7N_3$ (minor regioisomer)

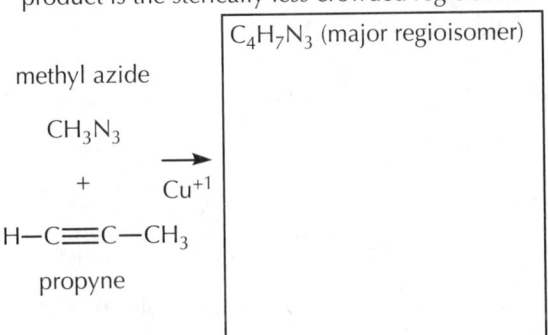

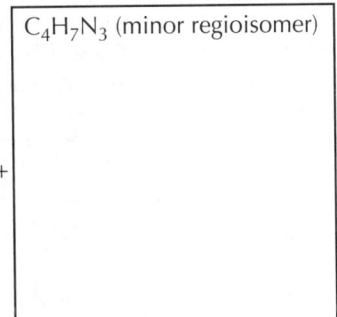

(c) Does the major regioisomer have an aromatic ring?

circle one: yes no

(d) Does the minor regioisomer have an aromatic ring?

circle one: yes no

B. The following parallel sequences result in diastereomeric products (*JCS Chem Comm*, **1987**, 1513). Hint: The format of the question contains significant clues to solving it.

2 equivalents $Li^{\oplus}$ $^{\ominus}NH_2$ → balanced equation → 1 equivalent CH_3-I followed by protonation → [ ]

H_2, Pd-$CaCO_3$, quinoline →

→ draw the structure that results in the stereochemistry given for the product + enantiomer

an intramolecular substitution catalyzed by strong acid →

\+ enantiomer

Na/NH_3 → include stereochemistry

→

\+ enantiomer

an intramolecular substitution catalyzed by strong acid → include stereochemistry + enantiomer

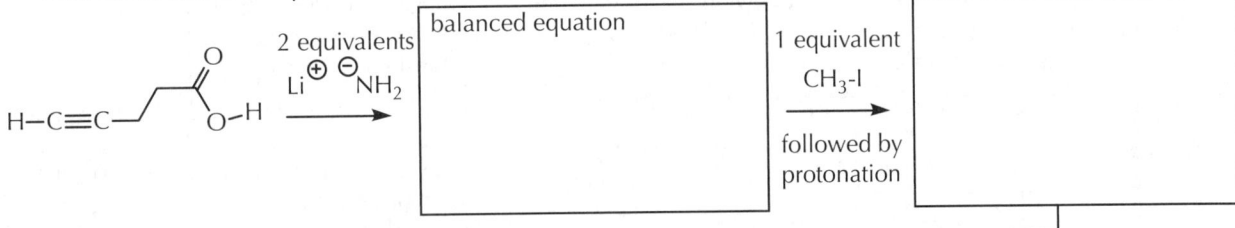

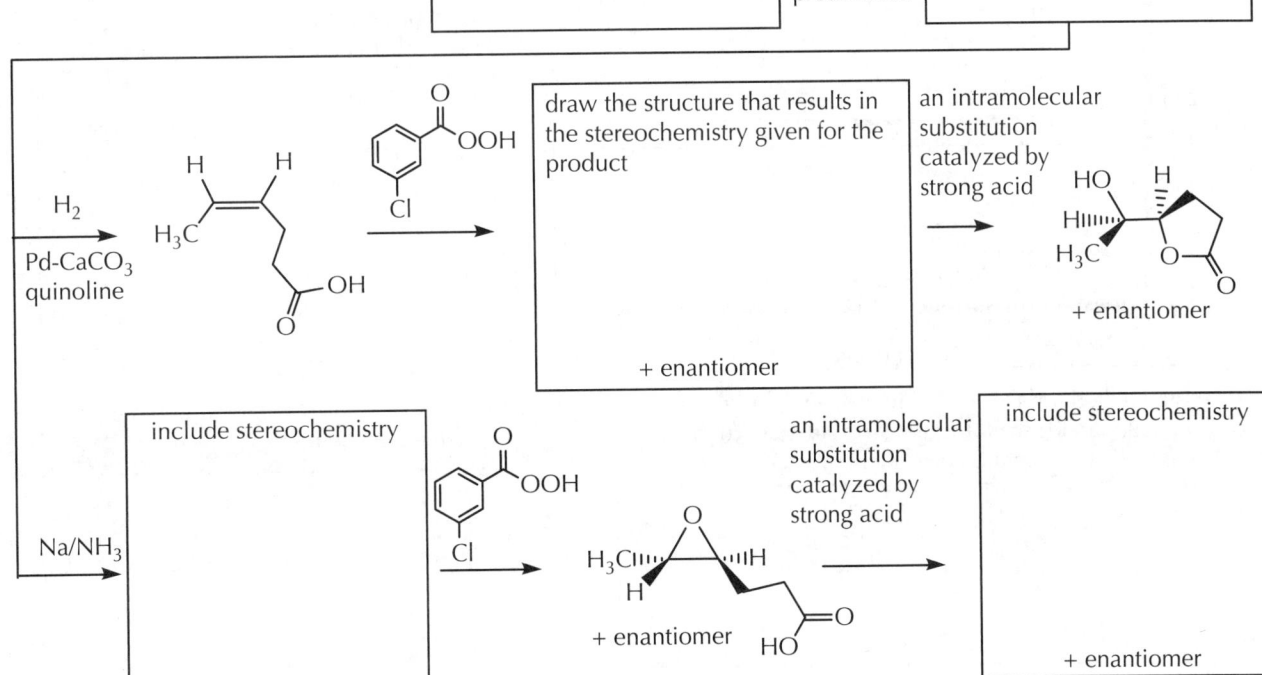

EQ 04.11

The following substitution reaction was studied in 1999 (JCS Dalton, **1999**, 3553).

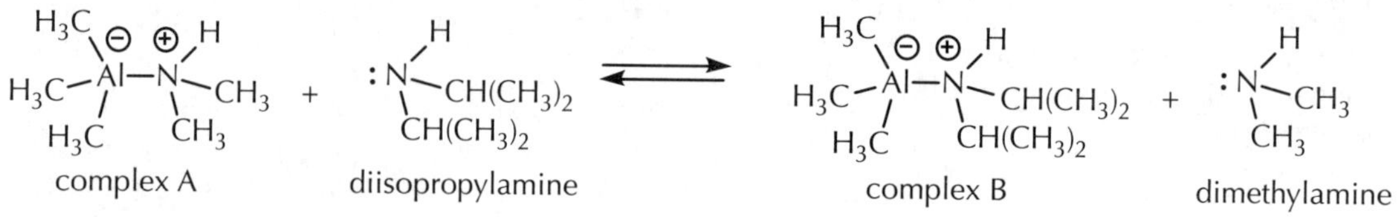

complex A + diisopropylamine ⇌ complex B + dimethylamine

EXPERIMENTAL INFORMATION

(1) The evidence favors an S_N1 mechanism, where complex A dissociates into dimethylamine and trimethylaluminum, and then diisopropylamine captures the open shell aluminum to give complex B.

(2) The activation energy to get from complex A to the open shell aluminum intermediate is 12 kcalmol^{-1} and the activation energy to get from the product, complex B, back to the open shell aluminum is 14 kcalmol^{-1}.

(3) The equilibrium favors complex A by 2 kcalmol^{-1}. The energy levels for compound A and the open shell aluminum intermediate are both shown in the energy diagram, below.

(a) Given this information, provide the curved arrow mechanism for the reaction of complex A to B. Show all of the involved molecules, nonbonding electrons, and charges.

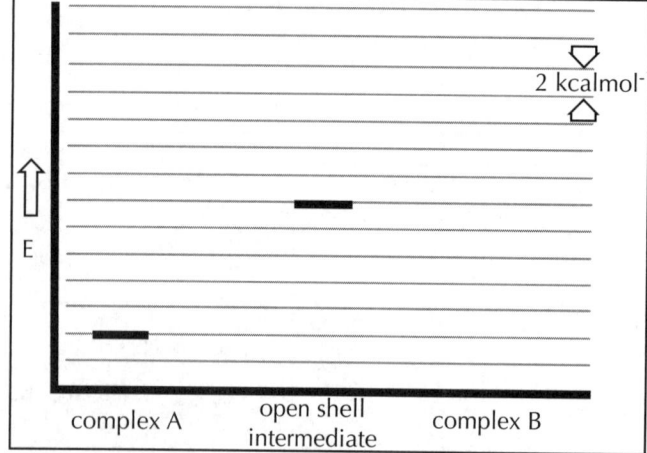

(b) Given this information, complete the following energy diagram; include the position of complex B and the required energy curves. Please note the scale (each spacing is 2 kcalmol^{-1}).

(c) What is the difference in energy between the two transition states? ☐ kcalmol^{-1}

(d) Using the nitrogen atom as the front atom and the aluminum as the back atom, draw the Newman projection for the most stable conformation of complex A. Exclude the atom labels "N" and "Al".

(e) Using dotted lines for partial bonds, and partial charges as needed, provide the transition state structure for the formation of complex B from the open shell intermediate.

(f) In just a few words, why is complex B less stable than complex A?

Exam Questions (Chapters 1–10) 871

EQ 04.12

A. In a direct analogy with molecular halogens (X_2), benzeneselenenyl bromide (PhSeBr) gives regioselective addition reaction products with the anti stereochemistry. (a) Based on this information, and the strcuture of the observed product, provide the structure of the ionic intermediate anticipated for this reaction. (b) The selenium atom can be oxidized with hydrogen peroxide (H_2O_2) and turned into a good leaving group. Thus, treating the oxidized compound with a base results in a final product with the molecular formula C_4H_5Br (draw it). (c) In principle, the same C_4H_5Br product could result from the regioselective addition of one equivalent of HBr to what compound?

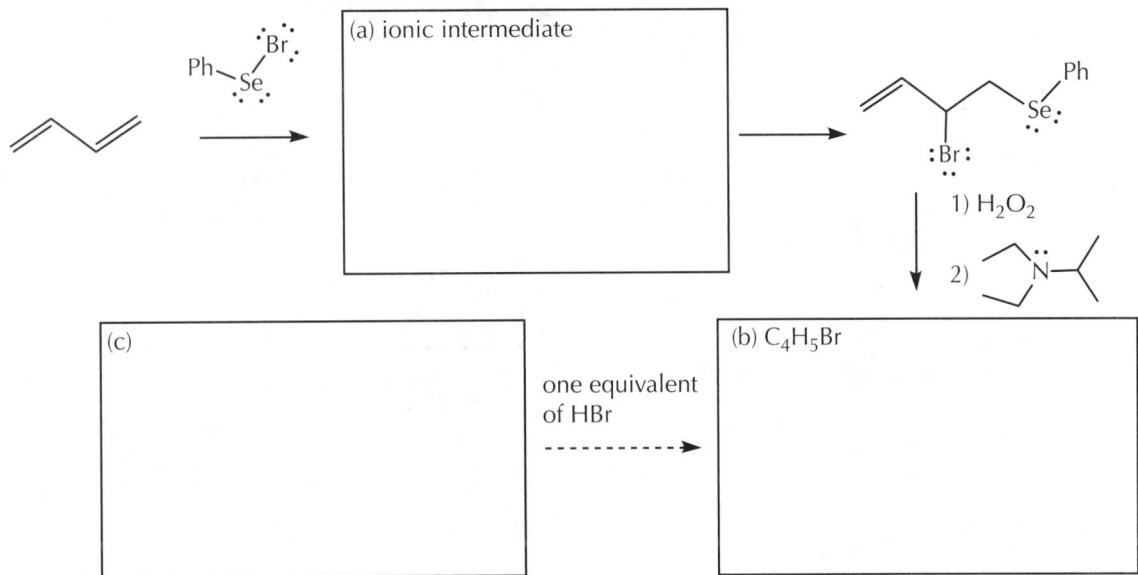

B. Complete the following reaction sequences. Separate experimental steps should be numbered.

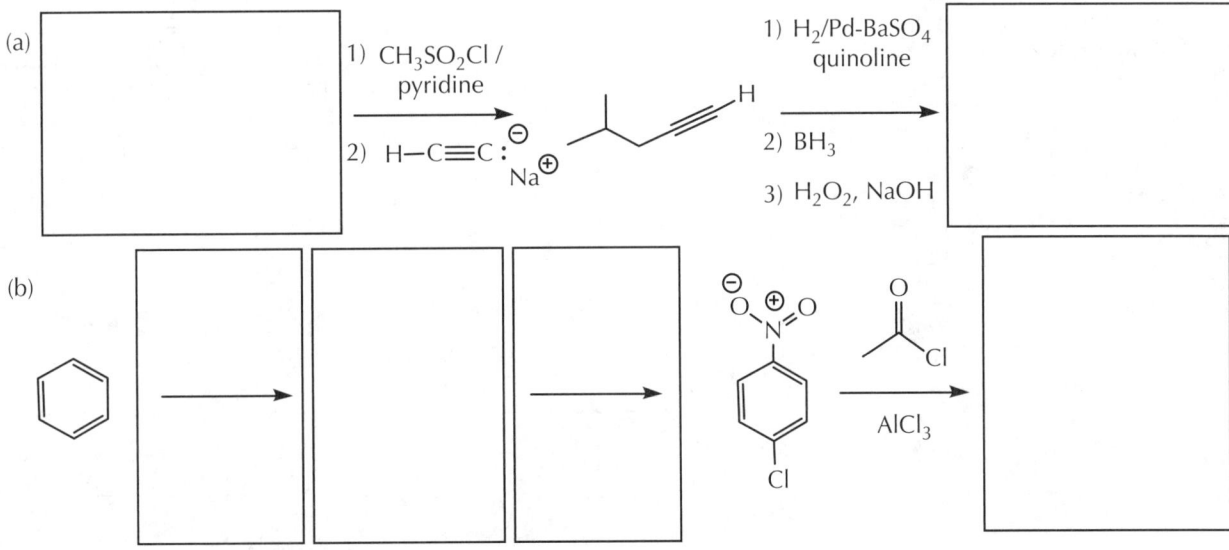

(c) Provide the starting material and reagent that give a balanced equation for the following reaction.

EQ 04.13

A. Picric acid was the first high explosive used in artillery shells (its use as an explosive dates back to 1885). However, picric acid can react with shell casings to form metal picrates, which are unstable and unsafe. A related compound, TNT, largely replaced picric acid as a high explosive between WWI and WWII.

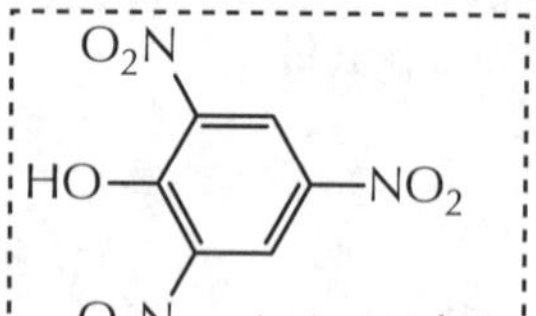

(a) Which of these two compounds (TNT or picric acid) has shorter carbon-nitrogen bonds, and provides a single, all closed shell atom resonance contributor that best explains your answer. Include all nonbonding electrons and formal charges.

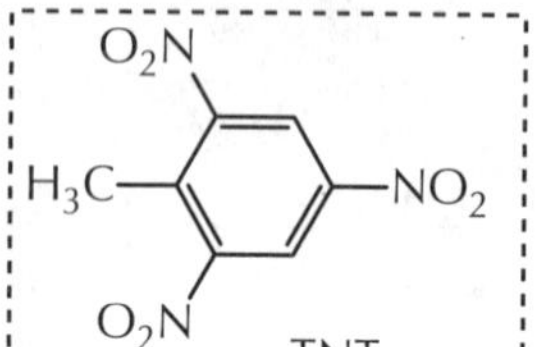

Compound with shortest carbon-nitrogen bonds (circle one):	resonance form
picric acid	
TNT	

(b) Complete the following transformation, which is used in the synthesis of picric acid.

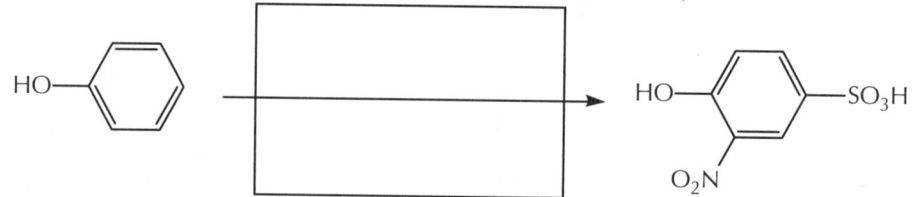

B. Draw the lowest energy chair conformation of compound Q. Clearly indicate the directionality for the H and non-H substituents on carbons 1–3 and 5 that are drawn on the structure below. You do not need to draw in the H atom substituents on carbons 4 and 6.

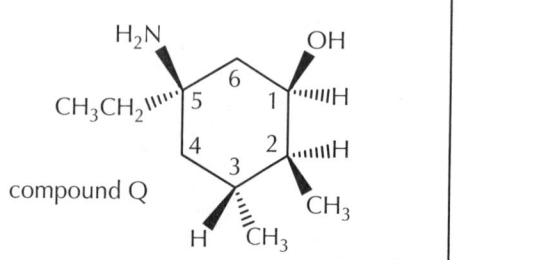

lowest energy chair form of compound Q

C. Answer the following questions about compound T.

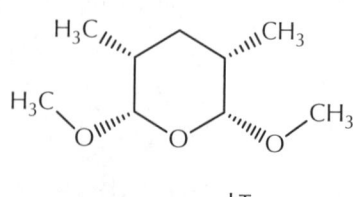

compound T

(a) How many atoms in compound T can act as H-bond acceptors?

(b) How many enantiomers does compound T have?

(c) How many signals appear in the ^{13}C-NMR spectrum of compound T?

(d) Estimate the pK_a of the conjugate acid of compound T.

EQ 04.14

Provide the reagents or structures, as necessary, to complete the following sequences of reactions. If more than one synthetic step is required, number them sequentially. You may use abbreviations for reagents (e.g., TsCl), but not acronyms (e.g., mCPBA—draw the structure instead). If any reaction produces a mixture of stereoisomers, draw one of them and write "+ enantiomer" or "+ diastereomer" in the box. For electrophilic aromatic substitution reactions, assume that only a single regioisomer will form.

(a)

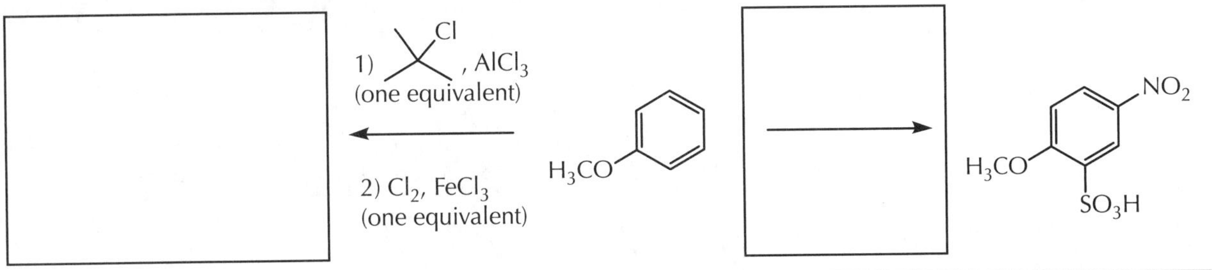

(b)

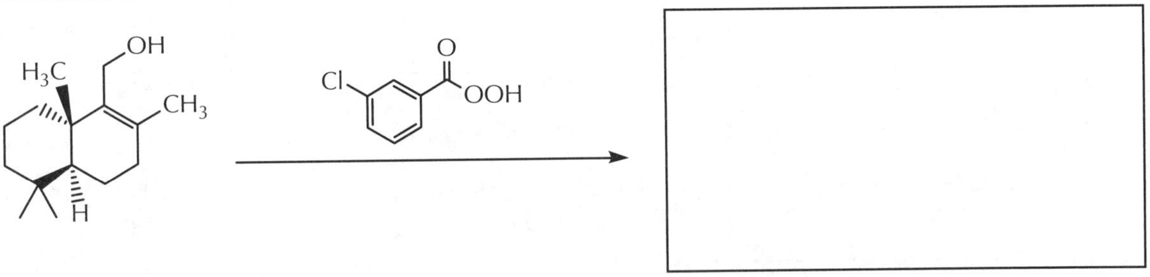

(c)

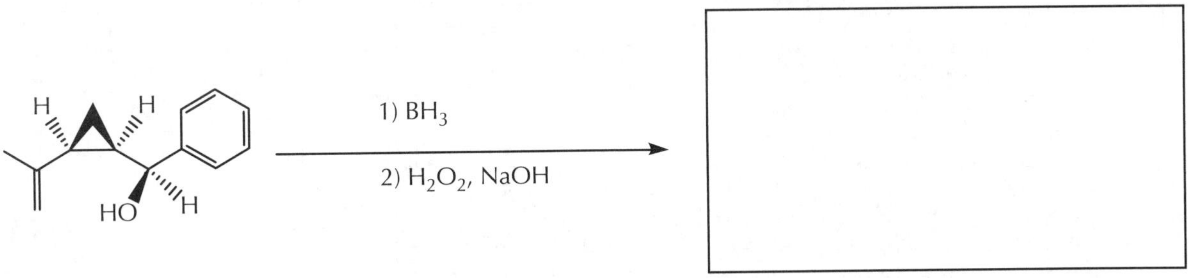

(d)

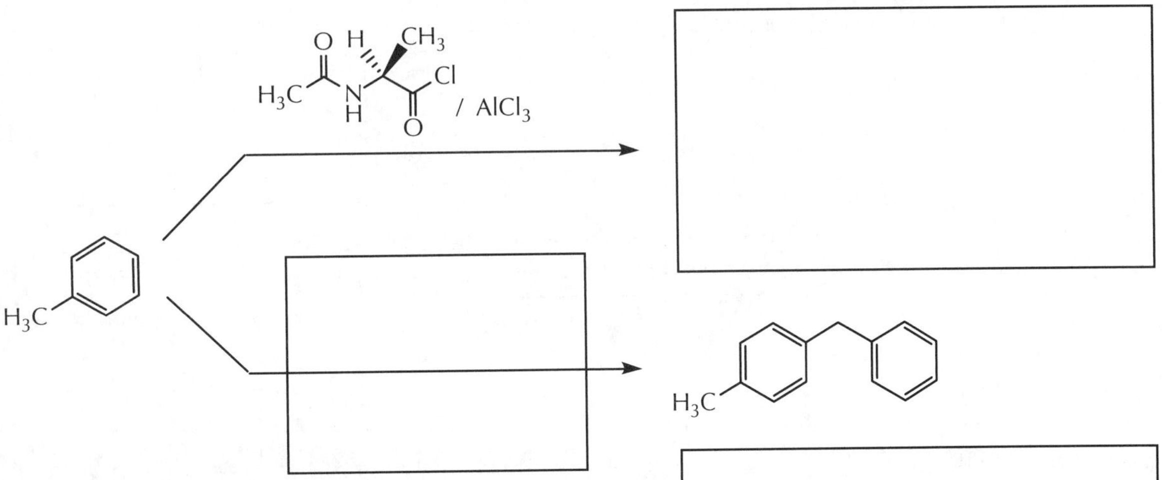

(e)

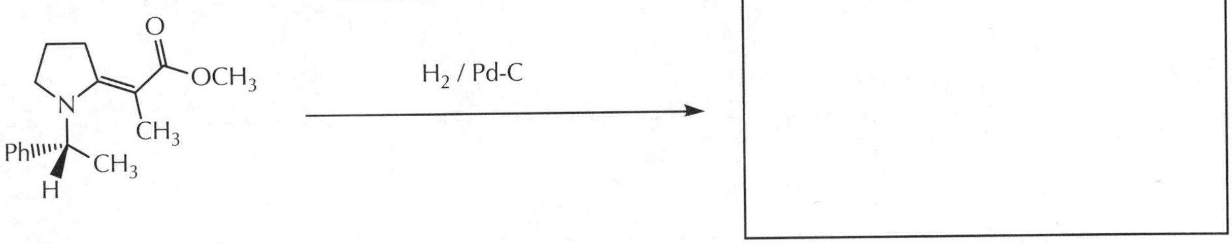

EQ 04.15

A. Check all of the boxes that apply to the following compounds.

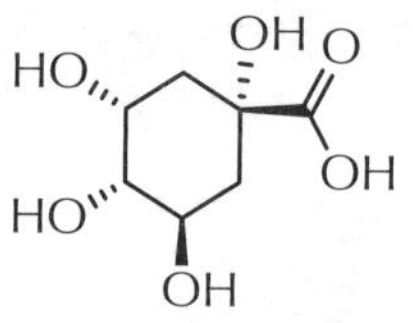

quinic acid
(plums, kiwifruit)

- ☐ is a *meso* compound
- ☐ has an enantiomer
- ☐ has an achiral diastereomer
- ☐ is chiral
- ☐ has an R stereocenter
- ☐ could be deprotonated by :N(CH$_2$CH$_3$)$_3$ with K$_{eq}$ >10^3

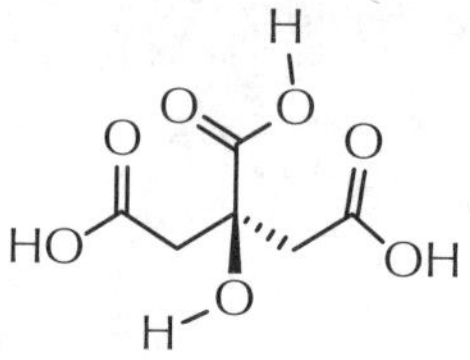

citric acid
(lemon, black currants)

- ☐ is a *meso* compound
- ☐ has an enantiomer
- ☐ has an achiral diastereomer
- ☐ is chiral
- ☐ has an R stereocenter
- ☐ could be deprotonated by :N(CH$_2$CH$_3$)$_3$ with K$_{eq}$ >10^3

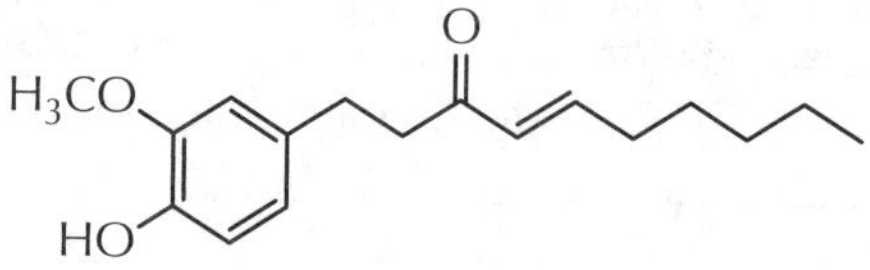

shogaol
(dried/cooked ginger)

- ☐ is a *meso* compound
- ☐ has an enantiomer
- ☐ has an achiral diastereomer
- ☐ is chiral
- ☐ has an R stereocenter
- ☐ could be deprotonated by :N(CH$_2$CH$_3$)$_3$ with K$_{eq}$ >10^3

B. (a) Draw the product from this reaction sequence. If a reaction results in a stereoisomeric mixture, draw one and indicate "+ enantiomer" or "+ diastereomer" in the box.

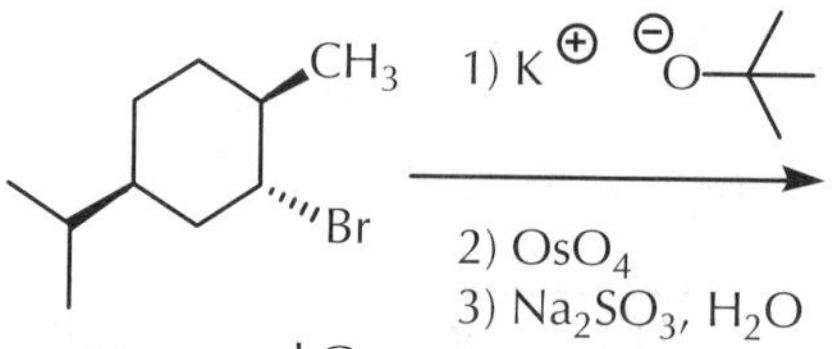

compound G

1) K$^\oplus$ $^\ominus$O-*t*-Bu
2) OsO$_4$
3) Na$_2$SO$_3$, H$_2$O

(b) Draw the most stable chair conformer for compound G, in which the isopropyl group is in the equatorial position. Include the hydrogen atoms for each substituent-bearing carbon in the ring.

C. For the elimination reaction, below:

(a) Draw the reactive conformation of alkyl bromide H (as a Newman projection), placing C$_1$ in the front, and then the resulting elimination product (compound W).

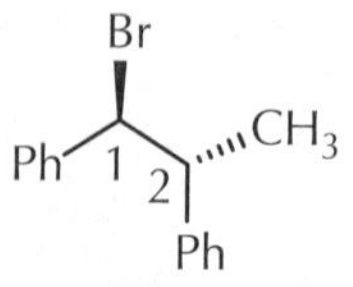

compound H

Newman projection of the reactive conformation for E2

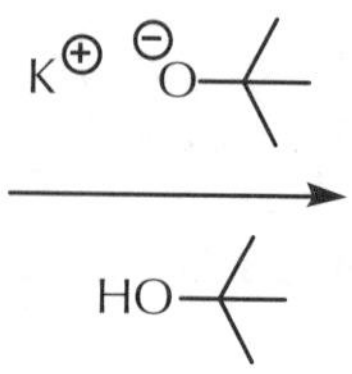

compound W

(b) Name compound W.

EQ 04.16

A. Any central atom with a tetrahedral array of other atoms can be a stereocenter. Assign the configuration (R or S) to the stereocenters in the following molecules. Reminders: Nonbonding electron pairs have an atomic number of "0," and isotopes are ranked according to atomic weight.

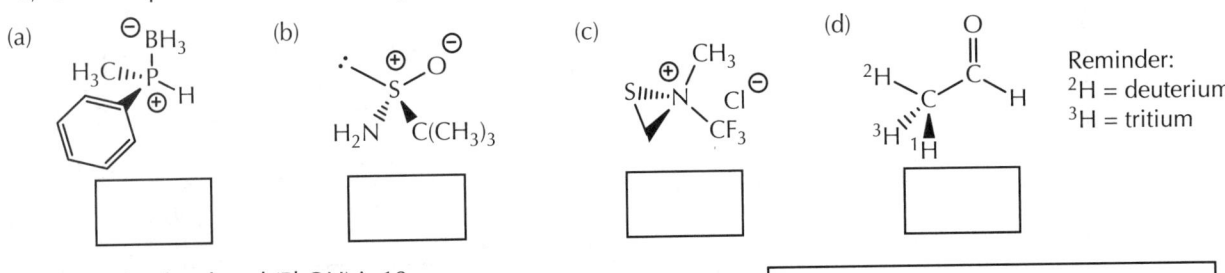

Reminder:
^2H = deuterium
^3H = tritium

B. The pK_a value for phenol (PhOH) is 10, and the pK_a for the phenolic proton in spirobacillene B is estimated to be 8. Provide the single most significant resonance contributor for the structure of the conjugate base of spirobacillene B that best explains this difference in acidity.

C. The bond indicated by the arrow in compound B has a significantly higher barrier to rotation than the corresponding bond in compound A. Provide a drawing for compound B that best explains this large difference in the ability to rotate that bond.

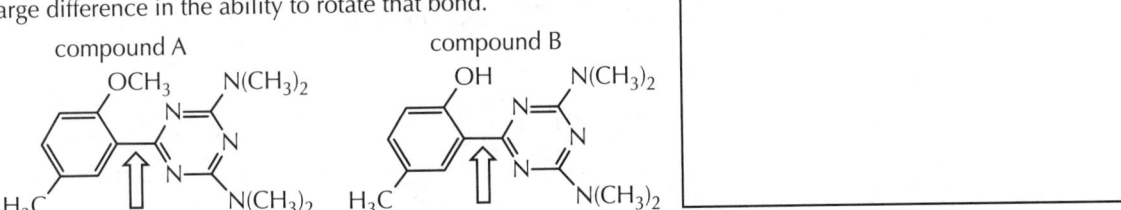

D. Given the following features:
   (a) molecular formula C_3H_3NO
   (b) all closed shell, uncharged atoms
   (c) hydrogen bond acceptor
   (d) not a hydrogen bond donor
   (e) 3 ^1H-NMR signals & 3 ^{13}C-NMR signals
   (f) an aromatic compound

Draw two structures that satisfy all of these criteria.

E. This molecule is known as a "superbase" because its conjugate acid has the remarkably high pK_a value (27) relative to those in comparable compounds, which tend to have values in the 8–13 range.

Draw the structure for the single best resonance contributor of this molecule's conjugate acid, and provide a few words that explain why this resonance contributor is so significant to its reactivity.

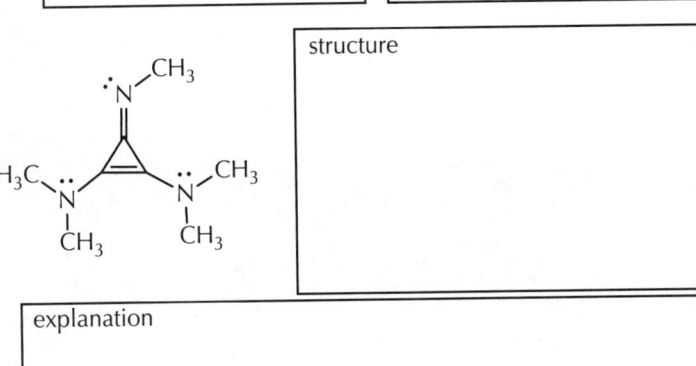

structure

explanation

EQ 04.17

A. Complete the following reactions by filling in the missing information. Be sure to number the steps in any sequential reaction.

(a)

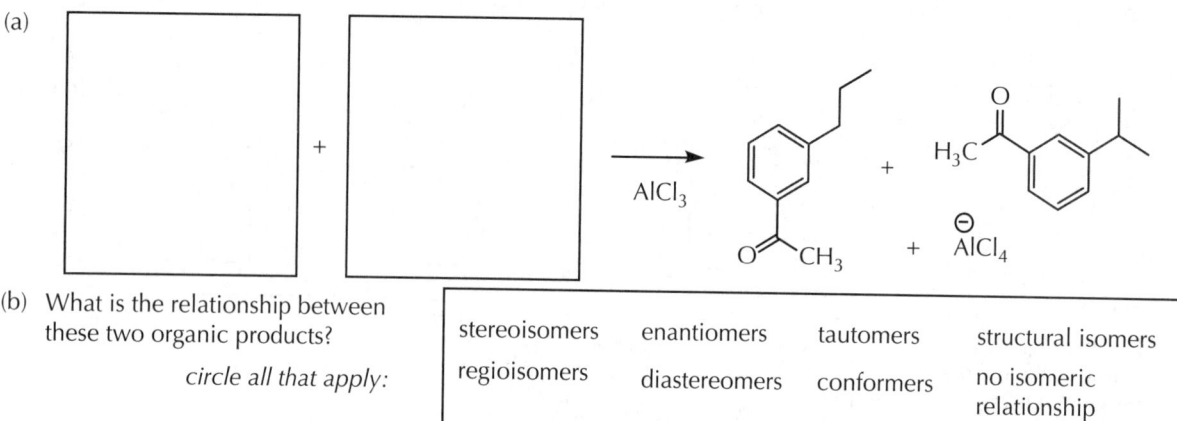

(b) What is the relationship between these two organic products?

circle all that apply: stereoisomers   enantiomers   tautomers   structural isomers   regioisomers   diastereomers   conformers   no isomeric relationship

B. Complete the following reactions by filling in the missing information. Be sure to number the steps in any sequential reaction (provide the major regioisomer when you have a choice).

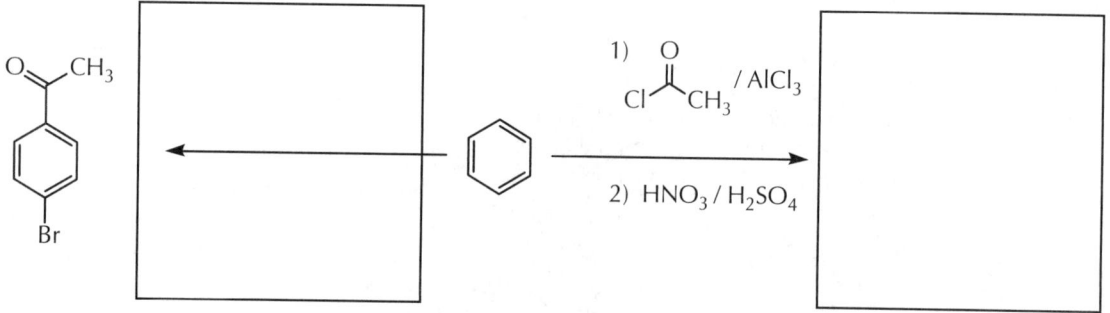

C. This reaction sequence was important in the synthesis of (±)-cis-pyrethrolone by Pattenden and co-workers in 1969 (J Chem Soc C, **1969**, 1016). Complete the missing information. If a reaction results in a stereoisomeric mixture, draw one and indicate "+ enantiomer" or "+ diastereomer" in the box.

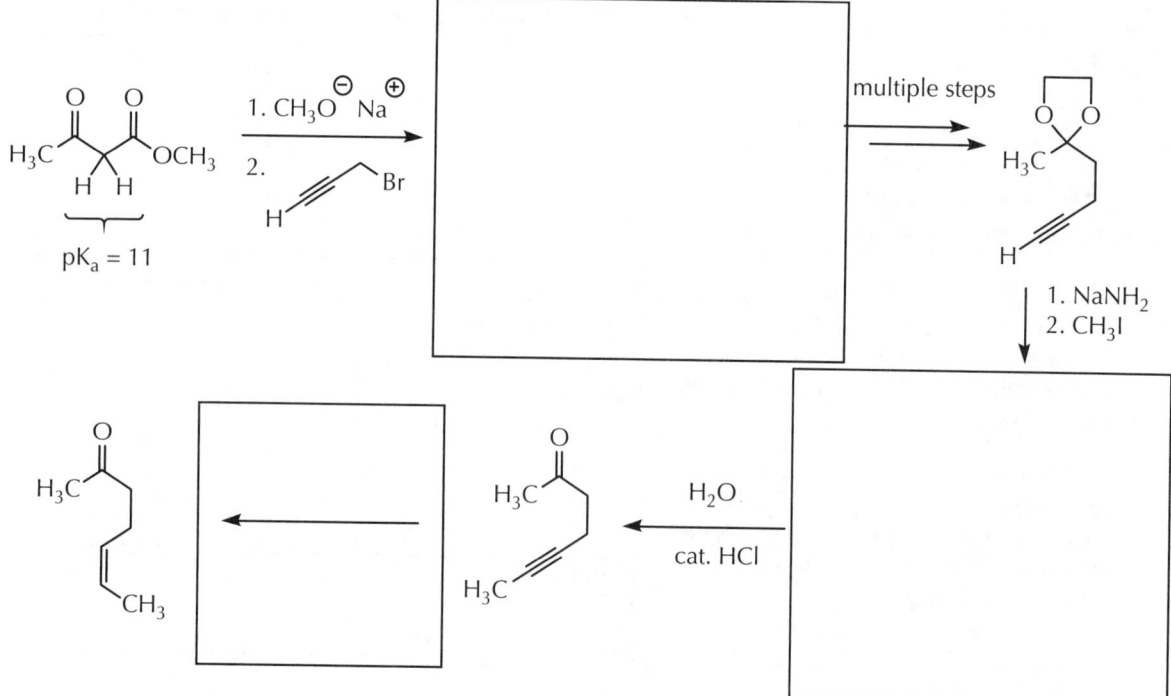

Exam Questions (Chapters 1–10)

EQ 04.18

A. Paraphrasing from *J Am Chem Soc*, **2016**, *138*, 15114:
In studying the competition between the intramolecular hydrogen bond that could form between the groups in compound N versus the intermolecular hydrogen bond that could form between compounds such as N and M, we found that proximity between the donor and the oxygen atom serving as the acceptor was the most significant factor. Indeed, the intramolecular hydrogen bond in compound N was 4 kcalmol^{-1} more stable than the intermolecular hydrogen bond formed between compounds N and M.

Given this information, draw compounds N and M in both boxes, and use a dashed line to represent the less stable and more stable hydrogen bonds.

compounds N, M, & the less stable hydrogen bond	compounds N, M, & the more stable hydrogen bond

B. Sulfonamides such as compound P are relatively weak Brønsted acids (pK$_a$ = 10). The following addition reaction requires the use of a strong acid catalyst (H–OTs) that, under the conditions of this reaction, provides the proton required for the first step and is regenerated in that last step. Given this information, and the scheme provided below, provide the complete, curved arrow mechanism (taken from *Org Lett*, **2006**, *8*, 4175). Also: Provide balanced equations in each step.

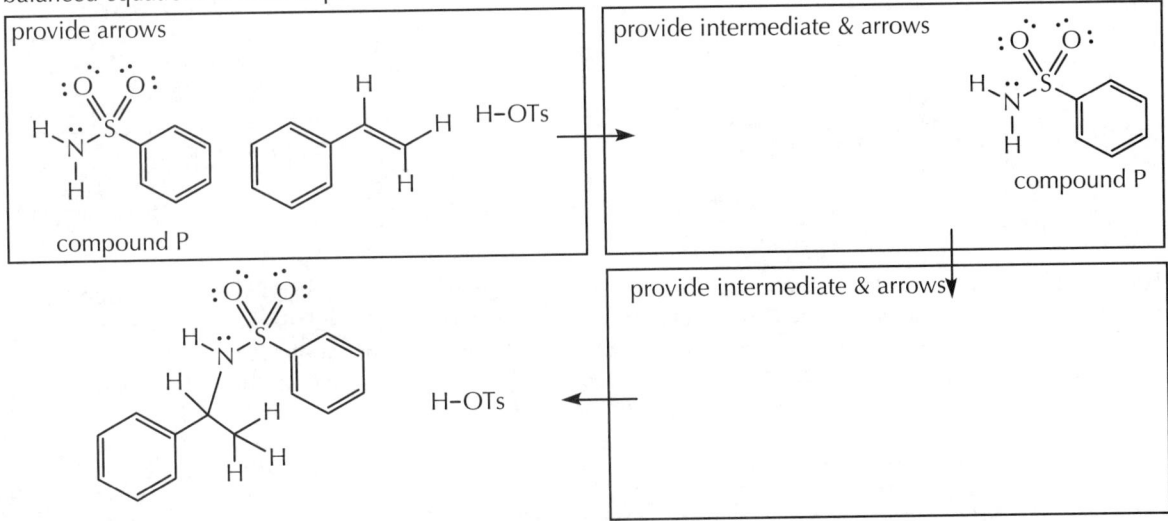

C. When the reaction in part B is carried out on a different starting material (below), the observed addition product clearly results from a rearrangement of the initially formed intermediate to its single most stable form. Given that information, what is the structure of the observed product?

the product has 10 ^{13}C-NMR signals

878 CHAPTER 10 Aromaticity and Electrophilic Aromatic Substitution Reactions

EQ 04.19

A. When compound H is heated in the presence of lithium bromide, an unusual transformation occurs through a two-step mechanism.

(a) Provide the curved arrow mechanism for the conversion of compound H to intermediate J, and also provide the structure of the product, compound K, resulting from the second step in the mechanism.

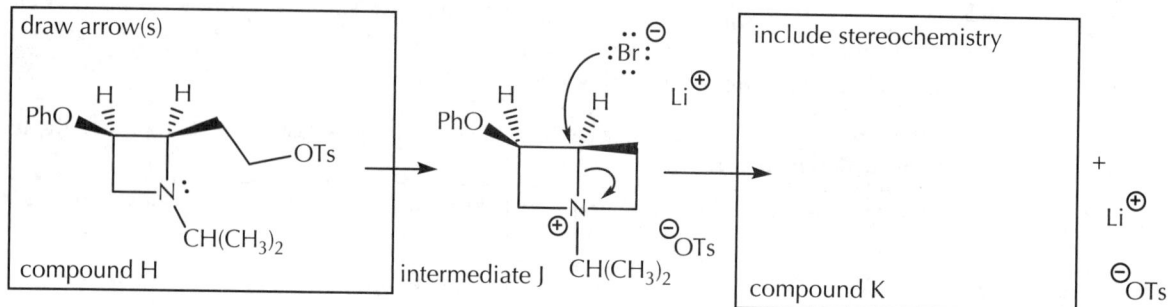

(b) The conversion of intermediate J to the product, compound K, is an example of what reaction (circle one)?

S_N1 reaction S_N2 reaction   E1 reaction   E2 reaction   electrophilic addition reaction

reduction reaction   complexation/decomplexation reaction   Brønsted acid/base reaction

B. For each of the following molecules provide the number of π electrons in the ring and indicate whether or not the ring is aromatic.

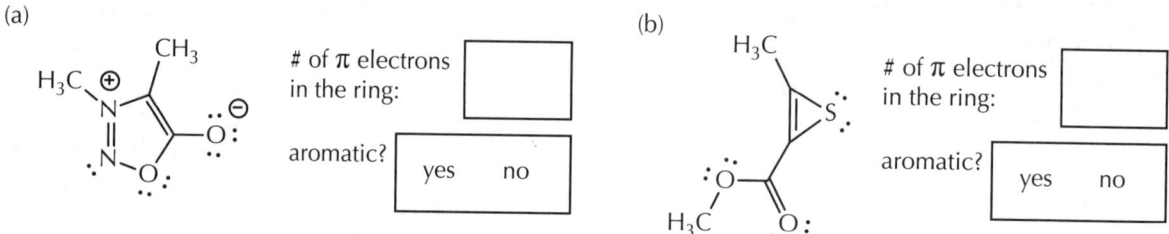

C. For the molecule shown below, assign the stereochemical configuration of the selected tetrahedral carbon stereocenters (R or S) and the alkene (E or Z) that are indicated by the arrows (note that you do not have to assign the configuration of every stereocenter in the molecule). If R/S or E/Z labels are not applicable because a carbon indicated by an arrow is not a stereocenter or the alkene cannot be assigned as E or Z, then circle "neither."

EQ 04.20

Chemists wanted to study the importance of an intramolecular hydrogen bond between the N-H bond of one substituent group and the dipolar PO bond of another. To do so, the following three compounds were prepared (compounds M, N, and O). Your objective is to select a compound that would place the two substituent groups in a position so that the potential intramolecular hydrogen bond would stabilize an otherwise destabilized conformation.

(a) Which of these compounds satisfies the requirement? (circle the letter)

(b) Draw its single chair conformation, including the intramolecular hydrogen bond (⅋⅋⅋⅋⅋⅋⅋) stabilizing an otherwise unfavored chair conformation. Include all of the atoms shown in the structural drawing, but not more.

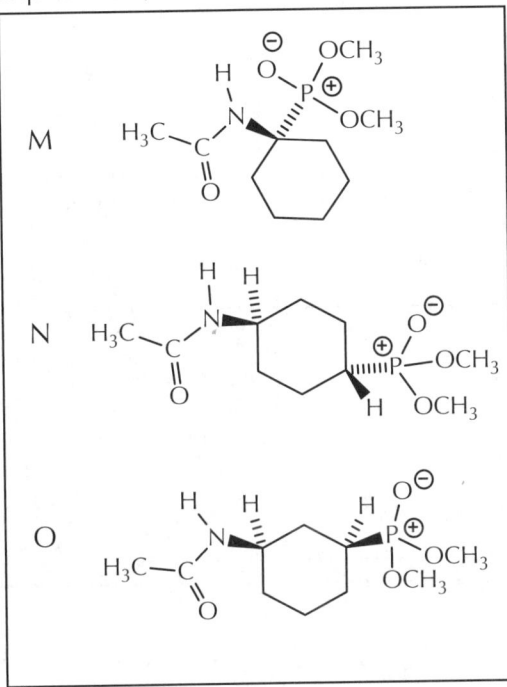

(c) Answer the following about compounds M, N, and O.

  (i) Number of chiral diastereomers it has? M ☐ N ☐ O ☐

  (ii) Number of (R) stereocenters it has? M ☐ N ☐ O ☐

  (iii) Number of enantiomers it has? M ☐ N ☐ O ☐

  (iv) Number of achiral diastereomers it has? M ☐ N ☐ O ☐

(d) Using the description of the study in part (a), circle the choices that make this a correct statement. This does not depend on the other answers you wrote.

> If the CH_3 in the group attached to the nitrogen was replaced by a CF_3 group, the NH would be a [circle one: better worse] hydrogen bond [circle one: acceptor donor], and cause the intramolecular hydrogen bond to be [circle one: stronger weaker].
>
> Consequently, the resulting chair conformation containing the hydrogen bond would be [circle one: more less] stable when the molecule with the CH_3 group is used, compared with the one with the CF_3 group.

EQ 04.21

The mechanisms for some impressively complex transformations can be understood as a sequence of more familiar, fundamental steps. Here is an example of such a transformation, published in early 2012 (*Org Lett*, **2012**, *14*, 5250).

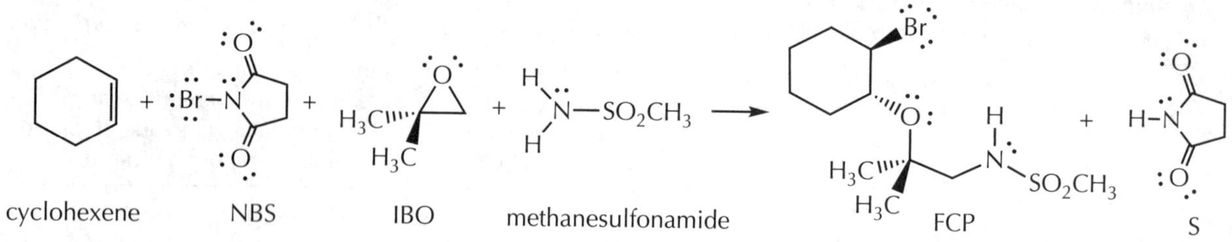

cyclohexene + NBS + IBO + methanesulfonamide → FCP + S

Provide the curved arrow mechanism for each step in the proposed sequence, based on the text description for the process. You should draw exactly those things described in the text. Please note that these texts refer to the molecules above. NOTE: the mechanism involves a bromonium ion which you have seen before as the three-membered charged intermediate involved in the addition of Br_2 to alkenes. Draw structures; do not use abbreviations.

(a) Cyclohexene reacts with NBS to give a single, ionic intermediate: the cation is the bromonium ion derived from cyclohexene (include the stereochemistry), and the anion is the conjugate base of S.

(b) The anion from part (a) [conjugate base of S] undergoes a Brønsted acid-base reaction with methanesulfonamide to give S and the conjugate base of methansulfonamide.

(c) The cation from part (a) [bromonium ion derived from cyclohexene] undergoes an anti ring opening reaction of the bromonium ion using the oxygen atom of IBO as the nucleophile, in order to give a single new cationic intermediate with a positively charged oxygen atom. Stereochemistry should be shown clearly, and be consistent with the anti ring opening.

(d) Finally, the new cationic intermediate formed in part (c) undergoes reaction with the conjugate base of methanesulfonamide [formed in part (b)] in order to give the observed product FCP, through an opening of the positively charged three-membered ring.

EQ 04.22

A. Provide a complete curved arrow mechanism for the most straightforward conversion of compound F to G. Carefully consider the reaction conditions when choosing acids or bases to draw in your mechanism (i.e., use the actual acid and the actual base that are involved in the transformation).

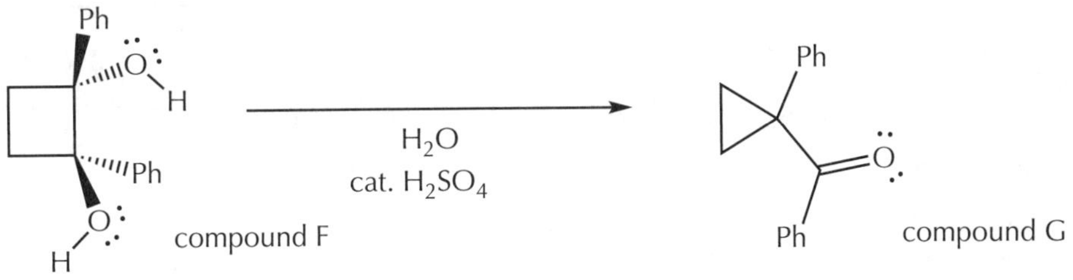

compound F → compound G (H$_2$O, cat. H$_2$SO$_4$)

B. Provide the products of the following reaction.

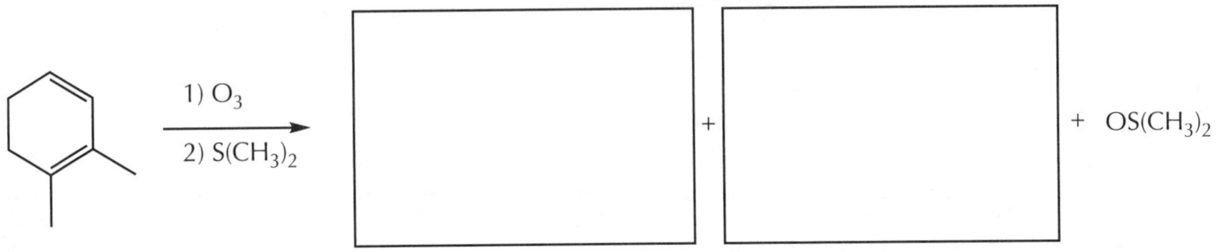

1) O$_3$
2) S(CH$_3$)$_2$

_____ + _____ + OS(CH$_3$)$_2$

C. Draw a complete Lewis structure, including nonbonding electrons and/or formal charges if/as necessary, for a C$_5$H$_5$O$^\oplus$ isomer that fulfills the following criteria:

- all atoms are closed shell
- the molecule has an aromatic ring
- there are no H-bond donors in the molecule

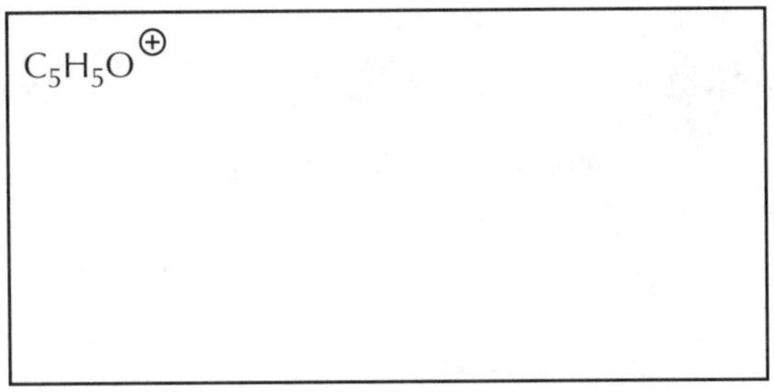

C$_5$H$_5$O$^\oplus$

EQ 04.23

A. Varubi is the name of a medical drug being used to treat breast and skin cancers. The active ingredient is rolapitant hydrochloride. What is the structure of the anticipated reaction product from two sequential bromination reactions? Reminder: The trifluoromethyl (CF$_3$) group is a strong ring deactivator.

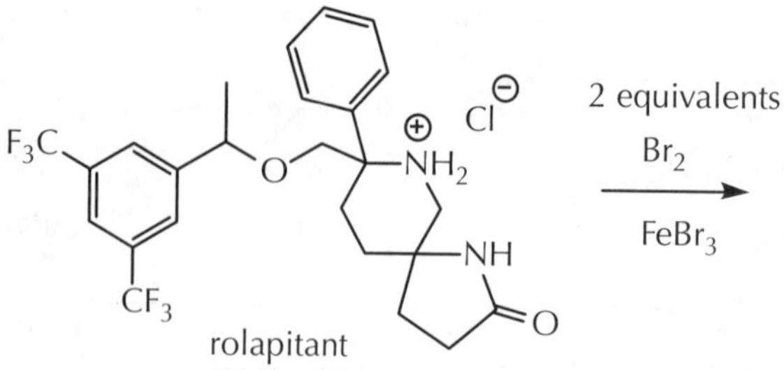

rolapitant
(HCl salt)

2 equivalents
Br$_2$
FeBr$_3$

B. You may have noticed that we rarely include fluorine atoms when talking about the halogens. This is because the chemistry of fluorine is often an exception to the generalizations that are true for the other halogen atoms. For example, here is a quote from 2006 (*J Org Chem,* **2006**, *2*, 19):

"It is well established that fluorine atoms on adjacent carbon atoms prefer to adopt a gauche conformational relationship rather than an anti conformational relationship…"

Using Newman projections for the C-C bond bearing the two fluorine atoms, draw the three staggered conformations for the molecule shown here.

(i) Account for the general statement about the fluorine atoms when evaluating the stability of the strutures.

(ii) You may show the view for the Newman projection from the right or left.

(iii) The gauche steric interaction between two -CO$_2$CH$_3$ groups is 3.5 times greater than the gauche steric interaction between a -CO$_2$CH$_3$ and an F atom.

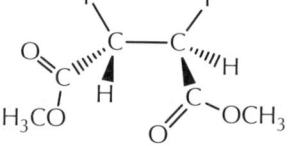

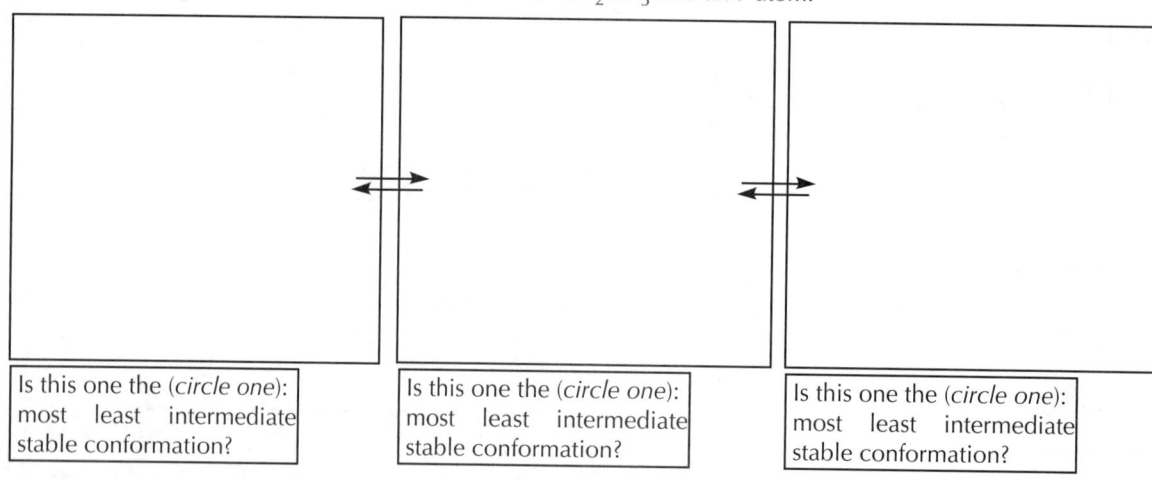

Is this one the (*circle one*): most   least   intermediate stable conformation?

Is this one the (*circle one*): most   least   intermediate stable conformation?

Is this one the (*circle one*): most   least   intermediate stable conformation?

C. Draw the structure for:

(*R*)-2-chloro-3-methylcyclohexa-2,4-dien-1-ol

EQ 04.24

A. The molecule shown below is synthesized by a mollusk called a "sea hare." This mollusk excretes this compound in a slime that acts as a natural defense mechanism against carnivorous fish!

(a) Identify and label all sites that have stereochemical configuration.

(b) Put a circle around the shortest bond.

(c) Put a square around the longest bond.

(d) For each of the atoms indicated with an arrow, provide the atom's most reasonable hybridization as well as its electronic (VSEPR) geometry, and its observable geometry (shape).

Cl
O2
H
O1
Br
C
H

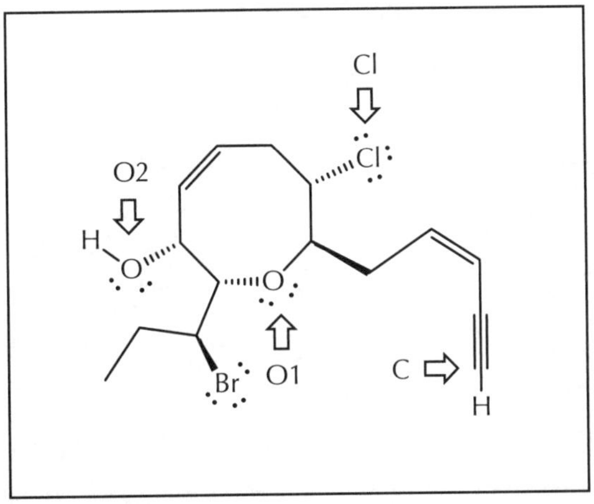

Atom	Hybridization	Electronic (VSEPR) geom.	Obs. geom. (shape)
C			
Cl			
O1			
O2			

B. Each of the following transformations yields one major product (i.e., no mixtures of stereoisomers or regioisomers). Draw this product for each.

(a)

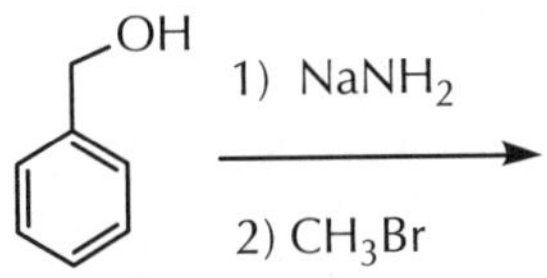

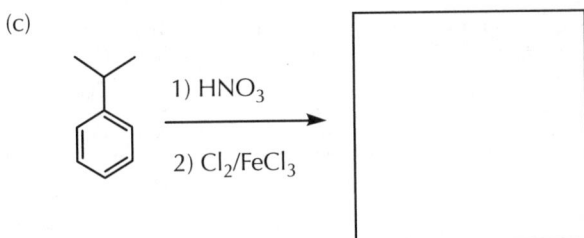

OH
1) NaNH2
2) CH3Br

(b)

1 equiv. OH
H2SO4 (cat.)

(c)

1) HNO3
2) Cl2/FeCl3

(d)

CH3NH2
H3C
Cl
H

(e)

OH
1) CH3ONa
2)
H3C
O
S
Cl
O
O
K

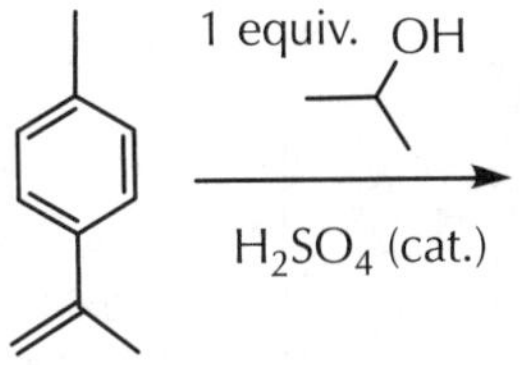

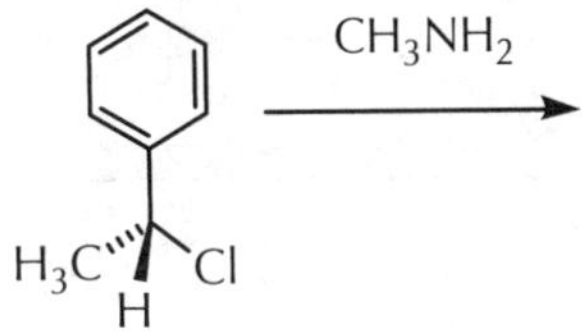

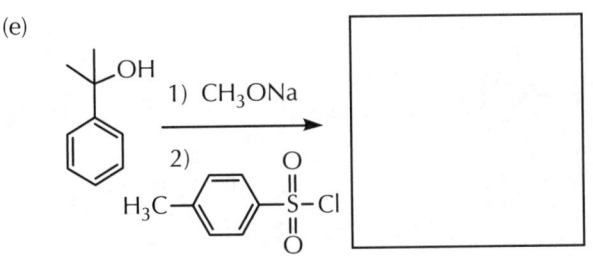

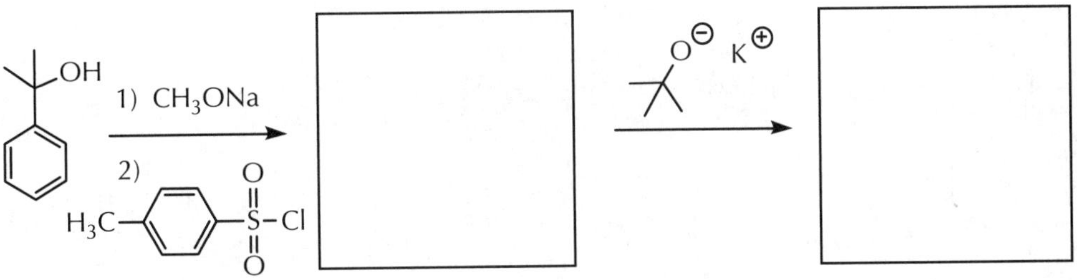

EQ 04.25

A. Compound W undergoes a rapid isomerization to form toluene using catalytic acetic acid.

(a) Provide a complete, stepwise, curved arrow mechanism for this process.

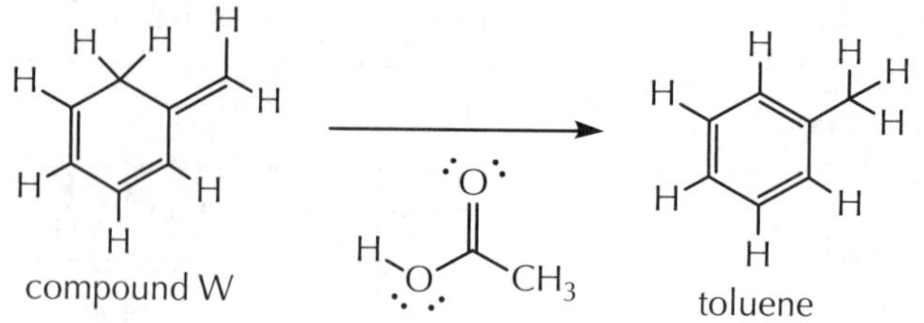

compound W   acetic acid   toluene

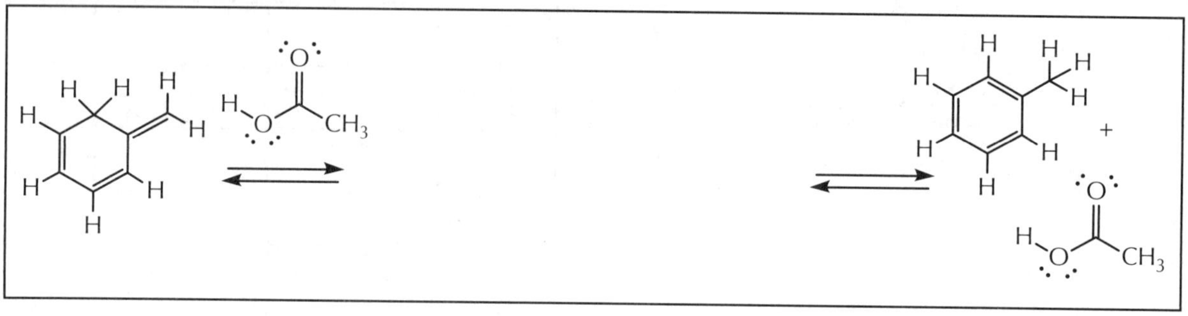

(b) There is a constitutional (structural) isomer of compound W that also, when treated with acetic acid, isomerizes to give toluene. The experimental data for this compound is provided; what is its structure?

mol. formula: C_7H_8

^{1}H-NMR spectrum:
4 signals (1:1:1:1)

^{13}C-NMR spectrum:
5 signals

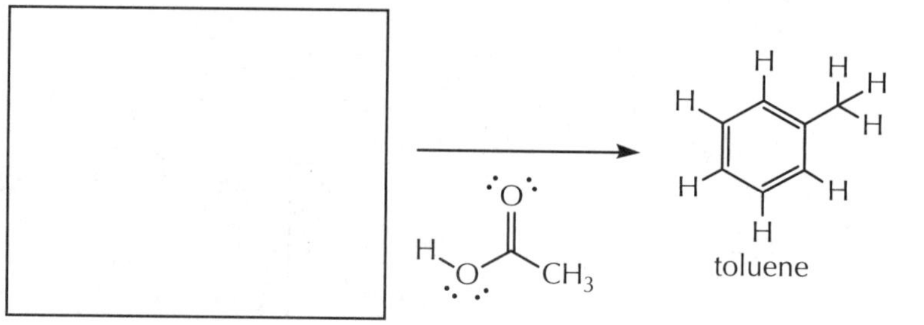

toluene

B. The following combination of reagents is observed to result in an electrophilic aromatic substitution. No other reagents except the ones shown here are required. The NCS (N-chlorosuccinimide) reagent provides an electrophilic chlorine atom, analogous to the combination of $Cl_2/AlCl_3$ and the mechanism does NOT proceed through a chloronium ion. Draw the complete, stepwise, curved arrow mechanism for this reaction (*ACS Med Chem Lett*, **2015**, 6, 584).

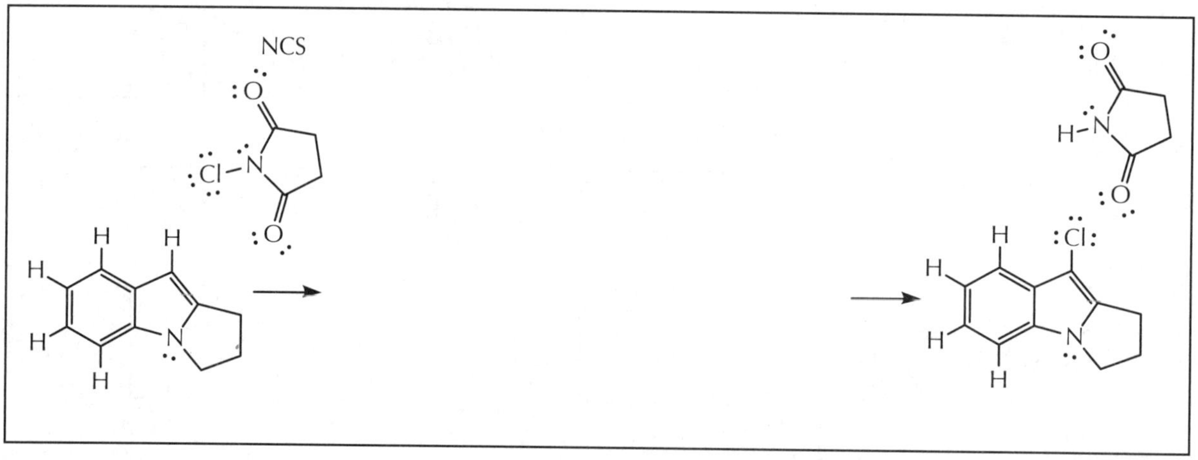

EQ 04.26

A. The length of the carbon-nitrogen bond in compound P is different than the length of the carbon-nitrogen bond in compound Q. Indicate which of these molecules has the shortest carbon-nitrogen bond, and provide a single structural drawing that best explains your answer.

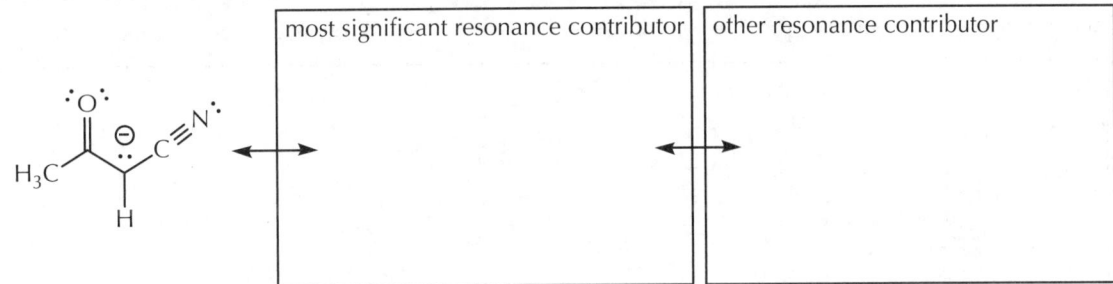

The molecule with the shortest carbon-nitrogen bond is:	drawing that best explains your answer
compound P compound Q circle one	

B. Draw two other resonance contributors of the anion shown below that have all closed shell atoms with no formal charges greater than +/– 1 on any single atom. Be sure to include nonbonding electron pairs and formal charges as necessary. Place the most significant resonance contributor in the left-hand box.

most significant resonance contributor	other resonance contributor

C. Nitrosonium tetrafluoroborate ($NO^{\oplus} BF_4^{\ominus}$) is an ionic compound that has found many applications as a reagent for organic synthesis. Draw a complete Lewis structure for nitrosonium tetrafluoroborate. Be sure to include nonbonding electron pairs and formal charges as necessary in your structure. Note that your Lewis structure should contain all closed shell atoms, with no formal charges greater than +/– 1 on any atom.

Lewis structure for $NO^{\oplus}$ $BF_4^{\ominus}$

D. Cyanamide is a useful reagent that is used for the synthesis of many important heterocyclic molecules. Provide a three-dimensional orbital representation for the most significant resonance contributor of cyanamide (i.e. the one shown below). Use lines, dashes, and wedges for sigma bonds and to indicate the directionality of localized nonbonding electron pairs. Draw overlapped p orbitals for pi bonds. Be sure to carefully consider whether nonbonding electrons are localized or delocalized when assigning hybridization/geometry.

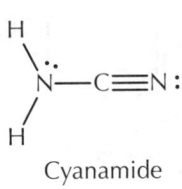

Cyanamide

886 CHAPTER 10 Aromaticity and Electrophilic Aromatic Substitution Reactions

EQ 04.27

(a) As reported in *J Org Chem,* **2016**, *81*, 10912, compound K reacts with N-bromosuccinimide (NBS, an electrophilic bromine source) and gives the following observed product ("PMB" and "TBS" are abbreviations for groups that do not participate in the reaction).

The mechanism is proposed to take place in three steps:

- Step 1: compound K reacts with NBS to give a bromonium ion and the conjugate base of succinimide
- Step 2: an intramolecular reaction occurs between the alcohol and the bromonium ion
- Step 3: deprotonation of the cationic oxygen atom, which gives compound L and the succinimide byproduct

According to this information, provide the complete, stepwise, curved arrow mechanism for this reaction.

(b) The stereoisomer of the actual product is shown here. Assign the "R" or "S" configuration to each of the indicated stereocenters (as above, the abbreviated groups can be ignored), and assign the "E" or "Z" configuration to the double bond.

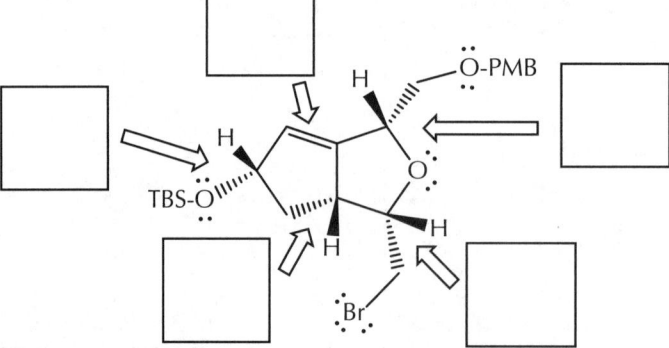

EQ 04.28

A. Provide a brief explanation, in words, for the following observations.

(a) The reaction of tropyllium bromide to give the tropyllium cation is faster than might be expected for a carbocation, while the formation of the anion is slower than expected.

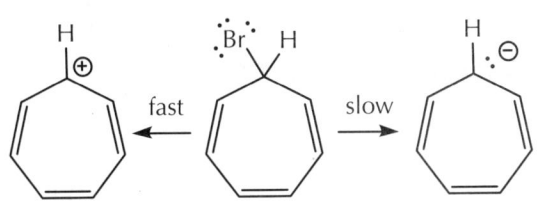

(b) Cl$_3$CCH$_2$CH$_2$OH has a significantly lower pK$_a$ value than CH$_3$CH$_2$CH$_2$OH.

B. Provide a single drawing that best explains the following observations.

(a) The indicated bond in compound W is significantly shorter than the corresponding bond in compound Y.

use a drawing:

(b) The rate of S$_N$2 substitution using sodium chloride as a nucleophile is 10,000 times slower in CH$_3$OH as compared with DMSO [(CH$_3$)$_2$S=O].

use a drawing:

C. Draw an optically active stereoisomer of: cyclodec-7-en-5,9-diyne-1,2,3,4-tetraol (*J Org Chem*, **2003**, *68*, 9379).

888 CHAPTER 10 Aromaticity and Electrophilic Aromatic Substitution Reactions

EQ 04.29

A. Complete the following energy diagram by providing the structures for the three intermediates in this reaction (B, C, and D). Although curved arrows are not required, you might still find it useful to think of this as a mechanism problem. Every answer space should show a balanced equation in your structures (i.e., account for all of the atoms). From: *Annalen der Chemie*, **1918**, *417*, 263.

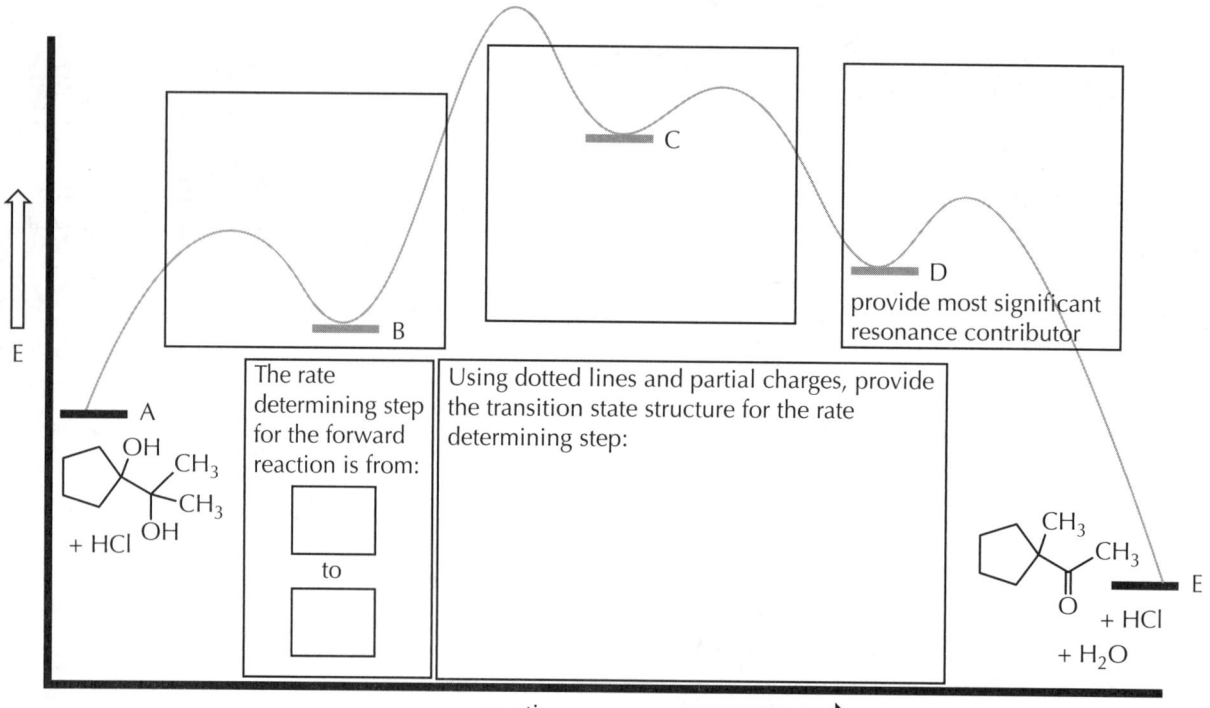

B. In the drawing for one of the resonance contributors of compound TDAC ($C_7H_{12}N_2$), all of the Hs are shown, and no atom bears a formal charge.

(a) What is the hybridization of each of these atoms?

N1:
N2:
C:

(b) Provide a resonance contributor with all closed shell atoms, including all nbe and charges.

(c) The three-membered ring in TDAC is:

(mark one)

☐ aromatic
☐ antiaromatic
☐ nonaromatic

(d) Provide the most significant, all closed shell atom resonance contributor for the conjugate acid of TDAC

EQ 04.30

A. Each of the following reactions gives two major products that share the same molecular formula. Draw these two isomeric products, showing all stereochemistry (if any) clearly, and define their relationship.

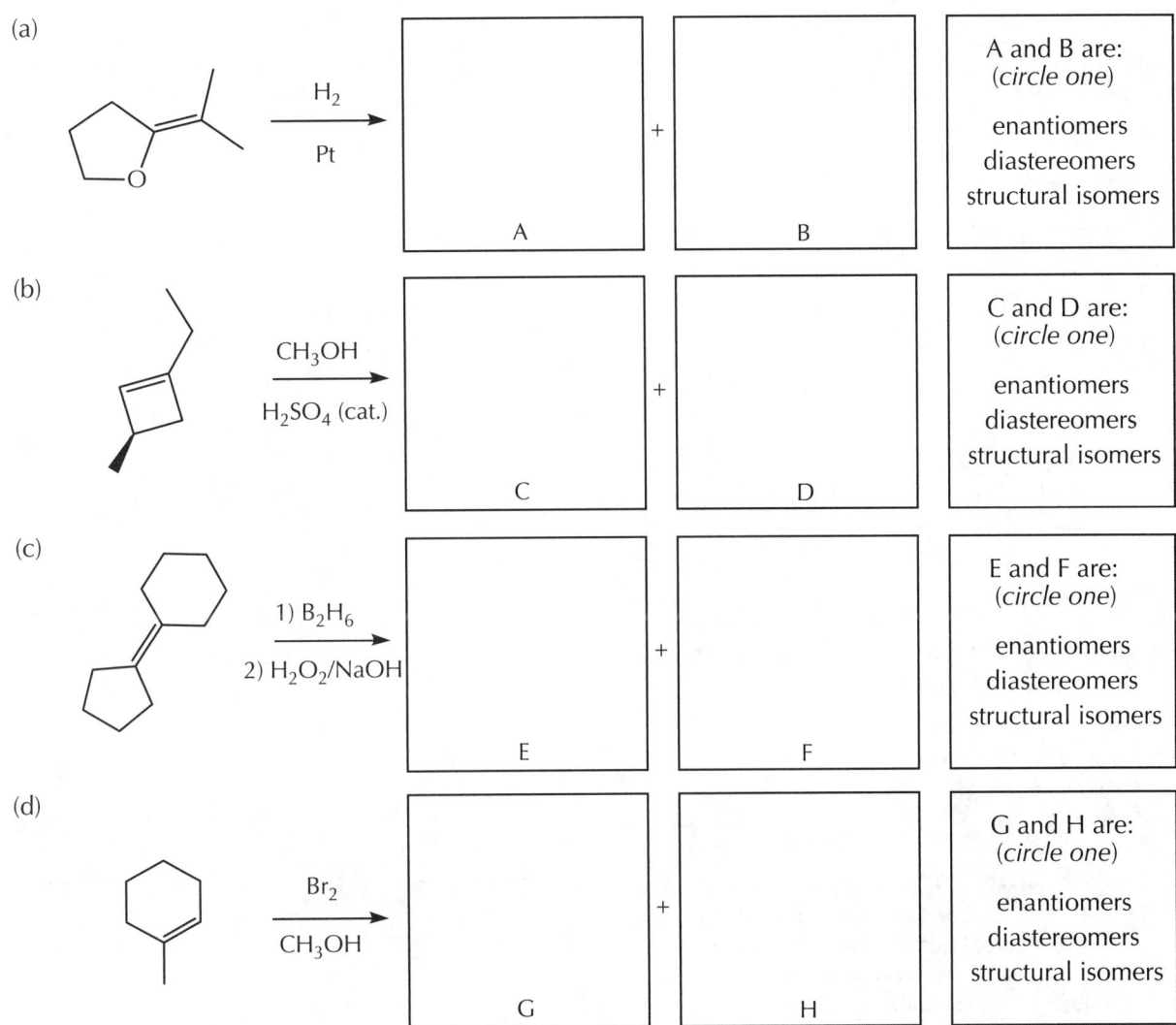

B. Provide the missing compounds in the proton transfer reaction shown below. Show only the most significant resonance contributor for the missing compounds. Also, estimate the K_{eq} for the reaction.

CHAPTER 10 Aromaticity and Electrophilic Aromatic Substitution Reactions

EQ 04.31

A. Provide the curved arrow mechanism for this single-step bimolecular reaction.

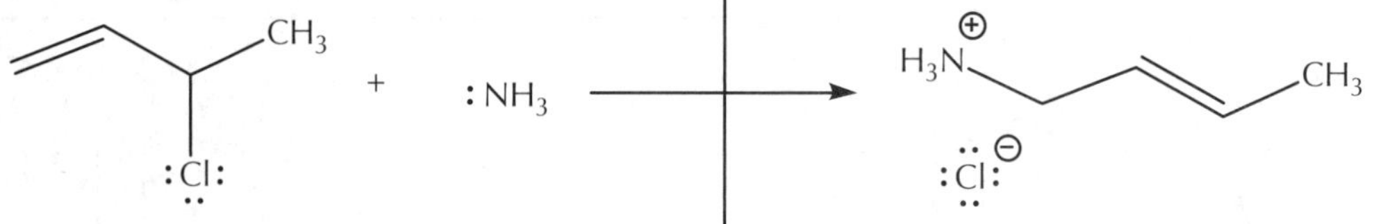

B. How many optically active diastereomers does each of these molecules have?

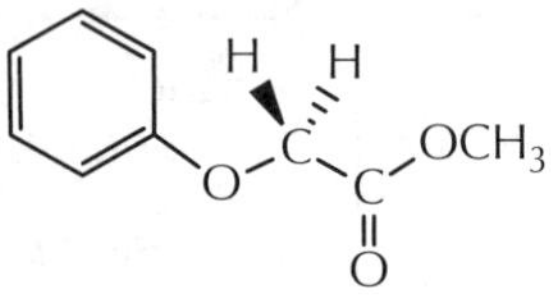

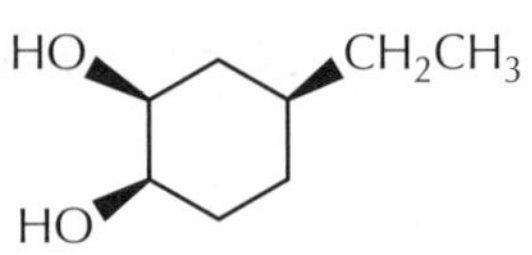

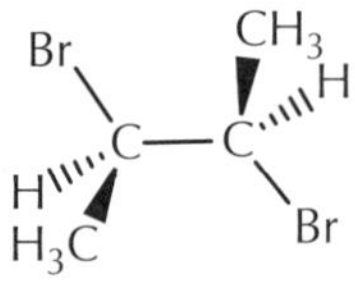

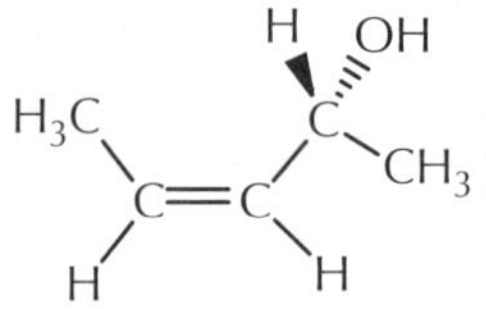

C. The experimentally determined pK_a values for serine are 16, 9.15, and 2.21. What base could be used to deprotonate the most acidic proton in serine ($K_{eq} > 10^2$) without causing significant deprotonation of the other acidic protons (keeping $K_{eq} < 10^{-3}$)?

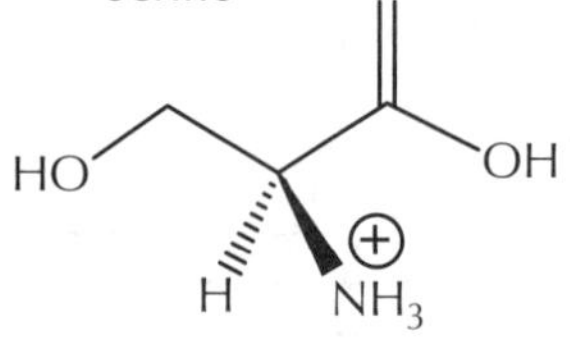

D. (a) The connectivity of methyl isocyanate is H_3CNC. Draw a Lewis structure of a resonance structure of H_3CNC where all of the atoms are closed shell. You should include all formal charges and nonbonding electron pairs.

Label each of the non-hydrogen atoms by pointing to the atoms with an arrow and indicating their expected hybridization.

(b) Draw a 3-dimensional orbital picture of methyl isocyanate. Use lines, dashes, and wedges to indicate sigma bonds and the directionality of nonbonding electron pairs, p orbitals for lone, unhybridized p orbitals and overlapping p orbitals for pi bonds.

EQ 04.32

A. Please indicate the relationship between the following pairs of molecular structures by placing an "X" in the appropriate space (or spaces, as needed).

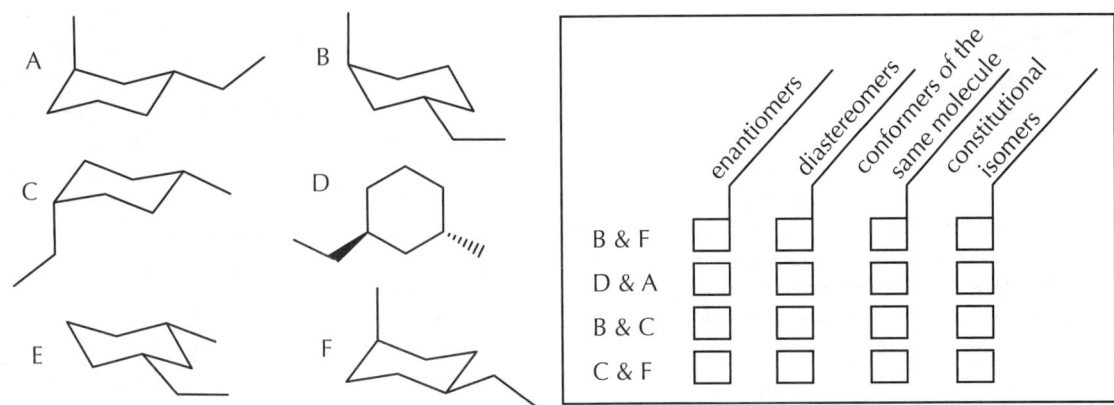

B. One of the resonance contributors for diaminocarbene is shown here. What are the anticipated hybridizations, electronic and observable geometries for each of the three indicated atoms?

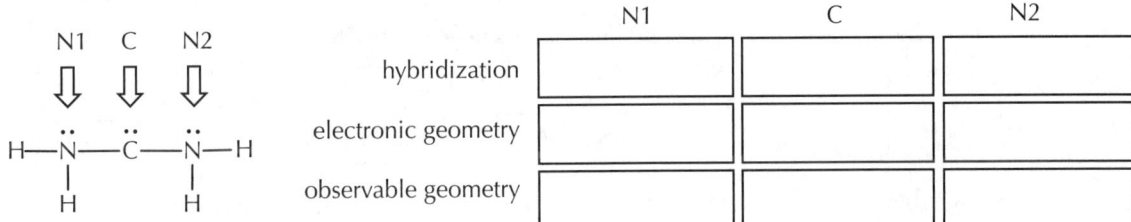

C. Provide the requested information about G, H, and I.

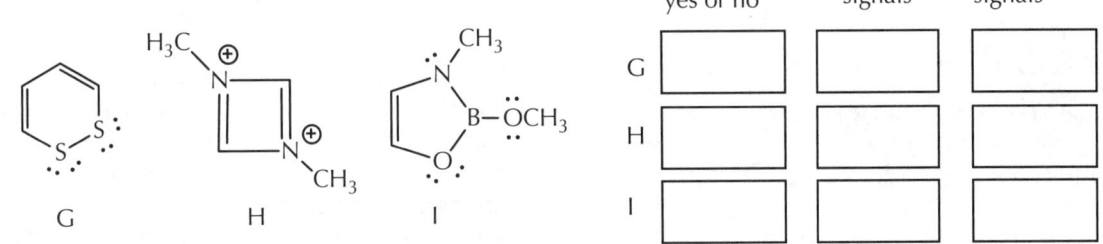

D. Provide the numerical value of the anticipated equilibrium constant (not just the exponent, but it can be an exponential expression).

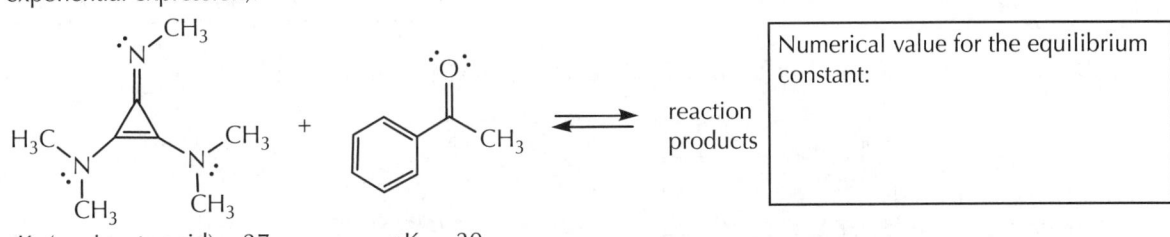

E. Which sites can be completely ($K_{eq} > 10^3$) deprotonated if this compound is combined with excess NaH?

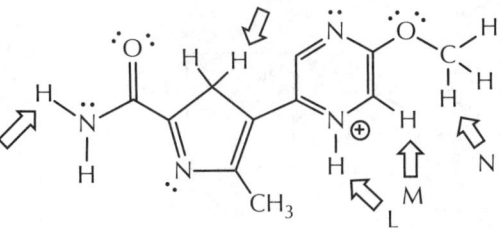

Completely deprotonated by NaH?

J yes no	K yes no
L yes no	M yes no

N yes no

EQ 04.33

A. Given these:

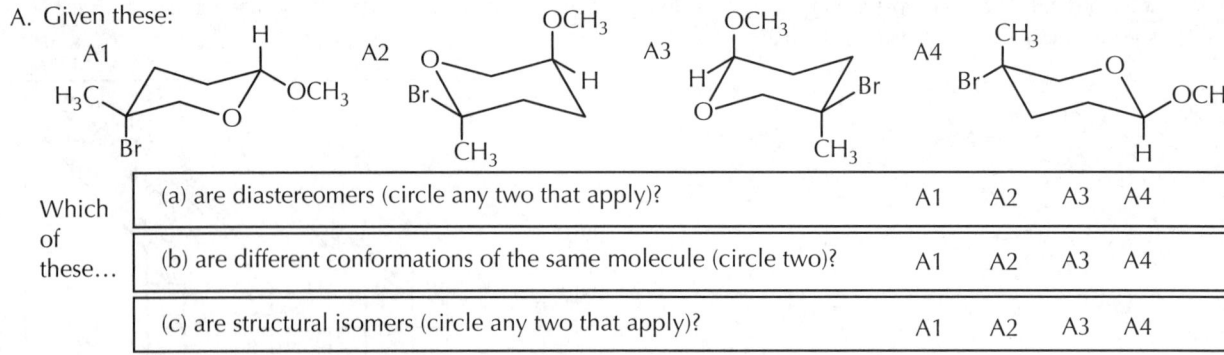

Which of these...	(a) are diastereomers (circle any two that apply)?	A1	A2	A3	A4
	(b) are different conformations of the same molecule (circle two)?	A1	A2	A3	A4
	(c) are structural isomers (circle any two that apply)?	A1	A2	A3	A4

B. How many chiral stereoisomers exist for each of the following compounds?

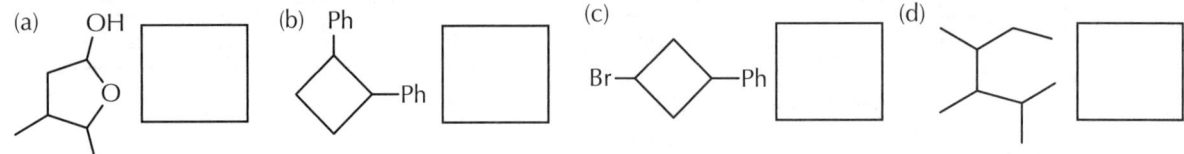

C. Including stereoisomers, how many organic products are there for each of these?

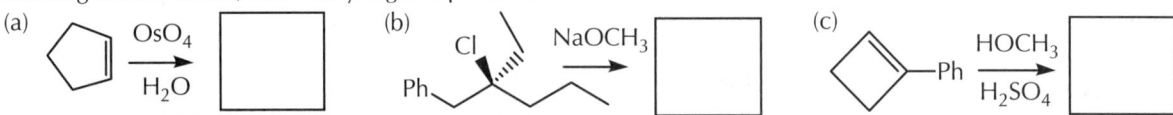

D. Including the one shown, how many all closed shell atom resonance contributors for each of these?

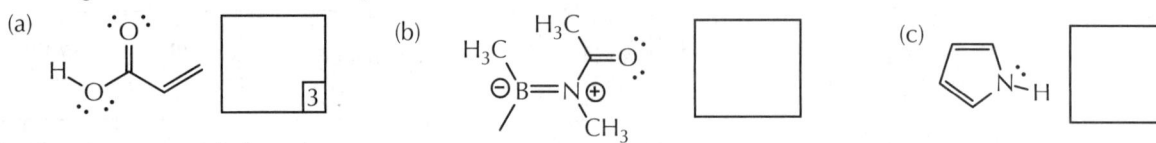

E. Predicted outcome (circle one):

(a) KCN + $(CH_3)_2CHCH_2Cl$ 　　　S_N2　E2　$S_N1/E1$　Acid-Base　Syn Addition　Anti Addition

(b) $NaNH_2$ + H–C≡C–⟨⟩　　　　　S_N2　E2　$S_N1/E1$　Acid-Base　Syn Addition　Anti Addition

(c) cyclopentene + Br_2, H_2O　　　S_N2　E2　$S_N1/E1$　Acid-Base　Syn Addition　Anti Addition

(d) (E)-2,3-diphenyl-2-butene + H_2/Pd-C　S_N2　E2　$S_N1/E1$　Acid-Base　Syn Addition　Anti Addition

F. The overall molecular charge on the major component in solution at the given pH value

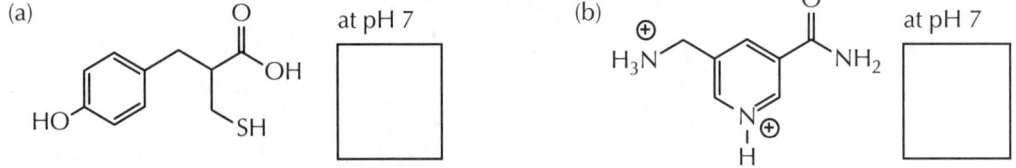

G. The number of ^{1}H-NMR or ^{13}C-NMR signals, as indicated?

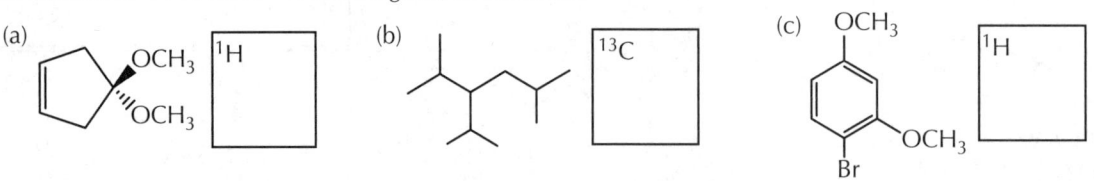

Exam Questions (Chapters 1–10)

EQ 04.34

A. Provide the reagents or structures as necessary to complete the following sequences of reactions. If more than one synthetic step is required, number them sequentially. You may use abbreviations for reagents (e.g., TsCl), but not acronyms (e.g., mCPBA—draw the structure instead).

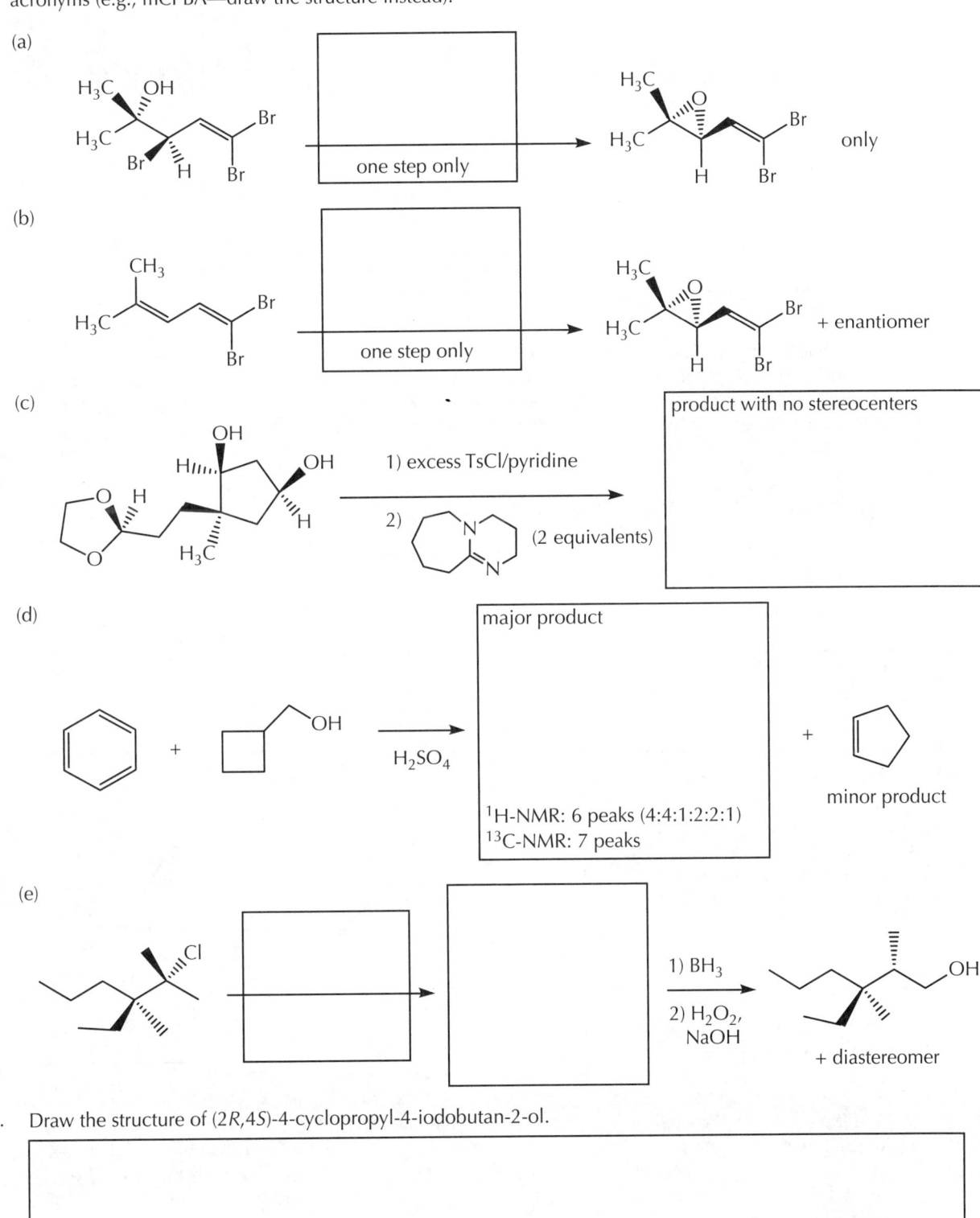

B. Draw the structure of (2R,4S)-4-cyclopropyl-4-iodobutan-2-ol.

894 CHAPTER 10 Aromaticity and Electrophilic Aromatic Substitution Reactions

EQ 04.35

A. (a) Draw the other chair conformation for this molecule.

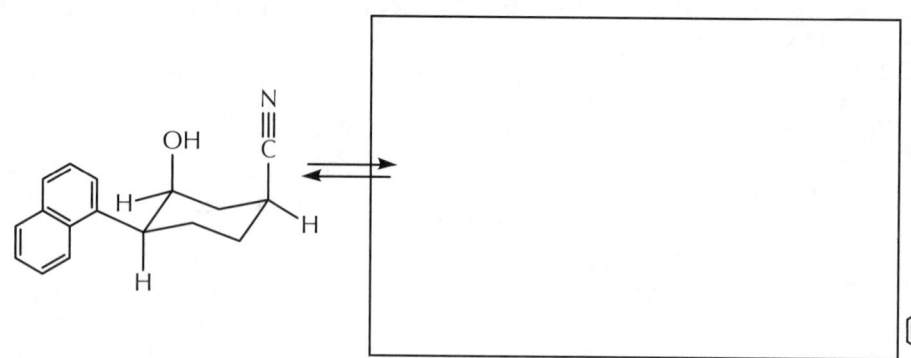

useful info:
K_{EQ} for cyclohexyl-X

X =	K_{EQ} $\dfrac{\text{X-equatorial}}{\text{X-axial}}$
CN	1.40
OH	2.80
naphthyl	2000

(b) As shown, the K_{EQ} is expected to be (circle one):   >1   =1   <1

B. Using titanium tetrachloride as a Lewis acid, compound A undergoes fast ionization to give the carbocation (shown below) at room temperature, which is an exceptionally low temperature for carbocation formation. This carbocation then behaves as a good electrophile in an electrophilic aromatic substitution reaction. (a) What is the reason that this carbocation forms so rapidly at such a low temperature? (b) Draw the ultimate reaction product (*J Org Chem*, **2005**, *70*, 5215).

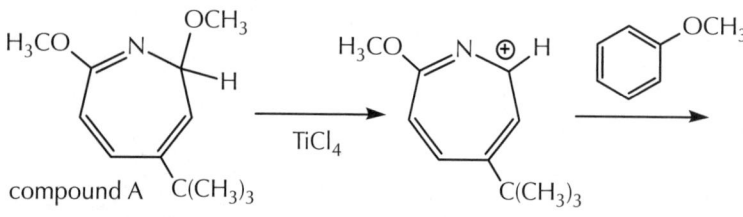

(b) EAS reaction product

(a) Why does the carbocation form so easily?

C. This energy diagram was used to describe two competing reaction pathways from compound S, giving possible reaction products X and Y through intermediates T and R, respectively (*J Org Chem*, **2005**, *70*, 9417).

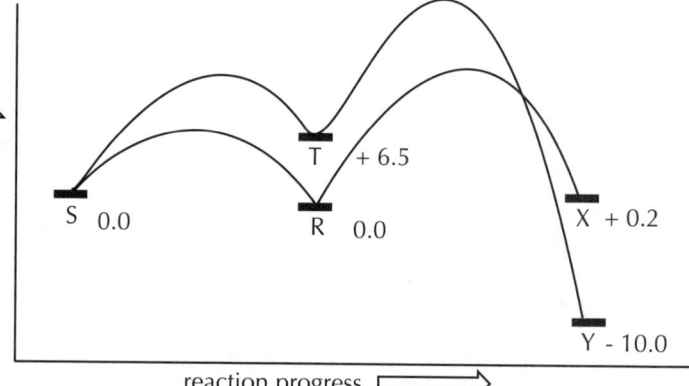

Complete the following by adding the letter or letters as appropriate.

(a) What is the major product if all steps are reversible?

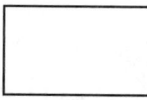

(b) What is the major product if all steps are irreversible?

(c) What is the rate-determining step in forming Y?

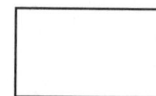

(d) If the S to T step is the only irreversible step, then what is the product?

(e) $K_{EQ} = 1$ between which 2 compounds?

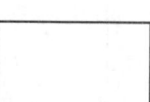

EQ 04.36

A. (a) Complete the following transformation (*Org Lett,* **2000**, *2*, 3893).

(b) What is the IUPAC name for compound K?

(c) What is the hybridization at each of the indicated carbon atoms?

(d) Draw a 3-dimensional orbital picture for compound K using lines, dashes, and wedges to indicate the sigma bonds, and overlapped p orbitals for the pi bonds (you may leave the methyl groups indicated as "CH$_3$" in your drawing.

B. (a) Serine is a naturally occurring amino acid. Our bodies also synthesize it when needed. Provide a base (from the pK$_a$ table) that could be used to deprotonate the most acidic proton in serine (K$_{EQ}$ > 10^4) without causing significant (> 50%) deprotonation of the other acidic protons? The observed pK$_a$ values in serine are 2.2, 9.5, and 16.1.

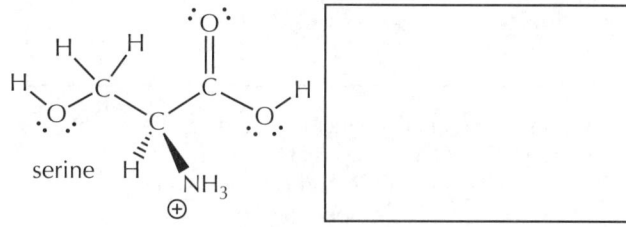

(b) What is the main form of serine in an aqueous buffer solution of pH = 10?

(c) Serine exists as a 1:1 mixture of a positively charged form and an uncharged form at what pH value?

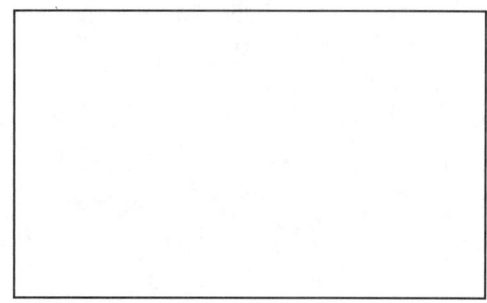

EQ 04.37

Compound J reacts to form the product compound K.

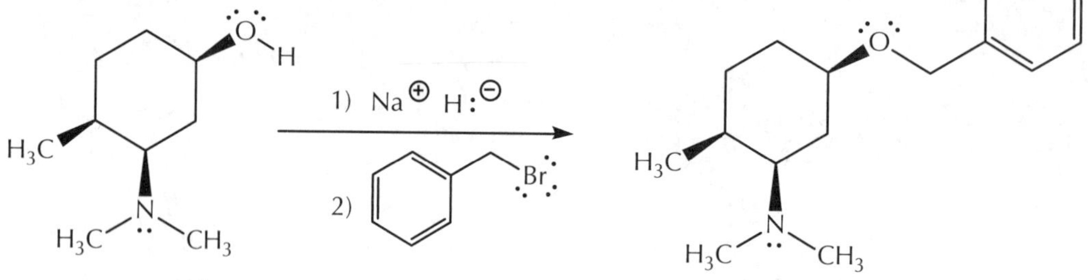

−ΔG° (A-Value) for the axial-to-equatorial chair-chair interconversion of these monosubstituted cyclohexanes.

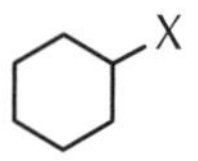

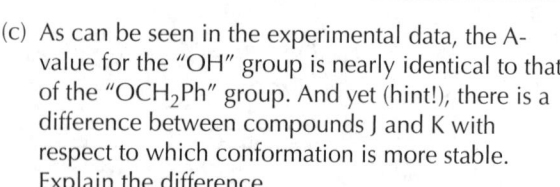

X = OH	0.87 kcal/mol
X = OCH$_2$Ph	0.91 kcal/mol
X = CH$_3$	1.70 kcal/mol
X = N(CH$_3$)$_2$	2.10 kcal/mol

Other information:

CH$_3$ / N(CH$_3$)$_2$ gauche interaction = 1.10 kcal/mol

OH / N(CH$_3$)$_2$ (diaxial) hydrogen bond = 2.50 kcal/mol

(a) Given the experimental data shown above, draw the most stable chair form of compound J.

(b) Given the experimental data shown above, draw the most stable chair form of compound K.

(c) As can be seen in the experimental data, the A-value for the "OH" group is nearly identical to that of the "OCH$_2$Ph" group. And yet (hint!), there is a difference between compounds J and K with respect to which conformation is more stable. Explain the difference.

(d) Circle and clearly label all stereocenters as R or S on the following drawing of compound J.

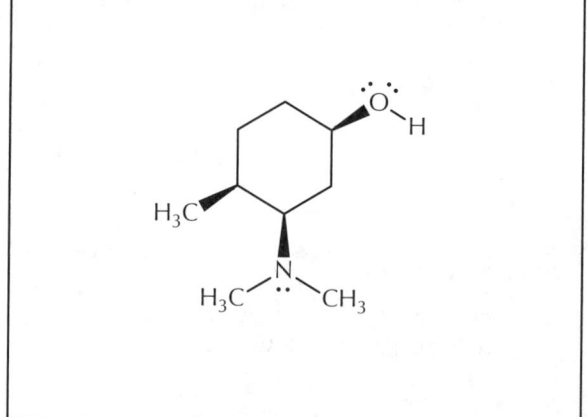

EQ 04.38

A. The number of ^{1}H-NMR and ^{13}C-NMR signals?

(a) 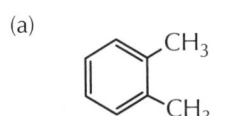 ^{1}H ☐   (b) 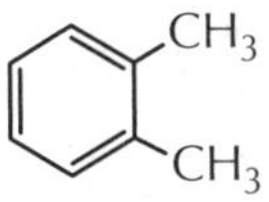 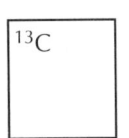 ^{13}C ☐   (c) 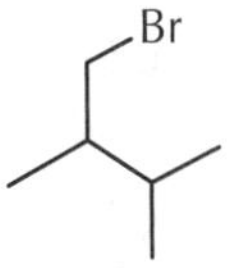 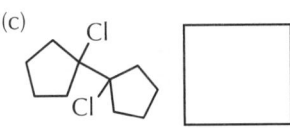 ^{13}C ☐

B. How many chiral stereoisomers exist for each of the following compounds?

(a) 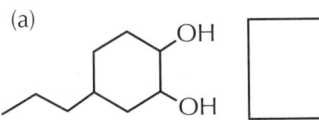 ☐   (b) 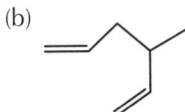 ☐   (c) 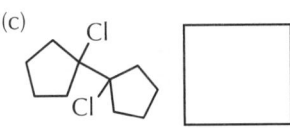 ☐

C. How many major organic products are there for each of these reactions, including stereoisomers?

(a) 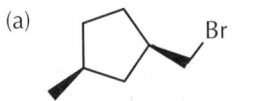 (CH$_3$)$_3$COK ☐   (b) 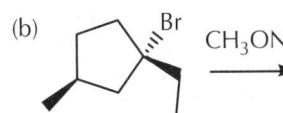 CH$_3$ONa ☐

D. Given these:

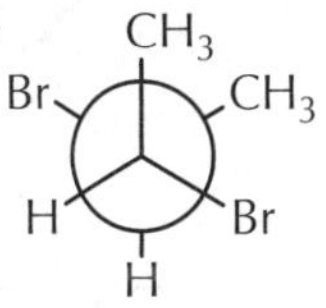

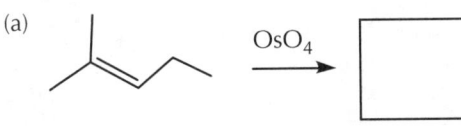

Which of these...	(a) are enantiomers (circle any correct pair)?	D1 D2 D3 D4
	(b) are different conformations of the same molecule (circle any correct pair)?	D1 D2 D3 D4
	(c) are diastereomers (circle any correct pair)?	D1 D2 D3 D4

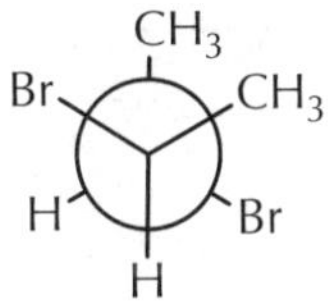

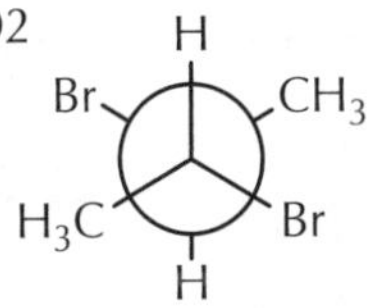

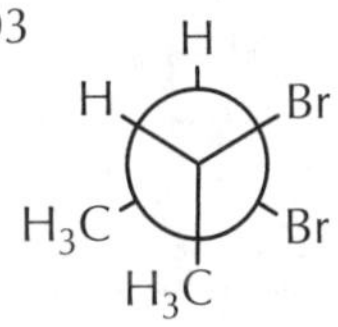

E. Including stereoisomers, how many major organic products are there for each of these?

(a) 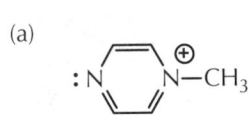 OsO$_4$ ☐   (b) 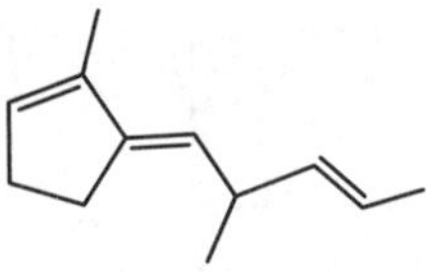 1) O$_3$ 2) Zn ☐

F. Is the ring aromatic, antiaromatic, or is the assessment not applicable?

(a) [pyrimidinium with N-CH$_3$] circle one: Aromatic / Antiaromatic / not applicable

(b) 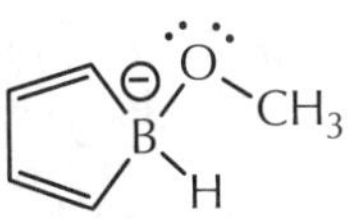 [borole with O-CH$_3$ and H] circle one: Aromatic / Antiaromatic / not applicable

(c) 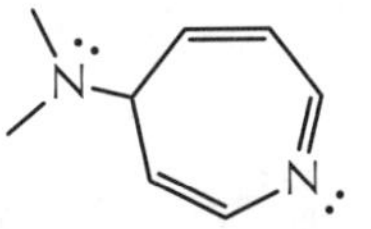 [azepine with N(CH$_3$) and N] circle one: Aromatic / Antiaromatic / not applicable

(d) 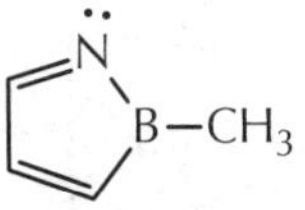 [azaborole B-CH$_3$] circle one: Aromatic / Antiaromatic / not applicable

G. Is the reaction expected to be S$_N$2, E2, S$_N$1/E1, or no reaction (NR)?

(a) H–C≡CNa + Ph–CH(Br)–CH$_3$ circle one: S$_N$2   E2   S$_N$1/E1   NR

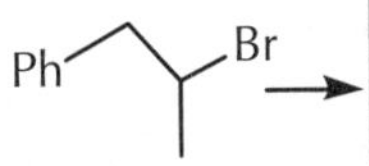

(b) 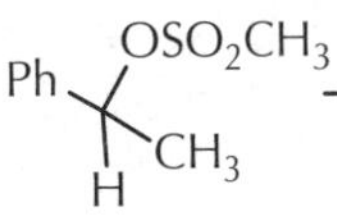 Ph–CH(OSO$_2$CH$_3$)–CH$_3$ + CH$_3$OH   circle one: S$_N$2   E2   S$_N$1/E1   NR

EQ 04.39

Wormwood and Asphodel

"Potter!" said Snape suddenly. "What would I get if I added powdered root of asphodel to an infusion of wormwood?"

Harry Potter and the Sorcerer's Stone 1997

Most wormwood species contain the toxic compounds α-thujone and β-thujone, which can cause convulsions. They are also the components of absinthe that are believed to cause hallucinations (*Proc Natl Acad Sci,* **2000**, *97*, 4417).

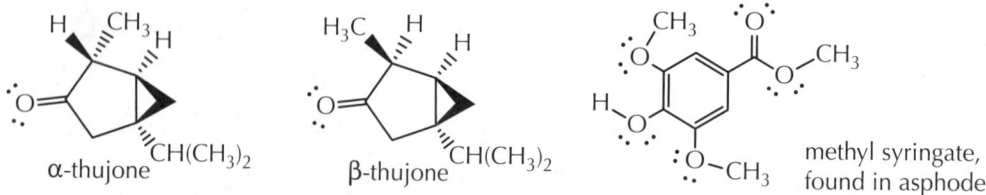

(a) Assign the stereochemical configurations to these two stereocenters in α-thujone.

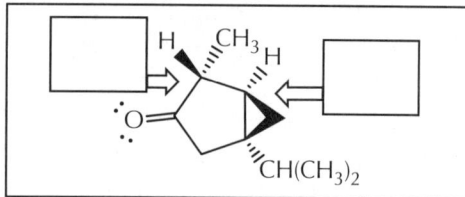

(b) What is the stereochemical relationship between α-thujone and β-thujone?

(c) Naturally occurring α-thujone has a specific rotation of -15 degrees. What can be said about the specific rotation of β-thujone?

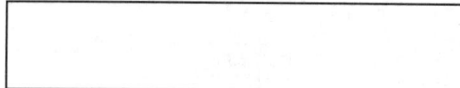

(d) The conjugate base of methyl syringate has a single significant resonance contributor in which the negative charge is delocalized to the oxygen atom of the C=O group; draw this resonance contributor.

(e) (i) How many ^{1}H-NMR signals are expected for methyl syringate?

(ii) How many ^{13}C-NMR signals are expected for methyl syringate?

(f) Complete the following three reactions, all of which were reported in a synthesis of α-thujone (*Chem Comm,* **2016**, *52*, 1170).

provide the balanced equation

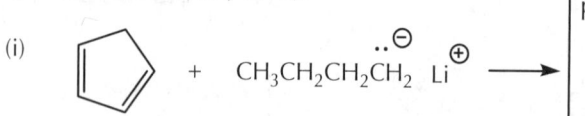

racemic

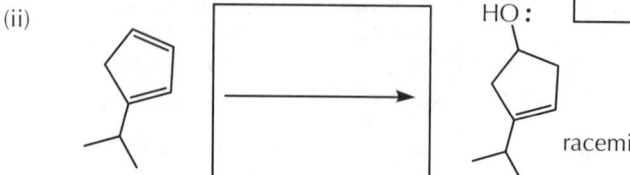

+ NaBr

EQ 04.40

A. The following questions refer to this molecule: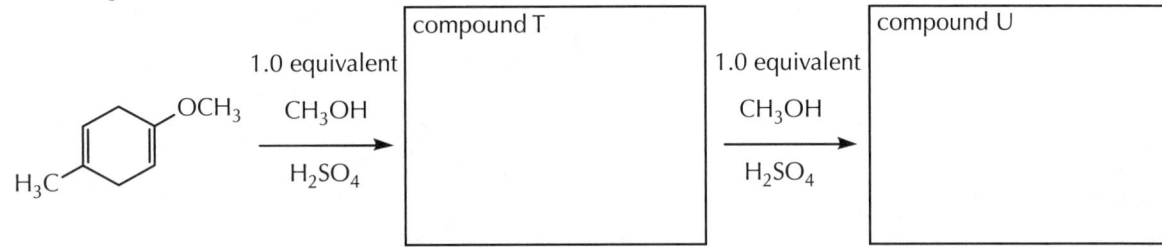

(a) Keeping in mind that "CH₃O" is named with the prefix designation "methoxy," name this molecule.

(b) When this molecule reacts with methanol (CH₃OH) with catalytic sulfuric acid, the initially formed product can undergo reaction with a second equivalent of methanol. What are these products?

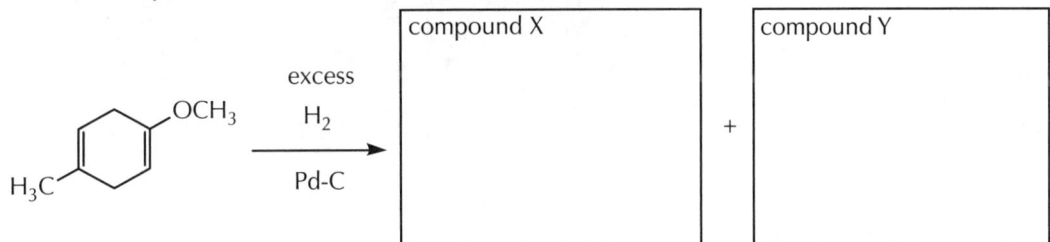

| compound T | compound U |

(c) When this molecule undergoes reaction with excess hydrogen with catalytic palladium-on-carbon, two stereoisomeric products are formed. What are these products?

| compound X | compound Y |

(d) (i) How many signals in the ¹H-NMR spectrum of 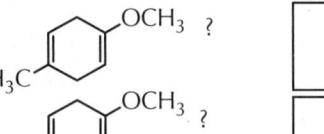 ?

(ii) How many signals in the ¹³C-NMR spectrum of 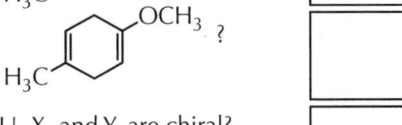 ?

(iii) How many of these compounds, from among T, U, X, and Y, are chiral?

(iv) How many of these compounds, from among T, U, X, and Y, are *meso*?

B. (a) How many aromatic rings are present in compound Z (shown on the right)? (*ACS Med Chem Lett,* **2014**, *5*, 1082)

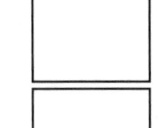

compound Z

(b) Draw the di-cation (+2 charge) that results when compound Z is protonated twice with a strong acid. The two protonation sites necessarily result in the two most stable postive charges.

CHAPTER 10 Aromaticity and Electrophilic Aromatic Substitution Reactions

EQ 04.41

Based on the information given below, complete the following reaction scheme by providing the structures for compounds C, D, E, F, and G. It is useful to work nonlinearly, particularly starting from the places where you have structural information, rather than trying to step through these in sequence. Please note that molecular formulas are provided in all of the answer spaces - double check your structures for accuracy.

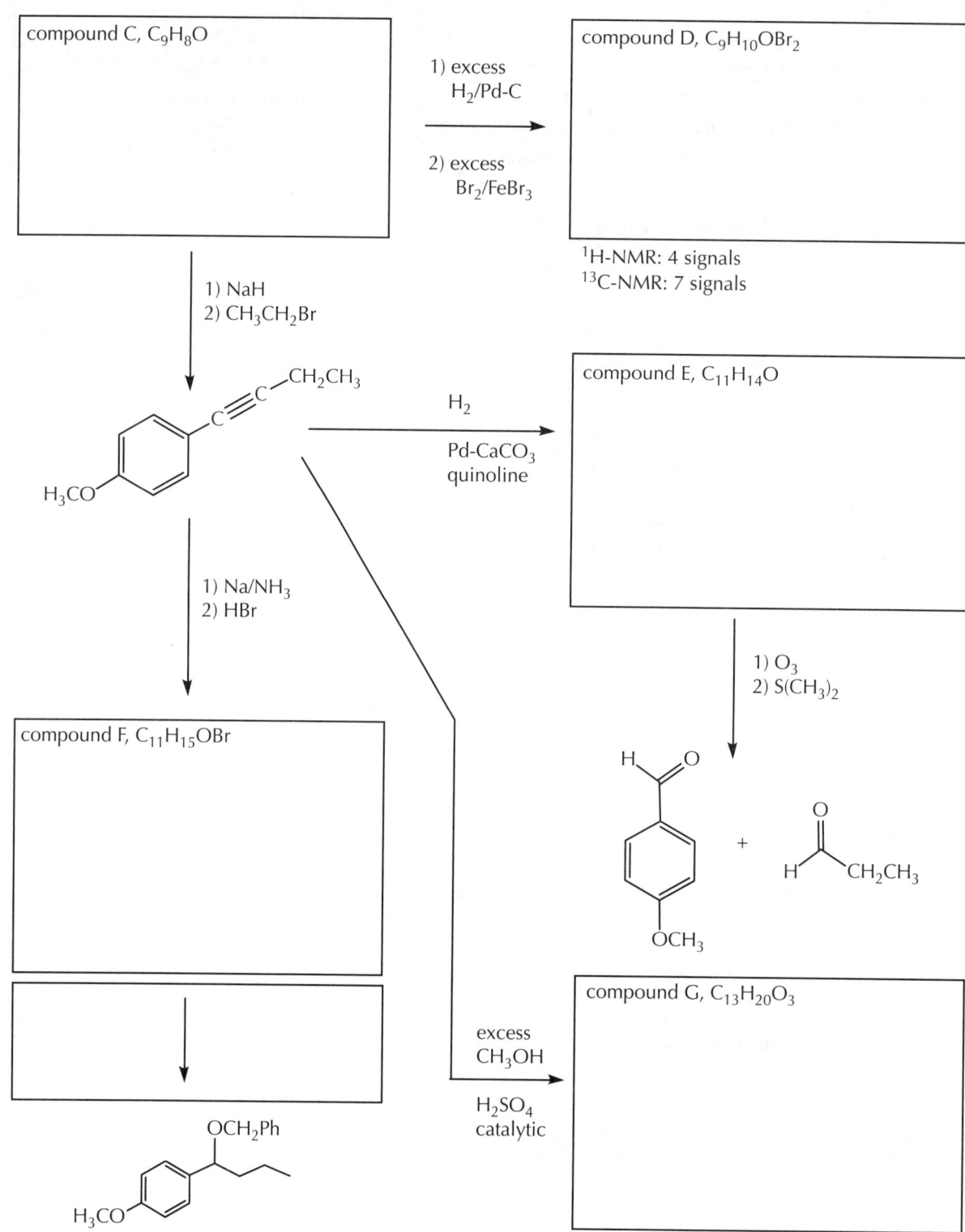

EQ 04.42

A. Given these molecules, answer the following (note, there may be more than one correct pairing):

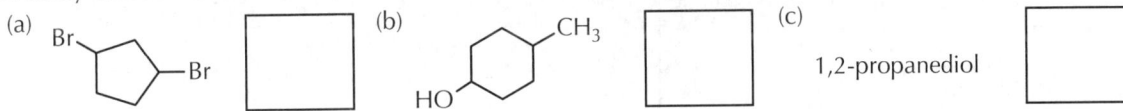

(a) Which of these are enantiomers (circle two)?	A1 A2 A3 A4
(b) Which are conformations of the same molecule (circle two)?	A1 A2 A3 A4
(c) Which of these are diastereomers (circle two)?	A1 A2 A3 A4

B. Given the reaction between CH₃SNa under each of the following conditions:

B1	B2	B3	B4
PhCH₂Br in DMSO	PhCH₂Cl in CCl₄	PhCH₂Cl in DMSO	PhCH₂Br in CCl₄

(a) Which reaction is fastest (circle one)?	B1 B2 B3 B4
(b) Which reaction is slowest (circle one)?	B1 B2 B3 B4

C. How many chiral stereoisomers exist for each of the following compounds?

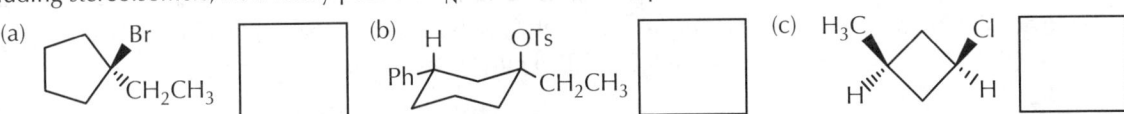

D. Including stereoisomers, how many possible S_N1 and E1 reaction products are there for each of these?

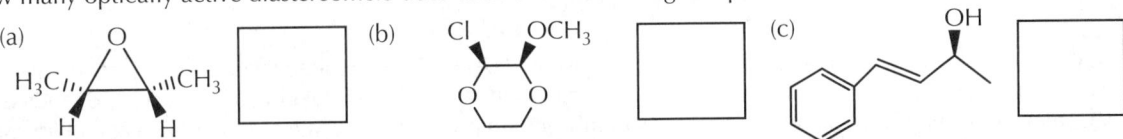

E. How many optically active diastereomers does each of the following compounds have?

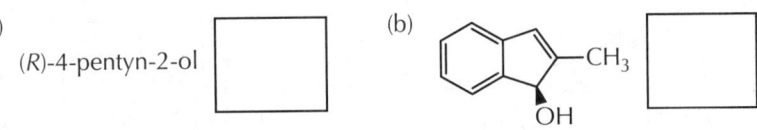

F. How many stereoisomers are produced when a completely regioselective hydroboration/oxidation reaction is carried out on:

(a) (R)-4-pentyn-2-ol

(b) [indene-CH₃/OH structure]

G. How many stereoisomers are produced when the completely regioselective addition reaction of trifluoroacetic acid (pK_a = 1, see below) is carried out on:

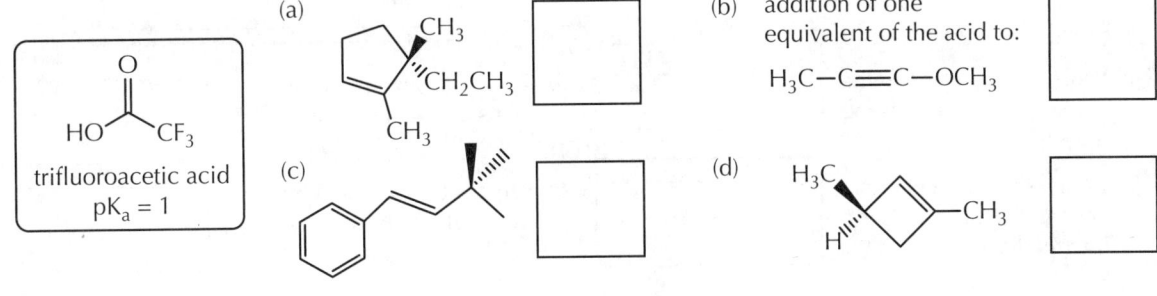

(b) addition of one equivalent of the acid to:

H₃C—C≡C—OCH₃

902 CHAPTER 10 Aromaticity and Electrophilic Aromatic Substitution Reactions

EQ 04.43

A. The acidities of compounds W, X, and Y differ significantly. Indicate which molecule is most acidic, provide a single Lewis structure that best explains your answer, and a brief (10 words or less) written explanation.

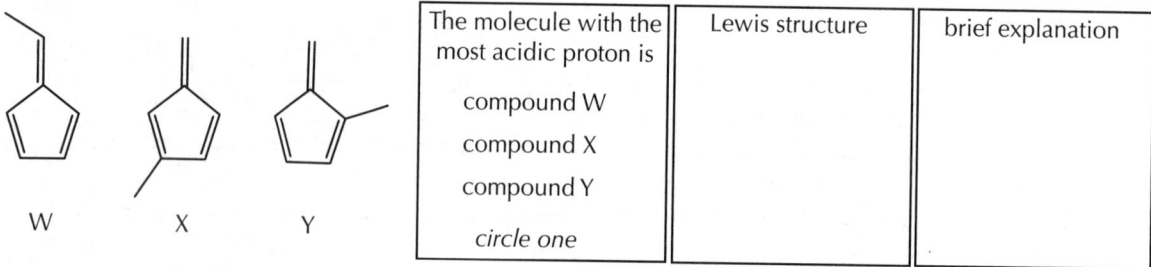

The molecule with the most acidic proton is	Lewis structure	brief explanation
compound W		
compound X		
compound Y		
circle one		

B. Nitromethane is a relatively strong acid with a pK_a value of 10.

(a) Provide a three-dimensional orbital representation for the most significant resonance form of the conjugate base of nitromethane. Use lines, dashes, and wedges for sigma bonds and to indicate the directionality of localized nonbonding electron pairs, and indicate formal charges if/as necessary. Draw overlapped p orbitals for pi bonds. Be sure to carefully consider whether nonbonding electrons are localized or delocalized when assigning hybridization/geometry.

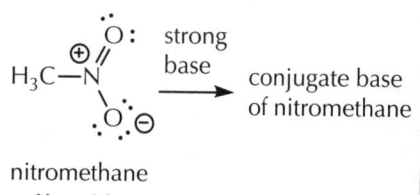

nitromethane
$pK_a = 10$

strong base → conjugate base of nitromethane

3D orbital drawing of conjugate base of nitromethane: most signifcant resonance contributor

(b) Provide the equilibrium constant for the reaction between the conjugate base of nitromethane and hydrofluoric acid. Your answer should be in the form $K_{EQ} = 10^x$ (fill in the box for x).

conjugate base of nitromethane + H–F ⇌ nitromethane ($pK_a = 10$) + F$^\ominus$ $K_{EQ} = 10^{\boxed{}}$

C. The hydrolysis of an ester to a carboxylic acid is illustrated below. These reactions are typically conducted under either aqueous acid or aqueous base conditions. A recent theoretical (computational) study was conducted on the mechanism of hydrolysis of methyl formate to formic acid in pure (pH = 7) water. An energy diagram for this reaction is shown below. On the basis of this energy diagram, answer the following questions.

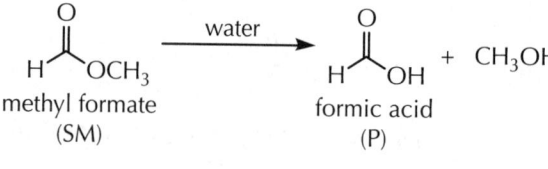

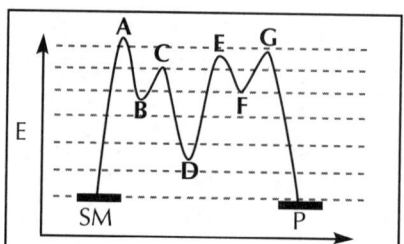

(a) Which of the letters in the energy diagram represent transition states?

(b) How many mechanistic steps are required for conversion of SM to P?

(c) Which of the letters in the energy diagram represent intermediates?

(d) What is the rate-determining step in the conversion of SM to P?

EQ 04.44

A. Under special experimental conditions, NMR spectra can be recorded for carbocations. There are 5 carbocations represented here along with three ¹H-NMR spectra. Match each ¹H-NMR spectrum with one of the carbocations.

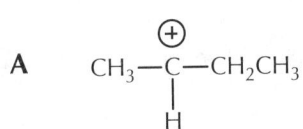

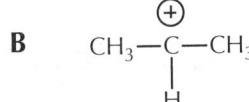

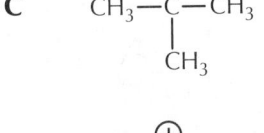

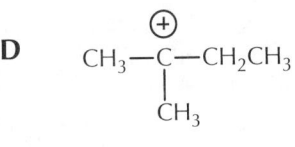

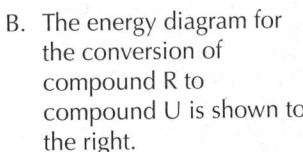

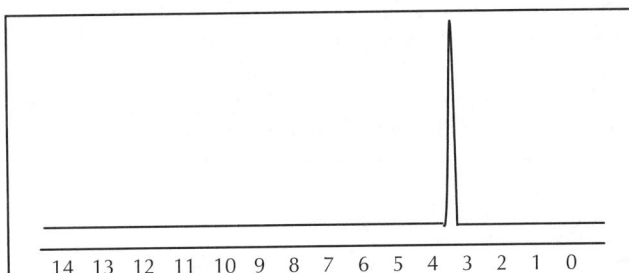

(a) Place the letter of the carbocation corresponding to this spectrum:

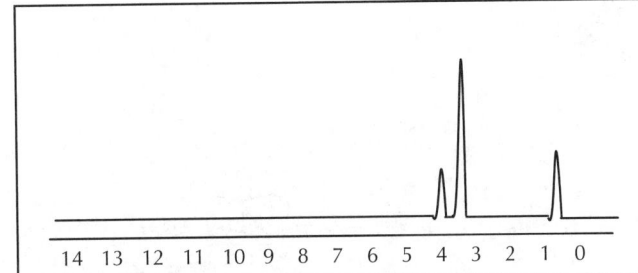

(b) Place the letter of the carbocation corresponding to this spectrum:

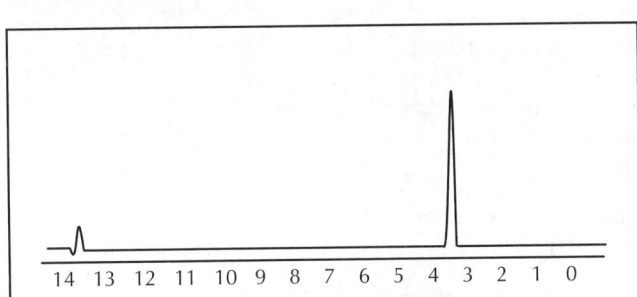

(c) Place the letter of the carbocation corresponding to this spectrum:

B. The energy diagram for the conversion of compound R to compound U is shown to the right.

Place the letter or letters corresponding to the points on this diagram in response to the following questions.

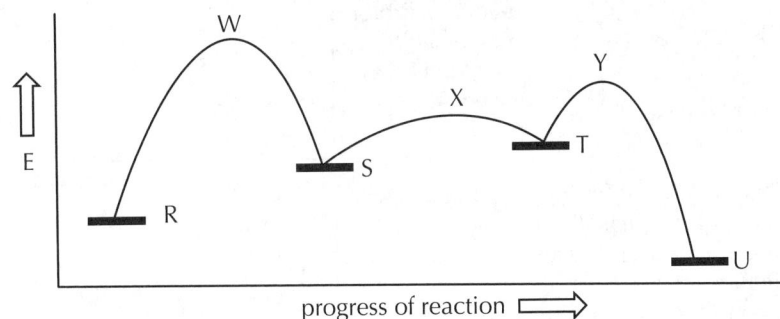

(a) How many mechanistic steps are shown in the conversion of R to U?

(b) How many transition states are shown for the conversion of R to U?

(c) What is the rate-determining step for the conversion of R to U?

(d) If all steps are reversible, which compound will be present at the highest equilibrium concentration?

(e) Which step, among all forward and reverse steps, has the lowest energy of activation?

904 CHAPTER 10 Aromaticity and Electrophilic Aromatic Substitution Reactions

EQ 04.45

A. Although halonium ions are well known for I, Cl, and Br, they had not been observed for F until early in 2013 (*Science*, **2013**, *340*, 57-60). When tosylate A was dissolved in H_2O, two substitution products were observed. The mechanism proceeds in four steps: (Step 1) A carbocation is formed. (Step 2) A fluoronium ion (which is not a 3-membered ring) forms. (Step 3) Nucleophilic attack of the fluoronium ion by water. (Step 4) Proton transfer to form the neutral product.

(a) Draw the complete curved arrow mechanism from tosylate A to product 1. Note that an incomplete scaffold for each intermediate is provided, fill in the needed substituents (with stereochemistry).

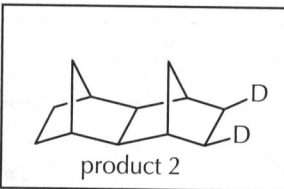

(b) Draw the other product that is predicted to form through this mechanism. An incomplete scaffold is provided, be sure to include stereochemistry.

(c) Products 1 and 2 are: (circle one)
enantiomers
diastereomers
structural isomers

B. The compound shown below is commonly called "Proton Sponge" because it is a strong base. This base was the subject of a study aiming to understand the impact of molecular structure on basicity (*Org Lett*, **2013**, *15*, 902).

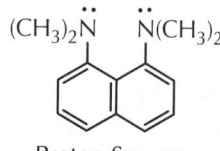

Proton Sponge
conj. acid pK_a = 12
(di-acid pK_a = -9)

(a) Draw the anticipated structure of Proton Sponge that is present in highest concentration at pH = 5.

(b) Proton Sponge's basicity is explained with the observation that its conjugate acid may be stabilized by a very strong intramolecular interaction. Draw Proton Sponge's conjugate acid and show this interaction.

(c) A structurally related compound is shown below.

(i) Draw a circle around the most basic position on the molecule.

(ii) Draw a square around the most acidic proton.

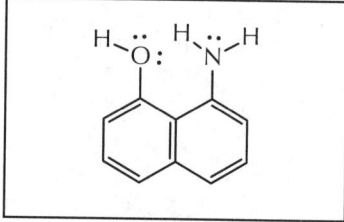

C. Draw a circle around the molecules that contain at least one aromatic ring.

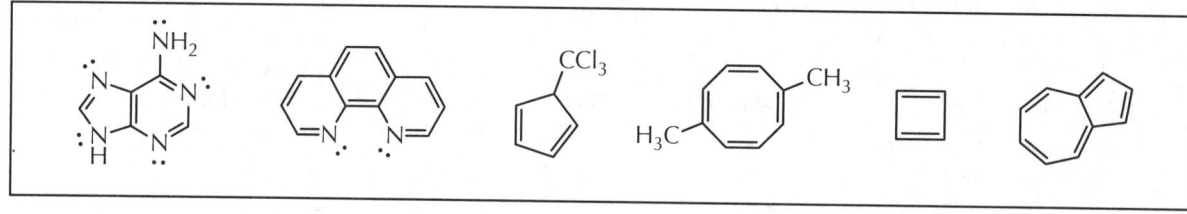

EQ 04.46

Complete the following as needed. Include stereochemistry as appropriate to your answer.

(a)
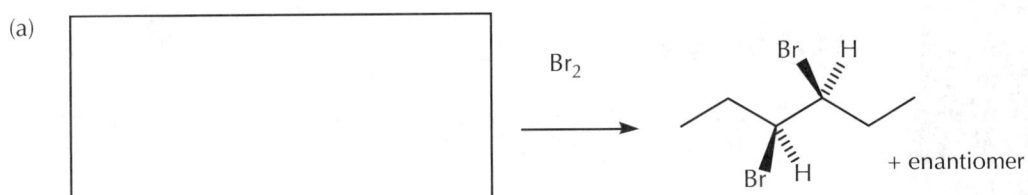

(b) Provide the complete IUPAC name for this compound:
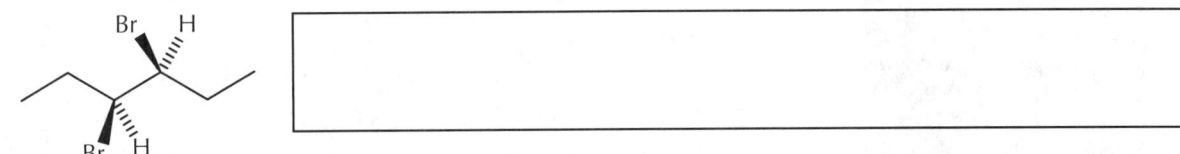

(c) Given the following pK_a values, what are the two structures in equilibrium at pH = 6.10?
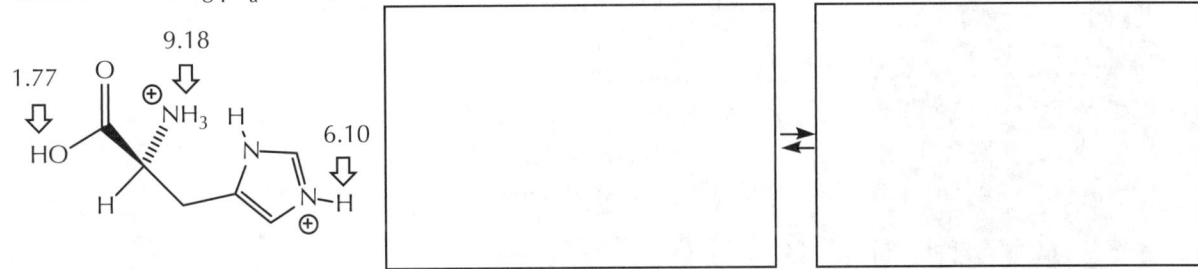

(d) Given the following data, what is the structure of the product in this reaction?
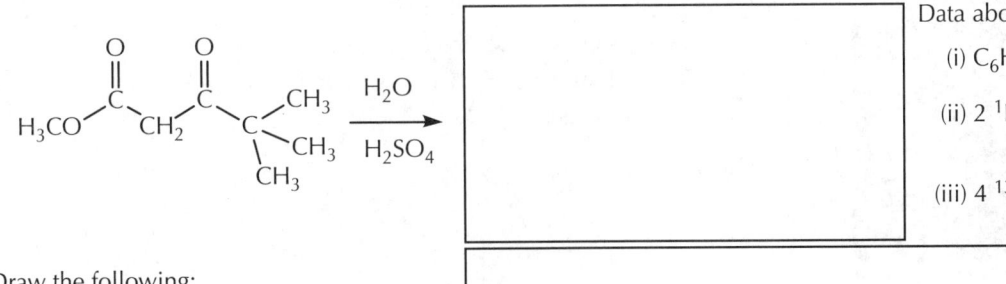

Data about the product:
(i) $C_6H_{12}O$
(ii) 2 ^{1}H-NMR signals in a 1:3 ratio
(iii) 4 ^{13}C-NMR signals

(e) Draw the following:

(3E, 2R)-4-methyl-3,5-hexadien-2-ol

(f) The following alcohol, in strong acid, gives an initially formed carbocation that undergoes rearrangement to give a new carbocation. The new carbocation, in turn, undergoes an intramolecular electrophilic aromatic substitution reaction. What are the structures of the two carbocations?

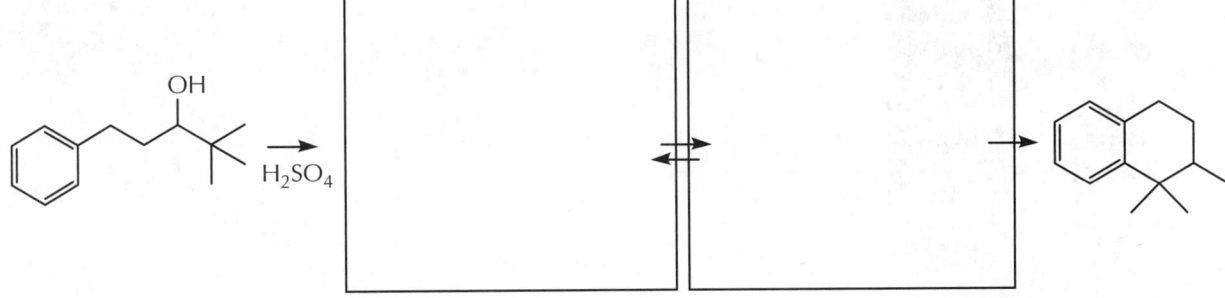

APPENDIX 1

Useful Expectations from General Chemistry

Preface

Introduction

A1.1 Atoms, Ions, and Molecules

    A. Atoms and the Periodic Table
    B. Molecular Structure: Valence Bond Model
    C. The Quantum World

PRACTICE QUESTIONS

A1.2 Periodic Trends

    A. Electronegativity and Covalent Bond Polarity
    B. Atomic Size and Bond Length
    C. Oxidation-Reduction

PRACTICE QUESTIONS

A1.3 Chemical and Physical Properties

    A. Rate (Kinetics) and Stability (Thermodynamics)
    B. Equilibrium (K_{EQ}), Enthalpy (ΔH), and Entropy (ΔS)
    C. Phase Changes and Mixing

PRACTICE QUESTIONS

APPENDIX 1

Useful Expectations from General Chemistry

Preface

For many years, students have begun their study of organic chemistry after taking a full year of general chemistry as a prerequisite. The natural assumption is clear: This requirement is based upon the perceived need for a year's worth of subject matter, which usually follows a year (or more) of high school chemistry. As it turns out, the amount of actual subject matter knowledge needed for learning organic chemistry is far less than what appears in a typical year of general chemistry. The background information is critical, but it is not beyond what is covered in a good high school course and does not account for a year of study.

In 1989, the University of Michigan removed a college-level course in general chemistry as a prerequisite for its organic chemistry courses, relying instead on an assessment of students' background knowledge and their own confidence about starting their university-level study of chemistry with the organic chemistry courses. For over 30 years, the vast majority of large numbers of first term, first-year students (1,000+ per year) have successfully completed the organic chemistry courses with "A" and "B" grades.

The University created a one-term general chemistry elective course for its engineering students and those students who wanted or needed a solid introduction to basic chemical principles. Although the amount of that subject matter necessary for the organic chemistry courses is limited, revisiting these more familiar introductory topics, combined with an experience with university-level classes, is useful for some students.

There is an old saying among educators: the problem in teaching a follow-up course is not that any group of ten students will have forgotten 25% of what they learned in the previous course, it's that all ten of them will have forgotten a different 25%.

The following is a list of typical questions that can be answered based upon learning in General Chemistry courses. Understanding that it might have been a while since you took General (or high school) Chemistry, if you can think for a few moments and then confidently write down (or just imagine) clear and coherent answers to these questions and believe that you can defend (justify) your answers in a conversation with others, then you probably have all the prior knowledge you need to move right into learning about organic chemistry.

Based upon the idea that every person forgets a different 25%, the remainder of the appendix is a reasonably complete refresher course in the General Chemistry concepts that are relevant in an organic chemistry course. You certainly do not need to work your way through the entire appendix, and you might locate and read those topics about which you want to get some reminders. Alternatively, you might enjoy a relatively short yet complete descriptive summary of the fundamental ideas about chemistry, where they came from, and how they fit into your study of organic chemistry.

The following five topics comprise a reasonably comprehensive list of the main ideas from a General (or high school) Chemistry background that are useful to be familiar with when studying organic chemistry, if not chemistry in general. These ideas will all be revisited in the main parts of the text, but not necessarily in any instructional detail.

APPENDIX 1 Useful Expectations from General Chemistry

Topic 1. *Using precise language and drawings to represent atomic and molecular entities (also called atomic and molecular species). Clear communication relies on shared symbols (words, drawings, sounds).*

For each of the following, use accurate representational forms to distinguish between the items in each set.
(a) bromide ion, molecular bromine, bromine atom
(b) ammonia, ammonium ion
(c) sodium ion, sodium atom
(d) molecular oxygen, oxygen atom
(e) hydrogen atom, molecular hydrogen, hydrogen cation (also called proton), hydrogen anion (also called hydride ion)

Topic 2. *Electron configurations and the relative stability of the closed shell electron configuration.* The periodic table is one of the most incredible tools ever constructed, in that it forms the foundation for the entire material world.

Write electron configurations for the monoatomic species in Topic 1 and identify which have closed shell configurations.

Topic 3. *Lewis structures.* The idea of a molecule as the organized assembly of atoms, following the rules of bonding, is just over 200 years old, while the model for bonding based on shared electron pairs only dates to the 1920s. Being able to draw molecular structures that include covalent and ionic bonding is a prerequisite skill for organic chemistry.

Representing covalent bonds with lines, using "dots" (i.e., ":") for nonbonding electrons and charge-separated ions for ionic bonds, draw Lewis structures for the following molecules:
(a) ammonium bromide
(b) hydrogen
(c) hydrogen chloride
(d) nitrogen triiodide
(e) ammonium tetrafluoroborate
(f) sodium fluoride
(g) potassium hydroxide
(h) lithium hydride
(i) lithium tetrachloroaluminate

Topic 4. *Phase changes and mixing.* Although it is automatically relevant in the laboratory, it is quite useful for you to have an accurate sense about the physical reality of chemical systems. We write "NaCl + H_2O" as though there is one water molecule combining with a single pair of ions in sodium chloride. Yet the reality of a what you should imagine as a grain of sodium chloride and a drop of water is quite different than any manipulation of the symbols "NaCl" and "H_2O" can accomplish. For Topic 4, you are reminded to imagine the physical reality of what a chemical system really means.

Represent the following processes using a series of 3 or 4 drawings that involve a collection of atomic or molecular species.
(a) dissolving sodium chloride in water
(b) evaporation of liquid water
(c) melting sodium
(d) mixing oxygen and nitrogen gases

(e) evaporation of gold
(f) the solubility of cesium fluoride in water is 100 times greater than sodium fluoride (why?)

Topic 5. *Inductive effects (electronegativity and electropositivity).* Consequences from the fundamentals of electrostatic attraction (opposite charges attract and provide stabilization) and electrostatic repulsion (like charges repel and provide destabilization) sit at the core of many principles in chemistry.

Differences in electronegativity are correlated with many different properties and trends among the elements.
(a) Identify all dipolar and nonpolar covalent bonds in the molecules named in Topics 1–4. Indicate the direction of the polarization in dipolar bonds according to the partial charges anticipated at each atom.
(b) Is iodide or fluoride a stronger reducing agent (why)?
(c) Can sodium or potassium be oxidized more easily (why)?

The five topics are embedded in the text of Appendix 1, which follows, along with the history and development of the ideas that you might encounter in a typical introduction to chemistry.

The full appendix is written to help students revisit topics from a typical introductory chemistry class that are foundational to beginning a study of organic chemistry. The appendix is not meant to replace a first course in chemistry and is written as a comprehensive review for students who have previously taken a high school or college-level introductory course. The content on these approximately 50 pages is a little more than a good high school course might cover, and about 2/3 what a college course might cover.

Introduction

This appendix serves to remind students about the foundational topics of introductory chemistry and how those ideas developed. And while you might not have every nuance on the top of your head, the topics should be familiar in a "oh, yeah… I generally remember that" way.

Even if you are quite confident about your background, you will also find it is useful to revisit the topics in this appendix because they are approached from a perspective that anticipates what you will need for learning organic chemistry.

APPENDIX 1 Useful Expectations from General Chemistry

A1.1 Atoms, Ions, and Molecules

A. Atoms and the Periodic Table

Ideas in chemistry are typically not taught or learned in the same order they were discovered. For example, a common starting point is learning the elements as they appear on the periodic table: a stepwise and incremental movement from position number 1 (defined as any element with one proton in its nucleus, with a space located at the 1s energy level that can hold electrons, or not) to spot number 2 (anything with two protons, also with the available 1s energy level), and so on, through number 118 (with 94 of them occurring naturally). It is quite reasonable, with the typical level of detail in this introduction, to wonder out loud: "How did they come up with all of that?"

The answer, hopefully unsurprisingly, is: "They did not do it all at once. It evolved over a couple of centuries."

In 2019, the "year of the periodic table" was celebrated to commemorate the 150th anniversary of Dmitri Mendeleev's 1869 sketch that ordered the known elements according to their chemical properties (i.e., the first periodic table). He also predicted, according to some gaps in the record, the existence of elements that had not yet been discovered. But Mendeleev knew nothing of subatomic particles and energy levels. Electrons, protons, and neutrons were not discovered until the time period between 1900 and 1930, and only near the end of this period did scientists have a model for the structure of atoms that involved these particles and their energy levels, defining the nature of the chemical bond, decades after Mendeleev's death in 1907.

Recreating the ignorance of the past is not possible (we know things now that cannot be unknown). But historical facts can often remind you that a mature idea, with all of its complexity, starts off simply and builds on itself through the contributions of many (many) people.

> There are ancient cathedrals, which, apart from their consecrated purpose, inspire solemnity and awe. Even the curious whisper reverberates through the vaulted nave; the returning echo seems to bear a message of mystery. The labor of generations of architects and artisans has been forgotten, the scaffolding erected for the toil has long since been removed, and their mistakes have been erased, or have become hidden by the dust of centuries. Seeing only the perfection of the completed whole, we are impressed as by some superhuman agency. But sometimes we enter such an edifice that is still partly under construction; then the sound of hammers, the reek of tobacco, the trivial jests bandied from workman to workman, enable us to realize that these great structures are but the result of giving to ordinary human effort a direction and purpose.
>
> Science has its cathedrals, built by the efforts of a few architects and many workers...
>
> Gilbert Lewis and Merle Randall, from the preface to *Thermodynamics and the Free Energy of Chemical Substances*, London: McGraw-Hill Publishing, **1923**.

The basic organization of the periodic table, one of those chemical cathedrals, gives you a reference point for four useful concepts that have built up over the years (Figure AP0101):

(a) the energy levels (s, p, d, f) and how they fill, based on the numerical order of the elements (1s, 2s, 2p, 3s, 3p, 4s, 3d, 4p...), from which the electron inventory (electron configuration) can be defined;

(b) an element's chemical identity is based upon its proton count (anything called "hydrogen" has a nucleus with 1 proton, whether it is a hydrogen atom, a hydrogen ion, or a hydrogen isotope);

(c) the electron configurations for the Group 18 elements correspond to points of relative stability,

(d) based on lack of reactivity, that we call "closed shell" electron configurations; and
the number of electrons above an atom's imbedded closed shell core, which define an atom's reactivity, that we call an atom's valence electrons.

General statement about numerical data: Throughout the text, and particularly in this appendix, there are summary tables and lists of numerical data. Unless specifically directed, it is quite unusual for anyone to ask you to memorize and repeat these data back to them. Instead, these data are the critical evidence from experimental results that are used to develop the important trends, generalizations, and takeaway messages (the concepts of chemistry).

Figure AP0101

General principles: the location of elements on the periodic table.

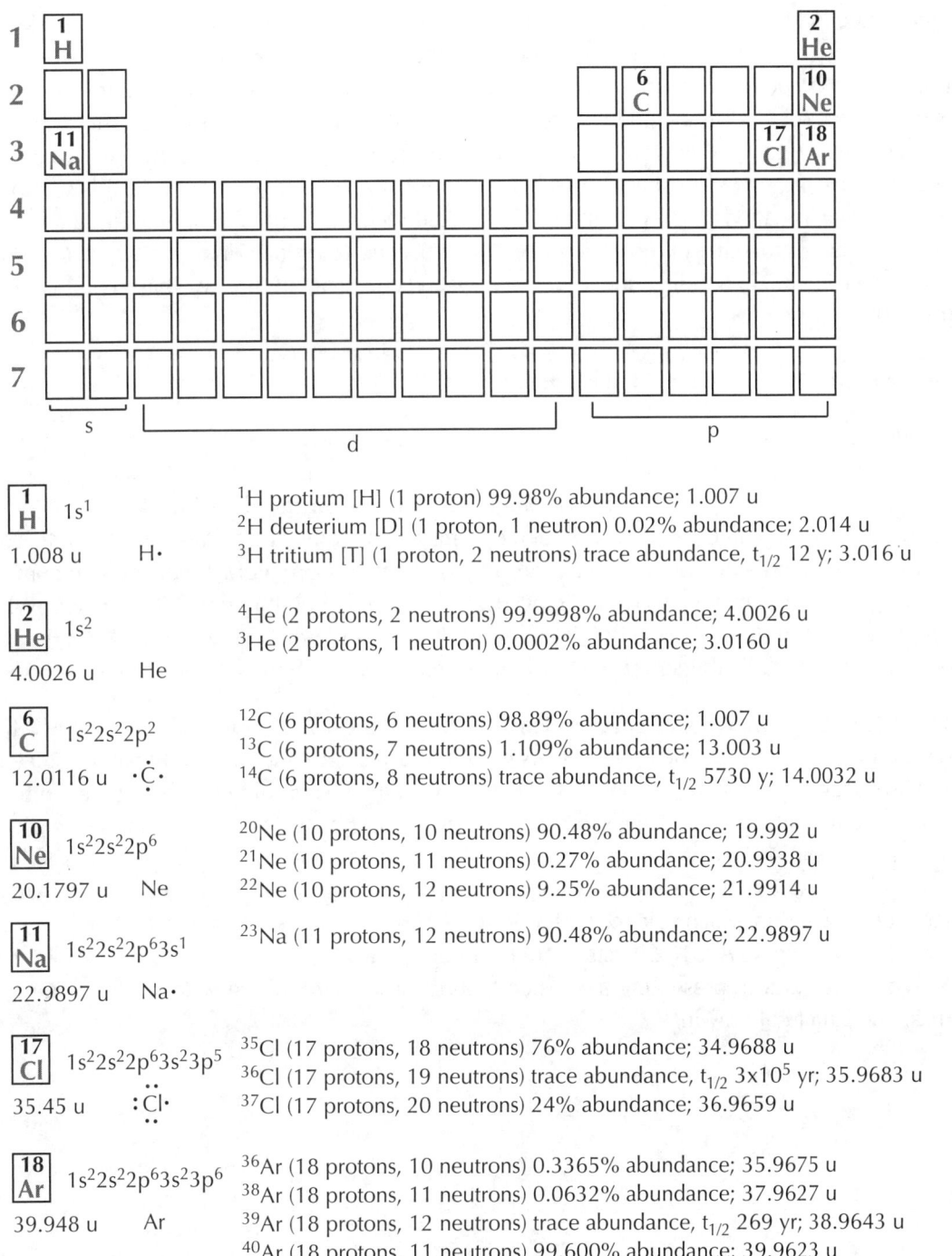

1 H	$1s^1$	^{1}H protium [H] (1 proton) 99.98% abundance; 1.007 u
1.008 u H·		^{2}H deuterium [D] (1 proton, 1 neutron) 0.02% abundance; 2.014 u
		^{3}H tritium [T] (1 proton, 2 neutrons) trace abundance, $t_{1/2}$ 12 y; 3.016 u

| 2 He | $1s^2$ | ^{4}He (2 protons, 2 neutrons) 99.9998% abundance; 4.0026 u |
| 4.0026 u He | | ^{3}He (2 protons, 1 neutron) 0.0002% abundance; 3.0160 u |

6 C	$1s^2 2s^2 2p^2$	^{12}C (6 protons, 6 neutrons) 98.89% abundance; 1.007 u
12.0116 u ·Ċ·		^{13}C (6 protons, 7 neutrons) 1.109% abundance; 13.003 u
		^{14}C (6 protons, 8 neutrons) trace abundance, $t_{1/2}$ 5730 y; 14.0032 u

10 Ne	$1s^2 2s^2 2p^6$	^{20}Ne (10 protons, 10 neutrons) 90.48% abundance; 19.992 u
20.1797 u Ne		^{21}Ne (10 protons, 11 neutrons) 0.27% abundance; 20.9938 u
		^{22}Ne (10 protons, 12 neutrons) 9.25% abundance; 21.9914 u

| 11 Na | $1s^2 2s^2 2p^6 3s^1$ | ^{23}Na (11 protons, 12 neutrons) 90.48% abundance; 22.9897 u |
| 22.9897 u Na· | | |

17 Cl	$1s^2 2s^2 2p^6 3s^2 3p^5$	^{35}Cl (17 protons, 18 neutrons) 76% abundance; 34.9688 u
35.45 u :Cl·		^{36}Cl (17 protons, 19 neutrons) trace abundance, $t_{1/2}$ 3x10^5 yr; 35.9683 u
		^{37}Cl (17 protons, 20 neutrons) 24% abundance; 36.9659 u

18 Ar	$1s^2 2s^2 2p^6 3s^2 3p^6$	^{36}Ar (18 protons, 10 neutrons) 0.3365% abundance; 35.9675 u
39.948 u Ar		^{38}Ar (18 protons, 11 neutrons) 0.0632% abundance; 37.9627 u
		^{39}Ar (18 protons, 12 neutrons) trace abundance, $t_{1/2}$ 269 yr; 38.9643 u
		^{40}Ar (18 protons, 11 neutrons) 99.600% abundance; 39.9623 u

Atomism, what the ancient Greek philosophers proposed as the existence of "natural minima" (the point at which a substance was indivisible), fell in and out of favor for a couple of thousand years, until John Dalton, in the early 1800s, helped establish an experimental basis for modern chemistry. The idea of a particle-based world was not the troublesome part of an atomic theory, rather that the existence of discrete particles meant accepting "the void," the existence of a nothingness in between the particles. The non-existence represented by the void is what gave scientists and philosophers, alike, their biggest hurdle, all the way through the mid-1800s. Many versions of the expression "Nature abhors a vacuum" exist in the history of science, beginning with *horror vacui*—attributed to Aristotle.

The modern idea of molecules took about a hundred years to develop after atoms were a generally accepted concept. Initially, every unique substance (water, sugar, oil) was imagined to be composed of a unique collection of atoms. By the early 1830s, however, pairs of substances that were clearly different were determined, experimentally, to be composed of the same collection of atoms. At this point, for the first time in history, people imagined that not only were substances composed of atoms, but that those atoms could have specific arrangements. The only way for these different arrangements to happen was to propose an organizing force that would hold the atoms in place. No one had any idea what this force was, but its existence would solve the problem about how to get fixed arrangements of atoms. The unknown force was called "chemical affinity," a made-up term with no definition other than "whatever it is that holds atoms together in molecules." Another hundred years later, in the 1930s, scientists settled on the idea that chemical bonding, the term that replaced chemical affinity, was based upon the valence electrons.

B. Molecular Structure: Valence Bond Model

Early models for understanding molecular structure were built upon a number of different ideas. In organic chemistry, two of these models stand out. The first one was called radical theory. Certain groups of atoms were observed to repeat from compound to compound, never occurring alone but always in combination with themselves or with other elements. These recurring groups were called "radicals" or "radicles" (from the German *radikal,* meaning fundamental or root). Using the modern atomic weights, two common examples of radicals were methyl (CH_3) and ethyl (C_2H_5), which in combination with the hydroxyl radical (OH) represented the structure of wood alcohol [(CH_3)OH] and grain alcohol [(C_2H_5)OH]. For about 20 years, the organic radicals (generically symbolized as "R") were thought to be as indivisible as atoms.

The second important model was called valence theory. By the 1850s, a more comprehensively atom-based view of molecules became more universal. When atoms combined with other atoms, they usually did so with specific and recurring numbers, initially called the "combining power" of the atom, and ultimately called its "valence" (from the same root as value or worth, coined in German as *valenz*). Carbon atoms had a combining power (valence, the number of other groups that would be connected) of four; nitrogen had a valence of three; oxygen, two; chlorine and hydrogen, one. There was no explanation for valence, but drawings in which valence was represented could be nonetheless used to imagine the structures of molecules. Figure AP0102 illustrates the "sausage" drawings used by Justus von Liebig in the late 1850s, with each lump representing a valence (connection) point, along with contemporary names, formulas, and structural drawings.

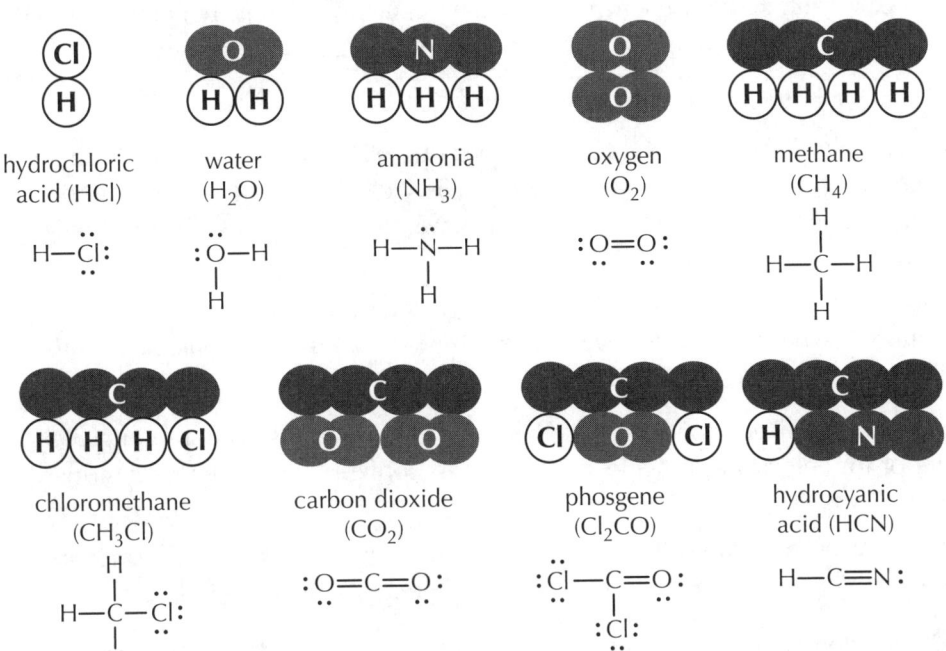

Figure AP0102
General principles: the use of valence in the 1850s.

Trying to understand the underlying cause(s) of valence consumed the attention of chemists throughout the end of the nineteenth century. In 1916, less than 20 years after electrons were discovered, German physicist Walther Kossel and American chemistry Gilbert Lewis independently published a pair of landmark publications called "On the Formation of Molecules as a Question of Atomic Structure" and "The Atom and the Molecule" (respectively). Their models are shown in Figure AP0103.

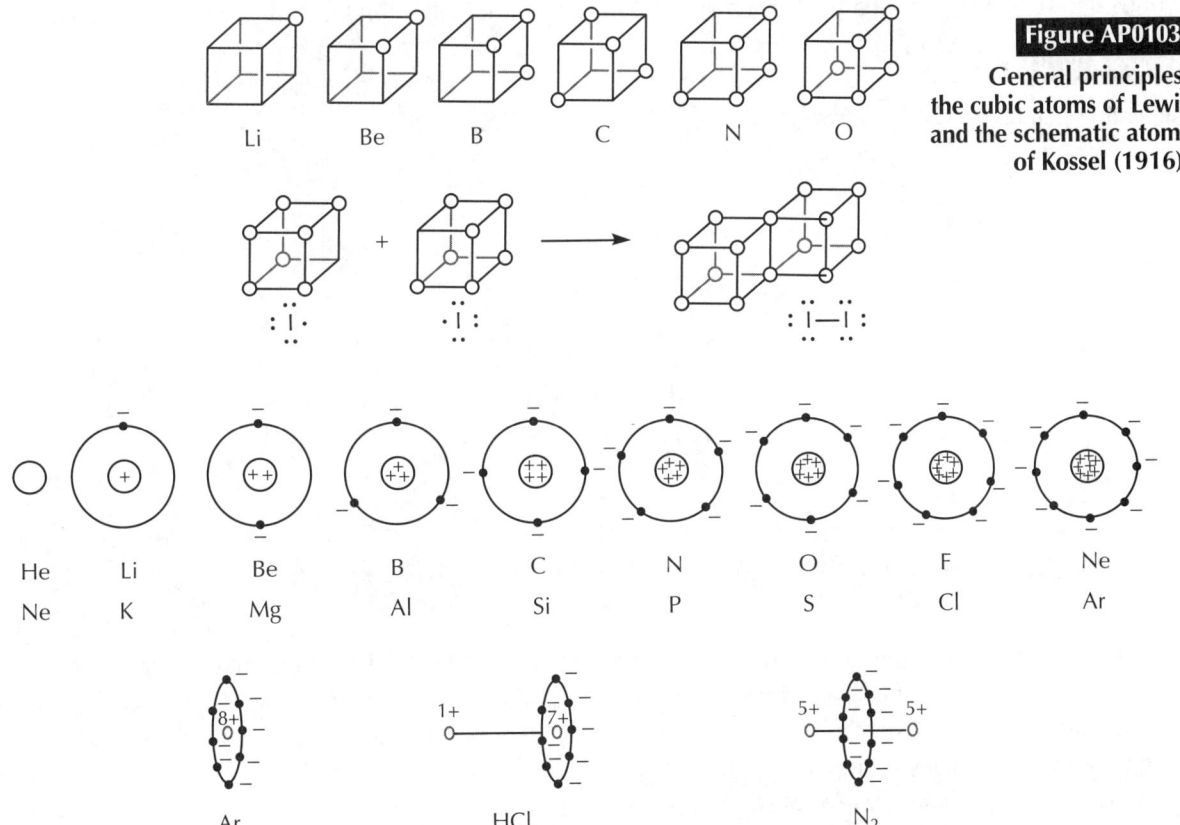

Figure AP0103
General principles: the cubic atoms of Lewis and the schematic atoms of Kossel (1916).

The proposals made by Lewis and Kossel were focused on how covalent and ionic bonding occurred in molecules, and they brought together a number of ideas that were emerging in chemistry:

(1) that bonds in molecules were 2 shared electrons;

(2) that, in general, a "rule of eight" governed bonding for atoms other than hydrogen and helium (what would be later called the octet rule), and that an 8-electron structure could be used to represent atoms; and that the driving force for bonding was achieving the 8-electron configuration of the nearest noble gas; and

(3) that, in addition to sharing, electrons could be lost or gained by atoms, resulting in atoms with a net charge (ions) because the positive charge of the kernel (nucleus) would be unequal to the negative charge provided by the electrons.

The bottom line (summary derived from Kossel and Lewis): Chemical bonds and ions form when atoms share, gain, or lose electrons to achieve the electron configuration of the nearest inert (noble) gas.

Walther Kossel, "Über Molekülbildung als Frage des Atombaus" [On the Formation of Molecules as a Question of Atomic Structure] *Annalen der Physik* (in German), **1916**, *354*(3), 229.

Gilbert N. Lewis, "The Atom and the Molecule" *J Am Chem Soc,* **1916**, *38*(4), 762.

In the same publications, they both linked the concept of valence (as combining power) with the drive for atoms to achieve the noble gas (closed shell) configuration using its valence electrons. They showed drawings of atoms using dots to represent electrons. Kossel used a more universally ionic model, while Lewis showed how shared 2-electron (covalent) bonds could give each atom its noble gas (closed shell) configuration. We still use these drawings today, calling them "Lewis dot structures" or the modified versions, "Lewis structures," in which covalent, shared electron pairs are represented as lines. Drawings with lines had been used for nearly 100 years, but Lewis redefined what the lines meant (Figure AP0104).

Figure AP0104

General principles: Lewis structures (1916).

Lewis also took on the question of multiple bonding (double and triple bonds). Although the cubic drawings captured the idea of 8 spaces for separate electrons, there were already some 50-year-old aspects of molecular structure that had been explained by tetrahedral atoms. Lewis imagined paired electrons sitting along alternating edges of the cube, which describes a tetrahedron. Lewis drew the modified cube but did not draw the actual tetrahedral structures (they were well known and considered common). Lewis's cubic structure with tetrahedral electron pairs and how those tetrahedra are imagined to be joined for single, double, and triple bonds, are shown in Figure AP0105.

On the other hand, the group of eight electrons in which the pairs are symmetrically placed about the center gives identically the model of the tetrahedral carbon atom which has been of such signal utility throughout the whole of organic chemistry. As usual, two tetrahedra, attached by one, two or three corners of each, represent respectively the single, the double and the triple bond.

Gilbert N. Lewis, "The Atom and the Molecule" *J Am Chem Soc,* **1916**, *38*(4), 762.

Figure AP1005

General principles: tetrahedral atoms and multiple bonds (1916).

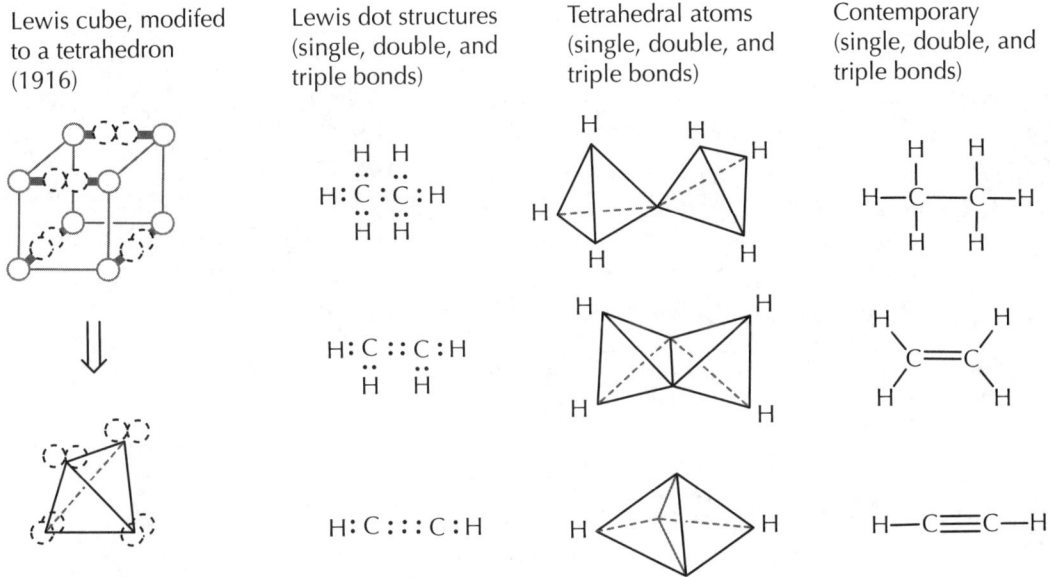

Kossel and Lewis connected the dots between ideas of valence, periodic properties, chemical affinity, molecular geometry, and the newly discovered science of subatomic particles, introducing the set of fundamental concepts that still drive all introductions to chemistry, today.

Valence electrons on atoms are the ones beyond the last closed shell configuration. Bonds form when atoms share, add to (gain), or lose electrons to achieve the nearest closed shell configuration.

Individual atoms, as they appear on the periodic table, have a formal charge of zero (0) because the number of negatively charged electrons equals the number of positively charged protons. Non-zero formal charges on atoms result when the number of electrons formally associated with an atom (nonbonding electrons plus half of the shared electrons).

In organic chemistry, whose original focus was on the molecules that appear in living organisms, the most common elements are only a narrow subset of the periodic table: H, C, N, O, S, and P. And so, studying the chemistry of what is called the "main group elements" dominates this subject (Figures AP0106 and AP0107).

Figure AP0106

Valence and bonding in the main group elements (rows 1 and 2).

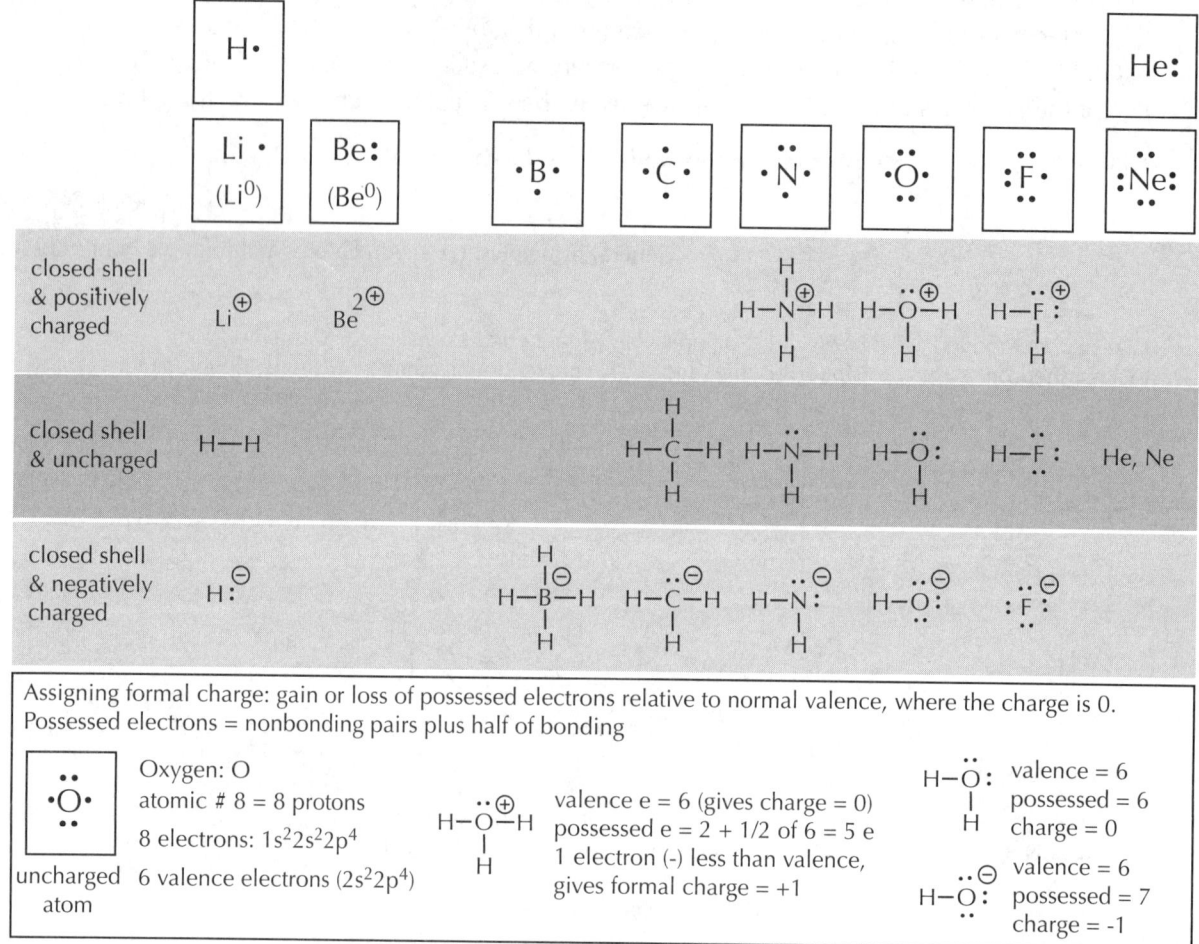

In the first row (H, He), the nearest closed shell electron configuration is that of helium (2 electrons in the "1s" level), also called the duet rule, typically achieved by H taking on 1 more electron through sharing or gaining.

In the second row (Li, Be, B, C, N, O, F, Ne), the nearest closed shell configuration is either that of helium (2 electrons, typically achieved by Li and Mg losing 1 or 2 electrons, respectively, giving the lithium cation, Li+, or beryllium dication, Be+2) or that of neon, called the octet rule, typically achieved by the remaining atoms taking on the required number of electrons through sharing or gaining to get a total of 8 electrons: 2 from the 2s sublevel plus 6 electrons from the 2p sublevel.

With atoms following a single rule (achieving a closed shell configuration), some broad generalizations about bonding result.

Compounds that include Group 1 and 2 metals are assumed to involve metal ions and to be ionic. Covalent compounds with uncharged, closed shell atoms have consistent and predictable bonding behavior: Hydrogen atoms have 1 shared electron bond; carbon atoms have 4; nitrogen atoms have 3 bonds, with 1 nonbonding electron pair (nbe); oxygen atoms have 2 bonds, with 2 nbe; and chlorine has 1 bond with 3 nbe. Any other situation for these atoms means, by definition, that the atom is open shell, or charged, or both. Although exceptions exist, atoms on the right-hand portion of the second row (B, C, N, O, F) rarely have formal charges other that -1, 0, or +1.

A shared electron pair bond does not have to arise from one electron being contributed from both atoms. An atom with a nonbonding electron pair can share that pair with an atom that needs two more to achieve its closed shell configuration. This type of bond, called coordinate covalent (also dative bond, coordinate bond), is commonly observed in the organic chemistry of boron, and when oxides form with individual oxygen atoms, as both of these atoms have 6 electrons and are in search of another pair to get to an octet.

Figure AP0107

Valence and bonding in the main group elements (row 3).

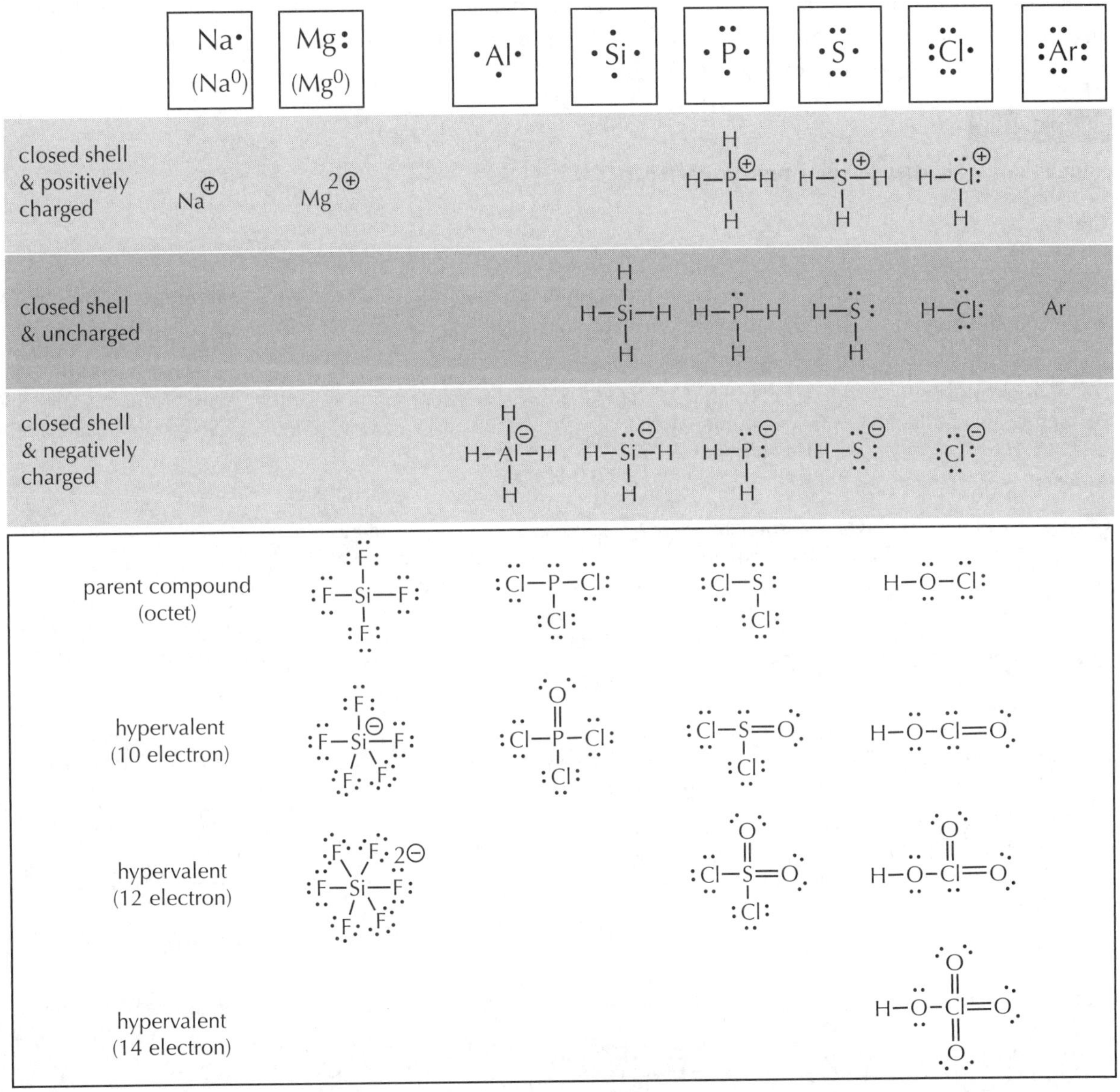

In the third row (Na, Mg, Al, Si, P, S, Cl, Ar), the nearest closed shell is either that of neon (sodium loses 1 electron to give the sodium ion, Na^+; magnesium loses 2 electrons to give the magnesium dication, Mg^{+2}, both of which are consistent with the octet rule), or that of argon, complete an octet of 2s + 6p electrons. For many chemical structures made up of the other atoms in this row, the octet generalization drives the formation of the molecules, and the compounds of aluminum mirror those of boron, silicon mirrors carbon, and so on.

918 APPENDIX 1 Useful Expectations from General Chemistry

Elements in the third row, unlike those in the second row, have an available 3d sublevel, and their coordinate covalent compounds, the commonly familiar ones being their oxides, are often drawn using structures with 10, 12, or 14 electrons associated with the central atoms, as opposed to using the drawings with high formal charges. The atoms with apparently over 8 electrons are said to be in a hypervalent state (see Figure AP0107 on the previous page for examples).

Knowing how to draw molecular structures cannot be calculated from a molecular formula. Familiarity with common examples (both structures and names), combined with experience and a willingness to play around with the pieces, are all important. So practice, practice, practice.

There are ions that recur numerous times in organic chemistry, and those shown in Figure AP0108 are useful to recognize and remember. These ions add to your baseline of examples for creating analogies when encountering more complex structures.

Figure AP0108

Common ions seen when learning organic chemistry.

Cations:

NH_4^+ ammonium
generic: ammonium

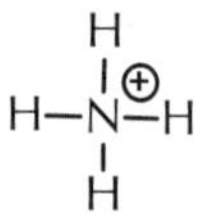

H_3O^+ hydronium
generic: oxonium

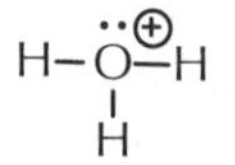

NO_2^+ nitronium

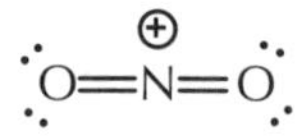

Fe^{2+} ferrous, also Fe(II)
Fe^{3+} ferric, also Fe(III)

Cr^{2+} chromous, also Cr(II)
Cr^{3+} chromic, also Cr(III)
Cr^{6+} chromium-6, also Cr(VI)

Cu^+ cuprous, also Cu(I)
Cu^{2+} cupric, also Cu(II)

Sn^{2+} stannous, also Sn(II)
Sn^{4+} stannic, also Sn(IV)

Anions:

CO_3^{2-} carbonate

HCO_3^- bicarbonate (hydrogen carbonate)

SO_4^{2-} sulfate

HSO_4^- bisulfate (hydrogen sulfate)

NO_3^- nitrate

NO_2^- nitrite

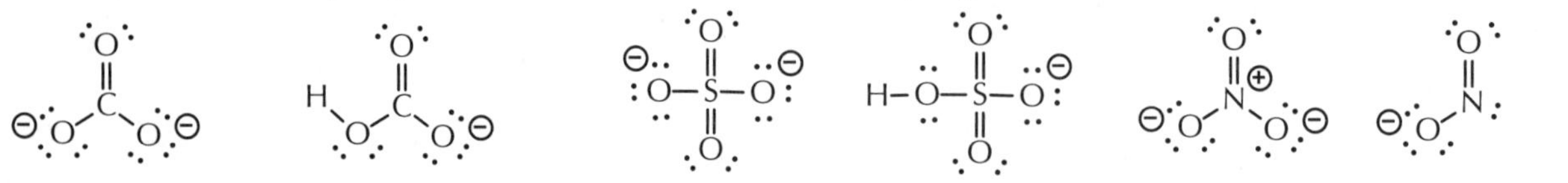

PO_4^{3-} phosphate (orthophosphate)

HPO_4^{2-} hydrogen phosphate

HPO_4^{2-} dihydrogen phosphate

$P_2O_7^{4-}$ diphosphate

$P_3O_{10}^{5-}$ triphosphate

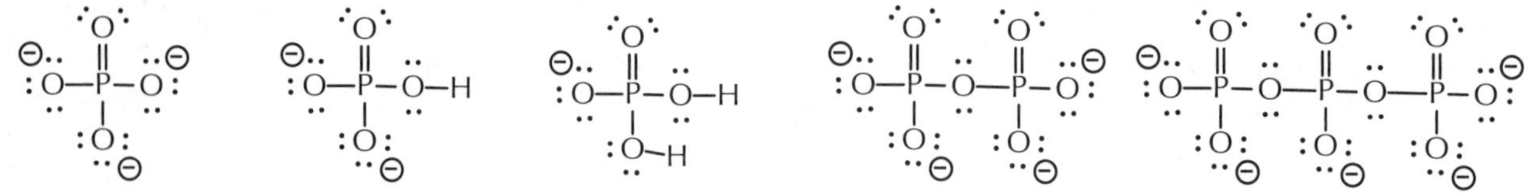

ClO^- hypochlorite

ClO_2^- chlorite

ClO_3^- chlorate

ClO_4^- perchlorate

CrO_4^{2-} chromate

$Cr_2O_7^{2-}$ dichromate

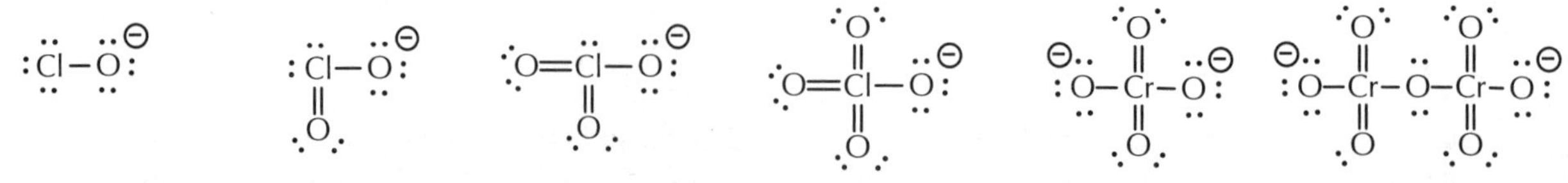

N_3^- azide

CN^- cyanide

HO^- hydroxide

HS^- sulfide

OCN^- cyanate

SCN^- thiocyanate

Three examples of drawing inferences from names and prior examples are shown in Figure AP0109.

Figure AP0109

Deducing molecular structures from the available information.

KMnO₄ (potassium permanganate)

⇒ ionic: K⁺ & MnO₄⁻
permanganate
(analogy: perchlorate)

K⊕ :Ö:
     ‖
:Ö=Mn−Ö:̈⊖
     ‖
    :Ö:

NH₄Fe(SO₄)₂ (ferric ammonium sulfate)

⇒ ionic: NH₄⁺ & Fe³⁺ & 2 SO₄⁻

         H   :Ö: :Ö:
         |⊕   ⊖.. ‖ ..⊖   ⊖.. ‖ ..⊖
     H−N−H   :Ö−S−Ö: :Ö−S−Ö:
         | ‖   ‖
         H   :Ö: :Ö:
                    Fe³⁺

PH₄I (phosphonium iodide)

⇒ ionic: PH₄⁺ & I⁻
phosphonium
(analogy: ammonium)

         H
         |⊕
     H−P−H   :Ï:⊖
         |
         H

C. The Quantum World

Quantum mechanics is an area of modern physics that started to develop in the mid-1800s, and mainly concerned itself with understanding how electromagnetic radiation (from x-rays to visible light to radio waves) interacts with matter.

In chemistry, by the early 1900s, the behavior of electrons in atoms and molecules was explained by energies that were not continuous, but by discrete energy bundles (quanta). Thus, when atoms or molecules absorb or emit energy, it appears that electrons can only exist with specific energy values.

Hydrogen atoms are the simplest to study, and the resulting ideas can be generalized.

First, though, a few cautions. The ways things operate at the atomic scale are quite different than at the familiar macroscopic scale of the material world. Although it was the first analogy used, electrons are not physical particles that orbit the nucleus like the planets around the sun. An electron associated with a proton exists with a single fundamental lowest energy (a lowest frequency), and we imagine the electron as a cloud of charge that is surrounding the proton like an atmosphere. The space inside where it is likely to find 90% of the electron's density is called its orbital. When we know its energy, precisely, we cannot locate where it is, only the space where it might be (Figure AP0110).

Figure AP0110

The planetary orbit model versus the orbital model for hydrogen atom.

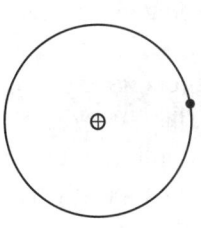
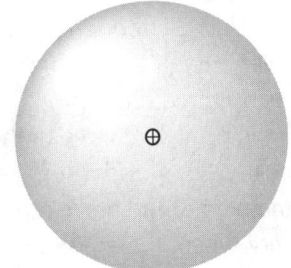

⊕ = nucleus (proton)
• = electron (as particle/orbit)

⊕ = nucleus (proton)
◯ = electron (as cloud of charge/orbital)

An unperturbed hydrogen atom is imagined to be a small point of positive charge (the proton) sitting within a cloud of negative charge (the electron). The spherical distribution is the most favorable for simple electrostatic reasons: It allows the greatest attractive charge/charge interaction between the negative and positive charges.

When the electron in a hydrogen atom absorbs energy, it only absorbs specific values, one of the best analogies for which is the waveforms of a vibrating string attached at its ends. Under a given set of conditions, a string suspended between two points has a finite set of vibrational frequencies associated with it (every string-based musical instrument is based on this principle).

As illustrated in Figure AP0111, the lowest vibrational frequency shows the longest waveform and the lowest energy. At the first overtone, the frequency is exactly twice the original, with half the wavelength, twice the energy, and one nodal point (a point where the string still exists but the value of its amplitude is zero). At the second overtone, the frequency is exactly tripled, the energy is three times higher than the original, and there are two nodes.

Figure AP0111
Vibrating string analogy for electron energy levels.

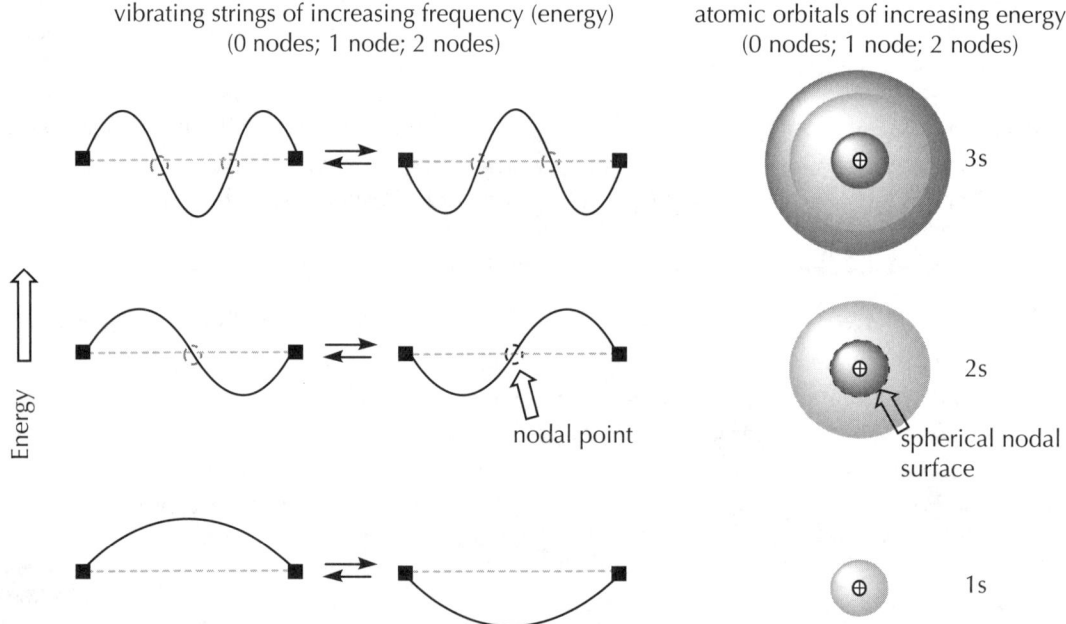

These wave properties are associated with electrons. The lowest energy state for an electron is surrounding the nucleus uniformly, with the lowest frequency (no node). By absorbing exactly the amount of energy, the electron goes from a lower energy (lower frequency) state and becomes a higher energy electron. Although it is hard to imagine in three-dimensional space, this higher energy electron is described by a sphere (larger = farther from the positive charge of the nucleus = less stable), and the higher "vibrational energy" of the sphere means there is a nodal surface with opposite amplitudes on either side of it. The vibrating string is the best analogy: At the first overtone, the string simultaneously has both positive and negative amplitude, and a spot where the amplitude is zero, but there is always a string present. We do not confuse the existence of the string with its properties. The same statement works for electrons.

The electron is less stable in the higher energy state than lower one, so the atom with the higher energy electron can spontaneously lose (emit) the specific energy it absorbed and return to the lower energy state. For hydrogen atoms, the energy is emitted as either infrared, visible, or ultraviolet light as a series of lines with quite specific frequencies. Seeing these specific few light lines, and measuring their frequencies, ultimately resulted in our picture of the atom as a set of specific energy levels that could be populated by electrons, and that the electrons filled these levels from the bottom up.

These drawings are complicated and imprecise, and so we more commonly use diagrams showing the energy levels. Hydrogen atoms have been studied intensely, and the frequencies of light emitted from high energy hydrogen atoms maps out a clear picture of seven major energy levels where the spacing is not even (Figure AP0112).

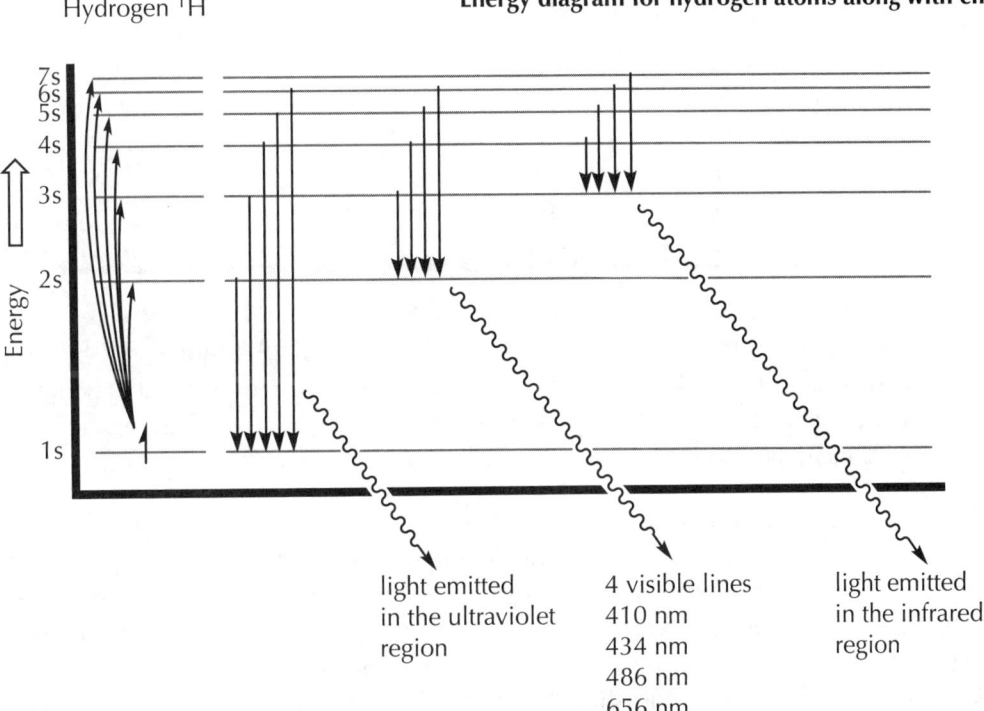

Figure AP0112

Energy diagram for hydrogen atoms along with emitted light.

As we move from hydrogen to other elements, the orbital picture becomes more complex, but at its root is the explanation of the assembly of the elements on the periodic table. In organic chemistry, we nearly exclusively focus on elements in the first and second rows, and the familiar assembly of the first 10 elements using electrons that can be in 1s, 2s, and 2p atomic orbital energy levels accounts for what we study the most.

The 1s level is a uniform sphere. The 2s level is its overtone, with a nodal surface (although, for our purposes in organic chemistry, we even ignore that node and just think of the 2s level as a larger sphere). Mathematically, the entire 2p level is also a sphere, but because any given orbital can only hold a maximum of two electrons, the 2p level is made up of the three mutually perpendicular p orbitals, with opposite amplitudes on either side of a nodal point in the center (Figure AP0113 on the next page).

Developing a quantum mechanical view of electrons in molecules necessarily accompanied the model used for atoms.

The basic question is reasonable: If we know what an electron does when it is captured by a single proton, what does it do when a pair of electrons is captured by a pair of protons that are sitting at bonding distance? The answer to this question gives the quantum mechanical view of the bonding in a hydrogen molecule.

Figure AP0113

Energy diagram and atomic orbitals for second row elements.

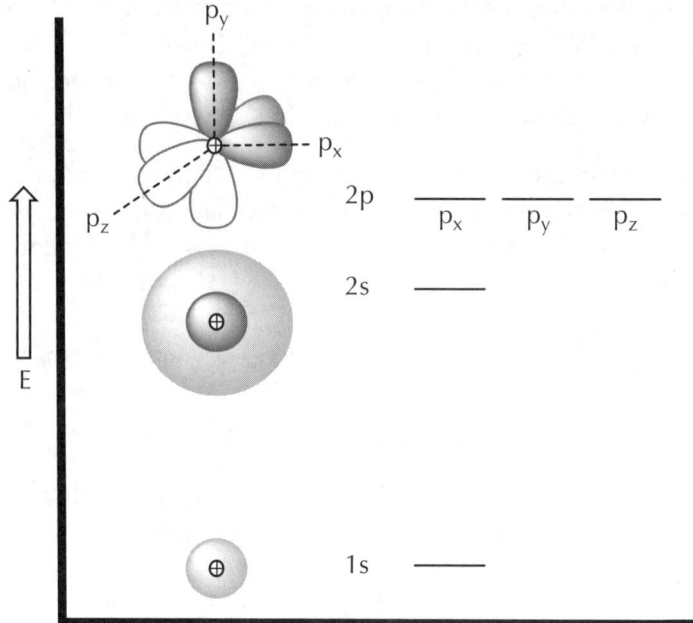

One specific question covers the most important topic for the bonding in molecules: Hydrogen atoms and helium atoms both possess 1s electrons; why, when two hydrogen atoms get within bonding distance, do they join with a result that is more stable than when they were apart, but when two helium atoms get within bonding distance, they do not join, but instead simply bounce off one another?

The explanation is the formation of molecular orbitals, a way for electrons to be distributed over more than one nucleus at the same time. When two nuclei get within bonding distance, the positive charges are repelling one another and need to be insulated. So, when two hydrogen atoms, each with an electron in a 1s orbital, combine to give a molecule of hydrogen, a new orbital forms that (a) encompasses both atoms, which means each electron can interact with 2 positive nuclei, and (b) places the bulk of the electron density between the two nuclei to shield them from their mutually repulsive force. This orbital is better for an electron to reside in, and so lower in energy. The diagram (Figure AP0114) illustrates how the 2 electrons in the new molecular orbital, because it is lower in energy than the atomic orbitals, creates a more stable situation for molecular hydrogen compared with the free atoms.

Figure AP0114

Forming a molecular orbital to create molecular hydrogen.

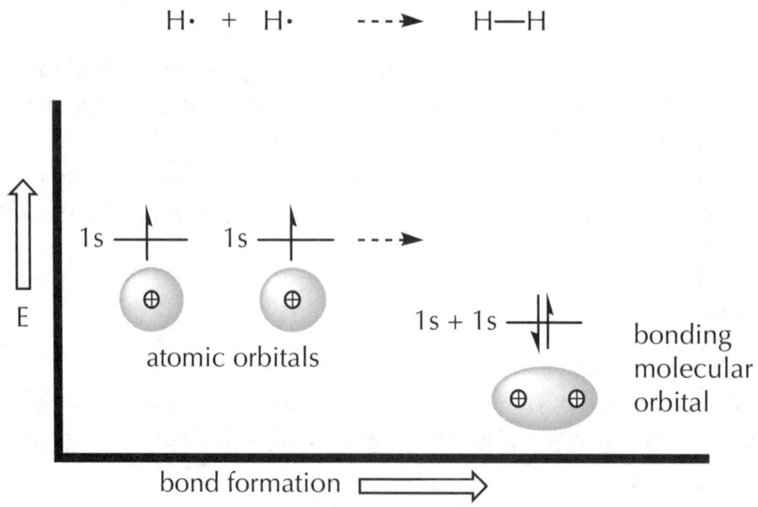

But this is not the whole story. If each of those atoms carried two electrons in their 1s atomic orbitals, the resulting molecular orbital also needs to exist, providing a place where two of the electrons are more stable. The result from the helium plus helium case, in which the atoms do not end up bonded, means that there is no net gain in stability when two atoms with 4 electrons come together, as opposed to the gain in stability when two atoms with 2 electrons combine (Figure AP0115).

Figure AP0115

Bonding and antibonding molecular orbitals in bond formation.

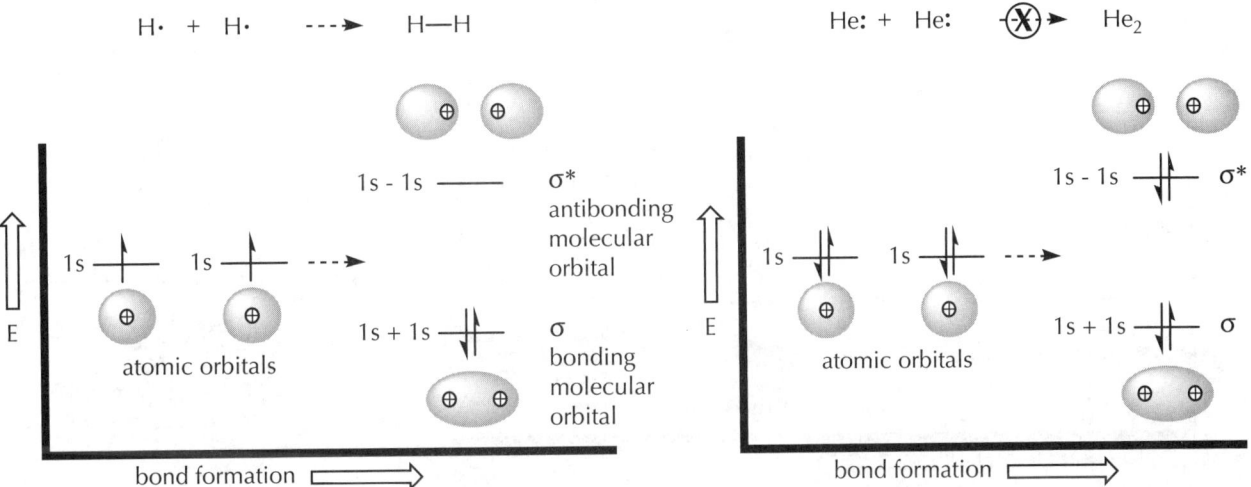

The best explanation for not forming diatomic helium, yet forming diatomic hydrogen, is that while 2 of the electrons go lower in energy (in diatomic helium), favoring the bond, the other two electrons go higher in energy (in diatomic helium), into a situation that disfavors the bond.

As a result, when two 1s atomic orbitals come within bonding distance, we imagine that two new molecular orbitals form, both of which encompass both nuclei. The lower energy molecular orbital, called the bonding molecular orbital, has no node and places the electron density between the nuclei, thus overcoming the internuclear repulsion. This bonding molecular orbital is also called a "sigma" (σ) bond. The higher energy molecular orbital, called the antibonding molecular orbital, has a node with no electron density in the center, and places the bulk of the electron density on the opposite sides of the bonding region. Holding two nuclei together in this situation is high energy because the two positive nuclei have no insulating effect from the electrons, and the nuclei, in fact, are being pulled apart by their attraction to the electrons. This antibonding molecular orbital is called "sigma star" (σ^*).

In second row elements, sigma bonds can also form when p orbitals are pointed towards one another along the internuclear axis (Figure AP0116).

Figure AP0116

Sigma bonding and antibonding molecular orbitals using p orbitals.

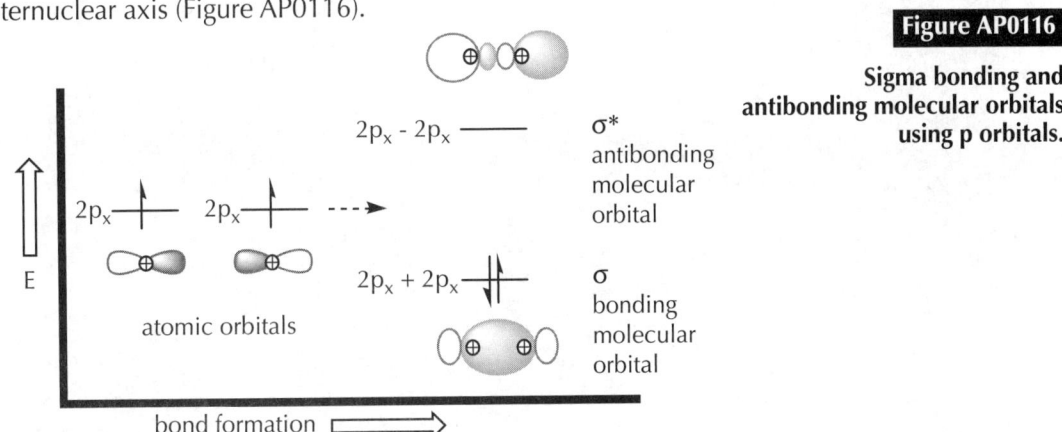

APPENDIX 1 Useful Expectations from General Chemistry

In the 1930s, the model for double and triple bonding changed from the all-tetrahedral atom view (Figure AP0105). In a double bond, one of them is a sigma bond, with a pair of electrons along the internuclear axis. The other is a pi bond (π), in which two parallel p orbitals share electron density above and below the sigma bond. A triple bond is constructed from a sigma bond and two pi bonds, the latter two derived from a set of mutually perpendicular pairs of p orbitals (Figure AP0117).

Figure AP0117
Double and triple bonding using sigma and pi bonds.

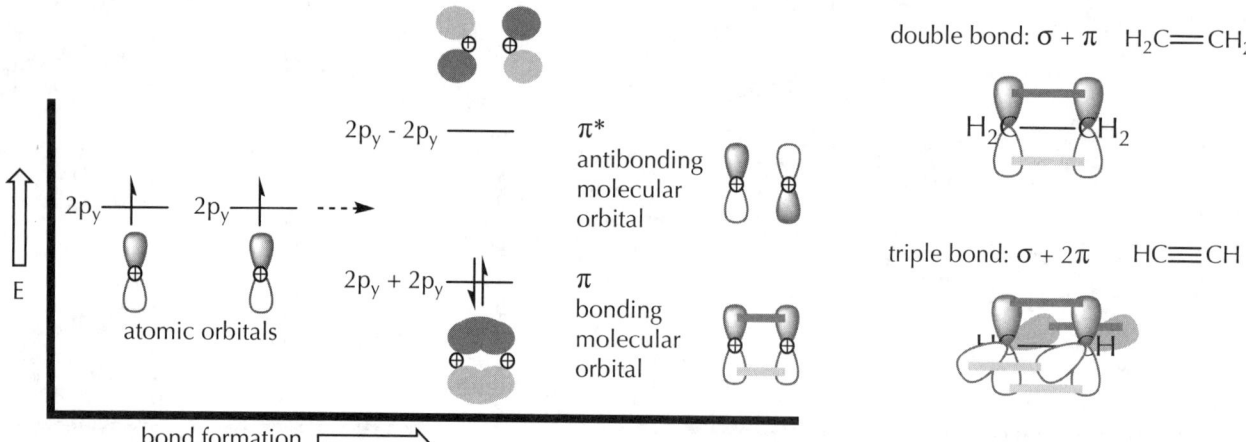

PRACTICE QUESTIONS

AP01.01 Using Lewis structures (lines for bonding electrons, dots for nonbonding/valence electrons), draw the common representations that distinguish between the following.

(a) bromide ion   bromine atom   bromine molecule   (b) sodium atom   sodium ion

(c) ammonia   ammonium ion   (d) oxygen atom   oxygen molecule

(e) hydrogen atom   hydrogen cation   hydrogen molecule   hydrogen anion

AP01.02 The following lists the monatomic examples from AP01.01. Provide the electron configurations for these atoms/ions and indicate whether or not this represents a closed shell configuration.

(a) bromine atom — closed shell?

(b) hydrogen anion — closed shell?

(c) sodium atom — closed shell?

(d) bromide ion — closed shell?

(e) oxygen atom — closed shell?

(f) hydrogen cation — closed shell?

(g) sodium ion — closed shell?

(h) hydrogen atom — closed shell?

AP01.03 Using Lewis structures (lines for bonding electrons, dots for nonbonding/valence electrons), draw the following. Use ion-separated drawings for ionic compounds.

(a) ammonium bromide

(b) sodium fluoride

(c) sodium hydrogen sulfate

(d) potassium carbonate

(e) ammonium tetrafluoroborate

(f) ferric chloride

(g) nitrogen triiodide

(h) lithium tetrachloroaluminate

(i) hydrogen chloride

(j) potassium hydroxide

(k) lithium hydride

(l) calcium dichromate

AP01.04 The following molecular structures contain atoms that are missing their nonbonding electron pairs and, in some cases, formal charges. Assuming that all of the atoms have closed shell electron configurations, complete the following structures.

(a) acetaminophen

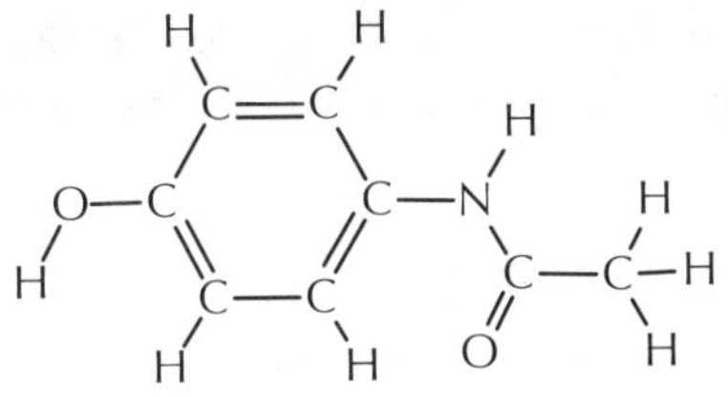

(b) caffeine

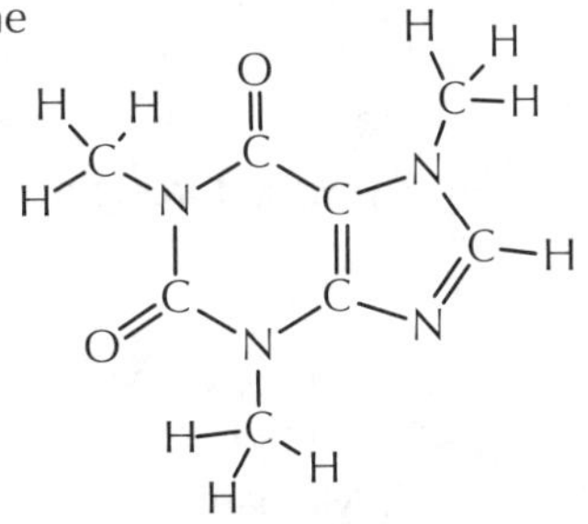

(c) sodium nitrite

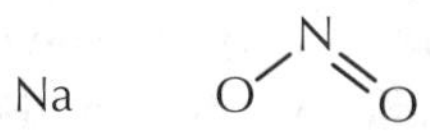

(d) ketamine

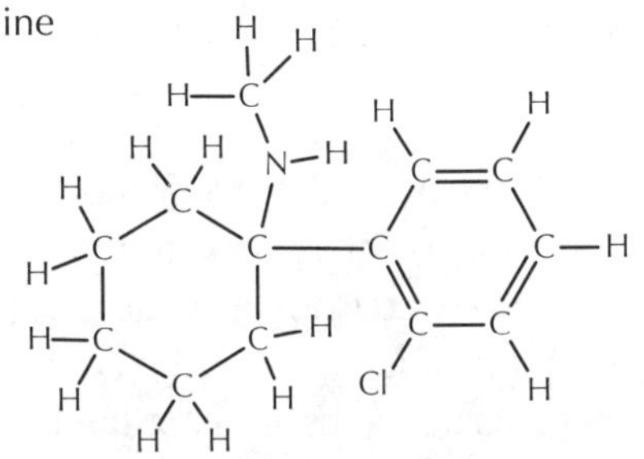

(e) nicotine

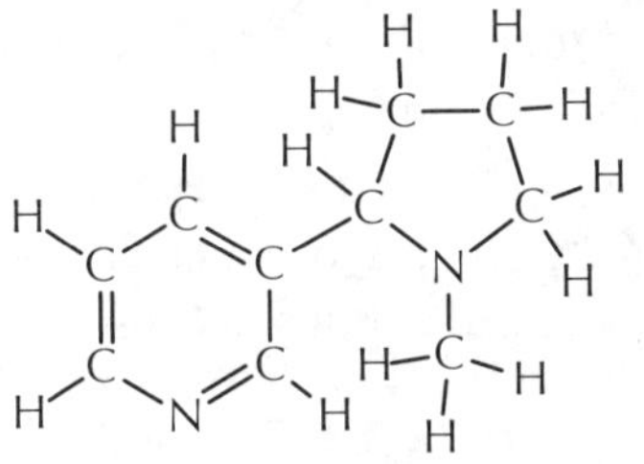

(f) acetylcholine picrate

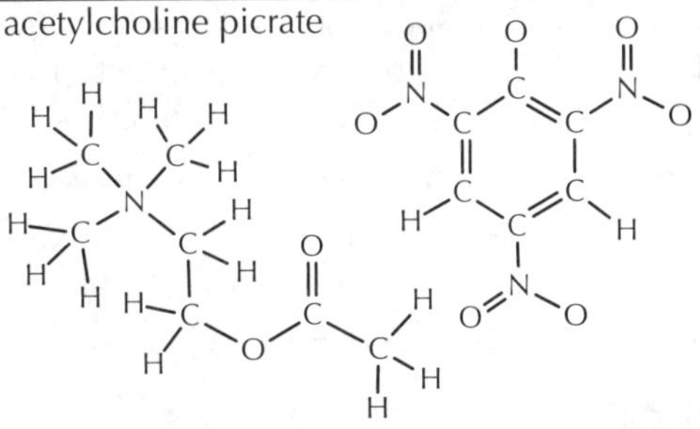

AP01.05 The model of bonding that involves atomic and molecular orbitals has proved to be useful for understanding and explaining the existence of compounds that the prior bonding models could not handle. Sticking with the "1s + 1s" system, in which a lower energy sigma (σ) bonding orbital and a higher energy sigma star (σ^*) antibonding orbital are formed, answer the following.

(a) In 1933, based on the molecular orbital model of bonding, Linus Pauling predicted the existence of the dihelium cation (He_2^+). Using an energy diagram as the explanation, what was the basis for this prediction?

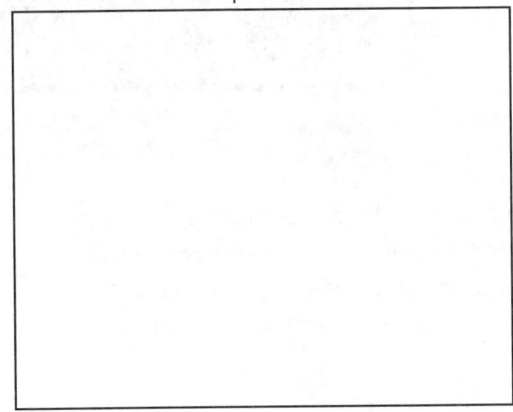

(b) The helium hydride cation (HeH^+) has been studied. Its bond length, 0.772 Å, compares quite favorably with the standard bond distance in molecular hydrogen (H_2; 0.74 Å). Why is this equivalence reasonable?

A1.2 Periodic Trends

A. Electronegativity and Covalent Bond Polarity

In the early 1800s, the first substantial batteries were being built, and so electricity was readily available for the first time. Curiosity about the effect of electricity on material substances was a natural outcome. When electricity was passed through water, two previously identified gasses were observed: oxygen and hydrogen.

Lots of familiar vocabulary is related to this experiment, which is called the electrolysis of water (electrolysis = to break apart or dissolve "lyse" with electricity). Electricity, which had an obvious flow, was originally treated as fluid. Michael Faraday named the pole where electricity "moved in" or "moved up" as the anode ("the up way"), and the other pole as the cathode ("the down way"), or where the electricity "moved out." Earlier, Benjamin Franklin had symbolized the cathode as "−" (exit) and the anode as "+" (entrance).

Thanks to the electrolysis experiments, scientists got their first real clue that some aspect of electricity was responsible for chemical affinity, because of its effect on water: (a) the hydrogen and oxygen gasses were strictly derived from the water, and (b) when the two gasses were combined, water (and only water) was re-formed. Indeed, it was the combination of these two gases, giving water as an outcome, that gave "hydrogen" its name (hydro-gen, the bringer of water).

As a consequence of the electrolysis of water, water was imagined to be built from two particles with opposite electrical charges that were attracted to the anode and cathode. These "charged, moving particles" were called ions, derived from the Greek verb "to go" (*ienai*). The oxygen gas showed up at the "+" anode, and so it was imagined that oxygen atoms were electrically negative ("electro-negative") and they were called anions, because of their attraction to the anode. The hydrogen gas, produced at the cathode, was imaged to be electrically positive ("electro-positive") ions, and called cations because of their attraction to the "−" cathode (Figure AP0118).

Figure AP0118
Terminology associated with the water electrolysis experiment.

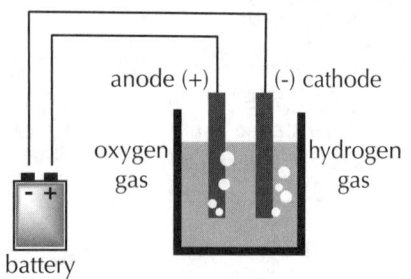
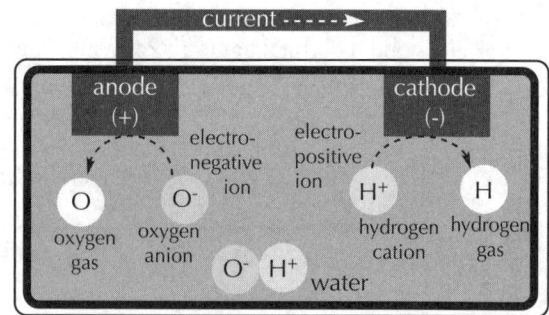

As noted on the figure, water was originally thought to be a binary compound, HO, composed on one oxygen ion and one hydrogen ion. Early tables of atomic weights list oxygen as about 8 g/mol rather than 16 g/mol for this reason. The fact that exactly twice as much volume of hydrogen gas than oxygen gas is formed in the electrolysis experiment was a key experimental observation that led to correcting the formula for water from HO to H_2O.

If you have any curiosity about the electrolysis experiment, you can take a standard 9V battery and two pieces of copper wire, hold the wires with your fingers to the two poles and dip the free ends into water. You will see bubbles of gas forming visibly on the end of at least one, if not both, of the wires. You can also use the graphite in pencils as electrodes (sharpen both ends of two pencils, connecting each one to your wires and dipping the other ends in the water).

A hundred years after Faraday, Lewis, in his publication about atoms and molecules, discussed electronegativity in more modern terms of bond polarity, where the influence of the chlorine atom on the electrons in, for example, a C-Cl bond shifted the electrons towards the chlorine atom (making it more electrically negative), and resulting in a polar covalent bond.

Lewis did not address the origin of electronegativity, which is simply a greater effective positive charge from the nucleus, that can better draw electrons to itself. The periodic trend, that the electronegativity of elements increases as you move to the right in a row of the periodic table, is consistent with the increasing amount of positive charge in the nucleus as you move from left to right.

In a column of the periodic table, predicting the trend is not so obvious. Moving down a column, the absolute size of the positive charge in the nucleus grows considerably, which should be favorable; on the other hand, bonding electrons on the larger atoms are farther from the nucleus, and the effect of the nuclear charge is masked by the inner shells of electrons. Experimentally, trends in reactivity that rely on bond polarity all favor fluorine atoms. It has minimal electrons to mask the nuclear charge and puts bonding electrons in greater proximity to the nucleus, so it ends up being the most electronegative atom. And the trend, that electronegativity decreases as you move down a column, is consistent throughout the periodic table.

In 1932, Linus Pauling defined a quantitative scale for electronegativity based upon the idea that the ionic character of an AB bond (A+B-) made it stronger than its purely covalent (A-B) character. His values are dimensionless and relative, but useful because they are highly internally consistent (Figure AP0119).

Figure AP0119

The Pauling electronegativity scale (partial).

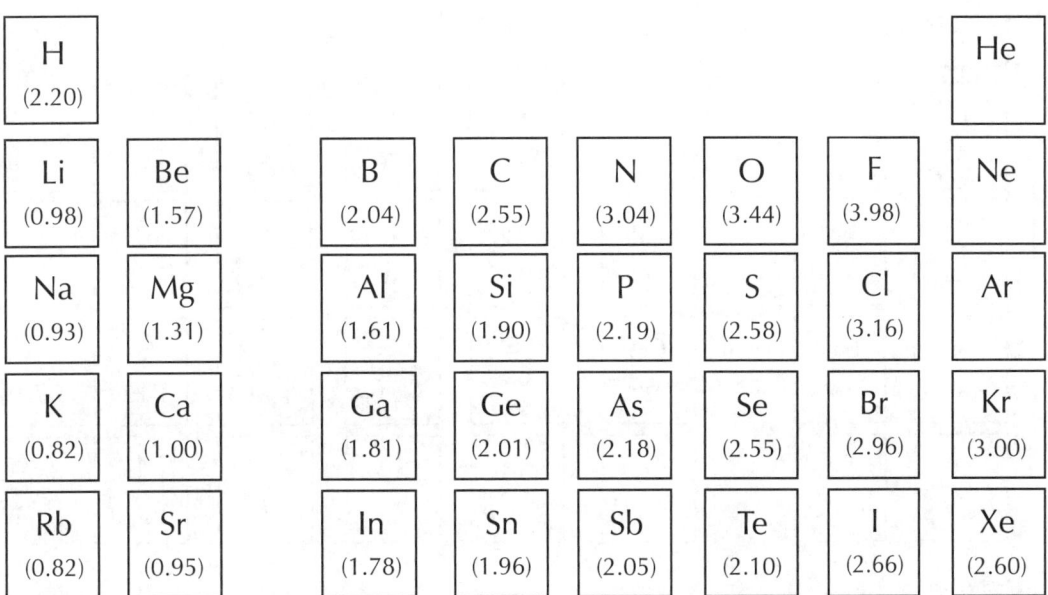

Linus Pauling "The Nature of the Chemical Bond. IV. The Energy of Single Bonds and the Relative Electronegativity of Atoms" *Journal of the American Chemical Society,* **1932**, *54*(9), 3570.

Although the left-to-right (row) and bottom-to-top (column) trends in electronegativity are useful for comparisons within rows and columns, the numerical scale is particularly useful for any diagonal comparisons, where neither of those generalizations is relevant, or if a relative ranking (e.g., O versus Cl, or N versus S) is called for (Figure AP0120).

Figure AP0120
Assigning partial charges to polar covalent bonds based on differences in electronegativities.

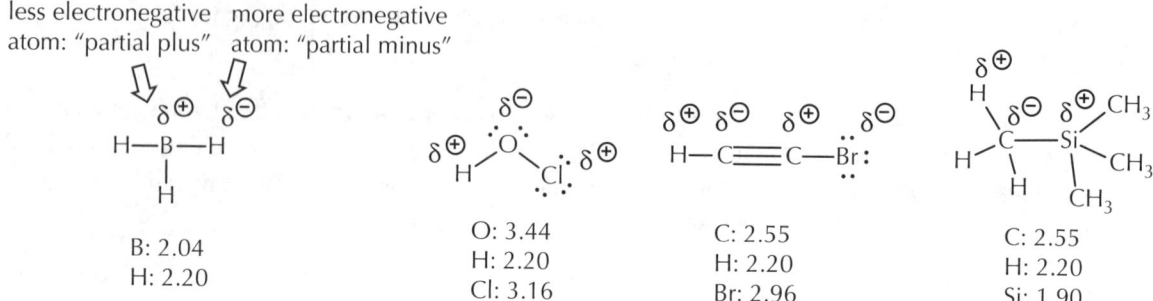

B. Atomic Size and Bond Length

Atoms do not have hard shells or boundaries, which means that the concept of atom size needs to be defined by the assumptions of the particular experiment being used to collect data. In general, the average size of an atom is derived from measuring a large number of covalent bond distances (nucleus to nucleus distances are measurable), and then inferring the size of the atoms, as given by an atomic radius, by looking at many different bond pairings.

The generalizations are consistent throughout the periodic table (Figure AP0121).

Figure AP0121
Trends in atomic size (radius).

H 0.37 Å							He 0.32 Å
Li 1.52 Å	Be 1.11 Å	B 0.88 Å	C 0.77 Å	N 0.70 Å	O 0.66 Å	F 0.64 Å	Ne 0.69 Å
Na 1.86 Å	Mg 1.60 Å	Al 1.43 Å	Si 1.17 Å	P 1.10 Å	S 1.04 Å	Cl 0.99 Å	Ar 0.97 Å
K 2.31 Å	Ca 1.97 Å	Ga 1.22 Å	Ge 1.22 Å	As 1.21 Å	Se 1.16 Å	Br 1.15 Å	Kr 1.10 Å
Rb 2.44 Å	Sr 2.15 Å	In 1.62 Å	Sn 1.40 Å	Sb 1.41 Å	Te 1.37 Å	I 1.33 Å	Xe 1.30 Å

Atom size (radius value) increases as you move down a column. As you go from row to row on the periodic table, electrons are being placed into new shells, which makes moving up and down in a column result in the largest size differences for adjacent elements: lithium (1.52 Å) to sodium (1.86 Å) to potassium

(2.27 Å) to rubidium (2.48 Å). And another example: oxygen (0.73 Å) to sulfur (1.03 Å) to selenium (1.19 Å) to tellurium (1.42 Å).

Atom size (radius value) decreases as you move across a row to the right. With all atoms in the same row adding electrons into the same shell, the rough scale of size for the atoms is about the same. As the positive nuclear charge increases as protons are added, left to right, the pull on those electrons increases (i.e., increasing electronegativity), and the size of the atom contracts. The difference in size from atom to atom, in a row, is not as dramatic as the change reported for adjacent atoms in a column. Looking at the 2p level: boron (0.80 Å) to carbon (0.77 Å) to nitrogen (0.74 Å) to oxygen (0.73 Å) to fluorine (0.72 Å) to neon (0.71 Å).

Because atom sizes are often derived from bond distance measurements, the trends in bond distance parallel those for atom size differences. The brief table of values shown here (Figure AP0122) summarizes the important generalizations about bond distance trends.

Figure AP0122

Trends in bond length.

H—C 1.09 Å	H—O 0.97 Å	C—C 1.54 Å	C—N 1.47 Å	C—O 1.43 Å	C—F 1.35 Å
H—Si 1.48 Å	H—S 1.35 Å	C=C 1.34 Å	C=N 1.28 Å	C—S 1.81 Å	C—Cl 1.74 Å
H—Ge 1.53 Å	H—Se 1.46 Å	C≡C 1.20 Å	C≡N 1.16 Å	C—Se 1.98 Å	C—Br 1.94 Å

All else equal, differences based on different rows are the most significant. All of the bonds between two second row elements, regardless of being single, double, or triple, are longer than a second row element bonded to hydrogen: HC and HO are both significantly shorter than any of the CC, CN, CO, or CF bonds. And the CS bond, for instance, which is the combination of a second row element with a third row element, is longer than all of those bonds between two second row elements.

For bonds between elements within the same row, bond order has an intermediate significance. As a group, the CC and CN triple bonds are shorter than the CC and CN double bonds which are, in turn, shorter than the CC, CN, CO, and CF single bonds.

Within the same row and the same bond order, the electronegativity compression shortens a bond length, and this is the least significant difference. The CN triple bond is shorter than the CC triple bond, the CN double bond is shorter than the CC double bond, and the CF single bond is shorter than CO, which is shorter than CN, which is shorter than CC. These generalizations hold elsewhere: C-Cl is shorter than C-S, H-S is shorter than H-Si, and C-Br is shorter than C-Se.

C. Oxidation-Reduction

In one of the original meanings of the term, oxidation meant a reaction in which the amount or fraction of oxygen in a compound increased ("to make the oxide"). The combination of iron metal (Fe) with oxygen to give rust (iron oxide, Fe_2O_3, among others) is a typical example. Reduction, also started out as its literal meaning: Metallic ores lost weight when heated (a reduction in weight) to produce the pure metal. Often, these ores were metal oxides, so the coupling of oxidation (gain of oxygen) with reduction (loss of oxygen) was established early in the history of modern chemistry (ca. the late 1700s).

The meanings of the two words have evolved with our change in understanding, but they remained linked together as complementary processes, often referred to as "redox."

In modern terms, a redox process involves the transfer of electrons between different atoms or molecules. A reducing agent provides an electron source, and in giving up its electron(s) it undergoes oxidation (loss of electrons), becoming more positive. An oxidizing agent is an electron destination, and in accepting electron(s) it undergoes reduction (gain of electrons), becoming more negative (Figure AP0123).

Figure AP0123
Oxidation-reduction.

	1700s	1800s	1900s
Oxidation	making an oxide (acidic taste)	> % oxygen or < % hydrogen	loss of electrons (increase in oxidation state)
	burning sulfur in moist air gives the oxide of sulfur	$S \xrightarrow{O_2} SO_3 \xrightarrow{H_2O} H_2SO_4$	$S^0 \longrightarrow S^{6+}$
Reduction	loss of mass (crude ore to refined metal)	< % oxygen or > % hydrogen	gain of electrons (decrease in oxidation state)
	heating iron ore with carbon to give iron	$Fe_2O_3 \xrightarrow[C,\ heat]{} Fe + O_2$	$Fe^{3+} \longrightarrow Fe^0$

The more easily an atom or molecule gives up its electrons, it is classified as a stronger reducing agent. And the more easily an atom or molecule takes on additional electrons, it is classified as a stronger oxidizing agent.

Some periodic trends can be predicted based on atomic structure. Sodium and lithium metal are both reducing agents because their chemistry is to transfer their single valence electrons away, when possible, to give monocations with closed shell electron configurations. The single 3s valence electron in sodium is farther away and less tightly held by the sodium nucleus than the single 2s valence electron is held by lithium. The 3s electron is more easily lost from sodium metal than the 2s electron is from lithium, and so sodium metal is a stronger reducing agent than lithium metal.

Seen in the complementary fashion, this trend means that the lithium cation is a stronger oxidizing agent (more readily gains and holds on to an electron) than the sodium cation, which is less likely to gain and hold on to an electron (as it gives it up more readily).

PRACTICE QUESTIONS

AP01.06 Using partial charge labels ($\delta^{\oplus}$, $\delta^{\ominus}$), indicate the direction of the polarity for each of the bonds indicated with an arrow in the following examples. Try predicting the direction, first, before looking up values, to see how well your instinct is.

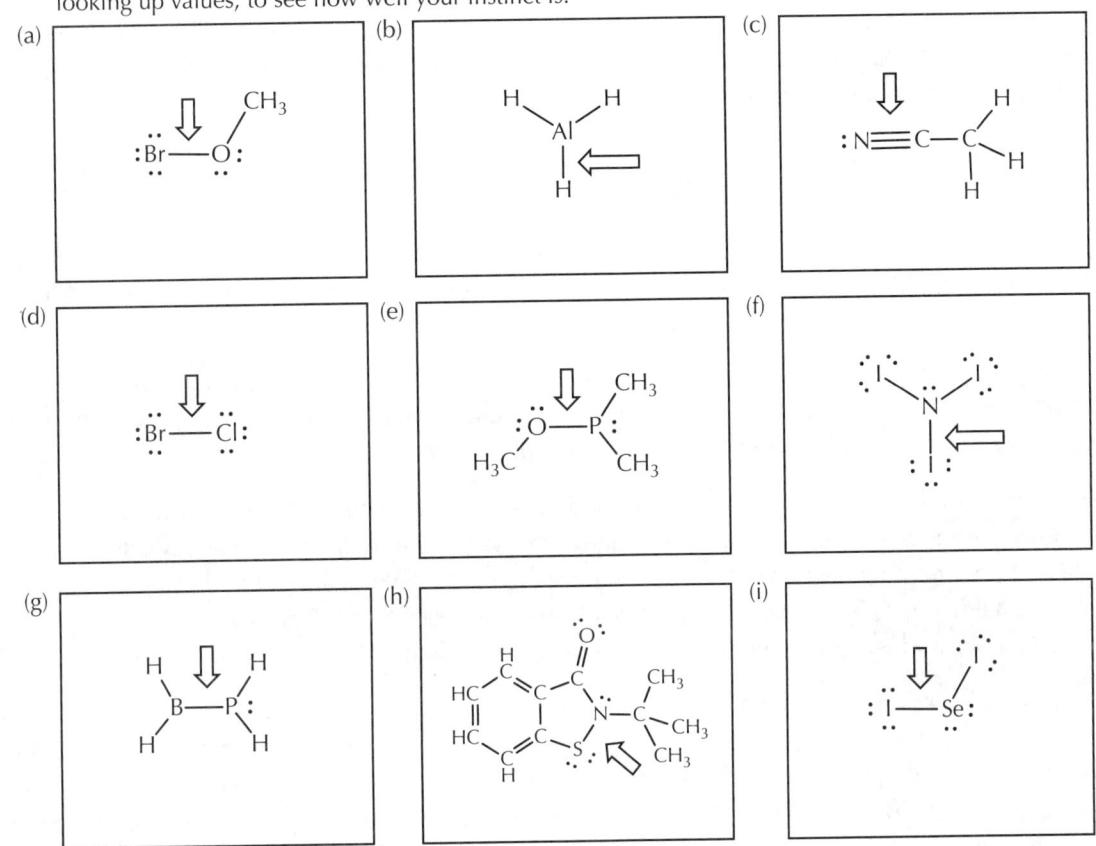

AP01.07 Using the generalizations about bond distances, rank the following bond from longest to shortest. In additon, rank the same bonds according to their polarity. Bonds with an electronegativity difference of less than 0.5 Pauling units are conventionally considered nonpolar covalent. Before you assess the polarity, predict how many of the marked bonds are nonpolar covalent.

AP01.08 Is iodide ion or chloride ion a stronger reducing agent? Why?

AP01.09 Can sodium metal or potassium metal be oxidized more easily? Why?

934   APPENDIX 1   Useful Expectations from General Chemistry

A1.3   Chemical and Physical Properties

A. Rate (Kinetics) and Stability (Thermodynamics)

For the most part, humans study chemistry based on the average behavior of incredibly large populations of molecules. Recall that a mole of water (6.02×10^{23} water molecules) occupies 18 mL of volume, which is barely over the amount delivered by a tablespoon.

Also, molecules are moving fast. The OH bonds in water vibrate on the order of 1014 times per second, and the water molecules are colliding about 109 times per second.

With so many molecules moving so quickly, and with them being too small to watch and record on film, the details of what goes on at the atomic and molecular level has always needed to be inferred from the measurements that could be carried out.

In the early years, measurements of mass and volume were the most common, and helped establish the incredibly important idea that atoms are conserved in a chemical process. Everything that goes into the front end of a reaction can be accounted for at the end.

Between about 1900 and 1935, as models for bonding and structure began to uncover the details of what molecules were, ideas also emerged about how structures varied based on the relative stability or instability of certain bonds or arrangements of atoms/molecules (thermodynamics), about the detailed (yet imagined) picture of what happens during bonding changes (called the reaction mechanism), and about how differences in structure and reaction conditions related to how fast chemical reactions take place (kinetics).

In the earlier discussion about bonding (Figure AP0115), moving the two electrons provided by two individual hydrogen atoms from their 1s orbitals into the lower energy level represented by the sigma (bonding) molecular orbital is a statement of going to greater thermodynamic stability. We say that it is favorable for a reaction to "go downhill in energy" and, as in this case, we draw the lower and more stable energy level of the sigma molecular orbital that is lower in energy than the atomic orbitals (Figure AP0124). Because energy needs to be conserved, the energy that is released when the bond is formed might be an emission of heat, light, or both. Alongside the energy diagram representing the orbital levels is another energy diagram, showing the change in overall energy as the interatomic distance decreases (the two atoms approach each other) and come to their ideal bonding distance (energy minimum). At shorter distances, the energy goes up dramatically as the two positive nuclei, too close to each other, have a strong and severe repulsion effect.

Figure AP0124

Relative thermodynamic stability in bond formation.

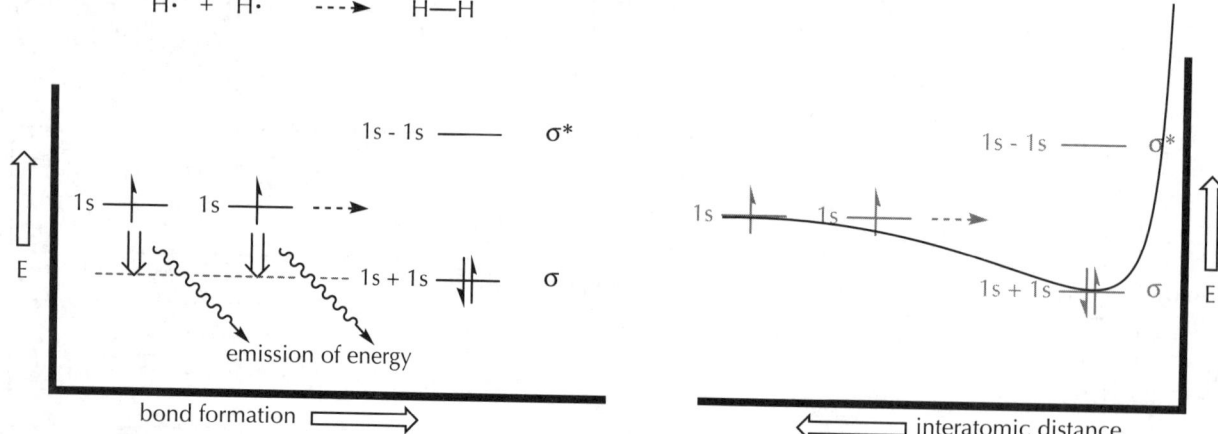

Reaction rates depend on many different experimental factors, but they all come down to a simple idea: What is the probability that the molecules have the proper proximity, orientation, and energy to carry out a reaction?

In the reaction to form hydrogen molecules from hydrogen atoms, the trillions of trillions of hydrogen atoms do not recognize or know how to seek out other hydrogen atoms, so they are also colliding with everything else around. As hydrogen molecules begin to form, hydrogen atoms are also colliding with them. A lone hydrogen atom (H•) can react with molecular hydrogen (H-H) to form a new molecule of hydrogen (H-H) and a new hydrogen atom (H•). The starting and ending points are the same atom and molecule, but in between there is a collision and an exchange in the bonding.

Not every collision has the proper orientation or the proper energy to cause a change to take place, so perhaps the two objects just bounce off one another. We imagine that the most effective collision takes place when the hydrogen atom collides with the molecular hydrogen from a direction directly opposite of its internuclear axis (Figure AP0125). There is great value, as you learn chemistry, to start thinking of atoms and molecules as real, physical objects for which the symbols are merely placeholders and not the objects themselves.

An energy diagram can be used to describe the change in energy taking place during the imagined collision. The molecular structure during the collision is less stable than the starting point, as the original hydrogen-hydrogen bond breaks and the new hydrogen-hydrogen bond forms. At some point during the collision, there is about half a bond on each side of the central hydrogen. We have no bonding theory to account for this as a stable structure, and we call this point of highest energy the transition state structure for the reaction.

Figure AP0125

Energy diagram for the hydrogen atom plus hydrogen molecule reaction.

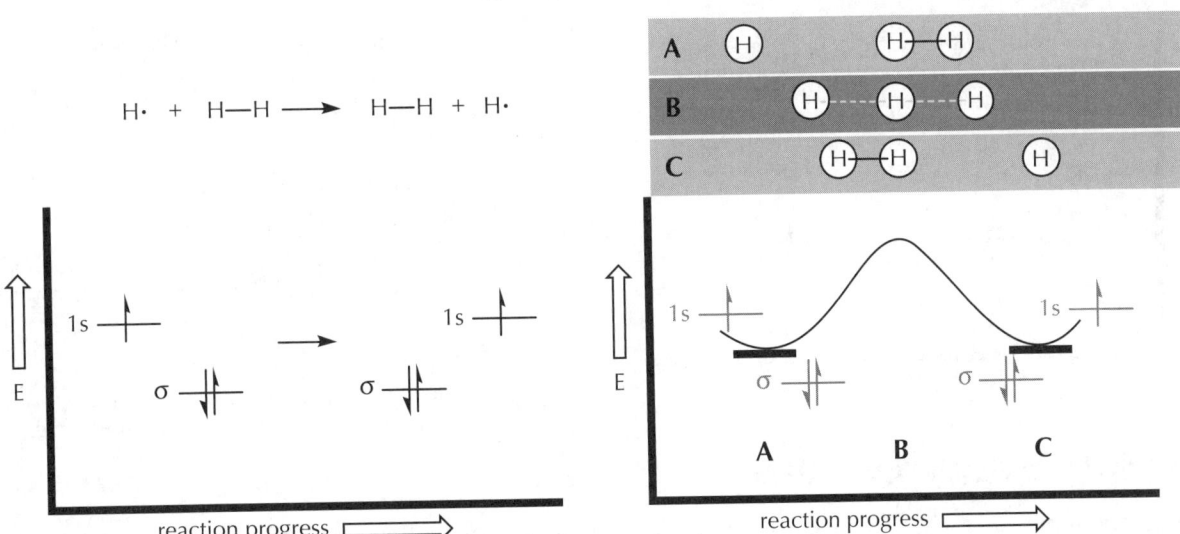

There needs to be enough available energy for the atoms and molecules to be moving around, colliding, as well as enough energy for the high energy structure (the transition state structure) to form, so that the process can occur. The energy difference between the starting point and the transition state structure is called the activation energy.

Note that this particular reaction would be impossible to study experimentally, because the two products are exactly the same as the starting materials. Scientists interested in chemistry theory, or chemical physics, use these simple reactions for creating computational models as a way to study the most fundamental aspects of chemistry.

APPENDIX 1 Useful Expectations from General Chemistry

The pathway for this collision is fixed (a bond breaks and a bond forms), but the experimental conditions under which the reaction occurs can influence greatly how fast the reaction takes place. Increasing concentration, pressure, and temperature all allow more collisions to take place per second, and so the odds of a productive collision go up, increasing the reaction rate. A higher temperature also means that more of the atoms and molecules will have enough energy to form the transition state structure. Mixing or stirring also increases the odds for collisions by allowing the proper two reagents to find one another. Many reactions can be productively catalyzed, forming the products at a faster rate. Catalysts generally modify the reaction pathway, creating lower activation energies. In this case, perhaps an open metal surface that would gather hydrogen molecules and hydrogen atoms to itself would cause a faster reaction because these two reactive partners would be able to find each other more readily.

Rate experiments are also useful because it can often be determined whether a collision between two different substances is required. In theory, because both partners are needed for the collision, doubling the concentration of the hydrogen atoms or the hydrogen molecules should double the rate, while doubling them both would increase the rate four-fold. In this current example, increasing the concentration of either the hydrogen atoms or the hydrogen molecules should cause the product to form faster, simply because the odds of a productive collision are higher. This relationship is usually expressed as a rate equation (Figure AP0126), which needs to be determined experimentally.

Figure AP0126

Rate equation for the forward reaction of a hydrogen atom plus a hydrogen molecule.

H· + H—H ⟶ H—H + H·

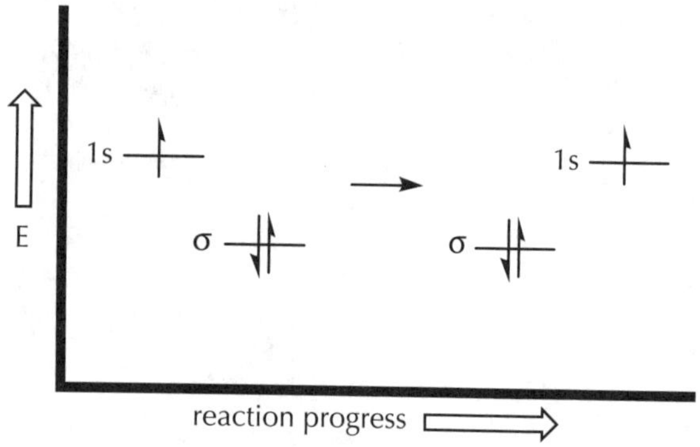

(a) double concentration H·, double the rate of reaction
  conclude: rate is proportional to the concentration of H·
  rate α [H·]

(b) double concentration H_2, double the rate of reaction
  conclude: rate is proportional to the concentration of H_2
  rate α [H_2]

(c) double concentrations of both H· and of H_2, quadruple the rate of reaction
  conclude: rate is proportional to the concentration of both, which implies a collision between the two of them

  Rate$_{forward}$ = k$_{forward}$ [H·][H_2]

  k = rate constant, identifies the fraction of the collisions that lead to productive outcomes

B. Equilibrium (K_{EQ}), Enthalpy (ΔH), and Entropy (ΔS)

The reaction between hydrogen atoms and molecular halogens have been studied experimentally (Figure AP0127).

Figure AP0127

Reaction between hydrogen atoms and halogen molecules.

H· + :Ẍ—Ẍ: ⟶ H—Ẍ: + ·Ẍ:

Enthalpy (ΔH) is commonly called the internal energy of a chemical system. Anything related to the structural stability of molecules falls in this category: bond formation energy, angle strain in the molecular structure, and/or different parts of the molecule interacting in a favorable (attractive) or unfavorable (repulsive) way. In this example, the bond formation energies have been measured, which means that the exact differences in energy between the two sides of the reactions are known. And from this information, an energy diagram can be constructed for both reactions (Figure AP0128).

Figure AP0128

Energy diagrams for the reaction between hydrogen atoms and halogen molecules.

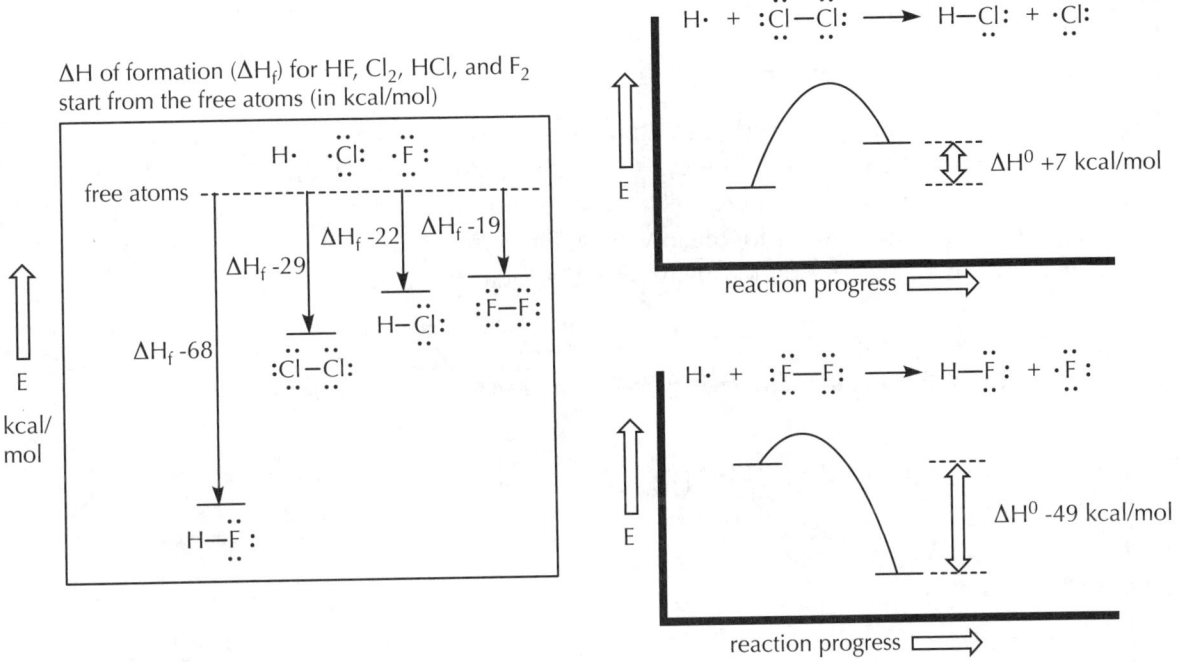

Equilibrium is the point at which the rate of a forward reaction equals the rate of the reverse reaction, at which point the ratio of the concentrations of the product to the reaction becomes constant and is called the equilibrium constant (K_{EQ}). The equilibrium constants for these two reactions are shown in Figure AP0129.

Figure AP0129

Equilibrium constants for the reactions between hydrogen atom and halogen molecules.

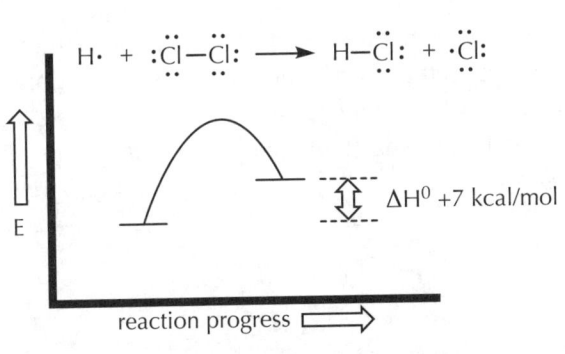

$Rate_{forward} = k_f [H\bullet][Cl_2]$

$Rate_{reverse} = k_r [Cl\bullet][HCl]$

Equilibrium: $Rate_{forward} = Rate_{reverse}$

$k_f [H\bullet][Cl_2] = k_r [Cl\bullet][HCl]$

$k_f / k_r = K_{EQ} = \dfrac{[Cl\bullet][HCl]}{[H\bullet][Cl_2]}$

$([Cl\bullet][HCl]) < ([H\bullet][Cl_2])$, so $K_{EQ} < 1$

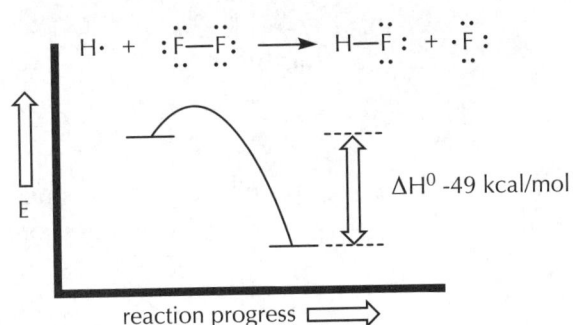

$Rate_{forward} = k_f [H\bullet][F_2]$

$Rate_{reverse} = k_r [F\bullet][HF]$

Equilibrium: $Rate_{forward} = Rate_{reverse}$

$k_f [H\bullet][F_2] = k_r [F\bullet][HF]$

$k_f / k_r = K_{EQ} = \dfrac{[F\bullet][HF]}{[H\bullet][F_2]}$

$([F\bullet][HF]) > ([H\bullet][F_2])$, so $K_{EQ} > 1$

Reading these energy diagrams needs to make sense. In the fluorine example, the energy difference, from left to right, is -49 kcal/mol, with the products being more stable than the starting materials. This is an exothermic (release of energy) reaction. At equilibrium, there will therefore be a higher concentration of the product relative to the starting material, and the numerical value of the equilibrium constant is going to be greater than 1.

In the chlorine example, the energy difference, from left to right, is +7 kcal/mol, with the starting materials being more stable than the products. This is an endothermic (input of energy) reaction. At equilibrium, there will therefore be a higher concentration of the starting materials relative to the product, and the numerical value of the equilibrium constant is going to be less than 1.

Returning to the earlier example (hydrogen atom plus hydrogen molecule): The starting materials and product are identical, and so equal in energy. The reaction is neither endothermic nor exothermic, as the energy difference is 0 kcal/mol. It should make sense that the equilibrium constant in this case would be 1, and because the rate expressions for the forward and reverse reactions are identical, the value of 1 for the equilibrium constant can be demonstrated with the equations (Figure AP0130).

Figure AP0130

Equilibrium constant for the reaction between hydrogen atoms and hydrogen molecules.

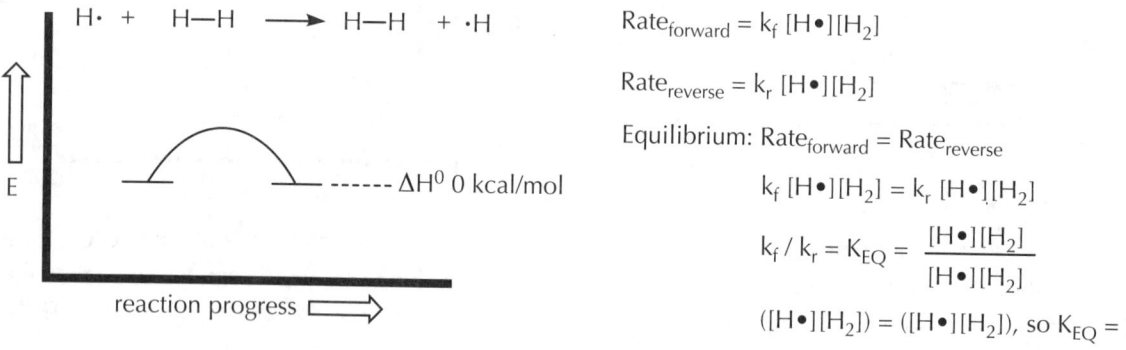

Entropy (ΔS) is commonly called the disorder of the system, although this is not a particularly meaningful term without a lot more definition. The lesson about entropy that is worth knowing about for organic chemistry is that a system that has the option of existing in a state with more particles is more favorable than a state with fewer particles. This idea will be illustrated by looking at what happens when you heat molecular chlorine, establishing the equilibrium between molecular chlorine and two chlorine atoms (Figure AP0131).

Figure AP0131

Equilibrium established by heating molecular chlorine.

$$:\ddot{Cl}-\ddot{Cl}: \rightleftharpoons :\ddot{Cl}\cdot + \cdot\ddot{Cl}: \qquad K_{EQ} = [Cl\bullet]^2 / [Cl_2]$$

As implied earlier, this is an endothermic process. The ΔH is +29 kcal/mol, meaning that the products (the two chlorine atoms) are less stable than the starting material (molecular chlorine). In this direction, the reaction would have a $K_{EQ} < 1$ (Figure AP0132).

A1.3 Chemical and Physical Properties

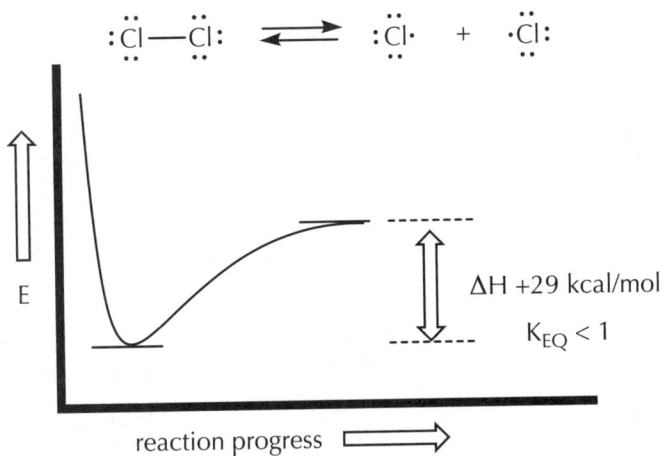

Figure AP0132

Enthalpy change for heating molecular chlorine.

The entropy change in the reaction is based upon comparing the number of arrangements for the total number of particles on both sides of the equation. To avoid the mathematics, a severe simplification will illustrate how to think about entropy.

Imagine that the reaction vessel only has one molecule of chlorine in it, and there are only four quadrants that the chlorine might occupy. That means there are four states (called microstates) available for the starting material. Now the reaction takes place, and there are two chlorine atoms in the same vessel, and those atoms can only be located in the same four quadrants. But the total number of arrangements (microstates) for these two chlorine atoms is far more than four; there are sixteen. All 20 of these microstates have the same energy and each of the 20 has the same likelihood, but there is a 16:4 ratio favoring the chlorine atoms relative to the chlorine molecule. This ratio, which says that seeing the two chlorine atoms is statistically favorable, is one way to express the idea that a system with two chlorine atoms has a higher entropy than the same system with one chlorine molecule (Figure AP0133).

Figure AP0133

Entropy change for heating molecular chlorine.

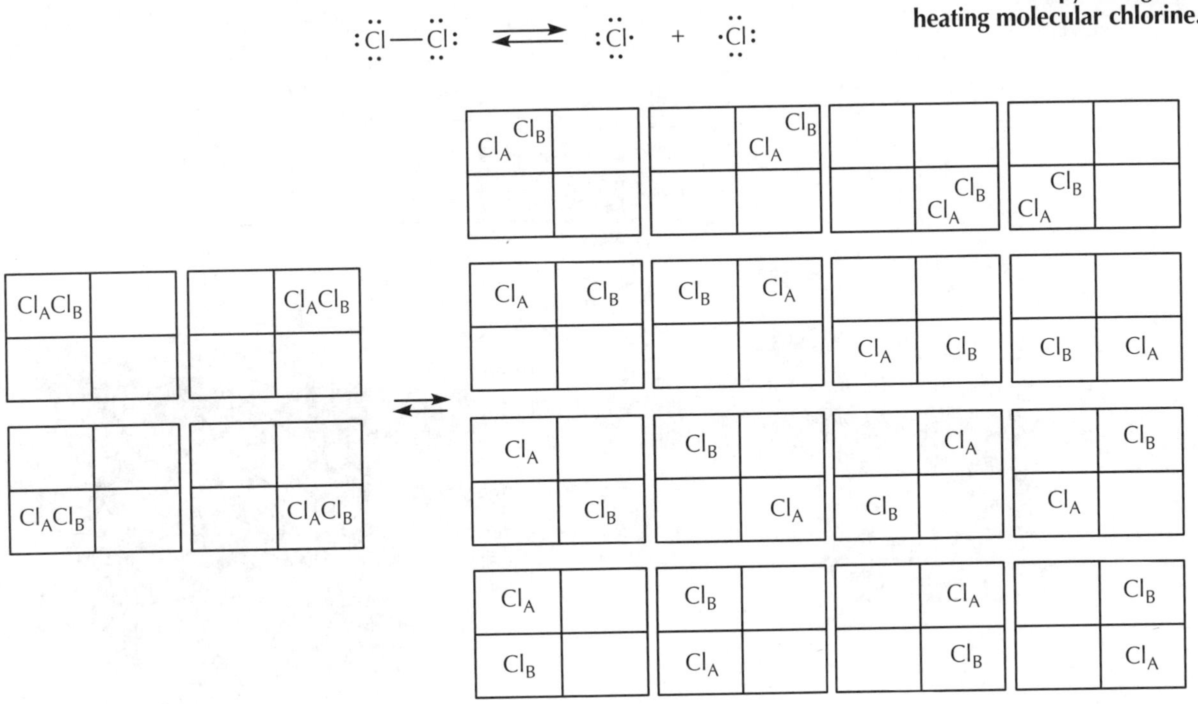

Based on the enthalpy change, the reaction should lie on the left ($K_{EQ} < 1$). Based on the entropy change, the reaction should lie on the right ($K_{EQ} > 1$). In this reaction, the enthalpy effect overcomes the entropy effect, and the observed $K_{EQ} < 1$. This "one way or the other" result is not a contradiction, but merely what happens when two effects are in opposition. The only way to resolve the balance is generally case by case, and by doing experiments to gather more data.

C. Phase Changes and Mixing

Covalent bonds hold atoms together to make individual molecules. Typically, groups of molecules can exist as solids, liquids or gases, the standard states of matter. The observed state of matter depends on the forces holding the molecules together versus the conditions under which the molecules exist apart, particularly according to temperature and/or being mixed with a solvent to make a solution.

States of Matter: solids. In a solid, the individual particles (atoms, ions, molecules) are packed together in a regular structure called a crystal lattice. A crystal lattice is a highly regular, three-dimensional arrangement of atoms, ions, or molecules. Generally speaking, the more spherical and well-packed the particles of a solid are, the more energy it takes to get them to separate from one another and start rolling around randomly (i.e., to melt and become a liquid).

Figure AP0134

Lattice structure and melting points for simple ionic compounds.

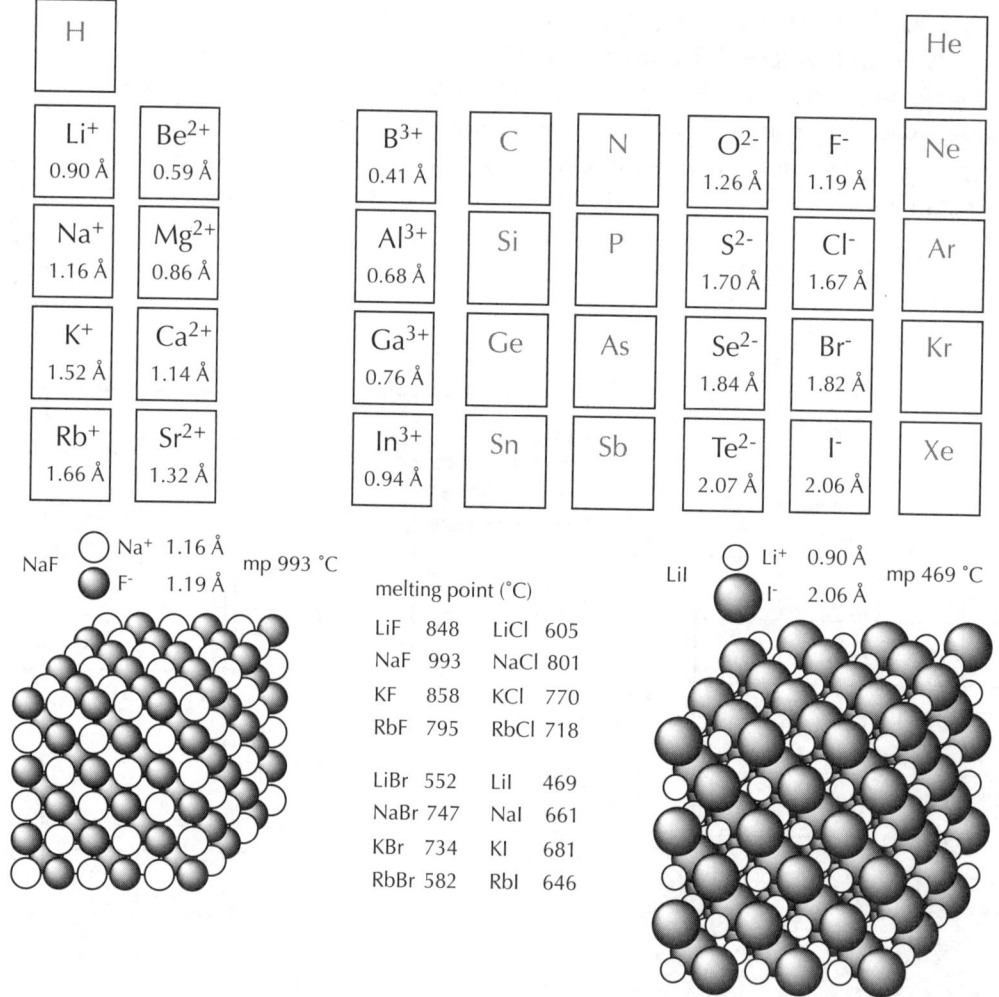

For simple ionic compounds, the relative match between the size of the cation and the anion matters the most. When the two ions are about the same size, the lattice structure is tightly packed and the ions are surrounded by their opposite charge, so the melting points are high. This attractive property is called having a high lattice energy. When the ions are mismatched in size, the lattice is looser and more easily disrupted (a lower lattice energy), consistent with their lower melting points. Some experimental data demonstrating these generalizations is shown in Figure AP0134. Look carefully at the melting points trends in the data table: The higher the melting point, the more equivalently sized the anion and cation, which results in better packing and thus better ion lattice energy (i.e., electrostatic attraction/stability).

Most organic molecules have highly irregular shapes, but the same principles hold. There are two sets of organic molecules in Figure AP0135, three with the molecular formula C_5H_{12}, and with the molecular formula $C_{10}H_{16}$.

Figure AP0135

Molecular shape and melting points in organic compounds.

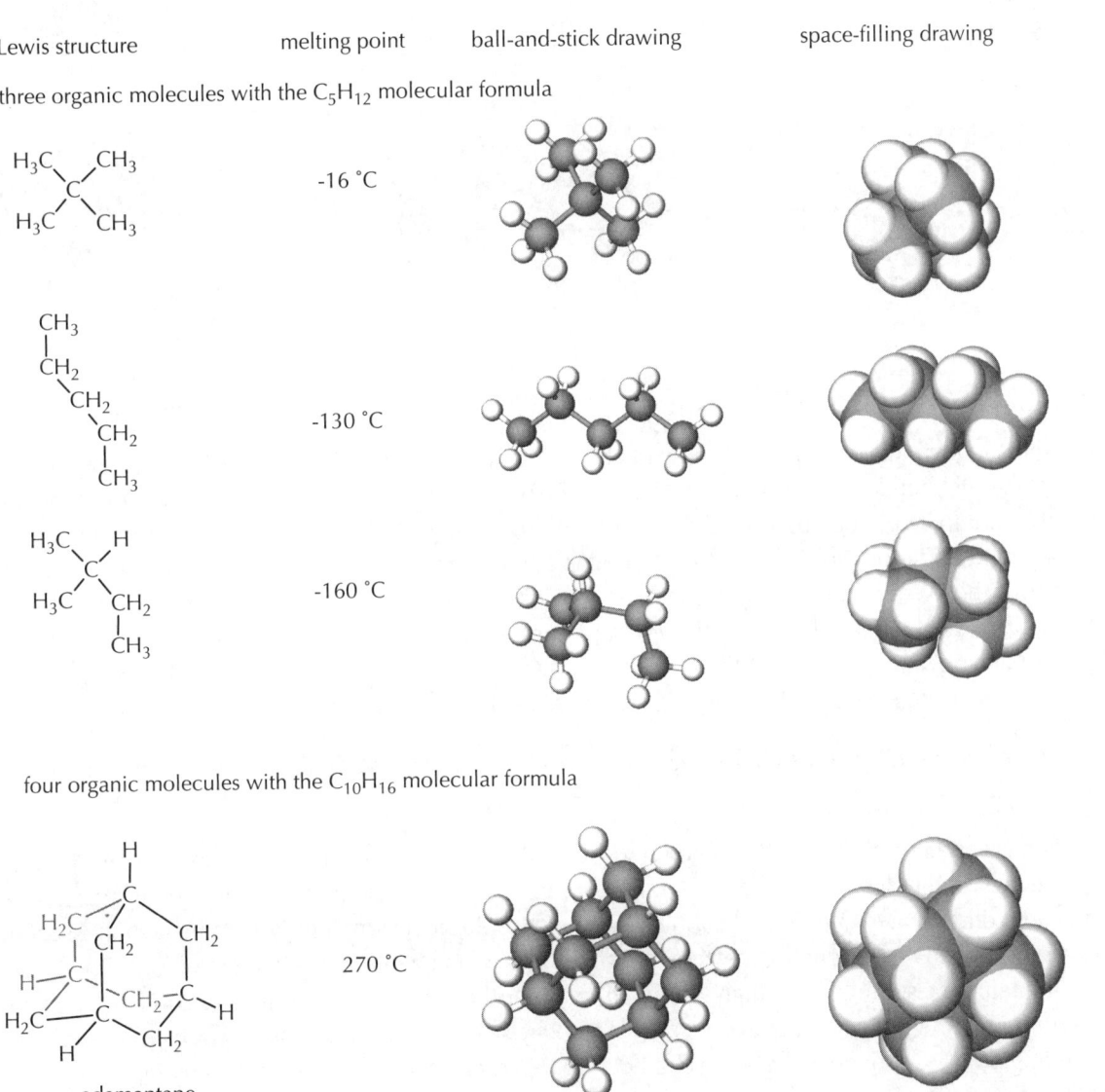

Remaining molecules continue on next page ...

Figure AP0135 continued . . .

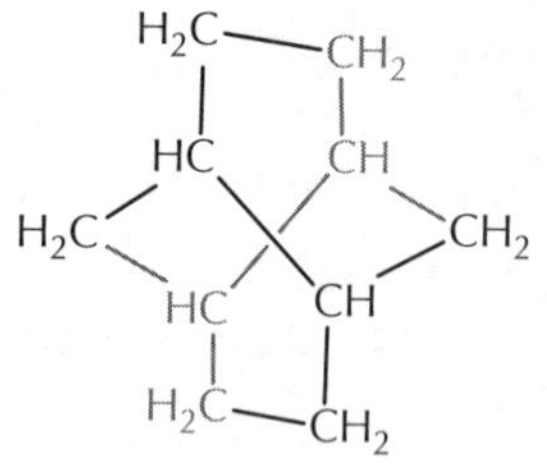

twistane   164 °C

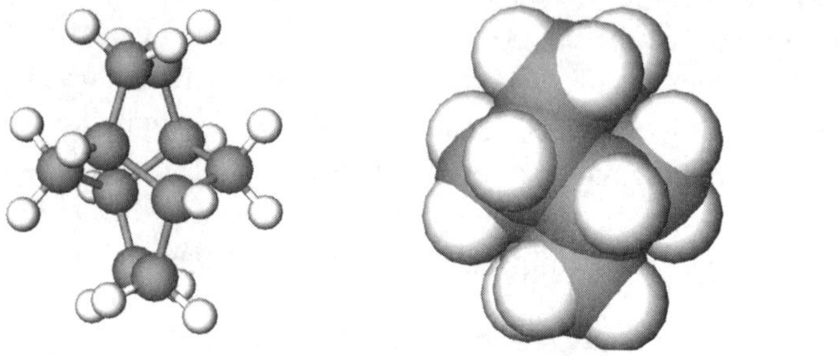

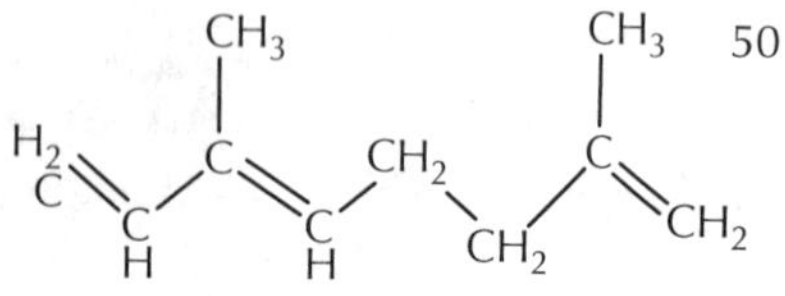

α-ocimene   50 °C

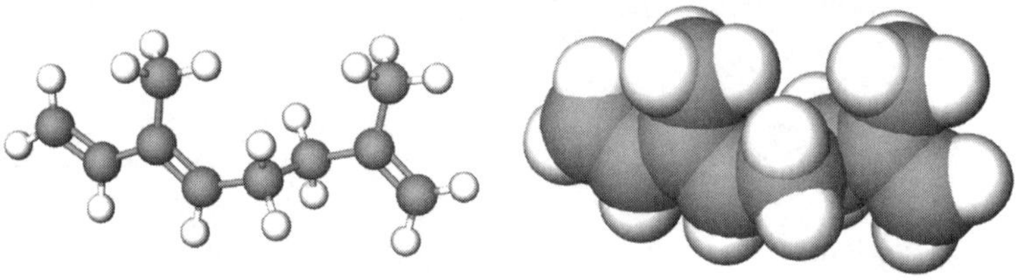

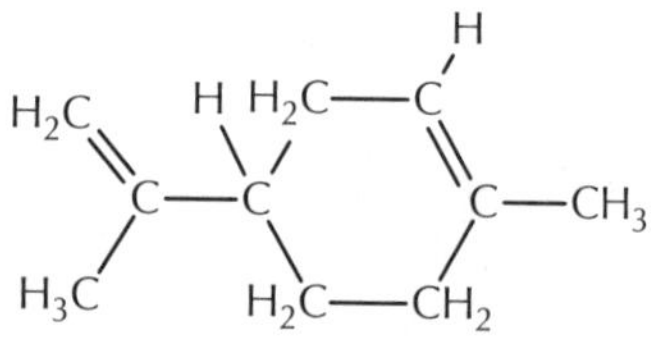

limonene   -74 °C

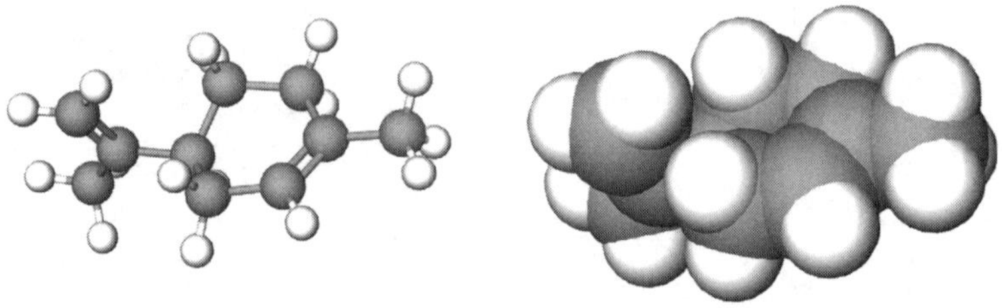

The first three of them have the same number of atoms (C_5H_{12}) but they differ in shape. The more spherical one packs into its solid lattice more readily, and it can spin in place as it is warmed up and not cause the crystal structure to be disrupted. It has the highest melting point (-16 °C) and so takes the most energy to disrupt the lattice. The next best shape for packing is a linear structure with no protruding atoms, which can stack together like bundles of sticks. That said, its melting point is dramatically lower (-130 °C). Breaking the linearity with random branching (lumps and bumps along the atomic chain) creates the worst situation for packing, and as you heat them, they more easily knock other molecules in the lattice out of position, as it starts to rotate, and so this one has the lowest melting point (-160 °C).

Predicting melting points, even with these generalizations, it not easy, particularly as the number of atoms increases.

The second group of organic structures shown in Figure AP0135 all have the $C_{10}H_{16}$ molecular formula. As you can see in the data, there is a wide range of melting points. The highly symmetrical structure of adamantane follows the generalization above; its close to spherical symmetry results in a small, smooth ball of atoms. When in the solid lattice, they pack well and, upon heating, can spin in place without disrupting the lattice, resulting in the high temperature to get them to separate from one another. Twistane has a symmetrical, compact structure, also. Not quite as spherical as adamantane, but still a relatively high melting point. Ocimene has a linear character to its structure, and has a significantly higher melting point than limonene, whose structure is a combination of shapes.

Note that these are fairly complex organic molecular structures, so you might want to revisit them after you have learned a bit more about how to read these structural drawings in Chapters 1 and 2.

A1.3 Chemical and Physical Properties

States of Matter: liquids. In a liquid, the arrangement of the molecules is disordered; they are mobile and in close contact. In liquids, shape does not matter for packing, because there is none, but it does matter for the surface area of the molecule and the amount of intermolecular contact that can take place. There are a number of important attractive intermolecular forces (noncovalent forces between molecules, collectivity called van der Waals forces) that account for the properties of liquids. The three most common types of van der Waals (intermolecular) forces are reviewed side-by-side in Figure AP0136.

Figure AP0136

Summary of intermolecular (van der Waals) forces.

London dispersion forces
induced dipole when atoms/molecules get in close proximity (increases boiling point)

He: 1.40 Å; -269 °C, 4K
Ne: 1.54 Å; -246 °C, 27 K
Ar: 1.88 Å, -186 °C, 87 K
Kr: 2.00 Å, -153 °C, 121 K

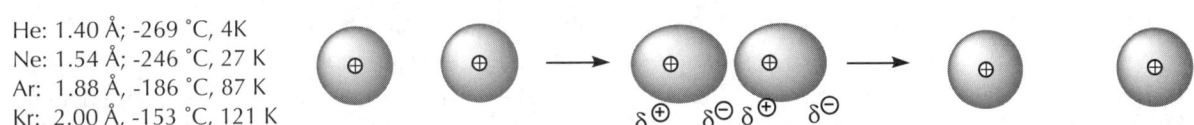

	boiling point		from **FIGURE AP0135**	boiling point		melting point
methane	-161 °C	CH_4				
ethane	-89 °C	CH_3CH_3	2,2-dimethylpropane	+10 °C	$CH_3C(CH_3)_3$	-16 °C
propane	-42 °C	$CH_3CH_2CH_3$	pentane	+36 °C	$CH_3CH_2CH_2CH_2CH_3$	-130 °C
butane	-1 °C	$CH_3CH_2CH_2CH_3$	2-methylbutane	+28 °C	$(CH_3)_2CHCH_3$	-160 °C

dipole/dipole forces
permanent dipoles attract (increases boiling point)

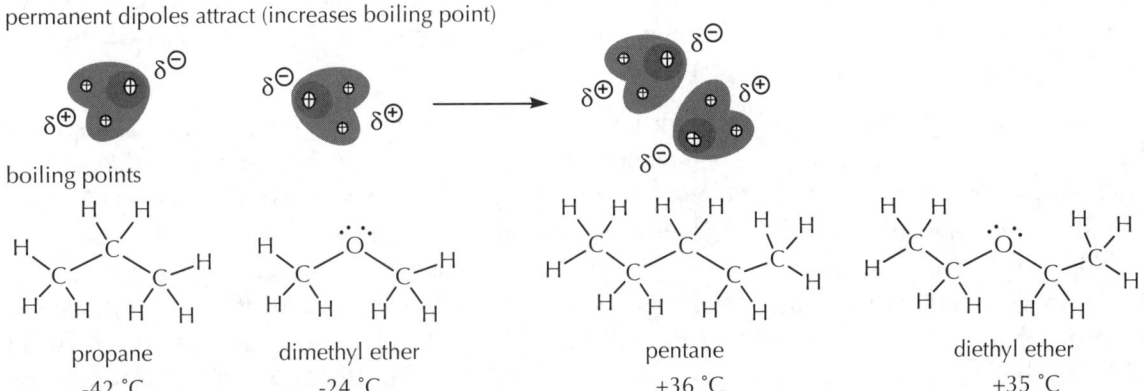

boiling points

propane   dimethyl ether   pentane   diethyl ether
-42 °C   -24 °C   +36 °C   +35 °C

hydrogen bonding
attraction between nonbonding electron pairs on N:, O:, and F: and the electron-deficient H of NH, OH, and HF

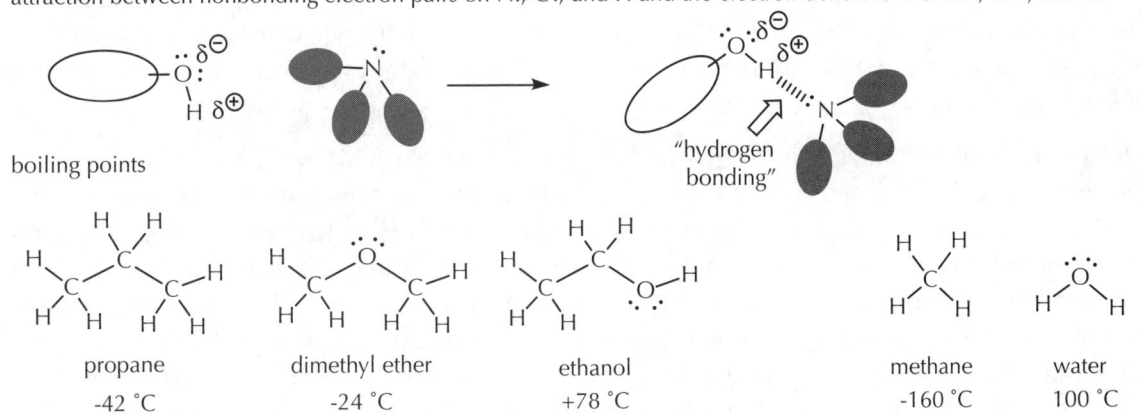

boiling points

propane   dimethyl ether   ethanol   methane   water
-42 °C   -24 °C   +78 °C   -160 °C   100 °C

(1) London dispersion forces– Intramolecularly (within a molecule) the repulsive forces between like charges (nuclei to nuclei, electrons to electrons) are counterbalanced by the attractive forces between unlike charges (nuclei to electrons). Because we imagine that the electron is not a particle, but rather a cloud of charge, the cloud is susceptible to distortion. As two molecules approach one another, we think that all of the charge/charge interactions are also taking place between the molecules. Consequently, the distortion of the electron clouds takes place and slight excesses of positive and negative charge build up, temporarily, as the two molecules get close enough. These are "induced dipoles" (i.e., separated partial charges that are caused by other nearby charges). And much like weak magnets, the net attractive forces provide cling between the molecules, keeping them in the liquid phase.

Even the smallest organic molecule (methane, CH_4, which is perfectly tetrahedral in shape occupying a 4 Å sphere) has a higher boiling point (-161 °C) than all of the noble gasses through krypton, which has about the same boiling point and almost four times the number of protons and electrons as methane. Helium is perfectly spherical (1.4 Å) and the most compact, and least susceptible to distortion of its electron cloud, and it has a boiling point of -269 °C (or 4 K, just 4 degrees above absolute zero). The data for the other noble gasses is consistent with their increased ability to at least temporarily attract one another as the number of more loosely held electrons increases (neon, 1.54 Å, -246 °C, 27 K; argon, 1.88 Å, -186 °C, 87 K; krypton, 2Å, -153 °C, 121 K).

As molecules increase in size, so does their surface area, and the amount of London dispersion forces increases dramatically. Linear organic molecules from 1–5 carbons in length, demonstrate a dramatic increase in boiling point (methane, -161 °C; ethane, -89 °C; propane, -42 °C; butane, -1 °C; pentane, +36 °C). And the three C_5H_{12} molecules that differ in shape (see Figure AP0135) also expose different surfaces: Tetrahedral structures have less contact with one another, and a lower boiling point, than more linear structures. Although both molecules are C_5H_{12}, the $CH_3C(CH_3)_3$ has a boiling point of +10 °C, while $CH_3CH_2CH_2CH_2CH_3$ (pentane, above) has a boiling point of +36 °C. The less branched but nonlinear $(CH_3)_2CHCH_2CH_3$ has a boiling point of +28 °C.

(2) Dipole-dipole forces– In the London dispersion forces, the formation of partial charges is temporary and variable. Many molecules, because of their constituent atoms and how they are arranged, have permanent dipoles. This is often due to differences in electronegativity. Propane ($CH_3CH_2CH_3$, 44.1 g/mL) has almost no permanent separation of change and a boiling point of -42 °C, while dimethyl ether (CH_3OCH_3), with nearly the same mass (46.1 g/mol), has a permanent dipole and a higher boiling point (-24 °C). Notice that differences in London dispersion forces have quite a large effect, proportionately, compared with this dipole-dipole effect. Pentane ($CH_3CH_2CH_2CH_2CH_3$), which also has no permanent dipole, has a boiling point of +36 °C. The analogously sized diethyl ether ($CH_3CH_2OCH_2CH_3$), with the same-sized permanent dipole as dimethyl ether, has a boiling point of +35°C. This second example is important because it tells you that the relative contribution from these different intermolecular forces will vary according to the molecular structures. No matter how much we might want a permanent dipole to define a molecular property, the experimental fact that pentane and diethyl ether have nearly identical boiling points tells us that London dispersion forces are the most significant contribution to the boiling point in these compounds.

(3) Hydrogen bonding– Although the word "bonding" is used here, the term "hydrogen bonding" refers to a strong intermolecular force and not to a shared electron bond. The use of the word "bond" is more metaphorical than literal, and mixing those different uses can lead to confusion. Hydrogen bonding is an intermolecular force that is part of the van der Waals collection; if you make that association, you can keep this use of the same word associated with its different meaning. Dimethyl ether (CH_3OCH_3) has exactly the same atoms and shape as ethyl alcohol (CH_3CH_2OH), so the London dispersion forces should be about the same. Experimentally, they both have permanent dipoles that are also about the same, but their boiling points are quite different. Dimethyl ether (-24 °C) has a significantly lower boiling point than ethyl alcohol (+78 °C).

A1.3 Chemical and Physical Properties

As a general observation, molecules with hydrogen atoms on the three most electronegative atoms from the second row (NH, OH, FH), called the hydrogen bond donors, are observed to have a strong intermolecular attraction with the nonbonding electron pairs associated with those same three atoms (N:, O:, F:), called the hydrogen bond acceptor. As seen in Figure AP0136, the intermolecular attraction, called a hydrogen bond, is customarily represented with a dashed line. Hydrogen bonding is mainly responsible for the exceptionally high boiling point (100 °C) for water (H_2O; 18 g/mol; 2.75 Å) relative to methane (-161 °C), which is about the same size (CH_4; 16 g/mol; 4 Å).

All of the van der Waals forces are attractive between molecules, and with a few exceptions, the London dispersion forces have the most significant effect on boiling point. One exception is in the case of small molecules where hydrogen bonding is significant (see methane versus water, above). Another exception is the case of molecules with highly spherical (symmetrical) shapes. Limonene and adamantane (see Figure AP0135) are composed of the same set of atoms arranged in significantly different shapes. Adamantane is nearly spherical and packs well in its crystal lattice, resulting in its high melting point (270 °C) relative to limonene (-74 °C). The boiling point of limonene (+176 °C) is 250 degrees higher than its melting point, indicative of its exposed surface and the opportunity for London dispersion forces. As near-spheres, mole-

Figure AP0137

Boiling versus sublimation.

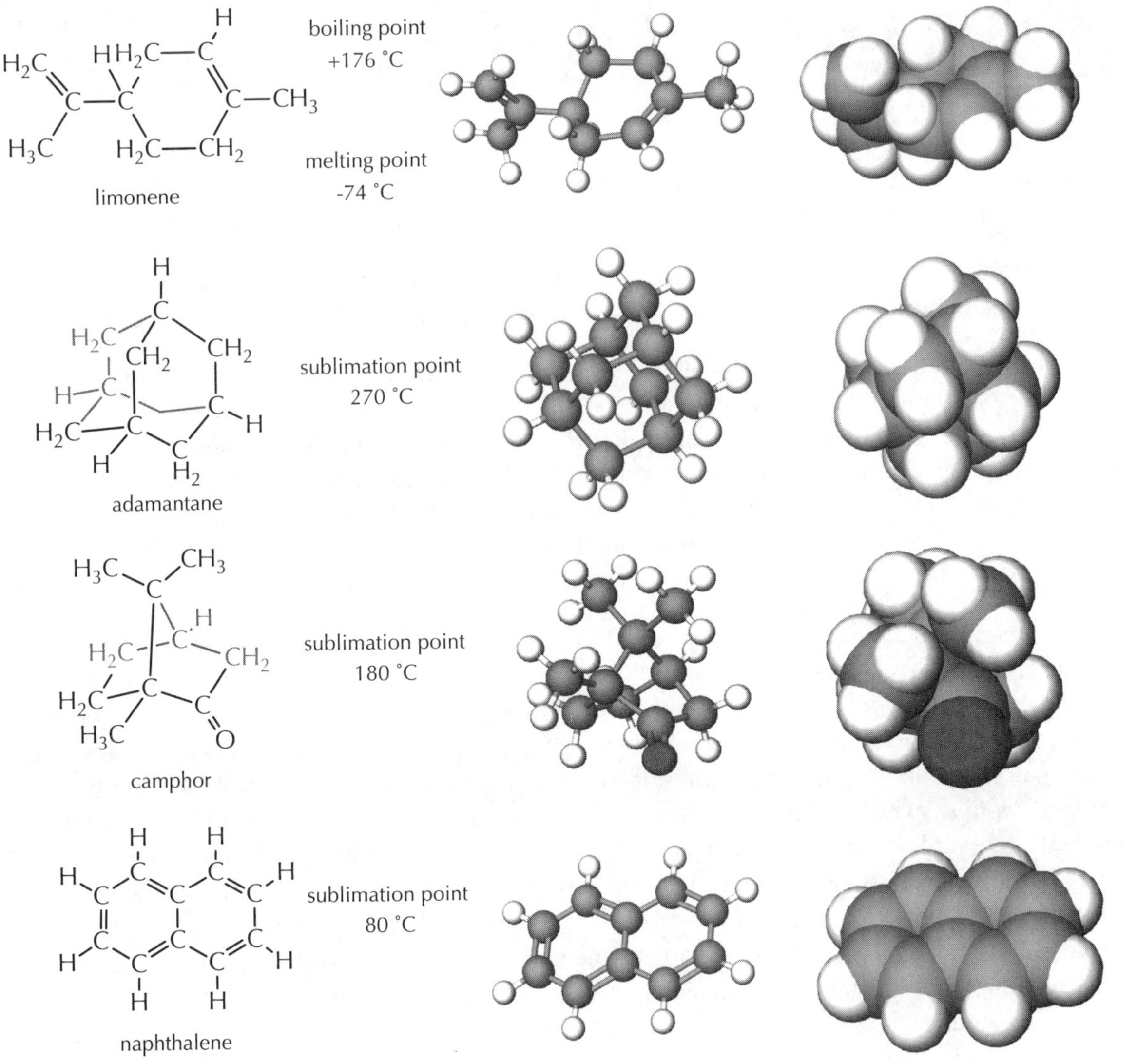

cules of adamantane are barely able to be in contact. As the solid is heated to 270 °C, it actually does not melt and go into a liquid state because the London forces are so low. Instead, at 270 °C, adamantane goes directly from the solid phase to the gas phase. Once the molecules start to move out of position in their crystal lattice, they separate from one another completely. Going directly from the solid phase to the gas phase is called sublimation, and it is characteristic of organic molecules with exceptionally high symmetry (Figure AP0137).

Mixing: liquids with liquids. Your intuition about mixing derives from your most common experience with objects: Mixing is inherently favorable. If you take a container with a layer of 20 blue marbles, on top of which is a layer of 20 white marbles, and then you shake the container, stop and look, and shake it again, stop and look, and so on, you never (ever) expect to see the two layers again (Figure AP0138).

Figure AP0138

Shaking a container of marbles creates a mixed arrangement.

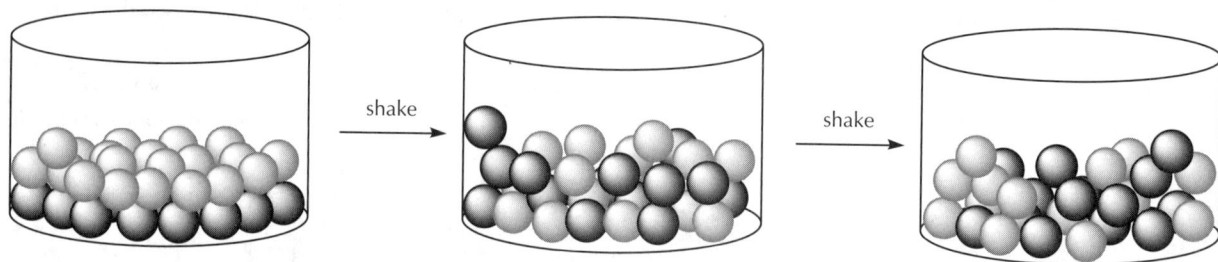

If you ask people why the marbles will always end up mixed, you are likely to get a variety of replies: Of course, it's mixed… mixing is better than unmixing… the mixed marbles have a higher entropy.

None of these answers is terribly satisfying as an explanation, because the unasked question is so much better: If one of the arrangements is better, then how do the marbles know what color they are?

That's a great question because the inherent stability of any given arrangement is exactly the same as any other. The marbles don't know what color they are! The arrangement ("mixed" versus "unmixed") is a value judgment made by us, the human observer. The answer to why the marbles will always end up mixed is that the number of total arrangements that look mixed to an observer exceeds the number that look unmixed, and from a purely statistical standpoint: The odds of producing an arrangement that looks mixed is so much greater than producing an unmixed arrangement that you just cannot live long enough until you shake the container and one of the unmixed arrangements shows up.

None of the mixed arrangements is any more or less likely than one of the unmixed ones. It is also just as unlikely that you will ever see the same mixed arrangement, again. But to your eyes, you are only sorting what you see into "mixed" from "unmixed," not taking an inventory of any specific arrangement. Enter your classroom or your bedroom. You have no fear at all that when you open the door, that the 21% of air made up of oxygen atoms has become unmixed and is all hovering at the ceiling (in which case you would asphyxiate upon only inhaling nitrogen atoms). A room filled with air is the same as the container of marbles, except that the ratio of "blue" (nitrogen) to "white" (oxygen) is 80:20. No one has to shake the room to get the nitrogen and oxygen molecules to mix, because we live at about 295 K (degrees above absolute zero) and that temperature represents a lot of thermal motion. The statistical probability of encountering a highly mixed arrangement of oxygen and nitrogen is so much higher than an unmixed one… not because the molecules know who they are, but simply because the number of mixed-up looking arrangements so far exceeds the unmixed ones.

This statistical probability of encountering a more mixed-up looking arrangement is a reasonable definition of entropy.

Using a highly simplified example, you can actually count the difference. Imagine a container with 4 chambers, so that you have a 2x2 grid of spots to occupy. Now you introduce two molecules of A (A1 and A2) and two molecules of B (B1 and B2). There are a total of 24 different arrangement for these 4 molecules in the 4 chambers. If we define "unmixed" as a layer of A above or below a layer of B, then there are 4 instances (microstates) that look unmixed (A over B), 4 others that look unmixed (B over A), and 16 that look mixed (Figure AP0139).

Figure AP0139

Simplified model for counting the options with two layers of marbles.

4 look unmixed (**A** over **B**)

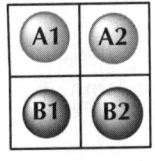 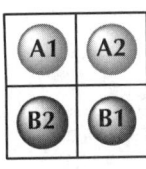 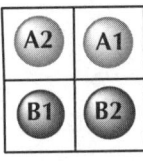 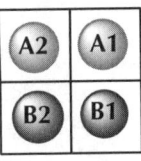

4 look unmixed (**B** over **A**)

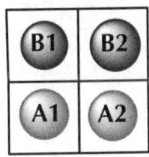 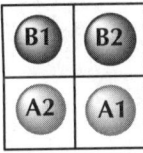 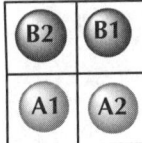 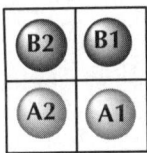

16 look mixed

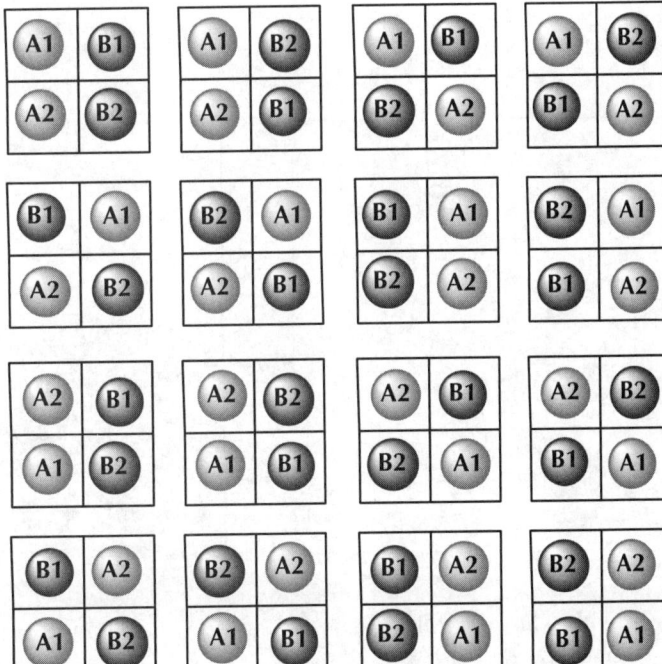

Even with only 4 objects in a two-dimensional grid, there is a 2:1 chance you will see a mixed-up arrangement when you shake this container compared with an unmixed one. If this was 3 x 10²³ molecules of water combined with 3 x 10²³ molecules of alcohol (all of which could sit easily in the three-dimensional space of a tablespoon), the probability of seeing an unmixed arrangement is so low as to be effectively zero.

There are times when two mixed liquids will spontaneously unmix. Based on the discussion in this section, that should seem somewhere between really weird and basically impossible. And yet, even after being vigorously shaken in a container, oil and water will spontaneously separate. The observation is quite general, too. There are many substances that are classified as "nonpolar" or "hydrophobic" (water-hating), and presumably prehistoric humans recognized the difference between materials that absorb water and materials that do not.

It was not until the 1940s that scientists began to propose molecularly based explanations for "the hydrophobic effect"—the spontaneously segregation of nonpolar molecules from water. This topic, which also turns out to have entropy as its origin, will be discussed in some detail during the later chapters in this book. If you are curious about it now, you have an available internet and you are likely carrying a mobile device with which you can connect to it... or you can skip to Chapter 17.2 in Book D.

Mixing: solids with liquids. Combining solids with liquids is also the process of either dissolving (mixing) or not dissolving (not mixing). Solubility of solids is a bit more complex than the solubility of liquids, because the strength of the interactions in the lattice also contributes to the observed solubility value. From a purely mixing standpoint, salt ought to dissolve (mix) with an oil such as octane ($CH_3CH_2CH_2CH_2CH_2CH_2CH_2CH_3$). But the molecules of octane are strictly nonpolar, and to get the ions to leave their lattice, where any given cation is surrounded by anions, the favorable charge/charge interaction would have to be compensated for (Figure AP0140). Therefore, salt is completely insoluble in octane.

Figure AP0140

Solubility: salt is insoluble in octane.

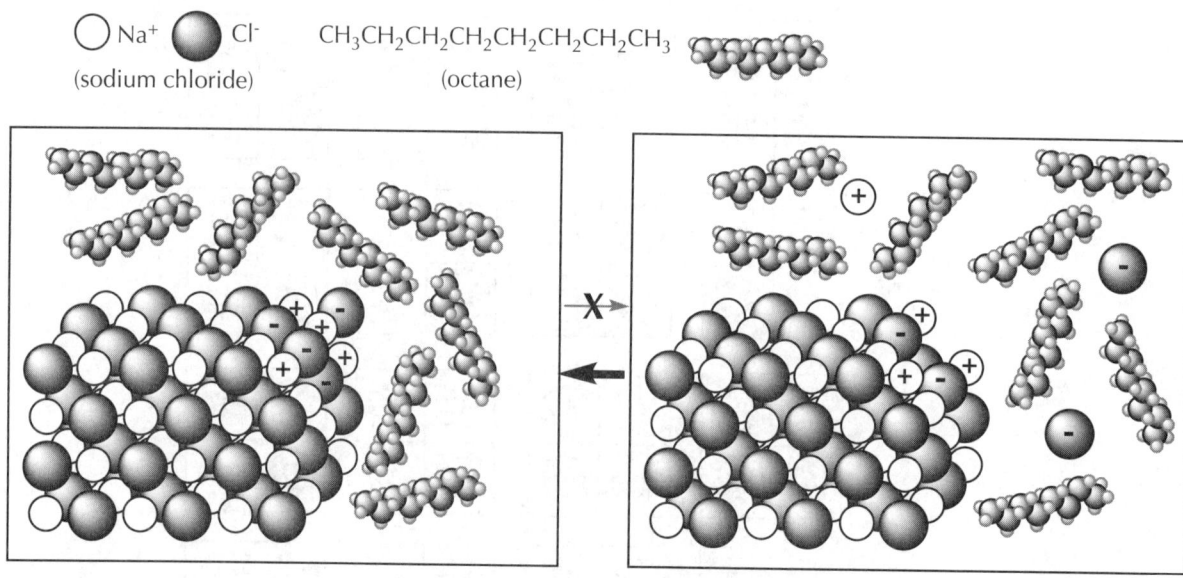

When salt is mixed with water, the dipolar nature of the water, and its ability to form hydrogen bonds with the anions, creates the usual situation where mixing is favorable. At some point, however, the water molecules start to get all tied up with the ions, and there are not enough of them remaining to interact. Here, you reach the point of saturation, where no more of the salt can dissolve without some also coming back to the lattice. At room temperature (about 20 °C), about 36 grams of sodium chloride can dissolve in 100 mL of water (Figure AP0141).

Figure AP0141

Solubility: dissolving salt in water.

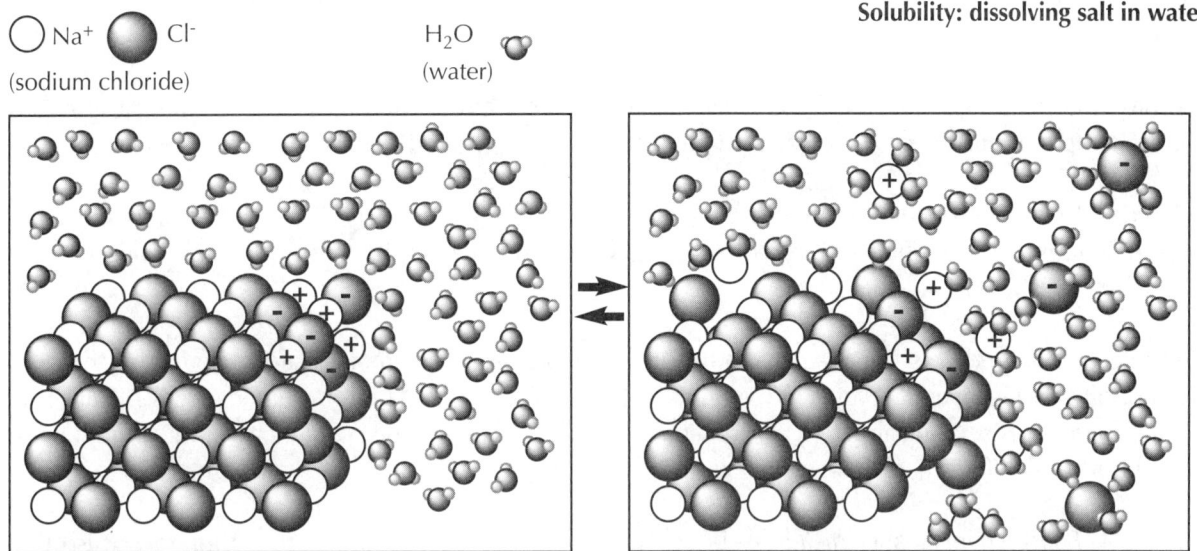

Solid organic compounds also exist in a crystal lattice that can be quite favorable. The solubility of naphthalene in octane is about 20 g/100 mL at room temperature. Thus, the stability of the crystal lattice for solid naphthalene is in competition with the favorable entropy of mixing, when the lattice is disrupted and the naphthalene dissolves. At the solubility value, the two competing effects are equal. Below that level, the mixing is favored, and the naphthalene will be completely dissolved. Above that level, and some of the naphthalene remains out of solution, in the lattice (Figure AP0142).

Figure AP0142

Solubility: dissolving naphthalene in octane.

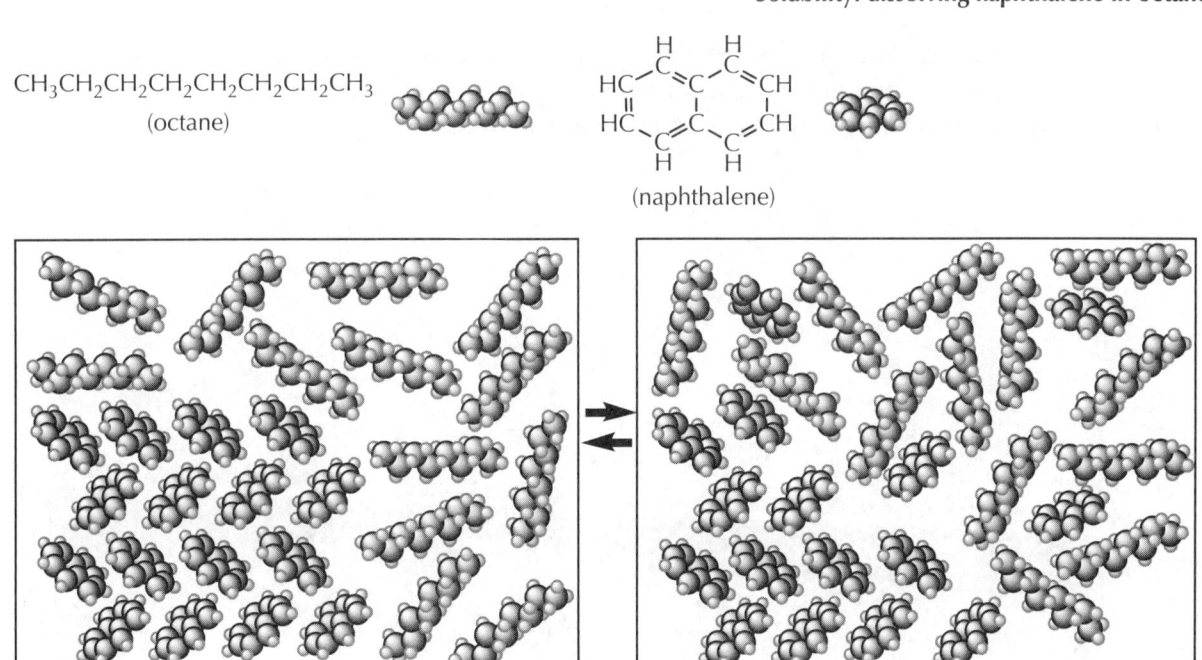

APPENDIX 1 Useful Expectations from General Chemistry

The principles taught in General (high school) Chemistry create an important foundation for your study of organic chemistry. Looking back at the preface for this appendix, the absolute list is not long, but all five of the topics are critical.

Topic 1. *Using precise language and drawings to represent atomic and molecular entities (also called atomic and molecular species).* Clear communication relies on shared symbols (words, drawings, sounds).

Topic 2. *Electron configurations and the relative stability of the closed shell electron configuration.* The periodic table is one of the most incredible tools ever constructed, in that it forms the foundation for the entire material world.

Topic 3. *Lewis structures.* The idea of a molecule as the organized assembly of atoms, following the rules of bonding, is just over 200 years old, while the model for bonding based on shared electron pairs only dates to the 1920s. Being able to draw molecular structures that include covalent and ionic bonding is a prerequisite skill for organic chemistry.

Topic 4. *Phase changes and mixing.* Although it is automatically relevant in the laboratory, it is quite useful for you to have an accurate sense about the physical reality of chemical systems. We write "NaCl + H_2O" as though there is one water molecule combining with a single pair of ions in sodium chloride. Yet the reality of a what you should imagine as a grain of sodium chloride and a drop of water is quite different than any manipulation of the symbols "NaCl" and "H_2O" can accomplish. For Topic 4, you are reminded to imagine the physical reality of what a chemical system really means.

Topic 5. *Inductive effects (electronegativity and electropositivity).* Consequences from the fundamentals of electrostatic attraction (opposite charges attract and provide stabilization) and electrostatic repulsion (like charges repel and provide destabilization) sit at the core of many principles in chemistry.

In addition, the appendix illustrates that experimental data are central to how these ideas were developed, and how the balance of opposing trends in a property can only be resolved by starting with the experimental measurement and imagining the different contributions to the property.

PRACTICE QUESTIONS

AP01.10 The following energy diagrams are all drawn on the same scale. Rank the anticipated magnitude of the five equilibrium constants for these reactions from largest to smallest. Provide a brief explanation for the rationale of the ranking.

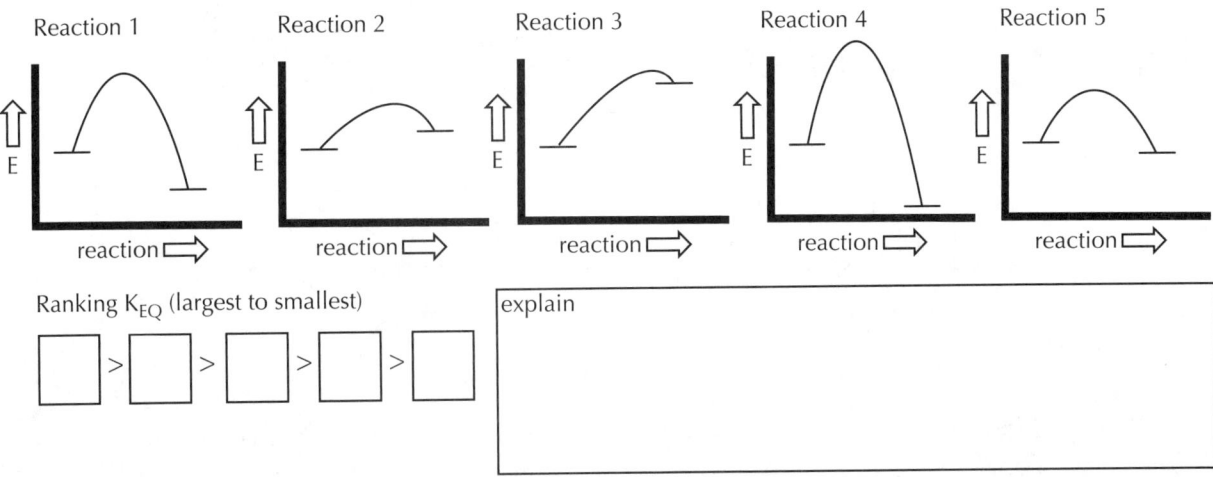

Ranking K_{EQ} (largest to smallest)

☐ > ☐ > ☐ > ☐ > ☐

explain

AP01.11 In the following general reaction, when compound A and B are in equilibrium, there is a 100:1 ratio of B to A.

(a) Which energy diagram is consistent with this result? Why?

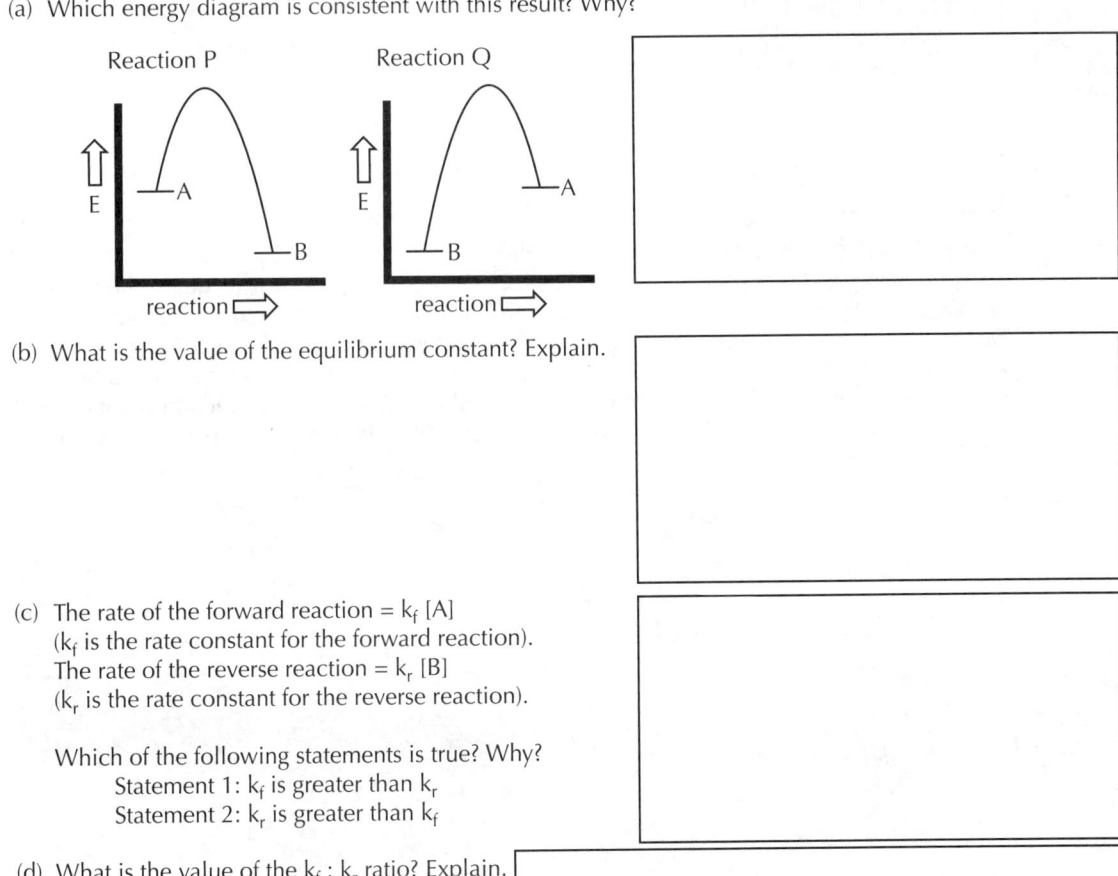

(b) What is the value of the equilibrium constant? Explain.

(c) The rate of the forward reaction = k_f [A]
(k_f is the rate constant for the forward reaction).
The rate of the reverse reaction = k_r [B]
(k_r is the rate constant for the reverse reaction).

Which of the following statements is true? Why?
  Statement 1: k_f is greater than k_r
  Statement 2: k_r is greater than k_f

(d) What is the value of the $k_f : k_r$ ratio? Explain.

AP01.12 The change in bonding energy in the following three reactions is nearly identical, yet there is a significant difference in the equilibrium constants, in the direction indicated below. Provide an explanation for the difference in magnitude of these equilibrium constants.

Reaction 1

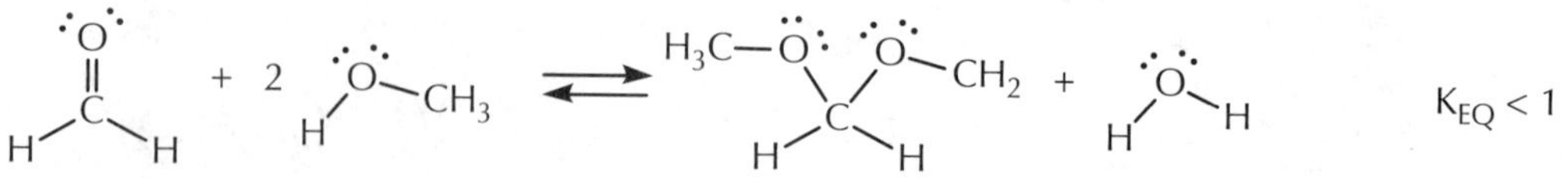

[formaldehyde] + 2 [methanol] ⇌ [dimethyl acetal of formaldehyde] + [water] $K_{EQ} < 1$

Reaction 2

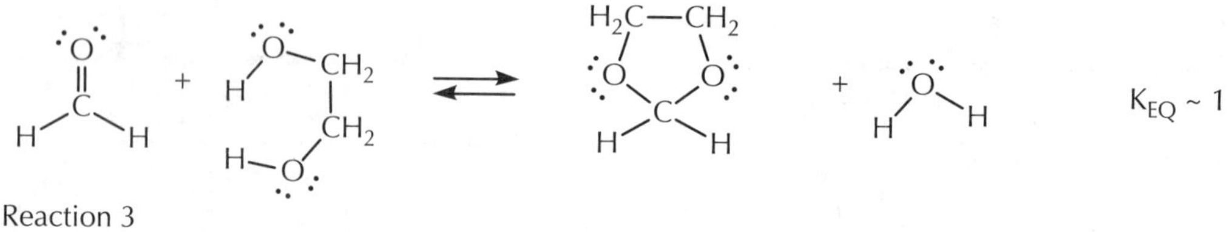

[formaldehyde] + [ethylene glycol] ⇌ [cyclic acetal (1,3-dioxolane)] + [water] $K_{EQ} \sim 1$

Reaction 3

[dimethyl acetal] + [ethylene glycol] ⇌ [cyclic acetal] + 2 [methanol] $K_{EQ} > 1$

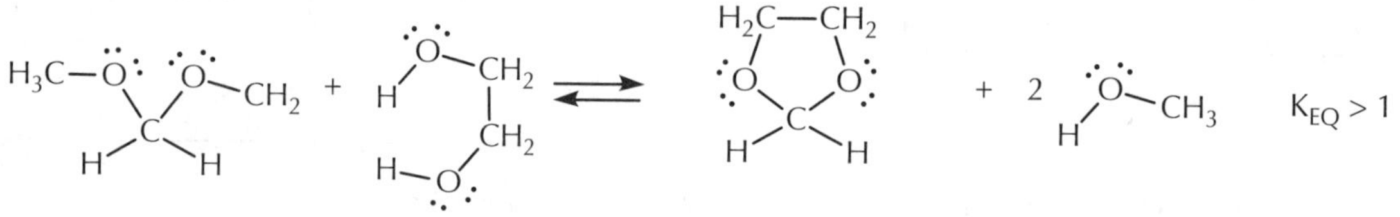

AP01.13 Triacetone triperoxide (TATP) is a crystalline solid that detonates quite easily, producing about 70% of the explosive force of TNT on a gram per gram basis. In most explosions, the decomposition is highly exothermic, producing heat along with, generally, a large number of super-heated molecules in the gaseous state that expand rapidly and create the shockwave of explosive force. In 2005, the outcome from the detonation of TATP was studied in detail. For every mole of the solid, four moles of small molecules form as a rapidly expanding ball of gas. Interestingly, however, the change in bond energy is zero; the reaction is not exothermic. What is driving this explosion of TATP? Explain briefly (nonbonding electrons on the oxygens are not shown for clarity).

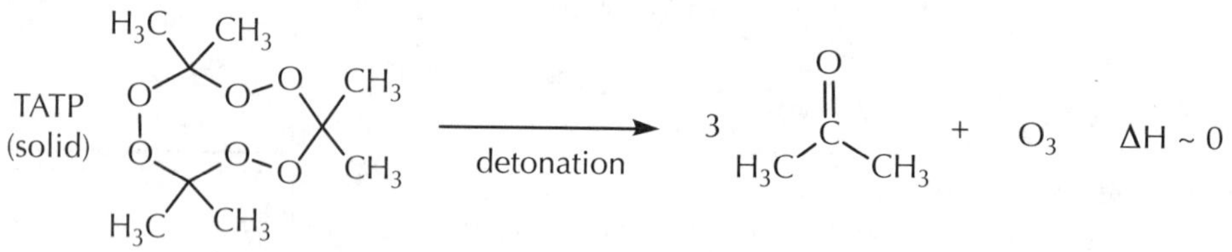

TATP (solid) $\xrightarrow{\text{detonation}}$ 3 acetone + O_3 $\Delta H \sim 0$

AP01.14 The two compounds shown below are named benzene and toluene. Three different representational forms are shown for each of them: a Lewis structure, a ball-and-stick model, and a space-filling model. One of them has a melting point of 6 °C and a boiling point of 80 °C, and the other one has a melting point of -95 °C and a boiling point of 111 °C. Assign which properties belong to which compound and provide an explanation for the difference.

benzene

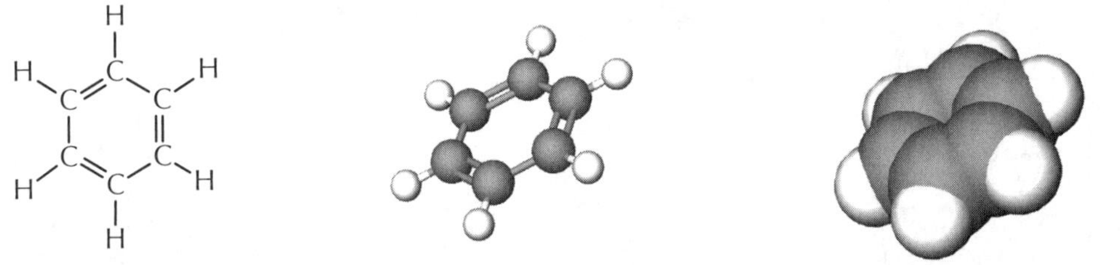

toluene

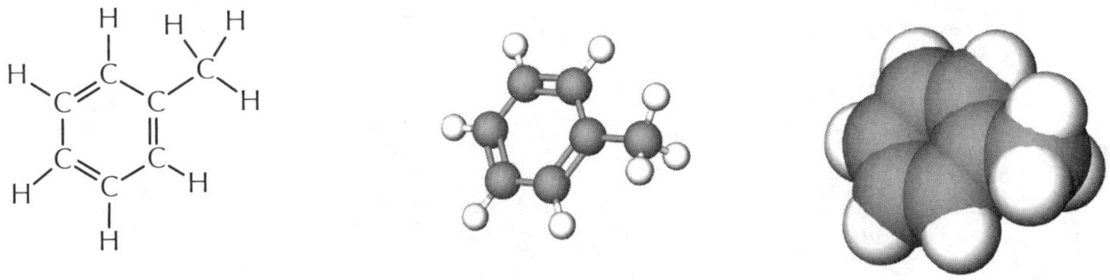

compound with
melting point -95 °C, boiling point 111 °C

compound with
melting point 6 °C, boiling point 80 °C

explanation

AP01.15 In the three spaces below, provide drawings for a group of about 40 sodium atoms: (a) in the solid phase, (b) as the solid is melting, and (c) in the liquid phase. Use a plain circle to represent the sodium atoms.

AP01.16 In the three spaces below, provide drawings for a group of about 20 sodium atoms: (a) in the liquid phase, (b) as the liquid is vaporizing, and (c) in the gas phase. Use a plain circle to represent the sodium atoms.

AP01.17 Use words and drawings to explain the following observations about the solubility of these ionic compounds in water.

compound		solubility in water (20 °C)
sodium fluoride	NaF	4 g/100 mL
cesium fluoride	CsF	573 g/mL
cesium iodide	CsI	85 g/mL

AP01.18 The following apparatus contains two chambers, one with 10 oxygen molecules (O_2) in it and the other with 10 nitrogen molecules (N_2).

a) The chambers are connected by a gateway that can be opened and closed. It is currently closed. Draw a picture of the arrangement of the molecules in both of the chambers using "O_2" and "N_2" to represent the molecules. The conditions in both of the chambers are identical, and both are under normal atmospheric temperature and pressure.

b) The gateway is opened. Within a short period of time, the overall arrangement has changed and appears to have stabilized. Draw a representation of the final state of these 20 molecules.

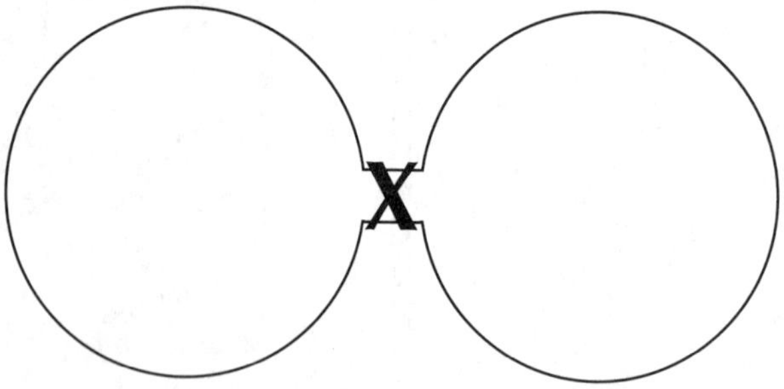

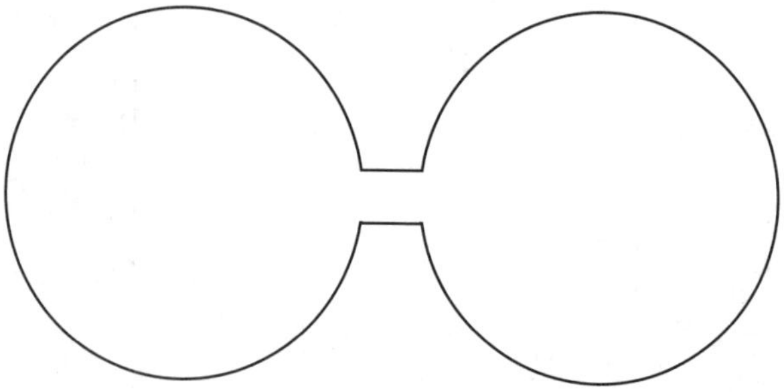

c) What is the origin of the energy that allows the arrangements to change?

d) You can wait for a lifetime and you will not see the initial arrangement again. Why not?

e) Molecules never stop moving between the chambers. How does the overall arrangement stabilize, then?

A1.3 Chemical and Physical Properties 955

AP01.19 The structures for two $C_{10}H_{16}$ compounds are given below (turpentine and camphene). One them has a melting point of -55 °C and a boiling point of +154 °C, while the other has a melting point of +52 °C and a boiling point of +159 °C. Which one is which, and why?

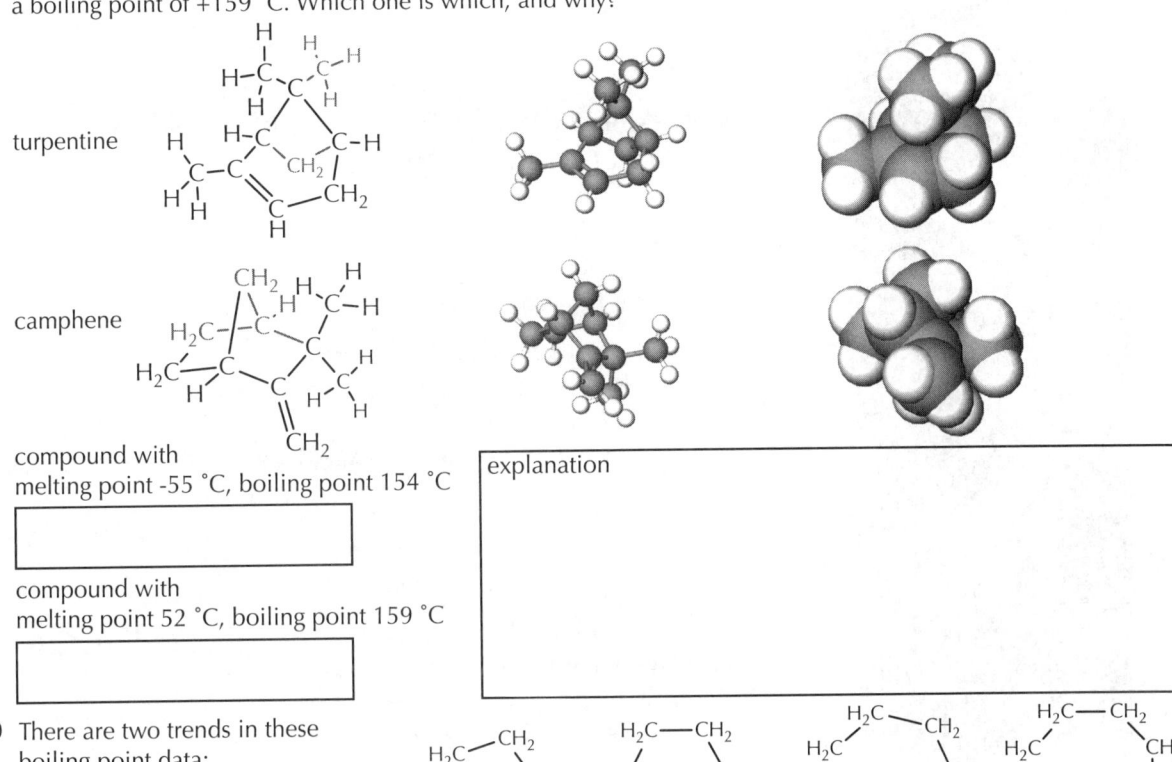

compound with
melting point -55 °C, boiling point 154 °C

compound with
melting point 52 °C, boiling point 159 °C

explanation

AP01.20 There are two trends in these boiling point data:

(1) as you increase the ring size, the boiling points systematically increase with each of the three families of compounds;

(2) the boiling point of the molecule with an NH in the ring is the highest, with the oxygen-containing ring usually next, although as you increase the ring size, all three boiling points move towards converging on the same value.

Explain the trends.

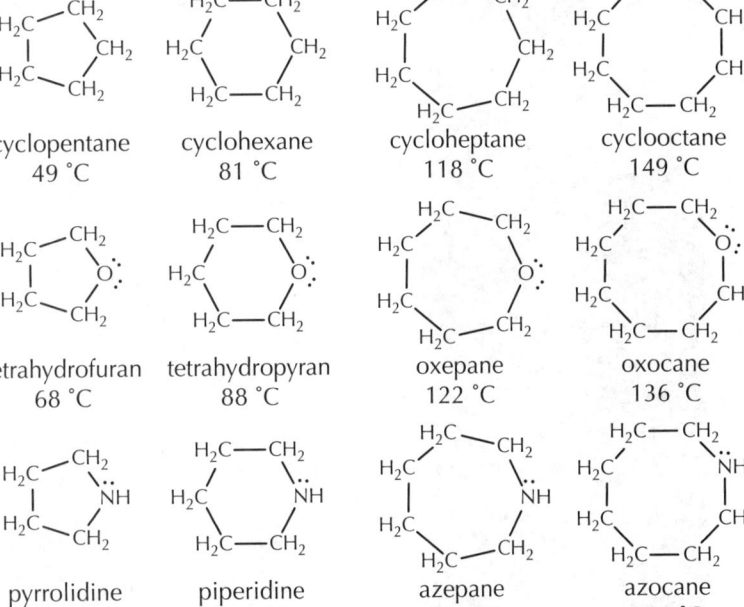

cyclopentane 49 °C
cyclohexane 81 °C
cycloheptane 118 °C
cyclooctane 149 °C

tetrahydrofuran 68 °C
tetrahydropyran 88 °C
oxepane 122 °C
oxocane 136 °C

pyrrolidine 87 °C
piperidine 106 °C
azepane 142 °C
azocane 155 °C

APPENDIX 2

Nomenclature of Organic Compounds I: Alkanes, Halides, and Alcohols

Introduction

A2.1 Acyclic Saturated Hydrocarbons
- A. Longest Carbon Chain (Root Name)
- B. Identifying Substituents and Numbering the Root Name
- C. Assembling the Name
- D. Three More Rules and the Four Allowed Branched Alkyl Group Names

PRACTICE QUESTIONS

A2.2 Cyclic Compounds, Halides, and Alcohols
- A. Cyclic Hydrocarbons
- B. Alkyl and Aryl Halides
- C. Alcohols
- D. Priorities When Numbering the Root Name

PRACTICE QUESTIONS

APPENDIX 2

Nomenclature of Organic Compounds I: Alkanes, Halides, and Alcohols

Introduction

"Of course 'irregardless' is a made-up word that was entered into the dictionary through constant use; that's pretty much how this racket works. All words are made-up: Do you think we find them fully formed on the ocean floor, or mine for them in some remote part of Wales?"

Kory Stamper, *Word by Word: The Secret Life of Dictionaries*, **2017**

Thor: "Where we have to go is Nidavellir."
Drax: "That's a made-up word."
Thor: "All words are made up."

Christopher Markus and Stephen McFeely, *Avengers: Infinity War*, **2018**

Historically, compounds were given names, usually by their discoverers, that reflected their origin or their properties. For example, the painkiller morphine was named for Morpheus, the Greek god of dreams. Cholesterol, the chief component of gallstones, got its name from the Greek words for bile and solid. And still today, giving short and descriptive names for complex molecular structures makes talking and writing about them easier. In 1962, Howard Whitlock reported the preparation of twistane. And in 1984, the discoverers of the soccer-ball shaped C_{60} molecule named it buckminsterfullerene, in honor of Buckminster Fuller, an architect who designed geodesic domes that resemble parts of this molecule's structure.

The structural formulas, along with both common and systematic names, of these four molecules are shown in Figure AP0201, not so you can memorize them but so you will have some idea why the systematic names are not always commonly used (although they are, in fact, valuable for computer indexing).

Figure AP0201
Complex molecules and their names.

morphine

(4*R*,4a*R*,12b*S*)-3-methyl-2,3,4,4a,7,7a-hexahydro-1*H*-4,12-methanobenzofuro[3,2-e]isoquinoline-7,9-diol

cholesterol

(3*S*,8*S*,9*S*,10*R*,13*R*,14*S*,17*R*)-10,13-dimethyl-17-((*R*)-6-methylheptan-2-yl)-2,3,4,7,8,9,10,11,12,13,14,15,16,17-tetradecahydro-1*H*-cyclopenta[a]phenanthren-3-ol

Complex names continue on next page . . .

Figure AP0201 continued . . .

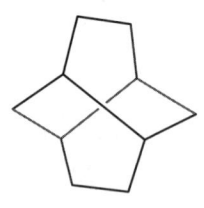

twistane

tricyclo[4.4.0.0^{3,8}]decane

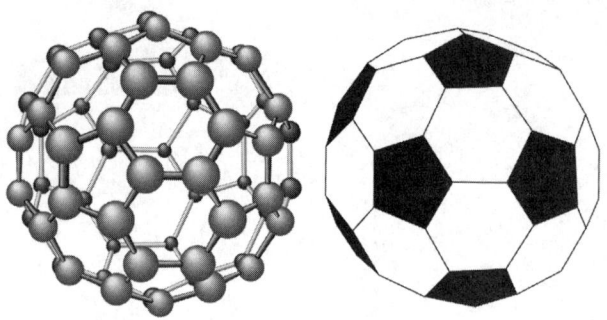

buckminsterfullerene

Since 1997: (C$_{60}$-I_h)[5,6]fullerene

Before 1997: hentriacontacyclo[29.29.0.0^{2,14}.0^{3,12}.0^{4,59}.0^{5,10}.0^{6,58}.0^{7,55}.0^{8,53}.0^{9,21}.0^{11,20}.0$^{13.18}$.0^{15,30}.0^{16,28}.0^{17,25}.0^{19,24}.0^{22,52}.0^{23,5}.0^{26,49}.0^{27,47}.0^{29,45}.0^{32,44}.0^{33,60}.0^{34,57}.0^{35,43}.0^{36,56}.0^{37,41}.0^{38,54}.0^{39,51}.0^{40,48}.0^{42,46}]hexaconta-1,3,5(10),6,8,11,13(18),14,16,19,21,23,25,27,29(45),30,32(44),33,35(43),36,38(54),39(51),40(48),41,46,49,52,55,57,59-triacontaene.

In 1787, as the first modern ideas about atoms and compounds were maturing, de Morveau, Lavoisier, Bertholet, and de Fourcroy coauthored *Méthode de Nomenclature Chimique* (*Method of Chemical Nomenclature*), in which they proposed the first truly systematic approach to naming chemical substances. Before this time, the same substance might have anywhere from 2–15 names, all granted by different people in different locations. Antoine Lavoisier, who gets much of the historical credit for the new nomenclature, proposed a system that relied on a substance's chemical makeup or its chemical properties, such as sodium chloride and ferric sulfate (for those whose constituents where known), and invented names such as oxygen (acid-bringer) and hydrogen (water-bringer) based upon the chemical properties of these new gases.

"We think only through the medium of words."

"The impossibility of separating the nomenclature of a science from the science itself, is owing to this, that every branch of physical science must consist of three things; the series of facts which are the objects of the science, the ideas which represent these facts, and the words by which these ideas are expressed. Like three impressions of the same seal, the word ought to produce the idea, and the idea to be a picture of the fact. And, as ideas are preserved and communicated by means of words, it necessarily follows that we cannot improve the language of any science without at the same time improving the science itself; neither can we, on the other hand, improve a science, without improving the language or nomenclature which belongs to it."

A. Lavoisier, *Elements of Chemistry* (translation by Robert Kerr, Edinburgh, **1790**, xiii) from Dover facsimile edition, **1965**.

A hundred years later, the number of organic compounds that were being prepared in the laboratories or isolated from natural sources was growing rapidly. The same situation had evolved: It eventually became impossible for chemists to learn the names randomly assigned to compounds by their discoverers, especially when the names showed no correlation to the structures.

In 1892, chemists from all over the world met for the first time, and continue to meet periodically, to decide on systematic rules for naming organic compounds. These rules, which are constantly evolving, are called the International Union of Pure and Applied Chemistry (abbreviated as IUPAC) rules. The IUPAC system of nomenclature was developed so that each organic compound would have a unique name that would allow the structural formula to be written from it.

Ideally, a chemist would have to learn only the systematic names of compounds. In the real world, however, there are three complications. First, even the IUPAC rules allow for some variations in the naming of compounds. Second, the correct IUPAC names for some compounds are so complicated and cumbersome that everybody continues to use the common, unsystematic names for them (see, e.g., morphine, twistane, etc.). Finally, many compounds had names before the IUPAC rules came into being, and these names are still used, especially by chemical supply houses and by industry, and they exist (and will always exist) in decades of published journal articles. To be literate about chemistry, you must be able to recognize many common names.

Therefore, while the emphasis in this book is on the systematic names of compounds, the common names of important compounds are also given and are used in cases where they are the ones used overwhelmingly by chemists in their daily work.

Just to give you an idea of how complex these naming problems can be, the discovery of the fullerenes (see Figure 0201) created a situation where it was nearly impossible to name a new molecule without a computer, and even then, the naming experts recognized a "good news, bad news" feature of naming these molecules:

> The limited availability of different isomeric fullerenes turns out to be a lucky circumstance. The naming of fullerenes according to these rules is unfeasible without a computer. Even then the task may become intractable, as the problem involved in deriving a unique name belongs to the so-called NP-complete class. This class of problems is characterized by nonpolynomial complexity which means that the time needed for performing the task increases with the size of the considered object faster than any polynomial function. Sooner or later, no program and no computer will be capable to handle the task properly.
>
> D Babic, AT Balaban, D J Klein, *Journal of Information and Computer Science,* **1995**, *35*, 515.

Ultimately, IUPAC needed to invent a new naming system, and one that humans could use conveniently, specific to this family of compounds (E W Godly, R Taylor, *Pure & Applied Chemistry,* **1997**, *69*(7), 1411). The post-1997 name is also shown in Figure AP0201.

In this appendix, we will begin naming with the simplest organic structures, namely, hydrocarbons. Most hydrocarbons (compounds of carbon and hydrogen only) occur in petroleum, a naturally occurring liquid found within the earth, which is proposed to have come from decomposed organic matter over millions of years.

Petroleum is a mixture of hundreds of different compounds. The hydrocarbons, and other organic compounds, are separated from one another through a process called refinement (hence, refineries), which is typically based on differences in boiling points. Gasoline, diesel fuel, and kerosene are all derived from petroleum.

Another source of organic compounds is coal tar, a waste product that results from purifying crude coal for use in industry (e.g., refined coal, called coke, is used for making steel). Coal tar is used for fuels and as asphalt. In the late 1700s, Germany was shipping crude coal to Great Britain at the start of the Industrial Revolution, and the barges returned to Germany with the coal tar as an industrial waste. Enterprising German scientists learned how to extract useful raw materials from the coal tar, particularly for making medicines, cosmetics, fertilizers, and dyes, and later for photographic materials, which resulted in the country's global dominance in the field of organic chemistry for more than a hundred years.

Petroleum, coal, and natural gas have accounted for about 80% of the energy production in the United States since their introduction in about 1900. As you learn to name these compounds, you should also remind yourself that there are large societal issues that surround hydrocarbons, in particular. Taking millions of years to produce, the reserves of petroleum are nonrenewable. Our rate of removal of petroleum from the earth, measured in seconds and minutes, far (far, far) exceeds its rate of replacement, measured in eons. And although quadrillions of BTUs of energy per year comes from combustion of these compounds, most of these carbon atoms end up as carbon dioxide, resulting in a change to our global atmosphere that, in turn, affects the global climate.

Thinking about alternatives to using fossil fuels and reducing the amount of carbon dioxide released from human industriousness is an ongoing concern in the 21st century. Solving the "carbon footprint" problem is an active area of research in many fields, and certainly no less so in chemistry, where every academic department includes many faculty members who are working in areas of "energy" and "sustainability."

A2.1 Acyclic Saturated Hydrocarbons

A. Longest Carbon Chain (Root Name)

As explained above, "hydrocarbon" means what the name says: a compound composed of only carbon and hydrogen. "Acyclic" means that there are no rings, and "saturated" means that there are only single bonds. As covered in detail in Chapter 1.4C, acyclic saturated hydrocarbons with uncharged and closed shell atoms all have a C_nH_{2n+2} molecular formula. Organic hydrocarbons with no rings or multiple bonds are classified as acyclic alkanes.

The names of the first four acyclic, straight-chain alkanes are methane (CH_4), ethane (CH_3CH_3), propane ($CH_3CH_2CH_3$), and butane ($CH_3CH_2CH_2CH_3$). These historical names are permitted under IUPAC rules. The systematic nomenclature of the other members of the series is based on a prefix that counts the number of carbon atoms in the chain, followed by the suffix "ane." The prefixes come from Greek or Latin words for the numbers. The names of the first twenty straight-chain alkanes are shown in Figure AP0202. These names are used in naming all other types of organic compounds derived from alkanes, so you should learn them.

Figure AP0202

Straight-chain alkanes (C_1–C_{20}).

CH_4	methane	$CH_3(CH_2)_9CH_3$	undecane
CH_3CH_3	ethane	$CH_3(CH_2)_{10}CH_3$	dodecane
$CH_3CH_2CH_3$	propane	$CH_3(CH_2)_{11}CH_3$	tridecane
$CH_3(CH_2)_2CH_3$	butane	$CH_3(CH_2)_{12}CH_3$	tetradecane
$CH_3(CH_2)_3CH_3$	pentane	$CH_3(CH_2)_{13}CH_3$	pentadecane
$CH_3(CH_2)_4CH_3$	hexane	$CH_3(CH_2)_{14}CH_3$	hexadecane
$CH_3(CH_2)_5CH_3$	heptane	$CH_3(CH_2)_{15}CH_3$	heptadecane
$CH_3(CH_2)_6CH_3$	octane	$CH_3(CH_2)_{16}CH_3$	octadecane
$CH_3(CH_2)_7CH_3$	nonane	$CH_3(CH_2)_{17}CH_3$	nonadecane
$CH_3(CH_2)_8CH_3$	decane	$CH_3(CH_2)_{18}CH_3$	eicosane

Most compounds are not straight-chain alkanes. Constructing the name begins with identifying the longest continuous chain of carbon atoms in the molecule. This will be the root name on which the name will be constructed. The goal in naming is to unambiguously represent the placement of the atoms in a structure with a word (a name) that allows you to communicate the structure without have to draw it and then point to it. Or, to take the name for a structure and accurately reproduce its structure.

There are three structural isomers for C_5H_{12} (Figure AP0203). One of them has all five carbons linked in a row, and this is the straight-chain hydrocarbon named pentane (historically, n-pentane for normal pentane). The other two structural isomers are called branched compounds because it is impossible to identify a straight line that connects all five carbons, as in pentane.

The longest continuous carbon chain that you can identify constitutes the root name. The other atoms or groups of atoms are the branches on that chain, and (for the purpose of naming) we consider these groups to have replaced, or substituted, one of the hydrogen atoms on that root chain. These groups are therefore called "substituent" groups.

Figure AP0203

Naming the three C_5H_{12} structural isomers: longest chain.

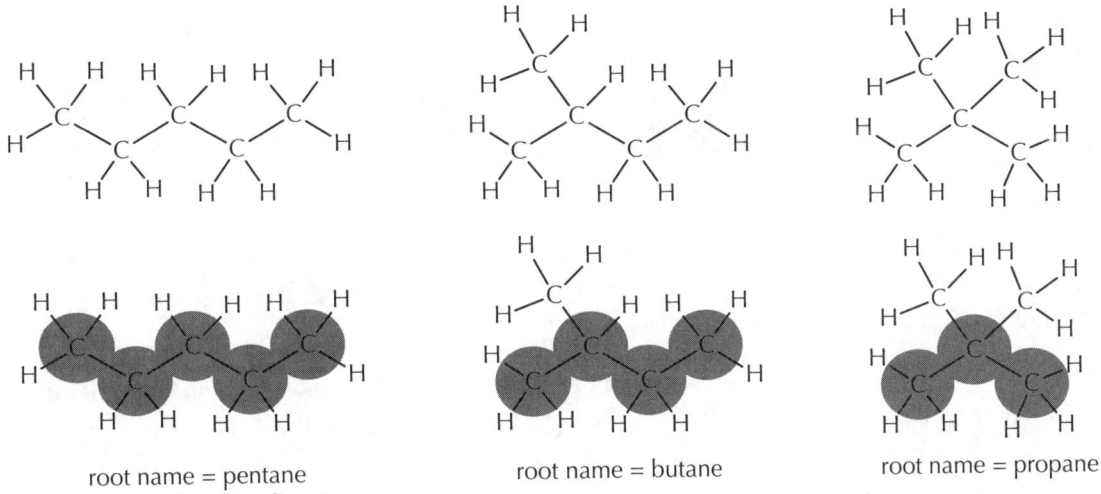

root name = pentane
(accounts for all carbons)

root name = butane

root name = propane

As shown in Figure AP0203, one of them has a longest continuous chain of four atoms, and so its root name is butane. The other has a longest continuous chain of only three atoms, and so its root name is propane.

B. Identifying Substituents and Numbering the Root Name

With "pentane," all of the atoms are accounted for, and that is its name. For the other two isomers, there are substituent groups hanging off the root chain that need to be accounted for (Figure AP0203). The isomer with the butane chain root has one substituent group (a CH_3 attached through the carbon) and the isomer with the propane chain root has two of these CH_3 substituent groups. A substituent group is named from its point of attachment, which means through the carbon atom. The one-carbon substituent group derived from the saturated compound (methane) is called a methyl group. The "meth" tells you that it is one carbon, and the "yl" is used to designate it as an attached group. A two-carbon substituent based on ethane is called an ethyl group, propane: propyl, butane: butyl, pentane: pentyl, and so on. Collectively, these alkane-based substituent groups are called alkyl groups, and they are historically abbreviated as "R-" groups (from the older German term, *radikal*).

The only thing remaining is to identify the location of the substituent groups (Figure AP0204). The location of the substituent is identified by numbering the carbon atoms of the root chain, which means (always) only one of two options: start numbering from one end or start numbering from the other. The rule for assigning numbers to the atoms of the root chain is called "first point of difference": The numbering of the root chain is done to give the first encountered substituent group the lowest number (and never, ever, done by adding the numbers, in any circumstance; adding numbers never appears in the rules for naming).

Figure AP0204

Naming the three C_5H_{12} structural isomers: numbering substituents.

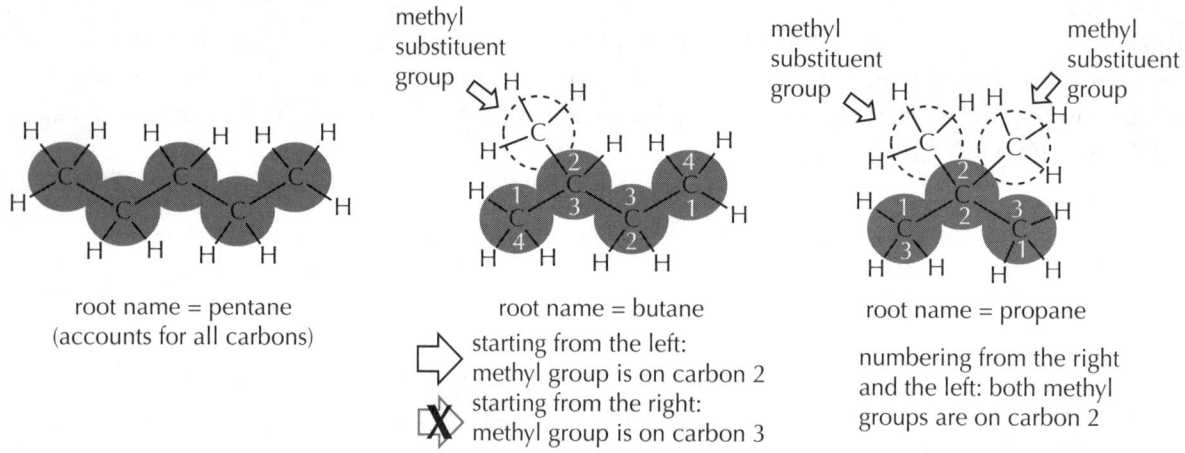

In the isomer with the 4-carbon root chain, numbering from one end places the attached methyl group at carbon 2, while numbering from the other end gives it the number 3. Because 2 is less than 3, the numbering of the root chain follows the first option.

In the isomer with the 3-carbon root chain, numbering from one end places both methyl groups on carbon 2, as does numbering from the other end. In this case, the results are identical.

C. Assembling the Name

In assembling the pieces into a name, alkyl groups are treated as prefixes on the root chain, so they appear in the front end of the name. The rules about punctuation and ordering prefixes are:

(a) Adjacent numbers and words are separated by a hyphen, adjacent numbers are separated by a comma, and adjacent words are not separated.

(b) Prefixes are ordered in front of the root name alphabetically.

(c) If there is more than one substituent of the same kind, the name of the alkyl group is given the additional combining prefix di (for two), tri (for three), tetra (for four), penta (for five), and so on, with the following provisions: (i) the combining prefix is not considered in the alphabetical ordering, and (ii) all numbers are retained.

Finally, the name can be constructed. As seen in Figure AP0205, the isomer with the butane root chain is named 2-methylbutane (historically, isopentane) and the one with the propane root chain is named 2,2-dimethylpropane (historically, neopentane). The standard way for representing the structure in a typed line is also shown: $CH_3CH_2CH_2CH_2CH_3$ for pentane, $(CH_3)_2CHCH_2CH_3$ for 2-methylbutane, and $(CH_3)_3CCH_3$ for 2,2-dimethylpropane.

Figure AP0205
Naming the three C$_5$H$_{12}$ structural isomers.

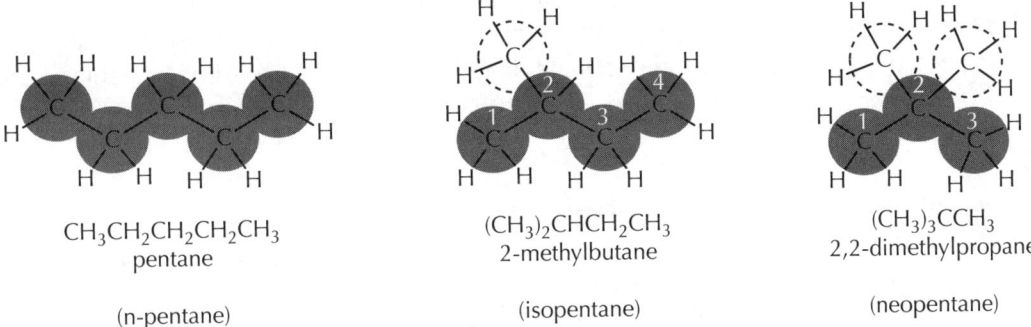

CH$_3$CH$_2$CH$_2$CH$_2$CH$_3$
pentane
(n-pentane)

(CH$_3$)$_2$CHCH$_2$CH$_3$
2-methylbutane
(isopentane)

(CH$_3$)$_3$CCH$_3$
2,2-dimethylpropane
(neopentane)

D. Three More Rules and the Four Allowed Branched Alkyl Group Names

There are 18 C$_8$H$_{18}$ hydrocarbons, and 17 of them can be named according to the rules above, with one additional rule needed for the last one: If there are multiple substituent groups and the numbering from one end versus the other yields the same first number, then the point of difference is extended to the second-encountered substituent. The stepwise construction of the names for four of these isomers, including the one that illustrates this additional rule, are shown in Figure AP0206.

Figure AP0206
Naming four of the 18 C$_8$H$_{18}$ structural isomers.

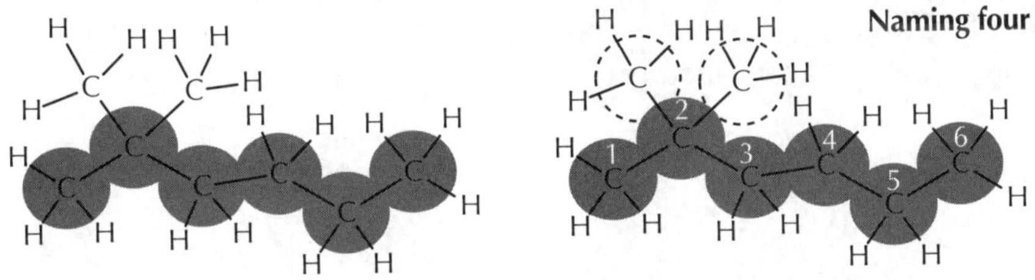

(CH$_3$)$_3$CCH$_2$CH$_2$CH$_2$CH$_3$
2,2-dimethylhexane

of note: there are 3 identical 6 carbon chains, you merely need to pick one

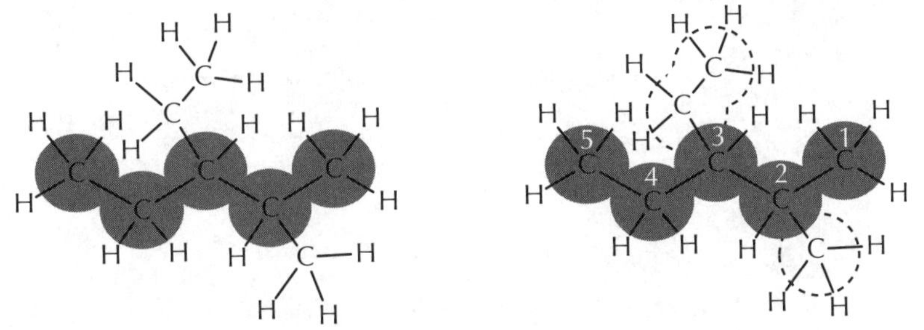

(CH$_3$CH$_2$)$_2$CHCH(CH$_3$)$_2$
3-ethyl-2-methylpentane

of note: first point of difference - numbering from the right gives the first group encountered at carbon 2, while from the right the first group is encountered at carbon 3; prefix groups appear in alphabetical order, not numerical

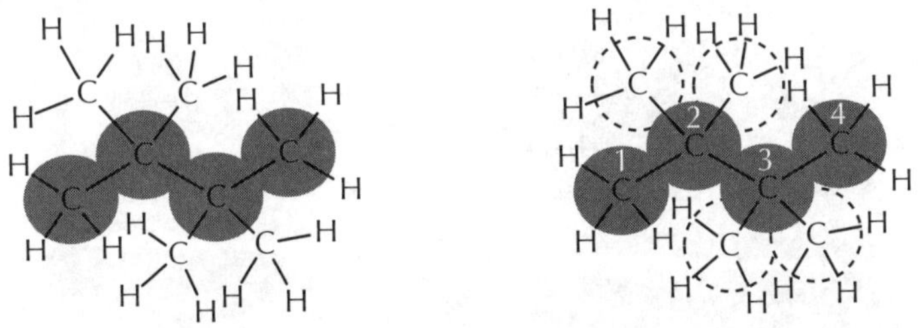

(CH$_3$)$_3$CC(CH$_3$)$_3$
2,2,3,3-tetramethylbutane

of note: the condensed formula can be deceptive; when you see this version, (CH$_3$)$_3$CC(CH$_3$)$_3$ it might not be obvious that the root name is butane, so drawing structures out is useful

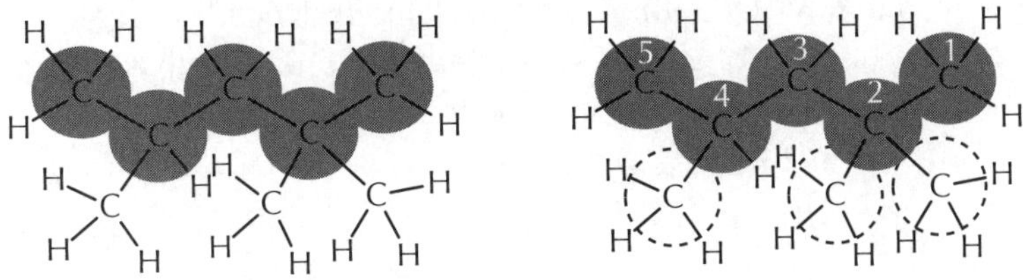

(CH$_3$)$_2$CHCH$_2$C(CH$_3$)$_3$
2,2,4-trimethylpentane

of note: the first encountered group from both the left and right is at carbon 2, so you need to use the second encountered group to determine a point of difference; from the left, the numbers would be 2,4,4- and from the right they are 2,2,4- (the second number determines the decision)

A caution here about online resources: Not everyone who publishes a website of chemistry notes and lessons actually knows the rules. Sometimes they learn them incorrectly and then pass them on, and there are no editors to monitor these resources. One of the common errors out there is that the sum of the numbers matters. It does not. Only the first point of difference. For instance, the molecule shown in Figure AP0207 is named 2,13,13-trimethylpentadecane (correct) and not 3,3,14- trimethylpentadecane (incorrect).

Figure AP0207
Follow the first point of difference rule; never add numbers.

2,13,13-trimethylpentadecane
not
3,3,14-trimethylpentadecane

of note: numbers are never added together to determine the root chain number assignments, the first point of difference rule is a one by one comparison (in this case, "2" versus "3" for the first encountered group settles the numbering of the chain)

The second of the remaining hydrocarbon rules (at least for this introduction) can be illustrated with two of the $C_{11}H_{24}$ isomers (Figure AP0208). Prefix-designated groups do not have any intrinsic priority for numbering (e.g., the number of atoms in the group does not matter). Sometimes, though, the first point of difference rule cannot distinguish between where to start the numbering because you get a tie. Ties are resolved in a wonderfully simple way: alphabetical order.

Figure AP0208
Naming two $C_{11}H_{24}$ structural isomers.

3-ethyl-6-methyloctane

of note: numbering from both the right and the left gives substituents at carbons 3 and 6, so number from the right is used because the "e" in ethyl comes before the "m" in methyl

3-ethyl-2,4,5-trimethylhexane

of note: numbering from both the right and the left gives substituents at carbons 2, 3, 4 and 5; the first substituent from both ends is 2-methyl, so the point of difference in the second encountered substituent (ethyl versus methyl) at carbons 3 and 4, favors the ethyl ("e") over the methyl ("m")

The third hydrocarbon rule addresses how to break a tie when there are different hydrocarbon chains of the same length. In the molecule shown in Figure AP0209, there are two different 10-carbon chains. The rule is to select the one that has the greatest number of simple substituents. In one case there are four simply named groups (3 methyls and an ethyl), while in the other case there are only three groups and one of them is itself branched, and not simply named.

Figure AP0209

Resolving the competition between equal chain lengths.

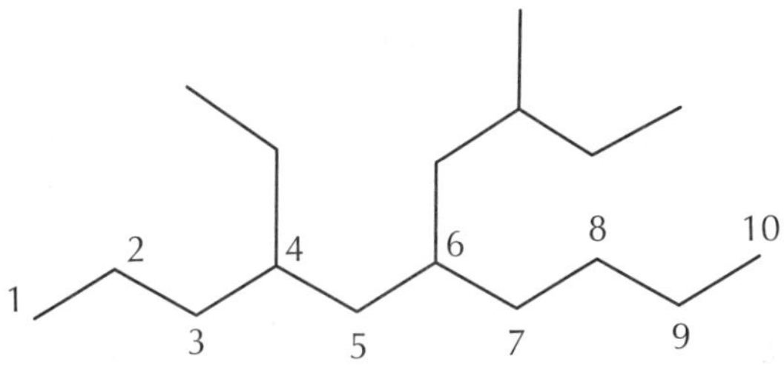

not the correct root chain

of note: there are two different 10-carbon chains; the one on the left (above) has two substituent groups at carbons 4 and 6 of the 10-carbon chain, and the substituent at carbon 6 is branched

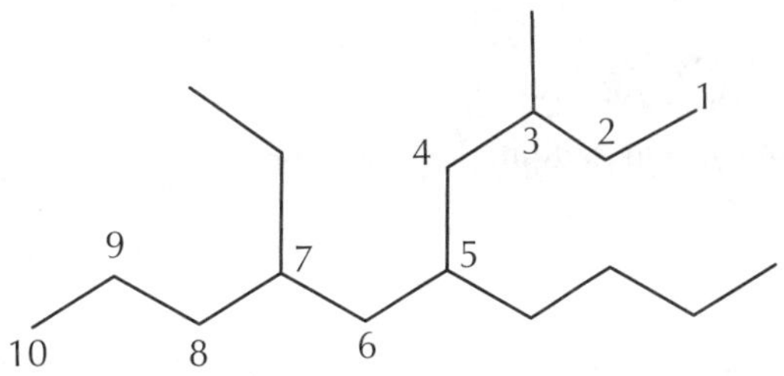

5-butyl-7-ethyl-3-methyldecane

of note: this 10-carbon chain has three substituent groups at carbons 3, 5, and 7 that can all be named simply; this is the selected root chain for building the name

Finally, as introduced in the previous example, substituent groups can certainly be branched, and there are rules to cover them. As with the 1- to 4-carbon hydrocarbons, the historical names for some of the most common branched substituent groups have survived and are recognized by the IUPAC rules. Four of the most commonly used branched substituent groups are shown in Figure AP0210. They are isopropyl, isobutyl, *sec*-butyl, and *tert*-butyl. These four can be treated the same way as any unbranched substituent group (methyl, ethyl) in naming. Isopropyl and isobutyl are single words and alphabetized on the letter "i," while the two hyphenated groups, *sec*-butyl, and *tert*-butyl, are alphabetized on the letter "b."

Figure AP0210

Four branched substituent groups with common names.

isopropyl group $-CH(CH_3)_2$

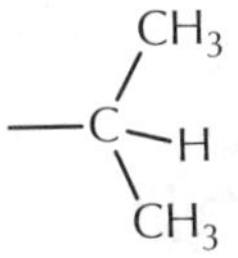

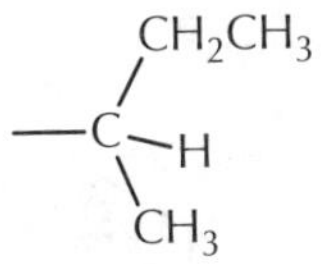

isobutyl group $-CH_2CH(CH_3)_2$

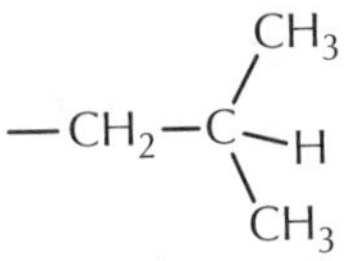

sec-butyl group $-CH(CH_3)CH_2CH_3$

tert-butyl group $-C(CH_3)_3$

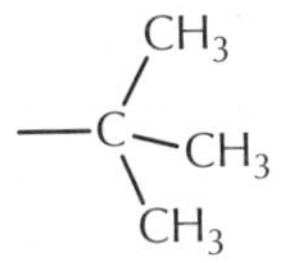

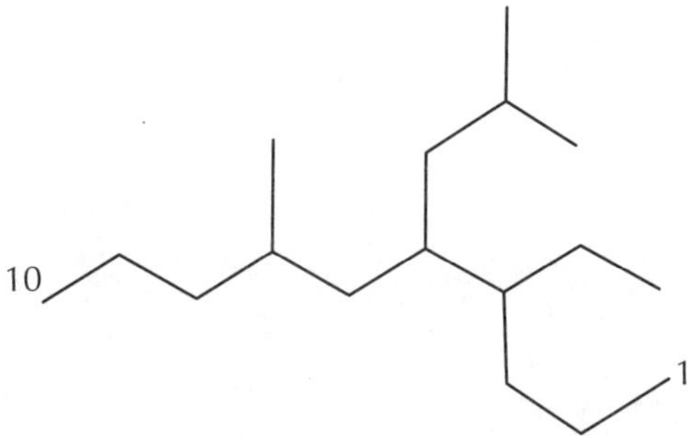

4-ethyl-5-isobutyl-7-methyldecane

of note: isobutyl (and isopropyl) are alphabetized on the letter "i"

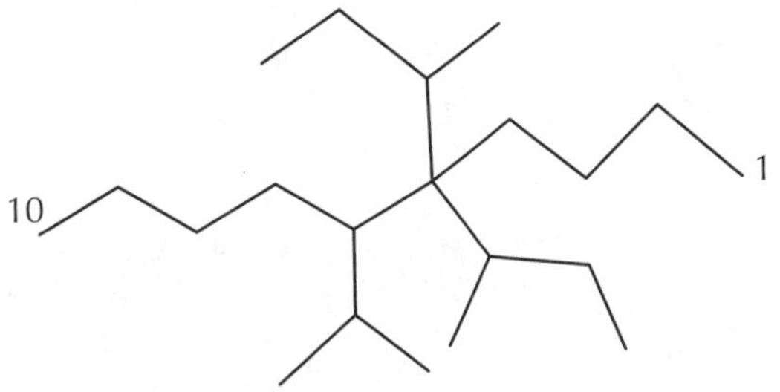

5,5-di-*sec*-butyl-6-isopropyldecane

of note: the first substituent encountered is at carbon-5 when starting from either end, so the final decision is on the second point of difference; note also the punctuation: the "*sec*" comes with a hyphen in "*sec*-butyl" and takes one with the combining term "di"

Three more examples that integrate all of these rules are provided in Figure AP0211, including some important notes about punctuation and handling the priority assignments for the hyphenated terms *tert*-butyl and *sec*-butyl.

Figure AP0211
Naming acyclic saturated hydrocarbons.

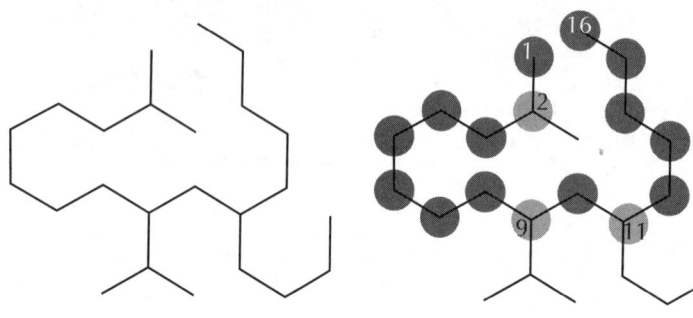

root: hexadecane

substituents:
2-methyl
9-isopropyl
11-butyl

11-butyl-9-isopropyl-2-methylhexadecane

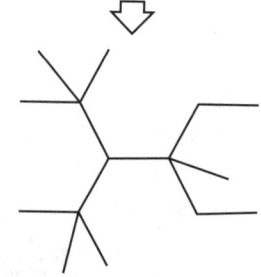

((CH$_3$)$_3$C)$_2$CHC(CH$_2$CH$_3$)$_2$CH$_3$

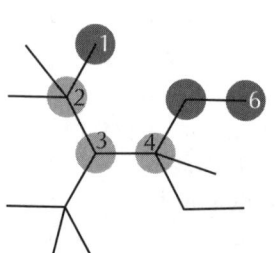

root: hexane

substituents:
2-methyl
2-methyl
3-*tert*-butyl
4-methyl
4-ethyl

3-(*tert*-butyl)-4-ethyl-2,2,4-trimethylhexane

to note: when the hyphenated branched groups *tert*-butyl and *sec*-butyl are alone (i.e., no di, tri, tetra), they are placed in parentheses

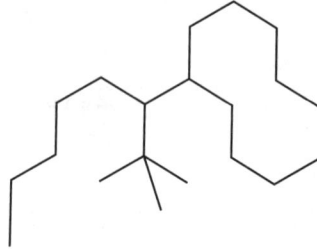

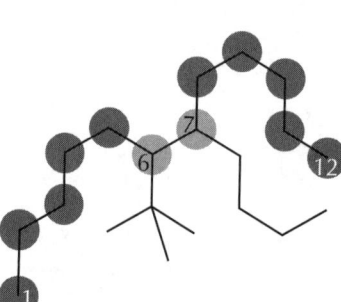

root: dodecane

substituents:
6-(*tert*-butyl)
7-butyl

6-(*tert*-butyl)-7-butyldodecane

to note: when a tie-breaker is needed between *tert*-butyl and butyl (or *sec*-butyl and butyl), the hyphenated group takes the priority; a tie between *tert*-butyl and *sec*-butyl goes to the *sec*-butyl for alphabetical order

PRACTICE QUESTIONS

AP02.01 Provide the IUPAC names for the following compounds.

(a)

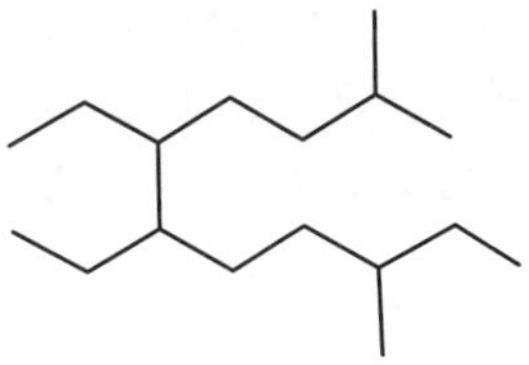

(b) $(CH_3CH_2CH_2)_3CCH_2CH_3$

(c)

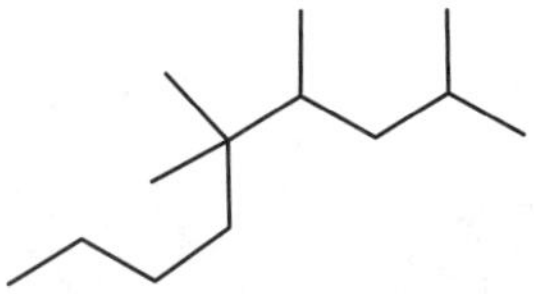

(d) $CH_3CH_2CH_2CH(CH_3)CH(CH_2CH_3)_2$

(e)

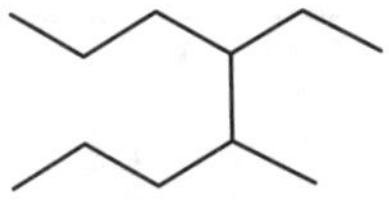

AP02.02 Draw the following compounds from their names.

(a) 3,7-diethyl-2-methyldecane

(b) 4-(*tert*-butyl)-5-isopropyloctane

(c) 6-ethyl-2,2,4-trimethyldodecane

(d) 2,2-dimethylheptane

(e) 6-isobutyl-2-methyldecane

(f) 3,3,4,4-tetraethylhexane

APPENDIX 2 Nomenclature of Organic Compounds I: Alkanes, Halides, and Alcohols

AP02.03 Provide the IUPAC names for the following compounds.

(a)

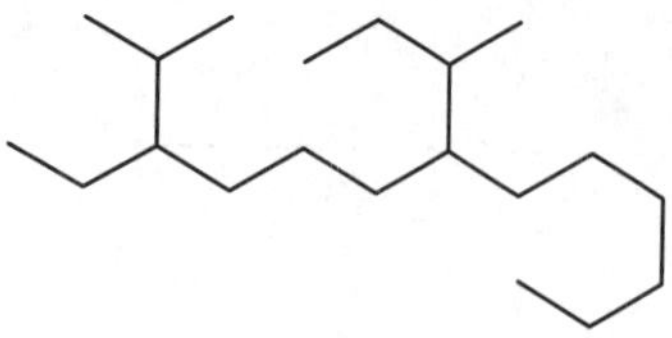

(b)

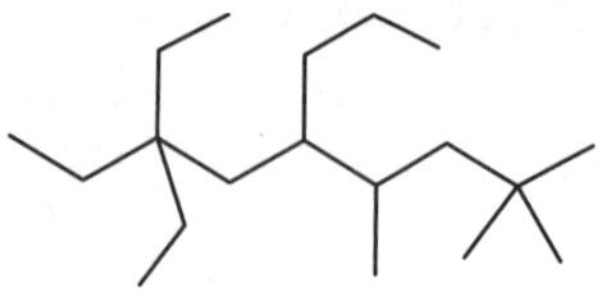

(c)

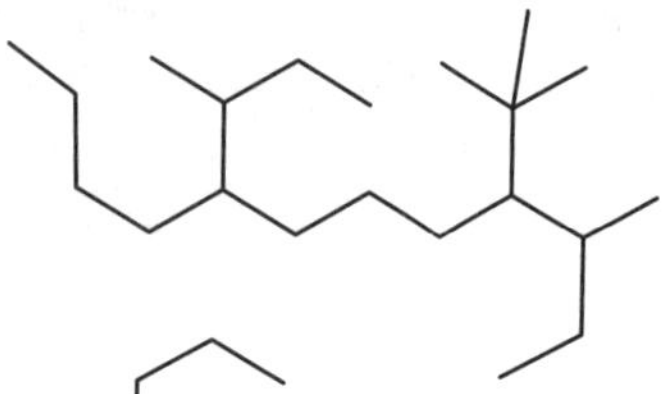

(d)

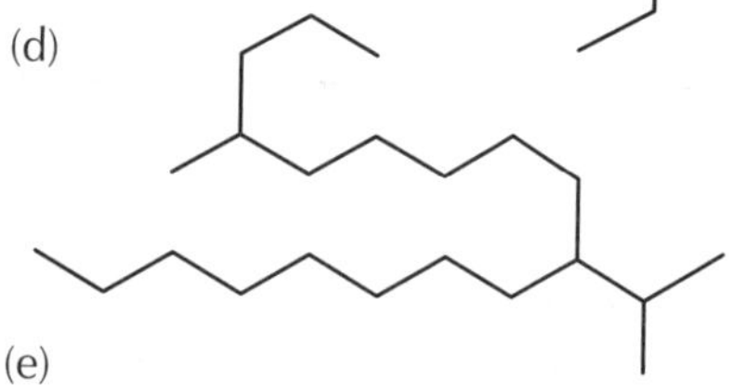

(e) $((CH_3)_3C)_4C$

AP02.04 Draw and name the following compounds from these descriptions (there may be more than one answer).

(a) a saturated acyclic hydrocarbon with 2 isopropyl groups that have the same number

(b) a trimethylheptane where the methyls have different numbers; then name it

(c) a dodecane with 4 different substituent groups

(d) a hexane with the maximum number of ethyl groups that can be a part of the structure

A2.2 Cyclic Compounds, Halides, and Alcohols

A. Cyclic Hydrocarbons

Monocyclic alkanes (cycloalkanes) are hydrocarbons that have the general formula C_nH_{2n} and in which some or all of the carbon atoms form a ring, accounting for the unit of unsaturation. A cycloalkane is named by combining the term "cyclo" with the name of the alkane having the same number of carbon atoms as are in the ring: cyclopropane, cyclobutane, cyclopentane, and so on (Figure AP0212)

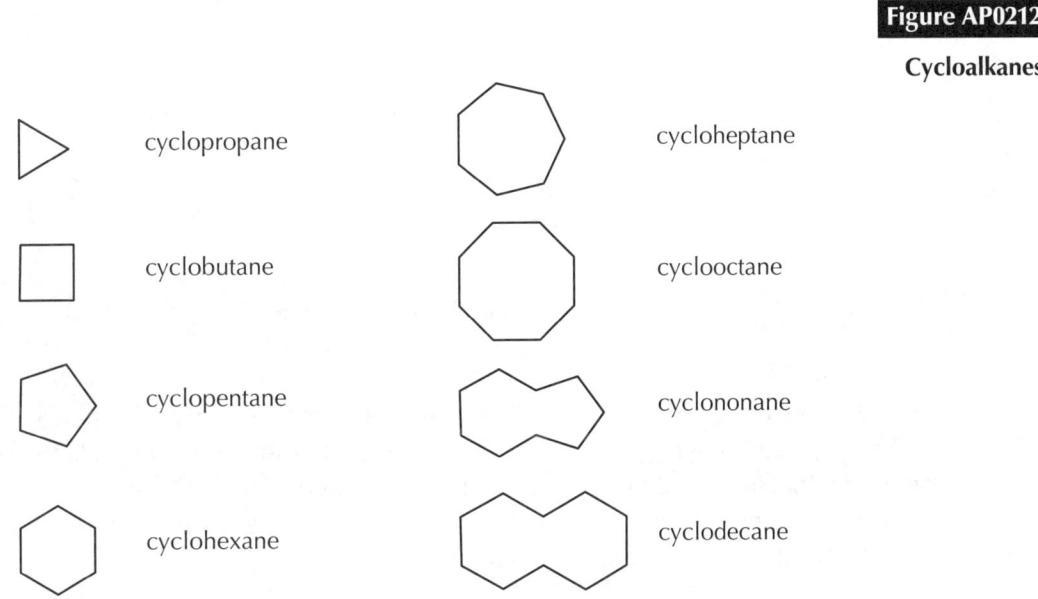

Figure AP0212

Cycloalkanes.

IUPAC rules allow for a bit of flexibility in naming substituted cycloalkanes, although the major generalization tends to work the best: Select the longest carbon chain with the greatest number of unbranched substituents, and number the atoms of the root chain by the first point of difference rule.

A few key examples of monosubstituted cycloalkanes are shown in Figure AP0213. The ring counts as an integral unit: You only count its atoms and never extend the count outside the ring. When the number of atoms in the substituent is less than the number of atoms in the ring, the ring is the root chain. In these cases, designating the substituent as "1" is superfluous because it defines the number 1 position. As you learn about naming, keep in mind that it is never incorrect to include the superfluous "1," so rather than worrying about when it is (or is not) needed, just include it until you are absolutely confident it is not needed.

When the number of atoms in the ring is less than the substituent group, then the ring is used as a substituent on the linear chain (and uses the substituent designation for an attached group, such as cyclopropyl, cyclobutyl, and cyclopentyl). When the ring and the substituent group have the same number of atoms, IUPAC gives the preference to using the ring as the root chain (although the alternative is always listed as a synonym). When the substituent on the ring is branched, and it is one of the simply named branched groups, then the ring can be the root name. Otherwise, the longest chain with the greatest number of simply named substituents is the linear chain, which is typically used as the root name. Even for these relatively simple molecules, IUPAC rules often allow multiple synonyms.

Figure AP0213

Naming monosubstituted cycloalkanes.

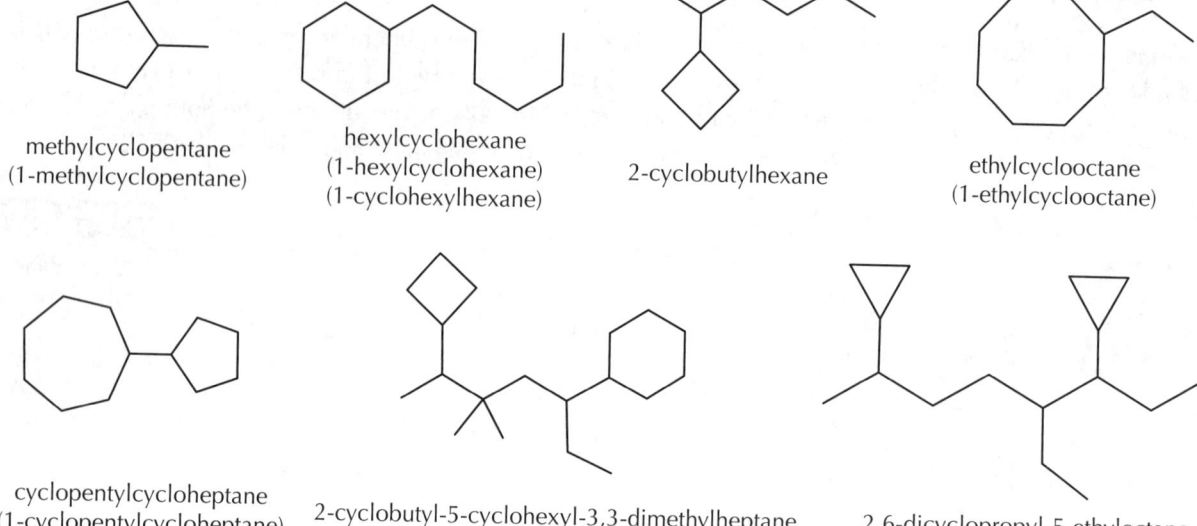

methylcyclopentane
(1-methylcyclopentane)

hexylcyclohexane
(1-hexylcyclohexane)
(1-cyclohexylhexane)

2-cyclobutylhexane

ethylcyclooctane
(1-ethylcyclooctane)

cyclopentylcycloheptane
(1-cyclopentylcycloheptane)

2-cyclobutyl-5-cyclohexyl-3,3-dimethylheptane

2,6-dicyclopropyl-5-ethyloctane

For disubstituted, trisubstituted, and above, cycloalkanes, the ring is usually the root chain (Figure AP0214). One of the substituent groups will get the number 1, and the point of difference rule dictates which one it is by looking at the numbering for the other groups. Disubstituted rings always result in a tie, which is broken by the alphabetical order tie-breaker rule. Some of the trisubstituted bonding patterns also need the tie-breaker to assign the numbering.

Figure AP0214

Naming disubstituted (and above) cycloalkanes.

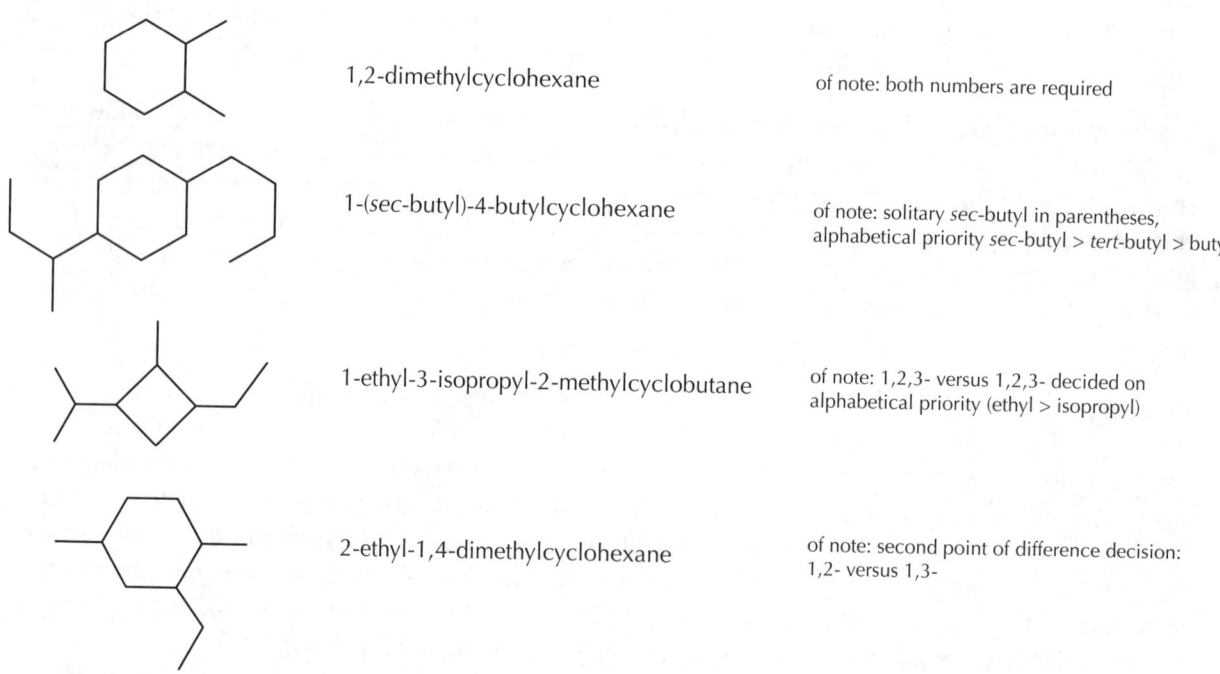

1,2-dimethylcyclohexane — of note: both numbers are required

1-(sec-butyl)-4-butylcyclohexane — of note: solitary sec-butyl in parentheses, alphabetical priority sec-butyl > tert-butyl > butyl

1-ethyl-3-isopropyl-2-methylcyclobutane — of note: 1,2,3- versus 1,2,3- decided on alphabetical priority (ethyl > isopropyl)

2-ethyl-1,4-dimethylcyclohexane — of note: second point of difference decision: 1,2- versus 1,3-

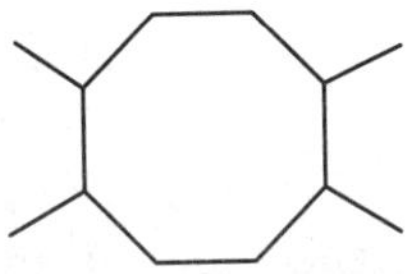

1,2,5,6-tetramethylcyclooctane — of note: all numbers are required

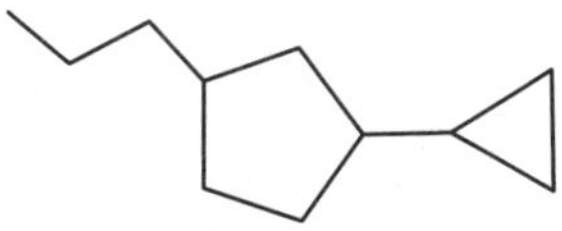

1-cyclopropyl-3-propylcyclopentane — of note: alphabetical tie-breaker, the "c" in "cycloprop" is integral to the root name

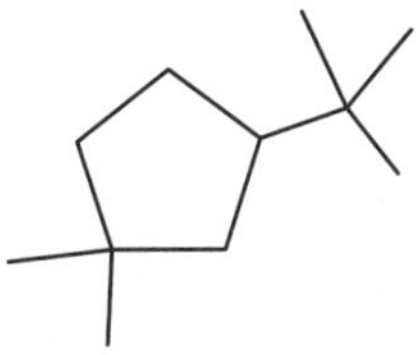

3-(*tert*-butyl)-1,1-dimethylcyclopentane — of note: second point of difference decision: 1,1- versus 1,3-

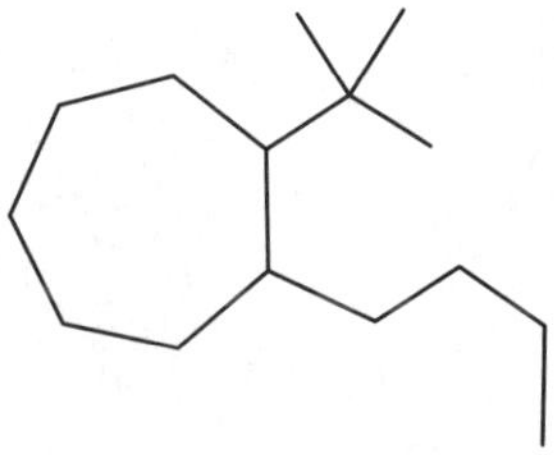

1-(*tert*-butyl)-2-butylcycloheptane — of note: solitary *tert*-butyl in parentheses, alphabetical priority *sec*-butyl > *tert*-butyl > butyl

There is one other ring that you should learn at this stage: benzene. This ring is treated in much the same way as saturated rings: It counts as a 6-carbon ring; the root name for the ring is "benzene;" and when it is used as a substituent group, It is named "phenyl." A few examples of compounds containing benzene rings are shown in Figure AP0215.

The monosubstituted benzene ring (phenyl group) is such a recurring feature in organic compounds that it has a commonly used abbreviation, namely, "Ph." This can be used regardless of whether the benzene ring is the root name or it is being used as a substituent group.

Figure AP0215

Naming hydrocarbons containing benzene rings.

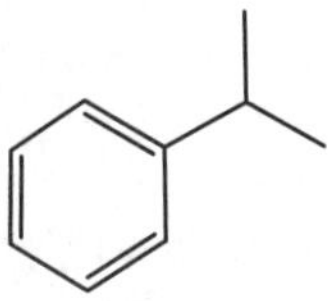

isopropylbenzene (1-isopropylbenzene)

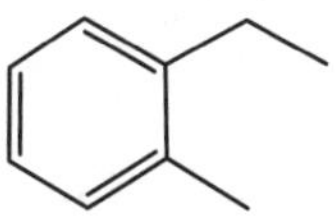

1-ethyl-2-methylbenzene

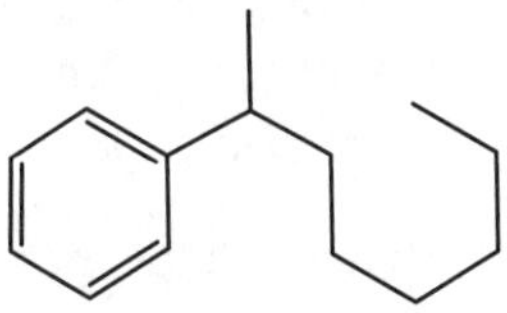

2-phenyloctane

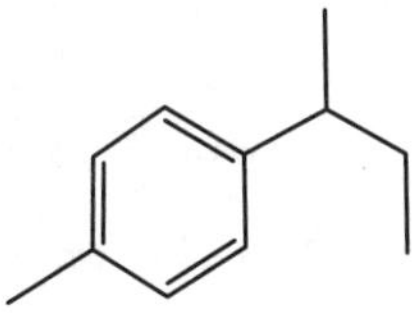

1-(*sec*-butyl)-4-methylbenzene

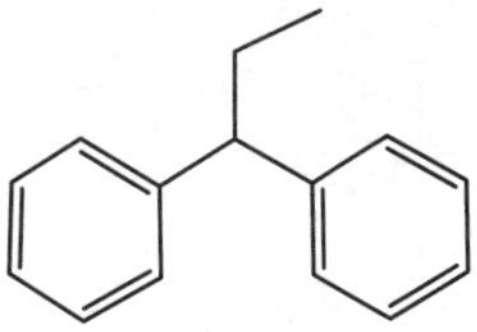

1,1-diphenylpropane

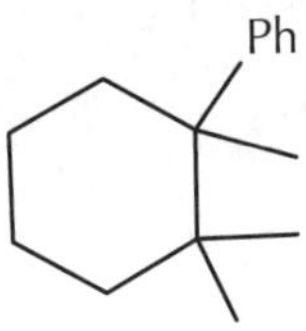

1,1,2-trimethyl-2-phenylcyclohexane

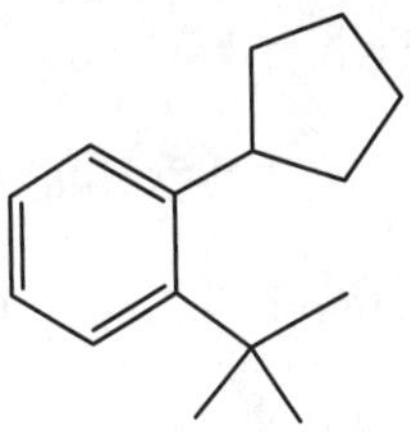

1-(*tert*-butyl)-2-cyclopentylbenzene

B. Alkyl and Aryl Halides

Halogen atoms (abbreviated as X) are treated as prefix-designated substituent groups, using the terms fluoro, chloro, bromo, and iodo in exactly the same way you use methyl, ethyl, propyl, and so on. There is no preference or priority for halogens versus alkyl groups in numbering, and so the same point of difference rule applies, with numbering ties being broken by the alphabetical order tie-breaker rule. Examples of naming alkyl halides (hydrocarbons with halogen atom substituents, RX), along with some examples that include benzene rings with attached halogen atom groups, generically called aryl halides (ArX), are shown in Figure AP0216.

Figure AP0216
Naming alkyl and aryl halides.

1,1-dichlorocyclohexane   5-bromo-2,5-dimethyloctane   1-iodo-3-methylbenzene   1-chloro-2-fluorocyclopentane

1,15,15-tribromo-8-isopropyl-13-phenylheptadecane

1,2-dibromo-4-chloro-1-methylcyclobutane
of note: no point of difference on bonding (1,1,2,4- versus 1,1,2,4-) and the third group gives the alphabetical tie-breaker (1,1,2-bromo versus 1,1,2-chloro)

C. Alcohols

Hydroxyl groups (OH) attached to alkyl or aryl groups give the class of molecules called alcohols. Unlike hydrocarbon groups and halogen atom groups, alcohols are usually named by changing the "e" ending of the name of the alkane to "ol" and using a number to indicate the position of the hydroxyl group. The preference is to place the number directly with the group suffix ($CH_3CH_2CH_2CH_2OH$ is named butan-1-ol), but a synonym can be created with group suffixes, as the number for the first numbered suffix can appear in front of the root chain name (1-butanol).

An alcohol owes its characteristic properties and reactivity to the hydroxyl group, so in naming an alcohol, the carbon chain is numbered so that the carbon atom bearing the hydroxyl group has the lowest possible number (Figure AP0217). When there is more than one hydroxyl group, normal first point of difference is used for the numbering, the combining terms are used (di, tri, tetra), and the "e" is brought back into the root name when the combining term begins with a consonant.

The unique name for the compound with a hydroxyl group on a benzene ring is phenol.

D. Priorities When Numbering the Root Name

So far, the decision about numbering the root chain has only been based upon the first point of difference rule, followed by the alphabetical order tie-breaker when point of difference creates a tie.

Figure AP0217
Naming alcohols.

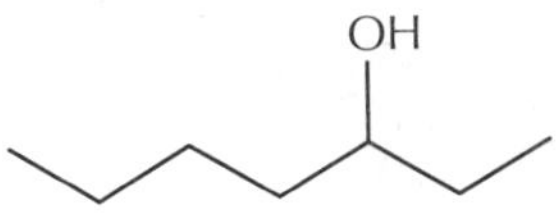

- heptan-3-ol (3-heptanol)
- cyclohexane-1,2-diol (1,2-cyclohexanediol) 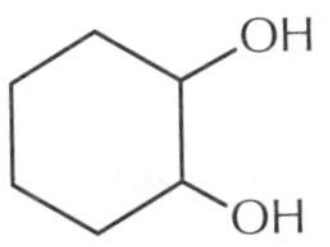
- 4-chloro-3-methylphenol 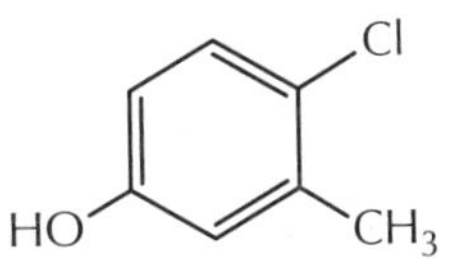
- 2,2,3-trimethylbutan-1-ol (2,2,3-trimethyl-1-butanol) 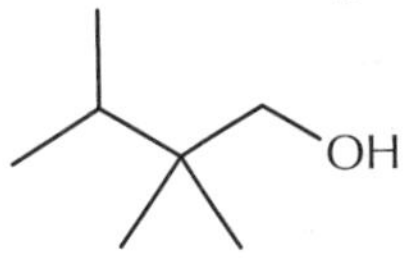

Although no prioritization for numbering exists for the prefix-designated substituent groups (methyl-, ethyl-, propyl-, phenyl-, chloro-, bromo-, etc., all use first point of difference), the IUPAC rules give suffix-designated substituent groups (so far, -ol) a higher priority for numbering than prefix-designated substituent groups.

When a molecule has both prefix- and suffix-designated groups, then the first attempt at numbering is based upon giving the suffix-designated group(s) the lowest point of difference and ignoring the prefix-designated groups completely in the numbering process.

If a decision based on the suffix-designated group can be made, then that fixes the numbering of the root chain. In the event of a numbering tie based upon the suffix group, then the previously used point-of-difference process for the prefix groups is used to break the suffix tie. And if that fails, then the alphabetical order tie-breaker is used (Figure 0218).

Figure AP0218
Naming compounds that need the substituent priority ordering for numbering.

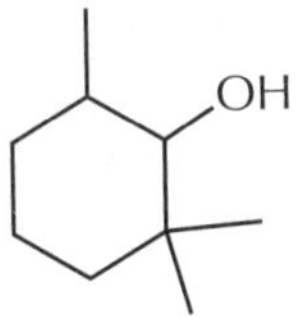

2,2,6-trimethylcyclohexan-1-ol
(2,2,6-trimethylcyclohexanol)
(2,2,6-trimethyl-1-cyclohexanol)

of note: the OH group defines carbon-1, and point of difference dictates how to number the rest of the ring (1,2,2...) versus (1,2,6...)

1-cyclohexyl-4-phenylbutan-2-ol
(1-cyclohexyl-4-phenyl-2-butanol)

of note: the OH group defines carbon-2 on the butane root chain containing 3 substituents

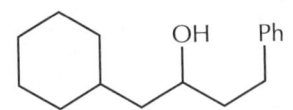

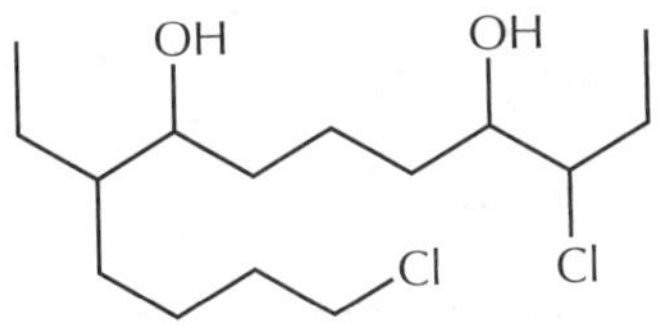

3,13-dichloro-9-ethyltridecane-4,8-diol
(3,13-dichloro-9-ethyl-4,8-tridecanediol)

of note: the two OH groups defines carbons 4 and 8 of the 13-carbon root chain

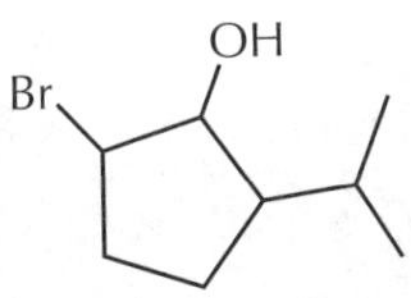

2-bromo-5-isopropylcyclopentan-1-ol
(2-bromo-5-isopropyl-1-cyclopentanol)

of note: the OH group defines carbon-1, and point of difference creates a (1,2,5-) versus (1,2,5-) tie, so the alphabetal order of bromo > isopropyl defines the numbering

In summary, there is a priority order for deciding which end of the root chain to start from for numbering: suffix groups (using point of difference) > prefix groups (using point of difference) > alphabetical order.

This is not the final set of priority rules that you will need. Looking ahead: (1) a molecule can only have one suffix-designated substituent group, so that situation will need to be resolved; and (2) some groups appear in the root chain itself (double bonds, triple bonds, heteroatoms), and so there are chain-designated groups, also. The rules for these situations appear in other appendixes.

PRACTICE QUESTIONS

AP02.05 Provide the IUPAC names for the following compounds.

(a) [cyclopentane with OH and CH₂CH₃ on same carbon]

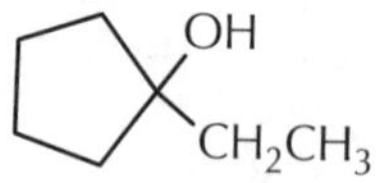

(b) [hexane chain with Br and methyl on C3, OH on C2]

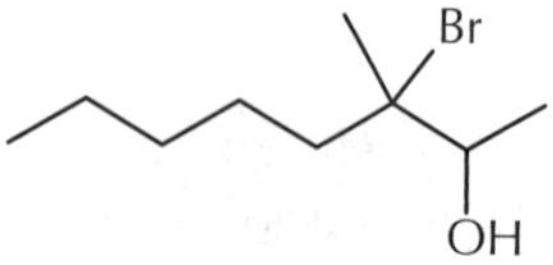

(c) Ph—[cyclohexane]—C(CH₃)₃

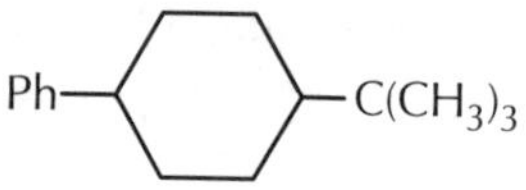

(d) Cl—CH₂CH₂CH₂—Br

(e) [pentane with I on C3, OH on C4, ethyl on C3 and C5... structure with I, OH, and ethyl branches]

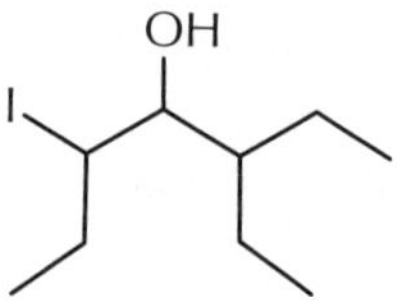

AP02.06 Draw the following compounds from their names.

(a) 2-methylhexan-2-ol

(b) 7-chloro-2,7-dimethyl-2-octanol

(c) isobutylcyclopentane

(d) 3-phenylheptane

(e) 4-(*tert*-butyl)tridecane

(f) 1,1-diiodocyclopropane

A2.2 Cyclic Compounds, Halides, and Alcohols

AP02.07 Provide the IUPAC names for the following compounds.

(a)

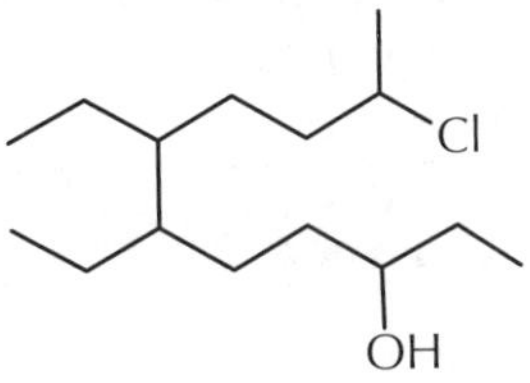

(b) (Br)₃CCH(CH₃)₂

(c)

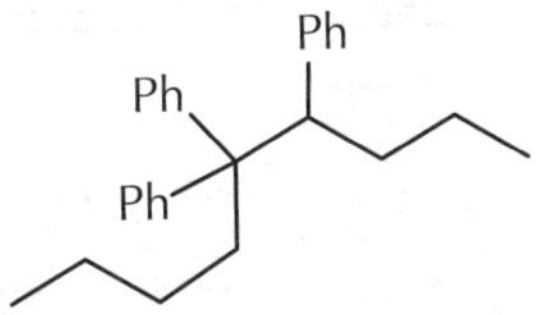

(d)

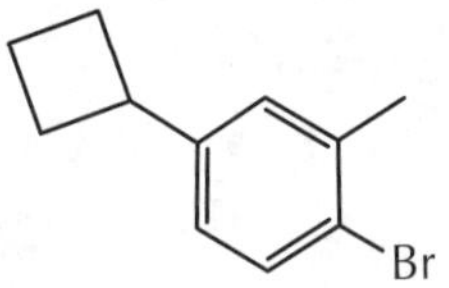

(e)

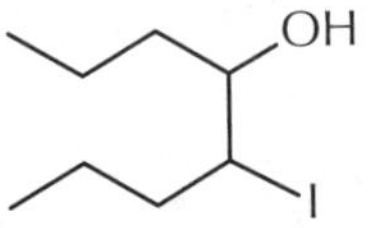

AP02.08 Draw the following compounds from their names.

(a) 1-chloro-4-ethyl-1,4-dimethylcyclohexane

(b) 2,3,5-tribromo-4-phenylhexane

(c) 2-cyclopentyl-1,4-dimethylcyclopentan-1-ol

(d) 2-phenylcyclooctane-1,2,4,6-tetraol

(e) 3,3-difluoro-4,4-diisopropylheptane

(f) 5-bromo-4-(sec-butyl)-2-(tert-butyl)phenol

AP02.09 Some of the following names are correct, and some are incorrect. Even for the incorrect ones, the intended structure can be deduced from the information given. In each case: (a) draw the molecule implied by the name; (b) decide if the name is correct or not; and (c), if the name is incorrect, correct it.

(a) 1-chloro-4-fluorobutane

(b) 1,1,1-triethyl-3-butaneol

(c) 1-methyl-1-ethyl-2-cyclobutanol

(d) 1-ethyl-3-propylbenzene

(e) 3-methyl-1-ol-cyclohexane

(f) 2-phenyl-3-pentanol

(g) 6-methylheptan-3-ol

(h) 1-methylcyclohexane-1,2,3-triol

A2.2 Cyclic Compounds, Halides, and Alcohols

AP02.10 Organic nomenclature is an expansive topic, and the introduction is always automatically limited and excludes many situations. For instance: complex branching can be handled in a few cases with common names (isopropyl, isobutyl, *tert*-butyl, *sec*-butyl), but there are many others. Two instances that you might want to be able to name are shown below along with a brief explanation of the naming system.

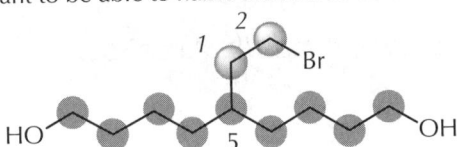

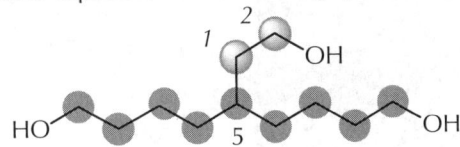

- longest chain with greatest number of substituents and suffix-designated groups root name = 1,9-nonanediol with a third group at position 5

- a 2-carbon attached group at position 5 on the root chain is an ethyl, and the ethyl is numbered from its point of attachment to the root chain; so the bromo group is at position 2 on the ethyl, and that entire substituent is called a "2-bromoethyl" group

The 2-bromoethyl group is attached to position 5 on the root chain, so the name is: 5-(2-bromoethyl)-1,9-nonanediol

- longest chain with greatest number of substituents and suffix-designated groups root name = 1,9-nonanediol with a third group at position 5

- a 2-carbon attached group at position 5 on the root chain is an ethyl, and the ethyl is numbered from its point of attachment to the root chain; so the hydroxyl group is named using the prefix designation "hydroxy" and the substituent is a "2-hydroxyethyl" group

The 2-hydroxyethyl group is attached to position 5 on the root chain, so the name is: 5-(2-hydroxyethyl)-1,9-nonanediol

Given this lesson, name the following:

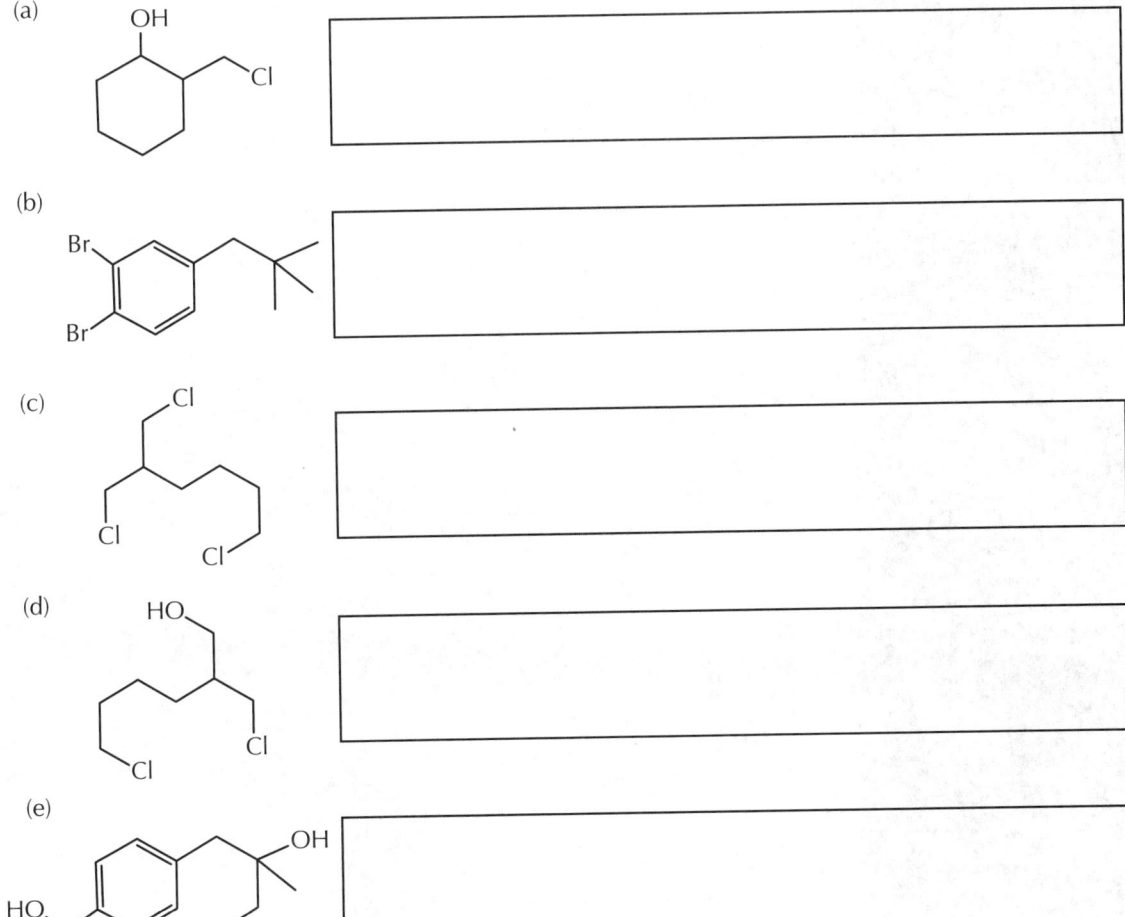

APPENDIX 3

Nomenclature of Organic Compounds II: Stereoisomers

Introduction

A3.1 Prioritizing Substituent Groups
    A. Cahn-Ingold-Prelog (CIP) Rules
    B. Point of Difference Comparisons
    C. Nonbonding Electron Pairs
    D. Duplicate and Phantom Atoms for Multiple Bonds

PRACTICE QUESTIONS

A3.2 Assigning Geometrical Labels
    A. Cis/Trans: Relative Stereochemistry
    B. Cis/Trans: Geometrical Label
    C. E/Z and R/S Geometrical Labels
    D. r/s and R_a/S_a Geometrical Labels

PRACTICE QUESTIONS

APPENDIX 3

Nomenclature of Organic Compounds II: Stereoisomers

Introduction

Stereoisomers are molecules with the same connectivity but which differ according to a fixed three dimensional geometry. The background and details about stereoisomerism are presented in Chapter 5. The usual rules for organic nomenclature only cover connectivity, so a set of supplemental rules is used to assign a geometrical label to a specific three-dimensional arrangement, which in turn becomes part of the name.

The naming rules for assigning a geometrical label usually relies on defining a priority relationship between a set of 2–4 substituents (either lone atoms or groups). The method for doing this prioritization was developed in the 1950s and 1960s, when determining the absolute spatial geometry of organic molecules was becoming commonplace and labels were needed. The naming rules were refined again in the 1980s before being adopted by IUPAC.

The prioritization rules were developed by Robert Sidney Cahn, Sir Christopher Ingold, and Vladimir Prelog, and they are called the Cahn-Ingold-Prelog (CIP) Rules. The rules are not based upon chemical principles, but rather on treating molecules as topological objects. Topology is a field of mathematics that concerns itself with the fixed geometrical relationships that exist in an object regardless of how it folds, bends, or twists.

> "Although the applications are chemical, the basic concepts are topological. It is, of course, because they are topological that they can be defined sharply, and also permanently, despite the ever-changing face of chemistry."
>
> RS Cahn, CK Ingold, V Prelog "Specification of Molecular Chirality" *Angewandte Chemie International Edition in English*, **1966**, *5*(4), 385.

> "We feel that, particularly for educational reasons, economy in coining new terms should be the order of the day."
>
> V Prelog, G Helmchen "Basic Principles of the CIP-System and Proposals for a Revision" *Angewandte Chemie International Edition in English*, **1982**, *21*(8), 567.

APPENDIX 3 Nomenclature of Organic Compounds II: Stereoisomers

A3.1 Prioritizing Substituent Groups

A. Cahn-Ingold-Prelog (CIP) Rules

The only goal in applying the CIP rules is to unambiguously rank atoms or groups of atoms that are connected to a source of molecular geometry.

The three-step procedure for generating a geometrical label is to (1) decide which groups need to be ranked because there is a source of stereoisomers (this procedure is covered in depth in Chapter 5, Book A); (2) using the CIP rules to rank the groups in question; (3) assign the geometrical label to the three-dimensional shape.

The basic CIP rules for step 2, which is the topic here, are simple:

(A) As in naming, point of difference is used; never (ever) add numbers.
(B) Ranking is based on atomic number.
(C) Ties in atomic number (isotopes) are broken by atomic weight.
(D) As a default, main group elements except hydrogen are treated as tetravalent.

A situation where ranking is needed is when there are four different atoms and/or groups attached to a tetrahedral, tetravalent atom. The example in Figure AP0301 is a basic illustration of assigning priorities to four atoms according to these rules. Generating the geometrical label is a completely separate exercise from assigning the priorities, so only the latter is of concern, here. The carbon atom is attached to a hydrogen atom, a chlorine atom, a bromine atom, and a fluorine atom. According to their atomic numbers, the CIP priority ranking is Br > Cl > F > H.

Figure AP0301

Assigning CIP priorities: four different atoms.

CIP ranking:

$Br_{35} > Cl_{17} > F_9 > H_1$

B. Point of Difference Comparisons

The most common situation in organic molecules is for at least two of the atoms directly attached to the center of interest to be the same atom, namely, carbon. Figure AP0302 illustrates how the first point of difference strategy works. The carbon of interest is attached to a hydrogen atom, a hydroxyl group, a methyl group and an ethyl group. The first analysis begins with comparing the four directly attached atoms (hydrogen, oxygen, carbon, and carbon). According to their atomic numbers, an incomplete ranking can be done: O > C = C > H, so the highest priority group is the hydroxyl group, and the lowest is the hydrogen atom. The procedure to break the tie and rank the methyl group versus the ethyl group also uses the first point of difference strategy.

To break the tie between the two carbon atoms (methyl versus ethyl), the sets of three atoms attached to both of these carbons are compared. Moving out from the original carbon of interest, the carbon of the

Figure AP0302

Assigning CIP priorities: methyl versus ethyl.

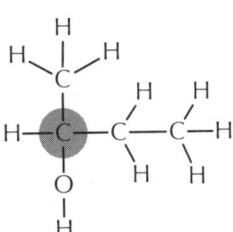

groups to prioritize:
"CH$_3$"
"CH$_2$CH$_3$"
"OH"
"H"

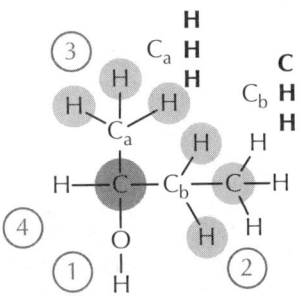

first level: O > C = C > H
conclude:
"OH" is priority 1
"H" is priority 4

break the tie: C$_a$ versus C$_b$
C$_a$: H = H = H; highest = H } first point of
C$_b$: C > H = H; highest = C } difference
second level: C (at C$_b$) > H (at C$_a$)
conclude:
"CH$_2$CH$_3$" is priority 2
"CH$_3$" is priority 3

methyl group is attached to three hydrogen atoms, while the comparable carbon of the ethyl group is attached to two hydrogen atoms and a carbon atom. Each of these subsets of three are ranked, independently, according to atomic number. Thus, the three hydrogens of the methyl group are a three-way tie (H = H = H), while the second carbon of the ethyl group has a higher priority than the two hydrogens (C > H = H).

Between these two subsets of three atoms, the atoms of highest priority are compared. The carbon atom from the ethyl group (the highest priority from that subset of three) has a higher atomic number than any hydrogen atom from the methyl group, which breaks the original tie between the two carbon atoms in the initial analysis. The ethyl group has a higher priority than the methyl group because of the single carbon versus hydrogen comparison in the second level of the analysis.

Note that atomic numbers are never added together in these analyses, and the comparisons are always one atom to one atom at a time.

The methyl versus ethyl case sets up the fundamental strategy for how the point of difference rule is applied. After the groups that need to be prioritized are identified, the first set of atoms is compared. If a full ranking can be done at that level (Figure AP0301), then you are done. If only a partial ranking can be done (Figure AP0302), the ties are broken by moving out to the next set of attached atoms and (a) ranking within each subset and then (b) comparing the highest-ranking atoms within each subset on a one by one basis until the tie is broken.

As long as the groups are different, they can be prioritized unambiguously through a systematic application of this strategy.

As you look at the four groups in the example in Figure AP0303 (on next page), you might be tempted to favor the substituent group containing the three bromine atoms as the automatically highest priority group. This conclusion is incorrect. The strategy is to move out systematically from the carbon atom of interest.

APPENDIX 3 Nomenclature of Organic Compounds II: Stereoisomers

Figure AP0303

Assigning CIP priorities: an extended example.

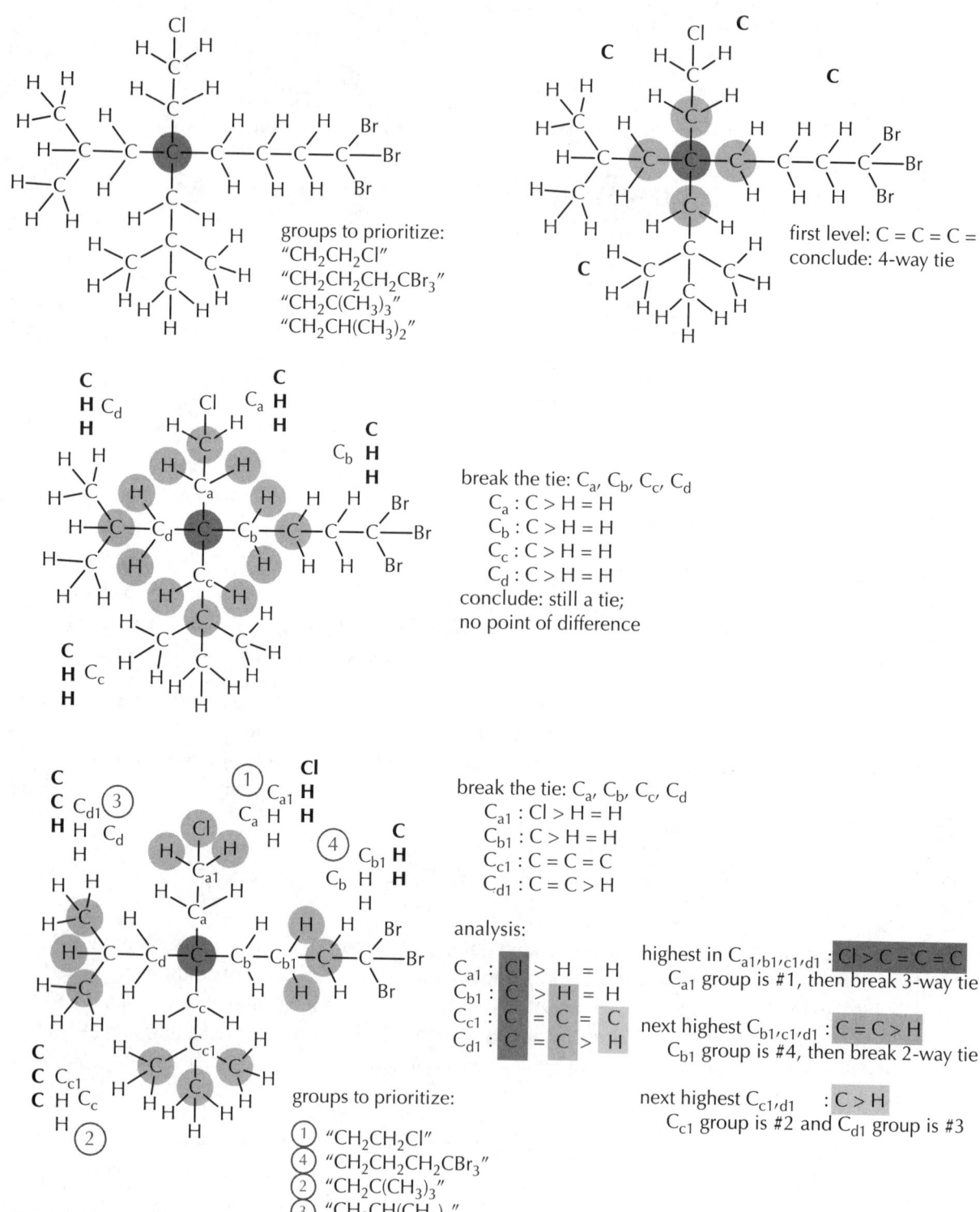

In Figure AP0303, the four directly attached atoms are all carbons, so after the first analysis there is a four-way tie (C = C = C = C). The four subsets of three next atoms attached to each of these carbons first need to be ranked, internally, and then the highest ranked atom from each subset is compared. The subsets, with their rankings, are (C > H > H), (C = C > H), (C = C = C), and (Cl > H > H). By comparing the highest ranked atom in each of the subsets, only the decision about the highest ranking of the original substituents can be decided (Cl > C = C = C). That decision reduces the analysis to the remaining three subsets. Now, the second highest ranked atom from the remaining three subsets gives (C = C > H), which means that the lowest ranking for the original subsets is decided. Finally, the lowest ranked atom from the remaining two subsets gives (C > H), and all four of the original substituents are now ranked. Notice that the substituent with the bromine atoms is actually in the lowest ranked subset.

C. Nonbonding Electron Pairs

All main group elements are treated as tetravalent, and so that means including nonbonding electron pairs in the ranking scheme. Nonbonding electron pairs, with no atom sitting on them, are given an atomic number of zero.

In Figure AP0304, the first analysis at the carbon of interest compares two oxygen atoms, a carbon and a hydrogen atom. The oxygen atoms are tied, but the third (C) and fourth (H) priorities can be assigned (O = O > C > H). Comparing the subsets on each of the oxygen atoms is required to break the tie. On the hydroxyl group (OH), the ranked subset (H > nbe = nbe) identifies hydrogen at the highest priority in that group, and on the methoxy group (OCH$_3$), the ranked subset (C > nbe = nbe) places the carbon as the highest priority. The one-to-one comparison of those two highest ranking items from each subset (C > H) results in assigning the methoxy group with the highest priority and the hydroxyl group with the second highest, then followed by the methyl group and the lone hydrogen atom.

Figure AP0304
Assigning CIP priorities: nonbonding electron pairs.

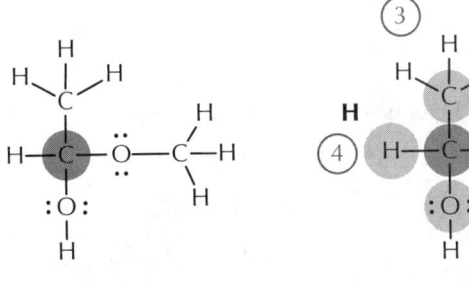
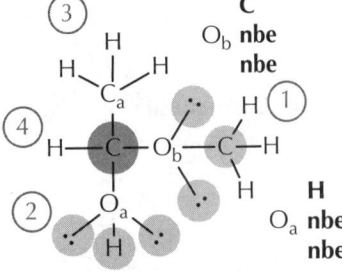

groups to prioritize:
"CH$_3$"
"OCH$_3$"
"OH"
"H"

first level: O = O > C > H
conclude:
"CH$_3$" is priority 3
"H" is priority 4

break the tie: O$_a$ versus O$_b$
O$_a$: H > nbe = nbe; highest = H ⎤ first point of
O$_b$: C > nbe = nbe; highest = C ⎦ difference
second level: C (at O$_b$) > H (at O$_a$)
conclude:
"OCH$_3$" is priority 1
"OH" is priority 2

D. Duplicate and Phantom Atoms for Multiple Bonds

The CIP rules for prioritizing double and triple bonds relies on two ideas, duplicate atoms and phantom atoms, which allow all atoms to be treated as tetravalent for the purpose of making comparisons.

Figure AP0305 (on the next page) lays out the procedure for some double and triple bonds.

Figure AP0305

Assigning CIP priorities: duplicate and phantom atoms used with double and triple bonds.

Duplicate atoms replicate the literal number of bonds, laying them out as singly bonded atoms. Duplicate atoms have their usual atomic number, but they are attached to three phantom atoms, all of which have an atomic number of zero. Sometimes you will see the duplicate atom written alone, with no superscripts, and sometimes you see the "000" superscript, which gives a visual reminder that these atoms are considered tetravalent.

In Figure AP0306, four substituent groups are attached to a carbon atom of interest: a 2-carbon group with a double bond (commonly called "vinyl" or, in IUPAC terms, "ethenyl"), an isopropyl group, a bromine atom, and a hydrogen atom. Ranking the four directly attached atoms to the carbon of interest allows for the highest ranking (Br) and the lowest (H), with a tie for the two intermediate positions (C = C). To break the tie, the next subsets of atoms need to be identified, internally ranked, and compared. The vinyl group is first expanded with its duplicate atoms, revealing that the isopropyl group and the vinyl group both have the same subsets, resulting in a tie: two carbons and a hydrogen (C = C > H), which means that the next subsets have to be inspected to break the tie.

Figure AP0306

Assigning CIP priorities: duplicate and phantom atoms used with a vinyl group versus an isopropyl group.

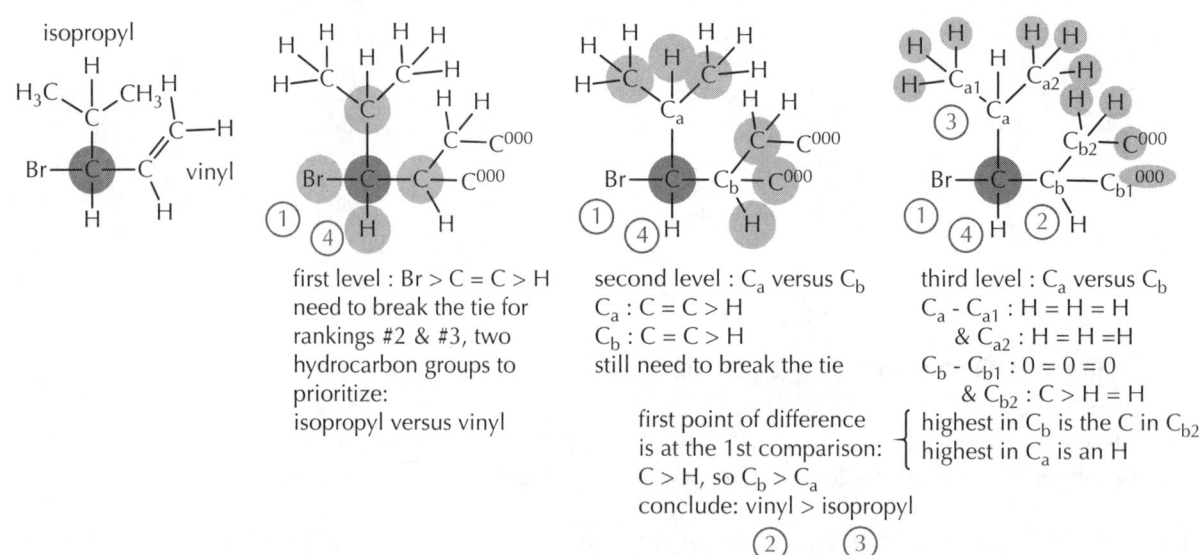

Both of the carbon atoms at the next level from inspecting the isopropyl group have three hydrogen atoms each. The two carbons (one real, one duplicate) at the same level in the vinyl group have subsets comprising three phantom atoms (on the duplicate carbon) and two hydrogen atoms and the second duplicate carbon atom (on the real carbon).

The two subsets on the isopropyl carbons (C_{a1} and C_{a2}) are the same, each with a three-way tie (H = H = H) and (H = H = H). The two subsets from the vinyl group are different. One of them (C_{b1}) has only phantom atoms (0 = 0 = 0) and the other (C_{b2}) has three atoms, one of which is the second duplicate carbon (C > H = H).

Thus, in the C_b substituent, the (C_{b2}) subset outranks the (C_{b1}) subset (C > 0), and is the one that first gets compared with the highest ranked subset from the C_a substituent (where, in fact, there is tie between C_{a1} and C_{a2}, so it does not matter which one is used first).

Notice how, for each decision, first point of difference is used to prioritize all choices: within C_{a1}, C_{a2}, C_{b1}, and C_{b2}; between C_{a1} and C_{a2}; between C_{b1} and C_{b2}. All of which arrives at one unambiguous answer: The C in C_{b2} outranks the H in either C_{a1} (or C_{a2}), breaks the tie, and gives the C_b substituent (vinyl) a higher priority than the C_a substituent, (isopropyl).

An instructive comparison exists if a vinyl group is now compared with a *sec*-butyl group (Figure AP0307). The expansion of the vinyl group is the same as in Figure AP0306. The directly attached atom for the *sec*-butyl group is the same as the isopropyl group: a carbon atom. And as in the isopropyl group, the *sec*-butyl, at the second level, has two carbon atoms and a hydrogen atom. At the third level out, the isopropyl group has two groups of three hydrogen atoms, while the *sec*-butyl group has one carbon with three hydrogen atoms and one carbon with a carbon atom and two hydrogen atoms.

Figure AP0307

Assigning CIP priorities: duplicate and phantom atoms used with a vinyl group versus a *sec*-butyl group.

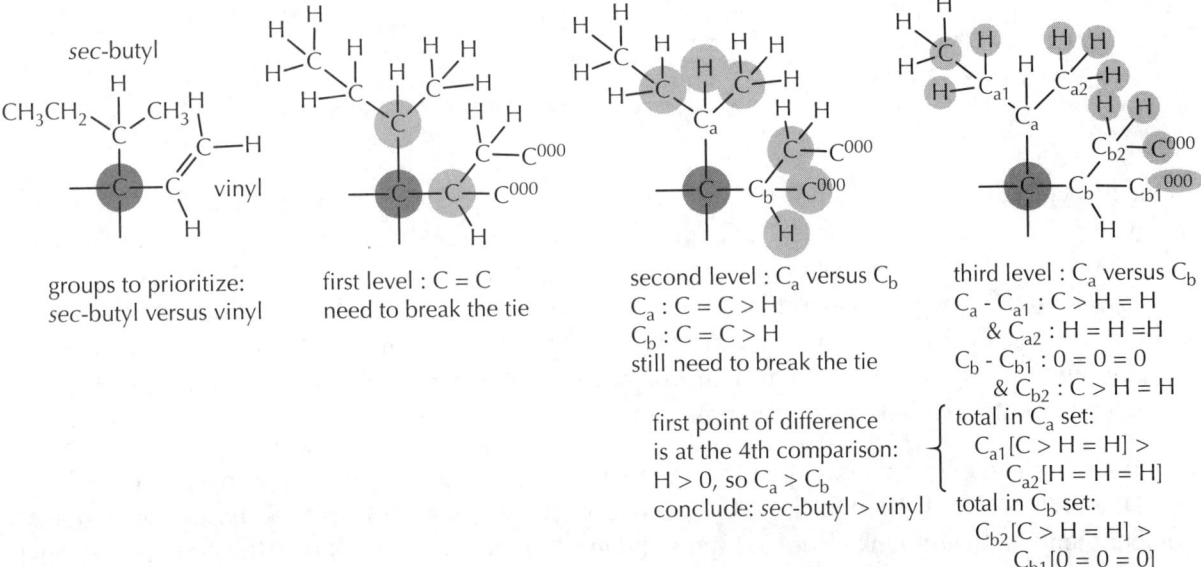

The highest ranked atom in the vinyl group at this level is the duplicate carbon in the (C > H = H) subset, followed by the three phantom atoms (0 = 0 = 0). On the *sec*-butyl group, the highest ranked atom is the carbon in its (C > H = H) subset, followed by the methyl group, with its three hydrogen atoms (H = H = H). The two subsets containing carbon atoms need to be compared first, resulting in a tie. But by going to the second subset, the tie is broken, as one of the hydrogen atoms (atomic number = 1) is ranked higher than a phantom atom (atomic number = 0). This gives the *sec*-butyl group a higher priority than the vinyl group.

Duplicate and phantom atoms are used in another common situation: cyclic versus acyclic substituent groups. In Figure AP0308, a cyclopropyl group is being compared with a carbon that has two ethyl groups on it.

Figure AP0308

Assigning CIP priorities: duplicate and phantom atoms used with a cyclopropyl group (comparison with two ethyl groups).

first level : C = C
need to break the tie

second level : C_a versus C_b
C_a : C = C = C
C_b : C = C = C
still need to break the tie

third level : C_a versus C_b
C_a - C_{a1} : C > H = H
& C_{a2} : C > H =H
C_b - C_{b1} : C > H =H
& C_{b2} : C > H =H
still need to break the tie

first point of difference is at the 4th comparison:
C > H, so C_b > C_a
conclude:
cyclopropyl is higher priority

fourth level : C_a versus C_b
C_a - C_{a3} : H = H = H
& C_{a4} : H = H =H
C_b - C_{b3} : C > H =H
& C_{b4} : C > H =H
total in C_a set:
[C > H = H] > [H = H = H]
total in C_b set:
[C > H = H] = [C > H = H]

At the first level there is a tie (carbon versus carbon). At the second level, there is also a tie: [(C > H = H) and (C > H = H)] are the two subsets for both of the substituents. Moving out to the third level, each of the carbon atoms from the ethyl groups have three hydrogen atoms each [(H = H = H) and (H = H = H)]. In the cyclopropyl group, the carbons at the third level have circled around and are attached to the first level carbon. The CIP rules are not set up so that you keep circling around these rings. Instead, when you circle around and "meet yourself" at an atom you have accounted for before, you split open the ring and use a duplicate atom with its three phantom atoms.

As a result of the duplicate atom expansion, the fourth level subsets for the cyclopropyl group are [(C > H = H) and (C > H = H)], and the highest ranked atom in one of these subsets (C) has a higher atomic number than the highest ranked atom at the same level in the ethyl groups (H). Thus, the cyclopropyl group has a higher priority than the group with the two ethyl groups.

As in the earlier example (Figure AP0306), the situation changes when the cyclopropyl groups is compared with a group that has two propyl groups instead of two ethyl groups (Figure AP0309).

At the third level, the tie is not broken. The two subsets from the cyclopropyl group [(C > H = H) and (C > H = H)] are exactly the same as from the two propyl groups [(C > H = H) and (C > H = H)]. In this case,

Figure AP0309

Assigning CIP priorities: duplicate and phantom atoms used with a cyclopropyl group (comparison with two propyl groups).

first level : C = C
need to break the tie

second level : C_a versus C_b
C_a : C = C = C
C_b : C = C = C
still need to break the tie

third level : C_a versus C_b
C_a - C_{a1} : C > H = H
 & C_{a2} : C > H = H
C_b - C_{b1} : C > H = H
 & C_{b2} : C > H = H
still need to break the tie

fourth level : C_a versus C_b
C_a - C_{a3} : C > H = H
 & C_{a4} : C > H = H
C_b - C_{b3} : C > H = H
 & C_{b4} : C > H = H
still need to break the tie

first point of difference is at the 1st comparison: H > 0, so C_a > C_b
conclude:
cyclopropyl is lower priority

fifth level : C_a versus C_b
C_a - C_{a5} : H = H = H
 & C_{a6} : H = H = H
C_b - C_{b5} : 0 = 0 = 0
 & C_{b6} : 0 = 0 = 0
total in C_a set:
 [H = H = H] = [H = H = H]
total in C_b set:
 [0 = 0 = 0] = [0 = 0 = 0]

the tie gets broken at the fifth level of comparison. The two carbons from the third level in the cyclopropyl group are duplicates, and they are both connected to three phantom atoms, giving a fifth level analysis for the cyclopropyl group of [(0 = 0 = 0) and (0 = 0 = 0)]. The two propyl groups, on the other hand, have two subsets of three hydrogen atoms each at the fifth level [(H = H = H) and (H = H = H)]. Hydrogen atoms outrank the phantom atoms, and so the substituent group with the two propyls has a higher priority than the cyclopropyl group.

In Chapter 5, the two other significant questions for understanding stereoisomerism that surround the CIP prioritization rules are answered in detail. First, how to identify an atom of interest as a source of stereoisomerism; and second, how to assign the geometrical labels. The second section of this appendix contains a summary of how geometrical labels are assigned, including the topics in Chapter 5 as well as a few closely related cases, which will illustrate some additional sources of stereoisomerism beyond the three cases in that chapter and how assigning their geometries follows quite directly from those three core cases.

APPENDIX 3 Nomenclature of Organic Compounds II: Stereoisomers

PRACTICE QUESTIONS

AP03.01 Use the CIP rules to prioritize the four substituents on the shaded carbon.

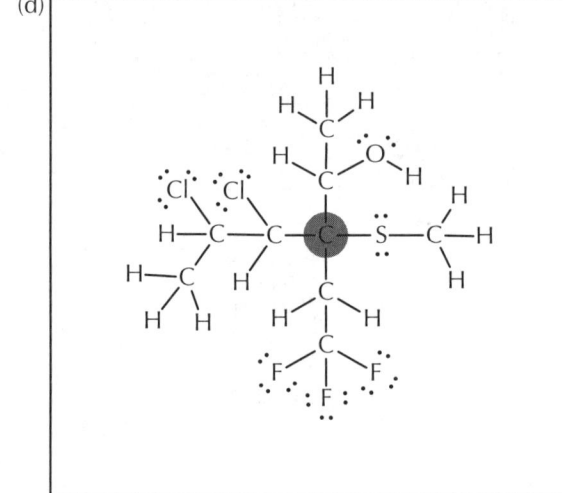

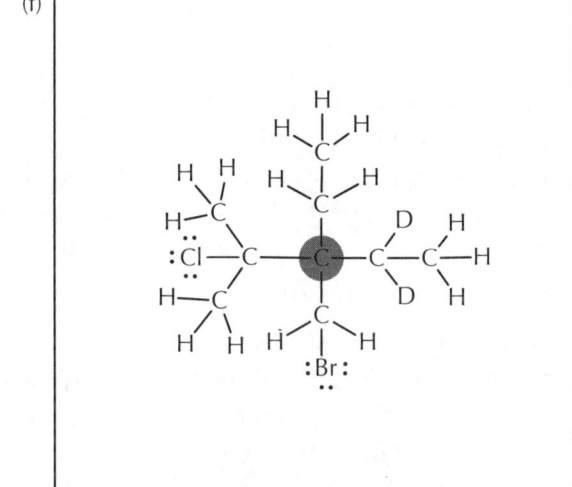

AP03.02 Use the CIP rules to prioritize the four substituents on the shaded carbon.

(a)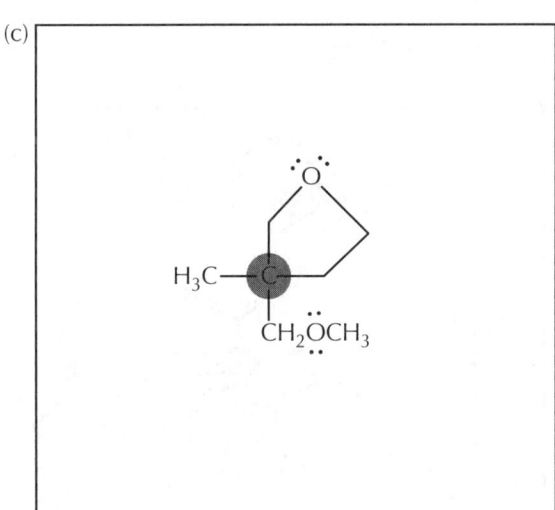

(b)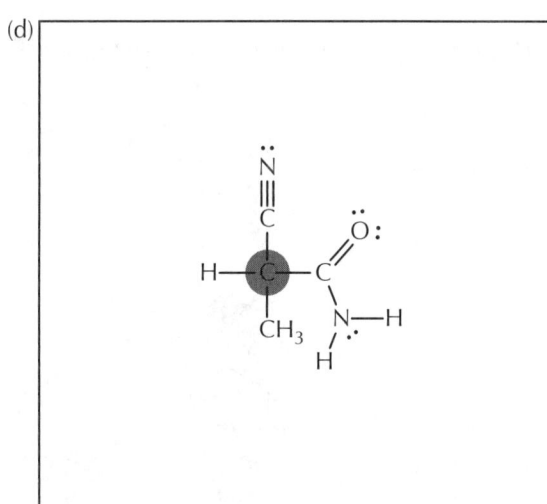

(c)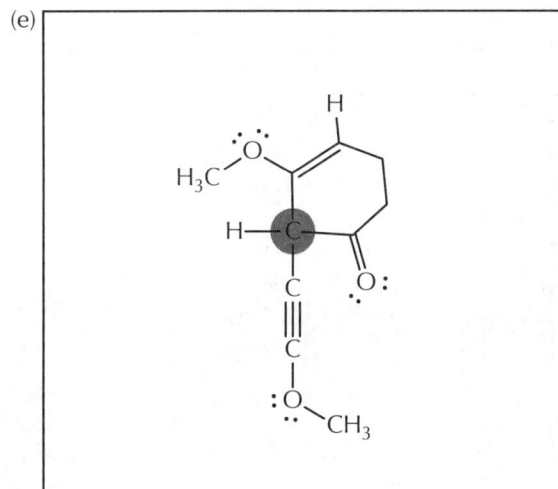

(d)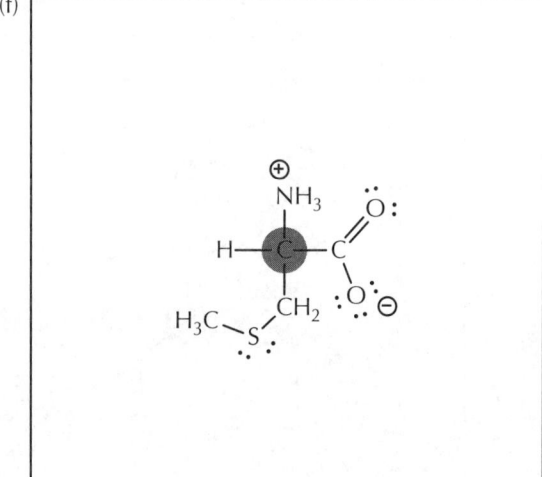

(e)

(f)

AP03.03 The following example highlights an important aspect of how the first point of difference rule works. How are these four isomeric substituents ranked? Be clear on the rationale.

AP03.04 Use the CIP rules to prioritize the four substituents on the shaded carbon.

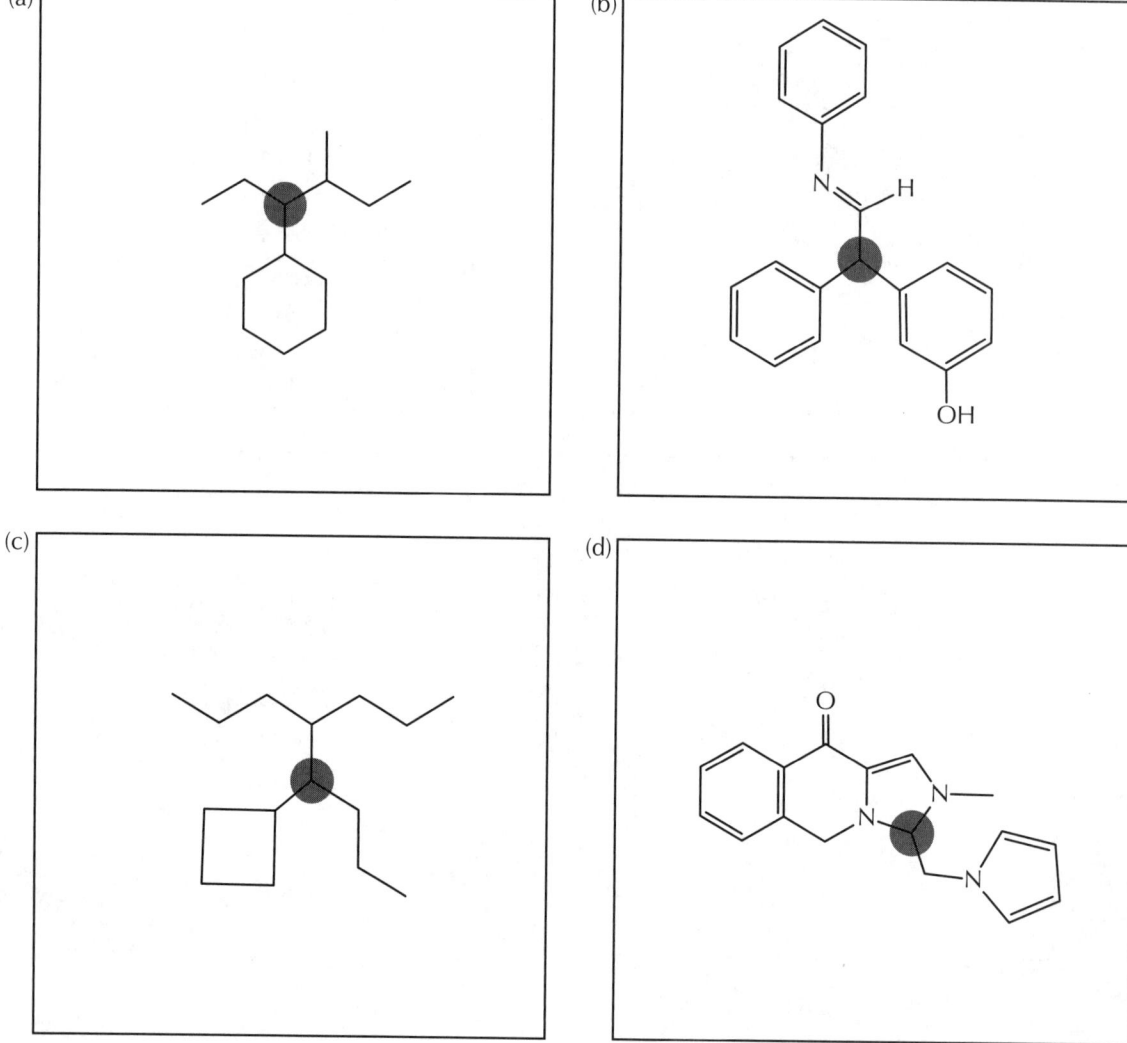

AP03.05 Use the CIP rules to prioritize the four substituents on the shaded carbon(s) in the following pharmaceuticals.

(a) warfarin (an anticoagulant)

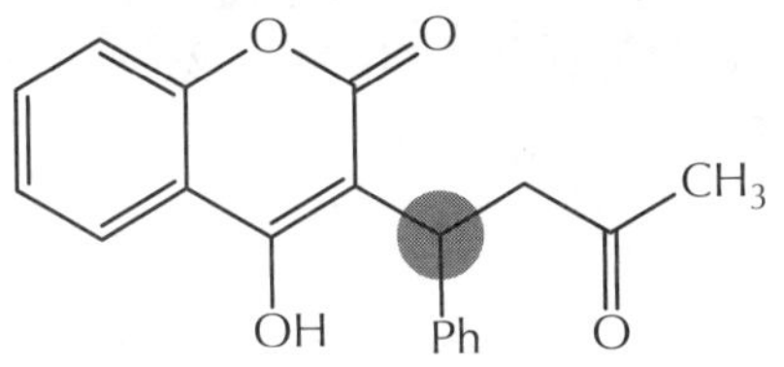

(b) ibuprofen (a nonsteroidal anti-inflammatory)

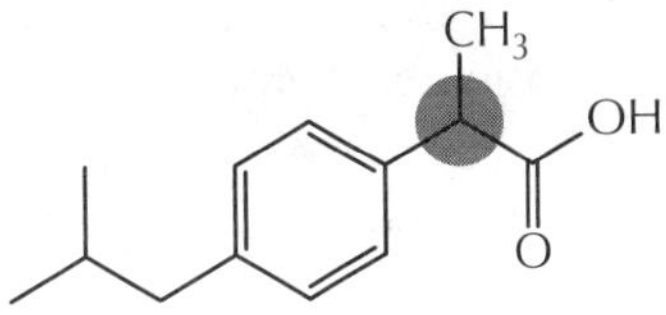

(c) pseudoephedrine (a decongestant)

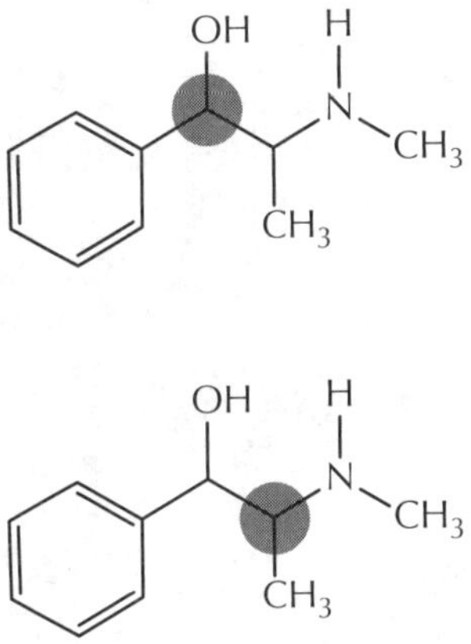

(d) zidovudine (AZT, an antiviral)

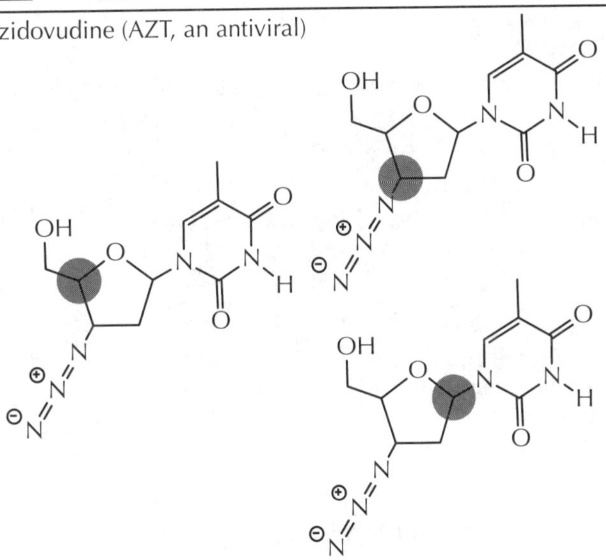

(e) estrogen (a human sex hormone)

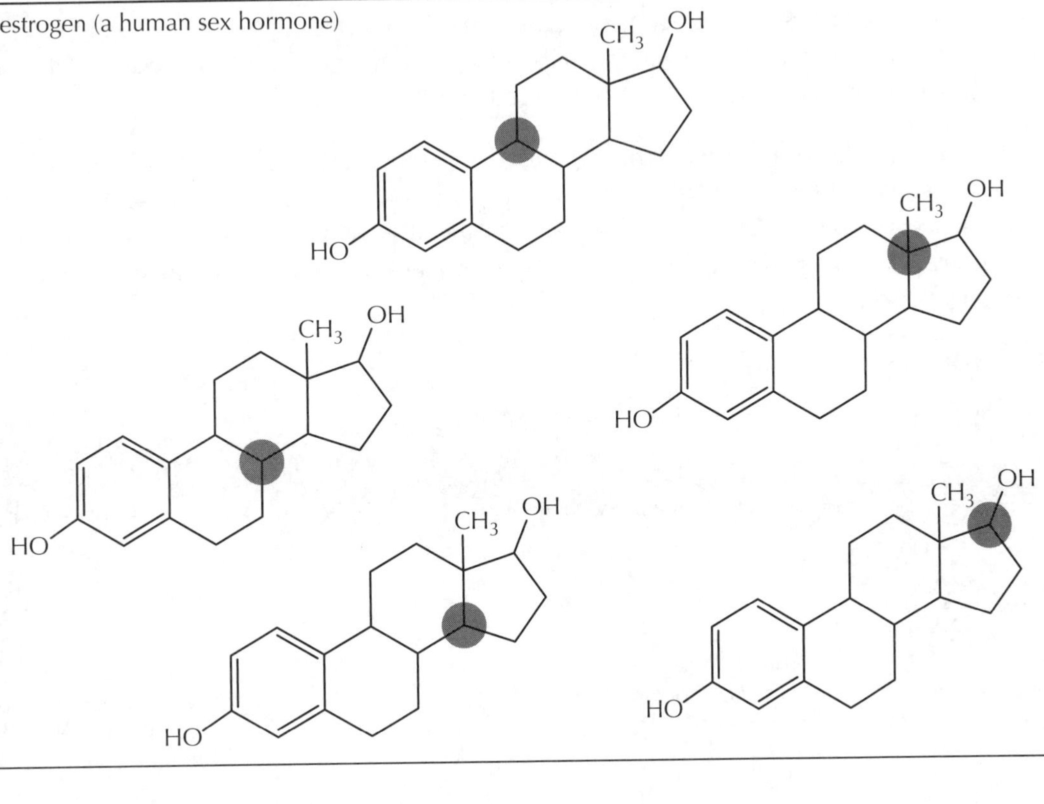

A3.2 Assigning Geometrical Labels

A. Cis/Trans: Relative Stereochemistry

Historically, the terms "cis" and "trans" have been used as both geometrical labels and as a way to describe the relative positions of groups attached to double bonds and rings. Except for a few cases, which will be detailed in Section B (below), the two terms are no longer used as geometrical labels but still used universally when the relative relationship between substituent groups is being described, in either writing or in conversation.

Examples of statements in which the relative relationship between substituent groups is being described are given in Figure AP0310.

Figure AP0310

Using "cis" and "trans" to describe relative stereochemistry.

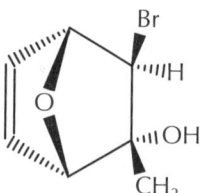

Br/CH₃ are cis
H/OCH₃ are cis
Br/OCH₃ are trans
H/CH₃ are trans
H/Br are on same atom

Cl/CH₃ are trans
H/OCH₃ are on the same atom
Cl/OCH₃ are cis
H/CH₃ are cis
H/Cl are trans

Br/CH₃ are cis
O/H are trans
OH/CH₃ are on the same atom
H/CH₃ are trans
Br/C=C are trans

B. Cis/Trans: Geometrical Label

In chemistry, the use of "cis" (on the same side) and "trans" (on opposite sides) started in the late 1800s, as commonly borrowed terms from Latin in English (e.g., transatlantic trip; cisalpine side of the Alps; and more recently in reference to cis- and trans-gendered people).

The IUPAC formally allows these two terms to be used as geometrical labels, in naming, for double bonds in which the geometry of a pair of hydrogen atoms is being compared. Examples of such molecules, and their names, are shown in Figure AP0311. These cases are all subsumed by the E/Z system, which can be used to name all double bond stereoisomers, but the historical and common use has kept cis/trans naming in common practice.

Figure AP0311

Using "cis" and "trans" to name double bond diastereomers.

trans-1-bromo-2-methoxyethene

cis-1,2-diphenylethene

3-chlorohex-2-ene cannot use "cis" or "trans" because a pair of Hs is not being compared

3-ethoxy-2-methylpent-2-ene has no other stereoisomer because the two groups on the left-hand end (methyl) are the same

Naming the two possible stereoisomers associated with oppositely-disubstituted even-sized rings is a common extrapolation of this "pair of hydrogens" principle (Figure AP0312). As with double bonds, there is a formal naming system that is used for all of these cases, and this is introduced in Section D. The detailed discussion about these stereoisomers appears in Chapter 5.3D.

Figure AP0312

Using "cis" and "trans" to name some disubstituted rings.

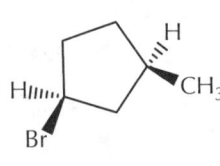 trans-4-ethylcyclohexan-1-ol

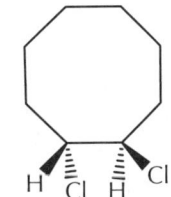 cis-1,3-dibromocyclobutane

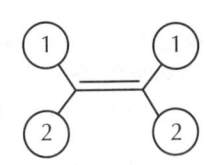 (1S,3R)-1-bromo-3-methylcyclopentane
although the Br/CH₃ relationship can be described as cis, the name is ambiguous because there are two cis isomers, this one and its enantiomer, the (1R,3S)- stereoisomer

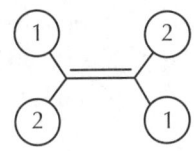 (1R,2R)-1,2-dichlorocyclooctane
although the Cl/Cl relationship can be described as trans, the name is ambiguous because there are two trans isomers, this one and its enantiomer, the (1S,2S)- stereoisomer

C. E/Z and R/S Geometrical Labels

Double bonds with pairs of different substituents at both atoms of the double bond create two and only two stereoisomers. After assigning the priorities for the two groups at each end, the isomer in which the two highest priority groups are cis is given the geometric label "Z" (from the German word *zusammen*, meaning "together"). The other stereoisomer, in which the two highest priority groups are trans, is given the geometric label "E" (from *entgegen*, meaning "opposite"). Examples are shown in Figure AP0313, and the process is also detailed in Chapter 5.3C.

Figure AP0313

Using "Z" and "E" to name disubstituted alkenes.

assign the CIP priorities at both carbons of the double bond; if they are cis, then the geometrical label is (Z)-

assign the CIP priorities at both carbons of the double bond; if they are trans, then the geometrical label is (E)-

(E)-1-bromo-2-methoxyethene

(Z)-1,2-diphenylethene

(Z)-1-chloro-3-ethoxy-2-methylpent-2-ene

(E)-3-chlorohex-2-ene

Tetrahedral atoms with four different substituents create two and only two stereoisomers. After assigning the priorities for the four groups, the geometrical label is defined by a specific point of view. The observer views the tetrahedral atom so that the lowest priority group is directly behind it (i.e., eclipsing it). The remaining three substituents describe the corners of a triangle. If the direction to move from priority 1 to 2 to 3 is clockwise (i.e., turning to the right), then the geometry is given the label "R" (from the Latin word, *rectus*, meaning "to the right"). And if the direction to move from priority 1 to 2 to 3 is counterclockwise (i.e., turning to the left), then the geometry is given the label "S" (from the Latin word *sinister*, meaning "to the left"). Examples are shown in Figure AP0314, and the process is also detailed in Chapter 5.3B.

Figure AP0314

Using "R" and "S" to name asymmetric tetrahedral atoms.

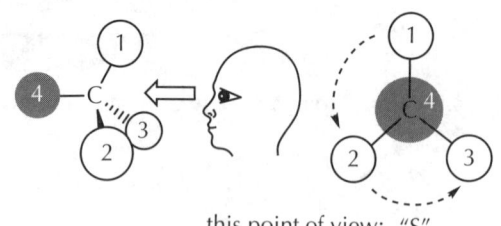

assign the CIP priorities to all four substituents at a tetrahedral stereocenter; using the viewpoint of an observer looking at the stereocenter with the lowest priority group behind the stereocenter, the 1-to-2-to-3 direction will be clockwise (turning right), and labeled "R," or counterclockwise (turning left), and labeled "S"

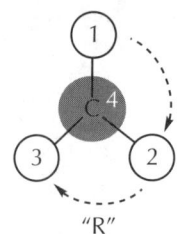

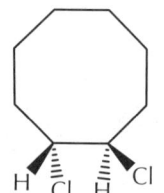 (1S,3R)-1-bromo-3-methylcyclopentane

(1R,2R)-1,2-dichlorocyclooctane

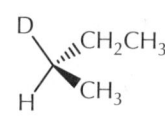 (4R,9R)-9-bromo-4-ethyltetradecan-4-ol

(R)-2-deuteriobutane

Many students find these visualizations challenging, particularly when the drawing of the tetrahedral atom shows the low priority substituent anywhere other than behind the plane of the writing surface. A common strategy used for assigning the "R" versus "S" label involves a right- and left-hand rule. If you imagine poking your thumb through the tetrahedral atom and along the bonding direction of the low priority group, then you look at the way your fingers curl and on which hand the curling of your fingers matches the 1-to-2-to-3 priority order of the three other groups. If the direction of turning from priority 1-to-2-to-3 follows the curl that the fingers on your right hand makes, then the geometrical label is "R." And if the 1-to-2-to-3 follows the curl of the fingers of your left hand, then the geometrical label is "S" (Figure AP0315).

Figure AP0315

Deciding "R" and "S" labels with your hands.

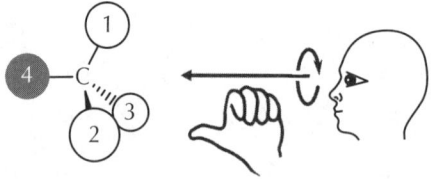

from this point of view, when the thumb of the left hand points at #4, the fingers on that hand curl from 1-to-2-to-3 ("S")

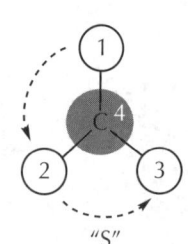

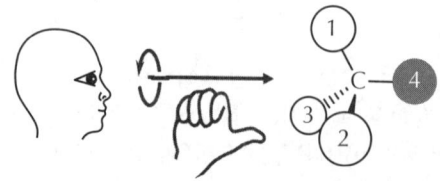

from this point of view, when the thumb of the right hand points at #4, the fingers on that hand curl from 1-to-2-to-3 ("R")

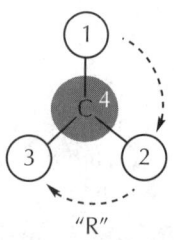

Sometimes, when applying the CIP rules for prioritizing substituents, the decision between cis versus trans, E versus Z, or R versus S needs to be made. According to the rules, when all else is equal, *cis*- has a higher priority than *trans*-, (Z)- has a higher priority than (E)-, and (R)- has a higher priority than (S)-.

D. r/s and R_a/S_a Geometrical Labels

As detailed in Chapter 5, molecules with two similar stereocenters have three stereoisomers, the pair of chiral (*R,R*)- and (*S,S*)- enantiomers along with the achiral (*R,S*)- (*meso-*) stereoisomer, a diastereomer of the other two. The 1,3-dimethylcyclohexane connectivity is an example of such a case, the general solution for which is described in Figure AP0316.

Figure AP0316

Stereochemical analysis for 1,3-dimethylcyclohexane.

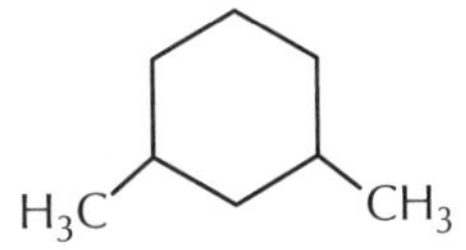

1,3-dimethylcyclohexane

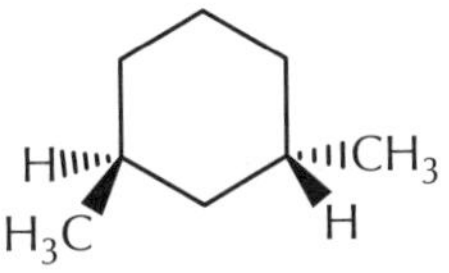

(1*S*,3*S*)-1,3-dimethylcyclohexane
chiral

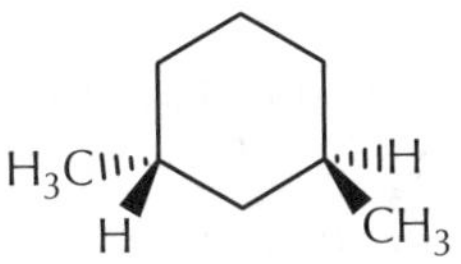

(1*R*,3*R*)-1,3-dimethylcyclohexane
chiral

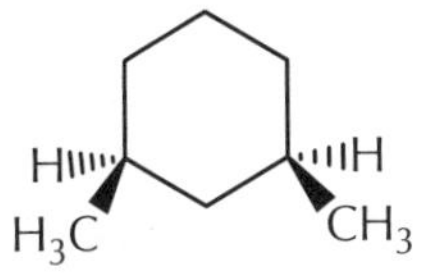

(1*R*,3*S*)-1,3-dimethylcyclohexane
achiral (meso)

Starting with 1,3-dimethylcyclohexane, there are three ways to create a new stereocenter by replacing a hydrogen atom with, for example, a bromine atom, at different locations on the ring. The easiest of these stereochemical analyses is shown in Figure AP0317 (1-bromo-2,4-dimethylcyclohexane). This connectivity represents the case of three completely different stereocenters, resulting in all eight possible stereoisomers, with all stereocenters able to be labeled as either "R" or "S."

Figure AP0317

Stereochemical analysis for 1-bromo-2,4-dimethylcyclohexane.

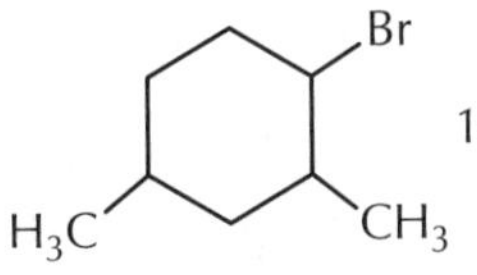

1-bromo-2,4-dimethylcyclohexane

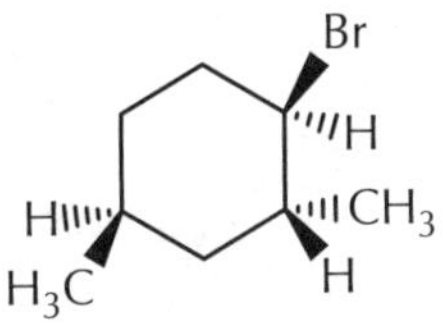

(1*R*,2*R*,4*S*)-1-bromo-2,4-dimethylcyclohexane

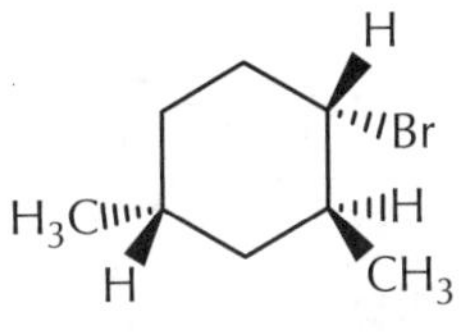

(1*S*,2*S*,4*R*)-1-bromo-2,4-dimethylcyclohexane

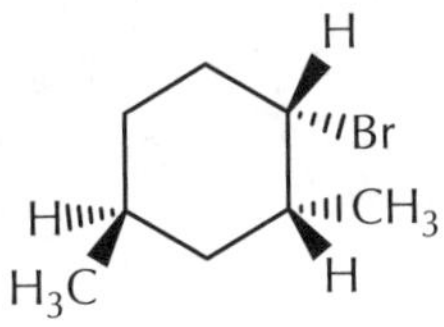

(1*S*,2*R*,4*S*)-1-bromo-2,4-dimethylcyclohexane

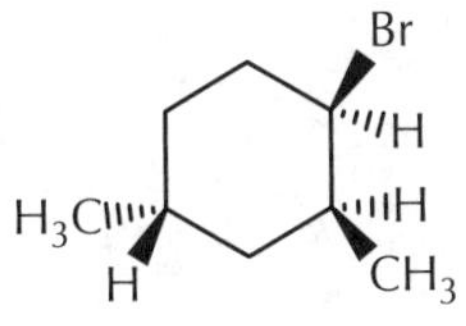

(1*R*,2*S*,4*R*)-1-bromo-2,4-dimethylcyclohexane

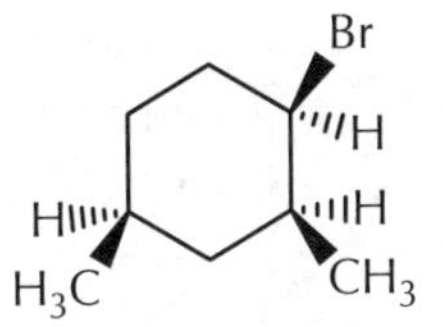

(1*R*,2*S*,4*S*)-1-bromo-2,4-dimethylcyclohexane

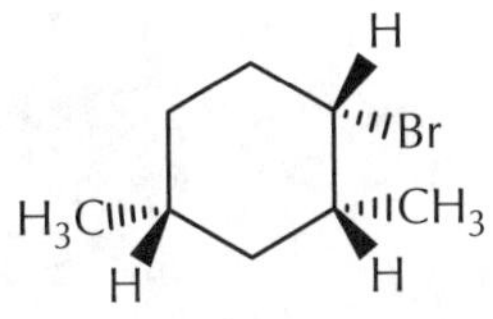

(1*S*,2*R*,4*R*)-1-bromo-2,4-dimethylcyclohexane

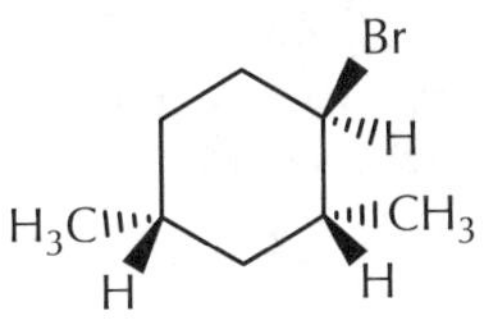

(1*R*,2*R*,4*R*)-1-bromo-2,4-dimethylcyclohexane

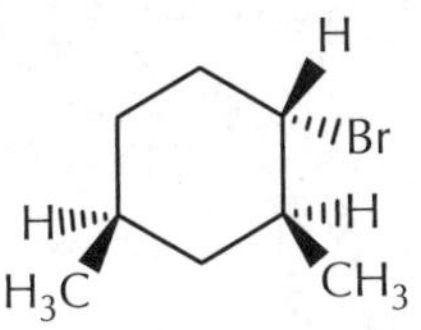

(1*S*,2*S*,4*S*)-1-bromo-2,4-dimethylcyclohexane

A second structural isomer (2-bromo-1,3-dimethylcyclohexane) is shown in Figure AP0318. In this case, the analysis gets more complicated. As in the simpler 1,3-dimethylcyclohexane example, the two stereocenters bearing methyl groups are similar, and so it is possible to have a *meso-* isomer when one of the centers is "R" and the other is "S." The carbon with the bromine atom is of interest because it is not a stereocenter when both of the methyl-bearing centers have the same configuration (both "R" or both "S"), because then the carbon with the attached bromine atom does not have four different groups attached to it. On the other hand, when the two methyl-bearing centers have the opposite configuration (one "R" and one "S"), two things are true: (1) the bromine-atom center technically has four different groups around it, and (2) any isomer with this relationship is *meso*.

Consequently, there are a total of four stereoisomers. The (1R,3R)- and (1S,3S)- isomers are chiral molecules, they are enantiomers, and the bromine-bearing carbon is not a stereocenter. And there are two (1R,3S)- isomers, both are achiral *meso* compounds, differing only according to the geometry at the bromine-atom center, where it is easy to see that the two stereoisomers are different. In one case, the bromo group is cis to the two methyl groups, and in other one it is trans to them.

The carbon with the attached bromine atom is called a pseudoasymmetric center: Both configurations are possible, and both molecules are achiral (*meso*).

Figure AP0318

Stereochemical analysis for 2-bromo-1,3-dimethylcyclohexane.

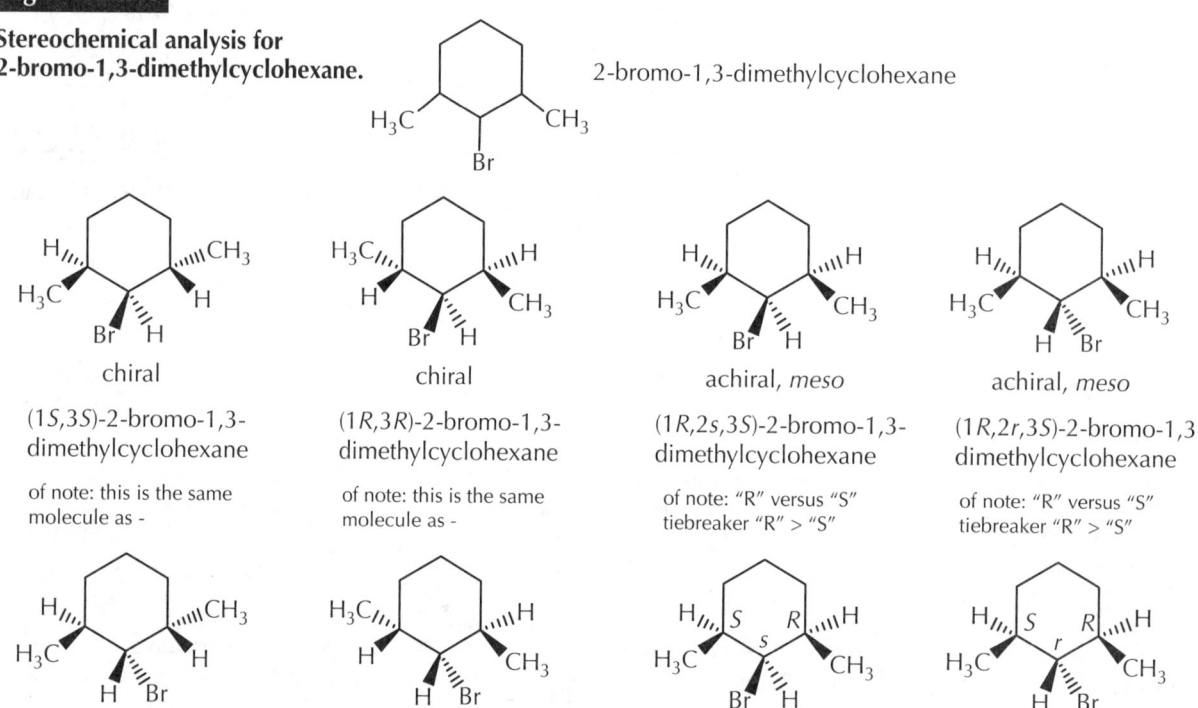

Pseudoasymmetric centers do not use the "R" and "S" labels because they are superimposable on (identical to) their mirror image. According to the CIP rules, the "R" geometry has a higher priority than the "S" geometry when there is a tie (thus, as above, the *meso*-1,3-dimethylcyclohexane is named (1R,3S)- and not (1S,3R)-).

For pseudoasymmetric centers, the labels "r" and "s" are used instead of "R" and "S," although exactly the same rules are used to make the assignment. In this example, the bromine-bearing carbon that would be assigned as "R" (based upon the CIP prioritization) is given the "r" label, and the "S" version gets the "s" label. The names of these two achiral, *meso* isomers, as shown in Figure AP0318, are (1R,2r,3S)- and (1R,2s,3S)-2-bromo-1,3-dimethylcyclohexane.

A3.2 Assigning Geometrical Labels

The result outlined in Figure AP0318 in generalizable. A *meso* compound with an added pseudoasymmetric atom has four stereoisomers: two achiral, *meso* compounds, (R,r,S)- and (R,s,S)-, which have a diastereomeric relationship. And they are both diastereomers with respect to the chiral (R,R)-/(S,S)- enantiomeric pair.

In the third structural isomer, 1-bromo-3,5-dimethylcyclohexane, the bromine-bearing carbon is also a pseudoasymmetric atom, and the result is exactly the same as the 1,2,3-isomer (Figure AP0319).

Figure AP0319

Stereochemical analysis for 1-bromo-3,5-dimethylcyclohexane.

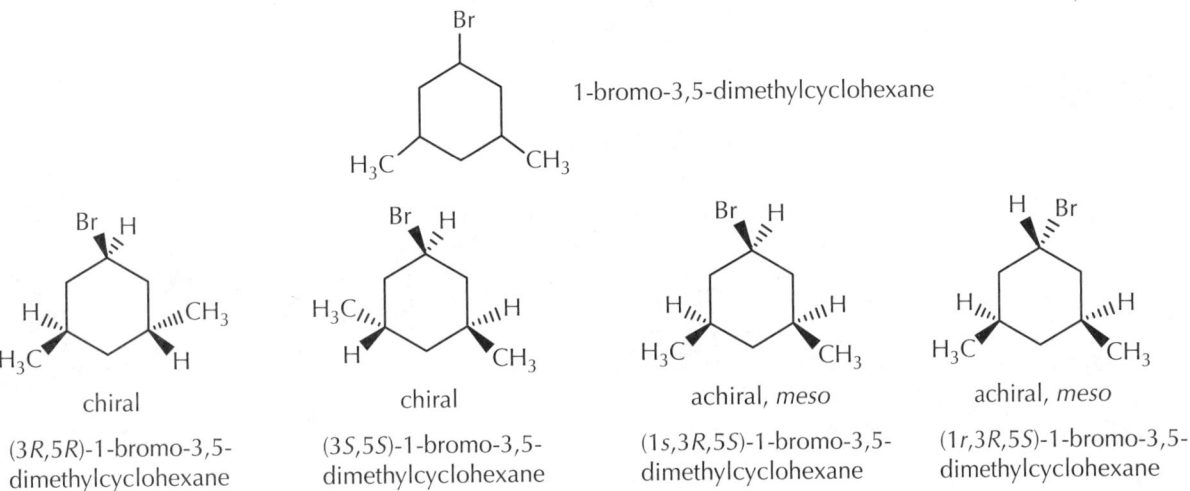

Another structural situation that uses the "r/s" labels is the oppositely-disubstituted even-sized ring, as in Figure AP0312. In the previous discussion of using "cis" and "trans" as geometrical labels, there was a limitation on comparing a pair of hydrogen atoms, and many examples of this case do not meet this requirement.

The IUPAC names for these molecules utilizes the "r/s" system combined with the system of duplicate and phantom atoms. IUPAC rules treat the two methyl-bearing carbons in 1,4-dimethylcyclohexane as pseudoasymmetric atoms. The analysis for the *cis*-1,4-dimethylcyclohexane isomer is shown in Figure AP0320, and is described below.

Figure AP0320

IUPAC naming for *cis*-1,4-dimethylcyclohexane (for one methyl).

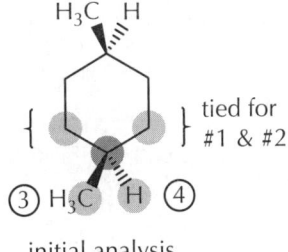

initial analysis

to break the tie, traversing the ring in both directions does not create a point of difference; upon reaching the original atom, two duplicate atoms with their phantom atoms are used at the ends of the open chain, turning the molecule into a *meso* isomer and allowing the "s" designation at the original (now pseudo-asymmetric) atom

The CIP rules for prioritizing the substituents at the two methyl-bearing carbons starts by comparing the hydrogen atom (fourth priority), the methyl group (third priority), and the two strands of the ring, which are tied for first and second priorities. Using the point of difference strategy, both pathways around the ring identify no difference at all, and you traverse a full trip around the ring until you get back to the methyl-bearing atom.

998 APPENDIX 3 Nomenclature of Organic Compounds II: Stereoisomers

Recall from the examples with the cyclopropyl group in Figures AP0308 and AP0309: when you make it all the way around a ring while assigning priorities, you then split the ring open and using a duplicate atom to replace the original starting point for both strands of the ring. In these molecules, two things happen when you do this: (1) the substituent on the other side of the ring ends up becoming a stereocenter, able to be labeled as "R" or "S," when you open the ring and attached the duplicate atom, bearing its three phantom atoms; and (2) the molecule, treated in this way, mimics a *meso* compound, turning the original center into a pseudo-asymmetric atom that can be given an "r" or "s" label.

The same treatment can be done for the other methyl group (Figure AP0321), which means that the IUPAC name for the *cis*-1,4-dimethylcyclohexane isomer is (1s,2s)- 1,4-dimethylcyclohexane. Using the same analysis, the *trans*- isomer gets the (1r,2r)- label.

Figure AP0321

IUPAC naming for *cis*-1,4-dimethylcyclohexane (for the other methyl).

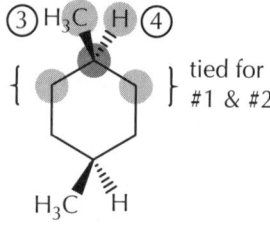

initial analysis

to break the tie, traversing the ring in both directions does not create a point of difference; upon reaching the original atom, two duplicate atoms with their phantom atoms are used at the ends of the open chain, turning the molecule into a *meso* isomer and allowing the "s" designation at the original (now pseudo-asymmetric) atom

For the same reason that the E/Z system is needed for double bonds when there is not a pair of hydrogen atoms for using cis/trans as labels, the oppositely-disubstituted even-sized rings need the r/s system. For instance, the two possible achiral diastereomers of 3-bromo-1-isopropylcyclobutan-1-ol cannot be unambiguously named with *cis*- and *trans*- labels.

The r/s analysis for both diastereomers of this cyclobutanol derivative is outlined in Figure AP0322.

Figure AP0322

IUPAC naming for the 3-bromo-1-isopropylcyclobutan-1-ol isomers.

(1r,3r)-3-bromo-1-isopropylcyclobutan-1ol

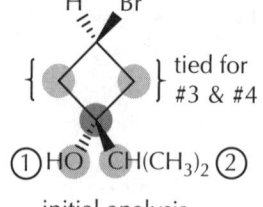

initial analysis

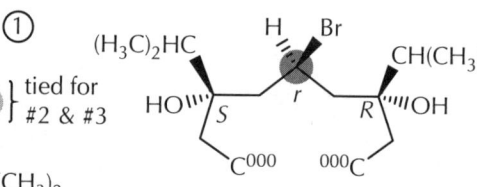

initial analysis

(1s,3s)-3-bromo-1-isopropylcyclobutan-1-ol

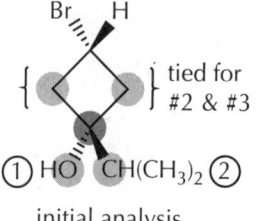

initial analysis

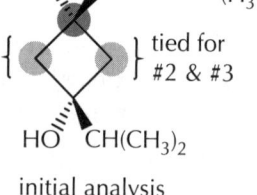

initial analysis

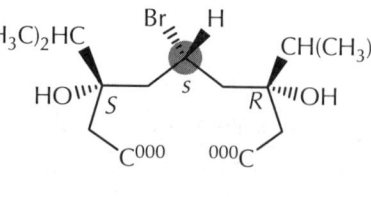

A3.2 Assigning Geometrical Labels

Chirality is not limited to tetrahedral stereocenters. In this final section, two examples of chiral axes are illustrated: cumulative double bonds and biphenyls.

Cumulative double bonds. The two sp² atoms of a double bond are coplanar, due to the pi bond, resulting in the possibility for E/Z diastereomers. When two double bonds share an atom, they are called cumulative double bonds, as in 1,2-propadiene (also called allene). The sp atom in the center of 1,2-propadiene means that the two sp² atoms at the ends are perpendicular and not coplanar (Figure AP0323). As you might anticipate, having three cumulative double bonds, with two sp atoms, moves the two sp² ends back into coplanarity, which will be true for having an odd number of double bonds in a cumulative system. The ends of an even number of cumulative double bonds will be perpendicular.

Figure AP0323

Geometry at the ends of cumulative double bond systems.

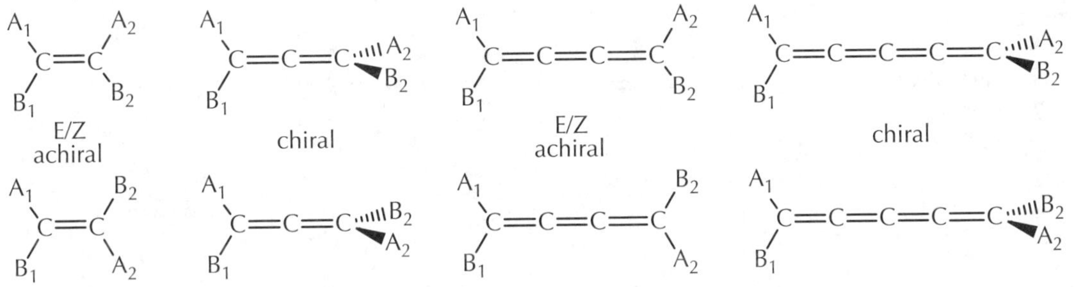

As in a simple double bond, cumulative systems with an odd number of double bonds will have "E" and "Z" diastereomers when the two ends have pairs of different substituents (Figure AP0324).

Figure AP0324

Stereoisomers in cumulative double bond systems: E/Z for odd numbers of double bonds.

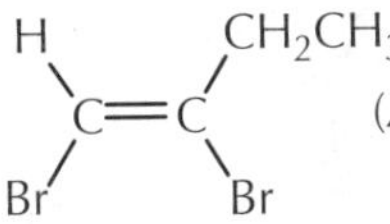

(Z)-1,2-dibromobut-1-ene

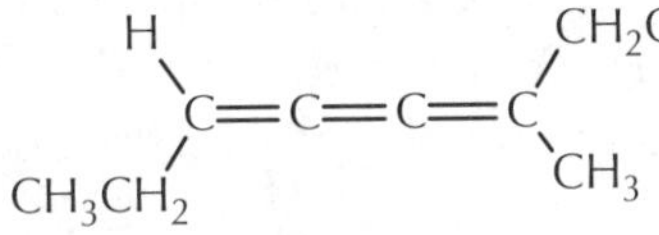

(E)-1-methoxy-2-methylhepta-2,3,4-triene

When there are different substituents at both ends of the even-numbered cumulative systems, on the other hand, enantiomers result. The two perpendicular ends on these even-numbered cases are treated as elongated tetrahedra (Figure AP0325).

Figure AP0325

Stereoisomers in cumulative double bond systems: R_a/S_a for even numbers of double bonds.

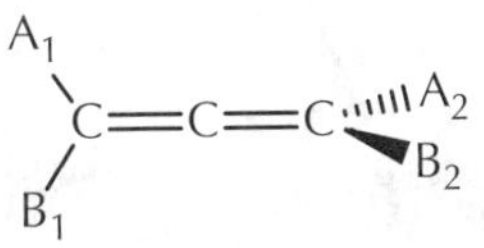

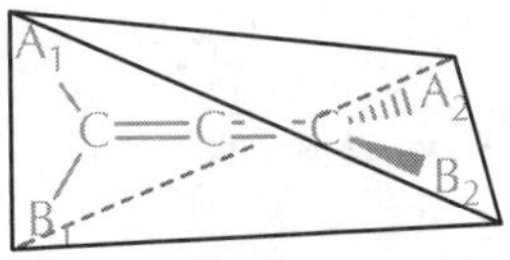

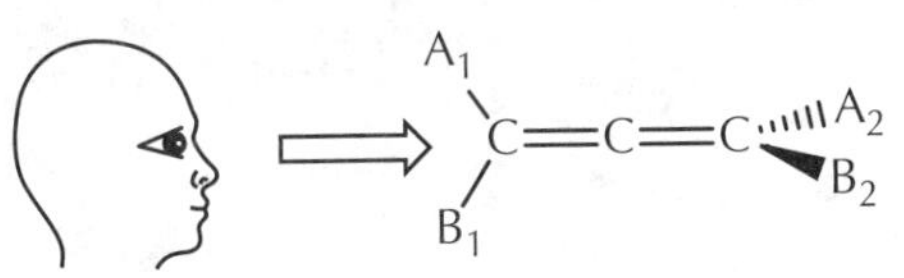

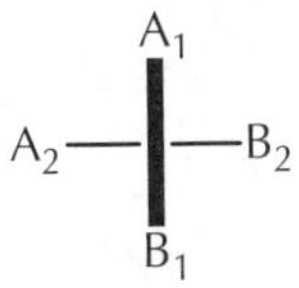

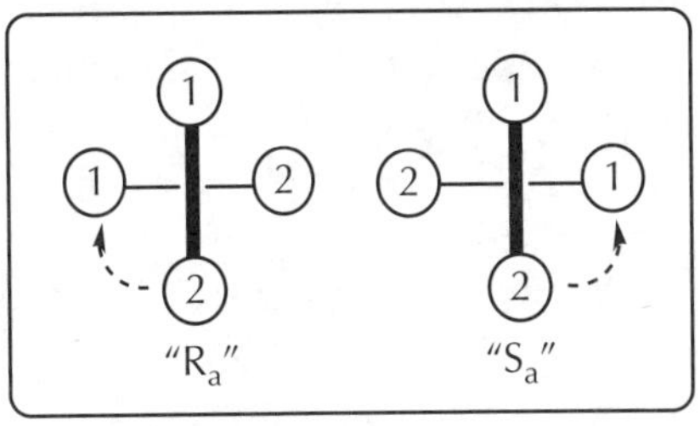

As in E/Z naming, the groups at both ends of the perpendicular system are prioritized using the usual CIP rules. Using a point of view that looks along the long axis of the molecule, analogous to a Newman projection.

Looking from either direction will give the same label. The two groups on the end closest to your point of view are perpendicular to the groups at the far end, and so the viewpoint can be depicted as a pair of perpendicular lines. At both the front and the back, the two groups can be prioritized. If the direction to move from the lower priority substituent in the front to the higher priority in the back is clockwise, then the geometry is labelled "R_a" ("R," as usual, meaning a turn to the right, and the "a" for axial). Comparably, if moving from the lower priority substituent in the front to the higher priority in the back is counterclockwise, then the geometry is labelled "S_a."

As is true in E/Z systems, there does not need to be four different groups, just two different groups at both ends, which means that there are two chiral stereoisomers for penta-2,3-diene (Figure AP0326).

Figure AP0326

Stereoisomers in cumulative double bond systems: the enantiomers of penta-2, 3-diene.

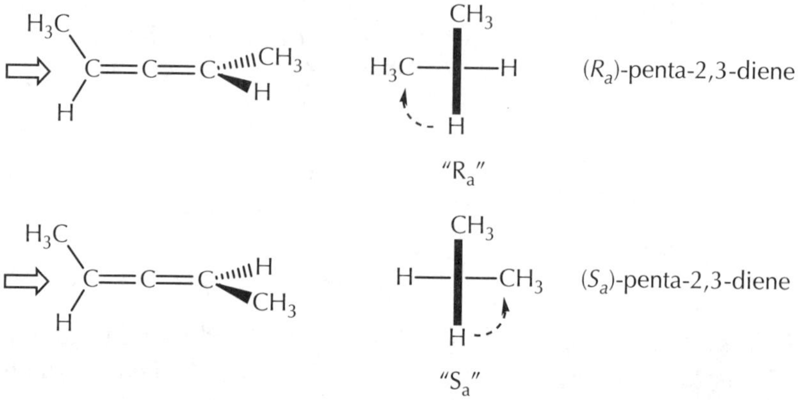

Biphenyls. Another structural case that results in axial chirality is that of biphenyls. In the parent compound, called biphenyl, the carbon-carbon bond between the rings is freely rotating, with a low activation energy, comparable to that in butane, of about 6 kcal/mol. The rings are not coplanar but are instead tipped at about a 45° angle. Although the coplanar rings would have the best delocalization, the sets of ortho hydrogen atoms are close enough to create steric hindrance (Figure AP0327). The shape derived from a +45° twist is enantiomeric to a -45° twist, and so these conformations are chiral and, as they are in butane, rapidly interconverting.

Figure AP0327

Chiral conformations of biphenyl.

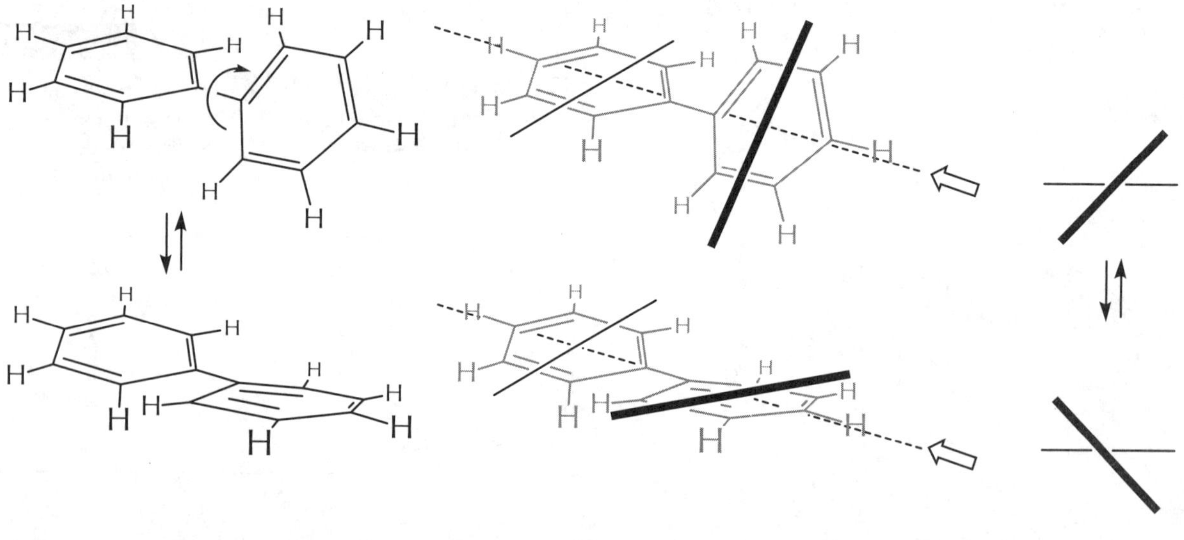

As the size of the groups in the ortho positions increases, the steric hindrance in the coplanar rings also increases, and it becomes increasingly difficult for the bond rotation to take place under normal laboratory conditions. In 1922, the first examples of chiral biphenyls based on these hindered bond rotations were isolated. For example, the enantiomers of the molecule shown in Figure AP0328 could be prepared. And looking from the ends, along the axis of the molecule, the "R_a" and "S_a" labels can be assigned.

Figure AP0328

Labeling the geometries of biphenyl enantiomers.

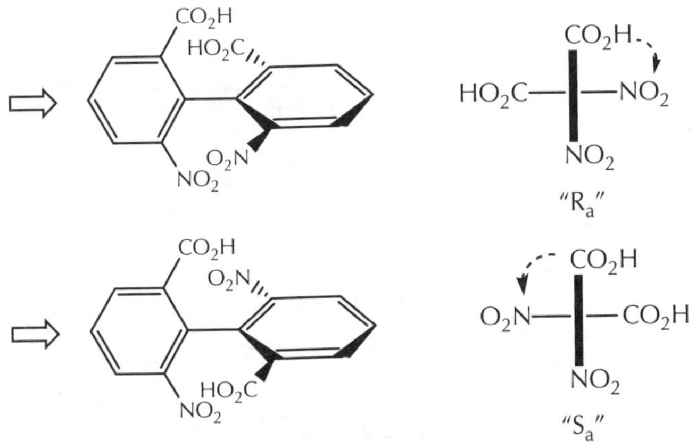

APPENDIX 3 Nomenclature of Organic Compounds II: Stereoisomers

PRACTICE QUESTIONS

AP03.06 Each of the following molecules requires one geometrical label in its name. Assign this label.

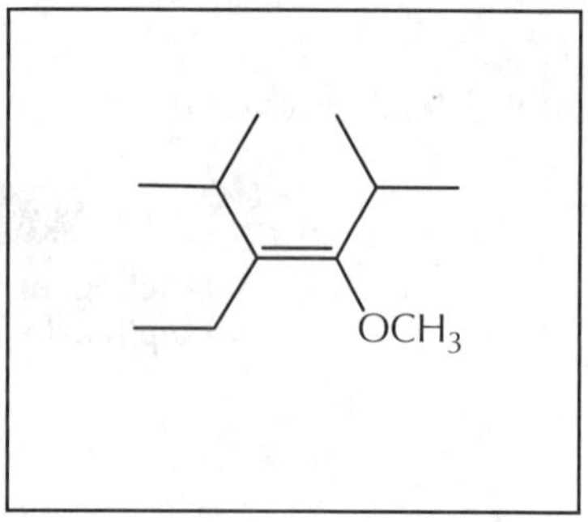

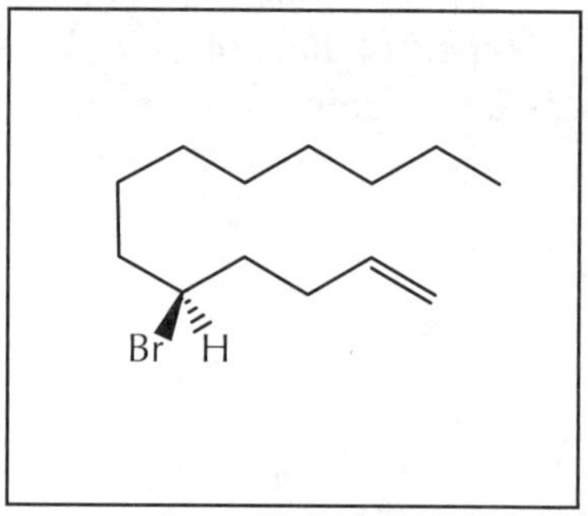

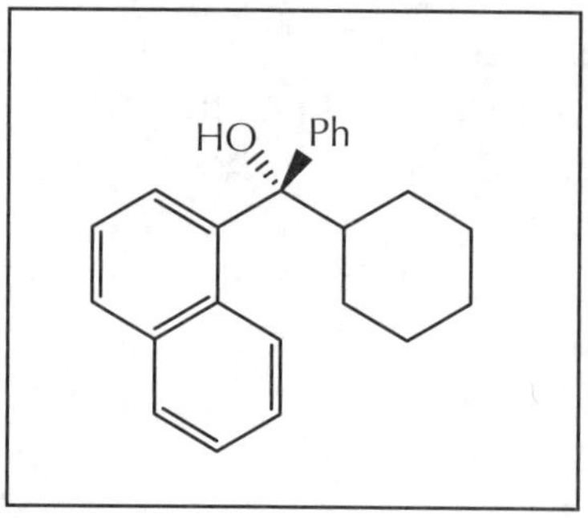

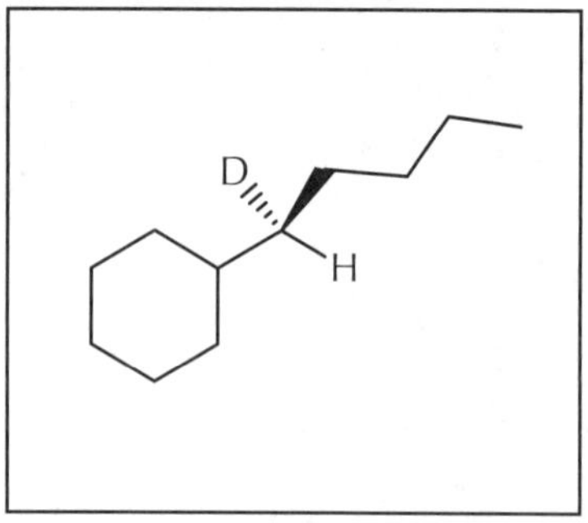

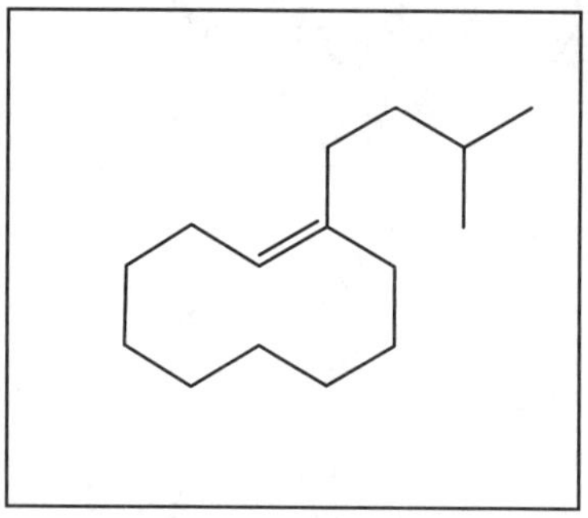

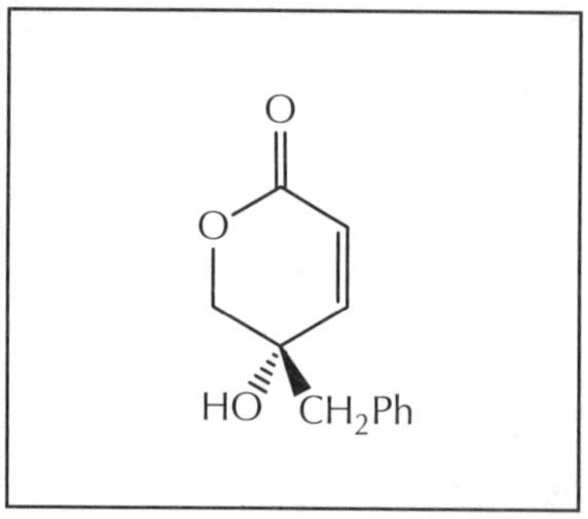

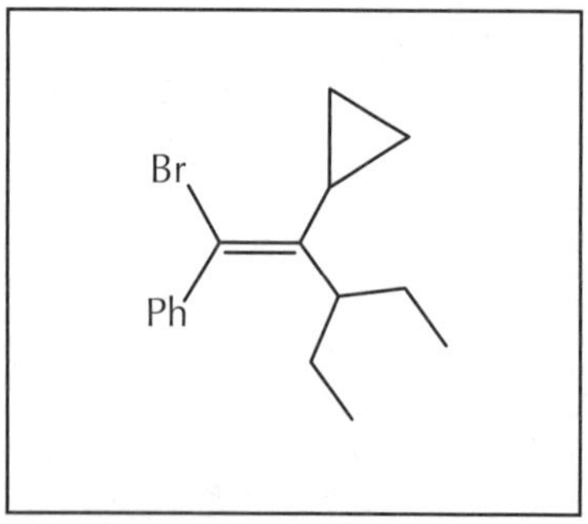

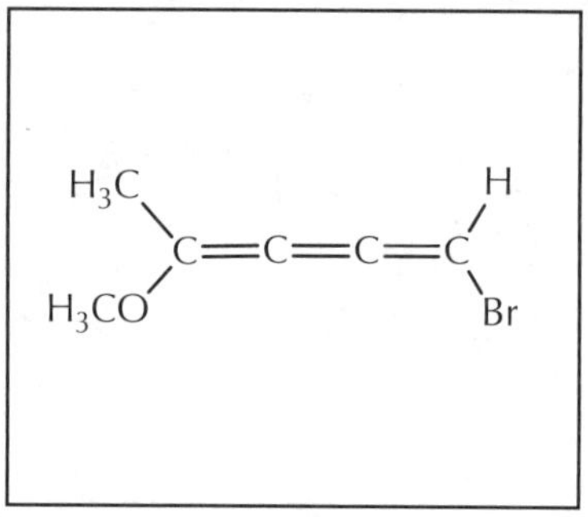

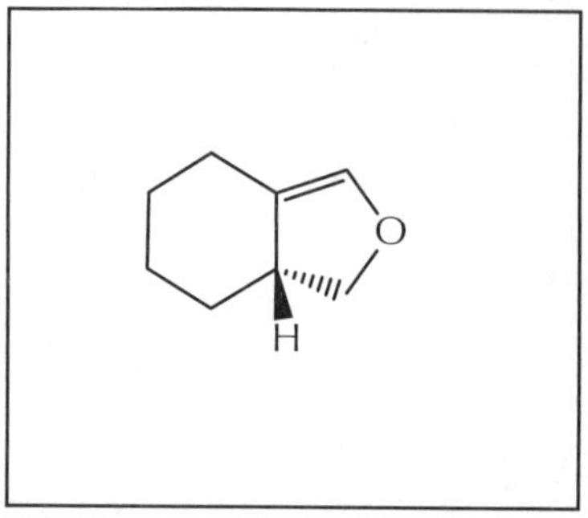

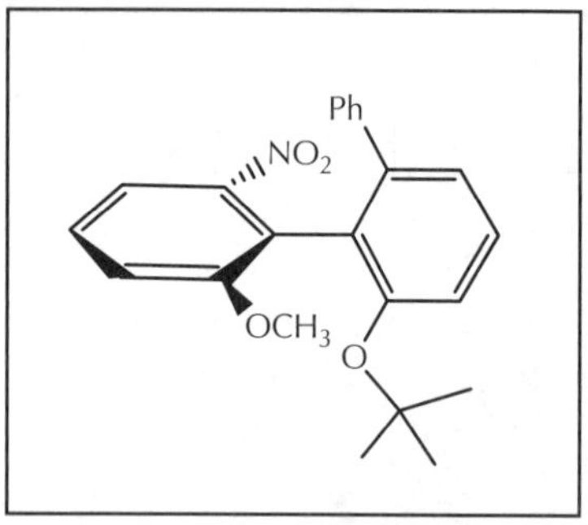

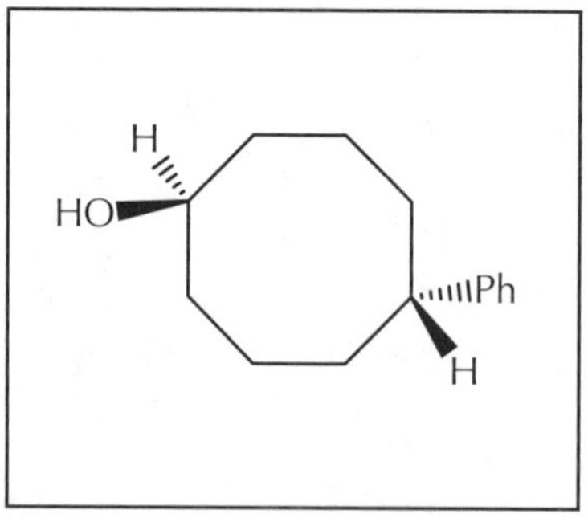

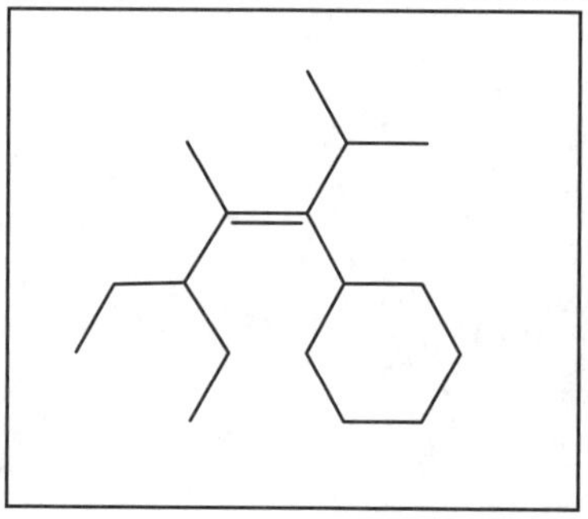

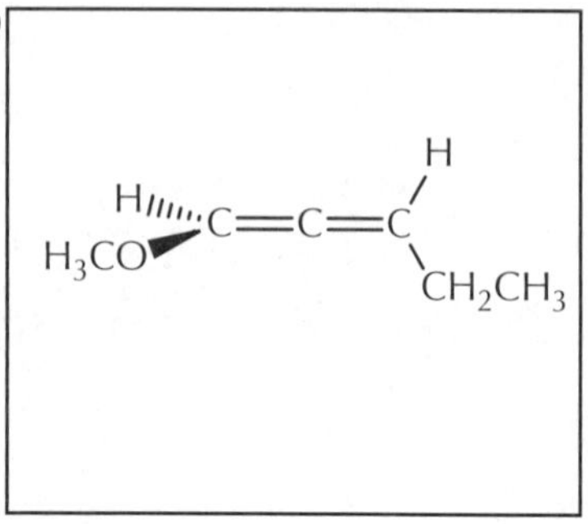

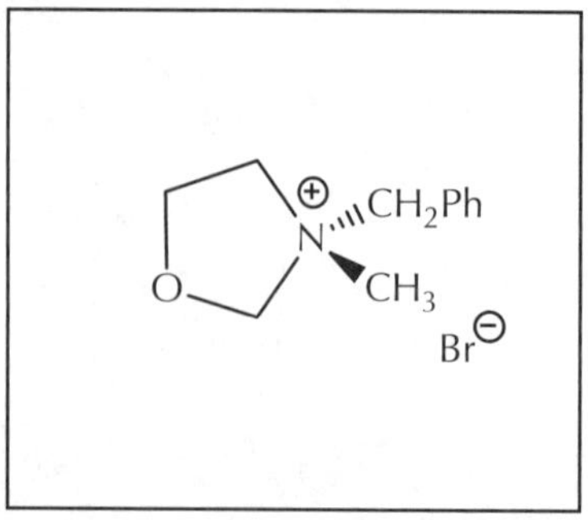

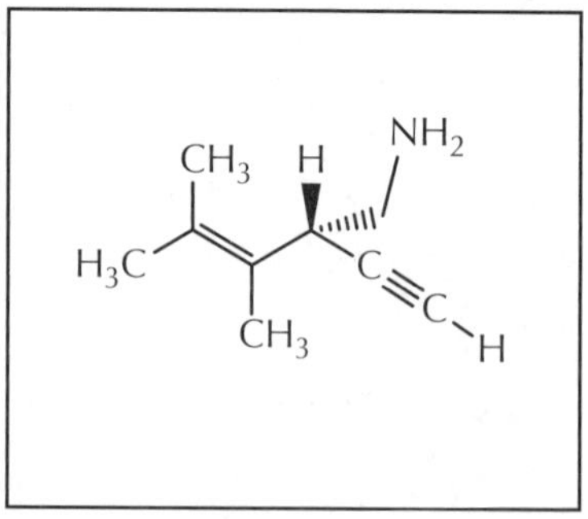

AP03.07 Each of the following molecules requires two geometrical labels in its name. Assign these labels.

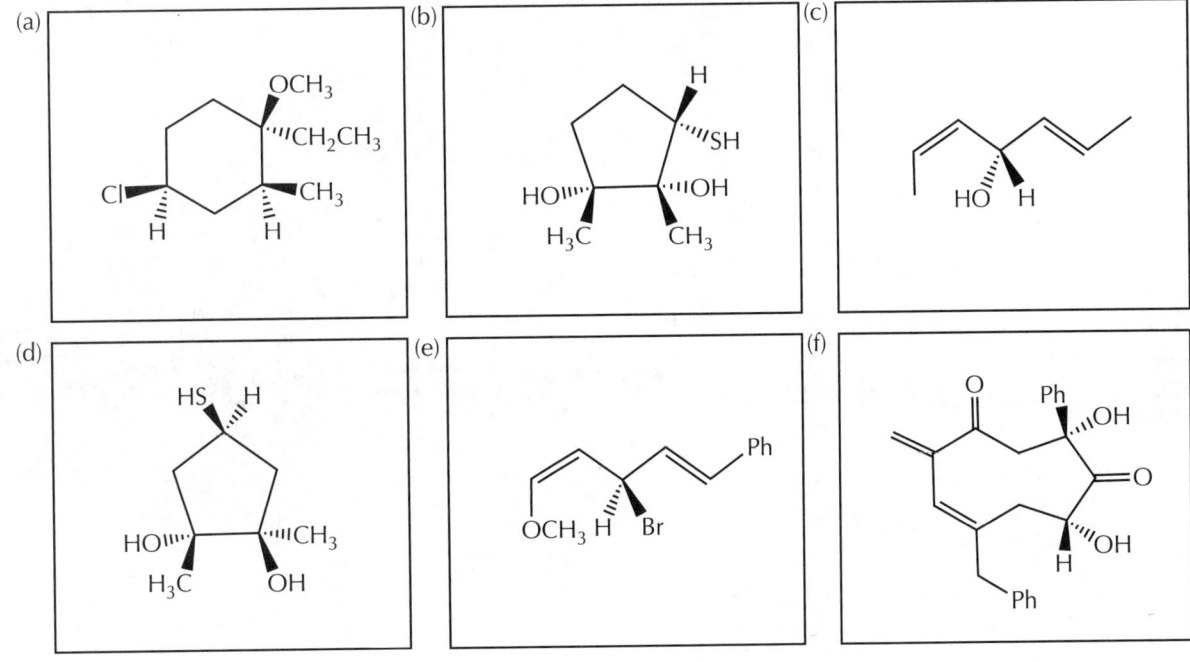

AP03.08 Each of the following molecules requires three geometrical labels in its name. Assign these labels.

AP03.09 Erythromycin is a macrolide antibiotic used to treat a variety of bacterial infections. It was descovered in 1952, and it is classified as one of the safest and most effective pharmaceuticals needed for global health care by the World Health Organization. It is also inexpensive, with a general wholesale price of 3–6 cents per pill for use in developing countries. Provide the 17 geometrical labels needed to describe the structure of erythromycin.

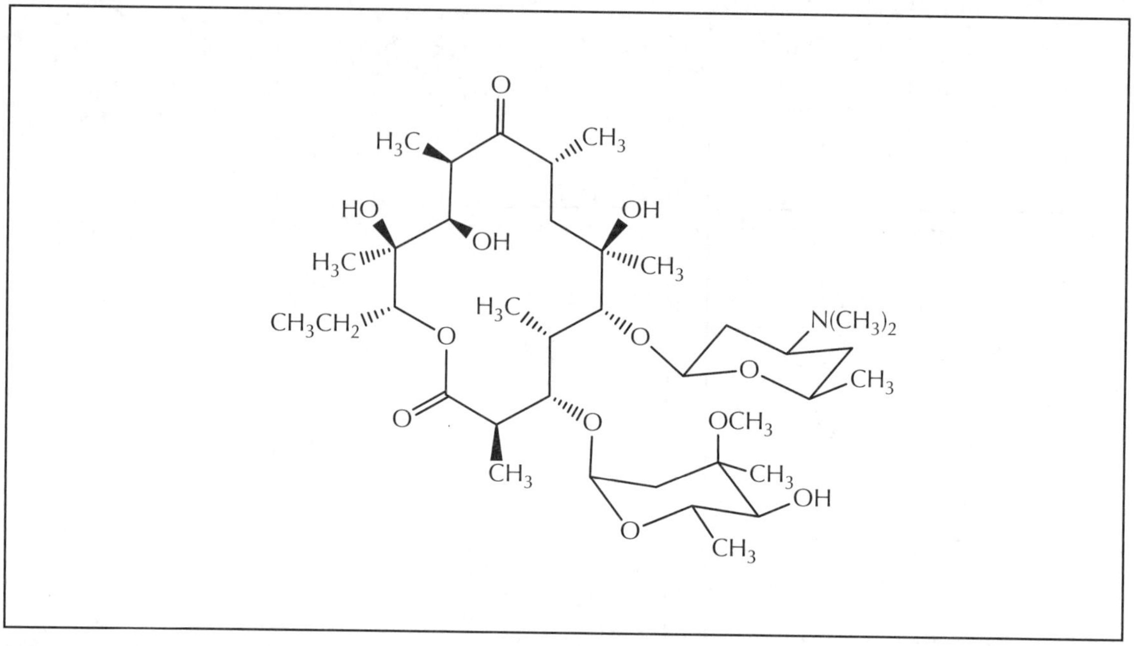

AP03.10 Vancomycin is another powerful, broad spectrum antibiotic discovered in the early 1950s. It is a macrolytic glycopeptide, given intravenously or by capsule. As with penicillins, it inhibits the formation of bacterial cell walls, causing the bacteria to rupture as they develop. Vancomycin is on the essential global medicines list compiled by the WHO. Depending on the form, the wholesale cost for doses in developing countries is anywhere between $1.50–6.00. Provide the geometrical labels for the indicated vancomycin stereocenters.

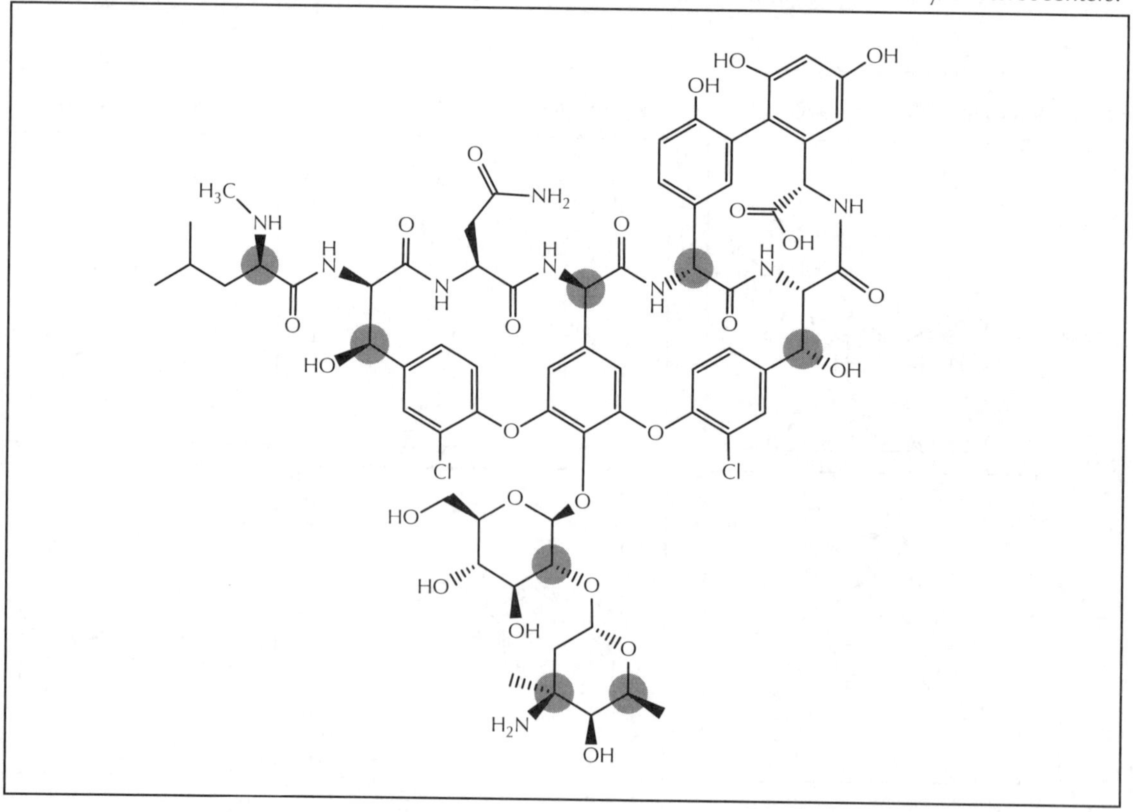

AP03.11 The pK_a for the α-hydrogens of carbonyl compounds is about 20, which means that under mild acid or base conditions, an equilibrium can be established between the original carbonyl-containing compound (called the "keto" form) and a structural isomer called the "enol" form. In many compounds, there is both an (E)- and a (Z)- enol possible. For the following enol forms, tell whether it is (E)- or (Z)-, or whether assigning stereochemical configuration is not applicable.

(a)

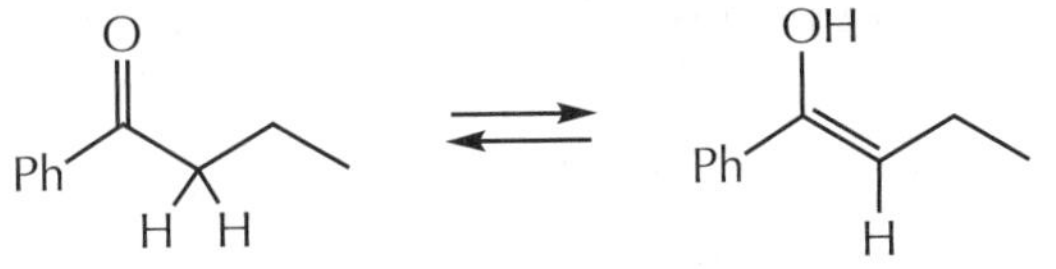

(b)

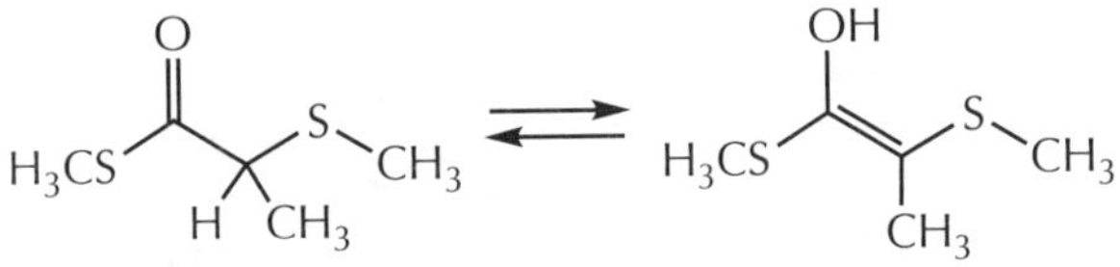

(c)

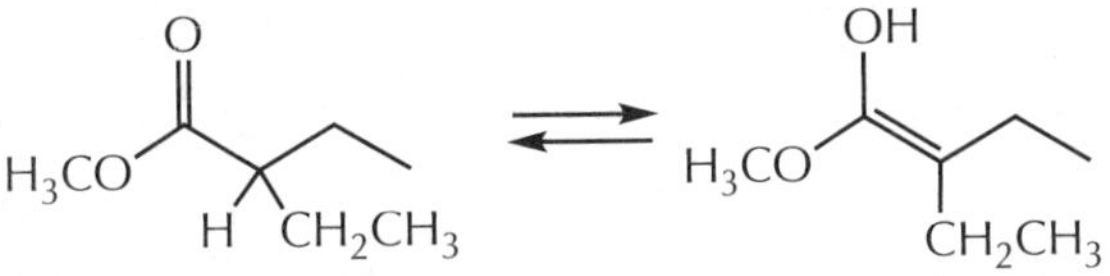

AP03.12 Axial chirality gives rise to R_a and S_a isomers. Draw the R_a isomer for the protonation reaction product from each of the following compounds.

(a)

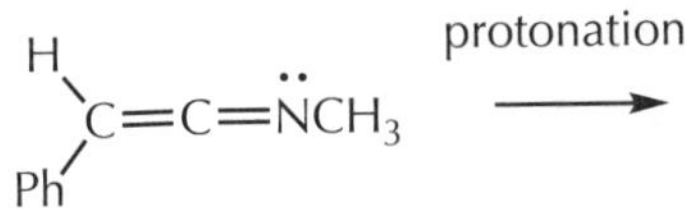

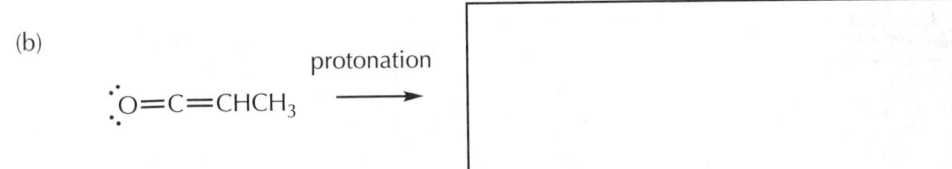

(b)

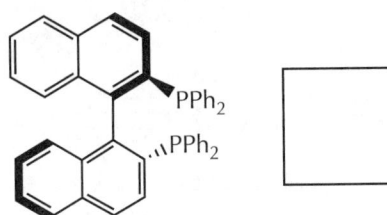

AP03.13 BINAP is the abbreviation for 2,2′-bis(diphenylphosphino)-1,1′-binaphthyl. It is a commonly used chiral compound with a 90° dihedral angle between the rings.

(a) Is this the R_a or S_a isomer?

(b) A chiral reducing agent (BINAL, shown here) has been prepared from a closely related derivative of BINAP, called BINOL. Do you predict that forming the ring with the aluminum hydride reduces or increases the dihedral angle between the rings? Why?

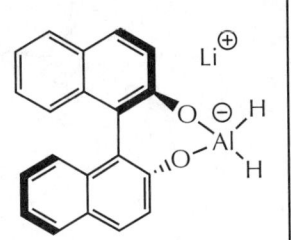

APPENDIX 4

Nomenclature of Organic Compounds III: Alkoxy Groups, Alkenes, and Alkynes

A4. Alkoxy Groups, Alkenes, and Alkynes
- A. Alkoxy Group Substituents and Ethers
- B. Chain Suffix: Alkenes and Alkynes
- C. Prioritizing Substituent Suffixes, Chain Suffixes, and Substituent Prefixes
- D. Geometrical Labels in Nomenclature

PRACTICE QUESTIONS

APPENDIX 4

Nomenclature of Organic Compounds III: Alkoxy Groups, Alkenes, and Alkynes

A4 Alkoxy Groups, Alkenes, and Alkynes

A. Alkoxy Group Substituents and Ethers

As the name implies, alkoxy groups are the combination of an alkyl group (methyl, ethyl, propyl, and so on.) with an oxygen atom to give methoxy (CH_3O-), ethoxy (CH_3CH_2O-), propoxy ($CH_3CH_2CH_2O$-), etc. A phenyl group attached to an oxygen is a phenoxy group (PhO-).

Alkoxy group are the principle units in the ether functional group, which is made up of two hydrocarbon groups attached to an oxygen atom (R-O-R). Ether is an ancient word, originally denoting a substance that occupied the open space beyond the moon (the *aether*, or "upper air," breathed by the gods) when the earth was believed to be the center of the universe. Later on, circa 1600–1800, it referred to the imagined matter that allowed electromagnetic radiation to be propagated. Its most familiar meaning comes from mid-1800s, as a highly volatile, sweet-smelling compound that was developed as the first generally useful anesthetic.

The original naming method for ethers was quite simple. For any given R_1-O-R_2, the name was "alkyl group #1_alkyl group #2_ether" (the alkyl groups listed in alphabetical order or combined with "di" if they were the same group. And so CH_3-O-CH_3 was named dimethyl ether, and CH_3CH_2-O-CH_3 was named ethyl methyl ether, and CH_3CH_2-O-CH_2CH_3 was named diethyl ether (and its common name, because it was used so widely, including as an anesthetic, is just "ether"). Common names for ethers, today, still use this construction, providing names such as methyl phenyl ether, cyclohexyl ethyl ether, and so on (Figure AP0401).

Figure AP0401

Common names for ethers.

dimethyl ether   ethyl methyl ether   diethyl ether (ether)   cyclohexyl ethyl ether   methyl phenyl ether (anisole)

Under IUPAC rules, an alkoxy group is treated as a prefix-designated substituent group, exactly as alkyl groups (methyl, ethyl, isopropyl), phenyl, and halogen atom groups (fluoro, chloro, bromo), with no additional rules. There are no internal priorities for prefix-designated substituent groups, so numbering the root chain, determined to be the longest carbon chain, follows from the first point of difference rule and then uses alphabetical order to break ties. In CH₃CH₂CH₂-O-CH₃ (methyl propyl ether), the longest chain is propane, and the methoxy group is at carbon-1, giving the IUPAC name 1-methoxypropane. Diethyl ether (CH₃CH₂-O-CH₂CH₃) is named ethoxyethane (or 1-ethoxyethane) using this system.

Examples of naming alkoxy groups using their prefix designation are shown in Figure AP0402.

Figure AP0402

IUPAC names for ethers using alkoxy group notation.

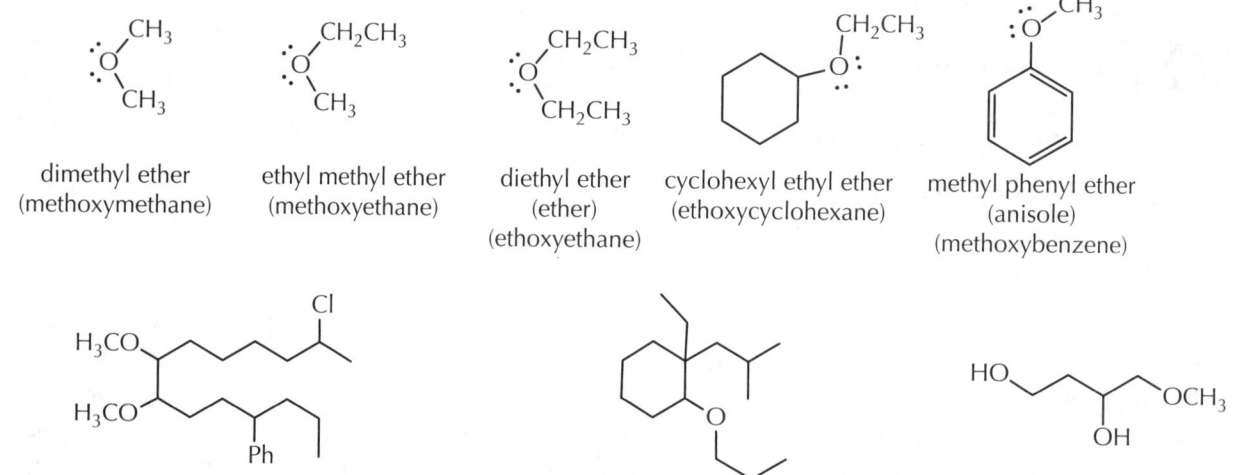

B. Chain Suffix: Alkenes and Alkynes

Double bonds and triple bonds are not substituent groups. They represent a structural modification of the carbon chain, and so chain suffixes are the major fourth piece in the construction of a name, in addition to the root chain name (meth, prop, but), prefix-designated substituent groups (methyl, ethyl, phenyl), and prefix-designated substituent groups (ol). In fact, you have been using the chain suffix for saturated alkanes (ane), already, in the construction of all of the root names seen so far (methane, ethane, decane).

For a double bond or a triple bond in the root chain, the first encountered atom of the multiple bond carries the number, and the presence of a double bond is designated by the chain suffix "ene," while a triple bond is designated with "yne."

With no other groups present, the lower numbered point of difference rule is used to decide which end of the root chain is carbon 1. Thus, a five-carbon chain with a double bond between the first two carbons is called pent-1-ene (or 1-pentene). This example and some other important molecules that illustrate the rules for chain suffixes (alone) are shown in Figure AP0403.

A summary list of what is illustrated:

(1) Carbon-carbon multiple bonds are designated by the chain suffixes "ene" and "yne."
(2) Numbering the chain follows the lowest number, first point of difference rule.

(3) More than one double bond or triple bond uses the standard combining terms di, tri, tetra, and so on; when a combining term is used, the root name recovers the "a" in the carbon counting term for ease of pronunciation (pent-1-ene versus penta-1,3-diene).
(4) Ene and yne chain suffixes are ordered in alphabetical order (ene before yne).
(5) There is no internal priority within chain suffixes for which group gets the lowest number, so first point of difference is used as it is with multiple prefixes; as always, an unresolvable tie is broken by alphabetical order (ene before yne).

Figure AP0403

Naming alkenes and alkynes.

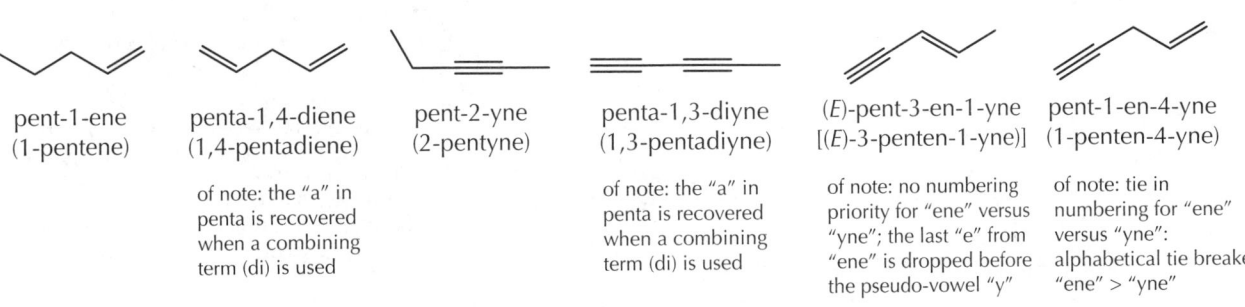

pent-1-ene
(1-pentene)

penta-1,4-diene
(1,4-pentadiene)

of note: the "a" in penta is recovered when a combining term (di) is used

pent-2-yne
(2-pentyne)

penta-1,3-diyne
(1,3-pentadiyne)

of note: the "a" in penta is recovered when a combining term (di) is used

(E)-pent-3-en-1-yne
[(E)-3-penten-1-yne)]

of note: no numbering priority for "ene" versus "yne"; the last "e" from "ene" is dropped before the pseudo-vowel "y"

pent-1-en-4-yne
(1-penten-4-yne)

of note: tie in numbering for "ene" versus "yne": alphabetical tie breaker "ene" > "yne"

C. Prioritizing Substituent Suffixes, Chain Suffixes, and Substituent Prefixes

Chain suffixes (ene, yne) fit into the prioritization rules for numbering the root chain in between suffix-designated substituents (only ol, so far) and prefix-designated substituents (methyl, chloro, phenyl, ethoxy). Thus, when two or all three of these features are in a structure, the numbering is first based upon the group suffix (ol). If there is a distinctive decision for the position of the suffix, then the numbering of the chain is established, and the name can be constructed. If the numbering analysis using the group suffix does not give a distinct result (i.e., no difference from the two ends of the chain), then the chain suffixes (ene, yne) are used to decide. And if those fail, then the prefix-designated substituents are used. And if those fail, then the final tie breaker is the alphabetical order of the prefixes at the first point of difference.

A series of examples is shown in Figure AP0404 that raise a variety of combinations and places where ties need to be broken.

Figure AP0404

Naming molecules with combinations of substituent suffixes, chain suffixes, and substituent prefixes.

(E)-6-bromohex-2-ene
[(E)-6-bromo-2-hexene]

of note: chain suffix > prefix for numbering

(E)-hex-4-en-1-ol
[(E)-4-hexen-1-ol]

of note: substituent group suffix > chain suffix for numbering; drop the "e" in "ene" before the vowel in "ol"

(Z)-1-methoxyhepta-1,6-diene
[(Z)-1-methoxy-1,6-heptadiene]

of note: point of difference decision based upon the methoxy group because the chain suffixes (yne) are tied; recover the "a" in hepta because of the consonant in diene

Naming molecules continue on next page . . .

Figure AP0404 continued ...

2-methylnona-3,6-diyn-5-ol
(2-methyl-3,6-nonadiyn-5-ol)

of note: numbering based upon point of difference with the methyl group because neither the substituent suffix (ol) nor the chain suffixes (yne) provide a decision; the "a" is recovered in nona because of "diyne" but at the same time the "e" in diyne is dropped because of the vowel in "ol"

8-methylnon-1-ene-3,6-diyn-5-ol
(8-methyl-1-nonene-3,6-diyn-5-ol)

of note: numbering based upon point of difference with the double bond; there is no "a" in nona because of the first "e" in "ene"; the second "e" in "ene" stays because of the consonant in "diyne," but the "e" in diyne is dropped because of the vowel in "ol"

3,3-dichloro-5-phenylcyclohex-1-ene
(3,3-dichloro-5-phenyl-1-cyclohexene)
(3,3-dichloro-5-phenylcyclohexene)

of note: numbering based on point of difference with the substituent groups: the two atoms of the double bond must be numbered first, and the position with the chlorine atoms is closer than the position with the phenyl (the chlorine position cannot be carbon 2 because then both atoms of the double bond have not come first); the "1" is optional because it is defined by the double bond

1-methylcyclopenta-2,4-dien-1-ol
(1-methyl-2,4-cyclopentadien-1-ol)
(1-methyl-2,4-cyclopentadienol)

of note: the substituent group (ol) sets the numbering, and so the "1" is optional; the "a" in cyclopenta is recovered because of the "d" in "diene," while the end "e" in "diene" is lost because of the vowel in "ol"

5-methoxycyclopenta-1,3-diene
(5-methoxy-1,3-cyclopentadiene)

of note: the chain suffixes having higher numbering priority than the prefix substituent (methoxy); the "a" in cyclopenta is recovered because of the "d" in "diene"

(E)-5-methyl-1-phenylhexa-1,4-dien-3-ol
[(E)-5-methyl-1-phenyl1,4-hexadien-3-ol]

of note: the substituent group (ol) sets the numbering; the "a" in hexa is recovered because of the "d" in "diene," while the end "e" in "diene" is lost because of the vowel in "ol"

(Z)-1-chloro-9-phenylcyclodec-1-ene
[(Z)-1-chloro-9-phenyl-1-cyclodecene]

of note: both of the atoms of the double bond need to be numbered first, and the point of difference rules uses the chloro group to define carbon-1; the phenyl merely takes its number after that decision and does not play a role; a reminder here that starting with cyclooctene, both (E)- and (Z)-stereoisomers are possible

D. Geometrical Labels in Nomenclature

As described in detail in Chapter 5.2 and Appendix 3, the basic name of a molecule only describes the connectivity. When stereoisomers are possible, then geometrical labels are used, and these appear in front of the name in parentheses. The geometrical labels included in the text are *R* and *S*, *E* and *Z*, *cis* and *trans* (in special cases), *r* and *s*, and R_a and S_a.

When more than one geometrical label is part of a name, they are simply ordered numerically in front of the connectivity, in a comma separated string contained in one set of parentheses (Figure AP0405).

A4 Alkoxy Groups, Alkenes, and Alkynes

Figure AP0405
Naming molecules with multiple geometrical labels.

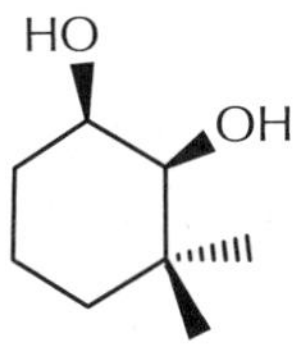

(1R,2S)-3,3-dimethylcyclohexane-1,2-diol

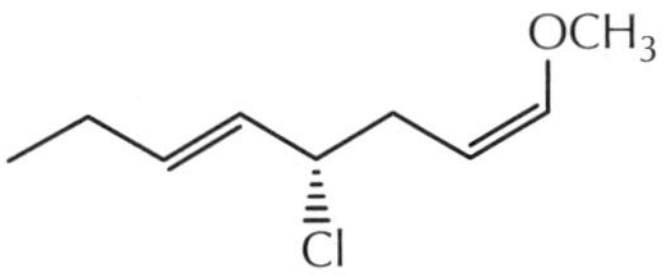

(1Z,4S,5E)-4-chloro-1-methoxyocta-1,5-diene
[(S,1Z,5E)-4-chloro-1-methoxyocta-1,5-diene]
one stereocenter means that the "4" with the "S" is optional, and if it is left off, the unnumbered "S" goes to the front

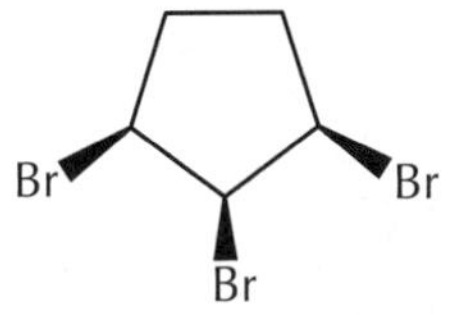

(1R,2s,3S)-1,2,3-tribromocyclopentane
(R) > (S) for tie-breaking, both in naming and priority assignment

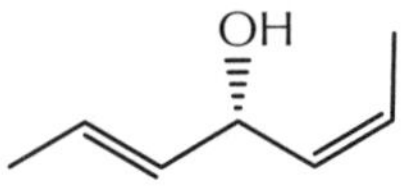

(2Z,4S,5E)-hepta-2,5-dien-4-ol
[(S,2Z,5E)-hepta-2,5-dien-4-ol]
one stereocenter means that the "4" with the "S" is optional, and if it is left off, the unnumbered "S" goes to the front;
(Z) > (E) for tie-breaking, both in naming and priority assignment

APPENDIX 4 Nomenclature of Organic Compounds III: Alkoxy Groups, Alkenes, and Alkynes

PRACTICE QUESTIONS

AP04.01 Provide the IUPAC names for the following compounds.

(a)

(b)

(c)

(d)

(e)

AP04.02 Draw the following compounds from their names.

(a) 1-isopropylcyclohex-1-ene

(b) (E)-penta-2,4-dien-1-ol

(c) 4,4-diphenyl-1-butyne

(d) (Z)-1-ethoxyprop-1-ene

(e) 5-methylhex-4-en-1-yn-3-ol

(f) 4-bromo-1-methylcyclobut-1-ene

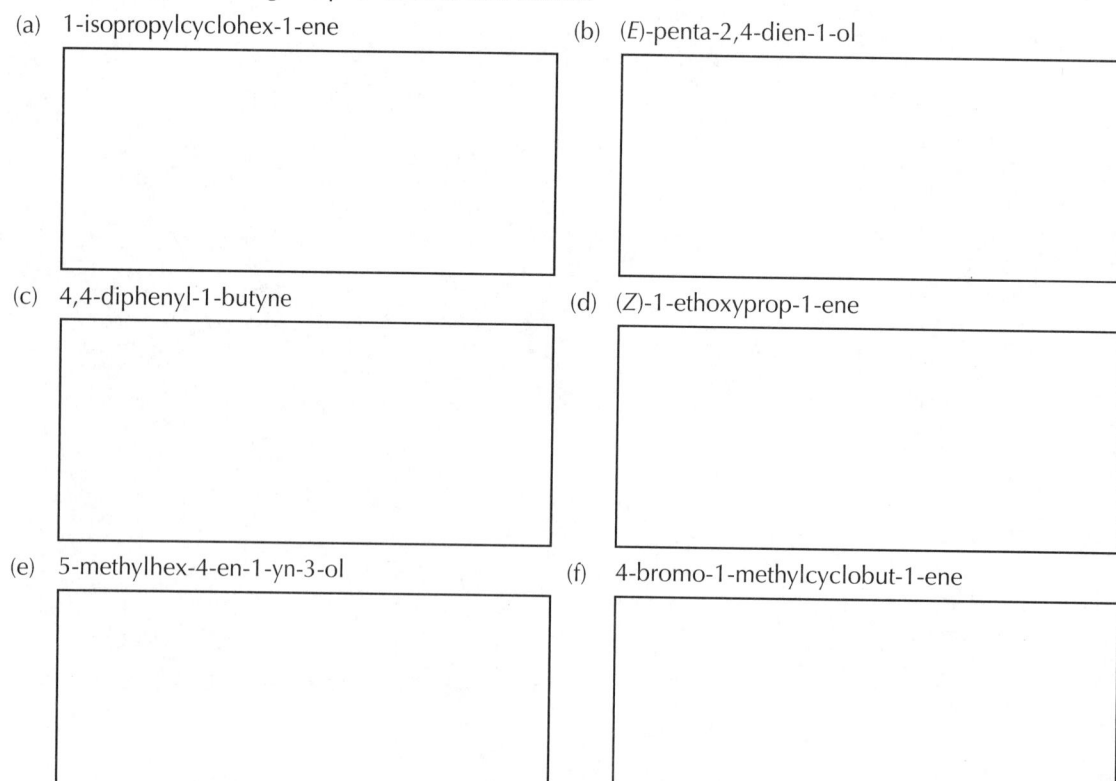

AP04.03 Provide the IUPAC names for the following compounds.

(a)

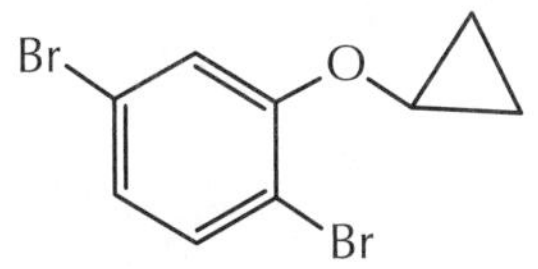

(b)

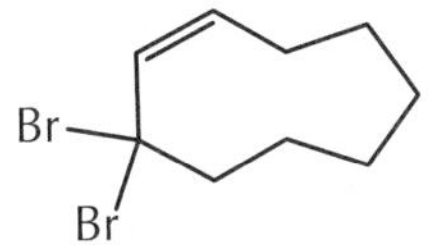

(c)

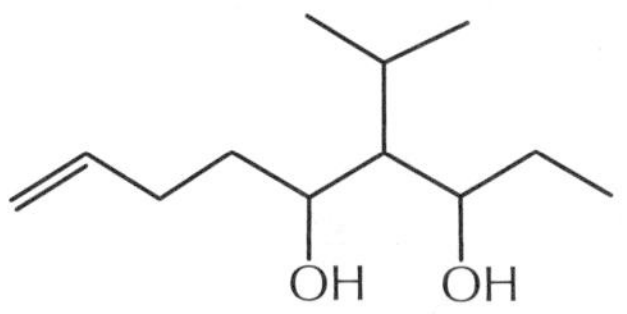

(d)

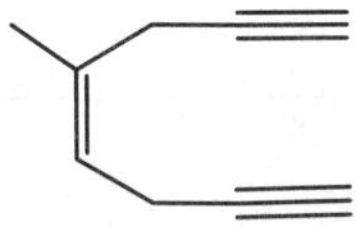

(e)

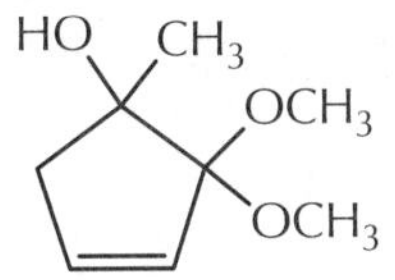

AP04.04 Draw the following compounds from their names.

(a) (2Z,4E)-3-methoxyhexa-2,4-diene

(b) 3,3-dichloro-6-ethoxycyclohepta-1,4-diene

(c) 2,5-dimethoxyhexa-1,5-dien-3-yne

(d) (Z)-2-bromo-1-cyclohexylhept-1-en-3-yne

(e) 1-isopropoxycycloprop-2-en-1-ol

(f) 7-phenylhepta-1,4,6-triyn-3-ol

AP04.05 Provide the IUPAC names for the following compounds.

(a)

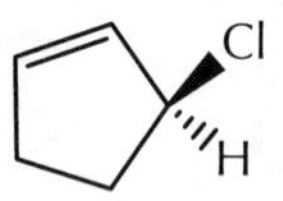

(b)

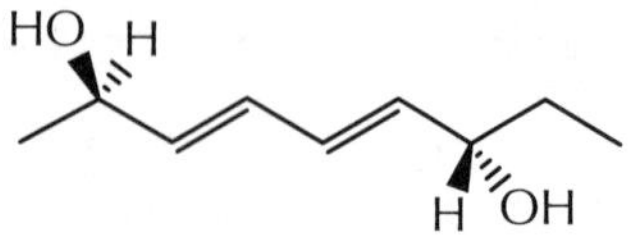

(c)

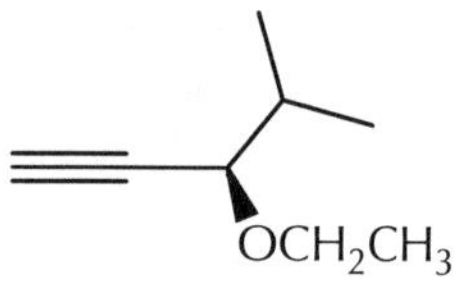

AP04.06 Draw the following compounds from their names.

(a) (S)-2-methoxybut-3-en-2-ol

(b) (3S,4R)-3,4-dichlorocyclohex-1-ene

(c) (1S,2S)-1-phenylpent-4-yne-1,2-diol

AP04.07 Provide the IUPAC names for the following compounds.

(a)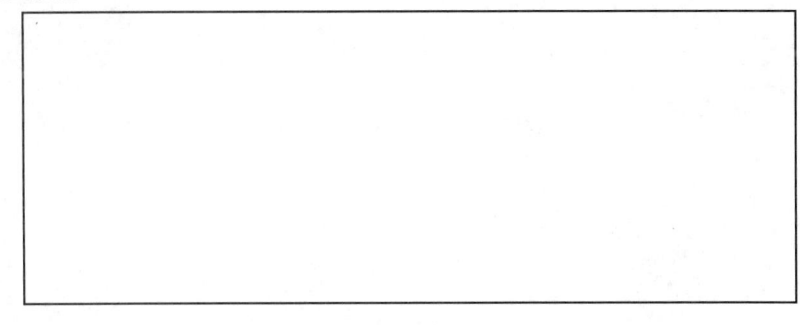

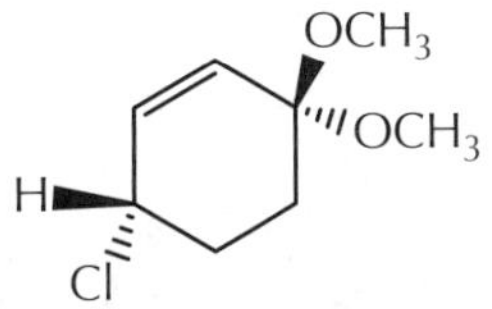

(b)

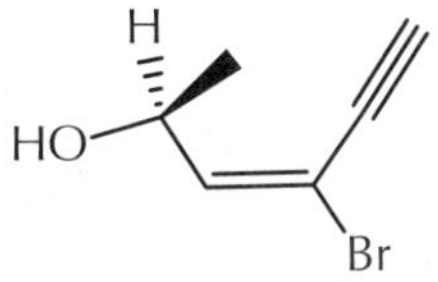

(c)

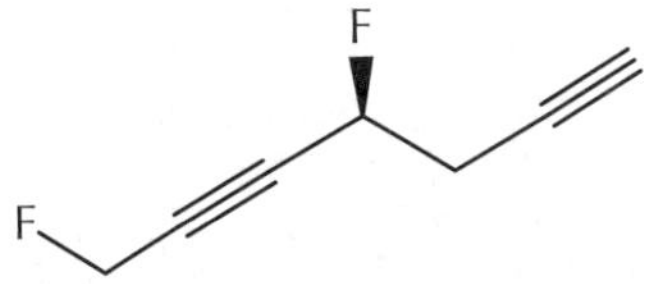

AP04.08 Draw the following compounds from their names.

(a) (2Z,4R,5Z)-4-bromo-4-methylocta-2,5-diene

(b) (1R,2Z,4S,7Z)-4-chlorocyclodeca-2,7-dien-1-ol

(c) (1R,4S)-4-(cyclopentyloxy)cyclopent-2-en-1-ol